The Structure
and Rheology of
Complex Fluids

TOPICS IN CHEMICAL ENGINEERING

A Series of Textbooks and Monographs

Series Editor
Keith E. Gubbins, North Carolina State University

Associate Editors
Mark A. Barteau, *University of Delaware*
Edward L. Cussler, *University of Minnesota*
Klavs F. Jensen, *MIT*
Douglas A. Lauffenburger, *MIT*
Manfred Morari, *ETH*
W. Harmon Ray, *University of Wisconsin*
William B. Russel, *Princeton University*

Receptors: Models for Binding, Trafficking and Signalling
D. Lauffenburger and J. Linderman

Process Dynamics, Modeling, and Control
B. Ogunnaike and W. H. Ray

Microstructures in Elastic Media
N. Phan-Thien and S. Kim

Optical Rheometry of Complex Fluids
G. Fuller

Nonlinear and Mixed Integer Optimization: Fundamentals and Applications
C. A. Floudas

An Introduction to Theoretical and Computational Fluid Dynamics
C. Pozrikidis

Mathematical Methods in Chemical Engineering
A. Varma and M. Morbidelli

The Engineering of Chemical Reactions
L. D. Schmidt

Analysis of Transport Phenomena
W. M. Deen

The Structure and Rheology of Complex Fluids
R. Larson

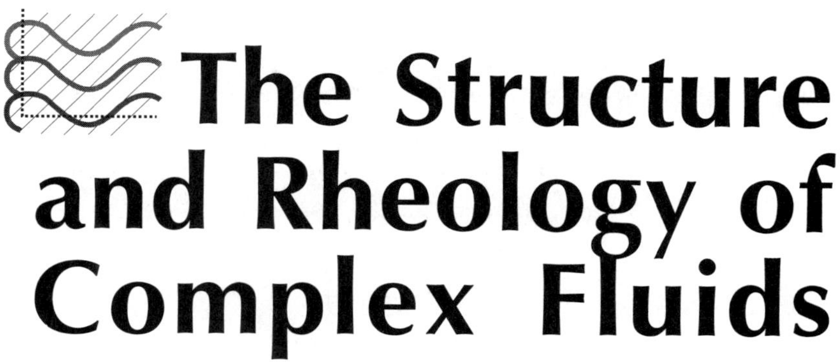

The Structure and Rheology of Complex Fluids

Ronald G. Larson

University of Michigan, Ann Arbor

New York • Oxford
OXFORD UNIVERSITY PRESS
1999

Oxford University Press

Oxford New York
Athens Auckland Bangkok Bogotá Buenos Aires Calcutta
Cape Town Chennai Dar es Salaam Delhi Florence Hong Kong Istanbul
Karachi Kuala Lumpur Madrid Melbourne Mexico City Mumbai
Nairobi Paris São Paulo Singapore Taipei Tokyo Toronto Warsaw

and associated companies in
Berlin Ibadan

Published by Oxford University Press, Inc.
198 Madison Avenue, New York, New York 10016

Library of Congress Cataloging-in-Publication Data

Larson, Ronald G. (Ronald Gary), 1953–
    The structure and rheology of complex fluids / Ronald G. Larson.
        p.   cm.—(Topics in chemical engineering)
    Includes bibliographical references and index.
    ISBN 0-19-512197-X (cloth)
    1. Complex fluids.   2. Rheology.   I. Title.   II. Series: Topics
chemical engineering (Oxford University Press)
    QD549.2.C66L37   1998
    530.4'2—dc21   98-19940
                                                                  CIP

Printing (last digit): 9 8 7 6 5 4 3 2 1

Printed in the United States of America
on acid-free paper

To Rachel, Emily, Andrew, and Eric

CONTENTS

PART II POLYMERS, GLASSY LIQUIDS, AND POLYMER GELS 105

PART IV LIQUID CRYSTALS AND SELF-ASSEMBLING FLUIDS 441

Chapter 10 Liquid Crystals 443

PREFACE

Every elementary school pupil is taught that there are three states of matter: solid, liquid, and gas. To these might be added "plasma," which exists at extremely high temperatures, or the "Bose–Einstein condensate," at extremely low ones. However, one needn't resort to superfrigid or superhot extremes to find states of matter that challenge the ordinary division into solids, liquids, and gases. Such everyday substances as mayonnaise or window glass will do the trick.

While it is true that the classical definition of "liquid" as opposed to "solid" provides a basis for deciding whether a substance is a solid or a liquid, the classical definition is inadequate not only for many everyday purposes, but for engineering ones as well. According to the classical definition, a "fluid" is any substance that deforms continuously under the application of an arbitrarily small shearing stress. By this definition, mayonnaise is a solid, since it holds its shape against gravity, while window glass is a liquid (!), since it creeps, albeit ever so slowly, against forces even much weaker than gravity. Clearly, for many workaday uses, in the home and on the factory floor, the classical definition is inadequate.

This book deals with the thick, rubbery, gooey, and pasty substances that defy the classical definitions of solids and liquids. These substances are often called "complex fluids," which I define as "substances that flow at modest stresses." By "flow," I mean a smooth deformation on a humanly accessible time scale. A piece of slate or ceramic does not "flow"; it deflects under modest loads, and it fractures under great ones. Peanut butter, on the other hand, "flows" rather smoothly when a child pushes a butter knife across it. Hence peanut butter is a "complex fluid," while a brick is not. Although the packed ices of glaciers "flow" on geological time scales, and metals creep under large loads by defect motion, and "even the mountains flow before the Lord," we mortals usually prefer to call these solids. Clearly, there are ambiguous cases: A piece of window glass seems solid enough at room temperature and ordinary periods of time, but heat it up a bit, or wait a few decades, and its solidity becomes dubious. Reiner (1949) expressed well the dilemma: "Strictly defined rheological divisions belong to ideal abstract bodies and not to real materials. If we say that concrete is a liquid, every builder will laugh at us. . . . If we say that glass is a solid, the theoretical physicists will consider us to be simple and crude." The definition of a "complex fluid" might therefore seem as plastic as the materials it is meant to define! Nevertheless, so many materials fall well within the rough definition given above that "complex fluids" are worthy of a title—and indeed, a book—of their own.

"Complex fluids" include polymeric liquids and melts, suspensions of colloidal particles, micellar solutions, and liquid foams. Complex fluids are distinguished from simple crystalline solids and simple liquids in that they usually possess molecular or structural length scales much larger than atomic. Complex fluids are used in foodstuffs, pharmaceuticals, and cosmetics, and they are intermediates in the manufacture of fibers, films, packaging, and so on. The rheological (i.e., flow) properties of complex fluids often dictate the uses to which they can be put; in fact, the choice of a particular complex fluid for a given application is often driven by the rheological properties that are needed in the application. This is certainly true for foods or personal care products, where, for example, the rheological properties of ketchup, ice cream, shampoo, or toothpaste can determine the success or failure of the product. Further examples in the cosmetic and health care industries could be multiplied. Even when rheological properties are not critical in the final product, they often influence the processability of the material as it is shaped into final form. This is certainly true in the plastics and ceramics industries.

In the chapters that follow, common and not-so-common complex fluids are grouped into classes, specifically:

> Polymers
> Glassy liquids
> Polymer gels
>
> Particulate suspensions
> Particulate gels
> Electro- and magnetoresponsive suspensions
> Foams, emulsions, and blends
>
> Liquid crystals
> Liquid-crystalline polymers
> Surfactant solutions
> Block copolymers

This book is organized as follows. In Chapter 1, the topic of complex fluids is introduced, along with methods of measuring rheological and other properties; simulation techniques are also discussed. In Chapter 2, the basic forces that produce the amazingly diverse structures of complex fluids are described. These include *excluded-volume, van der Waals, electrostatic, hydrogen-bonding*, and other forces. This, along with Chapter 1, completes Part I, "Fundamentals." Part II covers "Polymers, Glassy Liquids, and Polymer Gels"; it has a chapter each on "Polymers" (Chapter 3), "Glassy Liquids" (Chapter 4), and "Polymer Gels" (Chapter 5). Part III deals with "Suspensions" and covers "Particulate Suspensions" (Chapter 6), "Particulate Gels" (Chapter 7), "Electro- and Magnetoresponsive Suspensions" (Chapter 8), and "Foams, Emulsions, and Blends" (Chapter 9). Finally, Part IV deals with "Liquid Crystals and Self-Assembling Fluids," which includes chapters on "Liquid Crystals" (Chapter 10), "Liquid-Crystalline Polymers" (Chapter 11), "Surfactant Solutions" (Chapter 12), and "Block Copolymers" (Chapter 13).

The chapters need not be read in order. However, since Part I contains the basic material necessary for a thorough understanding of what follows it, the reader who is not already familiar with this material ought to at least scan the first two chapters. Besides these, the first chapters in each of Parts II–IV—that is, Chapters 3, 6, and 10—are the most essential. Having read these, any other chapter of the book should be accessible to the scientifically literate reader. For example, the chapters on "Polymers" (Chapter 3) and "Liquid Crystals" (Chapter 10) should provide enough background for the reader to tackle "Liquid-Crystalline Polymers" (Chapter 11) or "Block Copolymers" (Chapter 13).

REFERENCE

Reiner M (1949). *Deformation and Flow. An Elementary Introduction to Theoretical Rheology*, HK Lewis & Co, London.

Ronald G. Larson
University of Michigan, Ann Arbor

ACKNOWLEDGMENTS

This book is the product of efforts spread over a 5-year time span, from 1993 through 1997. During this period, I worked at three different institutions: Bell Laboratories in Murray Hill, New Jersey, the Isaac Newton Institute of Cambridge University, Cambridge, England, and the University of Michigan, Ann Arbor. I am indebted to each of these great institutions for providing the support, infrastructure (especially library resources), and, most of all, the intellectual stimulation, without which my exertions would surely have come to naught. I thank Xina Quan and Department 11145 at Bell Laboratories, Tom McLeish, Ken Walters, and Anthony Pearson at the Isaac Newton Institute, and Ralph Yang at the University of Michigan for leadership and support. The members of the Society of Rheology and the international rheology community were the source of many ideas and much inspiration. I am particularly indebted to the following individuals for critiquing portions of the manuscript: Bob Anderssen, Lynden Archer, Bob Butera, Michael Cates, Gerry Fuller, S. Hyde, Julie Kornfield, Greg McKenna, Kalman Migler, David Pine, Bill Russel, and Mohan Srinivasarao. I am also grateful to the students of the 1997 "Complex Fluids" course at the University of Michigan, especially Paul Suding, who helped me make several of the chapters more readable to students. Cattaleeya Pattamaprom was of great help in preparing the Index. My wife, family, friends, and numerous colleagues were a constant source of encouragement and strength. A.M.D.G.

R.G.L.

The Structure and Rheology of Complex Fluids

PART I

FUNDAMENTALS

Chapter 1

INTRODUCTION TO COMPLEX FLUIDS

1.1 COMPLEX FLUIDS VERSUS CLASSICAL SOLIDS AND LIQUIDS

A simple substance such as water below its freezing point is a hard three-dimensional crystalline solid, and above its freezing point it is a low-viscosity Newtonian liquid. In the liquid state, the mechanical properties of such a substance are specified by its shear viscosity η, which is of course temperature- and pressure-dependent.

It has been known for a century or more, however, that some condensed-phase materials are neither simple liquids nor simple crystalline solids, and thus they do not fall readily within the classical scheme of materials classification. These "complex fluids" possess mechanical properties that are intermediate between ordinary liquids and ordinary solids. Specification of a viscosity or an elastic modulus does not even begin to describe the mechanical properties of such a substance. In many cases, the relationship between stress and deformation for a complex fluid is nonlinear, is unknown, or is under dispute. The number of recognizably different kinds of complex fluids has gradually increased throughout this century, so that it is now possible to find materials that possess to an intermediate degree almost any of the properties that distinguish classical solids and liquids.

Perhaps the most important distinction between classical solids and classical liquids is that the latter quickly shape themselves to the container in which they reside, while the former maintain their shape indefinitely. Many complex fluids are intermediate between solid and liquid in that while they maintain their shape for a time, they eventually flow. They are "solids" at short times and "liquids" at long times; hence, they are *viscoelastic*. The characteristic time required for them to change from "solid" to "liquid" varies from fractions of a second to days, or even years, depending on the fluid. Examples of complex fluids with long structural or molecular relaxation times include *glass-forming liquids, polymer melts and solutions*, and *micellar solutions*.

There are also complex fluids that change from solid-like to liquid-like, or vice versa, when subjected to a modest deformation. Complex fluids of this type include particulate and polymeric *gels*. Some fluids change to solids when an electric or magnetic field is applied; these are *electrorheological* and *magnetorheological suspensions*. A classical liquid or solid, on the other hand, does not change character in response to a weak field unless it is extremely close to a phase transition temperature.

Another distinction between classical crystalline solids and liquids is that the former can be *anisotropic* (i.e., their mechanical properties can depend on the orientation of the deformation with respect to the crystallographic axes), while classical liquids are *isotropic* (i.e., their properties are the same in all directions). *Liquid crystals* are complex fluids that flow like liquids, but whose mechanical properties are anisotropic like those of crystals. Just as there are many kinds of crystalline symmetry, there are also many different symmetries of liquid crystalline phases.

Finally, there are complex fluids that are intermediate between solid and liquid in more than one of the ways listed above. *Liquid crystalline polymers (LCPs)* are both viscoelastic and liquid crystalline. Ordered *block copolymers* are viscoelastic and anisotropic. *Glassy polymers* possess long viscoelastic time scales both because they are glassy and because they are polymeric. *Filled polymer melts* possess the properties of both polymer melts and suspensions.

In this book, we review the most basic distinctions and similarities among the rheological (or flow) properties of various complex fluids. We focus especially on their *linear viscoelastic* behavior, as measured by the frequency-dependent *storage and loss moduli G' and G''* (see Section 1.3.1.4), and on the *flow curve*—that is, the relationship between the shear viscosity η and the shear rate $\dot{\gamma}$. The storage and loss moduli reveal the mechanical properties of the material at rest, while the flow curve shows how the material changes in response to continuous deformation. A measurement of G' and G'' is often the most useful way of mechanically characterizing a complex material, while the flow curve $\eta(\dot{\gamma})$ shows how readily the material can be *processed*, or shaped into a useful product. The moduli are usually more readily predicted by molecular theories than are other rheological properties.

In this first chapter, some examples are given of complex fluids (Section 1.2). Then, the most common methods of measuring their rheological properties are presented along with a qualitative description of how these properties reflect to varying degrees their "solid-like" or "liquid-like" character (Section 1.3). Next, we present the mathematics needed to describe velocity gradients, deformations, and stresses in complex fluids (Section 1.4). Then, we discuss how deformation imposed externally on a complex fluid can produce homogeneous *flow*, and we also discuss how that flow can be disrupted by *slip* or *yielding* (Section 1.5). Nonrheological methods of probing the structure of complex fluids at rest or under flow are then mentioned (Section 1.6). Some methods of calculating on the computer the structures of complex fluids under flow are discussed briefly (Section 1.7). Finally, a general-purpose expression for the stress tensor is given, which shows how the microstructure and microstructural forces control the bulk stresses in a complex fluid (Section 1.8).

1.2 EXAMPLES OF COMPLEX FLUIDS

1.2.1 Foods

Many delicious foods are complex fluids. Their rheology, along with their flavor and texture, determines their popularity; they are therefore gastrorheological fluids! *Mayonnaise* is a good example (Kurti and This-Benckard 1994). It contains vegetable oil, vinegar or lemon juice, and egg yolk. Droplets of vegetable oil in the vinegar or lemon juice are stabilized by lecithin, which is a natural surfactant contained in egg yolk (see Fig. 1-1). Thus, mayonnaise is an *emulsion*; that is, it contains one liquid dispersed in another. Mayonnaise holds its shape (more or less) against gravity, but flows smoothly under slight forces. It therefore has a low *yield stress*, a stress below which it will not flow. The magnitude of the yield stress is controlled by the size of the droplets (around 10 μm for good mayonnaise) and by the surface tension of the droplet surfaces, as discussed in Chapter 9.

Another even more delicious complex fluid is *ice cream*, which is a partially frozen *emulsion* of cream, milk, sugar, and flavoring and is *foamed* with 40–50% by volume air (Kilara and Sharkasi 1994). Foams are discussed in Chapter 9. A quick way to make ice cream is to pour an equal volume of liquid nitrogen into the ice cream ingredients while stirring with a wooden spoon (Kurti and This-Benckhard 1994). The liquid nitrogen both cools and foams the ingredients. The quick cooling keeps ice crystals small and the texture smooth. (If you wish to try this method, be sure to observe safety precautions: Use a metal bowl, wear gloves and safety glasses, and keep observers away from any liquid nitrogen splashes.)

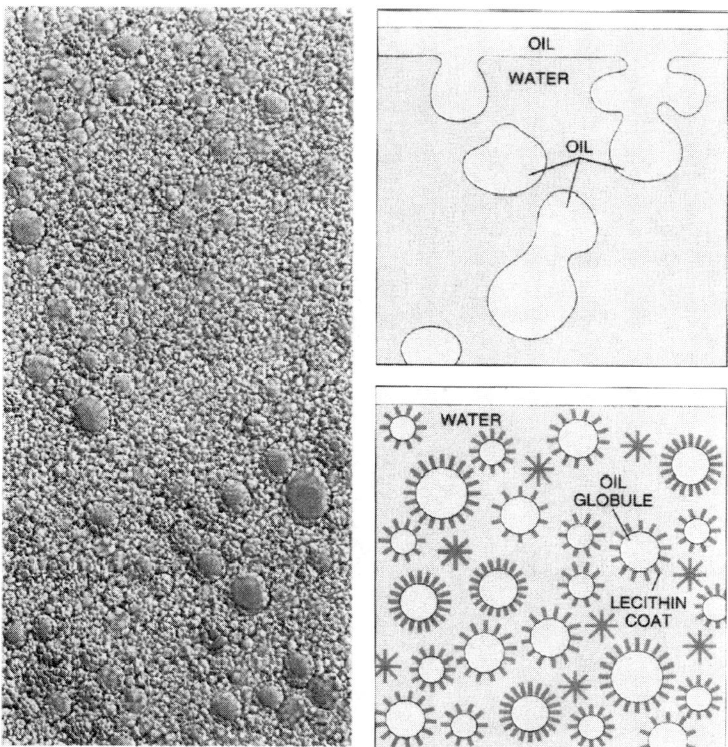

Figure 1-1   Mayonnaise (**left**, × 100) is a suspension of oil droplets in water. The tendency of oil and water to phase-separate (**top right**) is suppressed in mayonnaise by lecithin molecules from egg yoke that coat the oil droplets. Lecithin is a natural surfactant that can dissolve in either oil or water. (From Kurti and This-Benckhard 1994, with permission.)

Mayonnaise and ice cream both have a yield stress, which is produced in these foods by droplets and air bubbles that must deform if flow is to occur. Another food with a yield stress is *mustard*; it is neither an emulsion nor a foam, but a *suspension* or paste, containing particles 30 μm or so in diameter that attract each other and form a weak network (Gerhards and Schubert 1993). Mustard is made by simply grinding mustard seeds, together with vinegar, salt, spices, and water, into a mash. The grinding releases oils that impart to mustard its distinctive flavor. The rheology of particulate suspensions is covered in Chapter 6.

Another good example is *cheese*. Cheese is made from milk, which is itself a complex fluid containing *micelles* (small 0.04- to 0.3-μm spheres; see Chapter 12) of the protein casein (Sharkasi and Kilara 1994; deKruif 1992; Damodaran and Paraf 1997). By adding a weak acid and the appropriate enzyme (rennet), the micelles are destroyed and the proteins then *gel*, or congeal together to form a solid mass, or *curd*, which separates from the remaining liquid. Figure 1-2 shows electron micrographs of mozzarella cheese, in which stretching the curd orients the protein fibers. Ordinary *gelatin* is also a gel, produced from a biological protein called *collagen* mixed with a high volume fraction of water. Gels and gelation are discussed in Chapter 5.

Chocolate is not only delicious, but has remarkable mechanical properties. In the solid state, it can be "cold-extruded"—that is, forced to flow through a capillary die under pressure (Beckett et al. 1994). This process transforms the chocolate from a brittle solid to a plastic, malleable one that can be reshaped at will. This plastic state lasts for several minutes, after which the chocolate reverts to its brittle form, possibly because of recrystallization of sugars that had been melted under the stresses of extrusion. Not only does solid chocolate become plasticized under flow, but molten chocolate can be solidified by electric fields (Browne 1996)! In the high-temperature molten state, chocolate is an emulsion, the droplets of which line up into columns under an electric field, producing a rigid structure. Fluids that are solidified by electric fields are called *electrorheological fluids*; their structure and properties are described in Chapter 8. For more on food rheology, see Dickenson (1991).

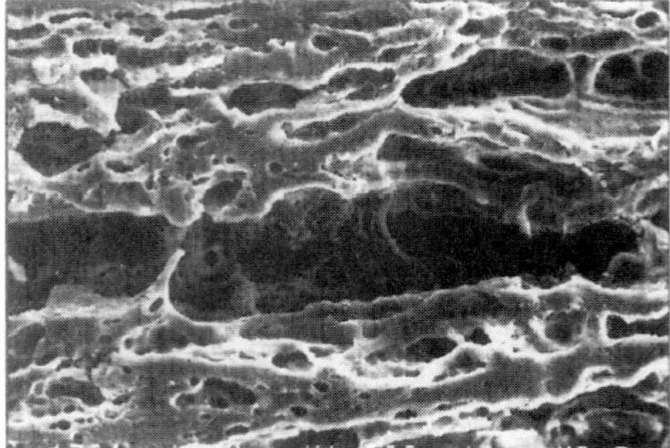

Figure 1.2   Low (**left**) and high (**right**) magnifications of scanning electron micrographs of mozzarella cheese samples. The protein fibers have an orientation roughly parallel to the stretch direction. (From Sharkasi and Kilara, MRS Bulletin, July 1994, p.47, reprinted with permission.)

1.2.2 Biofluids

Humans and other organisms are filled with complex fluids. *Blood* is a suspension containing around 40% by volume blood cells, which are flexible puckered disks roughly 10 μm in diameter, suspended in clear plasma, which is itself a viscoelastic fluid containing interacting protein macromolecules. The viscoelastic properties of blood determine the pumping load on the heart, for example, and affect the performance of artificial heart valves. At modest shear rates, the flow and orientation properties of red blood cells are similar to those of rigid disks (see Fig. 1-3), while at higher shear rates their flow behavior resembles that of fluid droplets (Goldsmith 1986). The rheological properties of suspensions of rigid and fluid particles are covered in Chapters 6 and 9, respectively. At low shear rates, red blood cells often stack up up like poker chips into so-called rouleaux (Goldsmith and Turrito 1986). The viscosity of blood therefore depends on the size of these aggregates, which, in turn, is shear-rate-dependent. The behavior of solutions of aggregated particles is described in Chapter 7.

There are many other biological fluids whose rheological properties are of significance. These include *mucin*, which is contained in stomach lining and lungs and whose rheological properties have been implicated as a contributor to asthma (Zayas and King 1990), *synovial fluid* (which lubricates joints), *spittle*, and other, even more disgusting fluids (Gabelnick and Litt 1973).

Other creatures possess even more remarkable complex fluids. Spiders, for example, synthesize and spin at ambient temperatures incredibly tough dragline silk fibers, with a breaking energy (10^5 J/kg) greater than any synthetic or natural substance (Tirrell 1996; Simmons et al. 1996; Grubb and Jelinski 1997; Termonia 1994). Silkworms also make very tough fibers, using special organs to synthesize a liquid-crystalline silk-precursor solution (Viney et al. 1994; Willcox et al. 1996). They extrude it, they coat it with water-soluble

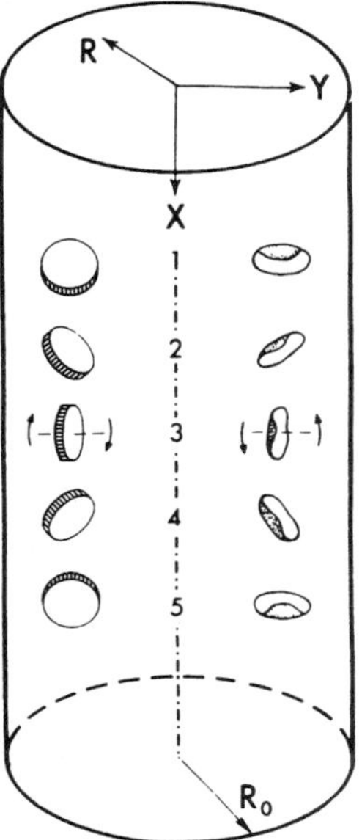

Figure 1.3   The "tumbling" motion of a red blood cell (**right**) as it is sheared in a Poiseuille flow down a tube at shear rates < 20 sec^{-1} is similar to that of a rigid disk (**left**). The tumbling rate is slowest in location 3. (From Goldsmith 1986, reprinted with permission from Academic Press.)

glue that binds fibers together, and finally they use their mouths as spinnerets to draw fibers whose properties are still unmatched by humans.

1.2.3 Personal Care Products

Shampoo, nail polish, lipstick, deodorant, and *toothpaste* are all commonplace products whose rheology is carefully tuned for customer satisfaction. Toothpaste, for example, must flow out of the tube only when squeezed, must flow under modest force, and must stop flowing immediately after it has been applied to the brush so that it doesn't sink into the bristles. Thus, the ingredients of toothpaste include not only gentle abrasives such as silica powder, fluoride for combating tooth decay, and flavoring and foaming agents, but also polymers, such as carboxymethylcellulose, for control of rheology (Pader 1993). Polymers are covered in Chapter 3.

Nail polish must be thick enough to cling to the bristles of the applicator brush and yet fluid enough to form a glassy smooth surface on finger nails. In addition, pigments for coloring and pearlescence must be prevented from settling; this is typically accomplished by adding carefully formulated clay powders such as montmorrillonite (Remz 1993). Structured suspensions of this kind that can support a small or large yield stress are described in Chapter 7.

Many shampoos and conditioners are designed to flow readily from the tube or bottle into one's hand while having enough "body" that they do not quickly drip. This is accomplished by careful control of the surfactant concentration as well as the addition of polymeric "thickeners" such as carbomers (very high molecular weight polyacrylic acid) (Lochhead 1993). Very thick "mousses" can be formulated using polymeric "associative thickeners," such as hydrophobically modified hydroxyethylcellulose (see Fig. 1-4). Other products requiring carefully designed rheological properties include cremes, lotions, and deodorants. The rheological properties of these products have been greatly improved even within the last 10 years. For a thorough discussion of the rheology of cosmetics and toiletries, see Laba (1993).

1.2.4 Electronic and Optical Materials

Many computers, such as laptops, have flat-panel screens or *displays* that contain *liquid crystals* (Bahadur 1990). Liquid crystals are composed of small semirigid molecules that are elliptical or oblong in shape and spontaneously *orient*, forming anisotropic fluids (Collings 1990). Liquid crystals commonly used in flat-panel displays have orientational order but do not have long-range positional order. The direction of the preferred orientation in a liquid crystal can be switched by application of an electric field, and since the optical properties (such as birefringence) are strongly orientation-dependent, the liquid crystal acts as an optical switch. The speed of switching in a liquid crystal is controlled by its viscous and elastic properties, as well as by its dielectric susceptibility. Displays can also be made in which the liquid crystal is *emulsified* as roughly $1\text{-}\mu\text{m}$ droplets in a polymer gel. Much more about liquid crystals and about their viscous and elastic properties can be found in Chapter 10.

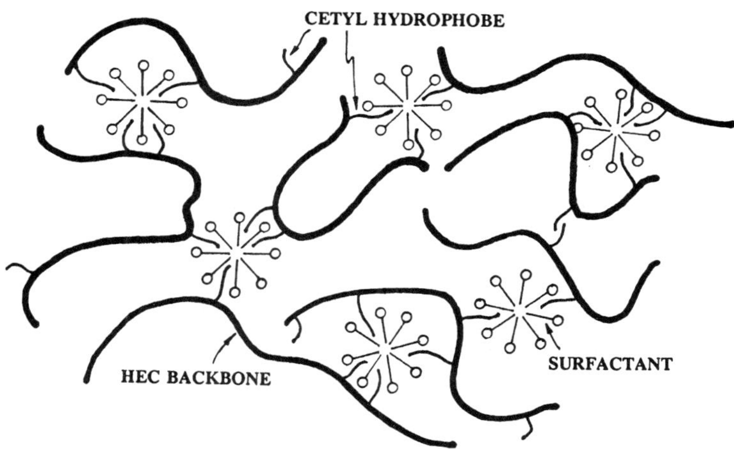

Figure 1.4 Network structure formed by an aqueous surfactant solution mixed with an associative thickening polymer hydroxyethylcellulose (HEC) with cetyl (16-carbon alkyl) side branches. (Reprinted from Clarke 1993, by courtesy of Marcel Dekker, Inc.)

When electronic devices are mounted onto circuit boards, their conducting leads must be soldered into place. A convenient way to do so is to first lay down small spots, or "pads," of *solder paste* onto the circuit board by screen printing. The components can then be placed onto the boards, with wire leads immersed into the solder-paste pads; the 50-μm lead solder balls in the paste are melted by heating, which allows them to flow together to provide a conductive bond between the wire leads and the circuit board. Coalescence of the molten solder balls is promoted by the acid in the viscous suspending medium, or "flux," which dissolves oxidized lead from the surfaces of the solder balls. The solder paste must have a yield stress high enough that the solder pads do not run together, but not so high that the paste cannot be forced through the screen during printing (Morris and Wojcik 1990; Trease and Dietz 1972).

1.2.5 Polymers

Solid plastic products, parts, and packaging, whether injection-molded polycarbonate compact laser disks, extruded polyethylene cable jacketing, high-strength fibers spun from Kevlar liquid crystalline polymer, soda bottles containing polyethylene terephthalate, or injection-molded polyurethane car bumpers, are all processed in a fluid state, either molten or in solution. In this state, these polymers are *viscoelastic* complex fluids, whose rheological properties determine the ease and expense of processing and, to some extent, the final properties of the manufactured parts. For example, the pressure drop required for injection molding depends on the melt viscosity. The attainable extrusion speed, as well as the shape of extruded parts, depends on elastic forces generated when the molecules are stretched out by the flow. "Molded-in" stresses can lead to warping of injection-molded parts or undesired residual birefringence in compact disks. Polymers are also increasingly important in bioengineering applications, such as encapsulants for drug delivery (Langer 1995), or in tissue engineering (Langer 1995; Hubbell 1996). The flow properties of such biocompatible polymers are especially important in applications in which artificial tissue, such as cartilage grown in a water-soluble polymer matrix, is injected through a small orifice into the body. The rheological properties of various kinds of polymers are discussed in Chapters 3, 5, 11, and 13.

1.2.6 Other Examples

Other rheologically complex fluids include oil-field fluids such as drilling muds and mobility control agents (Alderman et al. 1988; Maitland 1991), "waxy" crude oils which can form gels that clog pipelines (Wardhaugh and Boger 1991), pulp fiber used in producing paper (Forgacs et al. 1958; Biermann 1996), paints (Cohu and Magnin 1995), freshly mixed cement (Struble and Sun 1995), asphalt (Lesueur et al. 1996), and many others. These examples show both the diversity and the ubiquitous nature of complex fluids, and they also show the importance of characterizing and understanding their rheological properties.

1.3 RHEOLOGICAL MEASUREMENTS AND PROPERTIES

In a gross sense, a rheological measurement tells one how "hard" or "soft" a material is, or it indicates how "fluid-like" or "solid-like" it is. These characteristics of a material depend on the time scale at which the material is probed. For example, a ball of Silly Putty will bounce like an elastic solid if suddenly dropped, or it will flow like honey if left to rest on a table top. A "rheometer" measures the rheological properties of a complex liquid as a function of rate or frequency of deformation. For liquids, the simplest devices impose a *shearing flow* on the liquid and measure the resulting stresses, or alternatively, impose a shearing stress and measure the resulting shearing rate.

1.3.1 Shearing Flow

1.3.1.1 Flow Geometries

The simplest geometries for imposing a shearing flow are depicted in Fig. 1-5, from Middleman (1977). Either *drag flow* or *pressure flow* is imposed in the following geometries: (1) the sliding plate (or plane Couette geometry), (2) the concentric cylinder (or circular Couette geometry), (3) the cone and plate, (4) the parallel disks (or plate and plate), (5) the capillary (or circular Poiseuille geometry), (6) the slit (or plane Poiseuille geometry), and (7) the axial annulus. The depictions in Fig. 1-5 are not to scale; for example, the angle between the cone and plate in this figure is enormously exaggerated; typically the cone is a very shallow one so that the angle between it and the plate is no more than 5–10 degrees. The capillary and slit typically have lengths much greater than their diameters.

It may seem odd that such different geometries, some with curved streamlines and some

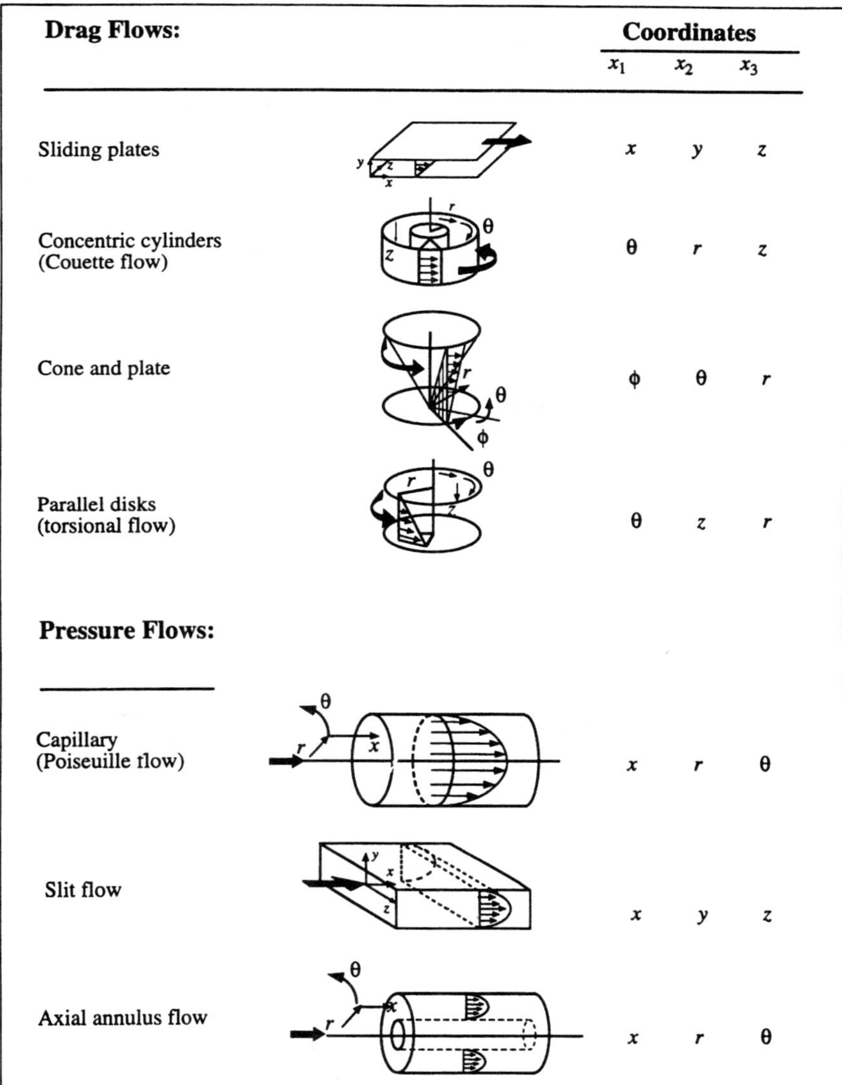

Figure 1.5   Geometries for producing shearing flows. (Adapted from Macosko, *Rheology Principles, Measurements, and Applications*, Copyright © 1994. Reprinted by permission from John Wiley & Sons.)

with straight ones, produce the same kind of flow. The key similarity in flows generated in these geometries is that they can be *viscometric*, meaning that each fluid element experiences a steady simple shearing flow with a shear rate that is constant in time. A more precise definition of a viscometric flow can be given once the velocity gradient tensor is introduced in Section 1.4.1.

There are advantages and disadvantages of each geometry. Geometries 1–3 admit nearly uniform shearing flows, while in 4–7, the shear rate varies within the geometry, and so data analysis is less straightforward. Geometries 1–4 readily allow time-dependent data to be gathered. Finally, one can measure *normal stress differences* (described in Section 1.4.3) with geometries 3 and 4, and with some modifications also with geometry 6. Because geometry 3, the cone and plate, is the only one that readily allows all of the above types of measurement, it has become the method of choice for many purposes. However, the other geometries are useful for special cases. For example, the plate-and-plate geometry is useful if the gap needs to be uniform—for example, for fluids for which the apparent rheological properties show a gap dependence (like liquid crystals, or fluids that slip at the wall). The capillary and the slit geometries, and to a lesser extent the parallel-disk geometry, readily allow attainment of high shear rates. The circular Couette device, as well as the pressure-flow devices, are valuable for liquids that are too runny to be held between cone-and-plate or plate-and-plate fixtures. For a thorough review of various shearing-flow rheometers, and their respective advantages and disadvantages, see Macosko (1994).

> • Problem 1.1 and Worked Example 1.2, just after this chapter, explore the degree of uniformity of the shear rate in the circular Couette and cone-and-plate geometries.

1.3.1.2 Steady-State Shear Viscosity

In each geometry, the steady shear rate imposed on the fluid depends on a driving velocity and the dimensions of the geometry. For the sliding-plate device, the shear rate $\dot{\gamma}$ is the velocity V of the moving plate (the other plate being held stationary), divided by the gap h between the two plates; hence $\dot{\gamma} = V/h$. In the cone-and-plate geometry, $\dot{\gamma} = \Omega/\tan\alpha$, where Ω is the steady angular rotation speed of the cone or plate (whichever is rotating), and α is the cone angle, which is usually less than or equal to about 0.10 radians.

The *shear stress* σ is the force that a flowing liquid exerts on a surface, per unit area of that surface, in the direction parallel to the flow. The *shear viscosity* η is then defined as

$$\eta = \sigma/\dot{\gamma} \tag{1-1}$$

After a steady shearing flow has been imposed on a fluid for a suitable period of time, the shear stress often (but not always) comes to a steady state, $\sigma(\dot{\gamma})$, which depends on the imposed shear rate $\dot{\gamma}$. The ratio of the steady shear stress σ to the shear rate $\dot{\gamma}$ is then the *steady-state* shear viscosity $\eta(\dot{\gamma})$.

1.3.1.3 Transient Shear Viscosity

If, as is often the case, the stress reaches a steady value only after a transient period of steady shearing starting from a state of rest, then the instantaneous stress $\sigma(\dot{\gamma}, t)$ during the transient start-up period, divided by the steady shear rate $\dot{\gamma}$, is the *transient start-up viscosity*

$$\eta^+(\dot{\gamma}, t) = \sigma(\dot{\gamma}, t)/\dot{\gamma} \tag{1-2}$$

The superscript "+" indicates that the shear rate was increased from zero at time $t = 0$. This definition is only meaningful if the fluid at time $t = 0$ is in a well-defined state, usually a stress-free state, and if this starting state and the transient viscosity are reproducible from one run to the next. The superscript "+" is sometimes omitted. Measurements of $\eta^+(\dot{\gamma}, t)$ give information about rates of structural rearrangement within a deforming complex fluid.

The *creep test* is a related way of obtaining time-dependent rheological information. In it, a constant shear *stress*, rather than a constant shear rate, is imposed on the material, and the shear rate is measured as a function of time until a steady shear rate is obtained. The creep test is especially useful for measuring the *yield stress* σ_y, since if the imposed stress is below σ_y the steady-state shear rate will be zero.

1.3.1.4 Storage and Loss Moduli

Another way to explore rates of structural rearrangement within a complex fluid, one that does not significantly deform the fluid's microstructure, is to impose *small-amplitude oscillatory shearing*. This kind of deformation can be achieved in a cone-and-plate geometry by rotating the cone about its axis with an angular velocity that oscillates sinusoidally, $\Omega(t) = \Omega_0 \cos(\omega t)$, where ω is the *frequency* of oscillation, in units of radians per second. The shear *rate* is then also a sinusoidal function of time, $\dot{\gamma}(t) = \Omega/\tan \alpha = \Omega_0 \cos(\omega t)/\tan \alpha$, and so is the shear *strain* γ, which is the time integral of the shear rate, $\gamma = (\Omega_0/\omega) \sin(\omega t)/\tan \alpha$. The ratio (Ω_0/ω) is the amplitude of the angular deflection of the cone, and $\gamma_0 = (\Omega_0/\omega)/\tan \alpha$ is the *strain amplitude* imposed on the fluid. In a plane Couette geometry, the strain amplitude is the amplitude of the displacement of the sliding plate, divided by the gap.

If the strain amplitude γ_0 is small enough (typically $\gamma_0 \ll 1$) that the fluid structure is not much disturbed by the deformation, then the stress measured during the oscillatory deformation is controlled by the rates of spontaneous rearrangements, or *relaxations* present in the fluid in the quiescent or equilibrium state. The shear stress $\sigma(t)$ produced by a small-amplitude deformation is proportional to the amplitude of the applied strain γ_0 and is itself *sinusoidally varying* in time. The maxima and minima of the sinusoidally varying stress signal are not necessarily coincident with the maxima and minima in the strain, however. In general, the sinusoidally varying stress can be represented as

$$\sigma(t) = \gamma_0[G'(\omega) \sin(\omega t) + G''(\omega) \cos(\omega t)] \tag{1-3}$$

The term proportional to $G'(\omega)$ is *in phase with the strain* and is called the *storage modulus*, while the term containing $G''(\omega)$ is *in phase with the rate of strain* $\dot{\gamma}$ and is called the *loss modulus*. The storage modulus represents storage of elastic energy, while the loss modulus represents the viscous dissipation of that energy. The ratio G''/G', which is called the *loss tangent* $\tan \delta$, is high ($\gg 1$) for materials that are *liquid-like*, but is low ($\ll 1$) for materials that are *solid-like*. The *complex modulus* G^* is defined by $G^* \equiv G' + iG''$, where i is the imaginary unit, $i \equiv \sqrt{-1}$. The magnitude of G^* is given by $|G^*| \equiv (G'^2 + G''^2)^{1/2}$; it is the amplitude of the stress waveform, divided by the strain amplitude. (The magnitude of G^* is sometimes also represented by just G^*.) The so-called "complex viscosity" is defined by $\eta' - i\eta'' \equiv \eta^* \equiv G^*(\omega)/i\omega$; its magnitude is $|G^*(\omega)|/\omega$. The regime of small-amplitude straining, in which the stress can be represented by Eq. (1-3), is called *the linear viscoelastic regime*.

1.3.1.5 Types of Rheological Response

Differences between solid-like and liquid-like complex fluids show up in all three of the shearing measurements discussed thus far: the shear start-up viscosity $\eta^+(\dot{\gamma}, t)$, the steady-state viscosity $\eta(\dot{\gamma})$, and the linear viscoelastic moduli $G'(\omega)$ and $G''(\omega)$. The start-up stresses $\sigma = \dot{\gamma}\eta^+(\dot{\gamma}, t)$ of prototypical "liquid-like" and "solid-like" complex fluids are depicted in Fig. 1-6. For the "liquid-like" fluid the viscosity instantaneously reaches a steady-state value after inception of shear, while for the "solid-like" fluid the stress grows linearly with strain up to a critical shear strain, above which the material "yields," or flows, at constant shear stress.

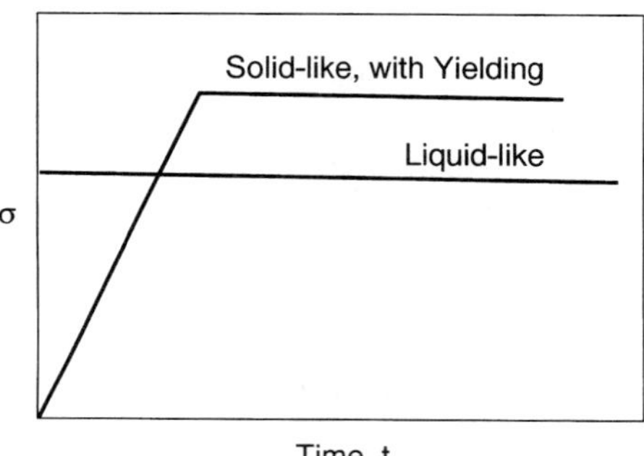

Figure 1.6   Illustration of transient shear stress $\sigma(t)$ for "liquid-like" and "solid-like" materials.

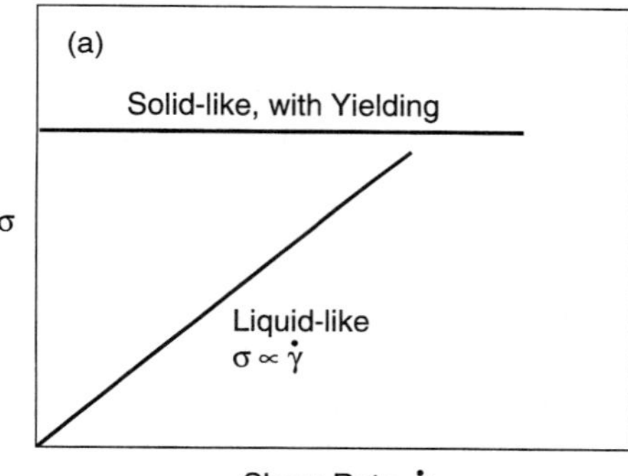

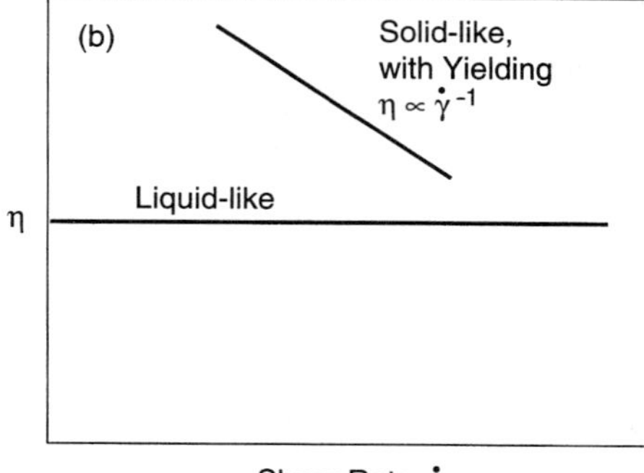

Figure 1.7   Illustrations of shear-rate-dependent (a) shear stress $\sigma(\dot{\gamma})$ and (b) shear viscosity $\eta(\dot{\gamma}) \equiv \sigma/\dot{\gamma}$ for prototypical "liquid-like" and "solid-like" materials.

Figure 1-7 shows schematic curves of the steady-state shear stress and shear viscosity versus shear rate for "solid-like" and "liquid-like" complex fluids. For a "solid-like" complex fluid, the steady-state shear stress is independent of shear rate (Fig. 1-7a), and so the shear viscosity *decreases* with increasing shear rate as $\eta(\dot{\gamma}) \propto \dot{\gamma}^{-1}$. A decreasing shear viscosity with increasing shear rate is referred to as *shear thinning*. For the "liquid-like" complex fluid, there is no shear thinning; the viscosity is a constant (Fig. 1-7b). This means that the shear stress increases linearly with shear rate, $\sigma(\dot{\gamma}) \propto \dot{\gamma}$.

The storage and loss moduli G' and G'' for our prototypical "liquid-like" and "solid-like" fluids are shown in Fig. 1-8. For the "liquid-like" fluid, the storage modulus is much lower than the loss modulus, and it scales with frequency as $G' \propto \omega^2$, the loss modulus is linear in frequency, $G'' \propto \omega$. The low-frequency "liquid-like" region in which G' and G'' obey these power laws is called the *terminal region*. For the "solid-like" fluid, $G' \gg G''$, and G' is nearly frequency-independent.

Real complex fluids often show behavior intermediate to the "solid-like" and "liquid-like" prototypes. Figure 1-9 shows the steady shear viscosity versus shear rate, for a polyethylene polymer melt (Laun 1978). At low shear rate, $\eta(\dot{\gamma})$ is nearly constant like the "liquid-like" fluid, but at high $\dot{\gamma}$ the viscosity drops off rapidly with $\dot{\gamma}$, more like that of the "solid-like" fluid. However, the drop in η is not as steep a power law as that of the prototypical "solid-like" fluid, in which $\eta \propto \dot{\gamma}^{-1}$. The low-shear-rate plateau viscosity is referred to as the *zero-shear viscosity* η_0. (Figure 1-9 also shows the shear-rate dependence of the first normal stress coefficient Ψ_1, discussed below.) Figure 1-10 shows the behavior of the transient shear stress $\sigma(\dot{\gamma}, t)$ for this same polymer melt. The "solid-like" growth in σ with time at short times is clearly evident, as is the plateau value at long times. Figure 1-11 shows $G'(\omega)$ and $G''(\omega)$, again for the same polyethylene melt. (The frequency in Fig. 1-11 is multiplied by a "shift factor" a_T which at the temperature 150°C is unity and can be ignored). At low frequencies the melt is "liquid-like," with $G' \ll G''$, while more nearly "solid-like" behavior, with $G' > G''$, is found at high frequencies.

Note that at the temperature 150°C, the transition from "liquid-like" to "solid-like" behavior occurs at frequency $\omega_c \sim 1 \ \mathrm{sec}^{-1}$, as indicated by the crossover of G' and G'' roughly at this frequency. For this melt, a corresponding crossover in the steady-state shear-viscosity curve from a "liquid-like" plateau in η to a "solid-like" shear thinning region occurs at a shear rate $\dot{\gamma}_c$ roughly in the vicinity of 1 sec^{-1} (see Fig. 1-9). Thus, the crossover shear rate $\dot{\gamma}_c$ in steady shearing is about equal numerically to the crossover frequency ω_c

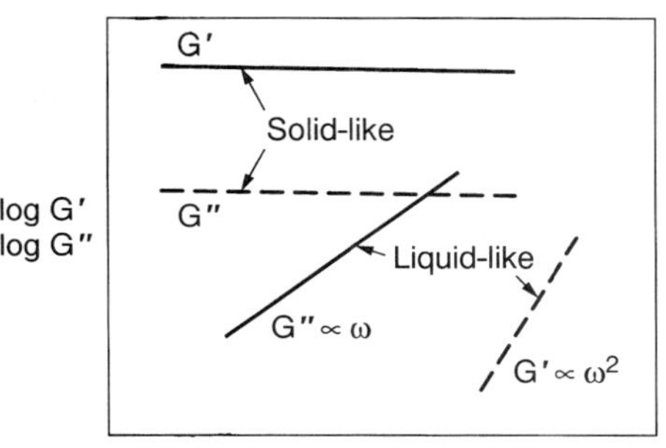

Figure 1.8   Illustrations of frequency-dependent storage and loss moduli G' and G'' for prototypical "liquid-like" and "solid-like" materials.

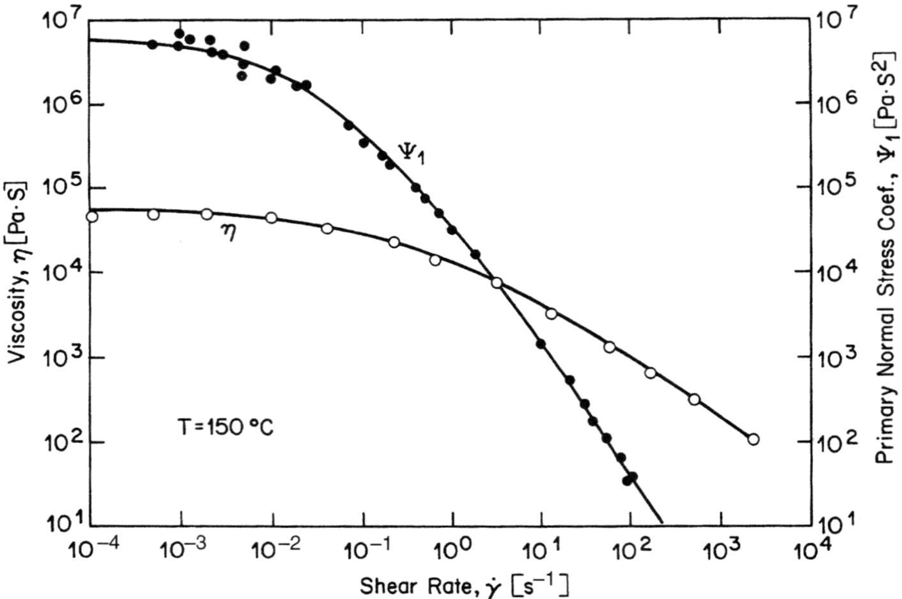

Figure 1.9   Steady-state shear viscosity η and first normal stress coefficient Ψ_1 versus shear rate for a low-density polyethylene melt, "Melt I." (From Laun 1978, reprinted with permission from Steinkopff Publishers.)

in oscillatory shearing. This is no coincidence; both $\dot{\gamma}_c$ and ω_c are approximately equal to the inverse of the fluid's *characteristic relaxation time* τ, which is roughly the longest time required for the elastic structures in the fluid to relax. Thus, $\tau \sim \dot{\gamma}_c^{-1} \sim \omega_c^{-1}$. One also can define a *characteristic modulus G* as the modulus $G' = G''$ at the crossover.

A handy, though very crude, rule of thumb is that the order of magnitude of the zero shear viscosity η_0 is given by the product of the characteristic relaxation time and the characteristic modulus; that is,

$$\eta_0 \sim G\tau \tag{1-4}$$

This rule of thumb goes back to Maxwell (1867), who said that a viscous fluid with viscosity η_0 can be thought of as a relaxing "solid" with modulus G that relaxes in a time period τ; hence, $\eta_0 \sim G\tau$. Another handy rule is that the characteristic modulus of a liquid is roughly equal to $\nu k_B T$, where ν is the number of "structural units" per unit volume. For a suspension of spheres, ν is the number of spheres per unit volume, while for a small-molecule liquid, ν is the number of molecules per unit volume; thus, $\nu = \rho N_A/M$, where ρ is the fluid density, M is the molecule's molecular weight, and N_A is Avogadro's number. Hence, for a small-molecule liquid with density $\rho = 1\text{g/cm}^3$, $M = 100$ g/mol, and $T = 300$ K, we estimate $G \approx 2.4 \times 10^7 \text{Pa} = 24$ MPa.

For many polymer melts and solutions, the shear viscosity as a function of shear rate, $\eta(\dot{\gamma})$, is almost identical to the complex viscosity $\eta^*(\omega)$ as a function of frequency,

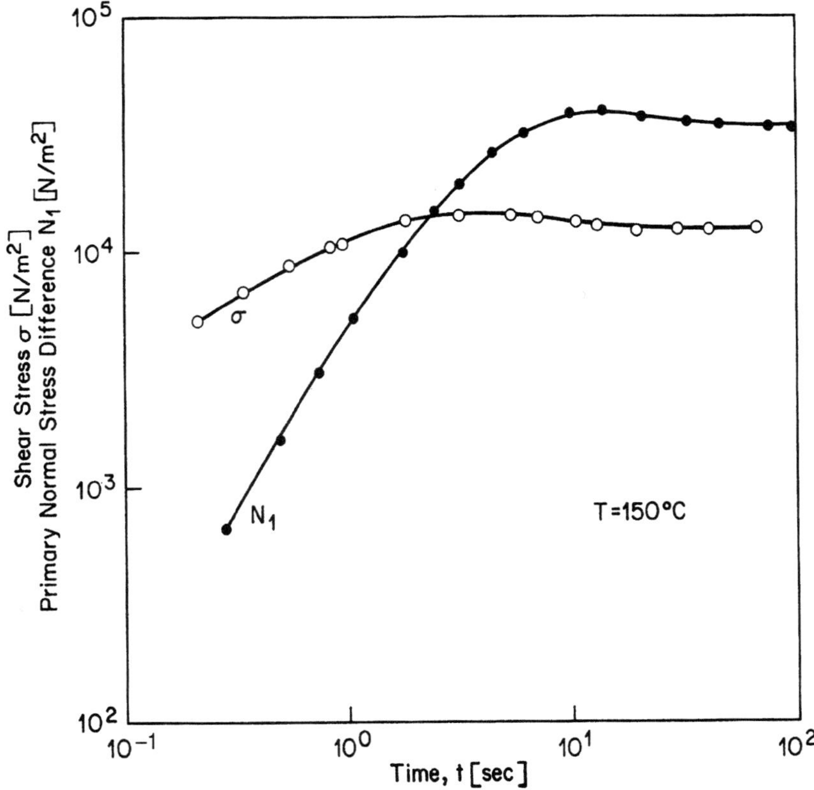

Figure 1.10   Transient shear stress σ and first normal stress difference N_1 after start-up of steady shearing for a low-density polyethylene melt, "Melt I," at a shear rate $\dot{\gamma} = 1$ sec^{-1}. (From Laun 1978, reprinted with permission from Steinkopff Publishers.)

an empirical finding known as the "Cox–Merz rule" (Cox and Merz 1958). This rule is often used in industry to estimate shear-viscosity curves from linear viscoelastic data. The conditions under which the "Cox–Merz rule" holds, and some reasons for its validity *for simple polymeric fluids*, are discussed in Dealy and Wissbrun (1990). The Cox–Merz rule is usually not reliable for more complex structured fluids, such as liquid crystalline polymers, concentrated colloidal dispersions, or gels, as discussed, for example, in Section 5.4.1.

Other complex fluids show behavior more complicated than that depicted in Figs. 1-9 to 1-11. For example, the shear viscosity for some fluids *increases* with increasing shear rate; this is called *shear thickening*. Some fluids show a region of shear thinning, followed at higher shear rates by shear thickening, followed at still higher shear rates by shear thinning again! Such behavior is encountered in the flow of dense suspensions, described in Chapter 6.

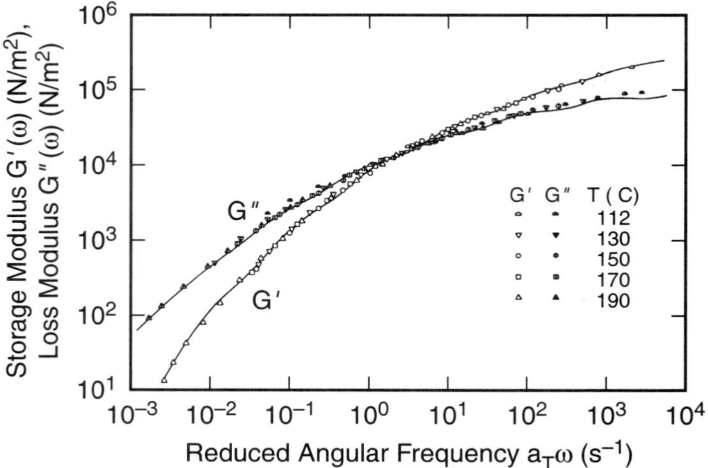

Figure 1.11 Storage and loss moduli for a low density polyethylene "Melt I." (These data were measured at several temperatures and shifted along the frequency axis by a "shift factor" a_T to form collapsed curves; see Section 3.5.2). The lines are empirical fits of Eqs. (3-25a) and (3-25b) to the data. (From Laun 1978, reprinted with permission from Steinkopff Publishers.)

1.3.1.6 Deborah Number, Weissenberg Number, and Peclet Number

A dimensionless quantity called the *Deborah number*, De, is defined as the fluid's "characteristic" relaxation time τ divided by a time constant t_f characterizing the flow (Reiner 1964). Thus, $\mathrm{De} \equiv \tau/t_f$. In an oscillatory shearing flow, for example, we might take t_f to be the inverse of the oscillation frequency ω, and then $\mathrm{De} = \tau\omega$. At high Deborah number, the flow is fast compared to the fluid's ability to relax, and the fluid will respond like a solid, to some extent. Thus, in an oscillatory shearing flow, when $\mathrm{De} \equiv \omega\tau \gg 1$ the complex modulus is "solid-like," while when $\mathrm{De} \equiv \omega\tau \ll 1$ a "liquid-like" terminal behavior is expected (see Fig. 1-11).

Other dimensionless groups similar to the Deborah number are sometimes used for special cases. For example, in a steady shearing flow of a polymeric fluid at a shear rate $\dot{\gamma}$, the *Weissenberg number* is defined as $\mathrm{Wi} \equiv \dot{\gamma}\tau$. This group takes its name from the discoverer of some unusual effects produced by *normal stress differences* that exist in polymeric fluids when $\mathrm{Wi} \gtrsim 1$, as discussed in Section 1.4.3. Use of the term "Weissenberg number" is usually restricted to steady flows, especially shear flows. For suspensions, the *Peclet number* is defined as the shear rate times a characteristic diffusion time t_D [see Eq. (6-12) and Section 6.2.2].

1.3.2 Extensional Flow

For a Newtonian liquid, the viscosity measured in a shearing flow can be used to predict the stress in other types of deformation. This, in general, is not so for complex fluids. In a steady *extensional flow*, for example, the rheological behavior of a complex fluid, especially one in which there are long polymer molecules, is often very different from that in shear. An extensional flow is one in which fluid elements are stretched or extended without being rotated or sheared. A *uniaxial extensional flow* can be generated by a device such as that depicted in Fig. 1-12. In this device, the ends of a cylinder of fluid floating on an oil bath or an air table are gripped by clamps that are fixed in place and rotate at a constant rate, thereby imposing a velocity V on one end of the cylinder between the clamps and $-V$ on the other (Meissner 1971). In this way, a *velocity gradient* $2V/L_0$ is imposed on each material element in the cylinder, where L_0 is the initial length of the cylinder. This velocity gradient is called the *extensional strain rate* $\dot{\varepsilon} \equiv 2V/L_0$. A velocity gradient is just the difference in velocity between one point and a second neighboring point, divided by the distance between those points.

The same velocity gradient can be produced by gripping the two ends of the cylinder and pulling these ends apart in such a way that the length of the cylinder grows exponentially in time, $L(t) = L_0 \exp(\dot{\varepsilon}t)$. In this way, both the velocity difference between the two ends, $\dot{L}(t) \equiv dL/dt = L_0\dot{\varepsilon} \exp(\dot{\varepsilon}t)$, and the distance of separation of these ends, $L(t) = L_0 \exp(\dot{\varepsilon}t)$, grow exponentially in time, and the ratio of the two, $\dot{L}(t)/L(t)$, which is the velocity gradient, is a constant, $\dot{\varepsilon}$. For polymer melts, it is difficult to grip the ends of the cylinder without creating flaws that can lead to sample rupture during the experiment. However, for sticky polymer solutions, this problem can be solved by merely bringing a flat plate into contact with the top of the cylinder and bringing another plate into contact with the bottom (see Fig. 1-13). When one disk is pulled away from the other at an exponentially increasing rate, the ends of the cylinder adhere to the plates, and a filament is created in which the velocity gradient is roughly constant, except near the regions of attachment to the plates (Tirtaatmadja and Sridhar 1993).

The instantaneous extensional stress $\overline{\sigma}(\dot{\varepsilon}, t)$ is the force $F(\dot{\varepsilon}, t)$ along the cylinder axis required to pull the cylinder ends apart, divided by the instantaneous cross-sectional area $A(\dot{\varepsilon}, t)$ of the cylinder; thus $\overline{\sigma}(\dot{\varepsilon}, t) \equiv F(\dot{\varepsilon}, t)/A(\dot{\varepsilon}, t)$. The time-dependent extensional viscosity, $\overline{\eta}(\dot{\varepsilon}, t)$, is then $\overline{\sigma}(\dot{\varepsilon}, t)/\dot{\varepsilon}$. If this viscosity reaches a time-independent value within the duration of the experiment, that value is called the *steady-state* extensional viscosity, $\overline{\eta}(\dot{\varepsilon})$.

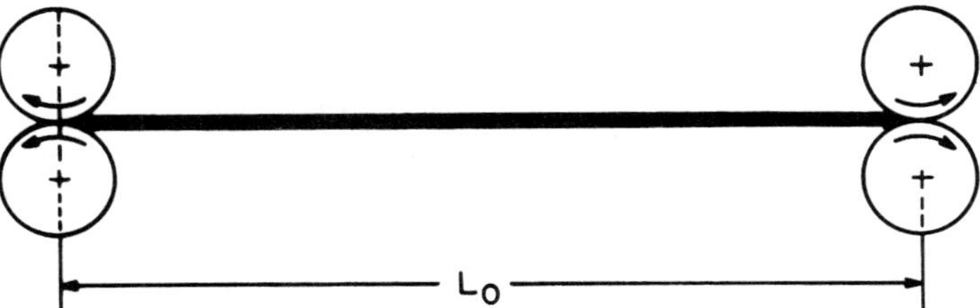

Figure 1.12   Rotating clamp device used by Meissner to impose a uniaxial extensional strain on a cylindrical filament of polymer of length L_0. Leaf springs in one of the sets of rotating clamps allow the extensional stress to be measured. (From Meissner 1971, reprinted with permission from Steinkopff Publishers.)

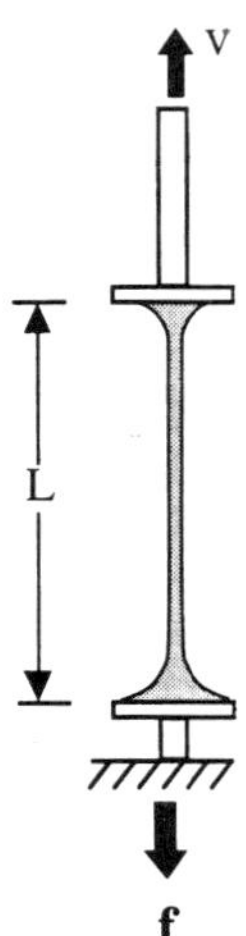

Figure 1.13   The stretching of a filament of viscoelastic liquid (shaded) sticking to two flat plates, one moving and the other attached to a force transducer. (From Macosko, *Rheology Principles, Measurements, and Applications*, Copyright © 1994. Reprinted by permission from John Wiley & Sons.)

Figure 1-14 compares the time-dependent extensional viscosity to the time-dependent shear viscosity, after onset of flow, at various shear and extension rates, for the same molten polyethylene described in Figs. 1-9 to 9-11 (Meissner 1972). This figure shows that the behavior of the extensional viscosity can be very different from that of the shear viscosity; the former *increases* while the latter *decreases* with increasing strain rate at a fixed time after inception of steady flow. Thus, while the shear viscosity is shear thinning, the extensional viscosity is *extension thickening*.

Many devices other than those mentioned above have been designed for measuring extensional viscosities. These include *crossed-slot* devices, *opposed jets*, and others described in Keller and Odell (1985) and Macosko (1994).

1.3.3 Mixed Flow

One might want to explore the behavior of a fluid in other flow fields, such as flows that are in some sense intermediate between shear and extension. One instrument for doing so is the *four-roll mill*, depicted in Fig. 1-15 (Fuller and Leal 1980, 1981). In the four-roll mill, a velocity field is generated by the rotation of four rollers in a container of liquid. By varying the rotation rate of one pair of rollers relative to that of a second pair, velocity fields ranging from planar extension to nearly simple shear can be produced near the stagnation point centered among the rollers. The four-roll mill is especially suitable for the study of fluids of low or modest viscosity, from which bubbles introduced during filling are easily removed. Because the desired flow is achieved only in one part of the device, the four-roll mill does not lend itself to direct stress measurements. However, for polymeric materials and some surfactant solutions, the stress in a small region of fluid around the stagnation point can be inferred from optical birefringence measurements, by using the *stress-optic rule* discussed in Section 1.6.3.

For a complex fluid, the stress depends not only on whether the flow is a shearing, extensional, or mixed type, but also on the whole history of the velocity gradient. Thus, the

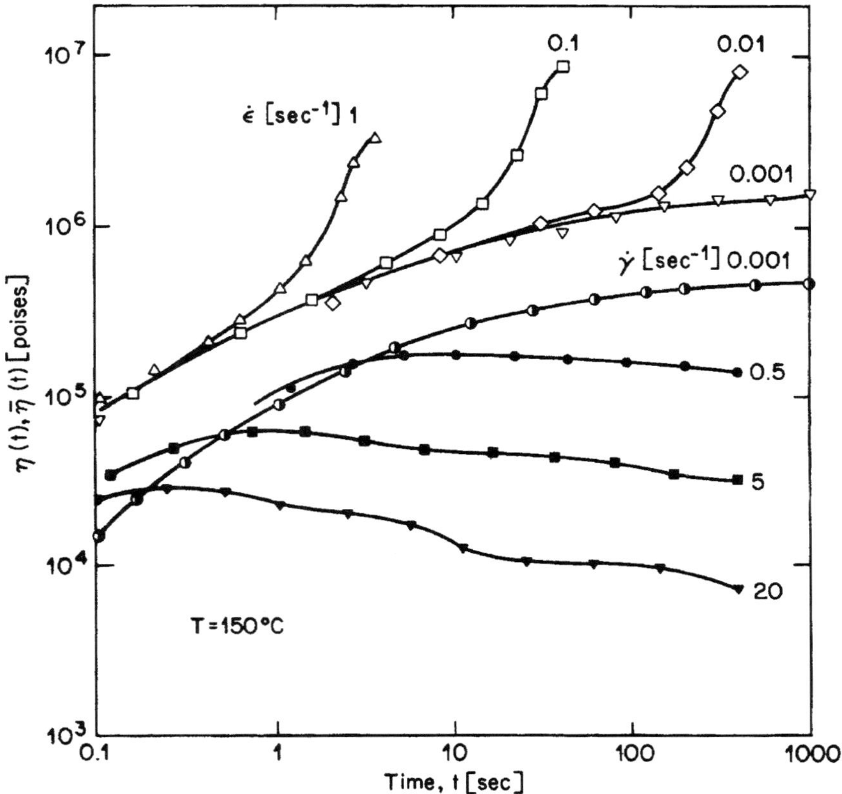

Figure 1.14 Uniaxial extensional viscosity $\bar{\eta}$ (open symbols) and shear viscosity η (closed and half-closed symbols) as functions of time after start-up of steady uniaxial extension or steady shearing for "Melt I." (From Meissner, J. Appl. Polym. Sci. 16:2877, Copyright © 1972. Reprinted by permission of John Wiley & Sons, Inc.)

characterization of the rheological properties of a complex fluid is a seemingly hopeless task, unless some theoretical guidance is provided, so that measurements in one flow history can be applied to those in another. In general, the goal is to develop a stress *constitutive equation*; this is a mathematical relationship allowing calculation of the stress tensor from a specified flow history. Examples of constitutive equations for various complex fluids are presented in Chapters 3 through 13.

1.4 KINEMATICS AND STRESS

In order to describe more general flow gradients, such as those generated near the stagnation point of a four-roll mill, some mathematics must be introduced; in particular, *tensors* (which can be represented as matrices) are needed, namely the *velocity gradient tensor*,

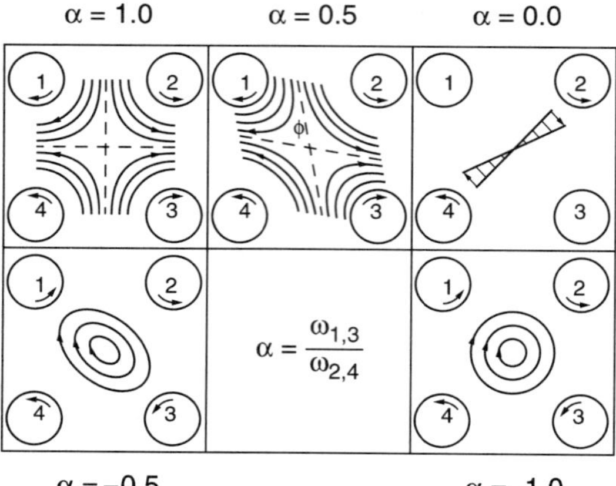

Figure 1.15   Flow fields produced by the four-roll mill. $\omega_{1,3}$ is the rotational velocity of rollers 1 and 3, and $\omega_{2,4}$ is that of rollers 2 and 4; α is the flow-type parameter [see Eq. (1-11)]. (From Fuller and Leal 1980, reprinted with permission from Steinkopff Publishers.)

the *deformation gradient tensor*, and the *stress tensor*. The reader should be warned that the sign and index conventions used in the definitions of these tensors are not universal (Bird et al. 1987a; Tanner 1985; Macosko 1994). The conventions used here are consistent with those in Larson (1988).

1.4.1 The Velocity Gradient Tensor

We define the vector $\mathbf{x} \equiv (x_1, x_2, x_3)$ to be a point in three-dimensional space with Cartesian coordinates x_1, x_2, and x_3. We also define $\mathbf{v}(\mathbf{x}) \equiv (v_1(\mathbf{x}), v_2(\mathbf{x}), v_3(\mathbf{x}))$ to be the velocity vector at point $\mathbf{x}$, where v_1 is the component of the velocity in the direction parallel to direction "1," and analogously for v_2 and v_3. Then the velocity gradient tensor $\nabla \mathbf{v}$ is given by the array

$$\nabla \mathbf{v} \equiv \begin{pmatrix} \dfrac{\partial v_1}{\partial x_1} & \dfrac{\partial v_2}{\partial x_1} & \dfrac{\partial v_3}{\partial x_1} \\[2ex] \dfrac{\partial v_1}{\partial x_2} & \dfrac{\partial v_2}{\partial x_2} & \dfrac{\partial v_3}{\partial x_2} \\[2ex] \dfrac{\partial v_1}{\partial x_3} & \dfrac{\partial v_2}{\partial x_3} & \dfrac{\partial v_3}{\partial x_3} \end{pmatrix} \tag{1-5}$$

The velocity gradient describes the steepness of velocity variation as one moves from point to point in any direction in the flow at a given instant in time. Thus if the velocity is $\mathbf{v}_0$ at some point in the flow, which we take to be the origin, and the velocity gradient there is $\nabla \mathbf{v}$, then at a nearby point, $\mathbf{x}$, the velocity is $\mathbf{v}_0 + \mathbf{x} \cdot \nabla \mathbf{v}$, where the dot represents the vector product. The transpose of $\nabla \mathbf{v}$ is frequently used in the literature; we therefore define this transpose by $\mathbf{K}$:

$$\mathbf{K} \equiv (\nabla \mathbf{v})^{\mathrm{T}} \tag{1-5a}$$

The *transpose* of a matrix is obtained by exchanging each row i with column i.

1.4.1.1 Shearing Flow

In a simple plane Couette shearing flow, as depicted at the top of Fig. 1-5, there is only one nonzero velocity component, namely v_1, and it varies only in direction 2, the direction orthogonal to the plates. Hence $\nabla \mathbf{v}$ is

$$\nabla \mathbf{v} = \begin{pmatrix} 0 & 0 & 0 \\ \dot{\gamma} & 0 & 0 \\ 0 & 0 & 0 \end{pmatrix} \tag{1-6}$$

where $\dot{\gamma} \equiv \partial v_1/\partial x_2$. The plane containing directions 1 and 2 will hereafter be called the "deformation plane" or "plane of deformation." (In the liquid crystal literature, this plane is often called "the shearing plane," but outside the liquid-crystal literature, "shearing plane" sometimes refers to the 1-3 plane.) For shearing flows generated by the other shearing geometries in Fig. 1-5, there is for each fluid element a frame that is possibly rotating and translating in which the velocity gradient tensor reduces to Eq. (1-6). Thus each geometry depicted in Fig. 1-5 in principle produces a *viscometric*, or simple shearing, flow.

1.4.1.2 Extensional Flow

The most general extensional flow is defined as a flow with a velocity gradient of the form

$$\nabla \mathbf{v} = \begin{pmatrix} \dot{\varepsilon}_1 & 0 & 0 \\ 0 & \dot{\varepsilon}_2 & 0 \\ 0 & 0 & \dot{\varepsilon}_3 \end{pmatrix} \tag{1-7}$$

where $\dot{\varepsilon}_i$ is the velocity gradient, $\partial v_i/\partial x_i$, in direction i.

The condition of incompressibility implies that

$$\mathrm{tr}(\nabla \mathbf{v}) = 0 \tag{1-8}$$

where the symbol "tr" means "trace"; it is just the sum of the diagonal elements of a tensor; that is,

$$\mathrm{tr}(\nabla \mathbf{v}) = \frac{\partial v_1}{\partial x_1} + \frac{\partial v_2}{\partial x_2} + \frac{\partial v_3}{\partial x_3} \tag{1-8a}$$

Thus, Eq. (1-8a) implies that for an incompressible *uniaxial extension*, in which a cylinder is extended axisymmetrically in direction 1 with a velocity gradient $\dot{\varepsilon}_1 \equiv \partial v_1/\partial x_1$, the velocity gradients in each of the other two directions are $\partial v_2/\partial x_2 = \partial v_3/\partial x_3 = -\dot{\varepsilon}_1/2$. The velocity gradient tensor for uniaxial extension with extension rate $\dot{\varepsilon}$ is therefore

$$\nabla \mathbf{v} = \begin{pmatrix} \dot{\varepsilon} & 0 & 0 \\ 0 & -\dot{\varepsilon}/2 & 0 \\ 0 & 0 & -\dot{\varepsilon}/2 \end{pmatrix} \tag{1-9}$$

For an incompressible *planar extensional flow*, $\dot{\varepsilon}_2 = -\dot{\varepsilon}_1$ and $\dot{\varepsilon}_3 = 0$; thus

$$\nabla \mathbf{v} = \begin{pmatrix} \dot{\varepsilon} & 0 \\ 0 & -\dot{\varepsilon} \end{pmatrix} \tag{1-10}$$

In Eq. (1-10), the third row and third column contain only zeros and have therefore been omitted. For an *equibiaxial extensional flow*, $\dot{\varepsilon}_1 = \dot{\varepsilon}_2$ and $\dot{\varepsilon}_3 = -2\dot{\varepsilon}_1$.

1.4.1.3 Mixed Flow

It is of interest to study flows of mixed or intermediate character that combine the characteristics of shearing and extensional flows. In two dimensions, the most general velocity gradient can be represented as

$$\nabla \mathbf{v} = \frac{1}{2} G \begin{pmatrix} 1+\alpha & 1-\alpha \\ -(1-\alpha) & -(1+\alpha) \end{pmatrix} \tag{1-11}$$

where α is a "flow-type" parameter, which is zero for simple shear and unity for planar extension; G is the shear rate $\dot{\gamma}$ if the flow is a shearing flow ($\alpha = 0$), and G is the extension rate $\dot{\varepsilon}$ if the flow is a planar extension. As mentioned in Section 1.3.3, a general 2-D mixed flow can be generated in a *four-roll mill* (see Fig. 1-15).

1.4.2 The Deformation Gradient and Finger Tensors

1.4.2.1 The Deformation Gradient Tensor

As discussed in Section 1.3, the stress in a complex fluid such as a polymer melt depends not only on the instantaneous velocity gradient, but also on the time period over which that velocity gradient has been imposed. Thus, the stress in a fluid element depends on the flow history, or *deformation history*, of that element. This deformation history is conveniently specified using the tensor $\mathbf{E}$, which is defined analogously to the velocity gradient tensor. Consider a point in the fluid that at some past time t' occupied position $\mathbf{x}'$, but that now resides at position $\mathbf{x}$. Then let $\delta\mathbf{x}'$ be a vector embedded in a small fluid element, with one end of the vector (the tail) at position $\mathbf{x}'$ and the other end (the vector's head) at position $\mathbf{x}' + \delta\mathbf{x}'$. Between times t' and t, the vector is stretched and rotated along with the fluid element in which it is embedded, and at time t the embedded vector is $\delta\mathbf{x}$. For example, if the fluid element is stretched in direction 1, the ratio $\lambda(t', t) \equiv \delta x_1/\delta x_1'$ is the *stretch ratio* in that direction; that is, it is the ratio of the length of the fluid element at time t to its length at time t'. For a general three-dimensional deformation, the tensor $\mathbf{E}$ is

$$\mathbf{E}(t', t) \equiv \frac{\partial \mathbf{x}}{\partial \mathbf{x}'} = \begin{pmatrix} \dfrac{\partial x_1}{\partial x_1'} & \dfrac{\partial x_2}{\partial x_1'} & \dfrac{\partial x_3}{\partial x_1'} \\[2ex] \dfrac{\partial x_1}{\partial x_2'} & \dfrac{\partial x_2}{\partial x_2'} & \dfrac{\partial x_3}{\partial x_2'} \\[2ex] \dfrac{\partial x_1}{\partial x_3'} & \dfrac{\partial x_2}{\partial x_3'} & \dfrac{\partial x_3}{\partial x_3'} \end{pmatrix} \tag{1-12}$$

The tensor $\mathbf{E}$ contains information not only on the stretching of the fluid element in each of its three dimensions, but also on the rotation of the fluid element. The inverse of this tensor, $\mathbf{F} \equiv \mathbf{E}^{-1} = \partial\mathbf{x}'/\partial\mathbf{x}$, is usually called the *deformation gradient tensor*. However, we will find it more convenient to work with $\mathbf{E}$. [Note that Eq. (1-12) assumes the index convention $(\mathbf{E})_{ij} = \partial x_j/\partial x_i'$. Elsewhere in the rheological literature, the convention used is often the transpose of this.] By definition, then, the embedded vector $\delta\mathbf{x}$ at time t is related to $\delta\mathbf{x}'$, the embedded vector at t', by

$$\delta\mathbf{x} = \delta\mathbf{x}' \cdot \mathbf{E} \tag{1-13}$$

where $\delta\mathbf{x}$ and $\delta\mathbf{x}'$ are understood to be row vectors for the purposes of vector multiplication. The column vector corresponding to $\delta\mathbf{x}$ is designated $(\delta\mathbf{x})^T$.

Figure 1-16 shows how the components of $\mathbf{E}$ are constructed for simple shear. Take an embedded vector oriented in direction 1, the flow direction, and of unit length at time t'; that is, $\mathbf{x}' = (1, 0, 0)$. Then E_{11}, E_{12}, and E_{13} are the components of this embedded vector after the deformation; $\mathbf{x} = (E_{11}, E_{12}, E_{13})$. Since shearing does not stretch or rotate lines that lie in the flow direction, $(x_1, x_2, x_3) = (E_{11}, E_{12}, E_{13}) = (1, 0, 0)$. Likewise E_{21}, E_{22}, and E_{23} are the coordinates of an embedded vector that at time t' is a unit vector oriented in direction 2. This vector is rotated and stretched by the shearing so that $(E_{21}, E_{22}, E_{23}) = (\gamma, 1, 0)$. Finally, $(E_{31}, E_{32}, E_{33}) = (0, 0, 1)$. Likewise, one can derive the tensor $\mathbf{E}$ for a general extensional deformation, as shown in Fig. 1-17.

For volume-preserving deformations,

$$\det(\mathbf{E}) = 1 \tag{1-14}$$

That is, the determinant of $\mathbf{E}$ is unity. For a uniaxial extensional deformation, incompressibility therefore implies that $\lambda_2 = \lambda_3 = \lambda_1^{-1/2}$, where λ_1 is the stretch ratio in the direction of elongation.

Using the chain rule of calculus, the tensor $\mathbf{E}$ can be related to the velocity gradient tensor $\nabla\mathbf{v}$, as follows:

$$\frac{\partial}{\partial t}\mathbf{E} = \frac{\partial\dot{\mathbf{x}}}{\partial\mathbf{x}'} = \frac{\partial\mathbf{x}}{\partial\mathbf{x}'} \cdot \frac{\partial\dot{\mathbf{x}}}{\partial\mathbf{x}} = \mathbf{E} \cdot \frac{\partial\mathbf{v}}{\partial\mathbf{x}} = \mathbf{E} \cdot \nabla\mathbf{v} \tag{1-15}$$

1.4.2.2 The Finger Tensor

As mentioned above, the tensor $\mathbf{E}$ contains information about both the stretching and the rotation of a material element. Yet, if a material element is rotated only and not stretched, no

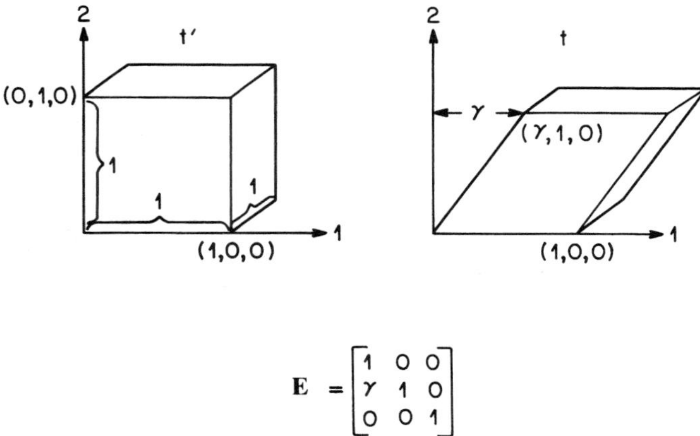

Figure 1.16 Definition of the deformation tensor $\mathbf{E}$ for the shearing deformation of a unit cube. (From Larson 1988, with permission.)

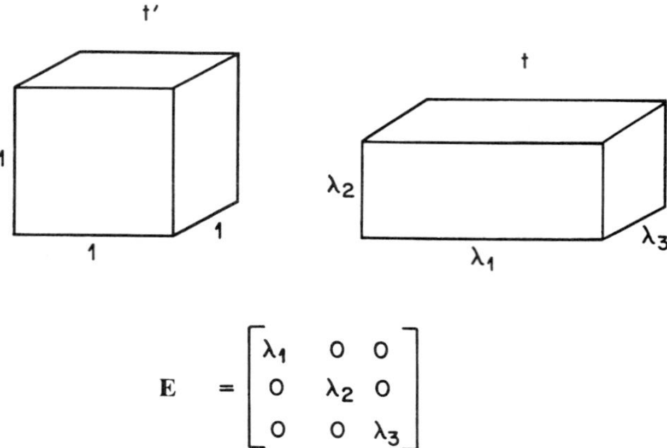

$$E \;=\; \begin{bmatrix} \lambda_1 & 0 & 0 \\ 0 & \lambda_2 & 0 \\ 0 & 0 & \lambda_3 \end{bmatrix}$$

Figure 1.17 Definition of the deformation tensor **E** for a general extensional deformation of a unit cube. (From Larson 1988, with permission.)

stress will be thereby produced. To remove from **E** information about solid body rotations that produce no stress, **E** is multiplied by its transpose to produce a new tensor, the "Finger tensor" **B**:

$$\mathbf{B} \equiv \mathbf{E}^T \cdot \mathbf{E} \tag{1-16}$$

By this definition, **B** is a *symmetric* tensor, whereas **E** can be asymmetric. The Finger tensor naturally arises when one considers the *length* of a deformed material line. From Eq. (1-13), the square of the length of an embedded vector $\delta\mathbf{x}$ following a deformation is

$$(\delta x)^2 = \delta\mathbf{x} \cdot \delta\mathbf{x} = \delta\mathbf{x}' \cdot \mathbf{E} \cdot \mathbf{E}^T \cdot \delta\mathbf{x}'^T \tag{1-17}$$

The tensor $\mathbf{E} \cdot \mathbf{E}^T$, called the Piola tensor (Astarita and Marrucci 1974), is closely related to **B**. In an extensional deformation, $\mathbf{E} \cdot \mathbf{E}^T$ is exactly equal to **B**. **B**, a symmetric tensor, contains information about the *orientation* of the three principal axes of stretch and about the *magnitudes* of the three principal stretch ratios, but *no information about rotations* of material lines that occurred during that deformation. Thus, for example, from the Finger tensor alone, one could not determine whether a deformation was a simple shear (which has rotation of material lines) or a planar extensional deformation (which does not). The Finger tensor $\mathbf{B}(t', t)$ *describes the change in shape of a small material element* between times t' and t, not whether it was rotated during this time interval.

For simple shear, **E** is given in Fig. 1-16; hence from Eq. (1-16), the Finger tensor for simple shear is

$$\mathbf{B} = \begin{pmatrix} 1+\gamma^2 & \gamma & 0 \\ \gamma & 1 & 0 \\ 0 & 0 & 1 \end{pmatrix} \tag{1-18}$$

where $\gamma = \gamma(t', t)$ is the shear strain accumulated between times t' and t. In steady shearing, $\gamma(t', t) = \dot{\gamma}(t - t')$, where $\dot{\gamma}$ is the shear rate.

For an extensional deformation for which $\mathbf{E}$ is given in Fig. 1-17,

$$\mathbf{B} = \begin{pmatrix} \lambda_1^2 & 0 & 0 \\ 0 & \lambda_2^2 & 0 \\ 0 & 0 & \lambda_3^2 \end{pmatrix} \tag{1-19}$$

where each λ_i is a function of t' and t: $\lambda_i(t', t)$ In a steady extensional flow, $\lambda_i(t', t) = \exp(\dot{\varepsilon}_i(t - t'))$.

From the definition of $\mathbf{B}$ given in Eq. (1-16), along with the relationship of Eq. (1-15), we find that

$$\dot{\mathbf{B}} \equiv \frac{\partial}{\partial t}\mathbf{B} = \nabla\mathbf{v}^T \cdot \mathbf{B} + \mathbf{B} \cdot \nabla\mathbf{v} \tag{1-20}$$

This identity is useful for relating integral and differential constitutive equations, as we shall see in Section 3.4.4. A thorough discussion of this and other relationships among kinematic tensors can be found in Astarita and Marrucci (1974).

1.4.3 The Stress Tensor

In Sections 1.3.1 and 1.3.2, we discussed the shear stress and the extensional stress in shearing flows and extensional flows, respectively. These are components of the three-dimensional *state-of-stress tensor* $\mathbf{T}$. The *i*th row of $\mathbf{T}$ is the force per unit area that material exterior to a unit cube exerts on a surface perpendicular to the *i*th coordinate axis (see Fig. 1-18). In general, if $\mathbf{F}$ is the force per unit area acting on a surface perpendicular to an arbitrary outward-directed unit vector, $\mathbf{n}$, then

$$\mathbf{F} = \mathbf{n} \cdot \mathbf{T} \tag{1-21}$$

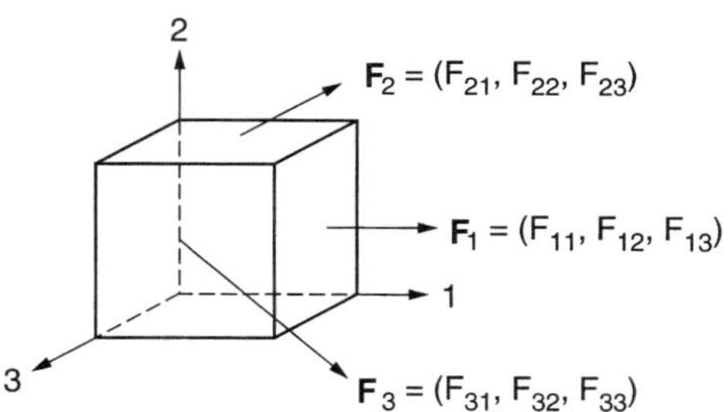

$$\text{The Stress Tensor, } \mathbf{T} = \begin{bmatrix} F_{11} & F_{12} & F_{13} \\ F_{21} & F_{22} & F_{23} \\ F_{31} & F_{32} & F_{33} \end{bmatrix}$$

Figure 1.18 The definition of the state-of-stress tensor in terms of force components acting on the faces of a cube. (From Larson 1988, with permission.)

The simplest state of stress is one in which the fluid element in the cube is under *hydrostatic pressure* only. In that case, $\mathbf{T} = -p\delta$, where δ is the unit tensor:

$$\delta = \begin{pmatrix} 1 & 0 & 0 \\ 0 & 1 & 0 \\ 0 & 0 & 1 \end{pmatrix} \tag{1-22}$$

By the convention used here, a positive pressure p is equivalent to a negative stress. Stresses that exist in addition to a hydrostatic pressure are expressed as the *extra stress tensor*, σ. Thus

$$\mathbf{T} = \sigma - p\delta \tag{1-23}$$

Tensors that are proportional to δ are sometimes called *isotropic* tensors. For an *incompressible material,* gradients of p, but not p itself, can affect fluid motion. Thus, a uniform isotropic tensor of arbitrary magnitude can be added to $\mathbf{T}$ (or σ) without consequence to the flow behavior. Adding such an isotropic tensor is equivalent to adding a constant to each *diagonal* component of the stress tensor. Thus, if the fluid is incompressible, σ *is determined only up to an additive isotropic tensor,* and the stress-free state is synonymous with the state of isotropic stress.

In all but the most unusual cases (Dahler and Scriven 1961), the principle of conservation of angular momentum for *isotropic materials* implies that σ is symmetric, that is, $\sigma_{ij} = \sigma_{ji}$. Isotropic materials are ones whose properties at rest are independent of direction; anisotropies arise only because of deformation. Isotropic materials include many highly disordered liquids, such as simple liquids, melts and solutions of ordinary flexible polymer molecules, glasses, disordered suspensions of rotund particles, and colloidal and polymeric gels. Anisotropic liquids have some degree of long-range orientational or positional order at rest and include liquid crystals, liquid-crystalline polymers, and ordered block copolymers.

In a shearing flow (Fig. 1-16) of an *incompressible isotropic liquid*, the stress tensor contains at least two nonzero components, $\sigma_{12} = \sigma_{21}$, as well as an isotropic pressure term. If the fluid is Newtonian, these are the only nonzero components of the stress tensor. However, a *non-Newtonian* liquid in general has other nonzero components of σ, namely the *normal stresses*, σ_{11}, σ_{22}, and σ_{33} (Weissenberg 1947). Since the stress tensor is only determined to within an additive isotropic tensor, only the normal stress *differences* $N_1 \equiv \sigma_{11} - \sigma_{22}$ and $N_2 \equiv \sigma_{22} - \sigma_{33}$ can be measured. Usually the first normal stress difference is *positive* in sign, while the second normal stress difference is *negative* and a factor of three or more smaller in magnitude than N_1. There are notable exceptions to this, however, as we shall see in Sections 6.4.4 and 11.3.3. In a cone-and-plate instrument, the first normal stress difference is the thrust per unit area of the plate that tends to push the plates apart if N_1 is positive but tends to pull them together if it is negative.

The time-dependent growth of N_1 after start-up of steady shearing for a polyethylene melt is shown in Fig. 1-10. Note that at steady state the first normal stress difference is *larger* than the shear stress at this particular shear rate. The normal stress differences usually are more shear-rate-dependent than the shear stress. In fact, if the isotropic liquid belongs to a fairly general class known as viscoelastic "simple" fluids with fading memory (Coleman and Noll 1961), then at low shear rates the normal stress differences depend quadratically on shear rate and the shear stress is linear in shear rate. For this reason, coefficients Ψ_1 and Ψ_2 are defined as

$$\Psi_1 \equiv \frac{N_1}{\dot{\gamma}^2}, \qquad \Psi_2 \equiv \frac{N_2}{\dot{\gamma}^2} \tag{1-24}$$

These coefficients, along with the shear viscosity $\eta \equiv \sigma_{12}/\dot{\gamma}$, often approach constant values at low shear rates; these are called the *zero-shear values*, η_0, $\Psi_{1,0}$, and $\Psi_{2,0}$. Figure 1-9 shows for a polyethylene melt that the zero-shear constant values of $\eta \rightarrow \eta_0$ and $\Psi_1 \rightarrow \Psi_{1,0}$ are approached at low shear rates. For a viscoelastic "simple liquid" with fading memory, the zero-shear values of the viscosity and first normal stress coefficient are related to the *zero-frequency* values of the dynamic moduli by

$$\eta_0 = \lim_{\omega \to 0} \frac{G''}{\omega}, \qquad \Psi_{1,0} = 2 \lim_{\omega \to 0} \frac{G'}{\omega^2} \tag{1-25}$$

At low shear rates, a viscoelastic "simple liquid" with fading memory satisfies the constitutive equation of a Newtonian liquid, that is,

$$\boldsymbol{\sigma} = 2\eta\, \mathbf{D} \tag{1-26}$$

where $\mathbf{D}$ is the *rate-of-deformation tensor* (i.e., the symmetric part of the velocity gradient):

$$2\mathbf{D} \equiv \nabla\mathbf{v} + (\nabla\mathbf{v})^T \tag{1-27}$$

For future reference, we also define the antisymmetric part of the velocity gradient, also called the *vorticity tensor*, as

$$2\boldsymbol{\omega} \equiv \nabla\mathbf{v} - (\nabla\mathbf{v})^T \tag{1-28}$$

The vorticity tensor is related to the angular velocity of the fluid element. For flows with no rotation, such as the extensional flow depicted in Figure 1-12, $\nabla\mathbf{v}$ is symmetric and the vorticity tensor is zero.

Analyzing the fluid flow behavior of complex fluids requires drastically different methods than those used for ordinary liquids like water or air. For Newtonian fluids, one simply replaces the stress tensor "$\boldsymbol{\sigma}$" with "$2\eta\mathbf{D}$" everywhere "$\boldsymbol{\sigma}$" appears in the momentum-balance equations. This yields the Navier–Stokes equations that contain the velocity field and pressure, but not the stress, as dependent variables. Then, after the geometry is defined, the Navier–Stokes equations are solved, along with the continuity and possibly other transport equations, using analytic methods or the computer. For complex fluids, however, the simple Newtonian constitutive law [Eq. (1-26)] fails, except possibly at low flow rates. Thus, although for complex fluids one still retains the momentum-balance equations (given in the Appendix for the case of negligible inertia), the familiar Navier–Stokes equations are not valid. To proceed, one must replace Eq. (1-26) with some new constitutive "law"—that is, one that is mathematically tractible and yet appropriate for the complex fluid whose flow one wishes to analyze. This new "law" must then be solved along with the momentum-balance and mass balance equations. The methods required to solve such equations depend on the mathematical structure of the constitutive law needed to represent the complex fluid (Crochet et al. 1984; Pearson and Richardson 1983; Keunings 1988).

> • Worked Example 1.3 and Problem 1.4 are designed to improve your understanding of the tensors $\nabla \mathbf{v}$, $\mathbf{D}$, $\boldsymbol{\omega}$, $\mathbf{E}$, and $\mathbf{B}$.

1.5 FLOW, SLIP, AND YIELD

1.5.1 Flow

A crude, but helpful, way to understand viscous flow of simple liquids and to estimate the viscosity η_0 is with *Eyring's absolute rate theory* (Eyring 1936; Bowden 1973; Argon 1975; Crist 1993). The idea is that even in the absence of macroscopic flow, each molecule undergoes "hopping" motions because of Brownian motion; under a shear stress σ the rate of "forward" hops (those in the direction favored by the imposed stress) is slightly greater than the rate of "reverse" hops, and, over time, the molecule drifts on average in a given direction with respect to its neighbors. Taking the rate of attempted hops to be Ω_0 and also taking the activation enthalpy barrier that the molecule must overcome to hop to a new site to be ΔH, the rate of hopping in the absence of an imposed stress is $\Omega_0 \exp[-\Delta H/k_B T]$. A typical molecular collision rate gives $\Omega_0 \approx 10^{13}$ sec^{-1}. Under an applied shear stress σ, the activation barrier for forward hops is lowered to $\Delta H - v^* \sigma$, where v^* is the *activation volume*, the size of the site involved in the hopping, which for small molecules should be roughly the molecular volume. The barrier to reverse hops is raised to $\Delta H + v^* \sigma$. Thus the shear rate $\dot{\gamma}$ under the stress σ is just the rate of forward hops minus the rate of backward hops, or

$$\dot{\gamma} = \frac{1}{2}\Omega_0 \left\{ \exp\left[\frac{-\Delta H + v^*\sigma}{k_B T}\right] - \exp\left[\frac{-\Delta H - v^*\sigma}{k_B T}\right] \right\} \tag{1-29}$$

The factor 1/2 enters because half the attempted moves are forward "hops" and the other half are backward ones. If the shear stress σ is small, so that $v^*\sigma \ll \Delta H$, then one readily finds that the shear viscosity $\eta_0 \equiv \sigma/\dot{\gamma}$ is

$$\eta_0 = \frac{k_B T}{v^*} \tau_0 \exp\left(\frac{\Delta H}{k_B T}\right) \tag{1-30}$$

where $\tau_0 \equiv 1/\Omega_0 \approx 10^{-13}$ sec. This expression for η_0 can be written as a product of a *modulus* $G = k_B T v$, where $v = 1/v^*$, and a relaxation time $\tau = \tau_0 \exp(\Delta H/k_B T)$. The exponential dependence of the relaxation time on inverse temperature is called an *Arrhenius* form; this form only holds at temperatures well above any *glass transition temperature*. The behavior of liquids near their glass transition is discussed in Chapter 4.

Rheometers, such as those mentioned in Section 1.3, are designed to impose deformation onto complex fluids by moving various bounding surfaces or by imposing a pressure drop across opposing surfaces of the material, with the expectation that the fluid will deform continuously and, in the simplest cases, homogeneously. The determination of viscosities or other rheological properties from forces measured in such instruments is based on the expectation that the flow within the geometry conforms to the one intended.

Sometimes, this expectation is not met. At high flow rates, there can be *hydrodynamic instabilities* that lead to secondary flows which ruin the rheological measurement. Such instabilities occur in Newtonian fluids, due, for example, to inertial effects, such as those in Poiseuille flow at Reynolds numbers exceeding 2000 (Drazin and Reid 1981). For some complex fluids, even at low Reynolds number there are instabilities that are driven by elastic effects (Larson 1992).

In addition to these impediments to rheological measurements, some complex fluids exhibit *wall slip*, *yield*, or a *material instability*, so that the actual fluid deformation fails to comply with the intended one. A material instability is distinguished from a hydrodynamic instability in that the former can in principle be predicted from the constitutive relationship for the material alone, while prediction of a flow instability requires a mathematical analysis that involves not only the constitutive equation, but also the equations of motion (i.e., momentum and mass conservation).

1.5.2 Wall Slip

In all simple-shear rheometers, there are solid boundaries against which fluid is intended to adhere, while neighboring fluid is set in motion by drag or pressure gradients. For simple small-molecule liquids like water, wall slip is usually negligible; that is, the "no-slip" boundary condition is valid, except in ultrathin geometries or near moving contact lines. However, if a solid block is confined between two plates, one of which is set in motion, one expects one or both surfaces of the block to *slip* with respect to its bounding surface. As we have seen (Section 1.3.1.5), a complex fluid can behave as a liquid at a low shear rate but behave as a solid at a higher rate. Hence, it should not be too surprising that, when sheared at rates where solid-like behavior prevails, complex fluids can slip against solid walls. As noted in Section 1.3, when a fluid is sheared steadily at a rate $\dot{\gamma}$ corresponding to a frequency $\omega = \dot{\gamma}$ at which the the storage modulus $G'(\omega)$ is much greater than the loss modulus $G''(\omega)$, the complex fluid will be "solid-like," and one should be on guard for possible slip phenomena.

In some cases the presence of slip is fairly obvious, as are its causes. For example, when an aqueous foam is sheared between smooth surfaces, the water in the foam can easily form a lubricating layer at the wall, leaving the bulk of the foam less sheared than intended (Yoshimura and Prud'homme 1988; Khan et al. 1988). Gelled colloidal suspensions are elastic materials that contain solvents capable of lubricating rheometer tool surfaces, and slip is a problem (Buscall et al. 1993; Persello et al. 1994). In these and other cases, slip can be counteracted in a number of ways, for example by using roughened rheometer surfaces (Khan et al. 1988; Buscall et al. 1993).

Other complex fluids, such as polymer melts, contain no solvent that can serve as a lubricant, and mechanisms for slip at or near a solid surface—and even the existence of wall slip—are less obvious (Denn 1990). Suspicion that slip may be occurring is aroused by observations of jumps, or abrupt slope changes, in curves of shear stress versus shear rate, or by oscillations in stress or pressure at fixed apparent flow rate, suggesting "stick-slip"—that is, alternating periods of stick and slip (Benbow and Lamb 1963; Blyler and Hart 1970; Vinogradov et al. 1972; Kalika and Denn 1987; Lim and Schowalter 1989; Piau et al. 1990; Hatzikiriakos and Dealy 1992). But molecular theories of slip for complex fluids such as polymer melts are only just now being developed (Brochard and de Gennes 1992). These theories are based on untested assumptions and contain rarely measured quantities (such as the density of polymers adsorbed onto a wall). Hence, experiments that clearly demonstrate the presence of wall slip and allow its velocity to be measured are very much needed.

1.5.2.1 Methods of Measuring Slip

1.5.2.1.1 GAP-DEPENDENT APPARENT SHEAR RATE. Indirect evidence of slip, as well as a measurement of its magnitude, can be extracted from the flow curve (shear stress versus shear rate) measured at different rheometer gaps (Mooney 1931). If slip occurs, one expects the slip velocity $V_s(\sigma)$ to depend on the shear stress σ, but not on the gap h. Thus, if a fluid is sheared in a plane Couette device with one plate moving and one stationary, and the gap h is varied with the shear stress σ held fixed, there will be a velocity jump of magnitude $V_s(\sigma)$ at the interfaces between the fluid and each of the two plates. There will also be a velocity gradient $\dot{\gamma}(\sigma)$ in the bulk of the fluid; thus the velocity of the moving surface will be $V = 2V_s(\sigma) + \dot{\gamma}(\sigma)h$. The *apparent* shear rate V/h will therefore be

$$\dot{\gamma}_{\text{app}} \equiv V/h = \frac{2V_s}{h} + \dot{\gamma} \tag{1-31}$$

A plot of $\dot{\gamma}_{\text{app}}$ against $1/h$ will then be a straight line with slope $2V_s$. This method has been used to measure the slip velocity for polyethylene melts in a sliding plate (plane Couette) rheometer by Hatzikiriakos and Dealy (1991). Analogous methods have been applied to shearing flows of melts in capillaries and in plate-and-plate rheometers (Mooney 1931; Henson and Mackay 1995; Wang and Drda 1996).

1.5.2.1.2 EFFECT OF SURFACE TREATMENT. A gap dependence of the flow curve is indirect evidence of slip. In addition, some role of the solid boundary in slip is clearly implicated in cases where one can show that changes in the wall material or surface treatments of it (such as a teflon coating) influence the magnitude of the apparent slip velocity (Ramamurthy 1986; Hatzikiriakos and Dealy 1991; Wang and Drda 1996).

1.5.2.1.3 PARTICLE TRACERS. A direct way to visualize slip near, although not at, solid surfaces is to seed the fluid with particles small enough and in dilute enough concentrations that they do not significantly disturb the flow, and yet large enough to be viewed with an optical microscope. Since the particles need to be around a micron or larger to be seen (unless fluorescent particles are used), the fluid should be rather viscous ($\gtrsim 10\text{Pa} \cdot \text{s}$) so that particle settling is minimized. Then, by viewing the motions of particles at or near solid boundaries, one can infer the existence of slip at the wall, or within a fluid layer whose thickness is no greater than the diameter of the particles. Galt and Maxwell (1964) appear to have been the first to use such a technique to infer slip of a polymer melt in a capillary tube, but they used rather large (50 μm) particles. Archer et al. (1995a) used smaller (1.5 μm) particles to infer sip of entangled polymer solutions sheared between two glass plates. They observed slip that increased rapidly as the solution was made more elastic either by increasing the polymer concentration or the molecular weight (Archer et al. 1995a, 1995b).

1.5.2.1.4 EVANESCENT WAVES. The most sophisticated method to measure slip velocities is a technique developed by Migler et al. (1993) that uses *evanescent-wave spectroscopy*. When a beam of monochromatic light is reflected from a surface or interface, the electric field associated with the light penetrates the interface, with the field strength decaying exponentially as a function of distance into the reflecting material. The decay constant of this exponential is a fraction of the wavelength of light; thus the penetration distance is around 0.1 μm. The intensity of the reflected light depends on the optical properties of the material in this fraction-of-a-micron-thick layer. In their method, Migler et al. (1993) used two interfering laser beams to bleach a sinusoidal "grating" into a dye-containing poly(dimethylsiloxane) melt confined between glass surfaces. When the sample was sheared by translating the top plate, motion of the grating in the layer of fluid near the bottom stationary plate could be detected by analyzing the time-dependent intensity of a beam reflected from the interface between the fluid and the bottom plate. Movement of the grating implied the existence of slip within the 0.1-μm layer and allowed the slip velocity to be measured.

1.5.2.1.5 LASER-DOPPLER VELOCIMETRY. Müller-Mohnssen et al. (1990) used laser-Doppler velocimetry to measure the velocities of small (0.15-μm diameter) tracer particles in flowing polyacrylamide solutions at various radial positions across a tube, as close as 1 μm from the wall. The measured velocity profiles showed an apparent finite slip velocity at the tube wall. Using total-reflection-microscope anemometry (a form of evanescent wave spectroscopy), the velocities as close as 0.15 μm from the tube wall were measured. These high-resolution measurements showed that the apparent slip at a resolution of 1 μm was caused by a *thin wall layer of fluid* in which the velocity gradient was much higher than in the bulk. Because of this thin high-shear layer, the bulk velocity gradient, when extrapolated to the wall, appeared to reach a nonzero value. The polymer was a polyelectrolyte whose charges were apparently repelled by the wall, leading to a near-wall layer denuded of polymer and therefore having a low viscosity. This low-viscosity layer could support a higher velocity gradient than the bulk and therefore acted as a lubricant, creating apparent wall slip. The existence of a polymer-denuded layer as a cause of slip had been proposed earlier by Cohen and Metzner (1985).

1.5.2.1.6 RHEO-NMR. Velocities near a wall, and indeed throughout the fluid, can be measured without tracer particles, and in opaque samples, by using nuclear magnetic resonance velocity imaging (Abbott et al. 1991; Rofe et al. 1996). Using "rheo-NMR," slip within about 10 μm of a wall can be inferred (Callaghan et al. 1996), as can the presence of yield surfaces and shear banding within the sample (Britton and Callaghan 1997). It is especially useful for soft solids, including foods, where the assumption of uniform shear can fail drastically. In such cases, uncritical analysis of ordinary rheology data, without the aid of velocity imaging, can lead to serious misinterpretations.

1.5.2.2 THE EXTRAPOLATION LENGTH

A useful way to think about slip, and its effect on rheological measurements, is to define an *extrapolation length b*. The extrapolation length is the distance from the fluid–solid

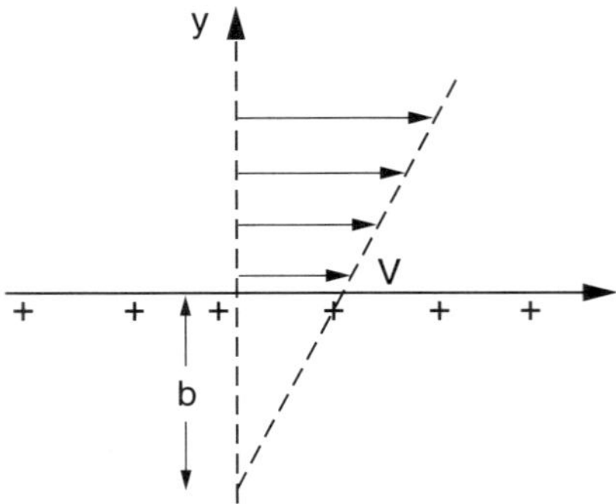

Figure 1.19 The slip extrapolation length b is the distance below the surface at which the velocity profile extrapolates to zero velocity; hence the slip velocity V_s is $b\dot{\gamma}$, where $\dot{\gamma}$ is the shear rate at the surface. (From Brochard and de Gennes, reprinted with permission. Langmuir 8:3033. Copyright 1992 American Chemical Society.)

interface that the velocity in the bulk extrapolates to zero (see Fig. 1-19) (Tolstoi 1952; de Gennes 1979; Brochard and de Gennes 1992). Thus b is defined as

$$b \equiv \frac{V_s}{\dot{\gamma}} \tag{1-32}$$

If, as is often assumed, the slip velocity is linear in the shear stress, and if the shear viscosity is shear-rate-independent, then the extrapolation length will also be shear-rate-independent.

It is possible for the extrapolation length to be *negative* if, for example, there is a layer of fluid so tightly bound to the interface that it does not move, and the point of zero velocity is moved from the wall *into* the fluid by one or more molecular layers.

From Eqs. (1-31) and (1-32), the apparent shear rate $\dot{\gamma}_{app}$ is

$$\frac{\dot{\gamma}_{app}}{\dot{\gamma}} = 1 + \frac{2b}{h} \tag{1-33}$$

where h is the gap. Thus, if the extrapolation length is small compared to the rheometer gap, $b/h \lesssim 0.01$, slip will have a negligible effect on rheological measurements.

1.5.2.3 Estimate of Extrapolation Length for Simple Liquids

A rough estimate of the slip velocity, or extrapolation length, can be obtained by modifying Eyring's rate equation [Eq. (1-29)], which gives the shear rate across a single layer of molecules. If, as is often the case, the molecules bind more strongly to each other than to the wall, then the activation enthalpy ΔH_w for movement of the molecules at the wall might be less than in the bulk. The activation volume v^* might also be somewhat less at the wall. Tolstoi (1952) related the activation enthalpy for molecular movement at the wall to the work of adhesion and work of cohesion, and thus he obtained a formula relating the extrapolation length to the contact angle of the liquid against the solid in a vacuum. We suppose that the effective shear rate, and hence the effective *viscosity* η_w for the first layer of molecules at the wall, can be computed from Eq. (1-29) by replacing ΔH and v^* with quantities modified by the wall–fluid interactions. A simple calculation of the extrapolation length then gives

$$b = a \left(\frac{\eta_0}{\eta_w} - 1 \right) \tag{1-34}$$

where a is the thickness of the wall layer—that is, the diameter of the molecules. In many cases, we expect that the "wall viscosity" η_w will not be many orders of magnitude different from the bulk viscosity η_0, and hence the extrapolation length is expected to be of the order of a few tens of molecular diameters at most. For small molecules, this is around 1–100 nm, which is consistent with experiments (Churaev et al. 1984; Blake 1990), and slip can be neglected for samples thicker than a micron or so. Flow in samples thinner than this is of interest in field of tribology (friction and wear), but much less so in the field of rheology.

For long polymer molecules, the appropriate molecular diameter a is likely to be the molecule's radius of gyration R_g, which can be as large as 0.1 μm. (The radius of gyration is defined in Section 2.2.3.2.) In addition, the viscosity ratio η_0/η_w might be huge because of molecular entanglements which are present in the bulk, but are suppressed at the wall, especially if the polymers adsorb only weakly (see Section 3.7.5.3) (Brochard and de Gennes 1992). Hence, extrapolation lengths b of order $1000 R_g$ (i.e., many microns, or even fractions of a millimeter) are possible. Thus, slip of polymers can be significant compared to bulk flow even when the gap is macroscopic (see Section 3.7.5.3).

> • Problem 1.5 exercises your understanding of wall slip and how it is measured.

1.5.3 Yield

When solids, or very viscous liquids, are subjected to stresses that are high, but not so high that they cause fracture, a process known as *yield*, or *plastic deformation*, can occur. Yield is especially important in metals, which, though they have a crystalline arrangement of atoms, are often *ductile*; that is, they deform irreversibly under high stresses, rather than fracturing (Hirsch 1975). Yield is also important in glassy solids (Bowdon 1973; Crist 1993) and in liquids near their glass transition points. Yielding occurs at stresses above the *yield stress* σ_y, which depends weakly on strain rate and temperature. At stresses below σ_y, the material deforms *reversibly*; that is, the material recovers its original shape when the stress is removed. Above σ_y, the material deforms *irreversibly* so that there is only partial recovery. Typical stress–strain curves for a glassy polymer that yields are shown in Fig. 1-20. While Fig. 1-20 is for uniaxial extension, analogous yield curves are obtained in shear (Crist 1993).

A simple mechanism for yield in a crystalline solid is depicted in Fig. 1-21 (Crist 1993). Under an increasing shear strain, each row of atoms is displaced from its equilibrium position (a) with respect to the neighboring row. Below a critical strain, if the stress is removed, the atoms spring back to their original positions. However, since the arrangement of atoms is spatially periodic, if the deformation continues, each row of atoms will eventually find itself back in registry with its neighboring rows, with each atom simply dispaced by one

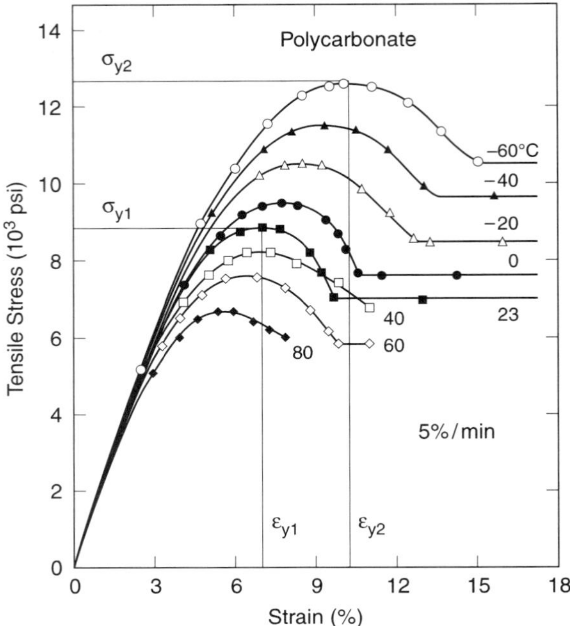

Figure 1.20   Stress–strain curves in extension for polycarbonate at various temperatures at a strain rate of 0.05 min^{-1}. (From Matsuoka 1992, reprinted with permission from Hanser Gardner Publications.)

interatomic distance from its original position relative to the layer below it (see Fig. 1-21c). The *strain energy* is therefore a periodic function of strain, oscillating between zero and a maximum value. The stress, which is the derivative of the energy with respect to strain, therefore also oscillates about zero, with a positive maximum value and negative minimum. The stress is zero at both the minimum and the maximum in strain energy (at strains $\gamma = 0$ and $\gamma = 0.5$ shown in Fig. 1-21a,b), and it is a maximum at the intermediate strain, $\gamma = 0.25$. This strain is the *yield strain* γ_y, and the stress maximum is the yield stress σ_y. If the macroscopic strain is applied slowly, then once the maximum in the stress is reached at $\gamma_y = 0.25$, there will be an instability whereby some layers slide forward, and others spring backwards to their initial positions with respect to the the underlying layer, with the result that all layers find themselves in registry with their neighbors, even though the strain is only $\gamma_y = 0.25$!

Thus, in this model, yielding involves *slippage* between some pairs of neighboring layers, with no relative movement at all between other pairs. Hence, the strain is not homogeneously distributed throughout the sample, but is *localized* to certain *slip planes*. In real crystalline materials, slippage often occurs in a much more complicated way than that depicted in Fig. 1-21, involving patterns of *dislocations* concentrated on slip planes. Mechanisms of slip analogous to that illustrated in Fig. 1-21 are involved in the flow of complex fluids such as colloidal crystals (Chapter 6) and block copolymers (Chapter 13).

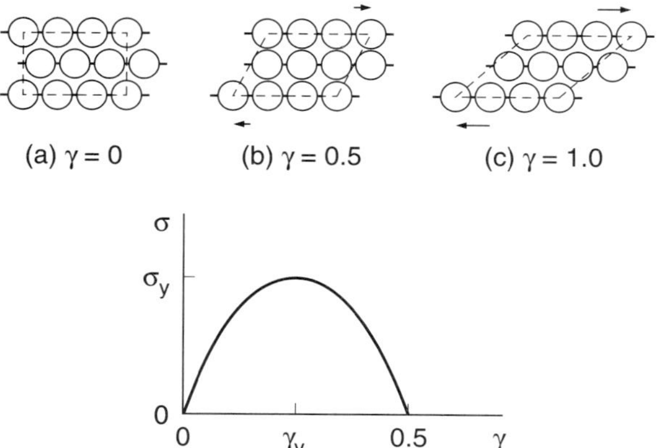

Figure 1.21   Illustration of idealized mechanism of yield. The structure shown in **(a)** represents the equilibrium positions of the atoms, and **(c)** is an equivalent equilibrium positioning with the top and bottom rows of atoms displaced by one lattice site to the right and left, respectively. As the layers are sheared over each other, the energy rises to a maximum in **(b)**, halfway between **(a)** and **(c)**. The stress, which is the derivative of the energy with respect to strain, is zero in **(a)**, **(b)**, and **(c)** and is maximally positive at $\gamma = 0.25$, halfway between **(a)** and **(b)**. (From Crist 1993, reprinted with permission from Wiley-VCH Verlag GmbH.)

Yield can occur even when the solid is a disordered one—that is, a *glass*. The stress–strain curve shown in Fig. 1-20 is that for uniaxial tension of polycarbonate, a glassy polymer. Note that the yield stress σ_y, which is the maximum in the curve, increases with decreasing temperature, and the yield strain γ_y is low, around 0.05–0.10. The yield stress is also a weak function of the rate of straining, generally increasing with increasing strain rate. Note also in Fig. 1-20 that at strains above the yield point, the stress drops somewhat and then levels out. The drop is known as *strain softening*; at still higher strains some materials show a reversal in slope and an increase in stress, which is called *strain hardening*.

A simple model for yield in *disordered* solids can be derived from the Eyring model, discussed in Section 1.5.1. If the enthalpy of a "hop" ΔH is very large, then flow is essentially impossible unless the applied shear stress is large, so that $v^*\sigma$ approaches ΔH in magnitude. Then the frequency of forward "hops" is much larger than that of reverse "hops," and the second term on the right side of Eq. (1-29) becomes negligible compared to the first term. Setting $\sigma = \sigma_y$, Eq. (1-29) then gives (Bowden 1973; Argon 1975; Crist 1993)

$$\sigma_y = \sigma_0 + \frac{k_B T}{v^*}\, \ln\left(\frac{2\dot{\gamma}}{\Omega_0}\right) \tag{1-35}$$

where σ_0 is the yield stress at zero temperature, $\sigma_0 = \Delta H/v^*$, and ln is the natural logarithm. (In this book, base 10 logarithms are designated by the term "log.") Since $\dot{\gamma} \ll \Omega_0$, this simple model predicts that the yield stress decreases with increasing temperature, but increases with increasing shear rate, in qualitative agreement with experiments. For polymer molecules, best-fit values of the activation volume v^* are generally the size of a few repeat units along the polymer chain. More detailed models of yielding, some of which show good agreement with experimental data for glassy polymers, have been reviewed by Crist (1993).

1.6 STRUCTURAL PROBES OF COMPLEX FLUIDS

While rheological measurements are wonderfully quantitative, they are usually poor qualitative probes of fluid structure. This is because in rheological experiments, the structural changes responsible for the measured relaxation behavior remain hidden. Thus, rheometry is often most useful when supplemented by other experimental methods that characterize fluid structure and flow-induced structural changes. Some of the most useful methods are microscopy, light, x-ray, and neutron scattering, and polarimetry.

1.6.1 Microscopy

Microscopy is the most direct way to obtain information on the structure of complex fluids. Optical microscopy can probe length scales down to around 0.5 μm (and even smaller if light is collected in the near-field with the sharpened tip of an optical fiber (Betzig and Trautman 1992). Electron microscopy can probe smaller length scales, as small as 0.5 Å for transmission (TEM) and 15 Å for scanning (SEM) modes (Sawyer and Grubb 1987; Reimer 1993; Goldstein et al. 1992), although the best resolution is usually not obtained for polymeric samples. Since molecular length scales, even polymeric ones, are usually in the submicron range, one cannot usually see them with an optical microscope, but one can see stress-producing supermolecular structures and textures, such as those present in emulsions, blends, foams, liquid crystals, and suspensions of large particles. Optical microscopy can readily be applied to such fluids under flow or deformation, by mounting a flow cell onto the stage of an optical microscope (Alderman and Mackley 1985; Larson and Mead 1992; Vermant et al. 1994; Mather et al. 1996).

Some biological macromolecules are very large and can be viewed directly by optical microscopy, often with the assistance of fluorescent dyes. Examples include DNA, actin filaments (which are part of the cellular cytoskeleton), and microtubules. Bustamante and coworkers (Smith et al. 1992, 1996), Chu and coworkers (Perkins et al. 1994), and Bensimon and coworkers (Strick et al. 1996) have pioneered the use of microscopy, combined with *micromanipulation*, to measure the mechanical properties of single DNA molecules. One tool that can exert forces on a single large molecule is an *optical trap*, or "optical tweezers," which consists of a focused beam of laser light, the momentum of which traps small particles whose index of refraction differs from that of the solvent medium (Perkins et al. 1994; Yin et al. 1995). Molecules attached to the small optically trapped particles can be dragged about or stretched by moving the focal position of the "tweezers." Some applications of "optical tweezers" are described in Chapter 3.

The force that optical tweezers can exert is modest, namely, several tens of piconewtons. Much larger forces (hundreds of piconewtons) can be exerted using a *micropipette*, which, as the name implies, generates a suction force through a pipette tip that is only a few microns in diameter (Evans et al. 1995, 1996; Smith et al. 1996). Yet another micromanipulation method uses *atomic force microscopy* (AFM), in which a sharp tip mounted on the end of a flexible cantilever is dragged along or pulled away from a surface, while simultaneously measuring the position of the tip and the force exerted on it (Bustamante and Keller 1995). In some special cases, one end of a molecule can be made to bind to the AFM tip while the other end remains tethered to the surface. In this way, Rief et al. (1997) have probed the elastic properties of a single polysaccharide molecule.

Another type of single-particle optical method is *interference microscopy* or *microrheology* (Mason et al. 1997; Gittes et al. 1997). In this method, the thermally driven motion of a probe particle is followed optically by measuring very precisely (on the scale of a few nanometers) the deflection of light rays using interference principles. In effect, one has a rheometer in which Brownian force acts as a motor and a beam of light as the transducer! The measurements of ensemble-average position $\langle \Delta r(t) \rangle$ as a function of time t can be converted to the complex modulus G^*. For example, $\tilde{G}(s) = k_B T/\pi a s \langle \Delta \tilde{r}^2(s) \rangle$, where a is the particle radius, s is the Laplace time, and the tildes represent Laplace-transformed functions (Mason and Weitz 1995). For a more complete description, see the original papers. This method is especially useful for nonhomogeneous samples, where one can in principle measure the complex modulus as a function of position within a transparent substance, such as a gel, protoplasm, or cytoskeleton of a single cell.

Electron microscopy cannot readily be carried out on liquids undergoing flow, because of the high vacuum of most (but not all) microscope chambers, the thinness required of the samples (100 nm for TEM), and the time required to obtain an image (seconds or longer). However, scanning electron microscopy has been used to observe *in situ* the deformation and fracture of solid polymers, using (for example) as displacement markers 20-nm dots burned by the electron beam into the sample surface (Corleto et al. 1996; see also Sharpe 1989). Furthermore, electron microscopy can be used to study flow-induced changes at the molecular level even in liquids, by using the time-consuming process of interrupting the flow, quenching the sample into a solid phase, microtoming, and mounting the sample on a grid to be placed in the microscope chamber. This method is especially useful for studying flow-induced changes in the structure of sluggish materials, such as polymer blends (Miles and Zurek 1988), liquid crystalline polymers (Hudson et al. 1993), block copolymers (Winey et al. 1993), and filled melts, as well as glassy materials that have been sheared to the point of yield. As we shall see in Chapter 12, ultrafast freezing and cryomicroscopy even permit the viewing of water-containing samples, such as surfactant solutions (Clausen et al. 1992).

1.6.2 Light, X-Ray, and Neutron Scattering

Scattering methods are among the most powerful and widely used in the study of complex fluids. Light, x-rays, and neutrons can be scattered (Lindner and Zemb 1991). For light scattering, length scales of 2000 Å to 100 μm are probed, while both x-rays and neutrons can reach smaller length scales, from 10 to 1000 Å for x rays and from 10 to 200 Å for neutrons.

In all scattering methods, one relies on a change in direction of propagation of the radiation due to interaction with inhomogeneities in matter. Figure 1-22 is a schematic of a beam of radiation scattered at an angle θ by inhomogeneities in the medium. If the scattering is *elastic*, the wavelength λ (in vacuum or air) of the incoming and outgoing beams are the same. The scalar *wavenumber* k is defined as $2\pi n/\lambda$, where n is the index of refraction of the medium, and λ/n is the wavelength in the scattering medium. The incoming *wavevector* $\mathbf{k}_i$ is $2\pi n/\lambda$ times a unit vector in the direction of propagation; the outgoing beam is defined analogously. Consider two rays that are scattered at the same angle from points in the medium separated by a distance vector $\mathbf{x}'$. From Fig. 1-22, there is a difference in path length of the two rays given by $\ell_2 - \ell_1$. From simple trigonometry, $k\ell_2 = \mathbf{k}_s \cdot \mathbf{x}'$ and $k\ell_1 = \mathbf{k}_i \cdot \mathbf{x}'$. Hence the difference in the phase of the two beams is

$$\Delta\delta = k\ell_2 - k\ell_1 = \mathbf{k}_s \cdot \mathbf{x}' - \mathbf{k}_i \cdot \mathbf{x}' = \mathbf{q} \cdot \mathbf{x}' \tag{1-36a}$$

where $\mathbf{q}$ is the *scattering vector* defined by

$$\mathbf{q} = \mathbf{k}_s - \mathbf{k}_i, \qquad |q| = \frac{4\pi n}{\lambda} \sin\left(\frac{\theta}{2}\right) \tag{1-36b}$$

If, for example, the scattering is caused by tiny particles separated from each other by a typical distance x', then there will be a peak in scattering at an angle θ for which the phase difference $\Delta\delta$ is equal to 2π. From Eq. (1-36a), this corresponds to a q vector of magnitude $2\pi/x'$. Hence, from Eq. (1-36b), $1/x' = (2n/\lambda) \sin(\theta/2) \approx \theta n/\lambda$ for small angles. Thus *there is an inverse relationship between scattering vector (or angle) and the structural length scale producing the scattering*. Since $\mathbf{q}$ is a vector quantity, scattering is an excellent way to determine not only the length scales but also the orientations of structures in complex fluids. The scattering intensity as a function of scattering wavevector $\mathbf{q}$ is called the *structure factor* $S(\mathbf{q})$. For a single-component material, it can be related to the Fourier transform $\rho_{\mathbf{q}}$ of the density pattern, $\rho(\mathbf{x})$, by $S(\mathbf{q}) = \langle \rho_{\mathbf{q}}\rho_{-\mathbf{q}}\rangle$, where $\langle \cdot \rangle$ denotes an average over space or time. For multicomponent samples, the structure factor is related to concentration as well as density fluctuations. Examples of the use of light scattering to infer structure are present throughout this book—for example, in Sections 3.6.2.2.2, 6.4.3, 7.2.5, 9.2.1, and 12.4.2.3.

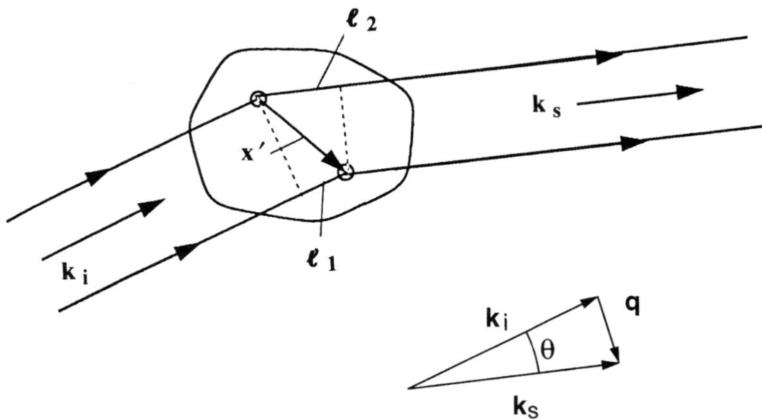

Figure 1.22   An incoming beam of radiation with wavevector $\mathbf{k}_i$ is deflected by an angle θ, so that its outgoing wavevector is $\mathbf{k}_s$. (From Fuller, Copyright © 1995 by Oxford University Press, Inc. Used by permission of Oxford University Press.)

Relatively large structures, of size ~ 0.1–$100\,\mu$m, are readily probed by the scattering of visible light. Smaller structures, down to molecular lengths, must instead be probed by x-ray or neutron scattering (Wignall 1996; Higgins and Maconnachie 1987; Glatter and Kratky 1982). The choice between scattering x-rays and neutrons hinges on practical concerns regarding the sources of contrast (electron versus nuclear density), the depth of penetration of the radiation through typical materials, and the availability of radiation sources. Neutron scattering for polymeric materials often requires selective deuteration (replacement of some of the hydrogen atoms by deuteriums) to achieve enough contrast, which, in turn, requires the synthesis of suitably deuterated polymeric materials.

For x-ray scattering, contrast is provided by differences in electron number density. For hydrocarbons, the mass density ρ is fairly insensitive to the ratio of hydrogen to carbon in the molecules, with ρ varying only modestly in the range 0.8–1.0 g/cm^3. Because of electroneutrality, the electron number density must be the same as the proton number density, so an ordinary hydrogen atom, which has no neutrons, adds more electron density per unit mass than does a carbon atom. Therefore, x-ray contrast is likely to exist between two species of hydrocarbons with differing ratios of hydrogen to carbon, even without any special labeling. However, the high x-ray absorptivity of most materials (especially those of high atomic weight) means that samples must be thin, and so must be the walls of containers or sample holders through which the x-rays must pass, unless the x-ray source is a particularly powerful one (such as a synchrotron source). Neutrons are absorbed much less readily than x-rays, and no special steps need be taken to keep neutron path lengths small. The time required to collect an x-ray or neutron scattering pattern (seconds to hours) is long enough that only rather slow dynamics can be followed in real time. The literature contains many descriptions of shearing cells designed to allow simultaneous scattering of light, x-rays and neutrons (Hadziioannou et al. 1979; Straty 1989; Safinya et al. 1991; Wu et al. 1991; Yanase et al. 1991; Hashimoto and Kume 1992; Koppi et al. 1992; Morrison et al. 1993; Navard 1994). Examples of the use of x-ray and neutron scattering are found in Sections 10.4.8 and 13.2.1.

1.6.2.1 Photon Correlation Spectroscopy

In light scattering, very fast ($\sim \mu$sec) processes can be followed by using *photon correlation spectroscopy*, also known as *dynamic light scattering* (Berne and Pecora 1990). This method relies on thermal noise to produce spontaneous position-dependent density or concentration fluctuations $\rho(x, t)$ that scatter coherent light. Since the scattering sites diffuse around randomly, the scattering amplitude varies in time. The time dependence of the scattering at a given wavevector q is controlled by the time required for a typical scattering site to diffuse a distance of order $2\pi/q$. Thus, information on length-dependent or diffusion rates can be obtained by correlating the instantaneous value of ρ_q with its value a time t later—that is, by $S(q, t) \equiv \langle \rho_q \rho_{-q} \rangle (t)$. $S(q, t)$ is called the *dynamic structure factor*, a generalization of the static structure factor defined in Section 1.6.2. The dynamic structure factor can be measured over a wide range of time scales from microseconds to seconds or longer. In dynamic light scattering, one therefore measures the relaxation rate as a function of the size of the microstructure. By comparing these results with stress-relaxation rates measured rheologically, one can gain insight into the microstructures that produce the stress. Examples are given in Sections 4.6 and 11.5.1.

1.6.2.2 Diffusing-Wave Spectroscopy

Traditional analyses of light scattering data assume that no photon is scattered from more than one scattering center; this assumption is *not* valid for samples that are visibly cloudy or milky. Unfortunately, many important complex fluids are milky, including foams, many emulsions, and some suspensions. However, Pine et al. (1988, 1990) showed that if the sample is *very turbid* so that each photon is scattered many times before exiting the sample, then each photon executes a many-step *random walk* as it scatters its way through the sample. The propagation of light in this limit is described by a diffusion equation with a diffusivity given by $c\ell^*/3$, where c is the speed of light in the medium and ℓ^* is the distance the light travels before its direction is randomized. This is the basis of *diffusing-wave spectroscopy (DWS)*, which is the extension of dynamic light scattering to multiply scattering samples. DWS has been used to measure the short-time diffusion coefficient of spheres in dense suspensions (Weitz and Pine 1993), the rates of rearrangement events in foams (Durian et al. 1991), the motions of magnetic particles in magnetorheological fluids (Ginder 1993), and the rate of particle aggregation during the formation of cheese (Horne 1989).

1.6.3 Polarimetry

Many complex fluids contain orientable molecules, particles, and microstructures that rotate under flow, and under electric and magnetic fields. If these molecules or microstructures have anisotropic polarizabilities, then the index of refraction of the sample will be orientation-dependent, and thus the sample will be *birefringent*. In general, the anisotropic part of the index of refraction is a tensor $\mathbf{n}$ that is related to the *polarizability* $\boldsymbol{\alpha}$ of the sample. The polarizability is the tendency of the sample to become polarized when an electric field is applied; thus $\mathbf{P} = \boldsymbol{\alpha} \cdot \mathbf{E}$, where $\mathbf{P}$ is the polarization and $\mathbf{E}$ is the imposed electric field. When the anisotropic part of the index of refraction is much smaller than the isotropic part (the usual case), the index-of-refraction tensor $\mathbf{n}$ can be related to $\boldsymbol{\alpha}$ by the Lorentz–Lorenz formula:

$$\mathbf{n} = \frac{(n^2 + 2)^2}{6n} \frac{4\pi}{3} \boldsymbol{\alpha} \tag{1-37}$$

where n is here the isotropic part of the index of refraction.

For simple *polymeric* liquids, the birefringence tensor $\mathbf{n}$ is often proportional to the stress tensor $\boldsymbol{\sigma}$, a relationship called the *stress-optic law* (Janeschitz-Kriegl 1983):

$$\mathbf{n} = C\boldsymbol{\sigma} \tag{1-38}$$

where C is called the *stress-optic coefficient*. [The molecular origins of Eq. (1-38) are discussed in Section 3.4.1.] Using the stress-optic law, one can measure stresses with a laser beam! This is useful, since laser beams can be made narrow and can easily be scanned through a sample, thus allowing the spatial dependence of the stress in a complex flow to be mapped out. If the polarization of the light is suitably time-modulated, two components of the birefringence tensor can be extracted nearly simultaneously with one beam of light, even when the birefringence is changing with time. Fuller (1995) and coworkers have developed these and related modulated polarimetry methods and applied them to the study of complex fluids. Examples of the use of birefringence can be found in Sections 3.4.1, 10.2.4.1, 10.2.6.2, 11.3.5, 11.5.2, and 13.3.1.2.1.

> • Worked Example 1.6 shows how the shear stress and the first normal stress difference can be measured using birefringence.

1.6.4 Linear Dichroism

Fuller (1995) and coworkers have also developed analogous methods to study orientation-dependent light attenuation, or *dichroism* in complex fluids. *Infrared dichroism* is the orientation-dependent absorption of infrared light due to bond stretching modes, while *conservative dichroism* is light attenuation due to anisotropic scattering. With infrared dichroism, one can selectively measure the degree of orientation of a single type of bond (the one that absorbs at the laser wavelength, such as a carbon–deuterium bond) in a sample containing many types of bonds. This is especially useful for analyzing the contribution of a single species to the total orientation or birefringence in a multicomponent blend. Conservative dichroism, on the other hand, is useful for studying structural orientation and relaxation in colloidal or liquid-crystalline polymer samples, as we shall see in Sections 6.3.1.1 and 11.3.5. For a thorough description of such methods, see Fuller (1995).

1.6.5 Polarized Light Scattering

While conservative dichroism measures the orientation-dependent loss of light due to scattering, the direction in which the light is scattered, as a function of the orientation of light's polarization, is probed in *polarized light scattering* (Ernst et al. 1990). In this combination of polarimetry and light scattering, the sample is placed between polarizers so that polarized light impinges on the sample. Scattering-induced changes in the polarization of the light are detected with the aid of the second polarizer—called the *analyzer*—through which the light passes after it has gone through the sample. Taking "H" (horizontal) and "V" (vertical) to be two directions normal to each other and to the propagation direction of the beam, the three most common types of polarized light scattering are "H–H," "V–V," and "H–V" (or "V–H"), where the first letter denotes the orientation of the slow axis of the "polarizer" (the one in front of the sample), while the second letter denotes the orientation of the analyzer. Usually either H or V corresponds to the flow direction. Polarized light scattering is especially useful in studying nonuniformly oriented fluids such as flowing liquid crystals, liquid-crystalline polymers, or block copolymers (see, for example, Section 12.3.1.3); it measures the length scales over which there is a change in the uniformity or direction of orientation.

1.6.6 Raman Scattering and Fluorescence Polarization

Ordinary methods of polarimetry, such as birefringence and dichroism, provide information about the *second moment* of the orientation distribution of the microstructural elements (molecules, particles, etc.). More complete information about the distribution function can be obtained using *Raman scattering* (Raman and Krishnon 1928), from which components of both the second and fourth moments can be extracted (Archer et al. 1992). In Raman scattering, a powerful laser beam (> 0.5 W) is polarized (with constant or time-varying polarization) and used to excite Raman vibrational modes of chemical bonds in the sample. The input beam can excite many different kinds of bonds (e.g., C–C, C–H, etc.), but the wavelength of the stimulated Raman output signal differs from one kind of bond to another. The intensity of the Raman output depends on the relative orientation of the chemical bond with respect to the polarization direction of the radiation. Using a spectrograph, the wavelength of a desired bond can be selected, and information on the second and fourth moments of its orientation distribution with respect to the polarization direction of the input beam can then be obtained. By modulating the polarization state of the input beam, two components of the second and fourth moment tensors can be thus obtained.

One great advantage of this Raman technique is that no sample labeling is required, and the orientation of essentially any bond in the sample can be studied. A drawback of the method is the feebleness of the stimulated Raman signal compared to the input intensity, necessitating use of a powerful laser, as well as long-time averages in steady-state measurements, or many repeat experiments in transient measurements, so that an adequate signal-to-noise ratio can be obtained. The method has been used to obtain the fourth moment of the bond orientation distribution in sheared polymer melts (Archer et al. 1992), as well as to probe changes in shear-induced bond rotation (Huang and Fuller 1996) that occur in a polymer melt as one approaches a glass transition. A much more detailed explanation of the method can be found in Fuller (1995). Examples of its use are found in Sections 4.3.2 and 10.2.4.1.

Polarized fluorescence arises from a different optical interaction with matter than does Raman scattering, but as in Raman scattering, the wavelength of the emitted light differs from that of the incident light. Also like Raman scattering, fluorescence polarization also allows one to measure both second and fourth moments of the orientation of specific bonds in a molecule. Monnerie (1987) has pioneered the application of fluorescence polarization methods to polymer dynamics.

1.6.7 Other Methods

Other experimental methods have proven their worth in the study of the dynamics of complex fluids. These include *dielectric spectroscopy*, which has a very wide frequency range ($\sim$ 15 decades). The complex dielectric constant is measured by mounting a sample between two parallel capacitor plates each of area A, applying an ac voltage V across the gap h between them at frequencies f ranging from 10^{-5} to 10^{10} Hz (Schönhals et al. 1991), and measuring the resulting complex impedance $Z \equiv V/I$, where I is the ac current produced in response to the imposed voltage. The complex dielectric constant $\varepsilon^*(f)$ is then given by $\varepsilon^* = (h/A)/(\epsilon_0 i 2\pi f Z)$, where $\varepsilon_0 = 8.8 \times 10^{-12} \mathrm{C^2 J^{-1} m^{-1}}$ is the permittivity of space. (The frequency f in Hz is related to the frequency ω in rad/sec by $\omega = 2\pi f$.) The frequency dependence of ε^* measures the rates of relaxation of electric dipoles borne by molecular segments. Hence, the dielectric spectrum typically probes the same molecular motions as those governing other relaxation phenomenon, including mechanical relaxation. Because of its wide frequency range, dielectric spectroscopy is useful for studying the dynamics of materials for which time–temperature superposition fails, such as glass-forming liquids (see Section 4.2). In certain polymer molecules, called "type A" by Stockmayer (1967), the dipole moments of the bonds along the backbone add up linearly, so that dielectric relaxation can be related to the relaxation of the end-to-end vectors of the molecules. An example is *cis*-polyisoprene (Adachi and Kotaka 1984; Watanabe et al. 1991; Patel and Takahashi 1992). Havriliak and Havriliak (1996) summarize the dielectric properties of many polymers.

Nuclear magnetic resonance (NMR) probes the environment an atom (carbon, hydrogen, deuterium) finds itself in; in particular, it measures lineshapes that indicate the reorientation speed of bonds to which the atom is attached (Slichter 1989, Tonelli 1989, 1996) Thus, it is sensitive to phase transitions that freeze or impede local molecular motion, such as (a) transitions to crystalline or liquid-crystalline states and (b) kinetic transitions such as the glass transition. It can also determine if there are multiple environments in a sample, such as small crystallites within an otherwise amorphous or liquid crystalline sample, and can quantify the percentages of each kind of environment. In glass-forming blends, it can determine the degree to which the individual components share a common glass transition temperature. NMR is usually sensitive, however, only to *local* dynamics and not to the motions or relaxations of large structures, such as the overall relaxation of a long polymer molecule. An example of the use of NMR is given in Section 12.3.4.2. The use of NMR to measure local velocity fields was discussed in Section 1.5.2.1.6.

In *differential scanning calorimetry*, or DSC, the temperature of a small specimen is made to rise or fall at a specified rate, with the heat flow (in or out of the material) needed to maintain the specified rate of temperature change carefully measured (Wunderlich 1990; Wen 1996). With off-the-shelf equipment, the method is simple, and phase changes reveal themselves as peaks or dips in heat flow, allowing the enthalpy of the transition to be quantified. Kinetic transitions, such as glass transitions, show up as broad peaks, and hysteresis in a transition is recognized by a shift in the transition temperature between heating and cooling.

Finally, techniques have been developed to measure *surface forces* generated by liquids in gaps as small as a few angstroms between very smooth surfaces, such as mica sheets (Tabor and Winterton 1969; Israelachvili 1992). Interference techniques are used to measure the gaps, and microcantilever springs measure forces at the piconewton scale. Alternatively, the surface force versus film thickness of thin, isolated, foam films can be measured by using the porous plate method (Mysels and Jones 1966; Exerowa et al. 1987; Bergeron and Radke 1992). In this method, a soap solution in a porous plate supplies liquid to a film bridging a hole drilled through the plate. By controlling the gas pressure in a cell containing the plate, the capillary pressure in the film can be set, and at equilibrium its thickness at that pressure can be determined interferometrically. Yet another method is to make a dilute emulsion of monodisperse ferrofluid droplets (see Section 8.4.4), which become magnetically polarized when a magnetic field is applied. The droplets then attract each other with a force that varies with the field. The equilibrium distance between neighboring droplets as a function of field can be measured accurately by light scattering, producing a force–distance curve (Mondain-Monval et al. 1996). With these techniques, static forces such as van der Waals, electrostatic, or steric, produced by various thin liquid layers, can be measured. As discussed in Chapter 2, these measurements have helped confirm our understanding of the basic forces controlling the structure of complex fluids. Shearing flows can also be imposed using modified versions of surface-forces devices, and stresses can be measured on very thin fluid layers, subjected to large or small confining pressures (Reiter et al. 1994). In this way, the gap dependence of G' and G'' can be measured; for polybutadiene of molecular weight 7000, it is found that departures from the bulk values of the moduli occur when the gap decreases below about 200 nm; below 20 nm, the melt even becomes "glassy" due to confinement effects (Luengo et al. 1997). Atomic force microcopy (AFM) can also be modified to provide measurement of friction forces during sliding of the AFM tip (Overney et al. 1994).

1.7 COMPUTATIONAL METHODS

Computational methods are increasingly valuable supplements to experiments and theories in the quest to understand complex liquids. Simulations and computations can be aimed at either molecular or microstructural length scales. The most widely used molecular-scale simulation methods are *molecular dynamics*, *Brownian dynamics*, and *Monte Carlo sampling*. Computations can also be performed at the continuum level by numerical solutions of field equations or by Stokesian dynamics methods, described briefly below.

1.7.1 Molecular Dynamics Simulations

Molecular dynamics (MD) is the most detailed molecular simulation method (Alder and Wainwright 1957; Allen and Tildesley 1989; Tildesley 1995). In it, Newton's equations of motion are solved for a large collection of molecules that interact with each other via *intermolecular potentials*. Thus, one solves a set of equations such as the following:

$$m_i \ddot{x}_i = \mathbf{F}_i \{\mathbf{x}_j\} \tag{1-39}$$

where $\mathbf{x}_i$ is the position of an atom (or, perhaps, a group of atoms, or molecule), $\dot{\mathbf{x}}_i$ is its velocity, $\ddot{\mathbf{x}}_i$ its acceleration, m_i is its mass, and $\mathbf{F}_i \{\mathbf{x}_i\}$ is the force acting on atom i, which is a function of the positions $\{\mathbf{x}_j\}$ of all atoms.

The thermal energy in an MD calculation is just the average kinetic energy of the atoms. Since long runs are required to obtain satisfactory averages of the fluid properties, small numerical errors in the integration of the equations of motion (1-39) can accumulate, producing gradual changes in the mean-square velocity, leading to *numerically generated heating or cooling*. To correct for this, the molecular velocities must be periodically rescaled upward or downward to bring the temperature back to the prescribed value. Examples of MD results are found in Figs. 3-28 and 4-24.

In MD calculations, the equations, such as (1-39), are *deterministic*; the pseudorandom motions characteristic of Brownian processes arise out of the cumulative effects of a huge number of uncoordinated intermolecular collisions (as is the case in real fluids). For fluids that contain large polymer molecules or colloidal particles in a small-molecule solvent, this approach is an inefficient one. Since the typical time step size taken in the simulation must be scaled to the motions of the fast-moving solvent molecules, most of the computation is consumed in calculations of the rapid, but uninteresting, jostling of solvent molecules, rather than in the more interesting, but lumbering, motions of the larger polymer molecules or particles.

1.7.2 Brownian Dynamics Simulations

Brownian dynamics is a more efficient approach for simulating such systems. In it, the solvent is treated as a viscous continuum which dissipates energy as macromolecules or particles move through it, and the Brownian motion of the macromolecules produced by random collisions with solvent molecules is mimicked by a stochastic force generated by pseudorandom numbers. The temperature of the system is set by the amplitude of the autocorrelation function of this imposed stochastic force, which is proportional to $k_B T$. Thus, in Brownian dynamics, no temperature drift occurs due to accumulated integration errors. A simple equation that might be integrated in a Brownian dynamics simulation is

$$\zeta_i (\dot{\mathbf{x}}_i - \mathbf{v}^s) = \mathbf{F}_i^e + \mathbf{F}_i^b \tag{1-40}$$

where $\mathbf{v}^s$ is the solvent velocity (which is controlled by the imposed flow field), ζ_i is a drag coefficient that represents viscous dissipation, $\mathbf{F}_i^e$ is an elastic or *conservative* force on a macromolecule or particle i produced, for example, by macromolecular deformation or interparticle interactions, and $\mathbf{F}_i^b$ is the Brownian force on macromolecule or particle i. $\mathbf{F}_i^b$ it is typically taken to have a "white-noise" spectrum, so that its magnitude at one time step is uncorrelated with that at the next time step. According to the fluctuation-dissipation theorem, the mean-square value of $\mathbf{F}_i^b$ at a given time step is given by (Reif 1965; Fixman 1978; Grassia and Hinch 1996)

$$\langle (\mathbf{F}_i^b)^2 \rangle = 2kT \zeta_i \delta$$

where δ is the unit tensor. One way to impose this condition is to choose a random vector $\mathbf{n}$ uniformly distributed over the interval $[-1, 1]$ and set $\mathbf{F}_i^b$ equal to $(6k_B T \zeta_i / \delta t)^{1/2} \mathbf{n}$, where δt is the time step. One can also impose *constraints* that, for example, keep bond lengths constant (Doyle et al. 1997). Body forces, such as gravity, or electric and magnetic fields, can easily be added. Brownian dynamics methods for polymer flow problems have been developed by Liu (1989), Zylka and Öttinger (1989), Grassia and Hinch (1996), and Shaqfeh and coworkers (Doyle et al. 1997). A method for handling hydrodynamic interactions between particles or between different parts of a polymer molecule has been developed by Ermak and McCammon (1978). For more details on polymer stochastic simulations, see Öttinger's (1996) book.

1.7.3 Monte Carlo Sampling

An even simpler simulation technique is the Metropolis Monte Carlo sampling method (Metropolis et al. 1953), which generates large numbers of configurations or microstates of the equilibrated system by stepping from one microstate to the next in small increments. Various schemes for moving polymer molecules on a lattice, for example, are described by Binder and coworkers (Kremer and Binder 1988; Paul et al. 1991). One can average quantities of interest over these microstates. To perform Monte Carlo sampling, one must define a system potential W that depends on the locations of all molecules or particles. Random test moves of one or more molecules change the potential by, say, ΔW. In the Metropolis method (the simplest Monte Carlo method), if $\Delta W < 0$, the move decreases the potential and is accepted; if $\Delta W \geq 0$, the move is accepted with probability

$P \equiv \exp(-\Delta W / k_B T)$. After an initial relaxation period consisting of many Monte Carlo moves, configurations typical of equilibrium are generated, and properties can be averaged over them. The effects of electric and magnetic fields on the average properties can be explored, and in some cases the effect of a flow field can be investigated, if the effect of the flow can be modeled by using an effective potential. Extensional flows, but not shearing flows, can sometimes be treated this way. Since Monte Carlo simulations are really sampling methods, they are not particularly well suited to simulations of time-dependent phenomena, although in some cases the sequence of Monte Carlo microstates can be mapped onto a progression in time (Binder 1994a). For reviews of the use of Monte Carlo and other methods in the simulation of the properties of polymeric materials, see Bicerano (1992), Colbourn (1994), and Binder (1994b, 1995). Results from Monte Carlo simulations are given in Figs. 3-21, 4-25, 8-13, 12-4, 12-20, and 13-12.

1.7.4 Numerical Solution of Composition-Field Equations

In some cases, one is interested in the structures of complex fluids only at the continuum level, and the detailed molecular structure is not important. For example, long polymer molecules, especially block copolymers, can form phases whose microstructure has length scales ranging from nanometers almost up to microns. Computer simulations of such structures at the level of atoms is not feasible. However, *composition field equations* can be written that account for the dynamics of some slow variable such as $\phi(\mathbf{x})$, the concentration of one species in a binary polymer blend, or of one block of a diblock copolymer. If an expression for the free energy f of the mixture exists, then a *Ginzburg–Landau type* of equation can sometimes be written for the time evolution of the variable ϕ with or without flow. An example of such an equation is (Ohta et al. 1990; Tanaka 1994; Kodama and Doi 1996)

$$\frac{\partial \phi}{\partial t} = \nabla \cdot \left[L(\phi) \nabla \left(-\kappa \nabla^2 \phi + \frac{d}{d\phi} f(\phi) \right) \right] - \nabla \cdot (\phi \mathbf{v}) \tag{1-41}$$

Here $\mathbf{v}$ is a flow velocity, $f(\phi)$ is the homogeneous free energy per unit volume, $L(\phi)$ is a composition-dependent "Onsager coefficient" or diffusivity, and κ is the coefficient of the gradient free energy term, which is related to the interfacial energy for inhomogeneous or multiphase liquids. For a polymer blend, $df(\phi)/d\phi$ has been approximated by the simple expression $-A \tanh\phi + \phi$ (Ohta et al. 1990). In the absence of flow ($\mathbf{v} = 0$), with ϕ small, $\kappa = 0$, and $L(\phi)$ constant, Eq. (1-41) reduces to a simple Fickian diffusion equation. More complex expressions for the free energies of block copolymers or blends can be included in Eq. (1-41). For a simple shearing flow, we set $\mathbf{v} = \dot{\gamma} y \mathbf{e}_x$, where $\mathbf{e}_x$ is a unit vector in the x, or flow, direction, and y is the gradient direction. Equation (1-41) can be solved numerically in two or three dimensions by finite-difference or finite-element methods, such as those used to solve other field equations for mass, momentum, or energy. Equation (1-41) assumes that the velocity field is a known one that is imposed externally. If the complex fluid consists of two or more phases whose effective viscosities differ widely, then it may be necessary to treat the velocity field itself as an unknown that satisfies a separate momentum balance equation [see, for example, Tanaka (1994)].

1.7.5 Suspension Simulations

Specialized algorithms have been developed to simulate the flow of suspensions, including suspensions of spheres (Brady and Bossis 1988; Doi and Chen 1989), ellipsoids, and cylinders (Yamane et al. 1994), as well as emulsions (Ohta et al. 1990) and foams (Khan and Armstrong 1986; Kraynik et al. 1991). Typically, the equations one solves are analogous to Eq. (1-40), with or without the Brownian term. By adding terms to describe electrostatic, magnetic, van der Waals, and excluded-volume forces, one can simulate the flow behavior of electro- or magnetorheological fluids, electrostatically stabilized suspensions, or gel-forming suspensions (see Chapters 7 and 8).

For dense suspensions of spherical particles, an especially accurate method called *Stokesian dynamics* has been developed by Bossis and Brady (1989). In Stokesian dynamics, one solves a generalized form of Eq. (1-40), in which the simple Stokes law for the drag force on sphere i, $\mathbf{F}_i^d = -\zeta(\dot{\mathbf{x}}_i - \mathbf{v}^s)$, is replaced by a more accurate tensor expression that accounts for the *hydrodynamic interactions*—that is, the disturbances to the solvent velocity field produced by the relative motions of the other spheres. The Stokesian dynamics method accounts for hydrodynamic interactions among widely separated spheres by a *multipole expansion*, as well as for closely spaced ones by a *lubrication approximation*. Results from this method appear in Figs. 6-8 and 8-8.

1.7.6 Boundary Conditions

In all of the methods discussed above, *periodic*, or "wrap-around," boundary conditions are often imposed on two- or three-dimensional boxes. With these boundary conditions, molecules on the right side of the box interact with an "image" of the molecules on the left side, and similarly the top interacts with an image of the bottom while the front interacts with an image of the back. In this way, one can avoid introducing walls or boundaries, and a limited number of molecules or particles might more fairly represent a macroscopic sample. Modifications of these boundary conditions have been devised to allow for the imposition of a uniform velocity gradient without introducing a wall or solid boundary.

1.8 THE STRESS TENSOR

Once the structure of a complex fluid has been simulated, computed, or derived by analytic theory, one would like to calculate the stress tensor $\boldsymbol{\sigma}$ and compare it to experimental stress measurements. The appropriate expression for the stress tensor depends on the type of complex fluid. However, if the idealized microstructure is built out of many small, point-like elements located at positions $\{\mathbf{x}_i, i = 1, 2, \ldots, N\}$, and on each such point a nonhydrodynamic force $\mathbf{F}_i$ is exerted by the rest of the microstructure, then $\boldsymbol{\sigma}$ can be obtained from the general Kirkwood (1949) formula (Doi and Edwards 1986):

$$\boldsymbol{\sigma} = -\frac{1}{V} \sum_{i=1}^{N} \mathbf{x}_i \mathbf{F}_i \qquad (1\text{-}42)$$

where V is the volume of the system. Eq. (1-42) quantifies the way in which microstructures and microstructural forces in a complex fluid control its stress. The microstructures and microstructural forces can be measured or inferred from the structural probes described in Section 1.6 and from the computational methods described in 1.7. Examples showing how Eq. (1-42) can be applied to polymers and electrorheological fluids are given in Sections 3.4.1 and 8.2.1.6. Specific expressions relating the stress tensor to the microstructure for polymers are given in Eqs. (3-11) and (3-68), those for suspensions of rigid particles are given in Eqs. (6-34)–(6-36), those for electrorheological fluids are given in Eq. (8-11), and those for nematic liquid crystals are given in Eq. (11-9). Derivations of these expressions can be found in Doi and Edwards (1986), Bird et al. (1987b), and Larson (1988).

1.9 SUMMARY

Complex fluids are amazingly diverse, ranging from foodstuffs, to biological materials, to plastic coatings. Nevertheless, such substances have shared characteristics, including (a) larger-than-atomic structural length scales ranging from nanometers to millimeters and (b) long time scales of milliseconds to years. Because of these ranges of length and time scales, the structures of complex fluids are responsive to ordinary "slow" flows, with shear rates in the range 10^{-3} to 10^3 sec^{-1}. The "liquid-like" or "solid-like" characteristics of complex fluids can be probed by oscillatory and steady shearing. Deformation and flow of complex flows are described using kinematic tensors, such as the velocity gradient and Finger deformation tensors.

Rheology provides an indirect, and hence limited, indicator of material structure. Rheological measurements are therefore more useful when combined with direct ways of interrogating the structure of the material under flow, such as microscopy, scattering, polarimetry, and others. In the last few years, such methods, combined with numerical simulations, have produced descriptions of flowing complex fluids that are breathtaking in detail. This is the golden age of complex fluids.

REFERENCES

Abbott SR, Tetlow N, Graham AL, Altobelli SA, Fukushima E, Mondy LA, Stephens TS (1991). *J Rheol* 35:773.

Adachi K, Kotaka T (1984). *Macromolecules* 17:120.

Alder BJ, Wainwright TE (1957). *J Chem Phys* 27:1208.

Alderman NJ, Mackley MR (1985). *Faraday Discuss Chem Soc* 79:149.

Alderman NJ, Gavignet A, Guillot D, Maitland GC (1988). SPE Paper 18035, 63rd Annual Technical Conference of the Society of Petrolium Engineers, Houston, Oct 2–5.

Allen MT, Tildesley DJ (1989). *Computer Simulation of Liquids*, Oxford University Press, New York.

Archer LA, Fuller GG, Nunnelley L (1992). *Polymer* 33:3574.

Archer LA, Larson RG, Chen Y-L (1995a). *J Fluid Mech* 301:133.

Archer LA, Chen Y-L, Larson RG (1995b). *J Rheol* 39:519.

Argon AS (1975). In *Polymeric Materials*, Baer E, Radcliffe VS (eds), American Society of Metals, Metals Park, OH, p 411.

Astarita G, Marrucci G (1974). *Principles of Non-Newtonian Fluid Mechanics*, McGraw-Hill, London.

Bahadur B (ed) (1990). *Liquid Crystals: Applications and Uses*, Vol 1, World Scientific, Singapore.

Beckett ST, Craig MA, Gurney RJ, Ingleby BS, Mackley MR, Parsons TCL (1994). *Trans I Chem E* C72:47.

Benbow JJ, Lamb P (1963). *Soc Petrol Eng Trans* 3:7.

Bergeron V, Radke CJ (1992) *Langmuir* 8:3020.

Berne BJ, Pecora R (1990). *Dynamic Light Scattering: with Applications to Chemistry, Biology and Physics*, Wiley, New York.

Betzig E, Trautman JK (1992), *Science* 257:189.

Bicerano J (ed) (1992). *Computational Modelling of Polymers*, Marcel Dekker, New York.

Biermann CJ (1996). *Handbook of Pulping and Papermaking*, 2nd ed, Academic Press, New York.

Binder K (1994a). *Prog Colloid Polym Sci* 96:7.

Binder K (1994b). *Adv Polym Sci* 112:181.

Binder K (ed) (1995). *Monte Carlo and Molecular Dynamics Simulations in Polymer Science*, Oxford University Press, New York.

Bird RB, Armstrong RC, Hassager O (1987a). *Dynamics of Polymer Liquids*, 2nd ed, Vol 1, Wiley, New York.

Bird RB, Curtiss CF, Armstrong RC, Hassager O (1987b). *Dynamics of Polymer Liquids*, 2nd ed, Vol 2, Wiley, New York.

Blake T (1990). *Colloid Surf* 47:135.

Blyler LL, Hart AC (1970). *Polym Eng Sci* 10:193.

Bossis G, Brady JF (1989). *J Chem Phys* 99:567.

Bowden PB (1973). In *The Physics of Glassy Polymers*, Haward RN (ed), Wiley, New York.

Brady JF, Bossis G (1988). *Annu Rev Fluid Mech* 20:111.

Britton MM, Callaghan PT (1997). *J Rheol* 41:1365.

Brochard F, de Gennes PG (1992). *Langmuir* 8:3033.

Browne MW (1996). *New York Times*, page C1.

Buscall R, McGowan JI, Morton-Jones AJ (1993). *J Rheol* 37:621.

Bustamante C, Keller D (1995). *Phys Today* Dec: 32.

Callaghan PT, Cates ME, Rofe CJ, Smeulders JBAF (1996). *J Phys II (France)* 6:375.

Churaev NV, Sobolev VD, Somov AN (1984). *J Colloid Interface Sci* 97:574.

Clarke MT (1993). In *Rheological Properties of Cosmetics and Toiletries*, Laba D (ed), Marcel Dekker, New York.

Clausen TM, Vinson PK, Minter JR, Davis HT, Talmon Y, Miller WG (1992). *J Phys Chem* 96:474.

Cohen Y, Metzner AB (1985). *J Rheol* 29:67.

Cohu O, Magnin A (1995). *J Rheol* 39:767.

Colbourn EA (ed) (1994). *Computer Simulation of Polymers*, Longmans, Harlow, New York.

Coleman BD, Noll W (1961). *Rev Mod Phys* 33:239.

Collings PJ (1990). *Liquid Crystals. Nature's Delicate Phase of Matter*, Princeton University Press, Princeton.

Corleto CR, Bradley WL, Brinson HF (1996). *J Mater Sci* 31:1803.

Cox WP, Merz EH (1958). *J Polym Sci* 28:619.

Crist B (1993). In *Structure and Properties of Polymers, Materials Science and Technology*, Vol 12, Thomas EL (ed), VCH Publishers, New York.

Crochet MJ, Davies AR, Walters K (1984). *Numerical Simulation of Non-Newtonian Flow*, Elsevier, New York.

Dahler JS, Scriven LE (1961). *Nature* 192:36.

Damodaran S, Paraf A (1997). *Food Proteins and Their Applications*, Marcel Dekker, New York.

Dealy JM, Wissbrun KF (1990). *Melt Rheology and Its Role in Plastics Processing*, Van Nostrand Reinhold, New York.

de Gennes PG (1979). *C R Acad Sci Paris* 288:219.

deKruif CG (1992). *Langmuir* 8:2932.

Denn MM (1990). *Annu Rev Fluid Mech* 22:13.

Dickenson E (ed) (1991). *Food Polymers, Gels, and Colloids*, Special Publication No. 82, Royal Society of Chemistry, Cambridge.

Doi M, Chen D (1990). *J Chem Phys* 90:5271.

Doi M, Edwards SF (1986). *The Theory of Polymer Dynamics*, Oxford University Press, New York.

Doyle PS, Shaqfeh ESG, Gast AP (1997). *J Fluid Mech* 334:251.

Drazin PG, Reid WH (1981). *Hydrodynamic Stability*, Cambridge University Press, Cambridge, England.

Durian DJ, Weitz DA, Pine DJ (1991). *Science* 252:686.

Ermak DL, McCammon JA (1978). *J Chem Phys* 69:1352.

Ernst B, Navard P, Hashimoto T, Takebe T (1990). *Macromolecules* 23:1370.

Evans E, Ritchie K, Merkel R (1995). *Biophys J* 68:2580.

Evans E, Bowman H, Leung A, Needham D, Tirrell D (1996). *Science* 273:933.

Exerowa D, Kolarov T, Khristov KHR (1987). *Colloid Surf* 22:171.

Eyring H (1936). *J Chem Phys* 4:283.

Fixman M (1978). *J Chem Phys* 69:1527.

Forgacs OL, Robertson AA, Mason SG (1958). In *Fundamentals of Papermaking Fibres*, British Paper and Board Makers Association, Kenley, Surrey, England.

Fuller GG, (1995). *Optical Rheometry of Complex Fluids*, Oxford University Press, New York.

Fuller GG, Leal LG (1980). *Rheol Acta* 19:580.

Fuller GG, Leal LG (1981). *J Non-Newtonian Fluid Mech* 8:271.

Gabelnick HL, Litt M (1973). *Rheology of Biological Systems*, C Charles Thomas, Springfield, IL.

Galt JC, Maxwell B (1964). *Mod Plastics* 42:115.

Gasvik KJ (1987). *Optical Metrology*, Wiley, New York.

Gerhards C, Schubert H (1993). *Rheology*, Dec: 256.

Ginder JM (1993). *Phys Rev E* 47:3418.

Gittes F, Schnurr B, Olmsted PD, MacKintosh FC, Schmidt CF (1997). *Phys Rev Lett* 79:3286.

Glatter O, Kratky O (1982). *Small-Angle X-Ray Scattering*, Academic Press, New York.

Goldsmith HL (1986). *Microvasc Res* 31:121.

Goldsmith HL, Turitto VT (1986). *Thromb Haemost* 55:415.

Goldstein JI, Newbury DE, Echlin P, Joy DC, Romig JAD, Lyman CE, Fiori C, Lifshin E (1992). *Scanning Electron Microscopy and X-Ray Microanalysis*, Plenum, New York.

Grassia P, Hinch EJ (1996). *J Fluid Mech* 308:255.

Grubb DT, Jelinski LW (1997). *Macromolecules* 30:2860.

Hadziioannou G, Mathis A, Skoulios A (1979). *Colloid Polym Sci* 257:136.

Hashimoto T, Kume T (1992). *J Phys Soc Jpn* 61:1839.

Hatzikiriakos SG, Dealy JM (1991). *J Rheol* 36:703.

Hatzikiriakos SG, Dealy JM (1992). *J Rheol* 36:845.

Havriliak S Jr, Havriliak SJ (1996). In *Physical Properties of Polymers Handbook*, Mark JE (ed), AIP Press, New York.

Henson DJ, Mackay ME (1995). *J Rheol* 39:359.

Higgins JS, Maconnachie A (1987). In *Neutron Scattering*, Skold K, Price DL (eds), Academic Press, New York.

Hirsch PB (1975). *Physics of Metal Defects*, Cambridge University Press, Cambridge, England.

Horne DS (1989). *J Phys D Appl Phys* 22:1257.

Huang K, Fuller GG (1996). *Macromolecules* 29:966.

Hubbell JA (1996). *MRS Bull* 21:33.

Hudson SD, Fleming JW, Gholz E, Thomas EL (1993). *Macromolecules* 26:1270.

Israelachvili JN (1992). *Intermolecular and Surface Forces*, 2nd ed, Academic Press, New York.

Janeschitz-Kriegl H (1983). *Polymer Melt Rheology and Flow Birefringence*, Delft University Press.

Kalika DS, Denn MM (1987). *J Rheol* 31:815.

Keller A, Odell JA (1985). *Colloid Polym Sci* 263:181.

Keunings R (1988). In *Fundamentals of Computer Modeling for Polymer Processing*, Tucker CL II (ed), Hanser Publishers, Munich.

Khan SA, Armstrong RC (1986). *J Non-Newt Fluid Mech* 22:1.

Khan SA, Schnepper CA, Armstrong RC (1988). *J Rheol* 32:69.

Kilara A, Sharkasi T (1994). *MRS Bull* July: 51.

Kirkwood JG (1949). *Rec Trav Chem* 68:649.

Kodama H, Doi M (1996). *Macromolecules* 29:2652.

Koppi KA, Tirrell M, Bates FS, Almdal K, Colby RH (1992). *J Phys II (Paris)* 2:1941.

Kraynik AM, Reinelt DA, Princen HM (1991). *J Rheol* 35:1235.

Kremer K, Binder K (1988). *Comp Phys Rep* 7:259.

Kurti N, This-Benckhard H (1994). *Sci Am* April: 66.

Laba D (1993). *Rheological Properties of Cosmetics and Toiletries*, Laba D (ed), Marcel Dekker, New York.

Langer R (1995). *MRS Bull* 20:18.

Larson RG (1988). *Constitutive Equations for Polymer Melts and Solutions*, Butterworths, New York.

Larson RG (1992). *Rheol Acta* 31:213.

Larson RG, Mead DW (1992). *Liq Cryst* 12:751.

Laun HM (1978). *Rheol Acta* 17:1.

Lesueur D, Gerard J-F, Claudy P, Letoffe J-M, Planche J-P, Martin D (1996). *J Rheol* 40:813.

Lim FJ, Schowalter WR (1989). *J Rheol* 33:1359.

Lindner P, Zemb T (eds) (1991). *Neutron, X-Ray, and Light Scattering*, Elsevier, New York.

Liu TW (1989). *J Chem Phys* 90:5826.

Lochhead RY (1993). In *Rheological Properties of Cosmetics and Toiletries*, Laba D (ed), Marcel Dekker, New York.

Luengo G, Schmitt F-J, Hill R, Israelachvili J (1997). *Macromolecules* 30:2482.

Macosko CW (1994). *Rheology Principles, Measurements, and Applications*, VCH Publishers, New York.

Maitland GC (1991). *Br Soc Rheol Bull* 33:78.

Mason TG, Weitz DA (1995). *Phys Rev Lett* 74:1250.

Mason TG, Ganesan K, van Zanten JH, Wirtz D, Kuo SC (1997). *Phys Rev Lett* 79:3282.

Mather PT (1996). *Liq Cryst* 20:527.

Matsuoka S (1992). *Relaxation Phenomena in Polymers*, Hanser, New York.

Maxwell JC (1867). *Philos Trans R Soc* A157:49.

Meissner J (1971). *Rheol Acta* 10:230.

Meissner J (1972). *J Appl Polym Sci* 16:2877.

Metropolis N, Rosenbluth AW, Rosenbluth MN, Teller AH, Teller E (1953). *J Chem Phys* 21:1087.

Middleman S (1977). *Fundamentals of Polymer Processing*, McGraw-Hill, New York.

Migler KB, Hervet H, Leger L (1993). *Phys Rev Lett* 70:287.

Miles IS, Zurek A (1988). *Polym Eng Sci* 28:796.

Mondain-Monval O, Leal-Calderon F, Bibette J (1996). *J Phys II (France)* 6:1313.

Monnerie L (1987). In *Developments in Oriented Polymers—2*, Ward IM (ed), Elsevier, New York.

Mooney M (1931). *J Rheol* 2:210.

Morris JR, Wojcik T (1990). *Soldering & Surface Mount Technology*, No. 5.

Morrison FA, Mays JW, Muthukumar M, Nakatani AI, Han CC (1993). *Macromolecules* 26:5271.

Müller-Mohnssen H, Weiss D, Tippe A (1990). *J Rheol* 34:223.

Mysels KJ, Jones MN (1966). *Discuss Faraday Soc* 42:42.

Navard P (1994). In *Rheo-Physics of Multiphase Polymer Systems: Characterization by Rheo-Optical Techniques*, Sondergaard K, Lyngaae-Jorgensen J (eds), Technomic, Lancaster, PA.

Ohta T, Nozaki H, Doi M (1990). *J Chem Phys* 93:2664.

Öttinger HC (1996). *Stochastic Processes in Polymeric Fluids*, Springer, New York.

Overney RM, Takano H, Fujihira M, Paulus W, Ringsdorf H (1994). *Phys Rev Lett* 72:3546.

Pader M (1993). In *Rheological Properties of Cosmetics and Toiletries*, Laba D (ed), Marcel Dekker, New York.

Patel SS, Takahashi KM (1992). *Macromolecules* 25:4382.

Paul W, Binder K, Heermann DW, Kremer K (1991). *Chem Phys* 95:7726.

Pearson JRA, Richardson SM (1983). *Computational Analysis of Polymer Processing*, Elsevier, New York.

Perkins TT, Smith DE, Chu S (1994). *Science* 264:819.

Persello J, Magnin A, Chang J, Piau JM, Cabane B (1994). *J Rheol* 38:1845.

Piau JM, El Kissi N, Tremblay B (1990). *J Non-Newt Fluid Mech* 34:145.

Pine DJ, Weitz DA, Chaikin PM, Herbolzheimer E (1988). *Phys Rev Lett* 66:3132.

Pine DJ, Weitz DA, Zhu JX, Herbolzheimer E (1990). *J Phys France* 51:2101.

Ramamurthy AV (1986). *J Rheol* 30:337.

Raman CV, Krishnon KS (1928). *Nature* 121:501.

Reif F (1965). *Fundamentals of Statistical and Thermal Physics*, McGraw-Hill, New York.

Reimer L (1993). *Transmission Electron Microscopy: Physics of Image Formation and Micro-analysis*, Springer-Verlag, Berlin.

Reiner M (1964). *Phys Today*, Jan: 62. Reiner quotes the prophetess Deborah in Judges 5:5, "even the mountains flowed before the Lord." Thus, solids are really liquids if viewed over a long enough period of time.

Reiter G, Demirel AL, Peanasky J, Cai LL, Granick S (1994). *J Chem Phys* 101:2606.

Remz HM (1993). In *Rheological Properties of Cosmetics and Toiletries*, Laba D (ed), Marcel Dekker, New York.

Rief M, Oesterhelt F, Beymann B, Gaub HE (1997). *Science* 275:1295.

Rofe CJ, de Vargas L, Perez-González J, Lambert RK, Callaghan PT (1996). *J Rheol* 40:1115.

Safinya CR, Sirota EB, Plano RJ (1991). *Phys Rev Lett* 66:1986.

Sawyer LC, Grubb DT (1987). *Polymer Microscopy*, Chapman and Hall, New York.

Schönhals A, Kremer F, Schlosser E (1991). *Phys Rev Lett* 67:999.

Sharkasi T, Kilara A (1994). *MRS Bull* July: 47.

Sharpe WN Jr (ed) (1989). *Micromechanics: Experimental Techniques*, American Society of Mechanical Engineers, New York.

Simmons AH, Michal CA, Jelinski LW (1996). *Science* 271:84.

Slichter CP (1989). *Principles of Magnetic Resonance*, 3rd ed, Springer-Verlag, New York.

Smith SB, Finzi L, Bustamante C (1992). *Science* 258:1122.

Smith SB, Cui Y, Bustamante C (1996). *Science* 271:795.

Stockmayer WH (1967). *Pure Appl Chem* 15:539.

Straty GC (1989). *NIST J Res* 94:259.

Strick TR, Allemand J-F, Bensimon D, Bensimon A, Croquette V (1996). *Science* 271:1835.

Struble LJ, Sun GK (1995). *Adv Cem Bas Mat* 2:62.

Tabor D, Winterton RHS (1969). *Proc R Soc Lond* A312:435.

Tanaka H (1994). *J Chem Phys* 100:5323.

Tanner RI (1985). *Engineering Rheology*, Oxford Univeresity Press, New York.

Termonia Y (1994). *Macromolecules* 27:7378.

Tildesley DJ (1995). *Faraday Discuss* 100:C29.

Tirrell DA (1996). *Science* 271:39.

Tirtaatmadja V, Sridhar T (1993). *J Rheol* 37:1081.

Tolstoi DM (1952). *Dokl Acad Nauk SSR* 85:1089.

Tonelli AE (1989). *NMR Spectroscopy and Polymer Microstructure: The Conformational Connection*, VCH Publishers, New York.

Tonelli AE (1996). In *Physical Properties of Polymers Handbook*, Mark JE (ed), AIP Press, New York.

Trease RE, Dietz RL (1972). *Solid State Technol* Jan: 39.

Vermant J, Moldenaers P, Picken SJ, Mewis J (1994). *J Non-Newtonian Fluid Mech* 53:1.

Viney C, Huber AE, Dunaway DL, Kerkam K, Case ST (1994). In *Silk Polymers: Materials Science and Biotechnology*, Kaplan D, Adams WW, Farmer B, Viney C (eds), American Chemical Society, Washington, DC.

Vinogradov GV, Malkin AY, Yanovskii YuG, Borisenkova EK, Yarlykov BV, Berezhnaya GV (1972). *J Polym Sci* A-2 10:1061.

Wang S-Q, Drda PA (1996). *Macromolecules* 29:2627.

Wardhaugh LT, Boger DV (1991). *J Rheol* 35:1121.

Watanabe H, Yamazaki M, Yoshida H, Kotaka T (1991). *Macromolecules* 24:5372.

Weissenberg K (1947). *Nature* 159:310.

Weitz DA, Pine DJ (1993). In *Dynamic Light Scattering: The Method and Some Applications*, Brown W (ed), Oxford University Press, Oxford.

Wen Y (1996). In *Physical Properties of Polymers Handbook*, Mark JE (ed), AIP Press, New York.

Wignall GD (1996). In *Physical Properties of Polymers Handbook*, Mark JE (ed), AIP Press, New York.

Willcox PJ, Gido SP, Muller W, Kaplan DL (1996). *Macromolecules* 29:5106.

Winey KI, Patel SS, Larson RG, Watanabe H (1993). *Macromolecules* 26:4373.

Wu X-L, Pine DJ, Dixon PK (1991). *Phys Rev Lett* 66:2408.

Wunderlich B (1990). *Thermal Analysis*, Academic Press, San Diego.

Yamane Y, Kaneda Y, Doi M (1994). *J Non-Newtonian Fluid Mech* 54:405.

Yanase H, Moldenaers P, Abetz V, van Egmond J, Fuller GG, Mewis J (1991). *Rheol Acta* 30:89.

Yin H, Wang MD, Svoboda K, Landick R, Block SM, Gelles J (1995). *Science* 270:1653.

Yoshimura AS, Prud'homme RK (1988). *J Rheol* 32:53.

Zayas JG, King M (1990). *Am Rev Respir Dis* 141:1107.

Zylka W, Öttinger HC (1989). *J Chem Phys* 90:474.

PROBLEMS AND WORKED EXAMPLES

Problem 1.1 Compute the shear rate profile in the Taylor–Couette (or circular Couette) geometry (see Fig. A1-1) for a Newtonian fluid with negligible inertia and negligible gravity and for

$\varepsilon \equiv (r_2 - r_1)/r_1 = 0.1$, where r_2 is the outer radius and r_1 is the inner radius. Show that the rate increases by 21% from the outer to the inner cylinder.

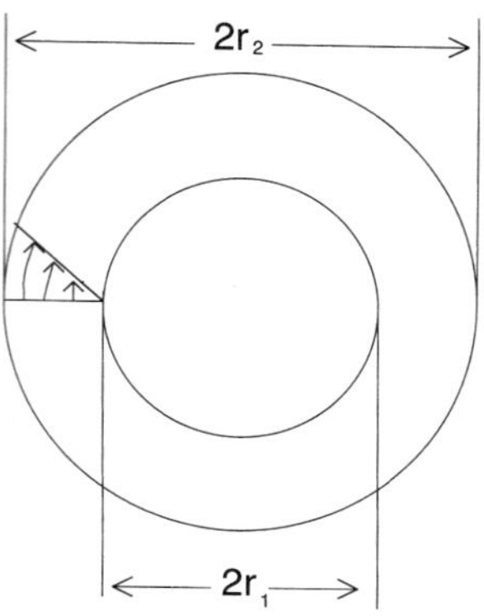

Figure A1.1

Problem 1.2 (Worked Example) Compute the shear-rate profile in a cone-and-plate geometry. For a cone angle α of 0.1 radians, what percentage increase occurs in shear rate as one migrates from the plate to the cone? (*Hint:* Look at the ϕ component of the momentum balance equation in spherical coordinates.)

ANSWER:

We work in spherical coordinates and define the polar angle as $\theta \equiv \pi/2 - \alpha$ (see Fig. A1-2). Now we go to the momentum-balance equations in spherical coordinates in the Appendix. By symmetry, derivatives with respect to r and ϕ are zero and Eq. (A-9) reduces to

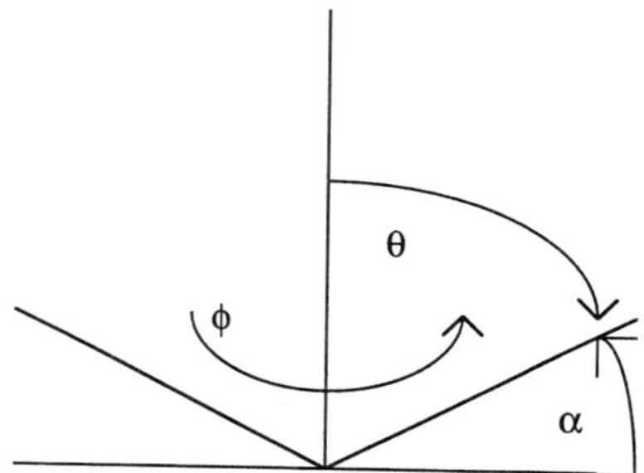

Figure A1.2

$$\frac{1}{r \sin \theta} \frac{\partial}{\partial \theta}(\sigma_{\theta\phi} \sin \theta) + \frac{\cot \theta \, \sigma_{\theta\phi}}{r} = 0 \qquad (A1\text{-}1)$$

This equation can be rearranged to

$$\frac{1}{\sigma_{\theta\phi}} \frac{d}{d\theta}(\sigma_{\theta\phi} \sin \theta) = -\cot \theta \sin \theta = -\cos \theta \qquad (A1\text{-}2)$$

where the partial derivative with respect to θ has been replaced by an ordinary derivative. Applying the product rule to the left side of Eq. (A1-2) gives

$$\frac{1}{\sigma_{\theta\phi}} \left(\frac{d\sigma_{\theta\phi}}{d\theta} \sin \theta + \sigma_{\theta\phi} \cos \theta \right) = -\cos \theta$$

Rearranging gives

$$\frac{d\sigma_{\theta\phi}}{\sigma_{\theta\phi}} = = -2 \cot\theta \, d\theta$$

This equation can be integrated to give

$$\sigma_{\theta\phi} = \frac{C_1}{\sin^2\theta} \tag{A1-3}$$

where C_1 is the constant of integration.

Now, we evaluate the shear stress $\sigma_{\theta\phi}$ on the plate, where the angle θ is $\pi/2$:

$$\sigma_{\theta\phi}\big|_{\theta=\pi/2} = C_1$$

Inserting this into Eq. (A1-3) gives

$$\sigma_{\theta\phi} = \frac{\sigma_{\theta\phi}(\theta = \pi/2)}{\sin^2\theta} \tag{A1-4}$$

On the cone, the angle θ equals $\pi/2 - \alpha$. Hence, from Eq. (A1-4)

$$\sigma_{\theta\phi}\big|_{\theta=\pi/2-\alpha} = \frac{\sigma_{\theta\phi}(\theta = \pi/2)}{\sin^2(\pi/2 - \alpha)} \tag{A1-5}$$

We now use a trigonometric identity to evaluate the denominator of Eq. (A1-5):

$$\sin^2(\pi/2 - \alpha) = [\sin(\pi/2)\cos(\alpha) - \sin(\alpha)\cos(\pi/2)]^2$$

$$= \cos^2(\alpha) = 1 - \sin^2\alpha \approx 1 - \alpha^2$$

Inserting this into Eq. (A1-4) gives

$$\frac{\sigma_{\theta\phi}(\theta = \pi/2 - \alpha)}{\sigma_{\theta\phi}(\theta = \pi/2)} = \frac{1}{1-\alpha^2} = 1.01 \qquad \text{for } \alpha = 0.1 \tag{A1-6}$$

Thus, the shear stress in the cone-and-plate rheometer varies by only 1% throughout the gap, if the cone angle is 0.1 radian (i.e., 5.7°).

Problem 1.3 (Worked Example) Consider a planar flow with the following velocity gradient tensor in the x–y plane:

$$\nabla\mathbf{v} = G\begin{pmatrix} 0 & -1 \\ 1 & 0 \end{pmatrix} \tag{A1-7}$$

where $G = 1$ sec^{-1}. What is the rate of strain tensor $\mathbf{D}$ and the vorticity tensor $\boldsymbol{\omega}$? Compute the tensor $\mathbf{E}$ for the deformation that occurs over the time interval 0 to t. Then compute the Finger strain tensor $\mathbf{B}$.

ANSWER:

Inserting Eq. (A1-7) into Eqs. (1-27) and (1-28) gives

$$\mathbf{D} = \begin{pmatrix} 0 & 0 \\ 0 & 0 \end{pmatrix}, \qquad \boldsymbol{\omega} = \begin{pmatrix} 0 & -1 \\ 1 & 0 \end{pmatrix}$$

To obtain $\mathbf{E}$, we use Eq. (1-15):

$$\frac{\partial}{\partial t}\mathbf{E}(0, t) = \mathbf{E}\cdot\nabla\mathbf{v} = \begin{pmatrix} E_{11} & E_{12} \\ E_{21} & E_{22} \end{pmatrix}\cdot\begin{pmatrix} 0 & -1 \\ 1 & 0 \end{pmatrix} = \begin{pmatrix} E_{12} & -E_{11} \\ E_{22} & -E_{21} \end{pmatrix} \tag{A1-8}$$

where, by definition, at time 0, **E** is the unit tensor:

$$\mathbf{E}(0,0) = \begin{pmatrix} 1 & 0 \\ 0 & 1 \end{pmatrix} \tag{A1-9}$$

We now break Eq. (A1-8) into component equations:

$$\frac{\partial}{\partial t} E_{11} = E_{12} \tag{A1-10}$$

$$\frac{\partial}{\partial t} E_{12} = -E_{11} \tag{A1-11}$$

$$\frac{\partial}{\partial t} E_{21} = E_{22} \tag{A1-12}$$

$$\frac{\partial}{\partial t} E_{22} = -E_{21} \tag{A1-13}$$

Taking a time derivative of Eq. (A1-10) and using Eq. (A1-11) gives

$$\frac{\partial^2}{\partial t^2} E_{11} = \frac{\partial}{\partial t} E_{12} = -E_{11}$$

Mindful of the initial condition in Eq. (A1-9), the solution of this second-order equation for E_{11} is

$$E_{11}(0, t) = \cos(t)$$

Then, using Eq. (A1-10),

$$E_{12}(0, t) = \frac{\partial}{\partial t} E_{11} = -\sin(t)$$

Now, taking a time derivative of Eq. (A1-12) and using Eq. (A1-13) to perform similar manipulations gives

$$E_{22}(0, t) = \cos(t), \qquad E_{21}(0, t) = \sin(t)$$

Thus, after assembling these pieces we obtain

$$\mathbf{E}(0, t) = \begin{pmatrix} \cos(t) & -\sin(t) \\ \sin(t) & \cos(t) \end{pmatrix} \tag{A1-14}$$

The Finger tensor, **B**, is given by Eq. (1-16):

$$\mathbf{B}(t, 0) \equiv \mathbf{E}^T(t, 0) \cdot \mathbf{E}(t, 0) = \begin{pmatrix} \cos(t) & \sin(t) \\ -\sin(t) & \cos(t) \end{pmatrix} \cdot \begin{pmatrix} \cos(t) & -\sin(t) \\ \sin(t) & \cos(t) \end{pmatrix}$$

$$\begin{pmatrix} \cos^2(t) + \sin^2(t) & -\cos(t)\sin(t) + \cos(t)\sin(t) \\ -\sin(t)\cos(t) + \sin(t)\cos(t) & \sin^2(t) + \cos^2(t) \end{pmatrix} = \begin{pmatrix} 1 & 0 \\ 0 & 1 \end{pmatrix} \tag{A1-15}$$

The "flow field" in Eq. (A1-7) is really just a solid-body rotation which rotates, but does not deform, the fluid element. As a result, the rate-of-strain tensor **D** is the zero tensor, and the Finger strain tensor is the unit tensor.

Problem 1.4 Compute the extensional viscosity $\bar{\eta}_b \equiv (\sigma_{xx} - \sigma_{zz})/\dot{\varepsilon}$ in a biaxial extensional flow for a Newtonian fluid with shear viscosity η and where $\partial v_x/\partial x = \partial v_y/\partial y = \dot{\varepsilon}$ is the rate of stretch. Show that $\bar{\eta}_b$ is six times the shear viscosity η.

Problem 1.5 Suppose you shear a polymer melt in a plane-Couette rheometer at a shear stress σ of 0.1 MPa and measure the apparent shear rate $\dot{\gamma}_{app} \equiv V/h$, where V is the velocity of the moving plate (the other is stationary) and h is the gap between the plates. At a gap h of 2 mm, you measure an apparent shear rate $\dot{\gamma}_{app}$ of 1.5 sec^{-1}; at $h = 1$ mm, you measure 2 sec^{-1}; and at $h = 0.5$ mm, you measure 3 sec^{-1}. What is the slip velocity V_s and the true shear rate $\dot{\gamma}$ at this shear stress? What is the extrapolation length b?

Problem 1.6 (Worked Example) Suppose you wish to measure a time-dependent first normal stress difference N_1 in a room-temperature polybutadiene "melt" after the start-up of steady shearing. You could do try to do this in a conventional cone-and-plate rheometer, if the transducer has a spring that permits small axial deflections. With this you could measure the axial force, which is directly related to N_1. However, such a spring makes the device *compliant* in the axial direction; that is, it permits small changes in the gap. Although these deflections are not a serious problem at steady state, if they are changing with time after start-up of shearing, they would, if not resisted, change the sample volume. Of course, they are resisted by a strong, counteracting suction which can ruin the transient measurements of N_1.

One convenient method for completely avoiding this problem is to measure the stresses with birefringence. Since birefringence measurements require no force transducers, the flow device can be made completely rigid, thus avoiding compliance. To obtain σ_{12} and $N_1 \equiv \sigma_{11} - \sigma_{22}$ using birefringence, one must measure the corresponding optical quantities, n_{12} and $n_{11} - n_{22}$, and convert them to stresses using the stress-optical law, Eq. (1-38). To obtain birefringence in the 12 plane, the light beam must propagate in direction "3," along the vorticity axis. A good geometry to do this is the circular Couette device shown in Fig. A1.1, where the beam is sent through the gap between cylinders, parallel to the cylinder axis. One can define a local coordinate system such that "1" is the local flow direction and "2" is the local gradient (or gapwise) direction.

Birefringence can be measured very simply using a *plane polariscope*, which is just two transparent polarizers (which can be obtained from antiglare sunglasses) between which the flow cell containing the sample is mounted. First, one sets the slow axes of the polarizers parallel to each other and measures the transmitted light intensity I_0. The best results are obtained using monochromatic light. Now cross the polarizers by rotating the slow axis of one of them until it is 90° relative to the other. Then, rotate them both, *keeping them crossed*. The intensity of light I will vary as a function of angle θ as you rotate the crossed polarizers:

$$\frac{I}{I_0} = \sin^2[2(\theta - \chi)] \sin^2\left(\frac{\delta}{2}\right)$$

At the angle $\theta = \chi$, the light intensity will fall to zero; χ is therefore called the *extinction angle*. At an angle $\theta = \chi + 45°$, the light intensity will be maximum, and I/I_0 is given by $\sin^2(\delta/2)$, where δ is the *retardation*—that is, the maximum phase shift of light between differing polarizations. (This occurs because the speed of light in the sample depends on the orientation of its electric vector relative to the dipoles in the sample. Thus, two beams with differing polarizations take slightly different times to pass through the sample.) The retardation δ, which depends on sample thickness d, is related to the sample's intrinsic birefringence Δn, by $\delta = 2\pi d \Delta n/\lambda$, where λ is the wavelength of light. Assuming that the sample is thin enough, that the birefringence is weak enough, and that $\delta/2$ is smaller than π, one can invert the sin function to obtain δ, and hence Δn. For more details on polarimetry see Gasvik (1987).

From these measured values of χ and Δn, along with the stress-optical coefficient C, calculate σ_{12} and N_1.

ANSWER:

In the original frame, where "1" is the flow direction, the birefringence tensor has the two-dimensional form

$$\mathbf{n} = \begin{pmatrix} n_{11} & n_{12} \\ n_{12} & n_{22} \end{pmatrix} \tag{A1-16}$$

If we consider a frame locked to the polarizers, the components of birefringence in the rotating frame, $\mathbf{n}'$, change with the angle θ through which we have rotated the polarizers. The effect of a rotation of frame can be described using the *rotation matrix* $\mathbf{Q}$:

$$\mathbf{Q} = \begin{pmatrix} \cos\theta & -\sin\theta \\ \sin\theta & \cos\theta \end{pmatrix} \tag{A1-17}$$

where θ is the counterclockwise angle through which we have rotated the polarizers. The birefringence tensor in the rotated frame is

$$\mathbf{n}' = \mathbf{Q}^T \cdot \mathbf{n} \cdot \mathbf{Q} \tag{A1-18}$$

The frame that extinguishes the light is one in which the tensor $\mathbf{n}'$ is diagonal, with component n'_{11} along direction 1', and n'_{22} along direction 2', where 1' and 2' are the directions into which directions 1 and 2 have been rotated. The component n'_{12} is zero when the polarizers have been rotated to the extinction angle, $\theta = \chi$.

Inverting Eq. (A1-18) to give $\mathbf{n}$ in terms of $\mathbf{n}'$, and carrying out the matrix multiplications, we obtain relationships between the components in the frame rotated to the angle χ and those in the unrotated frame:

$$2n_{11} = (n'_{11} + n'_{22}) + (n'_{11} - n'_{22})\cos(2\chi)$$

$$2n_{22} = (n'_{11} + n'_{22}) - (n'_{11} - n'_{22})\cos(2\chi) \tag{A1-19}$$

$$2n_{12} = (n'_{11} - n'_{22})\sin(2\chi)$$

We then find that

$$n_{11} - n_{22} = (n'_{11} - n'_{22})\cos(2\chi) = \Delta n \cos(2\chi)$$

$$n_{12} = \tfrac{1}{2}(n'_{11} - n'_{22})\sin(2\chi) = \tfrac{1}{2}\Delta n \sin(2\chi)$$

where $\Delta n = n'_{11} - n'_{22}$. From the stress-optical law, Eq. (1-38), we can convert these expressions into ones for σ_{12} and N_1:

$$N_1 = \sigma_{11} - \sigma_{22} = \; = C^{-1}\Delta n \cos(2\chi)$$

$$\sigma_{12} = \tfrac{1}{2}C^{-1}\Delta n \sin(2\chi)$$

The value of C for polybutadiene is around $+2.2 \times 10^{-9}\ \mathrm{m^2/N}$. Values of C for other melts are given in Janeschitz-Kriegl (1983) and Macosko (1994).

Chapter 2

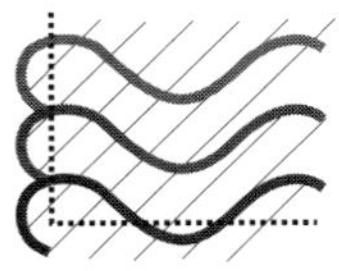

Basic Forces

2.1 INTRODUCTION

The structure of matter at the length scales greater than the atomic is governed by electromagnetic forces. The tendency toward electroneutrality, combined with thermal agitations and the discrete, or quantum, substructure of matter, produces an astonishing diversity of atomic arrangements. At the temperatures of interest to us, around 200–500 K, molecules composed of covalently bonded atoms can be regarded as indivisible units, and the electromagnetic forces that we need consider are those that the molecules exert on one another. The force F between two such molecules is often described using a *potential function* $W(r)$, which for spherical molecules separated by a distance r is given by

$$F = -\frac{dW}{dr} \tag{2-1}$$

A potential function can also be used to describe the force between a pair of colloidal particles. The electromagnetic forces that contribute to $W(r)$ can be grouped into several categories, namely *excluded volume* (or steric), *van der Waals, electrostatic, hydrogen bonding*, and *hydrophobic*.

2.2 EXCLUDED-VOLUME INTERACTIONS

When molecules or atoms are brought closer and closer together, their electron clouds eventually overlap, producing a very strong repulsion that increases so steeply with decreasing intermolecular distance that it easily overpowers all other forces. This strong, but short-ranged, repulsive interaction gives each molecule a reasonably well-defined size and shape that cannot be intruded upon by neighboring molecules. This *excluded-volume* force is largely responsible for determining the short-range structure of liquids and the crystallographic order of solids composed of small molecules, or of densely packed hard colloidal particles. The nature of the excluded-volume interaction depends, of course, on the shape and flexibility of the molecules or particles that interact. In what follows, we briefly consider the excluded-volume forces for the simplest cases, namely, hard spherical particles, hard nonspherical particles or molecules, flexible macromolecules, and semiflexible macromolecules.

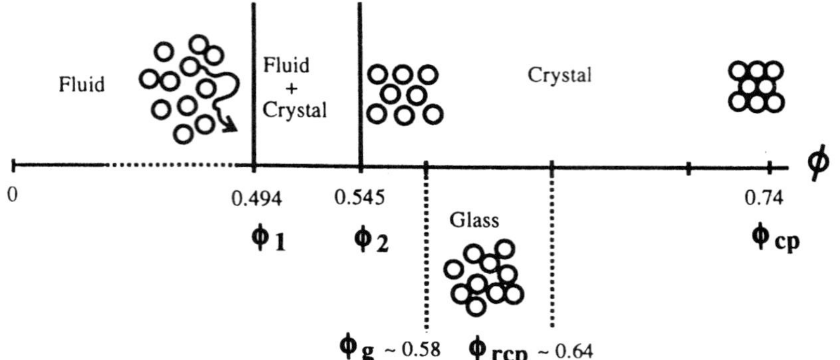

Figure 2.1   The hard-sphere phase diagram. Below volume fraction $\phi < \phi_1 = 0.494$, the suspension is a disordered fluid. Between $\phi_1 = 0.494$ and $\phi_2 = 0.545$, there is coexistence of this disordered phase with a colloidal crystalline phase with FCC (or HCP) order; the colloidal crystalline phase is the equilibrium one up to the maximum close-packing limit of $\phi_{cp} = 0.74$. Nonequilibrium colloidal "glassy" behavior can also occur between $\phi_g = 0.58$ and the limit of random close packing at $\phi_{rcp} = 0.64$. (From Poon and Pusey, fig. 5, with kind permission of Kluwer Academic Publishers, Copyright 1995.)

2.2.1 Hard Spheres

The equilibrium phase diagram for hard spheres (Fig. 2-1) has been determined by computer simulations and experiments on stable dispersions of monodisperse colloidal particles (Hoover and Ree 1968; Woodcock 1981; Poon and Pusey 1995). At volume fractions ϕ below 0.49, the spheres are disordered. At $\phi = \phi_1 = 0.49$, the spheres begin to order into a macrocrystalline structure, with close packing that can be considered a mixture of face-centered cubic (FCC) and hexagonally close packed (HCP) packings. In both of these packings, each sphere has 12 identically spaced nearest neighbors; the only difference between FCC and HCP is in the stacking sequence of the layers of spheres.

Amazingly, the hard-sphere crystallization transition is driven by entropy! At high packing densities, the ordering of the spheres onto a regular lattice gives each sphere greater room for positional fluctuations than would be the case for random packing at the same density, thus more than compensating for the entropic cost of the ordering (see Fig. 2-2). In the volume-fraction range $\phi_1 < \phi < \phi_2 = 0.545$, the disordered phase and the colloidal crystalline phase coexist. The colloidal crystalline phase can theoretically persist from ϕ_2 up to the concentration at the HCP limit, $\phi_{cp} = 0.7405$; this is the highest volume fraction that respects the hard-core diameter of the spheres.

In addition to these equilibrium phases, there is a metastable *glassy* disordered state that can exist at volume fractions above about 0.56 (Pusey and van Megan 1987). This phase exists because at such high densities the long-range Brownian motions of the spheres are suppressed by the crowding or "caging" effect of neighboring spheres, and critical nuclei needed to induce crystallization cannot form. Thus, if the concentration of spheres can be increased quickly enough (say, by centrifugation) so that the concentration regime where crystallization occurs is bypassed, one obtains a colloidal glass. The most densely packed

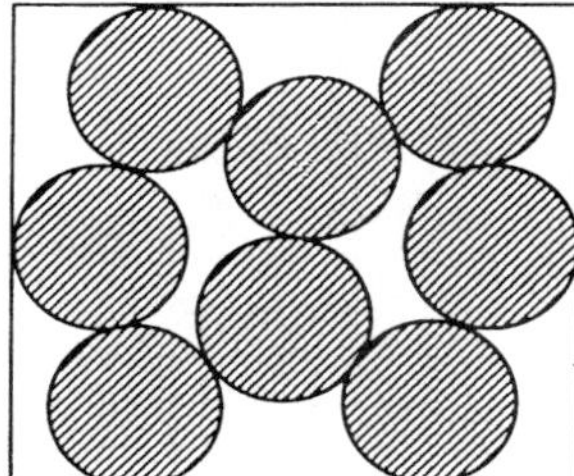 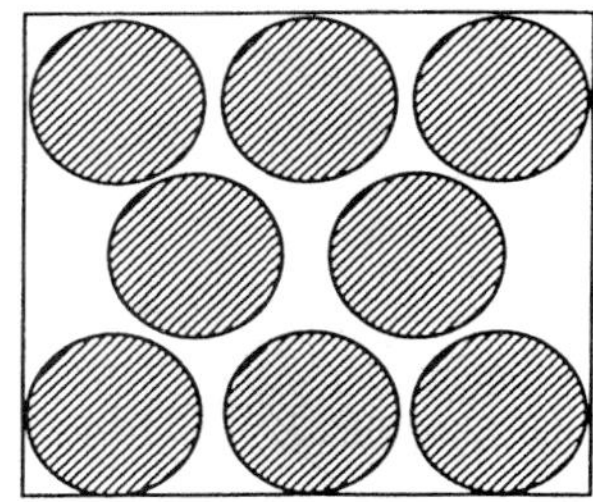

Figure 2.2. In **(a)** and **(b)**, the same number of spheres of the same size are packed into the same space. The disordered sphere packing in **(a)** can create more "free volume" by ordering into a regular packing in **(b)**, thereby creating volume entropy while losing configurational entropy (after Lekkerkerker, unpublished). (From Poon and Pusey, fig. 4, with kind permission of Kluwer Academic Publishers, Copyright 1995.)

state of a glassy suspension of hard spheres is "random close packing," for which $\phi = 0.64$. This concentration is 86% that of ordered close packing (see Fig. 2-1). This difference in maximum packing between the ordered and disordered states shows that the ordered state has more "free volume" than the disordered one, and it is the difference in entropy associated with this free volume that drives the ordering transition. (Interestingly, the density of liquids composed of spherical molecules or atoms at their melting point is also typically about 86% as high as the density of the crystal at 0 K (Lindemann 1910).

Even in the liquid state, with $\phi < \phi_1 = 0.49$, local order is not entirely absent. Liquid-state packing of hard spherical objects leads to correlations in molecular positions. For example, a hard spherical molecule in the liquid state is surrounded by, and is in near contact with, on average about nine *nearest neighbors*. There are also second and third nearest neighbors that are much more numerous and have much weaker positional correlation to the central sphere. The positional correlations that exist between pairs of molecules are described by the *radial distribution function*, $g(r)$; $g(r)$ is proportional to the probability of finding the center of mass of a second molecule a distance r away from the center of mass of a given central molecule. The normalization is chosen so that $g(r) = 1$ for molecules with no positional correlation. Figure 2-3 shows the theoretical radial distribution function for hard-sphere suspensions at different concentrations. Note that the largest peak is at nearest-neighbor contact, where $r/2a = 1$. At high concentrations ($\phi \gtrsim 0.4$), there are smaller peaks at next-neighbor packing "shells" located roughly at $r/2a \approx 2, 3$, and so on. In the colloidal crystal state ($\phi > 0.545$), these peaks become infinitely sharp and repeat out to infinite distances.

In the general case in which the phase might (or might not) have positional order, one can define an anisotropic pair correlation function, $g(\mathbf{x})$, where $\mathbf{x}$ is a position vector relative to a given molecule. The Fourier transform of the pair correlation function, namely,

$$S(\mathbf{k}) \equiv \int g(\mathbf{x}) \exp(i\mathbf{k} \cdot \mathbf{x})\, d\mathbf{x} \equiv \int_{-\infty}^{\infty} \int_{-\infty}^{\infty} \int_{-\infty}^{\infty} g(\mathbf{x}) \exp(i\mathbf{k} \cdot \mathbf{x})\, dx_1\, dx_2\, dx_3 \quad (2\text{-}2)$$

is called the *structure factor* with $\mathbf{k}$ the wavevector. $S(\mathbf{k})$ can be measured by scattering methods described in Section 1.6.2. An example is the structure factor for concentrated,

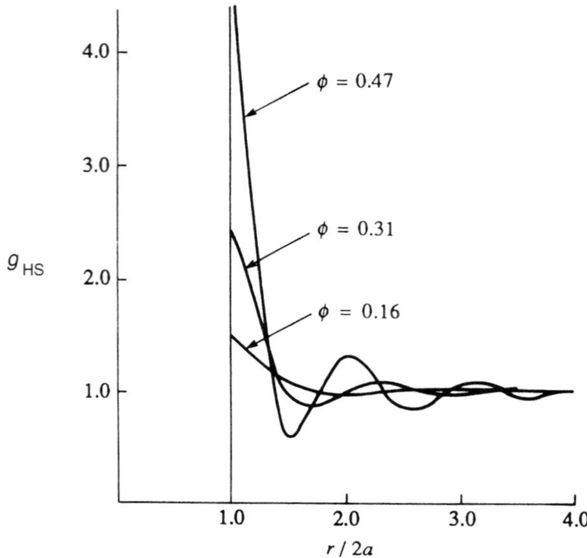

Figure 2.3. Radial distribution function $g_{HS}(r)$ for suspensions of hard spheres in the disordered state at various volume fractions ϕ, calculated from the Percus–Yevick equation. (From Russel et al. 1989, with permission of Cambridge University Press.)

but disordered, hard spheres, shown in Fig. 4-21. The rheological properties of hard-sphere suspensions in the liquid state are described in Section 6.2, and those in the glassy hard-sphere state are described in Section 4.6.

2.2.2 Rigid Nonspherical Particles or Molecules: The Nematic Phase

For molecules that are not spherical, packing and ordering transitions can occur that are more complex than those for spherical molecules. Perhaps the simplest nonspherical shape is a stiff, long cylinder (see Fig. 2-4). This shape is approximated by some synthetic and natural molecules and colloidal particles, such as tobacco mosaic virus, not-too-long molecules of poly(γ-benzyl glutamate), or akaganeite (βFeOOH) mineral particles (Frattini and Fuller 1986). The closest packing of cylindrical rods occurs when they are parallel to each other and packed hexagonally in the plane orthogonal to their axes; in this case, $\phi = 0.9069$. If the density of long ordered rods is decreased, a melting transition will occur in which the in-plane hexagonal order is lost, but the orientational order of the rod axes is partially preserved. This partially ordered state is called a *nematic*. States with partial order, including the nematic state, are common for stiff molecules of high aspect ratio. Another partially ordered state, called a *smectic A*, can occur in oriented solutions of hard or semiflexible rods when the ends of the rods congregate into *layers* that are orthogonal to the direction of rod orientation (Poniewierski 1992; Tkachenko 1996).

   The degree of orientational order in a nematic is described by an *orientational order parameter S. S* quantifies the degree to which molecular orientations are parallel to a

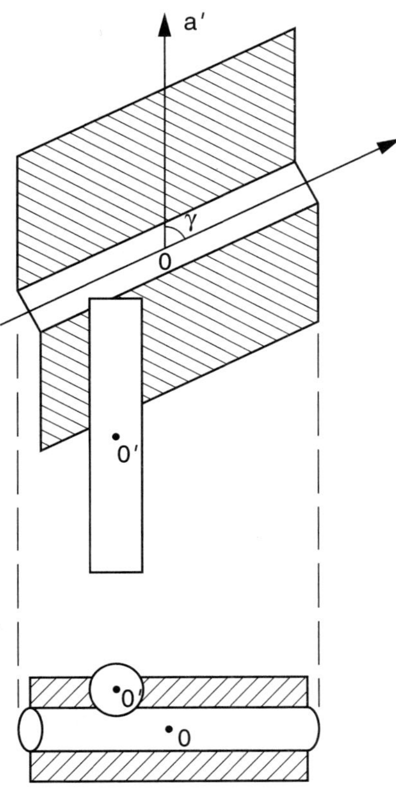

Figure 2.4. Packing of hard rod-like molecules. To avoid intersection, rod O excludes from the shaded region the center of mass of any other rod, such as O'. The volume of this region around rod O is proportional to the angle γ between the axes of the two rods. From this geometric argument, the *Onsager excluded-volume potential*, Eq. (2-5), is constructed. (From de Gennes and Prost, copyright © 1993 by Oxford University Press, Inc. Used by permission of Oxford University Press, Inc.)

common nematic axis. $S = 1$ corresponds to a perfectly parallel orientation of all rod-like molecules or particles, while $S = 0$ corresponds to a completely random, or *isotropic*, distribution of molecular orientations. At equilibrium, S can be defined precisely by

$$S = \tfrac{3}{2}\langle\cos^2\theta\rangle - \tfrac{1}{2} \tag{2-3}$$

where θ is the angle between a rod and the nematic axis, and the brackets "$\langle\ \rangle$" denote an average over all rods; that is,

$$\langle\cdot\rangle \equiv \int \cdot\,\psi(\mathbf{u})\,d u^2 \equiv \int_0^\pi \int_0^{2\pi} \cdot\,\psi(\mathbf{u})\,d\phi\,\sin\theta\,d\theta \tag{2-4}$$

where $\psi(\mathbf{u})\,d u^2$ is the probability that a rod's orientation lies between the unit vectors $\mathbf{u}$ and $\mathbf{u} + d\mathbf{u}$, and $\mathbf{u}$ can be represented as $\mathbf{u} = (u_1, u_2, u_3) = (\sin\theta\,\cos\phi, \sin\theta\,\sin\phi, \cos\theta)$, with θ the polar and ϕ the azimuthal angle. At the transition from the isotropic to the nematic state, S jumps from zero to a finite value, usually around 0.3 or higher.

2.2.2.1 Onsager Theory

The distribution of orientations can, in principle, be computed theoretically from a *nematic potential* that expresses the influence of one rod's orientation on that of its neighbors.

Onsager (1949) was the first to propose a theory from which one can obtain such a potential; his theory was derived for an ideal solution of long, perfectly stiff, hard rods interacting only by excluded-volume forces, at concentrations dilute enough that only pairwise interactions are significant. The Onsager potential, derived from a geometric excluded-volume argument illustrated in Fig. 2-4, is given by

$$V_{\text{nem}}(\mathbf{u}) = V_O(\mathbf{u}) = U_O k_B T \int \psi(\mathbf{u}') \sin(\mathbf{u}', \mathbf{u}) \, du'^2 \qquad (2\text{-}5)$$

where $\sin(\mathbf{u}', \mathbf{u})$ is the sine of the angle between $\mathbf{u}'$ and $\mathbf{u}$. The nematic parameter U_O is given by $U_O = 2\nu d L^2$, where ν is the number concentration of cylindrical rods and d and L are their diameter and length, respectively. One can also express U_O in terms of the volume fraction ϕ of rods in solution; using $\phi = \pi d^2 L \nu / 4$, we obtain $U_O = (8L/\pi d)\phi$.

From the potential $V_{\text{nem}}(\mathbf{u})$, one can obtain the rod orientation distribution function $\psi(\mathbf{u})$, and hence the order parameter S, by a *self-consistent* calculation. At equilibrium, the distribution function $\psi(\mathbf{u})$ is related to the potential $V_{\text{nem}}(\mathbf{u})$ by Boltzmann's equation:

$$\psi(\mathbf{u}) = C e^{-V_{\text{nem}}(\mathbf{u})/k_B T} \qquad (2\text{-}6)$$

where C is a normalization constant, $C^{-1} \equiv \int e^{-V_{\text{nem}}(\mathbf{u})/k_B T} \, du^2$. From Eqs. (2-5) and (2-6), one can solve simultaneously for the functions $\psi(\mathbf{u})$ and $V_{\text{nem}}(\mathbf{u})$.

For low concentrations ν—that is, small U_O—the only solution is the trivial one, $\psi = \text{const.} = 1/(4\pi)$, corresponding to the isotropic state. For a high enough value of U_O, there is in addition to the isotropic solution a stable nontrivial solution corresponding to a nematic state with $S > 0$. Of these two solutions, the one with the lowest free energy corresponds to the equilibrium state. As U_O increases, the lowest free-energy state changes from the isotropic to the nematic.

Onsager's potential is purely entropic; hence for a given rod diameter and length, the transition to the liquid crystalline state occurs at a concentration that is independent of the temperature T. And since the Onsager potential applies to a two-component system (rods + solvent), there is a biphasic range of concentrations over which the isotropic and nematic phases coexist. From numerical calculations, one finds that solutions with $U_O \leq U_{O,1} = 8.38$ are isotropic, while those for which $U_O \geq U_{O,2} = 10.67$ are nematic. These values correspond to volume fractions ϕ_1 and ϕ_2 of $3.3d/L$ and $4.1d/L$. Intermediate values of $U_O, U_{O,1} \leq U \leq U_{O,2}$ produce two-phase coexistence. At the lowest concentration for which the solution is completely nematic, the Onsager theory predicts that the order parameter is $S = 0.792$. S rises rapidly with increasing U_O, exceeding 0.95 at $U_O = 20$ (Lekkerkerker et al. 1984).

While these predictions of the Onsager theory for the dependence of S on U_O seem to be in qualitative agreement with experimental measurements for semirigid molecules such as poly(γ-benzyl-L-glutamate) (PBLG), the predicted values of $U_{O,1}$ and $U_{O,2}$ tend to be lower than the measured values (Abe and Yamazaki 1989). These deviations are probably due to the imperfect stiffness of these molecules. Evidence for this is presented in Fig. 2-5, which shows the values of ϕ_1 and ϕ_2, the volume fractions of molecules corresponding to the boundaries of the biphasic region. Figure 2-5 shows that for PBLG molecules, ϕ_1 and ϕ_2 decrease roughly as $1/L$ with increasing molecular length, in agreement with the Onsager theory, up to a length of around 600 Å, corresponding to about 400 monomers. For longer

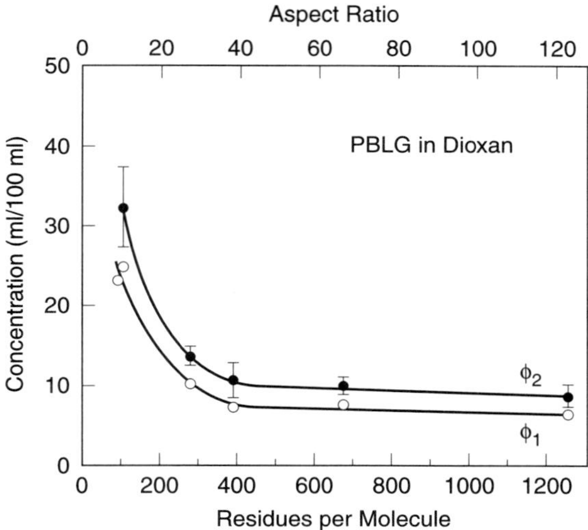

Figure 2.5 Critical volume fractions ϕ_1 below which a solution of poly(benzyl-L-glutamate) is fully isotropic ($\bigcirc$) and ϕ_2 above which the solution is fully liquid crystalline ($\bullet$) as functions of the number of peptide residues N in the chain. The molecular contour length is roughly 1.5N Å. (From Robinson et al. 1958, reproduced by permission of The Royal Society of Chemistry.)

molecules, ϕ_1 and ϕ_2 are roughly independent of molecular length, presumably because the longer molecules no longer behave as rigid rods (see Section 2.2.4).

Onsager's potential also applies to *oblate particles* or disks of high aspect ratio, if the parameter U_O is redefined as $4\pi \nu d^3$, where d is the disk diameter (de Gennes and Prost 1993). Solutions composed of disk-shaped particles or molecules can therefore also undergo a transition to a nematic phase, called a *discotic nematic*.

The Onsager theory and its extensions are valid only when the concentration is low enough that pairwise excluded-volume interactions are the dominant ones. Thus, these theories are not likely to apply to solvent-free bulk, or *thermotropic*, liquid crystalline phases, for which there are likely to be complex packing interactions and anisotropic *energetic* interactions, such as those produced by van der Waals forces.

2.2.2.2 Maier–Saupe Theory

An alternative nematic potential due to Maier and Saupe (1958, 1959, 1960) is perhaps more appropriate for thermotropic nematics. The Maier–Saupe potential is given by

$$V_{\text{nem}}(\mathbf{u}) = V_{\text{MS}}(\mathbf{u}) \equiv \text{const} - {}^3\!/_2 u_{\text{MS}} \mathbf{u}\mathbf{u} : \mathbf{S} \tag{2-7}$$

where u_{MS} is a phenomenological energy constant, and the order parameter *tensor* $\mathbf{S}$ is defined by

$$\mathbf{S} \equiv \langle \mathbf{u}\mathbf{u} \rangle - {}^1\!/_3 \boldsymbol{\delta} \tag{2-8}$$

with δ the unit tensor. At equilibrium, the order parameter tensor is related to the scalar order parameter S by

$$\mathbf{S} = S\,(\mathbf{nn} - \tfrac{1}{3}\delta) \tag{2-9}$$

where $\mathbf{n}$ is a unit vector, called the "director," parallel to the direction of average orientation, and S at equilibrium is obtained by a self-consistent solution of Eqs. (2-6) and (2-7). More generally, away from equilibrium where the tensor order parameter $\mathbf{S}$ may not be uniaxial [i.e., expressible by Eq. (2-9)], the scalar order parameter can be defined as

$$S \equiv \left(\tfrac{3}{2}\mathbf{S} : \mathbf{S}\right)^{1/2} \tag{2-10}$$

In their original theory, Maier and Saupe supposed that the molecular interactions responsible for the nematic state are anisotropic van der Waals interactions (discussed in Section 2.3), in which case u_{MS} should be temperature-independent. However, it is now recognized that shape anisotropy is also important, even for small-molecule thermotropic nematics. By making u_{MS} temperature-dependent, the Maier–Saupe potential can, in principle, accommodate both energetic and entropic effects. In fact, if the function $\sin(\mathbf{u}', \mathbf{u})$ in the purely entropic Onsager potential Eq. (2-5) is approximated by the expansion $1 - \tfrac{1}{2}\cos^2(\mathbf{u}', \mathbf{u}) + \ldots$, then to lowest order the Maier–Saupe potential (2-7) is obtained with $U_{\mathrm{MS}} = U_O k_B T/3$, where we have defined the dimensionless Maier–Saupe energy constant by $U_{\mathrm{MS}} \equiv u_{\mathrm{MS}}/k_B T$. Thus, the Maier–Saupe potential can be used as an approximation to describe orientational order in either *lyotropic* (solvent-based) or thermotropic nematics. For a thermotropic melt, the Maier–Saupe theory predicts a first-order transition from the isotropic to the nematic phase when $u_{\mathrm{MS}}/k_B T = U_{\mathrm{MS}} = U_{1,\mathrm{MS}} = 4.55$, and at this transition the scalar order parameter S jumps from zero to 0.43. S increases toward unity with further increases in U_{MS}. The spinodal point at which the isotropic phase is unstable to even small orientational perturbations occurs at $U = U_{\mathrm{MS}}^* = 5$ for the Maier–Saupe potential, and $U = U_O^* = 10.19$ for the Onsager potential. The two potentials are significantly different in their predictions of the dependence of the order parameter on U. For a given value of U/U^*, the Onsager potential predicts a higher-order parameter than does the Maier–Saupe potential.

2.2.2.3 Flory Lattice Theory

In addition to the Onsager and Maier–Saupe theories, there is also a *lattice theory* of Flory that predicts a transition to a nematic state (Flory 1956, 1984). By segmenting rods as depicted in Fig. 2-6, Flory computed the number of ways of packing these rods on the cubic lattice, thereby calculating the partition function and hence the free energy. Flory's theory is more appropriate to the densely packed state, while Onsager's theory becomes rigorous when the volume fraction of rods becomes small enough that only binary interactions between rods are important. For a solvent-free melt, the Flory theory predicts that molecules with an axial ratio L/d greater than 6.42 will form a nematic phase (see below); molecules with an aspect ratio smaller than this will produce an isotropic phase.

The rheology of isotropic suspensions and solutions of stiff molecules and particles is covered in Section 6.3. The rheology of lyotropic and thermotropic nematic liquid crystals composed of long, stiff molecules is described in Chapter 11.

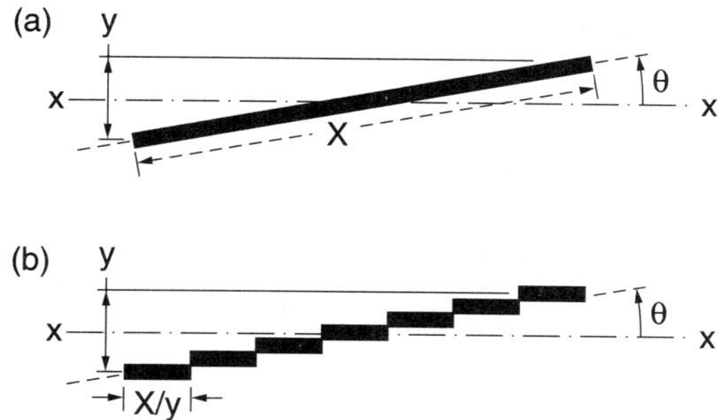

Figure 2.6 Illustration of Flory lattice theory for the thermodynamics of long stiff rods. In **(a)**, the rod of length X measured in units of the lattice spacing is oriented at an angle θ with respect to the preferred horizontal direction. In **(b)**, the rod is represented on a cubic lattice by subdividing the rod into y segments, each of length X/y. (From Flory 1984, reprinted with permission from Springer Verlag.)

> • Worked Example 10.5 of Chapter 10 shows how the Maier–Saupe potential can be used to predict the order parameter of a nematic.

2.2.3 Flexible Macromolecules

Molecules can be long without being stiff. Common examples of long, flexible molecules include hydrocarbon oligomers and flexible linear macromolecules such as polystyrene and polyethylene, which are composed of long sequences of identical chemical units called monomers (see Fig. 2-7). Synthetic polymers typically contain a few dozen to tens of thousands of monomers, or even more. (Oligomers are chains containing only a few to a few dozen monomers.) Materials made from flexible polymers are ubiquitous in everyday life, and they include polyethylene wrap for food, polyvinylchloride piping, polystyrene plastic cups, polyisoprene rubber tires, nylon fibers, and many others (see Section 1.2.5). The choice of a particular polymer for a given application is controlled to a large extent by the polymer's physical properties, which are tabulated in a handbook by Mark (1996), a book by Rodriguez (1996), and the *Modern Plastics Encyclopedia '95* (1994).

2.2.3.1 Polydispersity

The molecular weight M_0 of a typical monomer is around 100 daltons, so a polymer's *molecular weight M* ranges from a few thousand to a few million daltons, or perhaps even tens of millions of daltons. Biological macromolecules, especially DNA, can be truly enormous, with molecular weights exceeding a billion! Synthetic polymer molecules are almost always at least somewhat *polydisperse*—that is, possessing more than a single molecular weight. The average molecular weight and its distribution are typically characterized by *moments* of the distribution. The most important of these moments are the *number-averaged* and the *weight-averaged* molecular weights, M_n and M_w, defined by

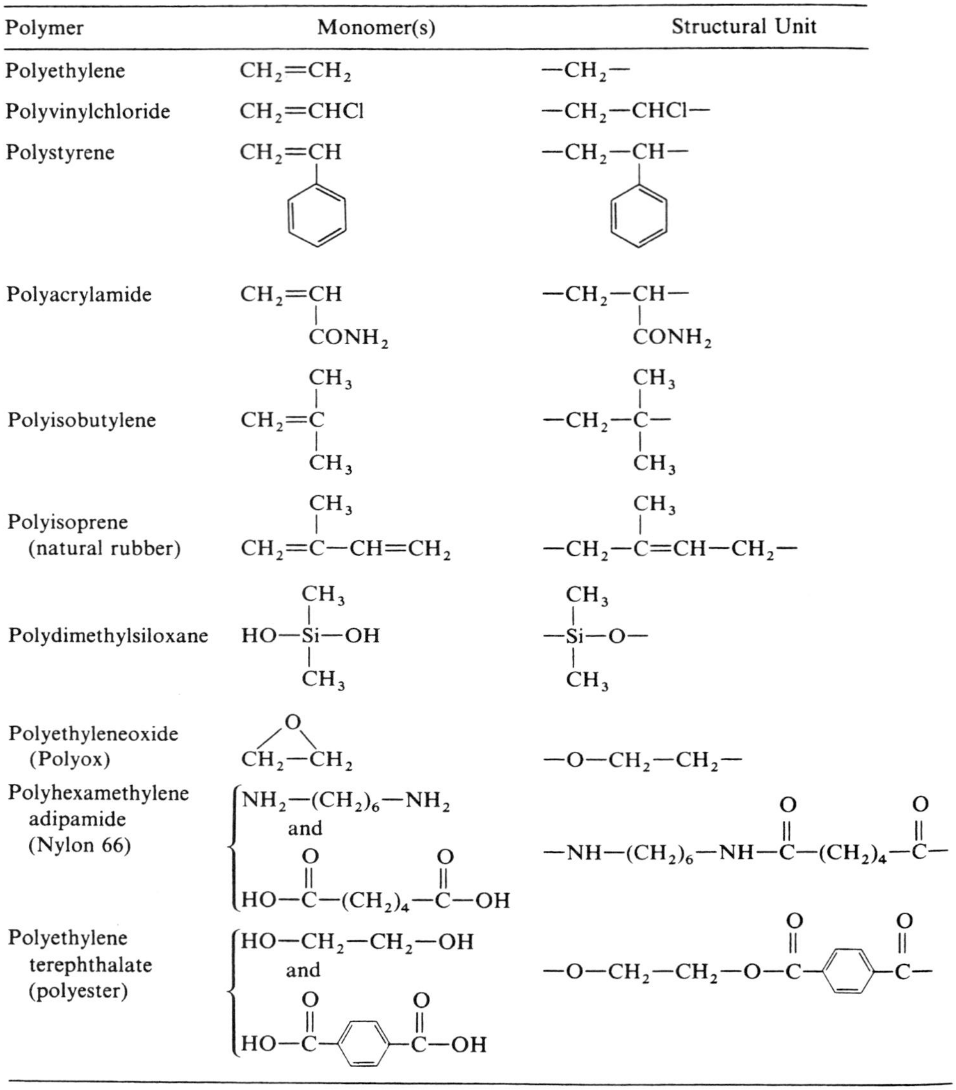

Polymer	Monomer(s)	Structural Unit
Polyethylene	$CH_2{=}CH_2$	$-CH_2-$
Polyvinylchloride	$CH_2{=}CHCl$	$-CH_2-CHCl-$
Polystyrene	$CH_2{=}CH$ (phenyl)	$-CH_2-CH-$ (phenyl)
Polyacrylamide	$CH_2{=}CH$, $CONH_2$	$-CH_2-CH-$, $CONH_2$
Polyisobutylene	$CH_2{=}C$ with two CH_3	$-CH_2-C-$ with two CH_3
Polyisoprene (natural rubber)	$CH_2{=}C(CH_3)-CH{=}CH_2$	$-CH_2-C(CH_3){=}CH-CH_2-$
Polydimethylsiloxane	$HO-Si(CH_3)_2-OH$	$-Si(CH_3)_2-O-$
Polyethyleneoxide (Polyox)	CH_2-CH_2 (epoxide)	$-O-CH_2-CH_2-$
Polyhexamethylene adipamide (Nylon 66)	$NH_2-(CH_2)_6-NH_2$ and $HO-C(O)-(CH_2)_4-C(O)-OH$	$-NH-(CH_2)_6-NH-C(O)-(CH_2)_4-C(O)-$
Polyethylene terephthalate (polyester)	$HO-CH_2-CH_2-OH$ and $HO-C(O)-C_6H_4-C(O)-OH$	$-O-CH_2-CH_2-O-C(O)-C_6H_4-C(O)-$

Figure 2.7 Some common polymers, the monomers from which they are synthesized, and the structural repeat unit in the polymerized chain. (From Bird et al. *Dynamics of Polymeric Liquids, Vol. 1: Fluid Mechanics*, Copyright © 1987. Reprinted by permission of John Wiley & Sons, Inc.)

$$M_n = \sum_{j=1}^{\infty} n_j M_j, \qquad M_w = \sum_{j=1}^{\infty} c_j M_j \qquad (2\text{-}11)$$

Here n_j is the number fraction, and $c_j = n_j M_j / \sum_k n_k M_k$ is the weight fraction of polymer molecules having molecular weight M_j. The *polydispersity* of the polymer can be characterized by the ratio M_w/M_n. M_w/M_n is unity for a perfectly monodisperse solution, and it becomes larger as the polydispersity increases. As an example, if two monodisperse fractions, one with twice the molecular weight of the other, are mixed in equal weight fractions, then $M_w/M_n = 1.125$. Thus, even values of M_w/M_n rather close to unity correspond to fairly broad molecular weight distributions.

> • Problem 2.1 and Worked Example 2.2, just after this chapter, will help you understand and use molecular weight distributions.

2.2.3.2 Random-Walk Statistics: The Freely Jointed Chain

Flexible molecules, such as those in Fig. 2-7, permit rotational motions of one bond about another, so that a combinatorially huge number of configurations is accessible (Flory 1969). On length scales of tens or hundreds of such monomers, the details of the distribution of allowed bond angles average out, producing in the melt a configuration distribution equivalent to that of a *random walk* (see Fig. 2-8). Because of the flexibility of these molecules, even in the densely packed melt state, they remain unoriented, or isotropic, at equilibrium.

Because of Brownian forces, each molecule in the molten state continually changes its configuration, but for long polymer molecules composed of hundreds or more monomers, the time-averaged mean-square distance $\langle R^2 \rangle_0$ separating one end of the molecule from the other obeys the random-walk formula

$$\langle R^2 \rangle_0 = n b_n^2 \tag{2-12}$$

where n is the number of backbone bonds composing the chain, and b_n is the length of an "effective" random-walk step. If each bond were a rigid link attached to its two neighboring

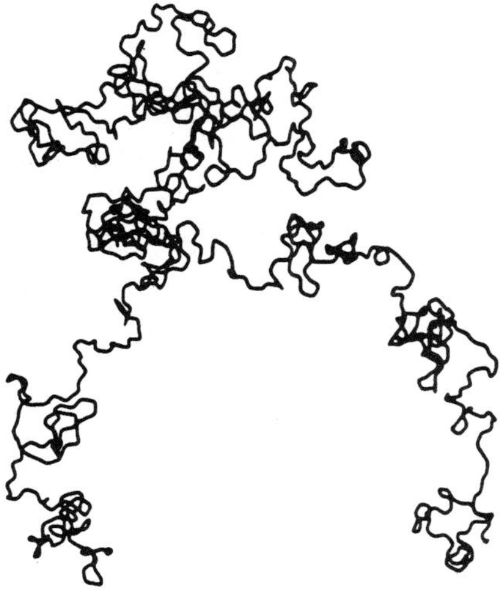

Figure 2.8   Random walk formed from 1000 links. (From Treloar, copyright © 1975 by Oxford University Press, Inc. Used by permission of Oxford University Press, Inc.)

monomers by completely flexible joints, then b_n would be the length of the rigid link. This simple model for a polymer molecule is called the *freely jointed chain*. Of course, the bonds in a real polymer chain do not have complete rotational freedom; because of bond angle restrictions there are correlations between the orientation of one bond and that of its near neighbors. Nevertheless, for a flexible chain, the orientation of a bond is uncorrelated with those of bonds that are many monomer units removed from it along the backbone of the chain. As a result, at a length scale larger than that of a monomer, the conformational properties of the chain are well represented by a random walk, as long as the effective step size b_n of the random walk is chosen so that Eq. (2-12) is satisfied. Since bond angle restrictions make the average coil size $\langle R^2 \rangle_0$ more expanded than would otherwise be the case, b_n is larger than the average bond length ℓ. Thus, for chains of many bonds one can write

$$\langle R^2 \rangle_0 = C_\infty n \ell^2$$

with the *characteristic ratio* C_∞ given by

$$C_\infty \equiv \frac{b_n^2}{\ell^2} \tag{2-13}$$

For a freely jointed chain, $C_\infty = 1$; but for real chains, bond angle restrictions lead to values of C_∞ in the range 5–10. For polystyrene, for example, $C_\infty \approx 9.6$ at 140°C. A tabulation of C_∞ for various polymers can be found by looking ahead to Table 3-3, as well as to Fetters et al. (1994; see also Flory 1969).

A quantity related to $\langle R^2 \rangle_0^{1/2}$ is the *radius of gyration R_g*, which is defined as the root-mean-square average distance separating a monomer from the center of mass of the chain; it can be expressed as

$$R_g^2 \equiv \left\langle \frac{1}{2} n^{-2} \left(\sum_{i=1}^{n} \sum_{j=1}^{n} (R_{ij})^2 \right) \right\rangle \tag{2-14a}$$

where R_{ij} is the spatial distance between monomers i and j. For a flexible chain, R_g is given by

$$R_g = \frac{\langle R^2 \rangle_0^{1/2}}{\sqrt{6}} \tag{2-14b}$$

• Worked Example 2.3 shows how to derive Eq. (2-14b).

In Eq. (2-12), the number of backbone bonds, n, was used in defining a random-walk step size b_n. Other choices are possible. If N, *the degree of polymerization*, is used instead, then one can define the "statistical segment length" b_N, or just b for short, using a formula analogous to Eq. (2-12), that is,

$$\langle R^2 \rangle_0 \equiv N b_N^2 = N b^2 \tag{2-15}$$

If j is the number of backbone bonds per monomer (often 2), then $N = n/j$, and the step size b is related to b_n by

$$b = b_n \sqrt{j} \tag{2-16}$$

If one can define a *fully extended length*, or contour length, L, of the molecule, then one can define yet another effective random-walk step size, or "Kuhn" length b_K, by

$$\langle R^2 \rangle_0 = L b_K \tag{2-17}$$

An effective freely jointed chain of contour length L would then contain $N_K \equiv L/b_K$ rigid links, each of length b_K, and N_K is the number of "Kuhn" steps in the chain (Semenov and Khokhlov 1988). Hence,

$$\langle R^2 \rangle = N_K b_K^2 \tag{2-18}$$

For ordinary flexible synthetic polymers, a reasonable definition for the contour length L is the maximum length to which the chain can be extended, say, in an aligned crystalline state. Since the bonds of the chain are not collinear, even when the chain is fully extended, L is somewhat less than $n\ell$; for tetrahedral (109°) bonding angles, one has

$$L = \sin\left[\tan^{-1}\left(\sqrt{2}\right)\right] n\ell \approx 0.82 n\ell \tag{2-19}$$

From Eqs. (2-12), (2-13), and (2-16)–(2-19), the three different random-walk step sizes b, b_n, and b_K can be related to the backbone bond length ℓ by

$$b_n = \sqrt{C_\infty}\,\ell, \qquad b = \sqrt{C_\infty j}\,\ell, \qquad b_K = \frac{C_\infty}{0.82}\ell \tag{2-20}$$

The numbers of such links are related to each other by

$$N = \frac{n}{j}, \qquad N_K = \frac{(0.82)^2}{C_\infty}n \tag{2-21}$$

2.2.3.3 Long-Range Excluded-Volume Effects in Solutions

The above calculations assume that the gross chain conformations are those of a random walk, which is the case in the melt. However, for an isolated polymer molecule in a dilute solution, the average conformation is affected by *excluded-volume* interactions between one part of the chain and another. Because the chain must avoid self-intersection, the conformation of the chain will be that of a *self-avoiding* walk, rather than a random walk, if the solution is *athermal*—that is, if all interactions are negligible except excluded volume. Self-avoiding walks lead, on average, to more expanded coil dimensions, since expanded configurations are less likely than contracted ones to lead to self-intersection of the chain. Thus, in an athermal solution, the mean-square end-to-end dimension of a polymer molecule scales as

$$\langle R^2 \rangle \sim N_K^{6/5} b_K^2 \tag{2-22}$$

instead of Eq. (2-18). Self-intersections are also forbidden in the melt, but in the melt a polymer molecule must also avoid intersecting other chains; if the chain were to expand to decrease the probability of self-intersection, it would correspondingly increase its probability of intersecting other chains. Hence, the chain in the melt has no tendency to expand beyond its random-walk dimensions.

2.2.3.4 Local Packing Effects in Melts

Although in the melt excluded-volume interactions have no *long-range* effects on overall polymer conformation, there are *local* packing effects associated with the requirement that the polymer segments fill space. These packing constraints influence the correlations between orientations of neighboring polymer segments. The local packing characteristics of a chain are defined not only by its effective segment length, but also by its effective "diameter" d. This "diameter" can be defined using a cylindrical segment model, in which the volume v of a cylindrical segment of length b_K is given by $v = \pi b_K d^2/4$. The total volume of the chain is therefore $V = N_K \pi b_K d^2/4$. The diameter d can then be obtained by matching this volume to the volume occupied by the real chain, $V = M_0 N/N_A \rho$, where M_0 is the molecular weight of a monomer, N is the number of monomers in the chain, N_A is Avogadro's number, and ρ the bulk density of the polymer. With this and Eqs. (2-20) and (2-21), we obtain

$$d^2 = \frac{4M_0}{0.82 j \pi \ell N_A \rho} \tag{2-23}$$

The length ℓ of a carbon–carbon single bond is about 1.54 Å. Thus, for polystyrene, with $M_0 = 104$ daltons, $\rho = 1.04$ g/cm^3, and $j = 2$, we obtain $d = 9.1$ Å. Since $C_\infty = 9.6$ for polystyrene (Table 3-3), from Eq. (2-20) the aspect ratio of the effective cylindrical link, b_K/d, is around 1.96. This ratio, b_K/d, is also known as the "axial ratio" ζ. For polyethylene ($C_\infty = 7.3$), $\zeta \equiv b_K/d = 2.67$, and, in general, $b_K/d \approx 1.5$–3 for flexible-chain macromolecules. Thus, in a gross sense, a flexible polymer chain can be modeled as a series of cylindrical links of aspect ratio 1.5 to 3, attached together at joints that permit completely free rotation. If the aspect ratio of the cylindrical link were to exceed a critical value of around 6.4, then Flory's lattice theory (in Section 2.2.2.3) suggests that the polymer melt would spontaneously produce nematic order.

Even when $b_K/d \approx 1.5$–3, there are short-range orientational correlations, or "orientational couplings" between segments of different chains that are packed close together. As a result, in a two-component mixture, long, easily oriented molecules induce partial orientation in surrounding fast-relaxing molecular species, such as shorter polymers or solvent molecules (Doi et al. 1989; Ylitalo et al. 1991; Deloche and Samulski 1981; Thulstrup and Michl 1982; Baljon et al. 1995; Gent 1969). In practice, such couplings need only be accounted for when the orientational order of individual components of a blend are measured separately—by infrared dichroism, for example.

A quantity closely related to b_K/d is the *packing length* p, defined by Fetters et al. (1994):

$$p = \frac{M}{\langle R^2 \rangle_0 \rho N_A} = \frac{v_0}{b^2}$$

where v_0 is the volume of a monomer. Since random-walk statistics imply that $\langle R^2 \rangle_0 \propto M$, the packing length p is a microscopic length that is independent of molecular weight. Using $M = M_0 N$ and Eqs. (2-15), (2-20), and (2-23), we find that p is related to the effective chain diameter d and Kuhn length b_K by

$$p = \frac{(\pi/4)d^2}{b_K}$$

A large value of p means that the chain segments are short and fat, so that within the radius of gyration of the chain relatively less room is left for other chains. The "packing length" is therefore related to the tendency of a chain of a given length to become *entangled* with other chains, as discussed in Section 3.7.

For more detailed information on the structure and morphology of polymers, see Flory (1953, 1969), Grosberg and Khokhlov (1994), and Painter and Coleman (1994). The rheology of flexible polymer solutions and melts is covered in Chapter 3.

2.2.4 Semiflexible Macromolecules

Some molecules, such as helical DNA, collagen, or PBLG, are stiffer than ordinary flexible polymers like polystyrene but are not rigid rods, and thus they are called "semiflexible" (Odijk 1986). Flexibility can be distributed either evenly or unevenly along the polymer backbone.

A uniform distribution of flexibility describes some molecules such as DNA and PBG; a simple model for such molecules is the flexible beam or "worm-like chain" (Saito 1967; Yamakawa 1971) which has a finite bending rigidity κ that is uniform along the chain contour. When we define a *contour variable s* as the distance along the chain contour from one of the ends and define $\mathbf{R}_s(s)$ as the curve in space defining the conformation of the worm-like molecule, the bending free energy is

$$F_{\text{bend}} = \frac{\kappa}{2} \int_0^L \left(\frac{\partial^2 \mathbf{R}_s}{\partial s^2}\right)^2 ds \tag{2-24}$$

The inextensibility of the worm-like chain implies that $|\partial \mathbf{R}_s/\partial_s| = 1$. For a flexible beam, the bending rigidity κ can be related to the Young's modulus E and the moment of inertia I by $\kappa = EI$. A circular beam of diameter d has a moment of inertia of $I = \pi d^4/64$. The orientation of a unit vector $\mathbf{u}(s) \equiv \mathbf{R}_s/|\mathbf{R}_s|$ parallel to the worm-like chain becomes uncorrelated with that of a unit vector $\mathbf{u}(s + \Delta s)$ when the distance Δs along the chain away from s is large enough. If $\mathbf{R}$ is the end-to-end vector of a worm-like chain, one finds that (Kloczkowski 1996)

$$\langle \mathbf{u}(0) \cdot \mathbf{R} \rangle = \lambda_p \left(1 - e^{-L/\lambda_p}\right) \tag{2-24a}$$

where $\lambda_p = \kappa/k_B T$ is the *persistence length* of the worm-like chain, and $\mathbf{u}(0)$ is a tangent unit vector at one end of the chain. The persistence length of DNA measured by light scattering, for example, is around 0.053 μm, around 150 base pairs, roughly consistent with the measured Young's modulus of $E = 3.5 \times 10^8$ Pa (Smith et al. 1996). (Under low-salt conditions, the effective persistence length of DNA is greater than this, because of electrostatic repulsion of charges along the molecule.)

For chain lengths $L \gg \lambda_p$, the orientation of the chain as a whole becomes uncorrelated with that at one end of the chain, and the gross conformational properties of an *isolated* worm-like chain are the same as those of the freely jointed one. The persistence length λ_p

is related to the Kuhn length b_K simply by $\lambda_p = b_K/2$. If, on the other hand, $L \ll \lambda_p$, the molecule is effectively a rigid rod. For arbitrary λ_p/L,

$$\langle R^2 \rangle_0 = 2\lambda_p L \left(1 - \frac{\lambda_p}{L} \left(1 - e^{-L/\lambda_p} \right) \right) \qquad \text{(worm-like chain)} \qquad \text{(2-24b)}$$

Nature produces the stiffest worm-like chains; the long persistence length of DNA ($\lambda_p = 0.053 \, \mu\text{m}$) is short compared to that of F-actin filaments, $\lambda_p \sim 17 \, \mu\text{m}$ (Ott et al. 1993); the latter are components of the cellular cytoskeleton. Self-assembled cellular microtubules are even stiffer than this, with λ_p on the order of millimeters!

In the second type of semiflexible polymer molecule, rigid units are interspersed with flexible ones. Some examples of molecules of this kind are given in Chapter 11. The freely jointed chain model might be a suitable model for such semiflexible polymers. If a persistently flexible molecule and a freely jointed molecule are characterized by the same values of L and λ_p (or, equivalently, of b_K and N_K), then the gross statistical measures of the coil dimensions of the two such *isolated* chains will be the same, despite the differences in the type of flexibility.

However, in a bulk nematic state, or in solutions concentrated enough to produce a nematic, it is likely that the average conformational properties of a semiflexible molecule depend on the type of flexibility. This is because in a nematic a free-energy penalty is paid for each unit length of chain that is not parallel to the nematic axis. Hence, the nematic potential favors sharp bends over gradual ones, because in the former less chain length is in orientations far from that of the nematic axis (de Gennes 1982; Warner et al. 1985). But for the worm-like chain, sharp "hairpin" bends are suppressed by the uniformly high bending rigidity of the chain, while such suppression is avoided when the chain contains flexible "joints"—that is, flexible spacers. Thus, molecules with flexible spacers between rigid links can accommodate themselves to the nematic state by forming folded conformations containing "kinks," while worm-like molecules experience "nematic stiffening"; that is, they are forced to straighten out to avoid the penalties imposed by both the nematic potential and the bending rigidity (Warner et al. 1985). Neutron-diffraction studies of the conformations of semiflexible polyester chains with 10- or 11-carbon flexible spacers indicate the existence of "hairpin" conformations in the nematic melt state with a frequency of only about 1 hairpin for every 12 spacers, implying a molecular weight of around 5000 between hairpins (Li et al. 1994). A similar conclusion has been inferred from the thicknesses of crystals grown from the nematic; these crystals are thin enough (as thin as 15 nm) that the polymer molecules must be folded to fit into them (Kent and Geil 1993). Studies of the conformations of worm-like chains in the nematic melt seem to have not yet been carried out.

As might be expected, molecular flexibility has a significant effect on the phase behavior of polymer melts and solutions, including the nematic–isotropic phase transition. As alluded to in Section 2.2.2.1, experimental deviations from the predictions of the Onsager theory for the concentration ν_2 for obtaining a fully nematic state can readily be attributed to the effects of flexibility. When the Onsager theory, which assumes completely rigid rods, is generalized to allow for flexibility, better agreement with experiment is obtained. The Onsager theory predicts that the minimum concentration ν_2 to form a fully liquid crystalline state is inversely proportional to the molecular length L. However, from Semenov and Khokhlov's generalized theory for worm-like flexible molecules, one finds that this prediction is only valid when $L \ll \lambda_p$. According to the Semenov–Khokhlov theory

(1988), for $L \gg \lambda_p$, the minimum volume fraction ϕ_2 needed for a fully nematic phase approaches a constant given approximately by $\phi_2 \approx 5.7\,d/\lambda_p = 11.4\,d/b_K$. Thus, for a long semiflexible molecule, the *axial ratio* $\zeta \equiv b_K/d$ of the Kuhn length to the diameter of the chain controls the concentration at which the nematic state is reached. For freely jointed chains, $\phi_2 = 4.86\,d/b_K$, a smaller value (Semenov and Khokhlov 1988; Flory 1978).

For example, PBLG molecules have a persistence length of around 120 nm (Jackson and Shaw 1991; Ookubo et al. 1976) in the solvent metacresol and a diameter d of 1.54 nm, so that the aspect ratio λ_p/d of a single persistence length of the molecule is around 78. According to the Semenov–Khokhlov theory, for long PBLG molecules the phase transition to a nematic should occur at a polymer volume fraction of $\phi_2 \approx 0.073$, not far from the experimental value of around 0.080. Further details can be found in Sato and Teramoto (1996).

The persistence length λ_p (or Kuhn length $b_K = 2\lambda_p$) is temperature-dependent. Figure 2-9, for example, shows that for hydroxypropylcellulose (HPC) in dimethylacetamide, the axial ratio ζ of the Kuhn length b_K to the molecular diameter d decreases with increasing temperature roughly as $\zeta \propto \exp(-\xi T)$ (Aden et al. 1984), with $\xi = 0.005$ and T in degrees K. The dependence of axial ratio on T can be represented by $\zeta \propto \exp(-\xi T)$ for other semiflexible molecules as well, with $\xi = 0.011$ for poly(hexylisocyanate) in toluene and tetrahydrofuran, and 0.0050 for tri-O-heptylcellulose in bromobenzene (Takada et al. 1995). According to the Flory theory, this decrease in ζ with temperature implies that the minimum concentration ν_2 for formation of a one-phase nematic (in this case a chiral nematic) increases with increasing temperature T, which is in agreement with experiments (Krigbaum 1985). When the axial ratio becomes small enough (at high temperatures), the concentration ν_2 should approach the concentration in the melt. Figure 2-9 indicates that at $T \approx T_{\mathrm{NI}}$, the isotropic–nematic transition temperature for *melts*, the axial ratios of HPC and

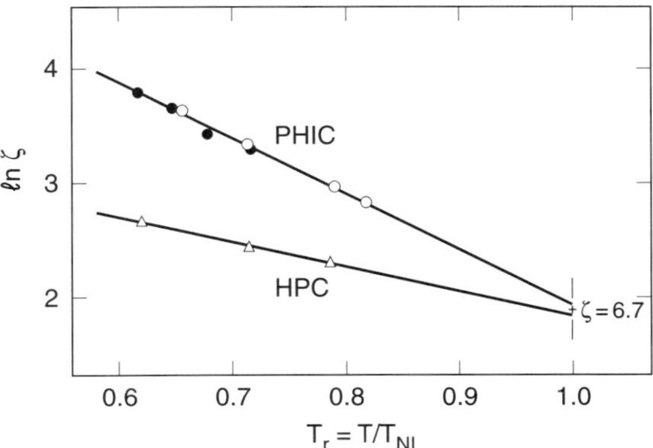

Figure 2.9 Temperature-dependence of the axial ratio $\zeta = b_K/d$ of Kuhn length b_K to molecular diameter d for poly(hexylisocyanate) (PHIC) in toluene ($\bigcirc$) and tetrahydrofuran ($\bullet$) and for hydroxypropylcellulose (HPC) in dimethylacetamide ($\triangle$). Here T_{NI} is the transition temperature from the nematic or chiral nematic to the isotropic phase in the melt. The lines fit the expression $\zeta \propto \exp(-\xi T)$, with $\xi = 0.011$ for PHIC and $\xi = 0.005$ for HPC. The "+" mark denotes Flory's predicted value of ζ at the nematic–isotropic transition in the melt state. (From Krigbaum 1985, with permission from the Royal Society of Chemistry.)

PHIC extrapolate to roughly 6.7, which is close to the value predicted by the Flory theory in the melt. This suggests that even for bulk HPC, the nematic–isotropic transition is driven primarily by excluded-volume, or packing, effects and only secondarily by anisotropic van der Waals interactions. The temperature dependence of the axial ratio could be incorporated into the Maier–Saupe potential by suitably adjusting the temperature dependence of the coefficient U_{MS}.

> • Problem 2.4 tests your understanding of semiflexibility in polymers.

2.3 VAN DER WAALS INTERACTIONS

2.3.1 Molecular Mixtures

While repulsive steric forces dominate the solid–liquid transition, attractive *van der Waals* forces control liquid–vapor and liquid–liquid phase transitions. Our discussion of these forces draws heavily from more complete discussions to be found in Israelachvili (1992) and Russel et al. (1989).

2.3.1.1 Dipolar Interactions: London, Keesom, and Debye Forces

Van der Waals forces arise from dipole–dipole interactions. In a dipolar molecule, there is a separation of positive from negative charge. For example, if the molecule contains charges $+q$ and $-q$ separated by a distance vector $\mathbf{r}$, then the molecule has a dipole moment of $\mathbf{u} \equiv \mathbf{r}q$. Dipolar moments are often measured in units of the *Debye* D, where $1\,\text{D} = 3.336 \times 10^{-30}$ coulomb-meter. While alkanes have no permanent dipole moments, organic molecules containing oxygen atoms usually do, because of the tendency of the oxygen atom to pull electrons away from a neighboring carbon.

The electric field created by a dipole ($\mathbf{u}_1$) either attracts or repels a neighboring dipole ($\mathbf{u}_2$) depending on their relative orientation—that is, the sign of $\mathbf{u}_1 \cdot \mathbf{u}_2$. In a liquid, where the molecules can rotate freely, the dipoles tend to align themselves with respect to each other to maximize attractive dipolar interactions and minimize repulsive ones. Of course, in an isotropic liquid, no permanent molecular orientations result from these interactions, only correlations between the orientation of one dipole and that of its neighbors. Because of these correlations, there is a net *attractive* force among molecules. This is the attractive force that holds most liquids together.

Even molecules such as alkanes (which are short oligomeric versions of polyethylenes; see Fig. 2-7) that possess no permanent dipole have *fluctuating* dipoles because of the motion of the electronic cloud around each molecule. Because the dipoles communicate with each other through their electric fields, they are able to correlate their fluctuating magnitudes and orientations to produce a net *attractive* force, called the *London* force. Thus, the electric field $\mathbf{E}$ produced by one temporary dipole can induce a dipole moment $\mathbf{u}_{\text{ind}}$ in another molecule, where $\mathbf{u}_{\text{ind}} = \alpha_0 \mathbf{E}$ and α_0 is the molecule's *polarizability*. Since these fluctuating dipoles also produce dispersion, or scattering, of light, the force is also called the *dispersion* interaction. Indeed, the magnitude of the dispersion force is closely related to the index of refraction of a material.

Liquids containing permanent dipoles have additional attractive interactions called *Keesom* forces, which are caused by the tendency of the permanent dipoles to align anti-parallel with each other. Finally, there are also *Debye* or *induction* interactions between permanent dipoles and fluctuating ones. The dispersion interactions are the most important of the three types, however, because they occur in all materials, and are usually stronger than the Keesom and Debye interactions, when the latter are present at all.

All three types of polarization interactions—Keesom, Debye, and London—are included in the following formula for the total van der Waals interaction potential between two spherical molecules separated by a distance r:

$$W(r) = -\frac{C}{r^6} \tag{2-25}$$

with

$$C = \frac{1}{(4\pi\varepsilon_0)^2}\left[(u_1^2\alpha_{02} + u_2^2\alpha_{01}) + \frac{u_1^2 u_2^2}{3k_B T} + \frac{3\alpha_{01}\alpha_{02}h\nu_1\nu_2}{2(\nu_1 + \nu_2)}\right] \tag{2-26}$$

Here u_1 and u_2 are the permanent dipole moments of molecule 1 and 2, respectively, the α_0's are their polarizabilities, the ν's are their ionization frequencies, $h = 6.625 \times 10^{-34}$ J·s is Planck's constant, and $\varepsilon_0 = 8.8 \times 10^{-12}$ C^{2}J^{-1}m^{-1} is the permittivity of free space. (The unit J is for joule $= 10^7$ erg, and C is for coulombs.) The dispersion term is the last of the three. If dispersion is dominant and the two molecules have the similar ionization frequencies, then we find that the dispersion interaction is proportional to the product of the polarizabilities of the two molecules:

$$W(r) \propto -\frac{\alpha_{01}\alpha_{02}}{r^6} \tag{2-27}$$

Note also in Eqs. (2-25) and (2-26) that since the α's are positive, the interaction is always attractive (C is positive) and falls off steeply, as the sixth power of the separation between molecules. An attractive interaction that falls off as the sixth power of the distance between the centers of mass of two molecules is often incorporated into intermolecular potentials, such as the famous Lennard-Jones 6-12 potential. This potential also includes a much steeper twelfth-power fall-off in repulsive interaction, which is used to approximate the "hard-core" excluded volume.

This steep fall-off of the van der Waals interactions with increasing molecular separation implies that the strongest van der Waals interactions a molecule experiences are those with its nearest neighbors. Since the strength of the interactions decreases as r^{-6}, while the number of interacting molecules increases roughly as r^2, the total strength of interactions with a second shell of neighbors twice as far away as the first is, crudely speaking, $2^4 = 16$ times weaker than that of nearest first-shell of interactions. Such considerations help to justify *lattice models* of liquids that neglect all but nearest-neighbor interactions. In these models, the hard-core excluded-volume interactions are imposed by requiring that no more than one molecule occupy any lattice cell, and nearest-neighbor van der Waals interactions are included by allotting a contact energy w_{ij} for each occurrence of a molecule of type i in a lattice cell adjacent to one occupied by a molecule of type j.

2.3.1.2 The Flory–Huggins Model

Using a simplified mean-field, or random-mixing, statistical mechanical analysis of such models, equations of state can be obtained for various types of liquid mixtures. For polymers, the equations are especially useful as qualitative guides for revealing the effect of molecular weight, composition, or temperature on phase behavior. For mixtures of two flexible polymer molecules, the *Flory–Huggins* lattice model (Flory 1941; Huggins 1941; Staverman and Van Stanten 1941) gives the following equation for the free energy ΔF of mixing:

$$\frac{\Delta F}{k_B T} = \frac{\phi_A \ln \phi_A}{N_A} + \frac{\phi_B \ln \phi_B}{N_B} + \chi \phi_A \phi_B \tag{2-28}$$

Here ϕ_A and ϕ_B are the volume fractions of polymers A and B, and N_A and N_B are the number of lattice sites occupied by an A and a B molecule, respectively. If N_B is set to unity, the above equation corresponds to polymer A dissolved in a small-molecule solvent B. $\chi k_B T$ is the energy cost per lattice site of moving an "A" polymer molecule from a reservoir of all A into a reservoir of all B. χ is thus related to the contact energies, w_{ij}, by

$$k_B T \chi = z_{\text{eff}} \left(w_{AB} - \tfrac{1}{2} w_{AA} - \tfrac{1}{2} w_{BB} \right) \tag{2-29}$$

where z_{eff} is the "effective coordination number" of a single site; it is the number of *intermolecular* contacts per lattice site. If molecules A and B occupy only single lattice sites, then $z_{\text{eff}} = z$, where z is the coordination number of the lattice, while if A is a polymer, then $z_{\text{eff}} < z$, since some of the contacts a chain makes are with itself.

Thus, χ is supposed to account for van der Waals interactions. For purely dispersive interactions between similarly sized molecules, using Eqs. (2-27) and (2-29), we expect that

$$k_B T \chi \propto -\alpha_{0A} \alpha_{0B} + \tfrac{1}{2} \alpha_{0A} \alpha_{0A} + \tfrac{1}{2} \alpha_{0B} \alpha_{0B} = \tfrac{1}{2} (\alpha_{0A} - \alpha_{0B})^2 \tag{2-30}$$

which is always positive, or zero. Hence in the liquid state, van der Waals interactions ordinarily produce a tendency for unlike molecule to demix, and the rule "like attracts like" holds. Equation (2-30) is the basis of a successful theory of mixing for small molecules that uses a "solubility parameter," δ (Hildebrand and Scott 1950; Barton 1993; Coleman et al. 1991), which is proportional to α_0. Using Eqs. (2-26) and (2-30) and absorbing $(h\nu)^{1/2} v_0^{-3/2}/\varepsilon_0$ into the definition of δ, where v_0 is a reference molecular volume, the χ parameter can be written as

$$\chi = \frac{v_0}{k_B T} (\delta_A - \delta_B)^2 \tag{2-31}$$

For polymers, χ is usually defined on a per monomer basis or on the basis of a reference volume of order one monomer in size. However, χ is usually not computed from formulas for van der Waals interactions, but is adjusted to obtain the best agreement between the Flory–Huggins theory and experimental data on the scattering or phase behavior of mixtures (Balsara 1996). In this fitting process, inaccuracies and ambiguities in the lattice model, as well as in the mean-field approximations used to obtain Eq. (2-28), are papered over, and contributions to the free energy from sources other than simple van der Waals interactions get lumped into the χ parameter. The temperature dependences of χ for polymeric mixtures are often fit to

$$\chi = \frac{A}{T} + B \tag{2-32}$$

where T is the absolute temperature. While plots of χ versus $1/T$ are indeed nearly linear for some polymer pairs over a range of temperatures, it is not uncommon for such plots to show curvature, or even local minima (Karis et al. 1995), implying that (2-32) is inaccurate.

Although the Flory–Huggins theory was derived from a lattice model in which units of polymer A and polymer B are the same size (i.e., they each occupy a single lattice cell), the theory is readily generalized so that it can apply to realistic cases in which the volumes of monomers A and B are unequal:

$$\frac{\Delta f}{k_B T} = \frac{\phi_A \ln \phi_A}{v_A N_A} + \frac{\phi_B \ln \phi_B}{v_B N_B} + \frac{\chi}{v} \phi_A \phi_B \tag{2-33}$$

where Δf is the free energy *per unit volume*, v_A and v_B are the volumes occupied, respectively, by A and B monomer units in the polymer, and v is a Flory–Huggins reference volume. The Flory–Huggins reference volume v can be arbitrarily set to either v_A or v_B; it is also sometimes taken to be the geometric mean of v_A and v_B: $v \equiv (v_A v_B)^{1/2}$. Note that the value assigned to χ depends on the choice of Flory–Huggins reference volume.

For a polymer in a small-molecule solvent, it is convenient to define $N \equiv N_A$ to be the polymer degree of polymerization and take $N_B = N_s = 1$ for the solvent. It is also convenient to use the solvent molecular volume v_s as the Flory–Huggins reference volume v and to define the ratio of the volume of a polymer molecule to that of a solvent by $x \equiv N v_p / v_s$, where v_p is the volume of a monomer of the polymer. Then Eq. (2-33) implies that a large polymer molecule ($x \gg 1$) will demix from a small-molecule solvent when $\chi > \chi_{\text{crit}} \approx 0.5 + x^{-1/2}$ (Flory 1953). For very large x, χ_{crit} approaches the value 0.5. When $\chi \gg \chi_{\text{crit}}$, high-molecular-weight polymers will remain in solution only if the composition is far from critical—for example, if the solution is dilute enough that polymer coils rarely overlap. In a dilute solution with a poor solvent, one polymer molecule will rarely find other molecules with which to form favorable polymer–polymer contacts; so, instead, as χ increases, isolated coils shrink in size in order to increase the number of self-contacts. Since χ is temperature-sensitive, the average coil size in dilute solutions is also sensitive to temperature. For some solvent–polymer pairs, there is a temperature called the *theta point* (Sundararajan 1996) at which the coil contracts sufficiently to cancel out the expansion produced by excluded volume forces, and a random-walk scaling of coil size with molecular weight is recovered [Eq. (2-12)]. For large N, the theta point corresponds to $\chi(T) = 0.5$. Thus, for $N \to \infty$, the theta temperature equals the critical temperature for phase separation. For $\chi < 0.5$, the polymer coil *expands* relative to random-walk dimensions, in theory to a size such that $R_g^2 \propto N^{6/5} b^2$, while for $\chi(T) > 0.5$, the coil *contracts*, in theory to a size $R_g^2 \propto N^{2/3} b^2$.

- Worked Example 2.5 derives the result $\chi_{\text{crit}} = 0.5 + x^{-1/2}$.

In polymer–polymer mixtures, both N_A and N_B in the Flory–Huggins equation (2-33) are large; hence phase separation is predicted to occur even for very small values of χ. For $N_A = N_B \gg 1$ and $v = v_A = v_B$, the spinodal point where phase separation first occurs in a 50/50 mixture of A and B is given by $\chi N = 2$. Hence, for typical polymer molecular weights, corresponding, say, to $N = 1000$, phase separation will occur at very small values of $\chi \approx 2 \times 10^{-3}$. Since, as we have discussed, χ is positive when van der Waals interactions predominate, phase separation is the norm for polymer–polymer mixtures, and *most polymer blends are immiscible*. The reason for this immiscibility is that the translational entropic terms (the first two terms on the right side) of Eq. 2-33 are very small for polymer molecules. Hence, even a very small positive χ term drives phase separation. Chemically distinct polymers of high molecular weight usually mix only when miscibility is promoted by special interactions, such as hydrogen bonding (see Section 2.5).

Bates and coworkers (Bates and Wignall 1986; Bates et al. 1988) have shown that even *isotopic substitution* of deuterium for hydrogen in polymers such as polystyrene can lead to phase separation of the deuterated polymer from its hydrogenous counterpart! The value of χ for interactions between a deuterated and a hydrogenated monomer is typically very small, around 10^{-4} (Londono et al. 1994), but this is large enough that chains composed of $\sim 10^4$ monomers ($M_w \approx 10^6$) will nevertheless phase separate. Since isotopic substitution is often introduced as a labeling method to provide species contrast in neutron scattering experiments, isotope-induced immiscibility can be a bothersome nuisance. But it also

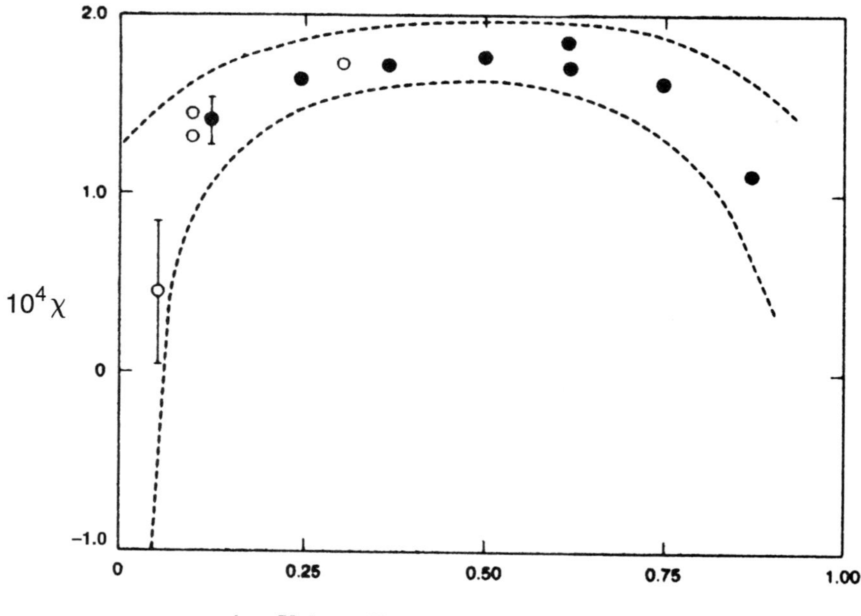

Figure 2.10 χ versus volume fraction of deuterated polystyrene in an isotopic blend of deuterated and hydrogenated polystyrenes at 160°C. The degrees of polymerization are: ($\bullet$) $N_H = 8700$, $N_D = 11,500$; and ($\bigcirc$) $N_H = 15,400$, $N_D = 11,500$. The dashed lines are the error limits. (Reprinted with permission from Londono et al., Macromolecules 27:2864. Copyright 1994, American Chemical Society.)

provides an excellent opportunity for probing miscibility and phase separation dynamics in polymer blends. Figure 2-10 shows the χ value measured by fitting neutron-scattering data to the Flory–Huggins theory for a mixture of deuterated and hydrogenated polystyrene as a function of composition. There appears to be a dependence of χ on composition, but this may be due to experimental error. Similar results were obtained by Schwahn et al. (1989).

If replacement of hydrogen by deuterium nuclei has an impact on polymer–polymer miscibility, then it is no surprise that even small chemical differences in monomeric units can lead to polymer–polymer immiscibility. Graessley et al. (1994) and Krishnamoorti et al. (1994) have studied the miscibility of various polyolefin polymers, including hydrogenated polyisoprene, poly(ethylenebutadiene), polypropylenes, and chains composed of various ratios of two randomly copolymerized four-carbon (C4) isomers, one linear and the other branched, as shown in Fig. 2-11a. These isomers differ in that one is a linear four-carbon chain, and the other a two-carbon chain with an attached two-carbon (or ethyl) branch. By *copolymerizing* various ratios of these monomers, polyolefins can be produced in which the percentage of vinyl content (monomers with two-carbon branches) is varied from nearly 0% to nearly 100%. Batches of polymer with differing degrees of ethyl branching can then be mixed together and the χ parameter measured (by neutron scattering) as a function of ethyl-branch content. A χ parameter of around 10^{-3}, high enough to produce phase separation for polymers of moderate molecular weight of around 10^5 daltons, is found between polymers whose vinyl content differs by as little as 10%!

As an equally dramatic example, consider ordinary polypropylene mixed with "head-to-head" polypropylene; the repeat units of these polymers are shown in Fig. 2-11b. These repeat units differ only in the placement of the methyl branches along the backbone. Yet the χ parameter, inferred from neutron scattering (Graessley et al. 1995) using a reference volume equal to the geometric mean of the two repeat units, is around 4.5×10^{-3} at 171°C. This is large enough that, according to the Flory–Huggins theory, pure polymers each of molecular weight 40,000 will phase separate if mixed in equal proportions. Small-molecule versions of these and similar molecules, containing 20 or fewer carbons, are common in gasoline, and are miscible in all proportions.

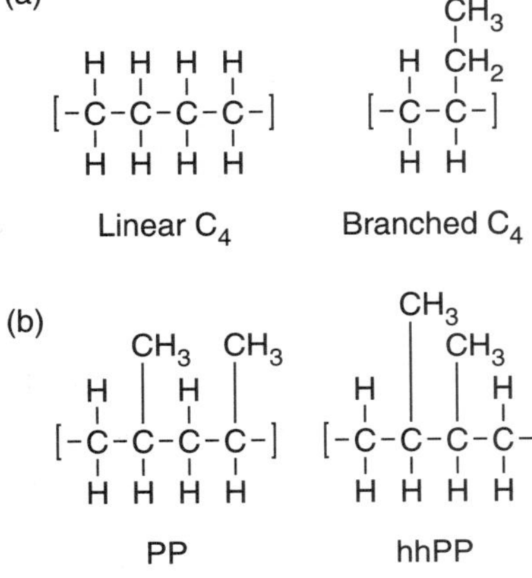

Figure 2.11 (a) Linear and ethyl-branched monomers that, when copolymerized together at various ratios, give polymers with a wide range of different ethyl branch content. By mixing together polymers with different levels of ethyl branching and measuring χ by neutron scattering, theories of polymer miscibility can be tested. (b) Repeat units for polypropylene and "head-to-head" polypropylene. (Reprinted with permission from Krishnamoorti et al., Macromolecules 27:3073. Copyright 1994, American Chemical Society.)

2.3.1.3 Theories for the χ Parameter

There have been recent efforts to predict, or at least rationalize, the χ parameters of these and other polyolefin–polyolefin blends. Bates et al. (1992) and Fredrickson et al. (1994) suggest that the χ parameter is correlated to a difference in *statistical segment length* of the polymer molecules, on a volume-normalized basis. The volume normalization is required because the definition of the "statistical segment length" depends on how the monomer unit is defined. Bates et al. (1992), for example, define

$$b_{v,\mathrm{A}}^2(T) \equiv \frac{b^2(T)v_{\mathrm{ref}}}{v_{\mathrm{A}}(T)} = \frac{6R_g^2(T)v_{\mathrm{ref}}}{N v_{\mathrm{A}}(T)} \tag{2-34}$$

where $v_{\mathrm{A}}(T)$ is the volume of monomeric unit A (which depends somewhat on temperature T), N is the number of such units in the chain, $R_g(T)$ is the (temperature-dependent) radius of gyration, and v_{ref} is a fixed reference volume, taken by Graessley et al. to be 100 Å^3. (The reference volume v_{ref} for the statistical segment length need not be the same as the Flory–Huggins reference volume v.) By this definition, the effective segment length $b_{v,\mathrm{A}}$ is independent of the definition of a monomer unit.

Fredrickson et al. (1994) have calculated a fluctuation correction to the Flory–Huggins free energy Δf, and find a positive entropic correction that is proportional to $\varepsilon \equiv b_{v,\mathrm{A}}/b_{v,\mathrm{B}}$, where the subscripts "A" and "B" denote the two different polymers in the blend. Thus, according to this theory, the differing conformational statistics of chains with differing b_v values produces a tendency to demix. Consistent with this prediction, Bates et al. (1992) find that the experimental χ values for binary blends of different polyolefin block copolymers can be correlated with the variation from unity in the ratio $\varepsilon \equiv b_{v,\mathrm{A}}/b_{v,\mathrm{B}}$. Graessley et al. (1995) also find from experimental values of χ for polyolefin mixtures a general correlation of the form $\chi \approx 0.21(\Delta b_v/\overline{b_v})^2$, where $\Delta b_v \equiv b_{v,\mathrm{A}} - b_{v,\mathrm{B}}$ is the difference in b_v values and $\overline{b_v}$ is a representative value of b_v arbitrarily set to 6 Å. The reference Flory–Huggins volume v for χ in this expression is set equal to v_{ref}, the reference volume for the statistical segment length.

Despite this general correlation between differences in statistical segment length and immiscibility in polymer blends, Graessley et al. (1995) have found significant exceptions and have noted that the experimental temperature dependences of χ are not consistent with its being controlled by entropic effects. Certainly, isotopic labeling produces negligible differences in statistical segment lengths; and at least for this case, differences in statistical segment length are not responsible for the observed nonzero χ value. The success of the correlation $\chi \approx 0.21(\Delta b_v/\overline{b_v})^2$ might therefore reflect a general correlation between statistical segment length and other factors that control polymer–polymer miscibility.

Graessley et al. (1994, 1995) have proposed an alternative scheme that correlates the χ values for polyolefin blends with differences in inferred values of the *solubility parameters* of the pure components [see Eq. (2-31)]. For the series of polyolefins they studied, values of δ for the pure components can be assigned in such a way that the χ parameters of the mixtures computed from Eq. (2-31) are in reasonable qualitative agreement with measured values. Direct estimates of the δ's can be obtained from the following formula derived from the van der Waals cohesive energy density (Hildebrand and Scott 1950; Krishnamoorti et al. 1996):

$$\delta \sim \left(\frac{T\alpha}{\kappa}\right)^{1/2} \tag{2-35}$$

where T is the absolute temperature, α is the thermal expansion coefficient, and κ is the isothermal compressibility of the liquid. From Eq. (2-35), one infers values of δ at 171°C equal to 15.8 and 16.6 MPa$^{1/2}$, respectively, for polypropylene and head-to-head polypropylene; their repeat units are depicted in Fig. 2-11b. The difference between these δ's is 0.8 MPa$^{1/2}$, which is about a factor of two larger than one infers from Eq. (2-31), using the measured χ value, $\chi \approx 4.5 \times 10^{-3}$.

Although the solubility-parameter scheme succeeds in systematizing the χ values for many polyolefin blends, there are several notable failures in which mixing is *irregular*. Irregular mixing means that no set of δ values can be assigned to the pure components such that the values of χ for the blends [obtained from Eq. (2-31)] match the experimental values (Graessley et al. 1995). In fact, in some cases *negative* values of χ are measured; such results are obviously inconsistent with Eq. (2-31), which predicts only nonnegative χ's.

There are also anomalous cases where the measured χ value has a *minimum* as a function of temperature. The existence of a minimum implies that for suitable molecular weights, the blend will first become miscible as the temperature is raised, but then with further increases in temperature will become immiscible again! The critical temperature at which the blend becomes miscible on heating is called an *upper critical solution temperature* (UCST), while the temperature at which it becomes immiscible on heating is called the *lower critical solution temperature* (LCST). The presence of both a UCST and an LCST in the same blend is a surprising, and not easily explained, phenomenon. Recently, Russel and coworkers (Karis et al. 1995) have found that with increasing temperature there is a disordering transition followed by an ordering transition at a higher temperature, in a *diblock copolymer* (see Chapter 13) of deuterated polystyrene and poly(n-butyl methacrylate). This observation implies that for this system also, χ has a minimum as a function of temperature.

In addition to these irregularities, Winey et al. (1996) have found that in random and alternating copolymers of styrene and methyl methacrylate, the *sequence distribution* of monomers along the backbone of the polymer strongly affects its miscibility with polystyrene and polymethylmethacrylate homopolymers, even when the overall ratio of styrene/methyl methacrylate in the copolymer chain is held constant. A strictly alternating sequence of monomers in the copolymer was found to be more miscible with the homopolymers than is a copolymer with a random sequence distribution. These results imply that simple group additive schemes, which have met with success in computing molecular miscibilities in small-molecule mixtures (Fredenslund et al. 1975; Guggenheim 1952), are not always adequate for polymers. Apparently energetic or packing effects that are too small to influence small-molecule mixing can easily overwhelm the low entropies of mixing of polymers. These observations suggest that no simple scheme will be able *in general* to predict polymer–polymer miscibility.

2.3.2 Suspensions

In *multiphase* complex fluids, it is useful to account for van der Waals forces on a continuum, rather than a molecular, basis. Consider, for example, two spherical particles of material A, each of radius a, approaching each other in a medium of substance B. Let r be the distance separating the centers of the spheres and let $D = r - 2a$ be the gap between them. Defining $x \equiv D/2a$, the van der Waals potential of interaction is found to be (Goodwin et al. 1986)

$$W_{\text{vdw}}(\text{spheres}) = -\frac{A_H}{12}\left\{\frac{1}{(x+1)^2 - 1} + \frac{1}{(1+x)^2} + 2\ln\left[1 - \frac{1}{(1+x)^2}\right]\right\} \quad (2\text{-}36)$$

where the sign and magnitude of A_H, the *Hamaker constant*, is controlled by van der Waals forces. If $D \ll a$ (i.e., $x \ll 1$), the *Derjaguin approximation* applies, and Eq. (2-36) is approximated simply by

$$W_{\text{vdw}}(\text{spheres}) \approx \frac{-A_H a}{12D} \tag{2-37}$$

For parallel, flat surfaces separated by a distance D, the potential per unit area is

$$W_{\text{vdw}}(\text{flat plates}) = \frac{-A_H}{12\pi D^2} \tag{2-38}$$

The value of the Hamaker constant can be estimated from the *Lifshitz theory* of dispersion forces in a continuum; in simplified form, one obtains (Israelachvili 1992; p. 184)

$$A_H = \frac{3}{4}k_B T \left(\frac{\varepsilon_A - \varepsilon_B}{\varepsilon_A + \varepsilon_B}\right)^2 + \frac{3h\nu_e}{16\sqrt{2}} \frac{(n_A^2 - n_B^2)^2}{(n_A^2 + n_B^2)^{3/2}} \tag{2-39}$$

where ε_i is the static dielectric constant of material i ($i = \text{A or B}$), n_i is the index of refraction of material i, and ν_e is the main ultraviolet absorption frequency, which typically has a value of $\sim 3 \times 10^{15}$ sec^{-1}. The first term in Eq. (2-39) includes the zero-frequency Keesom and Debye polarization terms, while the second term is the London dispersion contribution.

Several implications can be drawn directly from Eq. (2-39). First, A_H is always positive. Thus, the rule "like attracts like," inferred from Eq. (2-30) for molecular mixtures, should also hold at the continuum level. Second, when dispersion forces are dominant, the Hamaker constant is small when $n_A \dot{=} n_B$—that is when the dispersed phase (A) has an index of refraction close to that of the medium (B). These rules also apply to molecular mixtures. Nevertheless, small molecules with a significant difference in index of refraction often mix because of the large entropy thereby gained. But particles lose too little entropy on coagulation to resist doing so when there is an attractive van der Waals interaction, and so particle–particle clumping is the norm in suspensions, unless countermeasures are taken to stop it (see Section 7.1). Analogous considerations explain the prevalence of phase separation in polymer blends (see Section 2.3.1.2).

A final interesting observation is the existence of a frequency scale, $\nu_e \approx 3 \times 10^{15}$ sec^{-1} in Eq. (2-39). This is the frequency at which the electronic cloud around an atom fluctuates; it is therefore the rate at which the spontaneous dipoles fluctuate. Since the electromagnetic field created by these dipoles propagates at the speed of light $c = 3 \times 10^{10}$ cm/sec, only a finite distance $c/\nu_e \sim 100$ nm is traversed before the dipole has shifted. Since the dispersion interaction is only operative when these dipoles are correlated with each other, and this correlation is disrupted by the time lag between the fluctuation and the effect it produces a distance r away, the dispersion interaction actually falls off more steeply than r^{-6} when molecules or surfaces become widely separated. This effect is called the *retardation* of the van der Waals force. The effective Hamaker constant is therefore distance dependent at separations greater than 5–10 nm or so.

Figure 2-12 shows the curve of force versus distance measured between two crossed cylindrical surfaces of mica separated by aqueous electrolyte solutions. (The experimental data were corrected for weak electrostatic effects by subtraction of the electric-double-

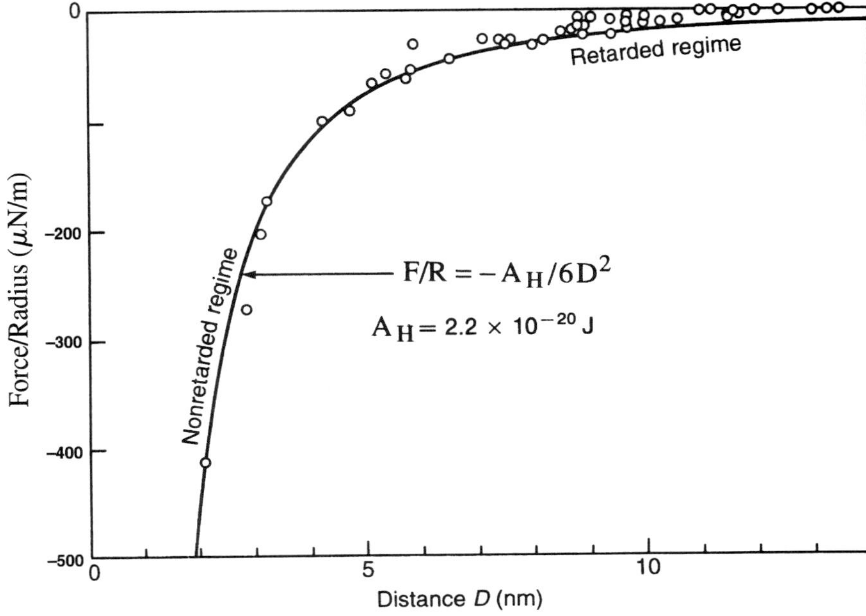

Figure 2.12 Van der Waals force F between two curved mica surfaces of radius $R \approx 1$ cm in water and electrolyte solutions. The line is the fitted van der Waals force with Hamaker constant $A_H = 2.2 \times 10^{-20}$ J. At distances D greater than 5 nm, the force is closer to zero than predicted because of retardation effects. (From Israelachvili and Adams 1978; and Israelachvili 1992, reprinted with permission from Academic Press.)

layer contribution.) Note that the force law $F = \partial W/\partial D = -A_H R/6D^2$ predicted for this geometry is well-obeyed for distances up to $D \approx 5$–10 nm. At distances of separation greater than this, the attractive force between the mica surfaces is reduced because of retardation. The value of the Hamaker constant obtained by fitting theory to this experiment in the nonretarded limit is $A_H \approx 2.2 \times 10^{-20}$ J, which is within 10% of the value computed from the Lifshitz theory. This value for A_H is a fairly typical one. For solids and liquids without hydrogen bonding, A_H can often be estimated from Γ, the surface energy (or surface tension) by

$$A_H \text{ (in joules)} \approx 2.1 \times 10^{-21} \Gamma \text{ (in mJ/m}^2) \tag{2-40}$$

2.4 ELECTROSTATIC INTERACTIONS

Electrostatic interactions are present in materials that contain ions. These include aqueous solutions of acids, bases, salts, polyelectrolytes (i.e., charged polymers), as well as colloidal suspensions of charged particles or droplets. Although fluids containing charged surfaces and mobile ions maintain electroneutrality overall, locally there are often charge imbalances.

2.4.1 The Poisson–Boltzmann Equation

Consider, then, a fluid containing ions that are nonuniformly distributed, producing a position-dependent electric field $\mathbf{E}$. Since an electric field is conservative, it is given by the gradient of a potential, $\mathbf{E} = -\nabla\psi$. Negative ions tend to collect at locations where this potential is positive relative to some datum, and positive ions collect where it is negative. At equilibrium, the number density n_i of ionic species i at each location is given by the *Boltzmann distribution:*

$$n_i = n_{0i} \exp\left(-z_i e\psi/k_B T\right) \tag{2-41}$$

where $-e = -1.6 \times 10^{-19}$ C is the charge of an electron, z_i is the charge valency of species i, and n_{0i} is the number density (number per unit volume) of species i at some reference location where ψ is taken to be zero.

Thus, Eq. (2-41) fixes the charged species distribution, once ψ is known. But the electric field (and hence ψ) is determined by the charge imbalance, according to Maxwell's equations, which for electrostatics reduce to the *Poisson equation:*

$$-\sum_i z_i e n_i = \varepsilon\varepsilon_0 \nabla \cdot \mathbf{E} = \varepsilon\varepsilon_0 \nabla^2\psi \tag{2-42}$$

where ε is the static dielectric constant of the medium, and ε_0 is the permittivity of free space. Combining Eqs. (2-41) and (2-42) gives the *Poisson–Bolzmann equation:*

$$\varepsilon\varepsilon_0 \nabla^2\psi = -\sum_i z_i e n_{0i} e^{-z_i e\psi/k_B T} \tag{2-43}$$

From this equation, the potential function can be determined, and therefore also the species distributions via Eq. (2-41), once suitable boundary conditions are given. One boundary condition is simply the reference position at which one chooses to set $\psi = 0$. The remaining boundary conditions are usually specifications of either the surface charge density σ (in coulombs/meter) or the surface potential ψ_s.

2.4.2 A Charged Surface and Its Double Layer

Consider a semi-infinite expanse of ionic solution, bounded on one side by a planar solid surface where σ is specified. Because of the condition of overall electroneutrality, the surface charge plus the bulk charge must sum to zero. From this and the Poisson and Boltzmann equations, one can derive (Israelachvili 1992)

$$\sigma^2 = 2\varepsilon\varepsilon_0 k_B T \sum_i (n_{si} - n_{\infty i}) \tag{2-44}$$

Here n_{si} is the density of species i in the solution next to the solid surface, and $n_{\infty i} = n_{0i}$ is the known density of species i in the bulk far from the surface, where $\psi = 0$. For each species, n_{si} can be obtained from ψ_s, the value of ψ at the solid surface, using the Boltzmann equation (2-41). Thus, Eq. (2-44) can be used to link the surface potential ψ_s to the density σ of charges bound to the solid surface.

> • Problem 2.6 helps you derive Eq. (2-44).

Consider, for example, a *symmetric electrolyte* which is one for which $z \equiv z_1 = -z_2$. An example is NaCl for which $z_1 = +1$ for sodium and $z_2 = -1$ for chloride. For a symmetric electrolyte, Eqs. (2-41) and (2-44) become

$$\sigma^2 = 2\varepsilon\varepsilon_0 k_B T n_{\infty 1} \left(e^{-ez\psi_s/k_B T} + e^{ez\psi_s/k_B T} - 2 \right)$$

$$= 4\varepsilon\varepsilon_0 k_B T n_{\infty 1} \left[\cosh\left(\frac{ez\psi_s}{k_B T} \right) - 1 \right] \tag{2-45}$$

Equation (2-45) gives the potential ψ_s at the surface in terms of the charge σ at the surface. Here $n_{\infty 1}$ is the bulk concentration of the cation, such as Na^+. The potential at a distance x from the surface is obtained by integrating the Poisson–Boltzmann equation, yielding

$$\psi(x) = \frac{2k_B T}{ez} \ln\left[\frac{1 + \gamma e^{-\kappa x}}{1 - \gamma e^{-\kappa x}} \right] \tag{2-46}$$

with

$$\gamma \equiv \tanh(ez\psi_s/4k_B T) \tag{2-47}$$

and

$$\kappa \equiv \left(\sum_i n_{\infty i} e^2 z_i^2 / \varepsilon\varepsilon_0 k_B T \right)^{1/2} \tag{2-48}$$

where for a 1:1 electrolyte, for example, the sum in Eq. (2-48) is over two species, with $z_1 = +1$ and $z_2 = -1$.

For a weak surface potential, $ez|\psi_s|/4k_B T \ll 1$, Eq. (2-46) reduces to the *Debye–Hückel equation*,

$$\psi(x) \approx \frac{4k_B T}{ez} \gamma \, \exp(-\kappa x) \approx \psi_s \, \exp(-\kappa x) \tag{2-49}$$

In this limit, the concentration of ions from the Boltzmann equation (2-41) can be expressed as

$$n_i \approx n_{0i}(1 - z_i e\psi/k_B T) = n_{0i}\left(1 - \frac{ez_i\psi_s}{k_B T} e^{-\kappa x} \right) \tag{2-50}$$

and Eq. (2-45) reduces to

$$\sigma \approx \varepsilon\varepsilon_0 \kappa \psi_s \tag{2-51}$$

According to Eq. (2-49) and (2-50), the potential, and hence the deviations of the ion concentrations from their bulk values, will decay exponentially as one moves away from the charged surface, with a decay length of κ^{-1}. This length scale κ^{-1} is called the *Debye length*, and it defines the characteristic distance from the surface over which ion concentrations are perturbed from their bulk values, and over which electrostatic forces are felt. Within roughly the distance κ^{-1}, ions with charge opposite that of the surface accumulate, and ions with the same charge as the surface are depleted.

The layer of thickness κ^{-1}, which contains enough charges to largely cancel out the surface charge, is called the *Gouy–Chapman diffuse double layer*. At room temperature, for a 1:1 electrolyte such as NaCl, using $e = 1.6 \times 10^{-19}$ C as the charge of an electron, $\varepsilon_0 = 8.854 \times 10^{-12} \mathrm{C}^2/\mathrm{Nm}^2$ as the permittivity of space, and $\varepsilon = 80$ as the dielectric constant for water, we obtain from Eq. (2-48) a Debye length κ^{-1} of

$$\kappa^{-1} = \frac{0.304}{\sqrt{[\mathrm{NaCl}]}} \quad \text{nanometers} \tag{2-52}$$

where [Nacl] is the molarity [moles per liter] of NaCl in the bulk. Thus Eq. (2-52) gives $\kappa^{-1} = 30$ nm for a 10^{-4} M solution and $\kappa^{-1} = 0.3$ nm for $[\mathrm{NaCl}] = 1$ M. *The thickness of the double layer is therefore sensitive to the electrolyte concentration.*

As an illustrative example taken from Russel et al. (1989), let us consider a 0.01 molar solution of sodium chloride in contact with a surface charged at a density of 5×10^{17} negative charges per square meter at room temperature, 298°K. Equation (2-52) gives $\kappa^{-1} = 3$ nm. The dimensionless surface potential $e\psi_s/k_BT$, obtained from Eq. (2-45), is -5.21, and Eqs. (2-46) and (2-49) give respectively the exact and the Debye–Hückel approximations for the potential as a function of distance from the surface. The results are plotted in Fig. 2-13. Note that since $-\psi_s/4k_BT > 1$, the Debye–Hückel approximation is not accurate. In this case, the Debye–Hückel approximation, Eq. (2-50), erroneously predicts (1) a negative concentration of anions close to the surface and (2) too low a concentration of cations.

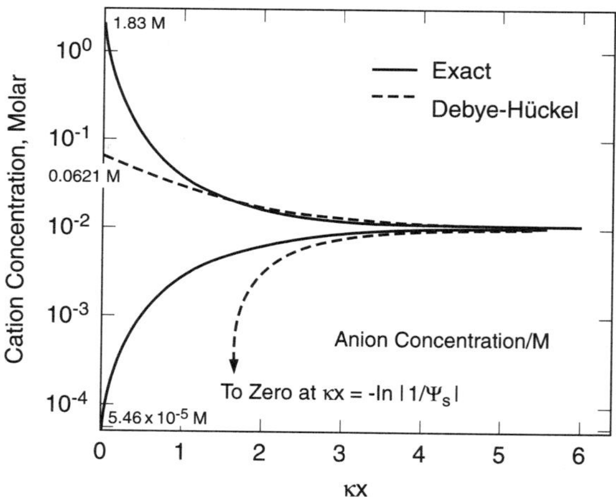

Figure 2.13   Calculated concentrations of monovalent cations and anions as a function of distance from a surface with a charge of $-8\ \mu\mathrm{C/cm}^2$ (5×10^{17} charges/m²), yielding a dimensionless surface potential of -5.21 (-133.9 mV). The bulk electrolyte concentration is 0.01 M, $T = 298$ K, and the solvent dielectric constant is 80, that of water. The Debye–Hückel theory [Eqs. (2-49) and (2-50)] fails close to the surface. The "exact" result is given by Eqs. (2-46) and (2-41). Ψ_s is defined as $e\psi_s/k_BT$. (From Russel et al. 1989, with permission of Cambridge University Press.)

In general, for highly charged surfaces, such as that in the preceding example, counterions are more concentrated near the surface than is predicted by the Debye–Hückel equation. At a density of around one charge per square nanometer on the surface, even in dilute aqueous electrolytes, the solution counterion concentration profile becomes so steep near the surface that there are even enough counterions within one counterion diameter of the surface to nearly balance the surface charge!

So far, we have assumed that the surface charge density is a known constant. How does a surface become charged? When a surface is exposed to an electrolytic liquid, surface groups spontaneously ionize, and solution ions can adsorb onto the surface, producing a net charge. Relatively immobile charges bound to the surface constitute what is called the *Stern layer,* which is only a couple of angstroms thick. This layer influences the surface charge density σ, which, in turn, controls the ion distribution in the double layer. However, if the surface charge density is not constant, but is influenced by the ion concentration in the double layer, then σ cannot be set equal to a constant, and the foregoing simple theory fails. This may happen, for example, when two charged surfaces are brought very close together in an electrolyte medium; some electrolyte ions may bind to the surface in order to reduce the surface charge, thereby reducing the repulsive interaction between the surfaces. This process is called *charge regulation.* One simple way to model the nonconstancy of surface charge is to assume constancy of the surface potential ψ_s, rather than the surface charge density σ.

A more complex, and perhaps more realistic, approach is to use a mass action expression for the surface charge. For example, if the surface charge is produced by the loss of protons from the surface, then some of these protons may reabsorb if the pH of the solution is lowered enough. In that case, the surface charge density might be controlled by a surface reaction such as

$$S^- + H^+ \leftrightarrow SH \qquad (2\text{-}53)$$

where S^- is the surface charge and SH is the neutralized surface group. If the dissociation constant for this reversible surface reaction is known, then the surface charge (which is proportional to $[S^-]$) can be determined self-consistently with the ion concentrations in the double layer (Israelachvili 1992; Russel et al. 1989). Multivalent ions, such as divalent cations, are especially prone to surface binding and can even *reverse* the surface charge, a fact that is of significance in colloidal stability (see Chapter 7). Multivalent counterions tend to congregate more densely at charged surfaces than do monovalent counterions, since a multivalent ion can deliver a greater neutralizing charge than can a monovalent ion for the same loss in entropy. Thus, relatively small concentrations of divalent cations can by themselves determine the surface potential ψ_s, even in the presence of a much greater concentration of monovalent cations.

2.4.3 The Force Between Two Charged Plates

Consider now two parallel flat surfaces, each with equal charge density σ, separated by a distance D. Although charged, there is no net electrostatic repulsion of the surfaces, since the counterions in the gap between the plates cancel the charges on the plates. However, because the counterions must be drawn into the gap, there is an entropy cost that becomes greater as the plates become closer together. This entropy cost produces a distance-dependent pressure $P(D)$ that tends to push the plates apart (Israelachvili 1992):

$$P(D) = k_B T \left[\sum_i n_{mi}(D) - n_{mi}(\infty) \right] \tag{2-54}$$

where $n_{mi}(D)$ is the density of species i at the midplane between the two surfaces separated by a distance D, and $n_{mi}(\infty)$ is the bulk concentration of i; or, equivalently, it is the concentration at the midplane when the plates are infinitely separated.

Although Eq. (2-54) above is a simple expression, obtaining the ion concentrations $n_{mi}(D)$ at the midplane between the plates requires, in general, a numerical solution to the Poisson–Bolzmann equation. However, using the Boltzmann equation (2-41) for a symmetric electrolyte at small dimensionless potential $ze\psi_m/k_B T$ and letting $\psi = 0$ in the bulk, so that n_{0i} in Eq. (2-41) equals $n_{mi}(\infty) = n_{\infty 1}$, we have

$$P = k_B T n_{\infty 1} \left(e^{-ze\psi_m/k_B T} + e^{ze\psi_m/k_B T} - 2 \right) \approx \frac{z^2 e^2 \psi_m^2 n_{\infty 1}}{k_B T} \tag{2-54a}$$

If the plates are greater than one Debye distance apart, an acceptable approximation is the *superposition approximation* in which the potential ψ at the midplane is taken to be the sum of the potentials from each of two isolated surfaces at $x = \frac{1}{2}D$. Evaluating ψ_m from twice $\psi(x)$ in Eq. (2-49) with $x = D/2$, we get

$$\psi_m \approx 2\psi(x) = \frac{8k_B T}{ez} \gamma \, \exp\left(-\frac{\kappa D}{2}\right)$$

Inserting this into Eq. (2-54a) gives for flat plates

$$P \approx 64 k_B T n_{\infty 1} \gamma^2 e^{-\kappa D} \tag{2-55}$$

for $\kappa D > 1$. Since $P = -\partial W_e/\partial D$, where W_e is the electrostatically induced potential per unit area, we obtain

$$W_e(\text{flat plates}) \approx 64 k_B T n_{\infty 1} \gamma^2 \kappa^{-1} e^{-\kappa D} \tag{2-56}$$

2.4.4 Interaction Potentials for Spheres

This result can be used to obtain the force F between two spheres of radius a, with $D \ll a$, by using the *Derjaguin approximation* that relates the force between two spheres to the potential between two flat plates; that is, $F \text{ (spheres)} = \pi a W_e \text{ (flat plates)}$. When this is integrated to obtain the potential, one obtains

$$W_e(\text{spheres}) \approx 64\pi k_B T a n_{\infty 1} \gamma^2 \kappa^{-2} e^{-\kappa D} \tag{2-57}$$

For low surface charge, using Eqs. (2-47), (2-48), and (2-51), this reduces to

$$W_e \approx 2\pi\varepsilon\varepsilon_0 a\psi_s^2 \, e^{-\kappa D} = \frac{2\pi a\sigma^2 e^{-\kappa D}}{\kappa^2\varepsilon\varepsilon_0}$$

Thus, as the two surfaces approach each other, the repulsive interaction increases exponentially, as long as $D > \kappa^{-1}$ (and $D \ll a$, so that the Derjaguin approximation holds).

The above expression has limited applicability, however, since it holds only when the gap D between particle surfaces is small compared to the particle radius, but still large compared to the Debye length. Fortunately, the superposition approximation can be generalized by linearly adding the potentials for two isolated spheres. This gives (Russel et al. 1989)

$$W_e = \frac{4\pi \varepsilon \varepsilon_0 a^2 \psi_s^2}{r} \exp(-\kappa D) \qquad (\kappa D \gtrsim 2) \qquad (2\text{-}58)$$

This potential is valid for arbitrarily widely separated spheres, but it breaks down when the Debye double layers begin to overlap significantly (Russel et al. 1989). However, for small separations, $\kappa D < 2$, the Derjaguin approximation may still be used, as long as $D < a$; that is, $\kappa a \gtrsim 2$. An analytic expression can then be obtained if the potential is small enough that the Poisson–Boltzmann equation can be linearized (Russel et al. 1989, p. 117). For small separations between particles, a choice must be made between a constant-potential or a constant-charge boundary condition. For a constant-potential boundary condition, one can write the approximate expression

$$W_e \approx 2\pi \varepsilon \varepsilon_0 a \psi_s^2 \ln[1 + \exp(-\kappa D)] \qquad (\kappa D \lesssim 2, \text{ constant } \psi_s) \qquad (2\text{-}59)$$

while for a constant-charge boundary condition we obtain

$$W_e \approx 2\pi \varepsilon \varepsilon_0 a \psi_s^2 \ln \left[\frac{1}{1 - \exp(-\kappa D)} \right] \qquad (\kappa D \lesssim 2, \text{ constant } \sigma) \qquad (2\text{-}60)$$

A useful composite potential, which reduces to Eq. (2-58) at large κD and to Eq. (2-59) at small D, is (McCartney and Levine 1969; Ogawa et al. 1997)

$$W_e \approx \frac{4\pi \varepsilon \varepsilon_0 a (r - a) \psi_s^2}{r} \ln \left[1 + \frac{a}{r - a} \exp(-\kappa D) \right] \qquad (\text{constant } \psi_s) \qquad (2\text{-}61)$$

Whatever the details of the electrostatic contribution to the interaction of two surfaces, at near contact the van der Waals (VDW) forces must begin to assert themselves. At close enough spacings, VDW forces overpower electrostatic forces, since W_{VDW} increases as a power law with decreasing separation distance D, while the electrostatic potential increases at most logarithmically. Since the van der Waals interactions are insensitive to electrolyte concentration, the total force between the two surfaces can be approximated by the sum of the van der Waals and the electrostatic contributions, a result leading to the *DLVO* (Derjaguin and Landau, 1941; Verwey and Overbeek, 1948) theory of surface interactions and colloid stability. Figure 2-14 shows the forces between two cylindrically curved mica surfaces separated by electrolyte solutions. At long distances, $\gtrsim 10$ nm, the force is dominated by an exponentially decaying electrostatic contribution, while at distances less than 5 nm, van der Waals interactions make a significant contribution. The lines in Fig. 2-14 are from the DLVO theory.

The long-range, repulsive nature of electrostatic interactions between particles of like charge can stabilize such suspensions, keeping the suspension in a fluid, rather than gelled, state. However, if the repulsive interactions are too strong, the particles will move to well-defined lattice points at which the separations between nearest-neighboring particles are as distant as possible; this leads to *colloidal crystalline ordering transitions* at low or modest

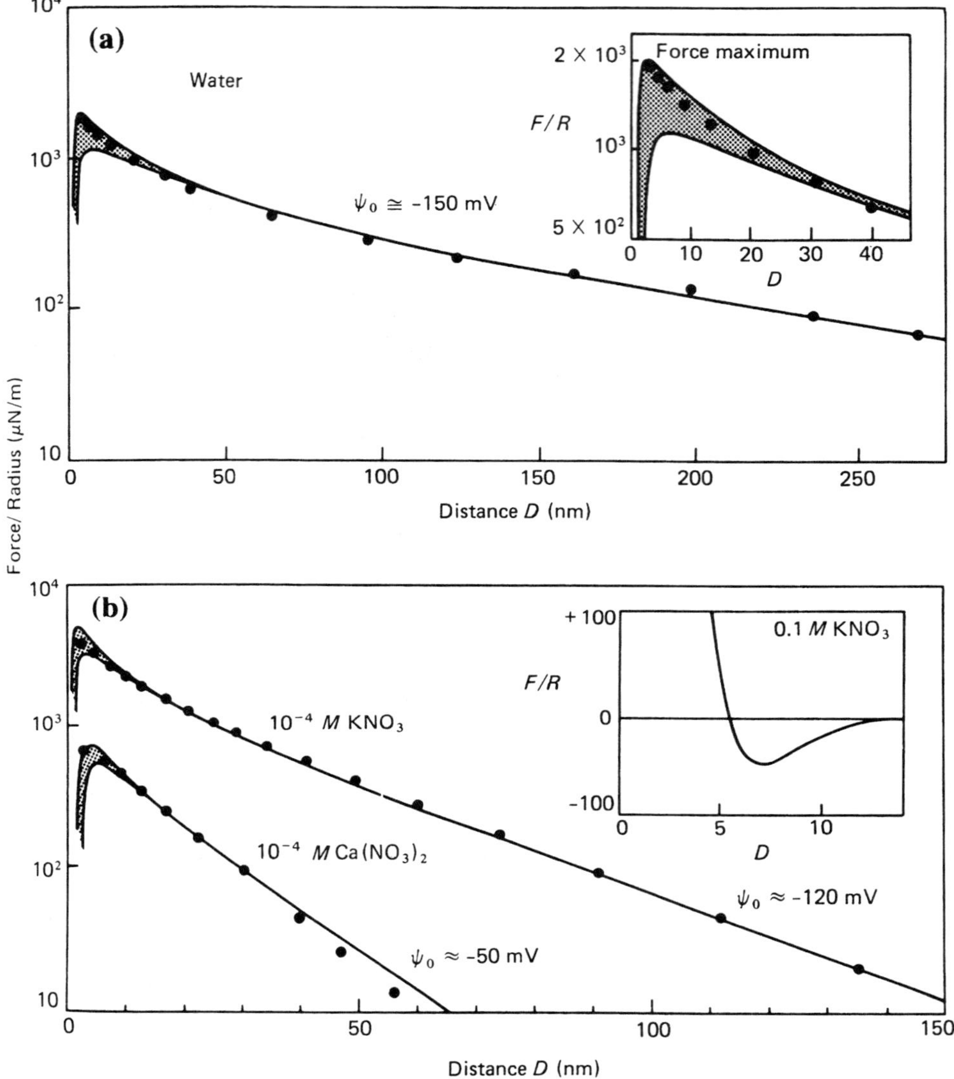

Figure 2.14   Measured electrostatic double-layer and van der Waals forces between two surfaces of curved mica of radius $R \approx 1$ cm in **(a)** water and **(b)** dilute KNO$_3$ and Ca(NO$_3$)$_2$ solutions. The lines are the predictions of the DLVO theory with a Hamaker constant of 2.2×10^{-20} J in the limits of constant surface charge and constant surface potential; here $\psi_0 = \psi_s$, the particle surface potential. (The lines for constant surface charge are slightly higher than those for constant surface potential at small D.) The inset in **(b)** is the measured force in 0.1 M KNO$_3$, which shows a force minimum at a distance of around 7 nm. Since this minimum in force occurs away from the deep minimum at the surface, it is called a *secondary minimum*. (From Israelachvili and Adams 1978; and Israelachvili 1992, reprinted with permission from Academic Press.)

particle volume fractions. These transitions, and the moduli of colloidal crystals, will be discussed in Section 6.4.

> • Problem 2.7 through 2.9 test your grasp of electrostatic forces.

2.5 HYDROGEN-BONDING, HYDROPHOBIC, AND OTHER INTERACTIONS

Other important forces, including hydration and solvation forces associated with the tendency of molecules or ions to bind tightly to surfaces, as well as hydrophobic and hydrophilic forces and other specific interactions, are present in complex fluids. These forces become important at separations of a couple of nanometers. As discussed in Chapter 7, such forces can influence the rheological properties of suspensions of particles in near contact.

Hydrogen bonding is a special case of a solvation force that acts between a hydrogen which is bound to an electron-withdrawing (i.e., electronegative) atom such as oxygen or fluorine and another electronegative atom, again such as oxygen or fluorine. Hydrogen bonds of importance to complex fluids typically have binding energies ΔE of 10–40 kJ/mol (Israelachvili 1992; Jeffrey 1997), making them stronger than van der Waals interactions (~ 1 kJ/mol), but much weaker than "permanent" covalent bonds (500 kJ/mol). The lifetime of a hydrogen bond can then be estimated as $\tau = \tau_0 \exp(\Delta E/RT)$, where $R = N_A k_B$ is the gas constant and τ_0 is the fundamental vibration period, $\tau_0 \sim 10^{-13}$ sec. Hence, at room temperature, hydrogen bonds last around $\tau \sim 10^{-11}$–10^{-7} sec. A small-molecule liquid of molecular weight $M \approx 50$ held together by hydrogen bonding should then have a viscosity η_0 of $\sim G\tau \sim \nu k_B T \tau = (\rho RT/M)\tau_0 \exp(\Delta E/RT) \sim 1$ cP to 100 P at room temperature. Hence, because of hydrogen bonding, the viscosities of water and glycerol, for example, are much higher than those of non-hydrogen-bonding liquids composed of molecules of similar size. Hydrogen bonding is a strong enough interaction to induce miscibility in polymer blends that are normally immiscible due to van der Waals or excluded-volume interactions, allowing, for example, the miscible blending of rod-like with flexible-coil polymers (Coleman et al. 1991; Painter et al. 1991).

The strong repulsion between water and hydrocarbons or other nonpolar molecules is legendary; everyone knows "you can't mix oil and water." Less well known, however, is that these repulsive interactions are *mainly entropic*, rather than energetic. The reason for this strange state of affairs is that each water molecule tenaciously holds onto its full complement of hydrogen bonds, even when forced to live next to a nonpolar neighbor with which it has no chance of hydrogen bonding. The water molecules around a hydrocarbon do this by orienting their tetrahedral bond structures to produce clathrate cages around the repulsive nonpolar neighbor. For most nonpolar inclusions, cages can be built that grant to each water its desired number of hydrogen bonds. While these clever cages allow most of the energetic penalty to be avoided, a high entropic price is paid for each cage. Hence, the free energy per unit surface area of contact between oil and water is high, 40–50 dyn/cm. The entropic nature of the strong immiscibility of oil and water is known as the *hydrophobic effect* (Israelachvili 1992; Kauzmann 1959; Tanford 1980). This surface free energy of oil–water contact is an important factor in the self-assembly of aqueous surfactant fluids, discussed in Chapter 12.

2.6 SUMMARY

The forces leading to microstructure formation in complex fluids are relatively few: Excluded-volume, van der Waals, and electrostatic forces are the main ones. In some fluids, hydrogen bonding, hydrophobic, or various solvation forces are also important. Simplified theories can account for the effects of these forces on fluid structure and, to some extent, on relaxation rates.

Crystalline or orientational orderings are mostly controlled by repulsive forces, such as excluded-volume forces. The crystalline transitions in ideal hard-sphere fluids and the nematic liquid crystalline transitions in hard-rod suspensions are convenient simple models of corresponding transitions in fluids composed of uncharged spherical or elongated molecules or particles. The transition from the isotropic to the nematic state can be described theoretically using the Onsager, Maier–Saupe, or Flory theories.

The conformations of long flexible polymer molecules in solution or in the melt can be described by statistical theories. In the melt or at special (theta) conditions in solution, polymer conformations on the large scale mimic those of random walks. In solution the excluded-volume effects tend to swell polymer coil dimensions, while in the melt the excluded-volume effects lead to local packing correlations, but no long-range deviation from random-walk statistics. Simple models of polymer flexibility include (a) the "freely jointed chain," which is a sequence of rigid rods connected by completely flexible joints, and (b) the "worm-like" chain, in which there is a uniform bending modulus along the chain's backbone. The freely jointed chain is often used to represent synthetic polymers with a backbone of carbon–carbon bonds, while the worm-like chain describes many biological polymers such as DNA and collagen.

Attractive van der Waals interactions produce liquid–liquid or liquid-gas phase separation. The van der Waals interactions between unlike molecules are almost always less attractive than those between like molecules. This leads to phase separation when the interactions are strong, or when the molecules are big enough that entropic differences between the mixed and unmixed states are small. Thus, phase separation is common in mixtures of unlike polymers. The Flory–Huggins lattice theory can account qualitatively for polymer–polymer phase separation by combining a simple expression for translational entropy, with a simple term for the interaction "energy" with a coefficient χ. In some cases, χ can be estimated from van der Waals cohesive energies. For suspensions of particles in a solvent, van der Waals interactions often lead to particle–particle clumping; the strength of the relative attractive forces between particles or surfaces can be quantified by the "Hamaker constant."

Electrostatic interactions in solutions containing charged particles and ions can be described using the Poisson–Boltzmann equation. A charged surface attracts counterions into a "double layer" of thickness defined by the "Debye length," which depends on counterion concentration and solvent dielectric constant. From simplified theories, expressions can be derived for the attractive interaction potential between charged spheres.

REFERENCES

Abe A, Yamazaki T (1989). *Macromolecules* 22:2145.

Aden MA, Bianchi E, Ciferri A, Conio G, Tealdi A (1984). *Macromolecules* 17:2010.

Baljon ARC, Grest GS, Witten TA (1995). *Macromolecules* 28:1835.

Balsara NP (1996). In *Physical Properties of Polymers Handbook*, Mark JE (ed), AIP Press, New York.

Barton AFM (1983). *Handbook of Solubility Parameters and Other Adhesion Parameters*, CRC Press, Boca Raton, FL.

Bates FS, Wignall GD (1986). *Macromolecules* 19:932.

Bates FS, Muthukumar M, Wignall GD, Fetters LJ (1988). *J Chem Phys* 89:535.

Bates FS, Schulz MF, Rosedale JH, Almdal K, (1992). *Macromolecules* 25:5547.

Bird RB, Armstrong RC, Hassager O (1987). *Dynamics of Polymeric Liquids, Vol 1: Fluid Mechanics*, Wiley, New York.

Coleman MM, Graf JF, Painter PC (1991). *Specific Interactions and Miscibility of Polymer Blends*, Technomic, Lancaster, PA.

de Gennes PG (1982). In *Polymer Liquid Crystals*, Ciferri A, Krigbaum WR, Meyer RB (eds), Academic Press, New York.

de Gennes PG, Prost J (1993). *The Physics of Liquid Crystals*, 2nd ed, Oxford University Press, New York.

Deloche B, Samulski ET (1981). *Macromolecules* 14:575.

Derjaguin BV, Landau L (1941). *Acta Physicochim URSS* 10:25.

Doi M, Pearson DS, Kornfield J, Fuller GG (1989). *Macromolecules* 22:1488.

Fetters LJ, Lohse DJ, Richter D, Witten A, Zirkel A (1994). *Macromolecules* 27:4639.

Flory PJ (1941). *J Chem Phys* 9:660.

Flory PJ (1953). *Principles of Polymer Chemistry*, Cornell University Press, London.

Flory PJ (1956). *Proc R Soc* A234:73.

Flory PJ (1969). *Statistical Mechanics of Chain Molecules*, Hanser, New York.

Flory PJ (1978). *Macromolecules* 11:1141.

Flory PJ (1984). *Adv Polym Sci* 59:1.

Frattini PL, Fuller GG (1986). *J Fluid Mech* 168:119.

Fredenslund A, Jones RL, Prausnitz JM (1975). *AIChE J* 21:1086.

Fredrickson GH, Liu AJ, Bates FS (1994). *Macromolecules* 27:2503.

Gent AN (1969). *Macromolecules* 2:262.

Goodwin JW, Hughes RW, Partridge SJ, Zukoski CF (1986). *J Chem Phys* 85:559.

Graessley WW, Krishnamoorti R, Balsara NP, Butera RJ, Fetters LJ, Lohse DJ, Schulz DN, Sissano JA (1994). *Macromolecules* 27:3896.

Graessley WW, Krishnamoorti R, Reichart GC, Balsara NP, Fetters LJ, Lohse DJ (1995). *Macromolecules* 28:1260.

Grosberg AY, Khokhlov AR (1994). *Statistical Physics of Macromolecules*, AIP Press, New York.

Guggenheim EA (1952). *Mixtures*, Clarendon Press, Oxford.

Hildebrand JH, Scott RL (1950). *The Solubility of Non-Electrolytes*, 3rd ed, Van Nostrand Reinhold, Princeton, NJ; reprinted by Dover Press, New York, 1964.

Hoover WG, Ree FH (1968). *J Chem Phys* 49:3609.

Huggins ML (1941). *J Chem Phys* 9:440.

Israelachvili JN (1992). *Intermolecular and Surface Forces*, 2nd ed, Academic Press, New York.

Israelachvili JN, Adams GE (1978). *J Chem Soc Faraday Trans I* 74:975.

Jackson CL, Shaw MT (1991). *Int Mater Rev* 36:165.

Jeffrey GA (1997). *An Introduction to Hydrogen Bonding*, Oxford University Press, New York.

Karis TE, Russell TP, Gallot Y, Mayes AM (1995). *Macromolecules* 28:1129.

Kauzmann W (1959). *Adv Protein Chem* 14:1.

Kent SL, Geil PH (1993). *Makromol Chem, Macromol Symp* 70/71:83.

Kloczkowski A (1996). In *Physical Properties of Polymers Handbook*, Mark JE (ed), AIP Press, New York.

Krigbaum WR (1985). *Faraday Discuss Chem Soc* 79:133.

Krishnamoorti R, Graessley WW, Balsara NP, Lohse DJ (1994). *Macromolecules* 27:3074.

Krishnamoorti R, Graessley WW, Dee GT, Walsh DJ, Fetters LJ, Lohse DJ (1996). *Macromolecules* 29:367.

Lekkerkerker HNW, Coulon Ph, van Der Haegen R, Deblieck R (1984). *J Chem Phys* 80:3427.

Li MH, Brûlet A, Cotton JP, Davidson P, Strazielle C, Keller P (1994). *J Phys II (France)* 4:1843.

Lindemann FA (1910). *Z Phys* 11:609.

Londono JD, Narten AH, Wignall GD, Honnell KG, Hsieh ET, Johnson TW, Bates FS (1994). *Macromolecules* 27:2864.

Maier W, Saupe A (1958). *Z Naturforsch* 13A:564.

Maier W, Saupe A (1959). *Z Naturforsch* 14A:882.

Maier W, Saupe A (1960). *Z Naturforsch* 15A:287.

Mark JE (1996) (ed). *Physical Properties of Polymers Handbook*, AIP Press, New York.

McCartney LN, Levine S (1969). *J Colloid Interface Sci* 30:345.

Modern Plastics Encyclopedia '95 (1994). Special issue of *Modern Plastics*, mid-November.

Odijk T (1986). *Macromolecules* 19:2313.

Ogawa A, Yamada H, Matsuda S, Okajima K (1997). *J Rheol* 41:769.

Onsager L (1949). *Ann NY Acad Sci* 51:627.

Ookubo N, Komatsubara M, Nakajima H, Wada Y (1976). *Biopolymers* 15:929.

Ott A, Magnasco M, Simon A, Libchaber A (1993). *Phys Rev E* 48:R1642.

Painter PC, Coleman MM (1994). *Fundamentals of Polymer Science*, Technomic Publishing, Lancaster, PA.

Painter PC, Tang W-L, Graf JF, Thomson B, Coleman MM (1991). *Macromolecules* 24:3929.

Poniewierski A (1992). *Phys Rev A* 45:5605.

Poon WCK, Pusey PN (1995). In *Observations, Prediction and Simulation of Phase Transitions in Complex Fluids*, Baus M, et al. (eds), Kluwer Academic Publishers, Hingham, MA.

Pusey PN, van Megan W (1987). *Phys Rev Lett* 59:2083.

Robinson C, Ward JC, Beevers RB (1958). *Discuss Faraday Soc* 25:29.

Rodriguez F (1996). *Principles of Polymer Systems*, 4th ed, Taylor & Francis, Washington, D.C.

Russel WB, Saville DA, Schowalter WR (1989). *Colloidal Dispersions*, Cambridge University Press, New York.

Saito N (1967). *J Phys Soc Jpn* 22:219.

Sato T, Teramoto A (1996). *Adv Polym Sci* 126:85.

Schwahn D, Hahn K, Streib, J, Springer T (1989). *J Chem Phys* 92:8383.

Semenov AN, Khokhlov AR (1988). *Sov Phys Usp* 31:988.

Smith SB, Cui Y, Bustamante C (1996). *Science* 271:795.

Staverman AJ, van Stanten JH (1941). *Recl Trav Chim Pays-Bas* 60:76.

Sundararajan PR (1996). In *Physical Properties of Polymers Handbook*, Mark JE (ed), AIP Press, New York.

Takada A, Fukuda T, Watanabe J, Miyamoto T (1995). *Macromolecules* 28:3394.

Tanford C (1980). *The Hydrophobic Effect*, Wiley, New York.

Thulstrup M, Michl J (1982). *J Am Chem Soc* 104:5594.

Tkachenko AV (1996). Phys Rev Lett 77:4218.

Treloar LRG (1975) *The Physics of Rubber Elasticity*, 3rd ed, Clarendon Press, Oxford.

Verwey EJW, Overbeek JTL (1948). *Theory of Stability of Lyophobic Colloids*, Elsevier, Amsterdam.

Warner M, Gunn JMF, Baumgärtner AB (1985). *J Phys A: Math Gen* 18:3007.

Winey KI, Berba ML, Galvin ME (1996). *Macromolecules* 29:2868.

Woodcock LV (1981). *Ann NY Acad Sci* 37:274.

Yamakawa H (1971). *Modern Theory of Polymer Solutions*, Harper's Chemistry Series, Harper and Row, New York.

Ylitalo CM, Kornfield JA, Fuller GG, Pearson DS (1991). *Macromolecules* 24:749.

PROBLEMS AND WORKED EXAMPLES

Problem 2.1   You mix 50% by mass of chains with molecular weight 100,000 with 50% by mass chains with molecular weight 200,000. What is the weight-averaged molecular weight of the mixture?

Problem 2.2 (Worked Example)   Consider a "box" distribution of molecular weights in which the weight fraction $c(M)$ of molecular weight M is a constant, c_1, over a range of molecular weights between a lower bound M_L and an upper bound M_H. There are no molecules with molecular weights greater than M_H or lower than M_L. Find the relationship between M_w/M_n and M_H/M_L. If $M_w/M_n = 2$, what is M_H/M_L?

ANSWER:

By extending the definition of the weight-averaged molecular weight M_w in Eq. (2-11) to a continuous distribution, we have

$$M_w = \int_{M_L}^{M_H} c(M)M \, dM = c_1 \int_{M_L}^{M_H} M \, dM = \frac{1}{2}c_1 \left(M_H^2 - M_L^2 \right) \tag{A2-1}$$

By definition, the weight fractions of all species add up to unity; hence

$$1 = c_1 \int_{M_L}^{M_H} dM = c_1 \left(M_H - M_L \right)$$

Thus,

$$c_1 = \frac{1}{M_H - M_L} \tag{A2-2}$$

Then, from Eqs. (A2-1) and (A2-2) we obtain

$$M_w = \frac{M_H^2 - M_L^2}{2 \left(M_H - M_L \right)} = \frac{1}{2} \left(M_H + M_L \right) \tag{A2-3}$$

The weight fraction $c(M)$ is related to the number fraction $n(M)$ by

$$c_1 = c(M) = \frac{n(M)M}{\int_{M_L}^{M_H} n(M)M \, dM} \tag{A2-4}$$

Let

$$c_2 \equiv \int_{M_L}^{M_H} n(M)M \, dM$$

Then, from Eq. (A2-4) we obtain

$$c_1 = \frac{n(M)M}{c_2}$$

which can be rearranged to

$$n(M) = \frac{c_1 c_2}{M} \tag{A2-5}$$

Also, by definition, the number fraction $n(M)$ must integrate to unity:

$$1 = \int_{M_L}^{M_H} n(M)\,dM = \int_{M_L}^{M_H} \frac{c_1 c_2}{M}\,dM = c_1 c_2 \ln\left(\frac{M_H}{M_L}\right)$$

Therefore

$$c_1 c_2 = \frac{1}{\ln\left(\dfrac{M_H}{M_L}\right)} \tag{A2-6}$$

By generalizing the definition of M_n in Eq. (2-11) to a continuous distribution and using Eqs. (A2-5) and (A2-6), we obtain

$$M_n = \int_{M_L}^{M_H} n(M)M\,dM = c_1 c_2 \int_{M_L}^{M_H} dM = c_1 c_2 (M_H - M_L) = \frac{M_H - M_L}{\ln\left(\dfrac{M_H}{M_L}\right)} \tag{A2-7}$$

From this expression for M_n and Eq. (A2-3) for M_w, we obtain

$$\frac{M_w}{M_n} = \frac{\frac{1}{2}(M_H + M_L)\ln\left(\dfrac{M_H}{M_L}\right)}{M_H - M_L} = \frac{\frac{1}{2}\left(\dfrac{M_H}{M_L} + 1\right)\ln\left(\dfrac{M_H}{M_L}\right)}{\dfrac{M_H}{M_L} - 1} \tag{A2-8}$$

For large M_H/M_L, we obtain $M_w/M_n \approx \frac{1}{2}\ln(M_H/M_L)$. Thus,

$$\frac{M_H}{M_L} \approx \exp\left(\frac{2M_w}{M_n}\right)$$

We then find that for $M_w/M_n = 2$, M_H/M_L is very large, namely $M_H/M_L \approx 45$!

Problem 2.3 (Worked Example)   Derive the formula for the radius of gyration R_g of a random-walk chain of n links, where n is large.

ANSWER:

Start with Eq. (2-14a), the definition of R_g. For a random-walk chain, each step of the walk is independent of the others. Hence, the ensemble average can be brought inside the summations. Thus,

$$R_g^2 \equiv \left\langle \frac{1}{2}n^{-2}\left(\sum_{i=1}^{n}\sum_{j=1}^{n}(R_{ij})^2\right)\right\rangle = \frac{1}{2}n^{-2}\left(\sum_{i=1}^{n}\sum_{j=1}^{n}\langle R_{ij}^2\rangle\right) \tag{A2-9}$$

Now, because the chain as a whole is a random walk, each subchain contained between monomers i and j can also be considered a random walk, and the random-walk formula (2-12) can be applied to obtain

$$\langle R_{ij}^2\rangle = |i - j|\,b_n^2$$

We can insert this into Eq. (A2-9) and convert the sums to integrals (because n is large):

$$R_g^2 = \frac{1}{2}n^{-2} \int_0^n \int_0^n |i - j| b_n^2 \, dj \, di \tag{A2-10}$$

To deal with the absolute value, we break the inner integral into two pieces:

$$\int_0^n \int_0^n |i - j| \, dj \, di = \int_0^n \int_0^i (i - j) \, dj \, di + \int_0^n \int_i^n (j - i) \, dj \, di$$

$$= \int_0^n \left(i^2 - \tfrac{1}{2}i^2 \right) di + \int_0^n \left[\left(\tfrac{1}{2}n^2 - \tfrac{1}{2}i^2 \right) - \left(ni - i^2 \right) \right] di \tag{A2-11}$$

The first of the two integrals above yields

$$\int_0^n \left(i^2 - \tfrac{1}{2}i^2 \right) di = \frac{1}{6}n^3$$

while the second one gives

$$\int_0^n \left[\left(\tfrac{1}{2}n^2 - \tfrac{1}{2}i^2 \right) - \left(ni - i^2 \right) \right] di = \int_0^n \left(\tfrac{1}{2}n^2 + \tfrac{1}{2}i^2 - ni \right) di = \frac{1}{6}n^3$$

Thus Eq. (A2-11) reduces to

$$\int_0^n \int_0^n |i - j| \, dj \, di = \frac{n^3}{3}$$

Hence, Eq. (A2-9) reduces to

$$R_g^2 = \frac{1}{2}n^{-2} \left(\frac{1}{3}n^3 \right) b_n^2 = \frac{nb_n^2}{6} = \frac{\langle R^2 \rangle_0}{6} \tag{A2-12}$$

Therefore, $R_g = \langle R^2 \rangle_0^{1/2} / \sqrt{6}$.

Problem 2.4   Suppose a worm-like chain with diameter $d = 1.5$ nm, persistence length $\lambda_p = 100$ nm, and length $L = 20$ nm is dissolved in a solvent. Estimate ϕ_2, the minimum volume fraction of polymer needed to form a wholly nematic phase. If L is increased to 500 nm, and d and λ_p are kept the same, what is ϕ_2?

Problem 2.5(a) (Worked Example)   From the Flory–Huggins free energy per unit volume, Eq. (2-33), show that the excess chemical potential of mixing of the solvent, $\mu_B - \mu_B^0$, in a polymer solution is given by

$$\frac{\mu_B - \mu_B^0}{k_B T} = \ln(1 - \phi_A) + \left(1 - \frac{1}{y} \right) \phi_A + \chi \phi_A^2 \tag{A2-13}$$

where μ_B^0 is the standard chemical potential of pure B and where $y \equiv N_A v_A / v_B$. For the solvent, you may take the "degree of polymerization" N_B to be unity. (In the text of Chapter 2, x is used for the quantity we here call y. We make this replacement here to avoid any confusion with the chi parameter χ.

ANSWER:

Since an A molecule occupies a volume $v_A N_A$ while a B molecule occupies a volume $v_B N_B$, in a volume V there are $n_A = \phi_A V / (v_A N_A)$ molecules of A and $n_B = \phi_B V / (v_B N_B)$ molecules of

B. We take the Flory–Huggins reference volume v to be the volume of the solvent molecule, $v = v_B$, and note that $N_B = 1$. Thus, we find

$$\phi_A = \frac{v_A N_A}{V} n_A, \qquad \phi_B = \frac{v_B N_B}{V} n_B = \frac{v}{V} n_B \tag{A2-14}$$

When we multiply the identity $1 = \phi_A + \phi_B$ by the total volume V and then substitute for ϕ_A and ϕ_B using Eq. (A2-14), we obtain

$$V = v_A N_A n_A + v n_B \tag{A2-15}$$

We can use this formula for V to express ϕ_A and ϕ_B in Eq. (A2-14) as

$$\phi_A = \frac{v_A N_A n_A}{v_A N_A n_A + v n_B}, \qquad \phi_B = \frac{v n_B}{v_A N_A n_A + v n_B} \tag{A2-16}$$

Now we can write the *extensive* free energy of mixing ΔF by multiplying Eq. (2-33) by V. Using Eq. (A2-14), we obtain

$$\frac{\Delta F}{k_B T} = n_A \ln \phi_A + n_B \ln \phi_B + \chi \phi_A n_B \tag{A2-17}$$

By definition, the chemical potential of mixing $\mu_B - \mu_B^0$ is obtained by taking the derivative of ΔF with respect to n_B, holding n_A constant. Thus,

$$\frac{\mu_B - \mu_B^0}{k_B T} = \frac{1}{k_B T} \frac{\partial \Delta F}{\partial n_B} \qquad \text{(holding } n_A \text{ constant)} \tag{A2-18}$$

where the partial derivatives with respect to n_B, here and in what follows, are always *taken with n_A held fixed*. Note from Eq. (A2-16) that ϕ_A and ϕ_B are functions of n_B and n_A. From Eq. (A2-17), we therefore obtain

$$\frac{1}{k_B T} \frac{\partial \Delta F}{\partial n_B} = n_A \frac{\partial}{\partial n_B} \ln \phi_A + \ln \phi_B + n_B \frac{\partial}{\partial n_B} \ln \phi_B$$
$$+ \chi \phi_A + \chi n_B \frac{\partial}{\partial n_B} \phi_A \tag{A2-19}$$

From Eq. (A2-14) and (A2-15), we obtain

$$\frac{\partial}{\partial n_B} \phi_A = -\frac{v_A N_A n_A}{V^2} \frac{\partial V}{\partial n_B}, \qquad \frac{\partial V}{\partial n_B} = v$$

These two results imply that

$$\frac{\partial}{\partial n_B} \phi_A = -\frac{v_A N_A n_A v}{V^2} = -\phi_A \frac{v}{V} = -\frac{\phi_A \phi_B}{n_B} \tag{A2-20}$$

We then also obtain

$$\frac{\partial}{\partial n_B} \ln \phi_A = \frac{1}{\phi_A} \frac{\partial \phi_A}{\partial n_B} = \frac{1}{\phi_A} \left(\frac{-\phi_A \phi_B}{n_B} \right) = -\frac{\phi_B}{n_B} \tag{A2-21}$$

Likewise, for derivatives of ϕ_B we get

$$\frac{\partial}{\partial n_B} \phi_B = \frac{v}{V} - \frac{v}{V^2} n_B \frac{\partial V}{\partial n_B} = \frac{v}{V} - \frac{v^2}{V^2} n_B = \frac{v}{V} - \frac{v \phi_B}{V} \tag{A2-22}$$

and then

$$\frac{\partial}{\partial n_B} \ln \phi_B = \frac{1}{\phi_B} \frac{\partial}{\partial n_B} \phi_B = \frac{1}{\phi_B} \left[\frac{v}{V} - \frac{v \phi_B}{V} \right] \tag{A2-23}$$

Substituting these expressions into Eq. (A2-19), we get

$$\frac{1}{k_B T} \frac{\partial \Delta F}{\partial n_B} = -\frac{n_A \phi_B}{n_B} + \ln \phi_B + \frac{n_B}{\phi_B} \left[\frac{v}{V} (1 - \phi_B) \right] + \chi \phi_A - \chi \phi_A \phi_B$$

$$= -\frac{n_A \phi_B}{n_B} + \ln \phi_B + \frac{n_B v}{\phi_B V} \phi_A + \chi (\phi_A - \phi_A \phi_B) \tag{A2-24}$$

Noting that $n_B v / \phi_B V = 1$ and also using Eq. (A2-14) for n_A and n_B, we obtain the following equation after a bit of rearrangement:

$$\frac{1}{k_B T} \frac{\partial \Delta F}{\partial n_B} = \ln \phi_B + \phi_A - \left(\frac{\phi_A / v_A N_A}{\phi_B / v} \right) \phi_B + \chi \phi_A (1 - \phi_B)$$

$$= \ln \phi_B + \phi_A - \frac{1}{y} \phi_A + \chi \phi_A^2$$

where we have used the definition for y given in the problem statement. Finally, going back to our expression for the chemical potential, Eq. (A2-18), we obtain

$$\frac{\mu_B - \mu_B^0}{k_B T} = \frac{1}{k_B T} \frac{\partial \Delta F}{\partial n_B} = \ln (1 - \phi_A) + \left(1 - \frac{1}{y} \right) \phi_A + \chi \phi_A^2 \tag{A2-25}$$

which was to be shown.

Problem 2.5(b) From Eq. (A2-25) in Problem 2.5(a), show that the critical value of χ at which a polymer solution will begin to phase separate is given by

$$\chi_c = \frac{1}{2} \left(1 + y^{-1/2} \right)^2 \approx \tfrac{1}{2} + y^{-1/2} \qquad \text{for large y} \tag{A2-26}$$

ANSWER:

First, we note that at the critical point we have

$$\frac{\partial \mu_B}{\partial \phi_A} = \frac{\partial^2 \mu_B}{\partial \phi_A^2} = 0 \tag{A2-27}$$

Noting that the derivatives of the standard chemical potential μ_B^0 are zero, from Eq. (A2-25) in Problem 2.5(a) we obtain

$$\frac{1}{k_B T} \frac{\partial \mu_B}{\partial \phi_A} = -\frac{1}{1 - \phi_A} + \left(1 - \frac{1}{y} \right) + 2 \chi \phi_A = 0 \tag{A2-28}$$

and

$$\frac{1}{k_B T} \frac{\partial^2 \mu_B}{\partial \phi_A^2} = -\frac{1}{(1 - \phi_A)^2} + 2 \chi = 0 \tag{A2-29}$$

where, when both these equations are satisfied, χ equals its critical value χ_c. From Eq. (A2-29), we can solve for ϕ_A:

$$\phi_A = 1 - (2\chi)^{-1/2} \tag{A2-30}$$

We substitute this expression for ϕ_A into Eq. (A2-28) and rearrange the result to

$$\chi - 2^{1/2}\chi^{1/2} + \frac{1}{2}\left(1 - \frac{1}{y}\right) = 0$$

This is a quadratic equation for $\chi^{1/2}$, which, when solved, gives

$$\chi^{1/2} = \frac{1}{\sqrt{2}} \pm \sqrt{\frac{1}{2y}}$$

The positive sign gives the appropriate answer, and we then obtain the following for the critical value of χ:

$$\chi \rightarrow \chi_c = \frac{1}{2}\left(1 + y^{-1/2}\right)^2 \approx \frac{1}{2}\left(1 + 2y^{-1/2}\right) = \frac{1}{2} + y^{-1/2} \tag{A2-31}$$

where the last steps assume that y is much less than unity.

Problem 2.6   From the Poisson and Boltzmann equations, derive Eq. (2-44) for the charge density σ on a flat plate bounded by an infinite expanse of electrolyte. [*Hint*: (a) First use electroneutrality to write $\sigma = -\sum_i \int_0^\infty z_i e n_i \, dx$, and use the Poisson equation to relate this to $(d\psi/dx)_s$, where s denotes the surface where $x = 0$. (b) Then get a formula for $\sum_i dn_i/dx$ in terms of $d^2\psi/dx^2$ and $d\psi/dx$ by differentiating the Boltzmann equation, and substituting the Poisson equation. Integrate this using the chain rule of calculus to get a term $[(d\psi/dx)_s]^2$, and link this to the result of part (a).]

Problem 2.7   Consider two uncharged silica spheres of radius $1 \ \mu$m immersed in water. The Hamaker constant is $A_H = 1.2 \times 10^{-20}$ J. If the spheres are brought to a distance of 10 nm of each other, what is the force in newtons pulling them together?

Problem 2.8   Suppose you place a charged surface into a 0.01 M NaCl electrolyte, and the surface potential is $\psi_s = 0.1$ V relative to the bulk potential of $\psi_\infty = 0$ V, where 1 V $= 1$ J/C. What is the concentration of Na$^+$ at the surface? (*Hint*: Use the Boltzmann equation.)

Problem 2.9   Compute the Debye screening length κ^{-1} for a particle with a weak surface potential of 0.005 V in an electrolyte of 0.0001 M NaCl.

PART II

POLYMERS, GLASSY LIQUIDS, AND POLYMER GELS

Chapter 3

POLYMERS

3.1 INTRODUCTION

Polymeric fluids are the most studied of all complex fluids. Their rich rheological behavior is deservedly the topic of numerous books and is much too vast a subject to be covered in detail here. We must therefore limit ourselves to an overview. The interested reader can obtain more thorough presentations in the following references: a book by Ferry (1980), which concentrates on the linear viscoelasticity of polymeric fluids, a pair of books by Bird et al. (1987a,b), which cover polymer constitutive equations, molecular models, and elementary fluid mechanics, books by Tanner (1985), by Dealy and Wissbrun (1990), and by Baird and Dimitris (1995), which emphasize kinematics and polymer processing flows, a book by Macosko (1994) focusing on measurement methods; and a book by Larson (1988) on polymer constitutive equations. Parts of this present chapter are condensed versions of material from Larson (1988). The static properties of flexible polymer molecules are discussed in Section 2.2.3; their chemistry is described in Flory (1953).

This chapter only deals with flexible, single-component, *homopolymers*—that is, polymers made of identical monomers. *Copolymers* containing more than one kind of monomer on a single chain are discussed in Chapter 13. The chemical structures of some of the most common flexible homopolymers are given in Fig. 2-7. The behavior of rod-like and semiflexible molecules is described in Chapter 11. Multicomponent polymer *blends* are discussed in Chapter 9.

Polymeric fluids often show strong viscoelastic effects, which can include shear thinning, extension thickening, viscoelastic normal stresses, and time-dependent rheology, phenomena defined in Chapter 1. These phenomena are common in polymeric liquids because the molecules are long and easily distorted, even in rather slow flows. For example, Fig. 3-1 shows a long DNA polymer molecule attached at one end to a small sphere held in a laser-optical trap and then stretched out in a series of slow constant-velocity flows, with the flow velocity increasing from left to right. At high velocity the polymer molecule is stretched to many times its undisturbed radius of gyration.

A solution or melt composed of such "stretchable" molecules can be highly "springy," especially in extensional flows (Tirtaatmadja and Sridhar 1993). The kinematics of an extensional flow are described in Section 1.4.1.2. From Eqs. (1-6) and (1-9), one can show that for a Newtonian fluid (for which $\boldsymbol{\sigma} = 2\eta\mathbf{D}$) the *Trouton ratio* $\mathrm{Tr} \equiv \bar{\eta}_u/\eta_0$ of the uniaxial extensional viscosity $\bar{\eta}_u$ to the zero-shear viscosity η_0 is numerically equal to 3. For polymers, Tr can be much higher than this. Figure 3-2, for example, shows Tr for a

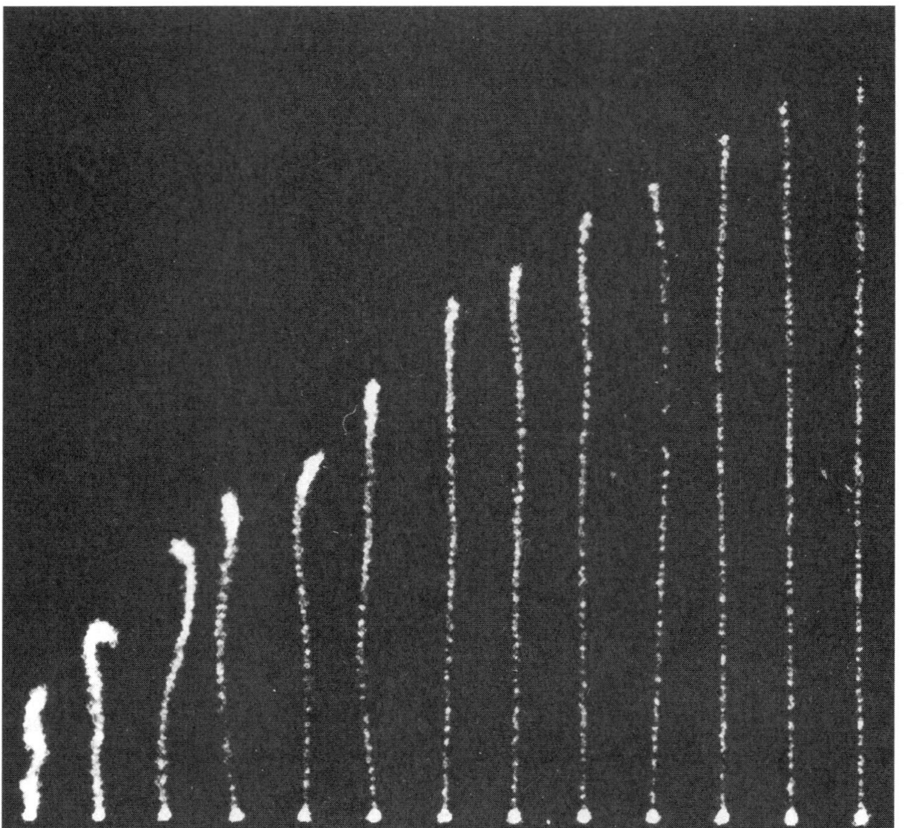

Figure 3.1 Images of a fluorescently labeled, 64.6-μm-long DNA molecule tethered at one end to a small (0.3 μm) sphere that is held in place by "optical tweezers," a focused beam of laser light (see Section 1.6.1). The solvent flows past the DNA and stretches it out at velocities of 1, 2, 3, 4, 5, 7, 10, 12, 15, 20, 30, 40, and 50 μm/sec, from left to right. (Reprinted with permission from Perkins et al, *Science*, 268:83 Copyright 1995, American Association for the Advancement of Science.)

dilute (0.185 wt%) solution of flexible polyisobutylene as a function of strain $\varepsilon = \dot{\varepsilon} t$ after start-up of steady extensional flow (Verhoef et al. 1997). Tr rises from a value of 3 to roughly 10^4 before a plateau value is reached!

At higher concentrations or in the melt state, fluids containing long polymer molecules become extremely viscous. Figure 3-3 is a famous plot by Berry and Fox (1968) of zero-shear viscosity data for several polymer melts, each as a function of polymer molecular weight M. Note that for each polymer there is a molecular weight, M_c, above which the shear viscosity rises more rapidly with molecular weight than at molecular weights below M_c. Below M_c, the zero-shear viscosity η_0 is found to depend roughly linearly on molecular weight, while above M_c it rises with a much higher power-law exponent, $\eta_0 \propto M^p$, where $p \approx 3.4$.

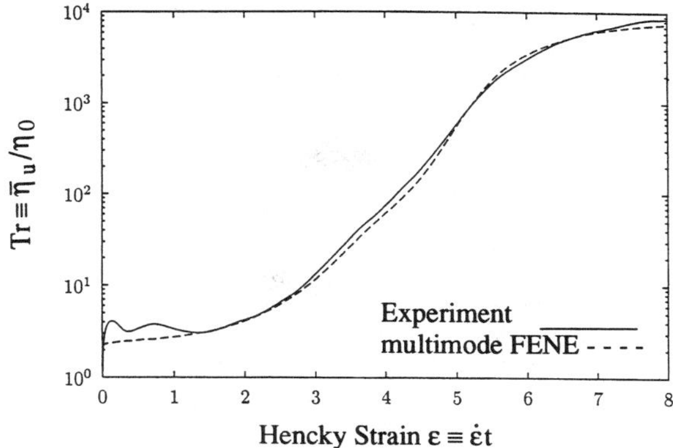

Figure 3.2   Trouton ratio, Tr, of uniaxial extensional viscosity $\bar{\eta}_u$ to zero-shear viscosity η_0 after start-up of steady uniaxial extension at a rate of $\dot{\varepsilon} \approx 1$ sec^{-1} for a "Boger fluid" consisting of a 0.185 wt% solution of flexible polyisobutylene ($M_w = 2.11 \times 10^6$) in a solvent composed mostly of viscous polybutene with some added kerosene (solid line). The dashed line is a fit of a "multimode" FENE dumbbell model, where each mode is represented by a FENE dumbbell model, with a spring law given by Eq. (3-56), without preaveraging, as described in Section 3.6.2.2.1. The relaxation times were obtained by fitting the linear viscoelastic data, $G'(\omega)$ and $G''(\omega)$. The slowest mode, with $\tau_1 = 5$ sec, dominates the behavior at large strains; the best fit is obtained by choosing for it an extensibility parameter of $B = 40{,}000$. The value of $B = 3L^2/\langle R^2 \rangle_0 = 3(0.82)^2 n/C_\infty$, predicted from the molecular characteristics, is around 20,000. This value is obtained using $n = M_w/28 = 75{,}000$ backbone bonds and $C_\infty = 6.8$ (see Table 3-3). (From Verhoef et al. (1997), with kind permission from Elsevier Science - NL, Sara Burgerhartstraat 25, 1055 KV Amsterdam, The Netherlands.)

The steep increase of viscosity with molecular weight for M above M_c is caused by *entanglements,* which are topological restrictions on molecular motion resulting from the fact that the chains cannot pass through each other. Because of entanglements, a long molecule surrounded by other such molecules cannot move very far in directions perpendicular to its own molecular contour (Edwards 1967). Therefore, molecular diffusion or relaxation is limited to a snake-like motion of the polymer molecule along its own contour. This slow snake-like motion is called *reptation* (de Gennes 1971). Observations of the motions of long DNA molecules entangled with other DNA molecules provide a dramatic illustration of this general kind of motion (Perkins et al. 1994a). Figure 3-4 shows that a single fluorescing DNA molecule stretched out in a contorted path in an entangled mesh of other, invisible DNA molecules, and held at one end by a laser-optical trap (see Section 1.6.1), relaxes by retracting its free end along the contorted path defined by its own contour. Thus, the entangled molecule appears to be confined to a *tube-like* region. For high molecular weights, motion is largely limited to reptation in a "tube," so relaxation is slow and the polymer melt viscosity is high.

Flexible polymer molecules are frequently modeled by beads-and-springs chains (Bird et al. 1987b), depicted in Fig. 3-5. The springs, when stretched, mimic the elastic forces in flexible polymer molecules, while the beads produce hydrodynamic drag. The simplest beads-and-springs model is the *elastic dumbbell* (Fig. 3-5b), which contains a single spring connecting two beads. The dumbbell model captures the simplest aspects of polymer flow

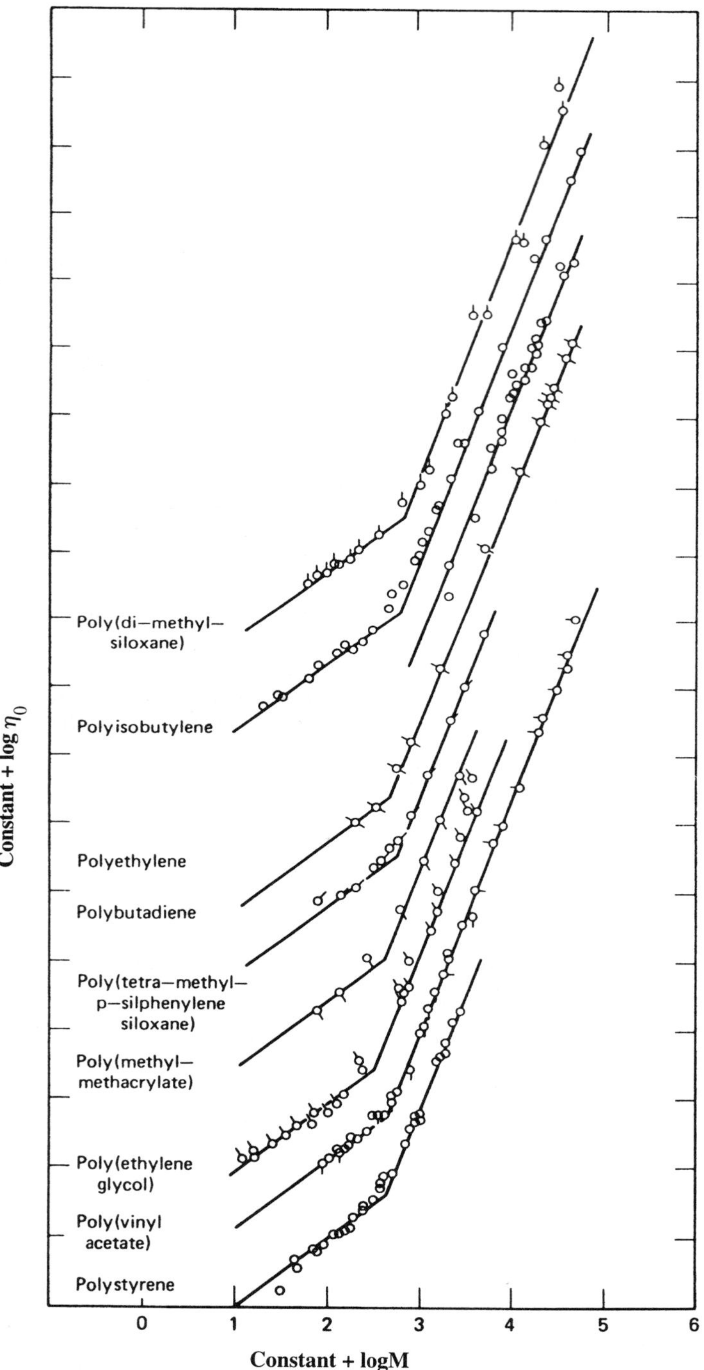

Figure 3.3 Relationship between zero-shear viscosity and molecular weight for several nearly monodisperse melts. For clarity, the curves are shifted relative to each other along both the abscissa and ordinate. (From Berry and Fox 1968, reprinted with permission from Springer Verlag.)

behavior, and it is readily generalized to chains with multiple beads and springs (see Section 3.4.5).

In any polymeric system, dilute or concentrated, the stress depends on the confor-

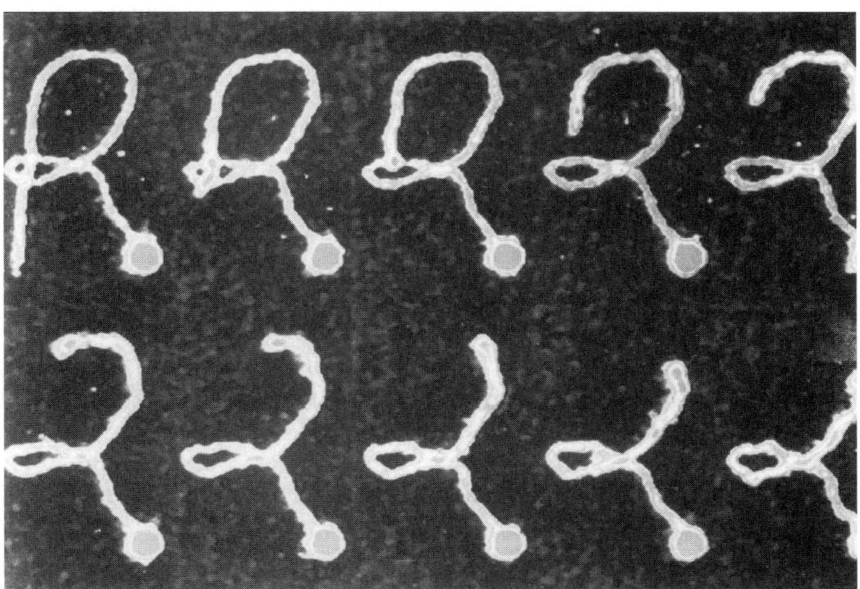

Figure 3.4   Time sequence of images showing retraction of one end of a fluorescing 60-μm-long DNA molecule entangled in a solution of other, non-fluorescing DNA molecules. The fluorescing molecule was attached at one end to a small sphere that was pulled through the solution using a laser-optical trap, to form the letter R. The free end then retracts through a "tube-like" region formed by the surrounding mesh of other, invisible DNA chains. (Reprinted with permission from the cover of *Science*, May 6, 1994; Copyright 1994, American Association for the Advancement of Science.)

(a)

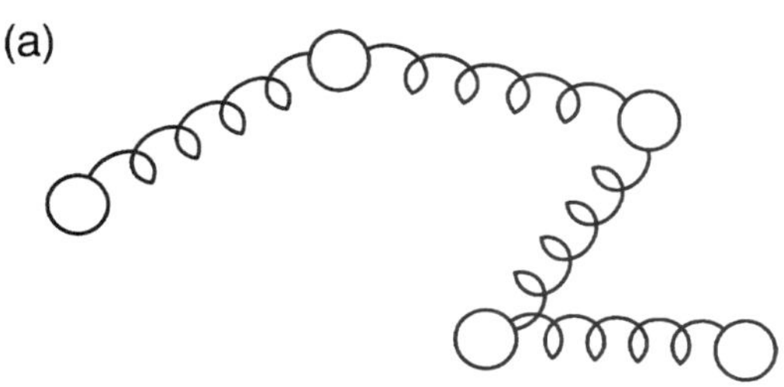

Figure 3.5   (a, b) Illustration of beads-and-springs models. Part (b) shows a two-bead flexible dumbbell. (From Larson 1988.)

(b)

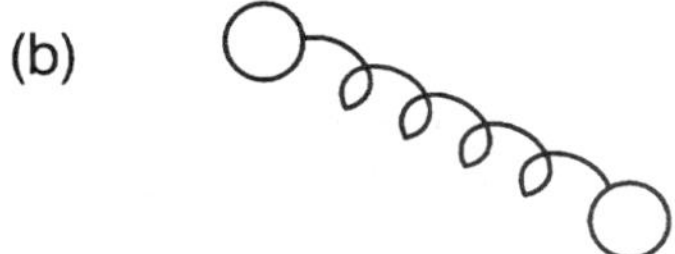

mations of the polymer molecules. As the orientation and degree of stretch of a molecule increases, so does its contribution to the stress. The total stress in a fluid element depends on contributions from all molecules in that element, each of which has its own conformation. Molecular theories for stress must therefore account for the *distribution of conformations* that results from a deformation. The starting point for developing such theories is to determine the distribution of conformations at equilibrium (i.e., in the absence of flow). This distribution function is described in Section 3.2. The distinction between dilute and entanglement polymers is described in Section 3.3. Section 3.4 describes simple dumbbell models, while Section 3.5 explains the principle of time–temperature superposition. Finally, Sections 3.6 and 3.7 cover the rheological properties of dilute and entangled polymer solutions and melts.

3.2 EQUILIBRIUM PROPERTIES

At equilibrium, the distribution of conformations in a solvent at the *theta* temperature (see Section 2.3.1.2), or in a concentrated solution, is given by a set of random walks or, equivalently, by the conformations of a *freely jointed chain* (see Section 2.2.3.2). If one end of the freely jointed chain with N_K links, each of length b_K, lies at the origin, then the probability, $\psi_0 dR^3$, that the other end lies at a position between $\mathbf{R}$ and $\mathbf{R} + d\mathbf{R}$ is approximately a *Gaussian function* (Flory 1969; Larson 1988):

$$\psi_0(\mathbf{R}) = \left(\frac{\beta}{\sqrt{\pi}} \right)^3 \exp(-\beta^2 R^2) \tag{3-1}$$

where $\beta^2 \equiv 3/(2N_K b_K^2)$ and $R^2 = \mathbf{R} \cdot \mathbf{R}$, where $\mathbf{R}$ is the end-to-end vector of the polymer chain. Equation (3-1) is only valid when R is a factor of three or so less than $N_K b_K$, the maximum extension of the freely jointed chain.

The second moment of the distribution function, ψ_0, is the average square of the end-to-end distance of the chain, $\langle R^2 \rangle_0$:

$$\langle R^2 \rangle_0 \equiv \int R^2 \psi_0(\mathbf{R}) \, dR^3 = \frac{3}{2} \beta^{-2} = N_K b_K^2 \tag{3-2}$$

where the integral $\int \cdot \, dR^3$ in Eq. (3-2) represents three nested integrals, one over each coordinate, dR_i, taken from $-\infty$ to $+\infty$. The notation $\langle \cdot \rangle_0$, where $\cdot$ represents any quantity, is an average of $\cdot$ over ψ_0:

$$\langle \cdot \rangle_0 = \int \psi_0 \cdot dR^3 \tag{3-3}$$

The subscript 0 denotes an equilibrium property.

Since N_K, the number of steps in the chain, is proportional to the polymer molecular weight, Eq. (3-2) implies that the root-mean-square end-to-end separation distance, $\langle R^2 \rangle_0^{1/2}$, of an undistorted coil, scales with molecular weight M according to a power law, $\langle R^2 \rangle_0^{1/2} \propto M^\nu$, with $\nu = 0.5$. Although this result applies in the melt state and for concentrated solutions, note that in dilute solutions with a good solvent the value of ν is as large as 0.6 because of *excluded-volume effects*, as discussed in Section 2.2.3.3.

3.3 INTRINSIC VISCOSITY AND OVERLAP CONCENTRATION

Figure 3-3 shows that for a polymer melt, the viscosity becomes especially sensitive to molecular weight once a critical molecular weight is exceeded. Similarly, if a high-molecular-weight polymer is dissolved in a solvent, the viscosity increases roughly linearly with polymer concentration at low concentrations, and then much more rapidly at higher concentration (Osaki et al. 1975). The threshold concentration above which a more rapid rise in viscosity is observed corresponds roughly to the concentration at which the polymer molecules begin to *overlap* and become *entangled* with each other. A polymer solution is considered dilute if polymer coils rarely overlap, because then the viscoelastic properties of the solution are governed by the behavior of a single polymer molecule in a bath of solvent.

The overlap concentration can be estimated as the concentration at which the number of coils per unit volume, v, times the volume pervaded by a single coil, R_g^3, is roughly unity, where $R_g = \langle R^2 \rangle_0^{1/2} / \sqrt{6}$ is the *radius of gyration* of the polymer coil. Now $v = cN_A/M$, where c is the mass per unit volume of polymer in solution and N_A is Avogadro's number. Therefore, in a theta solvent (a solvent at a temperature where polymer excluded-volume interactions are negligible; see Section 2.3.1.2), the *overlap concentration, c^**, is given by

$$c^* \equiv \frac{v^* M}{N_A} = \frac{R_g^{-3} M}{N_A} = \left[\frac{n^{1/2} \ell C_\infty^{1/2}}{\sqrt{6}} \right]^{-3} \frac{M}{N_A} \tag{3-4}$$

where n is the number of backbone bonds of the chain, ℓ is the average bond length, and C_∞ is the "characteristic ratio" for the polymer (see Section 2.2.3.2). For a polystyrene of molecular weight 10^6 in a theta solvent, according to this definition we have $c^* \approx 0.1$ g/ml. In a good solvent, overlap begins at a lower concentration.

Even below c^*, the viscoelastic properties of a solution vary somewhat from those of a dilute solution, due to occasional coil overlaps (Ferry 1980). To obtain true dilute-solution properties from measurements made at a series of low, but finite, concentrations, one can subtract from those measurements any solvent contribution and then *extrapolate* the results to zero concentration. The *intrinsic* steady-state shear viscosity, for example, is one such extrapolated quantity; it is defined as

$$[\eta] = \lim_{c \to 0} (\eta - \eta_s)/\eta_s c \tag{3-5}$$

where η_s is the solvent viscosity. Since the intrinsic viscosity depends on the coil dimensions, it can be used to determine the molecular weight of the polymer.

The intrinsic viscosity can be related to the overlap concentration, c^*, by assuming that each coil in the dilute solution contributes to the zero-shear viscosity as would a hard sphere of radius equal to the radius of gyration of the coil. This rough approximation is reasonable as a scaling law because of the effects of *hydrodynamic interactions* which suppress the flow of the solvent through the coil, as we shall see in Section 3.6.1.2. The Einstein formula for the contribution of suspended spheres to the viscosity is

$$\eta_0 - \eta_s = \eta_{p0} = 2.5 \eta_s \phi \tag{3-6}$$

where ϕ is the volume fraction of spheres, η_0 is the zero-shear viscosity of the solution, and η_{p0} is the zero-shear polymer contribution to the viscosity. From this and Eq. (3-5), we obtain

$$[\eta]_0 = \frac{\eta_{p0}}{\eta_s c} = \frac{2.5\phi}{c} \tag{3-7}$$

Now c^* is defined as the concentration at which the spheres begin to overlap. We expect this to occur roughly when $\phi \sim 0.40$. Thus from Eq. (3-7), we obtain

$$c^* \sim \frac{1}{[\eta]_0} \tag{3-8}$$

which, for theta solvents, gives a value about six times lower than Eq. (3-4) (see Section 3.6.1.1). Although c^*, as estimated in Eq. (3-8), is an adequate criterion for the onset of molecular interactions when the macromolecules are only slightly extended by flow, highly deformed molecules might entangle or otherwise interact at concentrations much lower than this.

3.4 ELEMENTARY MOLECULAR THEORIES

3.4.1 The Polymer Stress and Birefringence Tensors

From the configuration distribution function, $\psi_0(\mathbf{R})$, the elasticity of flexible polymer molecules can be predicted. Suppose the ends of the polymer chain are held fixed so that the end-to-end vector of the chain is $\mathbf{R}$. The number of internal configurations, Ω, of the chain that satisfies this constraint is $\Omega = c\psi_0(\mathbf{R})$, with c a constant. The entropy, S, is then (Wall 1942)

$$S = k_B \ell n(\Omega) = k_B \ell n(c\psi_0) = k_B[\ell n\, c + 3\ell n\, (\beta/\sqrt{\pi}) - \beta^2 R^2] \tag{3-9}$$

where we have used Eq. (3-1) for ψ_0. If the enthalpy contributed by the chain does not depend on its configuration, then the configurational free energy, W, is $-TS$, where T is the absolute temperature.

If we now pull on a chain end to increase the end-to-end distance $\mathbf{R}$, the force we need to exert to overcome the entropic "spring" force of the chain is

$$\mathbf{F}^s = \frac{\partial W}{\partial \mathbf{R}} = -T\frac{\partial S}{\partial \mathbf{R}} = 2k_B T \beta^2 \mathbf{R} \tag{3-10}$$

An increase in the separation distance, R, reduces the number of allowed configurations; the molecule resists this action with a force proportional to $\mathbf{R}$. *A Gaussian chain therefore acts like a Hookean spring.* The validity of the "entropic-spring" concept in describing low-frequency dynamics in polymeric materials has been directly established by experiments on single, long DNA molecules (Perkins et al. 1994b) and by molecular dynamics simulations of "pearl necklace" chains (Gao and Weiner 1992). Note in Eq. (3-10) that this force is proportional to the temperature, T; and since $\beta^{-2} \propto n$, it is inversely proportional to the molecular weight M of the polymer. Equation (3-10) breaks down, and the relationship between force and molecular extension becomes nonlinear when the chain is stretched out to greater than around 30% of its fully extended length (see Section 3.6.2.2.1).

To obtain the polymer contribution to the stress tensor σ^P for a fluid containing a large number N_s of springs, we define the position of one end of the spring i to be $\mathbf{r}_{i,1}$ and define the position of the other end to be $\mathbf{r}_{i,2}$. We then define $\mathbf{R}_i \equiv \mathbf{r}_{i,2} - \mathbf{r}_{i,1}$. We note that the spring exerts a force $-\mathbf{F}_i^s$ on point 2 and exerts a force $\mathbf{F}_i^s$ on point 1. Now we use the Kirkwood formula for the stress tensor, Eq. (1-42). The summation must be carried out over the $2N_s$ locations of the ends of all N_s different springs. This summation can be expressed as

$$\sigma^P = -\frac{1}{V} \sum_{i=1}^{N_s} \mathbf{r}_{i,1} \mathbf{F}_i^s + \mathbf{r}_{i,2}(-\mathbf{F}_i^s) = \frac{1}{V} \sum_{i=1}^{N_s} (\mathbf{r}_{i,2} - \mathbf{r}_{i,1}) \mathbf{F}_i^s = \frac{1}{V} \sum_{i=1}^{N_s} \mathbf{R}_i \mathbf{F}_i^s$$

We now define $\nu \equiv N_s / V$ to be the number N_s of springs per unit volume V. If this number is large, the sum in the above equation can be replaced by ν times an ensemble average; hence (Bird et al. 1987b)

$$\sigma^P = \nu \langle \mathbf{R}\mathbf{F}^s \rangle \tag{3-11}$$

Here $\langle \cdot \rangle$ without the subscript "0" is an average over the nonequilibrium distribution, ψ:

$$\langle \cdot \rangle \equiv \int \psi \cdot dR^3 \tag{3-12}$$

For Gaussian chains, using Eq. (3-10), we obtain

$$\sigma^P = 2k_B T \nu \beta^2 \langle \mathbf{R}\mathbf{R} \rangle \tag{3-13}$$

Thus, viscoelastic stresses are produced by *distortions of the distribution of polymer configurations*. For Gaussian distributions, the stress tensor σ^P is proportional to the second moment of the configuration distribution function. Mechanical stresses can therefore be interpreted in terms of *anisotropies in molecular orientations*. In simple shearing of ordinary polymer melts and solutions, for example, the first normal stress difference, $N_1 \equiv \sigma_{11} - \sigma_{22}$, is positive, implying that there is a higher degree of orientation in the direction of flow, the "1" direction, than in the "2" direction, which is the direction of the velocity gradient. The second normal stress difference, $N_2 \equiv \sigma_{22} - \sigma_{33}$, in melts and entangled solutions of flexible polymers is usually at least a factor of 3–10 smaller in magnitude than N_1 and negative in sign (Keentok et al. 1980; Larson 1988; Brown et al. 1995). The negative sign implies that the orientation in direction "2" is *depleted* of polymer orientations relative to the vorticity direction "3."

The relationship between stress and polymer orientation implies that there is a proportionality between the polymer stress tensor σ^P and the tensor $\mathbf{n}^P$ of optical anisotropy or *birefringence*. This proportionality is called the *stress-optic law* (Janeschitz-Kriegl 1983) and is expressed as (Section 1.6.3)

$$\mathbf{n}^P = C\sigma^P \tag{3-14}$$

The proportionality constant C is the *stress-optic coefficient*. From the freely jointed chain model, one can derive its value (Kuhn and Grün 1942; Treloar 1975):

$$C = \frac{2\pi}{45k_B T} \frac{(n^2 + 2)^2}{n} (\alpha_1 - \alpha_2) \tag{3-15}$$

where n is the isotropic part of the index of refraction, and α_1 and α_2 are, respectively, the axial and transverse polarizabilities of a chain link. Worked Example 1.6 at the end of Chapter 1 shows how the stress-optic law can be used to obtain stresses from optical measurements.

Experimental studies have nearly universally supported the validity of the stress-optic law for *single-phase* melts and entangled solutions that are well above their glass transition temperatures, as long as the chain deformation is not so large that the force–extension relationship $\mathbf{F}(\mathbf{R})$ becomes nonlinear (Wales 1976; Janeschitz-Kriegl 1983; Talbott and Goddard 1979; Cathey and Fuller 1990; Fuller 1995). In a uniaxial extension of a polystyrene melt, such a nonlinearity produces a breakdown of the stress-optical law only when the stress exceeds around 2 MPa (Muller and Froelich 1985). The validity of the stress-optic law for stress levels lower than this confirms that the stress is determined by the configurations of the polymer chains. Hydrodynamic drag serves to orient polymer molecules, and therefore affect $\langle \mathbf{RR} \rangle$, but otherwise makes no significant direct contribution to the stress in melts or concentrated solutions. For dilute polymer solutions, the stress-optic rule has also been found to hold, if the solvent contributions to the total stress and birefringence tensors are suitably subtracted (Amelar et al. 1991; Fuller 1995).

Deviations from the stress-optic law are found at high frequencies in oscillatory deformations for melts near their *glass transition temperatures* (Janeschitz-Kriegl 1983). The glass transition temperature is the temperature at which amorphous polymers become very sluggish and solid-like; the glass transition is discussed in detail in Chapter 4. Osaki and coworkers (Inoue et al. 1991, 1992) have, however, found that a modified version of the stress-optic rule can apply under oscillatory shearing of polymers near the glass transition if the stress and birefringence are each divided into two portions, a rubbery contribution and a glassy contribution:

$$\boldsymbol{\sigma} = \boldsymbol{\sigma}_r + \boldsymbol{\sigma}_g \qquad (3\text{-}16\text{a})$$

$$\mathbf{n} = \mathbf{n}_r + \mathbf{n}_g \qquad (3\text{-}16\text{b})$$

Each of these two contributions has its own stress-optic coefficient; that is, $\mathbf{n}_r = C_r \boldsymbol{\sigma}_r$ and $\mathbf{n}_g = C_g \boldsymbol{\sigma}_g$. C_r and C_g can differ from one another in magnitude and even in sign. (Negative values of the stress-optic coefficient are generally found for polymers in which the molecular polarizability is greater in directions transverse to the chain axis than it is along the chain axis. The rubbery contribution to the stress in polystyrene, for example, has a negative stress-optic coefficient, owing to its bulky phenyl side group.) Recent theory and simulations provide some justification for the stress and birefringence expressions of Osaki and coworkers (Gao and Weiner 1994; Osaki et al. 1995).

Breakdown of the stress-optic rule can also occur in multiphase or multicomponent mixtures, as well as in melts with crystalline domains. However, as in glassy polymers, for miscible blends, a revised stress-optic law can sometimes be recovered by breaking the stress and birefringence tensors into two components, one for each component in the blend (Zawada et al. 1994; Kannan and Kornfield 1994).

3.4.2 Rubber Elasticity Theory

From Eq. (3-13), polymer stresses in a flow or deformation can be calculated if the distribution function for that flow can be predicted. To make such predictions, molecular theories of polymer dynamics are required. Conceptually, the simplest theory is classical rubber elasticity theory, which applies to rubbers made up of cross-linked polymer chains, schematized in Fig. 3-6. Although cross-linked rubbers are the topic of Chapter 5, we present ideas from rubber elasticity theory here because they are the basis for much of the theory for polymer liquids.

The theory of rubber elasticity was developed by Wall (1942), Flory and Rehner (1943), James and Guth (1943), and Treloar (1943). Suppose that cross-link points are evenly spaced along each polymer chain a distance N_K random-walk steps apart. The portion of polymer between two neighboring cross-links is called a *strand*. We suppose that in the absence of deformation the probability, $\psi_0(\mathbf{R}')d R'^3$, that a strand has an end-to-end vector between $\mathbf{R}'$ and $\mathbf{R}' + d\mathbf{R}'$ is the same it would be without the cross-links; that is,

$$\psi_0(\mathbf{R}') = \left(\frac{\beta}{\sqrt{\pi}}\right)^3 \exp(-\beta^2 R'^2) \tag{3-17}$$

where $\beta^2 = 3/(2N_K b_K^2)$.

We now consider an extensional deformation of an incompressible rubber network (Fig. 3-7), where the stretch axes are oriented along the coordinate axes and where the stretch ratios λ_1, λ_2, and λ_3 are in directions 1, 2, and 3, respectively. For the example in Fig. 3-7, the deformation is a *uniaxial extension* that increases the length of the cylinder by a factor of λ_1 over its initial length. By volume conservation, the radius of the cylinder then shrinks to $\lambda_1^{-1/2}$ times the original radius. If the cross-link points are *convected with*

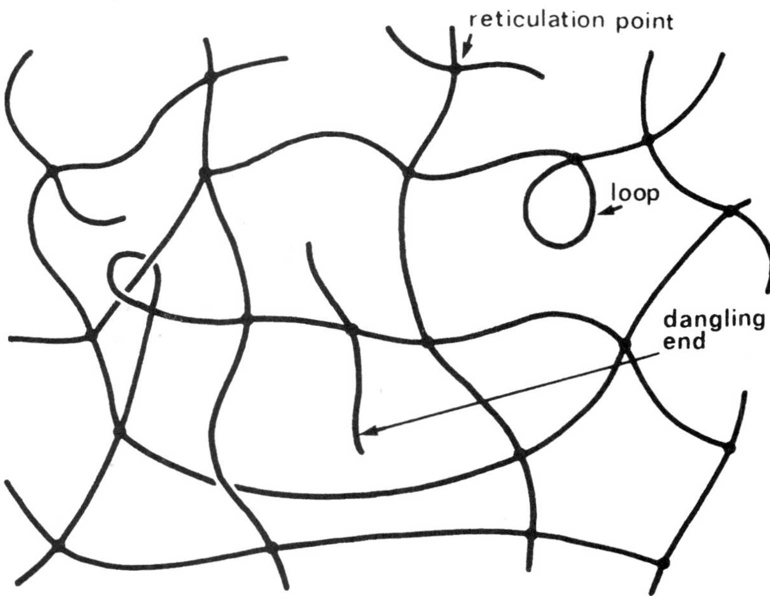

Figure 3.6   Schematic of a cross-linked polymer network. (Reprinted from Pierre-Gilles de Gennes, *Scaling Concepts in Polymer Physics.* Copyright © 1979, by Cornell University. Used by permission of the publisher, Cornell University Press.)

this macroscopic deformation (i.e., they move *affinely*), then each component, R_i, of **R** is given by

$$R_i = R'_i \lambda_i \tag{3-18}$$

For a volume-conserving deformation, the strands that before the deformation had end-to-end vectors between **R**′ and **R**′ + d**R**′ have end-to-end vectors between **R** and **R** + d**R** after the deformation. Therefore $\psi_0(\mathbf{R}')dR'^3 = \psi(\mathbf{R})dR^3$. Since $R'_i = R_i/\lambda_i$ and $dR^3 = \lambda_1\lambda_2\lambda_3 dR'^3 = dR'^3$ (since $\lambda_1\lambda_2\lambda_3 = 1$), we have

$$\psi(\mathbf{R}) = \psi_0\left(\frac{R_i}{\lambda_i}, \ i = 1, 2, 3\right) = \left(\frac{\beta}{\sqrt{\pi}}\right)^3 \exp\left[-\beta^2\left(\frac{R_1^2}{\lambda_1^2} + \frac{R_2^2}{\lambda_2^2} + \frac{R_3^2}{\lambda_3^2}\right)\right] \tag{3-19}$$

Each surface of equal probability is therefore deformed from a sphere to an ellipsoid with principal axes λ_1, λ_2, and λ_3 times the diameter of the undeformed sphere.

Inserting Eq. (3-19) into Eq. (3-13), the components of the stress tensor are

$$\sigma_{ii} = 2\nu k_B T \beta^2 \langle R_i R_i \rangle = 2\nu k_B T \beta^2 \int R_i^2 \left(\frac{\beta}{\sqrt{\pi}}\right)^3 \exp\left[-\beta^2 \sum_j \frac{R_j^2}{\lambda_j^2}\right] dR^3$$

The above is a triple integral, which can be rearranged into a product of three integrals, each over one of the coordinates dR_j. Two of the three integrals can be reduced to the form $(\lambda_j/\sqrt{\pi}) \int \exp(-x^2) \, dx = \lambda_j$, while the third has the form $(\lambda_i^3/\sqrt{\pi}) \int x^2 \exp(-x^2) \, dx = \frac{1}{2}\lambda_i^3$. Using $\lambda_1\lambda_2\lambda_3 = 1$, we obtain simply

$$\sigma_{ii} = \nu k_B T \lambda_i^2 \tag{3-20}$$

where ν is the number of strands per unit volume. This can be expressed in tensor form as

$$\boldsymbol{\sigma} = \nu k_B T \begin{pmatrix} \lambda_1^2 & 0 & 0 \\ 0 & \lambda_2^2 & 0 \\ 0 & 0 & \lambda_3^2 \end{pmatrix} = G\mathbf{B} \tag{3-21}$$

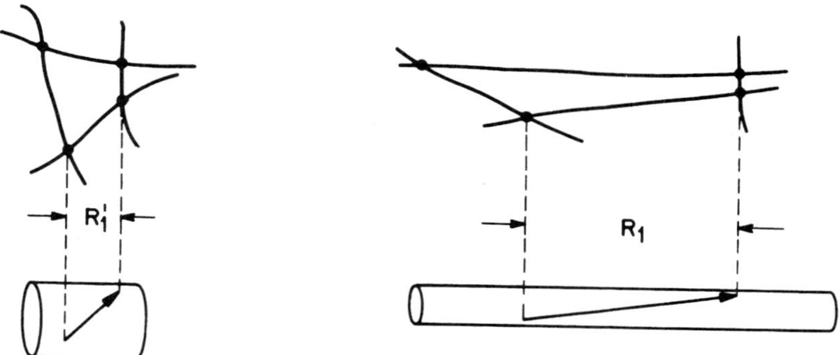

Figure 3.7   Cross-linked rubber network before (**left**) and after (**right**) a uniaxial extensional deformation. The stretching of the polymer strands connecting cross-link points is *affine*—that is, directly proportional to the macroscopic deformation. (From Larson 1988.)

where $G = \nu k_B T$ is the modulus, and $\mathbf{B}$ is the *Finger tensor* defined in Eq. (1-16). Equation (3-21) in the form $\boldsymbol{\sigma} = G\mathbf{B}$, applies not only to extensional deformations but also to shearing and to any other volume-preserving deformation.

Equation (3-21) describes a *purely elastic* material—that is, one that dissipates no energy during or after deformation. Hence, Eq. (3-21) predicts that in a small-amplitude oscillatory shearing deformation, $G'(\omega) = G = $ constant and $G''/G' = 0$. Only for cross-linked rubbers is this behavior found, and even then only at low frequency. Figure 3-8, for example, shows that the linear moduli for a cross-linked polyurethane rubber approximately satisfy these relationships only at low frequency, below about 10 sec^{-1}. At higher frequencies, G' becomes frequency-dependent, and G'' is about as large as G'. At these frequencies, the strands between cross-link points are not able to relax their conformations fast enough to keep up with the imposed deformation, and energy is dissipated. The next section describes a more sophisticated model that includes relaxation.

3.4.3 The Temporary Network Model

Green and Tobolsky (1946) proposed a theory that was inspired by rubber elasticity theory. The theory was originally intended for polymer melts, but it can also be applied to elastic rubbers at deformation rates high enough that elastic energy is dissipated. In this model, polymer chains are physically restrained by "temporary junctions" that spontaneously form and break. Thus, the melt is a "liquid" network in which the strands are not permanently bound to the same cross-links, but sporadically break loose, relax their configuration, and then become bound again to new junctions. The average concentration of physical junction points, ν, is constant; thus the strands "break" and "re-form" at equal rates. These temporary

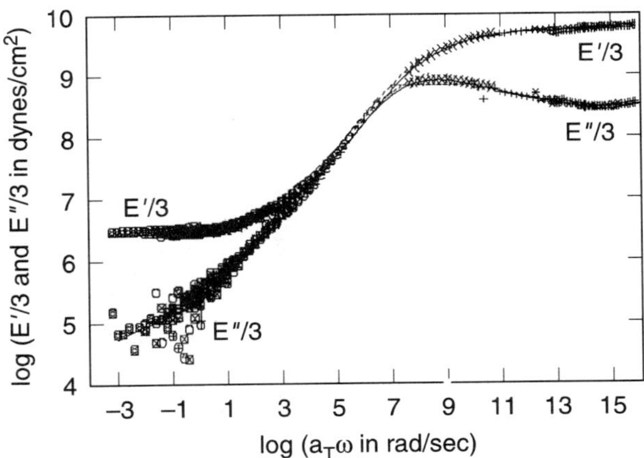

Figure 3.8 Storage and loss moduli versus reduced frequency in small-amplitude oscillatory deformation for a cross-linked polyurethane rubber, Sorbothane 70, at 20°C. The imposed deformation was uniaxial extension, yielding tensile moduli E' and E'' which were converted to shear moduli using $G' = E'/3$ and $G'' = E''/3$. The data at 16 different temperatures ranging from -81°C to 80°C were collapsed to a "master curve" using methods described in Section 3.5.2. (From Larson et al. 1996, reprinted with permission from Steinkopff Publishers.)

constraints or physical junction points might be physical bonds, such as hydrogen bonds or other weak bonds formed by physical polymer gels (see Chapter 5). In ordinary polymer melts, however, the "junction points" are thought of as *entanglements*.

Let $1/\tau$ be the probability, per unit time, that a strand breaks free of a "junction point" and relaxes. Then the probability, $P_{t',t}$, that the strand does *not* break free any time during the time interval t' to time t obeys the differential equation

$$\frac{d}{dt} P_{t',t} = -\frac{1}{\tau} P_{t',t} \tag{3-22a}$$

with $P_{t',t'} = 1$. Hence

$$P_{t',t} = e^{(t'-t)/\tau} \tag{3-22b}$$

When the material is deformed, each strand is stretched affinely until it breaks free from its junction. After it breaks free, it relaxes to a configuration typical of equilibrium. As often as a strand breaks free, another relaxed strand becomes entangled. The probability that a strand breaks free and becomes re-entangled in an interval of time between t' and $t' + dt'$ is dt'/τ. The probability that it survives without breaking from time t' to time t is $P_{t',t}$. The contribution $d\sigma$ to the stress from those strands that meet both of these conditions is, according to Eq. (3-21),

$$d\sigma = \frac{dt'}{\tau} P_{t',t}\, G\mathbf{B}(t',t) = G\frac{1}{\tau}e^{(t'-t)/\tau}\, \mathbf{B}(t',t)\, dt' \tag{3-23}$$

with $G = \nu k_B T$. The total stress produced by strands that became re-entangled at all past times, t', is then

$$\sigma = \int_{-\infty}^{t} m(t - t')\mathbf{B}(t',t)\, dt' \tag{3-24}$$

where $m(t - t') \equiv (G/\tau)\exp[(t' - t)/\tau]$. This assumes that there is only one relaxation process, with a single relaxation time, τ.

However, Eq. (3-24) can be extended to allow multiple relaxation *modes* simply by setting $m(t - t')$ equal to a sum of exponentials (Lodge 1956, 1968; Yamamoto 1956, 1957, 1958); that is,

$$m(t - t') = \sum_i \frac{G_i}{\tau_i} \exp[(t' - t)/\tau_i]$$

With this choice for $m(t - t')$, $G'(\omega)$ and $G''(\omega)$ are given by

$$G'(\omega) = \sum_i G_i \frac{\omega^2 \tau_i^2}{1 + \omega^2 \tau_i^2} \tag{3-25a}$$

$$G''(\omega) = \sum_i G_i \frac{\omega \tau_i}{1 + \omega^2 \tau_i^2} \tag{3-25b}$$

By choosing a discrete set of values of τ_i distributed over some interval and then fitting the above expressions to experimental data for G' and G'', the constants G_i can be obtained. Figure 3-9, for example, shows that by adding up the contributions from eight modes,

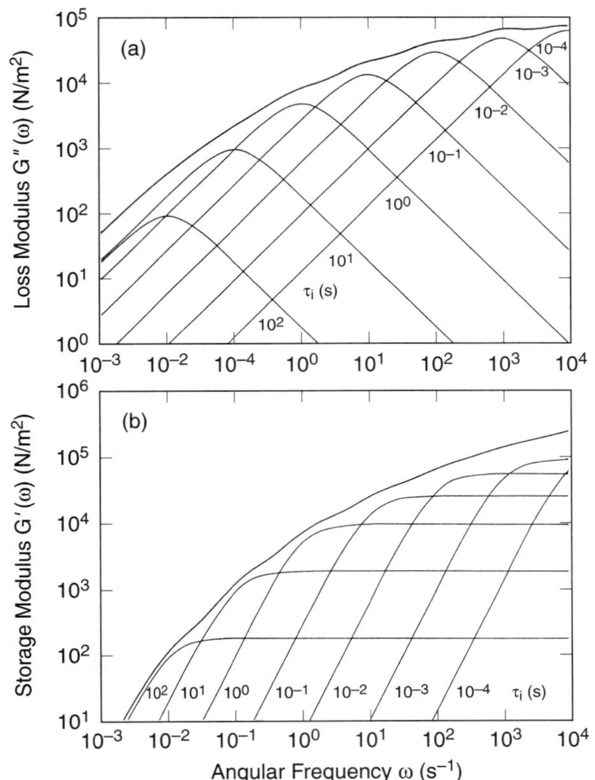

Figure 3.9 (a) Loss modulus G'' and (b) storage modulus G' versus frequency computed from Eqs. (3-25a) and (3-25b) for the eight modes given in Table 3-1 for Melt I, a polyethylene melt. Summing up over all the modes gives the "envelope" curves shown; these curves are reproduced in Fig. 1-11, where they are seen to represent accurately the linear data for this melt. (From Laun 1978, reprinted with permission from Steinkopff Publishers.)

arbitrary linear moduli G' and G'' can be fit over seven decades of frequency. In this case the moduli to be fit were taken from a polyethylene melt, "Melt I"; the constants providing the fit are given in Table 3-1 (Laun 1978). Equation (3-24) with multiple relaxation times is often called the "Lodge equation" for a "rubber-like liquid." The set of values of τ_i and G_i are referred to as the relaxation "spectrum."

To solve Eq. (3-24) for a particular flow history, one must compute the tensor **B** and carry out the integration. An example of how this is done for start-up of steady uniaxial extension is given in Worked Example 3.1 at the end of this chapter.

The temporary network model predicts many qualitative features of viscoelastic stresses, including a positive first normal stress difference in shear, gradual stress relaxation after cessation of flow, and elastic recovery of strain after removal of stress. It predicts that the time-dependent extensional viscosity $\overline{\eta}$ rises steeply whenever the elongation rate, $\dot{\varepsilon}$, exceeds $1/2\tau_1$, where τ_1 is the longest relaxation time. This prediction is accurate for some melts, namely ones with multiple long side branches (see Fig. 3-10). (For melts composed of unbranched molecules, the rise in $\overline{\eta}$ is much less dramatic, as shown in Fig. 3-39.) However, even for branched melts, the temporary network model is unrealistic in that it predicts that $\overline{\eta}$ rises to infinity, whereas the data must level eventually off. A hint of this leveling off can be seen in the data of Fig. 3-10. A more realistic version of the temporary network model

TABLE 3-1
Spectrum of Relaxation Modes for "Melt I"

Mode Number, i	τ_i (sec)	G_i (Pa)
1	10^3	1.00×10^0
2	10^2	1.80×10^2
3	10^1	1.89×10^3
4	10^0	9.80×10^3
5	10^{-1}	2.67×10^4
6	10^{-2}	5.86×10^4
7	10^{-3}	9.48×10^4
8	10^{-4}	1.29×10^5

allows the probability of junction breakage to increase at high stresses (Phan-Thien and Tanner 1977, 1978) [Eq. (3-81c)]. This keeps the stresses in extensional flows bounded; and in shearing flow it produces thinning of the shear viscosity and normal stress coefficients, as observed in experiments.

An example of the success of the temporary network model for a practical application is shown in Fig. 3-11. Here, the predictions of Eq. (3-24) are compared to experimental force-deflection data for "impact tests" in which a heavy flat-bottomed object is dropped onto a flat circular pad of dissipative Sorbothane rubber at various velocities and two different temperatures. Since the material is nearly incompressible under these conditions, the impact

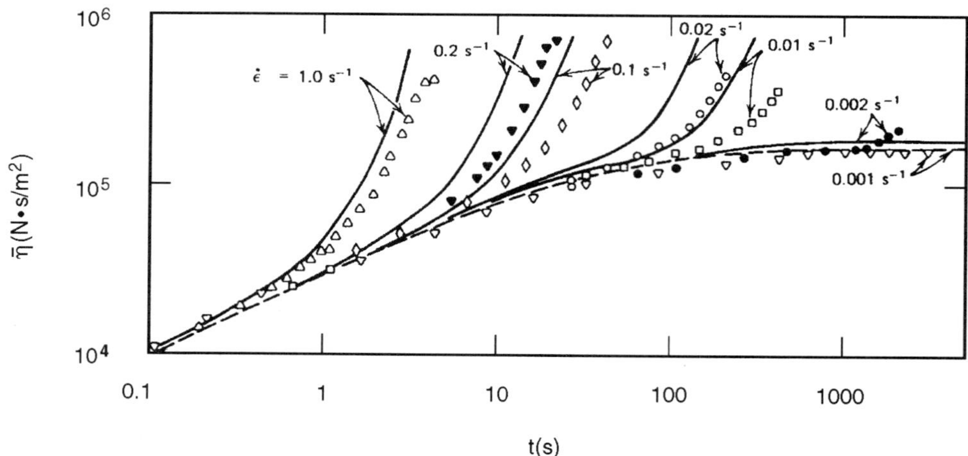

Figure 3.10   Predictions of the temporary network model [Eq. (3-24)] (lines) compared to experimental data (symbols) for start-up of uniaxial extension of Melt I, a long-chain branched polyethylene, using a relaxation spectrum fit to linear viscoelastic data for this melt. (From Bird et al. *Dynamics of Polymeric Liquids, Vol. 1: Fluid Mechanics*, Copyright © 1987. Reprinted by permission of John Wiley & Sons, Inc.)

produces compression in one direction and equal extensions in the other two; it is therefore an equibiaxial extensional deformation (see Section 1.3.2). Each curve in Fig. 3-11 shows hysteresis, as the force rises during compression of the pad and falls along a different path as the impacting object rebounds. The hysteresis reflects the dissipation of elastic energy during impact, and its magnitude is controlled by the viscoelastic relaxation spectrum. This spectrum is represented by the memory function $m(t-t')$ in Eq. (3-24), which was obtained by fitting the linear data given in Fig. 3-8 and using "time–temperature shifting" (described in Section 3.5). While Eq. (3-24) is nominally intended for an elastic liquid, it can be applied to a rubbery solid merely by including in the relaxation spectrum at least one relaxation time that is much longer than the duration of the deformation.

While empirically useful, the temporary network model gives no indiction of the relationship between the relaxation spectrum and the *molecular relaxation processes*. In the next sections, this deficiency is addressed by returning to the bead-spring models of Fig. 3-5.

3.4.4 The Elastic Dumbbell Model

In Fig. 3-5a, the polymer coil is modeled as a series of "beads" equally spaced along the polymer backbone and connected to each other by springs. The beads account for the viscous forces and the springs the elastic forces in the molecule; the portion of the chain represented by a single spring is called a *submolecule*. The bead-spring model is

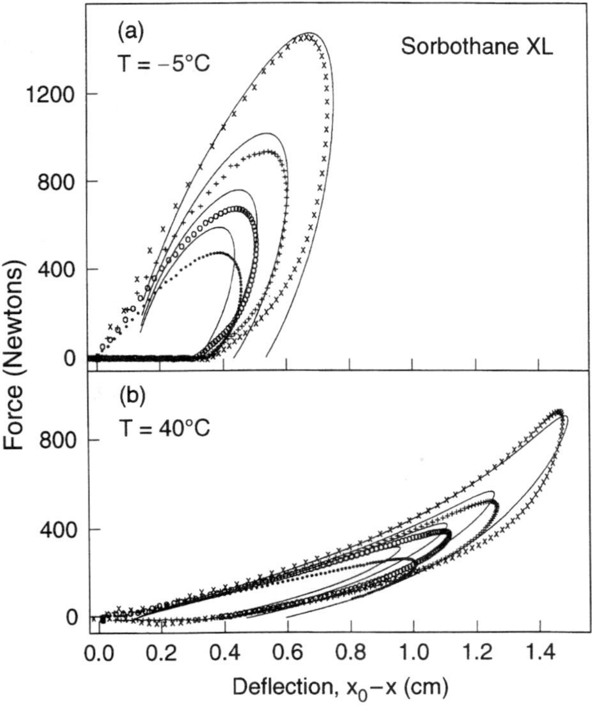

Figure 3.11 (a, b) Curves of instantaneous force against instantaneous sample deflection during impact of a heavy flat-bottomed mass against a flat pad of Sorbothane rubber, 2.59 cm thick, at two different temperatures. Each curve corresponds to a different impact, which lasted on the order of 1 msec. The different curves correspond to different velocities at initial contact. The symbols are experimental measurements and the lines are the predictions of Eq. (3-24) with the memory function fit to the data of Fig. 3-8. (From Larson et al. 1996, reprinted with permission from Steinkopff Publishers.)

appropriate if the internal motions of each submolecule are rapid enough that they remain nearly at equilibrium under flow. This is reasonable for the calculation of "coarse-grained" properties that involve portions of the deformed macromolecule that are longer than a single submolecule. Long-range changes in conformation controlled by highly cooperative, and therefore slow, relaxation processes are thus expected to be well-represented by the bead-spring model. Surprisingly, for dilute polymer solutions in the linear viscoelastic regime, not only are these slower relaxation processes well-described by beads-and-springs models, but so are much faster relaxation processes, even the relaxation of a single submolecule represented by one spring (Lodge et al. 1982) (see Section 3.6.1.2).

The slowest relaxation, which is that of the entire molecule, can be captured by the "elastic dumbbell model." The elastic dumbbell has just one spring with a bead at each end of it (Kuhn 1934) (see Figure 3-5b). For an elastic dumbbell in a solvent, there are three forces that each bead exerts on its surroundings: the viscous drag, the elastic spring force, and a random Brownian force. (The inertia of the beads can be neglected.) It can readily be shown that the center of mass of the dumbbell, which is the position halfway between the two beads, moves with the velocity of the solvent at this position, $\mathbf{v}_{cm}$. The configuration of the dumbbell is described simply by the end-to-end vector $\mathbf{R}$ obtained by subtracting the position of bead 1 from that of bead 2. The balance of the three forces gives

$$\frac{1}{2}\zeta(\dot{\mathbf{R}} - \mathbf{R} \cdot \nabla \mathbf{v}) + 2k_B T \beta^2 \mathbf{R} + k_B T \frac{\partial \ln \psi}{\partial \mathbf{R}} = \mathbf{0} \tag{3-26}$$

The first term in Eqn. (3-26), $\mathbf{F}^d = \frac{1}{2}\zeta(\dot{\mathbf{R}} - \mathbf{R} \cdot \nabla \mathbf{v})$, is the drag force; it is directly proportional to the difference between the rate of stretching of the spring, $\dot{\mathbf{R}}$, and that of the continuum fluid element containing the spring, $\mathbf{R} \cdot \nabla \mathbf{v}$. This term assumes that the velocity gradient, $\nabla \mathbf{v}$, is uniform over the length of the dumbbell. $\mathbf{F}^d$ is zero if the dumbbell is deformed *affinely*—that is, if the dumbbell deformation is the same as that of the macroscopic continuum. The factor of $\frac{1}{2}$ enters because the velocity of bead 2 is just $\mathbf{v}_{cm} + \frac{1}{2}\dot{\mathbf{R}}$ and that of the fluid element containing bead 2 is $\mathbf{v}_{cm} + \frac{1}{2}\mathbf{R} \cdot \nabla \mathbf{v}$. The difference of these two velocities times ζ gives the drag force in Eq. (3-26). For spherical beads, the friction coefficient ζ is given by Stokes' law, $\zeta = 6\pi \eta_s a$, where a is the bead radius and η_s is the solvent viscosity. Equation (3-26) neglects *hydrodynamic interaction*, which is the effect the motion of one bead has on the solvent velocity near the other bead. Hydrodynamic interaction is discussed in Section 3.6.1.

The second term of Eq. (3-26) is the elastic spring force, $\mathbf{F}^s$. The form assumed, $\mathbf{F}^s = 2k_B T \beta^2 \mathbf{R}$, is that of a linear, or Hookean, spring, given in Eq. (3-10). It is an appropriate expression when the molecule is stretched to no more than about a third of its maximum extension.

The third term is the Brownian force, $\mathbf{F}^b = k_B T \partial \ln \psi / \partial \mathbf{R}$. It represents the average force exerted by bead 2 as a result of random bombardments by the surrounding (mostly solvent) molecules.

If Eq. (3-26) is solved for $\dot{\mathbf{R}}$ and inserted into a probability balance equation,

$$\frac{\partial}{\partial t}\psi + \frac{\partial}{\partial \mathbf{R}} \cdot (\dot{\mathbf{R}}\psi) = 0 \tag{3-27}$$

the result is the *Smoluchowski equation*, which is an equation of change for the distribution function ψ:

$$\frac{\partial \psi}{\partial t} + \frac{\partial}{\partial \mathbf{R}} \cdot \left[\mathbf{R} \cdot \nabla \mathbf{v} \psi - \frac{4 k_B T \beta^2}{\zeta} \mathbf{R} \psi - \frac{2 k_B T}{\zeta} \frac{\partial \psi}{\partial \mathbf{R}} \right] = 0 \qquad (3\text{-}28)$$

Multiplying Eq. (3-28) by $\mathbf{RR}$ and integrating over the $\mathbf{R}$ coordinates gives the following terms (Larson 1988):

$$\int \frac{\partial \psi}{\partial t} \mathbf{RR} \, dR^3 = \frac{\partial}{\partial t} \int \psi \mathbf{RR} \, dR^3 = \frac{\partial}{\partial t} \langle \mathbf{RR} \rangle \qquad (3\text{-}29a)$$

$$\int \mathbf{RR} \frac{\partial}{\partial \mathbf{R}} \cdot (\mathbf{R} \cdot \nabla \mathbf{v} \psi) \, dR^3 = -\nabla \mathbf{v}^T \cdot \langle \mathbf{RR} \rangle - \langle \mathbf{RR} \rangle \cdot \nabla \mathbf{v} \qquad (3\text{-}29b)$$

$$-\frac{4 k_B T \beta^2}{\zeta} \int \mathbf{RR} \frac{\partial}{\partial \mathbf{R}} \cdot (\mathbf{R} \psi) \, dR^3 = \frac{8 k_B T \beta^2}{\zeta} \langle \mathbf{RR} \rangle \qquad (3\text{-}29c)$$

$$-\frac{2 k_B T}{\zeta} \int \mathbf{RR} \frac{\partial}{\partial \mathbf{R}} \cdot \frac{\partial \psi}{\partial \mathbf{R}} \, dR^3 = -\frac{4 k_B T}{\zeta} \delta \qquad (3\text{-}29d)$$

The right sides of the last three equations were obtained by integrating the left sides by parts.

Since the polymer contribution to the stress tensor is $\sigma^P = 2 \nu k_B T \beta^2 \langle \mathbf{RR} \rangle$, where ν is the number of dumbbells per unit volume, we have, after collecting terms and multiplying by $2 \nu k_B T \beta^2$,

$$\dot{\sigma}^P - \nabla \mathbf{v}^T \cdot \sigma^P - \sigma^P \cdot \nabla \mathbf{v} + \frac{1}{\tau} (\sigma^P - G \delta) = 0 \qquad (3\text{-}30)$$

with

$$G = \nu k_B T, \qquad \tau = \frac{\zeta}{8 k_B T \beta^2} \qquad (3\text{-}31)$$

The dot in $\dot{\sigma}^P$ denotes a time derivative. In flows that are not spatially homogeneous, this time derivative is the *substantial time derivative* defined by

$$\dot{\sigma}^P \equiv \frac{\partial}{\partial t} \sigma^P + \mathbf{v} \cdot \nabla \sigma^P$$

Equation (3-30) can be arranged into a more compact form,

$$\overset{\triangledown}{\sigma}{}^P + \frac{1}{\tau} (\sigma^P - G \delta) = 0 \qquad (3\text{-}32)$$

where the "$\triangledown$" above σ^P is the *upper-convected time derivative*, defined in general by

$$\overset{\triangledown}{\mathbf{X}} \equiv \dot{\mathbf{X}} - \nabla \mathbf{v}^T \cdot \mathbf{X} - \mathbf{X} \cdot \nabla \mathbf{v} \qquad (3\text{-}33)$$

Often, the transpose velocity gradient tensor $\mathbf{K}$ is used instead of $\nabla \mathbf{v}$; that is, $\mathbf{K} \equiv (\nabla \mathbf{v})^T$ [see Eqs. (1-5) and (1-5a)]. In terms of $\mathbf{K}$, Eq. (3-33) is

$$\overset{\triangledown}{\mathbf{X}} \equiv \dot{\mathbf{X}} - \mathbf{K} \cdot \mathbf{X} - \mathbf{X} \cdot \mathbf{K}^T \qquad (3\text{-}33a)$$

Equation (3-32) is the *upper-convected Maxwell equation*. Note that in a state of rest ($\dot{\sigma}^P = 0$), with no flow ($\nabla \mathbf{v} = 0$), the stress tensor is an isotropic tensor, $G \delta$.

It can be shown using Eq. (1-20) that the upper-convected Maxwell equation is equivalent to the Lodge integral equation, Eq. (3-24), with a single relaxation time. This is shown for the case of start-up of uniaxial extension in Worked Example 3.2. Thus, the simplest temporary network model with one relaxation time leads to the same constitutive equation for the polymer contribution to the stress as does the elastic dumbbell model.

For a dilute solution of dumbbells, there is also a contribution to the stress from the Newtonian solvent, namely

$$\boldsymbol{\sigma}^P = 2\eta_s \mathbf{D}$$

The total stress tensor is then

$$\boldsymbol{\sigma} = \boldsymbol{\sigma}^P + \boldsymbol{\sigma}^s \tag{3-34}$$

Equations (3-32)–(3-34) are equivalent to the so-called Oldroyd-B equation. The Oldroyd-B equation is a simple, but qualitatively useful, constitutive equation for dilute solutions of macromolecules (see Section 3.6.2). Refinements to the simple elastic dumbbell model, such as the effects of the nonlinearity of the force–extension relationship at high extensions, are discussed in Section 3.6.2.2.1.

• Worked Examples 3.1 and 3.2 (at the end of this chapter) show how calculations of stress in simple flows are carried out using the temporary network model and the elastic dumbbell model.

3.4.5 The Rouse Model

The derivation of the constitutive equation for the dumbbell model can also be extended to allow multiple beads and springs. From the force–balance equations for such a model, one obtains (Bird et al. 1987b; Larson 1988)

$$\zeta_b(\dot{\mathbf{R}}_i - \mathbf{R}_i \cdot \nabla \mathbf{v}) + k_B T \sum_{j=1}^{N_s} A_{ij} \left(2\beta_s^2 \mathbf{R}_j + \frac{\partial \ln \psi}{\partial \mathbf{R}_j} \right) = \mathbf{0} \tag{3-35a}$$

where

$$A_{ij} = \begin{cases} 2 & \text{if } i{=}j \\ -1 & \text{if } i{=}j{+}1 \text{ or } i{=}j{-}1 \\ 0 & \text{otherwise} \end{cases} \tag{3-35b}$$

$\mathbf{R}_j$ is the end-to-end vector of the jth spring in the chain, and N_s is the number of springs. The constant ζ_b is the drag coefficient for a single bead, and $\beta_s^2 \equiv 3/(2N_{K,s}b_K^2)$ is calculated for a single spring, with $N_{K,s}$ the number of Kuhn steps corresponding to the portion of the polymer comprising a single spring, $N_{K,s} = N_K/N_s$. Hence $\beta_s^2 = N_s\beta^2$, where $\beta^2 = 3/(2N_K b_K^2)$ characterizes the chain as a whole. Note that Eq. (3-35a) for spring i is coupled to the corresponding equations for the neighboring springs, $i-1$ and $i+1$, because beads attached to spring i feel forces from the neighboring springs. This implies that Eq. (3-35) is a matrix equation with off-diagonal matrix elements.

Rouse (1953) transformed this matrix equation into a set of *uncoupled* equations—that is, into N_s independent equations for the *normal modes*:

$$\zeta_b(\dot{\mathbf{Q}}_i - \mathbf{Q}_i \cdot \nabla\mathbf{v}) + 4k_BT \, \sin^2\left(\frac{i\pi}{2(N_s + 1)}\right)\left[2\beta_s^2\mathbf{Q}_i + \frac{\partial \ln \psi}{\partial \mathbf{Q}_i}\right] = \mathbf{0} \qquad (3\text{-}36)$$

where $\mathbf{Q}_i$ is a transformed end-to-end vector. The first normal mode, $i = 1$, corresponds to cooperative relaxation of the entire chain, while the other normal modes, $i = 2, 3, ...,$ correspond to successively higher harmonics, which involve increasingly more localized motion.

The normal-mode decomposition permits one to break the stress into a sum of contributions,

$$\sigma^p = \sum_{i=1}^{N_s} \sigma_i \qquad (3\text{-}37)$$

each contribution coming from one of the normal modes,

$$\sigma_i = 2\nu k_B T\beta_s^2 \langle \mathbf{Q}_i\mathbf{Q}_i \rangle \qquad (3\text{-}38)$$

Each σ_i obeys an upper-convected Maxwell equation

$$\overset{\triangledown}{\sigma}_i + \frac{1}{\tau_i}(\sigma_i - G_i\delta) = \mathbf{0} \qquad (3\text{-}39)$$

where (Ferry 1980)

$$G_i = G = \nu k_B T, \qquad \tau_i = \frac{\zeta_b}{16k_B T\beta_s^2 \sin^2(i\pi/2(N_s + 1))} \qquad (3\text{-}40)$$

The longer relaxation times, for which i/N_s is small, are given approximately by

$$\tau_i \approx \frac{\zeta_b(N_s + 1)^2}{4\pi^2\beta_s^2 i^2 k_B T} = \frac{\tau_1}{i^2}, \qquad G_i = G = \nu k_B T \qquad (3\text{-}41)$$

Since the total frictional drag acting on the molecule must be divided among $N_s + 1$ beads, the drag coefficient ζ_b that one must assign to each bead should be inversely proportional to the number of beads $N_s + 1$ used to describe the chain. Hence, $\zeta_b = N\zeta_0/(N_s + 1)$, where N is the number of monomers in the whole chain, and ζ_0 is the drag coefficient per monomer. Thus, Eq. (3-41) can be rearranged to a form that shows no dependence on N_s (Ferry 1980; Larson 1988):

$$\tau_i \approx \frac{\zeta_0 N^2 b^2}{6\pi^2 i^2 k_B T} = \frac{6(\eta - \eta_s)}{\pi^2 i^2 \nu k_B T} \qquad (3\text{-}42)$$

Here b is the statistical segment length (so that $\beta_s^2/N_s = \beta^2 = 3/(2Nb^2)$). The polymer contribution to the shear viscosity is $\eta - \eta_s$. The second equality in (3-42) is derived from the relationship

$$\eta - \eta_s = \nu k_B T \sum_{i=1}^{N_s} \tau_i \qquad (3\text{-}42a)$$

and the identity $\sum_{i=1}^{N_s} \tau_i \approx \tau_1 \sum_{i=1}^{\infty} 1/i^2 = \pi^2\tau_1/6.$

The Rouse model is the earliest and simplest molecular model that predicts a nontrivial *distribution of polymer relaxation times*. As described below, real polymeric liquids do in fact show many relaxation modes. However, in most polymer liquids, the relaxation modes observed do not correspond very well to the mode distribution predicted by the Rouse theory. For polymer solutions that are dilute, there are *hydrodynamic interactions* that affect the viscoelastic properties of the solution and that are unaccounted for in the Rouse theory. These are discussed below in Section 3.6.1.2. In most concentrated solutions or melts, *entanglements* between long polymer molecules greatly slow polymer relaxation, and, again, this is not accounted for in the Rouse theory. *Reptation* theories for entangled polymers are described in Section 3.7.

Melts of rather short polymer molecules, with molecular weights less than the *entanglement threshold* M_c (which is typically in the range $\approx$ 2000–30,000), seem to be described by the Rouse theory at low or modest frequencies (Ferry 1980). However, even for these liquids, there are *high-frequency* relaxation modes not described by the Rouse model; these include bond rotations and coordinated movements of several backbone atoms (Ferry 1980; Zhu and Ediger 1995) These motions are sometimes called "glassy modes" since they account for much of the relaxation that occurs in bulk polymers that have been vitrified—that is, cooled to the point of solidification into a glass. This occurs at the glass transition temperature, mentioned earlier. The behavior of liquids near the glass transition temperature is covered in Chapter 4.

3.5 LINEAR VISCOELASTICITY AND TIME–TEMPERATURE SUPERPOSITION

3.5.1 Distribution of Relaxation Times

The relaxation times, τ_i, and corresponding moduli, G_i, constitute what is called the *distribution* or *spectrum* of relaxation times. The relaxation spectrum given in Eq. (3-40) is a distinctive feature of the Rouse model that can be tested experimentally. A simple type of rheological experiment from which this spectrum can be obtained is *small-amplitude oscillatory deformation*, discussed in Section 1.3.1.4. In this test, at low frequencies, $\omega \ll 1/\tau_1$, the Rouse model predicts the usual terminal relaxation behavior $G' = G\omega^2\tau_1^2$, and $G'' = G\omega\tau_1$. More significantly, at higher frequencies, where ω is in the range $1/\tau_1 \ll \omega \ll 1/\tau_{N_s}$, the Rouse model predicts a power-law frequency dependence of G' and G'':

$$G' = G'' - \eta_s\omega \propto \omega^{1/2} \tag{3-43}$$

This prediction is a consequence of the spacing of relaxation times in the Rouse model, namely, $\tau_i \propto 1/i^2$ (see Eq. 3-41). In general, if $\tau_i \propto 1/i^p$ and $G_i = G$ is independent of i, then in the range $1/\tau_1 \ll \omega \ll 1/\tau_{N_s}$ we obtain

$$G' \propto G'' - \eta_s\omega \propto \omega^{1/p} \tag{3-44}$$

3.5.2 Time–Temperature Superposition

In Eq. (3-40), each τ_i and G_i depend on absolute temperature T. For τ_i, the most important temperature dependence is not the explicit $1/T$ dependence, but instead the indirect dependence of the friction coefficient ζ_0 on T. Note that this dependence is *the same for each τ_i*. This means that on a log–log plot, a change in temperature merely shifts the curves $G'(\omega)$ and $G''(\omega)$ along the frequency axis, without changing their shapes. The curves are also shifted equally along the modulus axis through the dependence of G_i on T and chain density ν (Eq. (3-40)]. The shift along the modulus axis is often small or negligible, however, since when T increases, ν decreases by thermal volume expansion, and this effect at least partially cancels the direct effect of T on modulus. Thus, the curves $G'(\omega)$ and $G''(\omega)$ measured at different temperatures *can be superposed* by shifting along the logarithmic frequency axis, with a minor shift along the logarithmic modulus axis, if desired.

This procedure, called *time–temperature superposition*, is not restricted to fluids obeying the Rouse model, but holds for many polymer melts and solutions, as long as there are no phase transitions or other temperature-dependent structural changes in the liquid (Ferry 1980). For amorphous single-component polymers, large deviations from time–temperature superposition are rarely found except near the glass transition temperature (Plazek 1965; Ferry 1980). Time–temperature superposition also works for dilute solutions, as long as there are not large temperature-dependent changes in solvent quality (Amelar et al. 1991). Fluids for which linear rheological data, such as G' and G'', superimpose at different temperatures are said to be *thermorheologically simple*.

As an example taken from Ferry (1980), consider in Fig. 3-12a the curves of the *storage compliance J'* measured over a frequency range of 2.5 decades, roughly 10–3000 cycles per second, for poly(n-octyl methacrylate) at 24 temperatures ranging from $-14.3°C$ to $129.5°C$. The storage compliance is obtained by imposing an *oscillatory stress* on the sample and measuring the resulting oscillatory strain. J' is then the component of the strain that is in phase with the stress, divided by the amplitude of the imposed stress. The storage compliance is related to the storage and loss *moduli G' and G''* by $J' = G'/(G'^2 + G''^2)$. In Fig. 3-12b, the data of Fig. 3-12a have been *time–temperature superposed* by shifting along the horizontal log frequency axis by an amount required to bring the data into superposition. (There has also been a small shift along vertical axis for reasons given above.)

The temperature-dependent *shift factor*, a_T, required to obtain this superposition is plotted in Fig. 3-12c. A curve such as Fig. 3-12b with data from multiple temperatures superposed onto a single line is called a *master curve*; its abscissa ωa_T is the *reduced frequency*. At the reference temperature T_0 to which the data have been shifted, one has by definition $a_T = 1$. Since $a_T \ll 1$ at temperatures well above T_0 and $a_T \gg 1$ when $T \ll T_0$, the master curve typically covers a much wider range of effective frequencies than are covered by experiments at any single temperature. Figure 3-12b, for example, covers some 11 decades of reduced frequency, even though each data set from which Fig. 3-12b was obtained spans only 2.5 decades of real frequency. Since most rheometers can only cover a few decades of frequency at any one temperature, time–temperature superposition is invaluable as a means of enormously extending the frequency range of such instruments, as well as greatly simplifying the task of characterizing polymeric liquids.

Time–temperature shifting is also extremely useful in practical applications, since it allows one to make predictions of material response for time scales either much longer

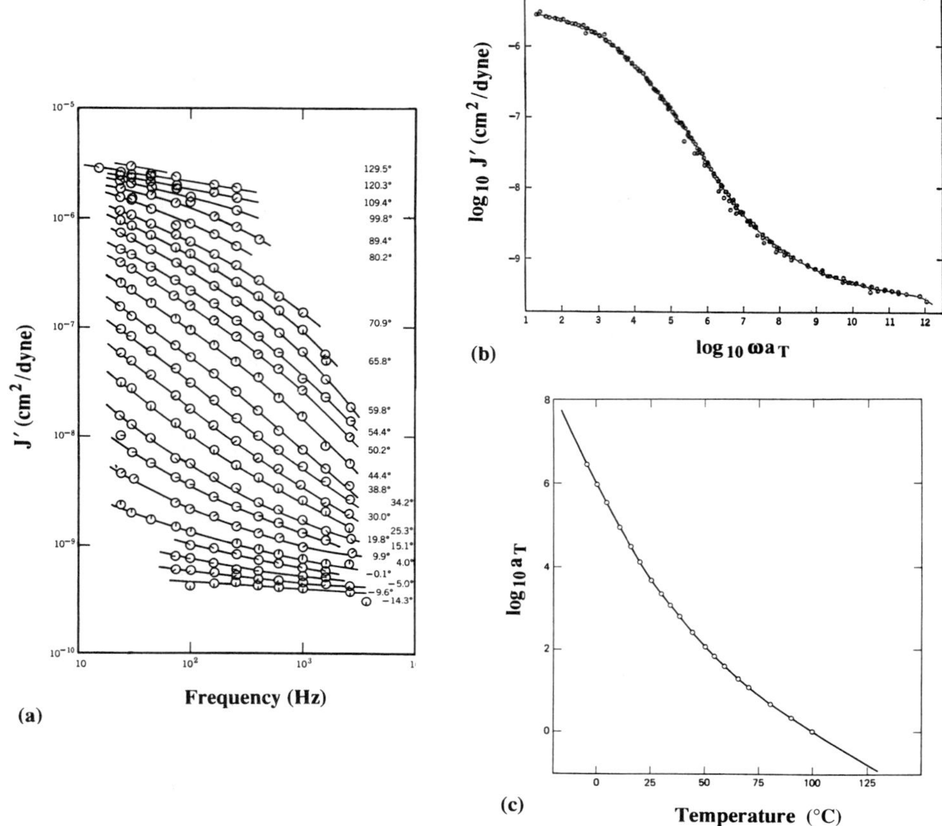

Figure 3.12 (a) Storage compliance J' of poly(n-octyl methacrylate) versus frequency measured at 24 temperatures, given in degrees Celsius. (b) "Master curve" of J' versus reduced frequency ωa_T obtained from the data of (a) by time temperature shifting. (c) Temperature-dependence of the shift factor a_T required to obtain the master curve of (b); the line is a fit of the WLF equation, (3-45a), with $T_0 = 373.2$ K, $c_1^0 = 7.60$, and $c_2^0 = 227.3$ K. (From Ferry *Viscoelastic Properties of Polymers*, Copyright © 1980. Reprinted by permission of John Wiley & Sons, Inc.)

or much shorter than one can measure by conventional rheometry. As an example, the predictions of impact forces at the temperatures −5°C and 40°C shown in Fig. 3-11 were obtained from Eq.(3-24) and a *single set* of relaxation times [extracted by fits of Eq. (3-25) to the data in Fig. 3-8)], suitably shifted using the value of a_T appropriate to the given temperature. Accurate prediction of these impact curves requires data corresponding to frequencies up to 10^5, much higher than can be obtained by ordinary rheometry at a single temperature. Using time–temperature superposition, data at *effective* frequencies much higher than this can easily be determined (see Fig. 3-8).

As a second simple example, consider how the zero-shear viscosity η_0 shifts with temperature. Since $\eta_0 = \lim_{\omega \to 0} G''/\omega$, we have

$$\eta_0(T) = \lim_{\omega \to 0} \frac{G''(T, \omega)}{\omega} = \lim_{\omega \to 0} \frac{G''(T_0, a_T\omega)}{\omega} = \frac{\eta_0(T_0)a_T\omega}{\omega} = a_T\eta_0(T_0)$$

The dependence of the shift factor a_T on temperature can often be fit to an empirical expression known as the *WLF (Williams–Landel–Ferry) equation* (Williams et al. 1955; Ferry 1980):

$$\log a_T = \frac{-c_1^0(T - T_0)}{c_2^0 + T - T_0} \tag{3-45a}$$

(Here and elsewhere, "log" represents a base 10 logarithm; "ln" represents a natural logarithm.) The line in Fig. 3-12c is the fit of the WLF equation to the shift factor data for poly(n-octyl methacrylate), obtained for the parameter values $T_0 = 373.2$ K, $c_1^0 = 7.60$, $c_2^2 = 227.3$ K. Defining $T_\infty \equiv T_0 - c_2^0$, this equation can be rewritten as

$$\log a_T = -\frac{c_1^0(T - T_0)}{(T - T_\infty)} \tag{3-45b}$$

where the temperature T_∞ is a material-dependent temperature that is typically around 50°C below the glass transition temperature. Tabulations of values of WLF parameters c_1^0, c_2^0, T_∞, and T_0 for various polymers can be found by looking ahead to Table 3-3 in Section 3.7 (Ferry 1980).

In the Rouse model, thermorheological simplicity holds because all relaxation processes, no matter how slow, involve relaxations of submolecules, each of which is controlled by the same drag coefficient ζ_0. The slower Rouse relaxation processes require the coordinated movement of more submolecules than do the faster processes, but since the degree of coordination does not depend on temperature, the rates of all modes change proportionately when the temperature is raised or lowered. For real polymeric materials, time–temperature shifting works for relaxation modes that involve motions of portions of the chain large enough to average out small-scale heterogeneities in the chain or in the viscous environment through which the chain moves. Fast relaxation modes involving no more than a few backbone bonds do not necessarily shift in accordance with time–temperature superposition. In rheological studies, these fast modes show up at low temperatures near the *glass transition temperature* T_g, and this is where thermorheological superposition is most likely to fail for simple homopolymers. Computer simulations and high-frequency spectroscopic studies have been used to gain an understanding of the details of the high-frequency, local relaxation modes of polymers for which time–temperature superposition fails (Helfand et al. 1980; Weber and Helfand 1983; Hall and Helfand 1982; Morawetz 1979; Zhu and Ediger 1995).

The relaxation spectrum can also be affected by changes in pressure, p, especially if these pressure changes are large—that is, hundreds of atmospheres (Ferry 1980; Tanner 1985). As with temperature, a simple shifting procedure can be used to account for pressure effects. Thus the shift factor $a_{T,p}$ is both temperature-and pressure-dependent.

• Worked Example 3.3 shows how useful the principle of time–temperature superposition can be.

3.6 THE RHEOLOGY OF DILUTE POLYMER SOLUTIONS

We now review the rheology of *dilute* polymer solutions. Diluteness is defined by the rarity of overlap or interference of one polymer coil with others. In the *dilute-solution limit*, the polymer contribution to the viscoelastic properties is just that of a single coil in an infinite bath of solvent, multiplied by the number of such coils in solution. One might suppose that the Rouse model might be accurate for such solutions, since in that model there are no interactions among polymer molecules. However, the Rouse model assumes that the solvent drag on each part of the polymer molecule is the same as it would be if the other parts of the polymer weren't present. Thus, it assumes that the coil is *freely draining*; that is, the effects of *hydrodynamic interactions* are neglected. Hydrodynamic interactions are the disturbances in the solvent velocity field created by the motion of one part of a polymer chain that then affect the drag exerted by the solvent on other parts of the same chain. Hydrodynamic interactions influence both the linear and nonlinear rheological properties of dilute solutions. As discussed in Section 3.6.1, the linear properties of dilute solutions are dramatically affected by hydrodynamic interactions; the intrinsic viscosity, relaxation time, and diffusivity depend more weakly on molecular weight, and the complex modulus depends more strongly on frequency than would be the case in the absence of hydrodynamic interaction. The effect of hydrodynamic interactions on the nonlinear rheology of dilute solutions is more subtle and is discussed in Section 3.6.2.2.2.

3.6.1 Linear Rheology

3.6.1.1 Molecular-Weight Scaling Laws

Because of hydrodynamic interactions, the Rouse model fails in its prediction of the scaling of the low-shear viscosity, η_0, with molecular weight. Note from Eqs. (3-41) and (3-42a) that for the Rouse model we obtain

$$\eta_0 = G \sum_{i=1}^{N_s} \tau_i + \eta_s \approx \nu k_B T \tau_1 \sum_{i=1}^{N_s} \frac{1}{i^2} + \eta_s$$

$$= \frac{\pi^2}{6} \nu k_B T \, \tau_1 + \eta_s$$

$$= \frac{\pi^2}{6} \frac{c N_A k_B T}{M} \tau_1 + \eta_s \qquad (3\text{-}46)$$

where $c = \nu M / N_A$ is the mass concentration per unit volume, N_A is Avogadro's number, and, from Eq. (3-42), $\tau_1 \approx \zeta_0 N^2 b^2 / 6\pi^2 k_B T$. Since $N \propto M$, the intrinsic viscosity at zero shear rate is proportional to the molecular weight,

$$[\eta]_0 \propto \eta_0 - \eta_s \propto M \qquad \text{(Rouse)} \qquad (3\text{-}47a)$$

Experimental results, however, fit a different power law:

$$[\eta]_0 = K M^a, \qquad a < 1 \qquad \text{(experiment)} \qquad (3\text{-}47b)$$

Since the exponent a is always less than unity, Eq. (3-47b) disagrees with the Rouse prediction. For theta solvents, a is 0.5. For *good solvents*, polymer–solvent contacts are favorable, so the polymer coil expands to contact as many solvent molecules as possible. Hence, a for good solvents is higher than 0.5, but it is always less than around 0.8 for flexible polymers (de Gennes 1979).

These scaling laws for the intrinsic viscosity can be explained by noting that in the limit of dominant hydrodynamic interaction, the drag on the polymer molecule as it moves through the solvent can be estimated crudely by assuming that the polymer coil acts like an impenetrable sphere. The drag force F^d on an undeformed polymer coil moving at a velocity V through the solvent is therefore roughly $F^d \sim 6\pi\eta_s RV$, where R is the coil radius (Kirkwood and Riseman 1948). A more precise calculation based on the *Zimm theory* (Section 3.6.1.2) gives

$$\frac{F^d}{V} \equiv \zeta_{\mathrm{coil}} = \frac{3}{8}(6\pi^3)^{1/2}\eta_s R = 5.11 R\eta_s = \frac{k_B T}{D} \tag{3-48}$$

where R is the *root-mean-square* separation of the ends of the molecule; that is, $R \equiv \langle R^2\rangle_0^{1/2}$. The *drag coefficient*, $F^d/V \equiv \zeta_{\mathrm{coil}}$, can be related to the diffusivity D of the molecule by $F^d/V = k_B T/D$. The longest relaxation time, in turn, scales as $\tau_1 \propto R^2/D$. The polymer contribution to the viscosity is proportional to $G\tau_1$, where G is the modulus, which is inversely proportional to the molecular weight. From these scaling rules and from the scaling for the coil radius $R \sim M^\nu$, with $\nu = 1/2$ for a theta solvent and $\nu = 3/5$ for a good solvent, the results in Table 3-2 are obtained (de Gennes 1979; Larson 1988).

In theta solvents, experimental results confirm the $M^{1/2}$ scaling predicted for $[\eta]_0$ in the limit of dominant hydrodynamic interaction. In good solvents, the measured exponent is generally slightly less than the value, 4/5, predicted for dominant hydrodynamic interaction. The scaling law, $[\eta]_0 = K_\theta M^{1/2}$, for a flexible polymer in a theta solvent is often used to obtain molecular weights from viscosity measurements. Here the coefficient K_θ depends on the chemical makeup of the polymer and, perhaps to a minor extent, on the solvent. It can be calculated from the elementary structural properties of the polymer using

$$K_\theta = \Phi[\langle R^2\rangle_0 /M]^{3/2} = \Phi\left(\frac{C_\infty \ell^2}{m_0}\right)^{3/2} \tag{3-49}$$

where Φ is the "universal hydrodynamic constant," $\Phi = 2.5 \times 10^{21}$ dl cm^{-3} mol^{-1}, ℓ is

TABLE 3.2
Predicted Scaling Laws in Dilute Solution

	Freely Draining Gaussian Chain (Rouse Theory)	Dominant HI Theta Solvent (Zimm Theory)	Dominant HI Good Solvent
Friction coefficient, F^d/V	M	$M^{1/2}$	$M^{3/5}$
Diffusivity, D	M^{-1}	$M^{-1/2}$	$M^{-3/5}$
Relaxation time, τ_1	M^2	$M^{3/2}$	$M^{9/5}$
Intrinsic viscosity, $[\eta]_0$	M	$M^{1/2}$	$M^{4/5}$

the backbone bond length, and m_0 is the polymer molecular weight per backbone bond. A tabulation of values of K_θ for various polymers can be found in Flory (1969). For polystyrene, $K_\theta \approx 8 \times 10^{-4}$ dl g^{-1} (g/mol)$^{-1/2}$. A more detailed discussion of these scaling laws can be found in de Gennes (1979).

• Problems 3.4 and 3.5 test your understanding of dilute-solution viscometry.

3.6.1.2 Complex Modulus: The Zimm Theory

The linear viscoelastic behavior of dilute high-molecular-weight polymers in good or theta solvents is typified by Fig. 3-13. In this figure, there is an *intermediate-frequency regime* which extends from log $\omega\tau_0 \sim 1$ to log $\omega\tau_0 = 2$ and beyond; the high-frequency regime, where $\omega \sim 1/\tau_{N_s}$, is off the scale of this figure. Experimentally, for a theta solvent, G' and $G'' - \omega\eta_s$ in the intermediate-frequency regime of Fig. 3-13 scale as $\omega^{2/3}$ rather than the predicted Rouse scaling, $\omega^{1/2}$. The failure of the Rouse model to predict the correct frequency dependence of the linear viscoelastic functions, G' and G'', and its failure to predict the correct molecular-weight dependence of $[\eta]_0$ can be corrected by accounting for hydrodynamic interaction (Zimm 1956), as described in what follows (Bird et al. 1987b; Larson 1988).

Suppose a force, $\mathbf{F}_c$, is exerted by a bead on the Newtonian solvent at the origin. This force sets the surrounding solvent in motion; away from the origin at a point $\mathbf{r}$ the solvent velocity, calculated from the Stokes equation, reaches the steady-state value

$$\mathbf{v}' = \mathbf{\Omega} \cdot \mathbf{F}_c$$

where $\mathbf{\Omega}$ is the *Oseen tensor* (Oseen 1910):

$$\mathbf{\Omega}(\mathbf{r}) = \frac{1}{8\pi\eta_s r}\left[\mathbf{\delta} + \frac{\mathbf{rr}}{r^2}\right] \tag{3-50}$$

and $\mathbf{\delta}$ is the unit tensor. The influence of a force applied by a bead in the solvent therefore decays as $1/r$ as the distance r from the bead increases. [In Brownian dynamics simulations (Section 1.7.2), the Oseen tensor is unsuitable because it loses positive definiteness when beads get close together. A better choice for simulations is the Rotne–Prager–Yamakawa tensor discussed in Zylka and Öttinger (1989)]. If the number of beads, $N_s + 1$, is large, then the longer of the relaxation times depend on a single dimensionless parameter, $h \equiv h^* N_s^{1/2}$, where $h^* \equiv \zeta_b/(12\pi^3)^{1/2} R_s\eta_s$, ζ_b is the friction coefficient for each bead, and R_s is the root-mean-square length of a spring at equilibrium (Amelar et al. 1991). At high molecular weights, $h \to \infty$; this defines the *nondraining limit*. In the nondraining limit, Zimm (1956) found a set of approximate normal modes. From these, the relaxation times are obtained; it turns out that their spacing is changed from $\tau_i \sim i^{-2}$ of Rouse to $\tau_i \sim i^{-3/2}$ of Zimm. At intermediate frequencies, G' and G'' therefore scale as $\omega^{2/3}$ rather than $\omega^{1/2}$ of the Rouse theory:

$$G'' - \eta_s\omega = \sqrt{3}\,G' \propto \omega^{2/3} \quad \text{(Zimm)} \tag{3-51}$$

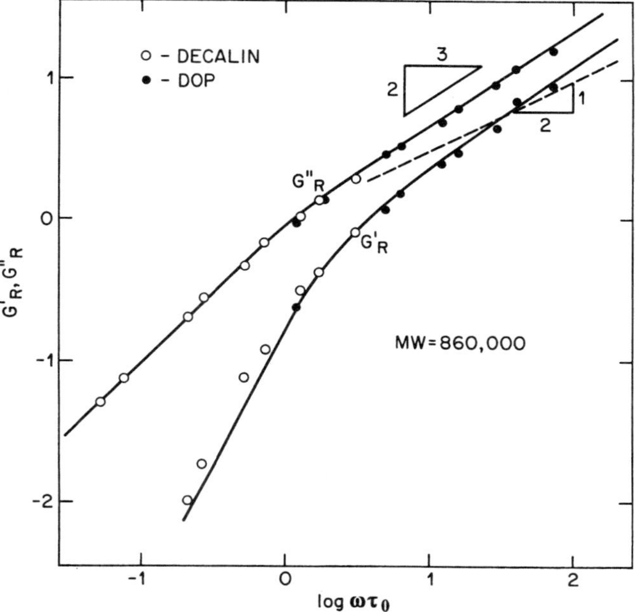

Figure 3.13 Linear viscoelastic data (symbols) for polystyrene in two theta solvents, decalin and dioctylphthalate, compared to the predictions (lines) of the Zimm theory with dominant hydrodynamic interaction, $h = \infty$. The *reduced* storage and loss moduli G'_R and G''_R are defined by $G'_R \equiv [G']M/N_A k_B T$ and $G''_R \equiv [G'']M/N_A k_B T$, where the brackets denote *intrinsic* values extrapolated to zero concentration, $[G'_R] \equiv \lim_{c\to 0}(G'/c)$ and $[G''_R] \equiv \lim_{c\to 0}[(G'' - \omega\eta_s)/c)$, and c is the mass of polymer per unit volume of solution. The characteristic relaxation time τ_0 is given by $\tau_0 = [\eta]_0 M\eta_s/N_A k_B T$. For frequencies $\tau_0\omega$ greater than 10, G'_R and G''_R are proportional to $\omega^{2/3}$, in agreement with the Zimm theory, and not the Rouse theory, which predicts $G' = G'' - \eta_s\omega \propto \omega^{1/2}$. (From Johnson et al. 1970, with permission of the Society of Polymer Science, Japan.)

Both the scaling with ω and the proportionality constant, $\sqrt{3}$, are confirmed by experimental data (see Fig. 3-13). The lines in Fig. 3-13 are proportional to G' and $G'' - \eta_s\omega$ computed in the nondraining limit. The agreement with data for a polystyrene of high molecular weight ($M = 860,000$) in theta solvents is excellent. In addition to its agreement with experimental data, the predictions of Zimm theory are supported by molecular dynamics simulations (Pierleoni and Ryckaert 1991; Dünweg and Kremer 1991).

In general, N_s and h cannot be considered infinite, especially for lower molecular weights or higher frequencies, and so the relaxation spectrum is in general governed by both h^* and N_s. This is *partial draining*. The sets of relaxation times, $\{\tau_i\}$, for various values of N_s and h^* have been tabulated by Lodge and Wu (1972). A value $h^* = 0.25$ has been found empirically to give good fits to data for theta solvents (Sahouani and Lodge 1992).

In a good solvent, where there are excluded-volume effects, G' and G'' can be fit to the Zimm theory simply by adjusting $h*$ downward, for finite N_s (Ferry 1980). Thus, as the solvent quality improves, the relaxation spectrum becomes more Rouse-like (since

$h^* = 0$ in the Rouse limit). For polystyrene in a good solvent, excellent fits to oscillatory birefringence data over a wide frequency range can be obtained when $h^* \doteq 0.15$ and the molecular weight of a single submolecule, M/N_s, is chosen to be around 5000 (Lodge et al. 1982; Sammler et al. 1990; Amelar et al. 1991). Sahouani and Lodge (1992) have noted, however, that the use of h^* as a fitting parameter for good-solvent systems is conceptually questionable and leads to a prediction of the molecular-weight dependence of the longest relaxation time that disagrees with the measured dependence. They discuss several other, preferable ways of accounting for excluded-volume effects within the bead-spring model.

3.6.1.3 High-Frequency Behavior

Above the intermediate-frequency regime, which spans from the inverse of the longest Zimm time, $1/\tau_1$, to the inverse shortest time, $1/\tau_{N_s}$, there is a high-frequency regime, $\omega \gtrsim 1/\tau_{N_s}$, which is beyond the range covered by Fig. 3-13. In this regime, individual submolecules of the bead-spring model are stretched, and a crossover from $G'' - \eta_s\omega \sim \omega^{2/3}$ to $G'' - \eta_s\omega = 0$ is predicted, where η_s is the solvent viscosity. Thus, at these high frequencies, the dynamic viscosity, $\eta' \equiv G''/\omega$, which is a measure of the viscous dissipation, is predicted to equal η_s, and hence there is no contribution to η' from the polymer! This occurs in the beads-and-springs model because at high frequency the deformation is too fast for even individual springs to relax, and so the polymer dissipates no energy. Experimentally, $\eta' = G''/\omega$ does approach a constant, η'_∞, at high ω (Massa et al. 1971; Ferry 1980; Morris et al. 1988), but η'_∞ usually does not equal the solvent viscosity η_s. Usually it is greater than η_s, but sometimes can be less than η_s (Landry 1985; Man 1984; Morris et al. 1988). The case $\eta'_\infty < \eta_s$, which occurs for polybutadiene in the solvent Aroclor, would seem to imply that the polymer has a *negative* viscous dissipation, which is a thermodynamic impossibility. Experiments by Lodge and coworkers (Morris et al. 1988; Lodge 1993), however, show that the relaxations in the solvent are affected by the presence of the polymer. Depending on its interactions with the solvent, a polymer can either increase or decrease the viscous dissipation in the surrounding solvent. In effect, the polymer changes the solvent viscosity. The viscosity of the solvent tends to go down when the polymer dissolved in it has a local relaxation rate (or bond reorientation rate) that is faster than that of the solvent. Roughly speaking, this occurs when the glass transition temperature of the polymer is lower than that of the solvent.

The still faster ($\sim$nsec) dynamics of local motions of polymer molecules in dilute solutions have been investigated by Ediger and coworkers (Zhu and Ediger 1995, 1997). They find that the rates of these local motions of a few bonds are *not* proportional to the solvent viscosity, unless the solvent reorientation rate is fast compared to the polymer local motion. Thus, for local motions (such as bond reorientations) of polymer molecules, Stokes' law of drag does not always hold.

3.6.2 Nonlinear Rheology

If a dilute polymer solution is subjected to a unidirectional or steady flow with a velocity gradient large enough to stretch out the polymer molecule, nonlinear viscoelastic effects are observed. The simple Hookean dumbbell model, described in Section 3.4.4, can predict qualitatively some of the nonlinear rheological behavior of dilute solutions. In a steady simple shear flow, the Hookean dumbbell model predicts that the shear viscosity η and first normal stress coefficient Ψ_1 are *shear-rate-independent* and are given by

$$\eta = \eta_p + \eta_s, \qquad \Psi_1 = 2\eta_p \tau \tag{3-52}$$

where $\eta_p = \nu k_B T \tau$ is the polymer contribution to the viscosity, ν is the number of polymer molecules per unit volume, and τ is the relaxation time. In an extensional flow, the Hookean dumbbell model predicts that the stress along the principal stretch direction increases without bound when the extension rate reaches the value $\dot{\varepsilon} = 1/2\tau$. The predictions of multiple-relaxation-time Rouse or Zimm models in steady-state flows are qualitatively similar to those of the Hookean dumbbell model. The behavior of actual dilute polymer solutions differs somewhat from that predicted by these models, as described below.

3.6.2.1 Experiments

The nonlinear rheological properties of dilute polymer solutions can be difficult to measure. One difficulty is inherent in diluteness; to obtain the *intrinsic* value of a rheological property in the dilute limit, one must measure that property at a series of low concentrations and then extrapolate toward zero concentration. Since the solvent contribution must be subtracted from these measurements before extrapolating, small experimental errors can be amplified. Another difficulty is that in low-viscosity solvents, the longest relaxation time of even long molecules tends to be small, often much less than 0.1 sec. Thus, to obtain properties in the highly nonlinear range, the shear rate must be high, around 100 sec^{-1} or higher. At such high shear rates, many rheological instruments suffer severe limitations due to inertia, viscous heating, meniscus instabilities, and so on. An exception is capillary rheometry, for which these effects are manageable, at least for shear rates less than around 10^4 sec^{-1} (Macosko 1994). However, from a capillary rheometer, one obtains only steady shear viscosities; normal stresses and time-dependent rheological properties are not probed. Specialty slit-flow devices, such as one designed by Lodge (1989), allow measurement of the first normal stress difference and the shear stress at very high shear rates ($\sim 10^6$ sec^{-1}), but systematic data from such instruments for dilute polymer solutions have not yet been published.

Many of the experimental difficulties listed above are avoided with special dilute solutions known as *Boger fluids* (Boger and Binnington 1977; Mackay and Boger 1987). These fluids are composed of low concentrations (typically around 1000 ppm) of a high-molecular-weight flexible polymer dissolved in a very viscous, yet Newtonian, liquid, such as low-molecular-weight polymer, or oligomer. Bianchi and Peterlin (1968) long ago reported rheological studies on such fluids. The high molecular weight of the polymer (usually in the millions of daltons), combined with the high viscosity of the solvent (typically tens or hundreds of poises), boosts the polymer relaxation time into the range of around 1 sec or so. With such long relaxation times, strongly nonlinear effects are brought into the range of shear rates of conventional torsional-flow rheometers. Boger fluids are also used as model fluids to investigate complex viscoelastic flows, such as those produced in two- or three-dimensional geometries, or to study viscoelastic flow instabilities (Nguyên and Boger 1979; Sridhar et al. 1986; Muller et al. 1993; Byars et al. 1994). As shown in the next two sections, the rheological properties of Boger fluids can be approximated to some extent by the elastic dumbbell model; hence, numerical simulations using the Oldroyd-B constitutive equation (which describes solutions of Hookean dumbbells) can usefully be compared to experimental flows of Boger fluids.

3.6.2.1.1 SHEARING FLOW. Figure 3-14 shows the shear-rate-dependent viscosities of polystyrenes of various molecular weights in a couple of low-viscosity solvents, decalin and toluene (Noda et al. 1968; see also Kotaka et al. 1966). Plotted is the intrinsic *relative* viscosity, $[\eta]/[\eta]_0$, against a dimensionless shear rate,

$$\beta^* \equiv \tau_0\dot\gamma \qquad (3\text{-}53)$$

where $\tau_0 = [\eta]_0 M \eta_s / N_A k_B T$ is a characteristic relaxation time. Since polymeric fluids possess a spectrum of relaxation times, different "characteristic" relaxation times can be defined. According to molecular theory, τ_0 is given by the sum over the spectrum of relaxation times, $\tau_0 = \sum \tau_i$. One can also choose simply the longest relaxation time τ_1 as the characteristic time. Using the Zimm spectrum, one finds $\tau_0 = 2.369\tau_1$. A relaxation time which can be defined from rheological data alone is the ratio $\tau_{\text{eff}} \equiv \Psi_{1,0}/2\eta_{p,0}$, where $\eta_{p,0}$ is the polymer contribution to the zero-shear viscosity. According to Zimm theory, $\tau_0 \approx 4\tau_{\text{eff}}$. The shear rate made dimensionless with one of these relaxation times is often called a "Weissenberg number," or sometimes "Deborah number."

Figure 3-14 includes data for polystyrene in both good (toluene) and theta (decalin) solvents. Note that the degree of shear thinning is greater in the good solvent than in the theta solvent. For both types of solvents, however, the degree of shear thinning is modest compared to that of polymer melts, for which η can fall by two or more orders of magnitude as the shear rate increases [see Fig. 1-9 (Laun 1978)].

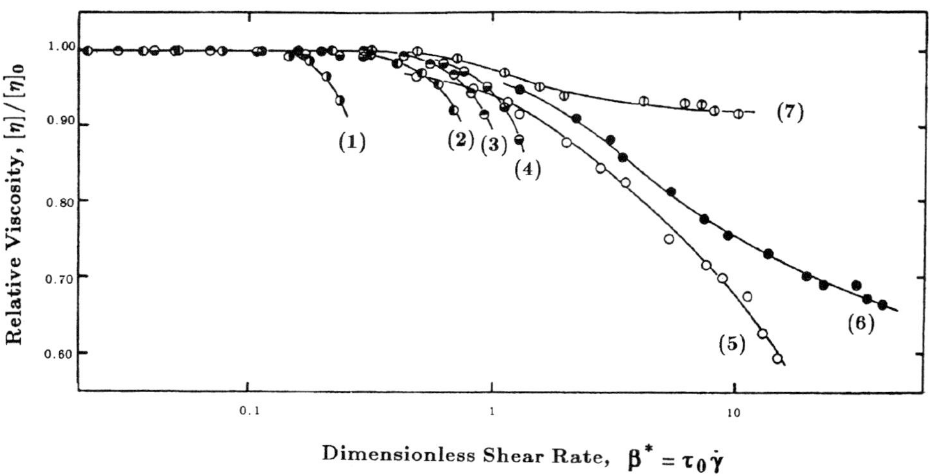

Figure 3.14   Curves of intrinsic relative shear viscosity versus dimensionless shear rate β^* for dilute solutions of poly(α-methylstyrene) with molecular weights of (1) 690,000, (2) 1,240,000, (3) 1,460,000, (4) 1,820,000, (5) 7,500,000, and (6) 13,600,000 in toluene (a good solvent), and (7) 13,600,000 in decalin (a theta solvent). (Reprinted with permission from Noda et al., Journal of Physical Chemistry 72:2890 Copyright 1968, American Chemical Society.)

Figure 3-15 shows linear and nonlinear viscoelastic data for a Boger fluid consisting of a dilute solution of polystyrene, described in the caption. The shear-rate-dependent first normal stress coefficient $\Psi_1(\dot{\gamma})$ is shear thinning at high shear rates, consistent with theoretical predictions of a FENE-type model described in Section 3.6.2.2.1. Note that the shear thinning in Ψ_1 begins at a shear rate almost two decades higher than that at which shear thinning begins in the linear viscoelastic curve $2\eta''/\omega$. Thus, Ψ_1 remains roughly constant even at values of N_1/σ_p as high as 20, where $N_1 \equiv \dot{\gamma}^2\Psi_1$, and $\sigma_p = \sigma - \eta_s\dot{\gamma} \approx 0.2\eta_0\dot{\gamma}$ is the polymer contribution to the shear stress. For entangled solutions or melts, shear thinning in Ψ_1 usually occurs when N_1/σ is near unity (Laun 1978). Since the ratio N_1/σ_p is a rough measure of the degree of stretch or aspect ratio of the macromolecule in the flow field, it is apparent that in some Boger fluids the coils can become highly extended (stretched by a factor of 20) before significant shear thinning occurs.

Figure 3-16 shows Ψ_1 versus dimensionless shear rate for a 1000 ppm solution of a high-molecular-weight polystyrene in a viscous two-component solvent. This solution, like the low-viscosity dilute solutions described in Fig. 3-14 and like the Boger fluid described in Fig. 3-15, is shear thinning at high shear rates. Note that at shear rates just below the high-shear-rate shear-thinning regime, there is a very weak *shear thickening* regime, so that Ψ_1 has a weak *maximum*. A slight upturn in Ψ_1 has also been reported by Mackay and Boger (1987) for a different Boger fluid, namely a high-molecular-weight polyisobutylene in polybutene.

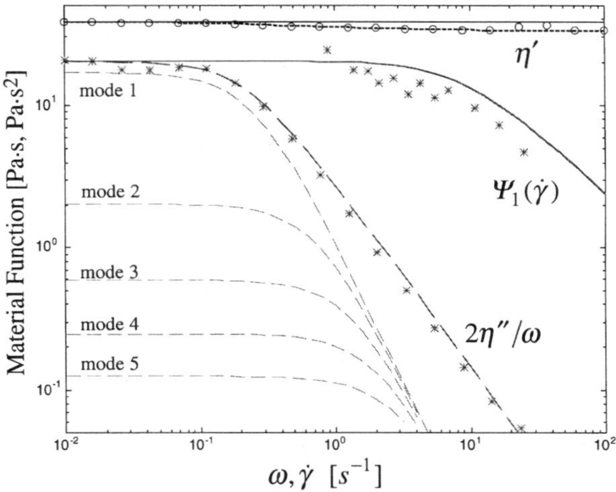

Figure 3.15 The frequency-dependent in-phase and out-of-phase components of the dynamic viscosity, η' and η'', in small-amplitude oscillatory shear, along with the shear-rate dependence of the first normal stress coefficient $\Psi_1(\dot{\gamma})$ for a 0.05 wt% solution of polystyrene of molecular weight 2.25×10^6 in a solvent of oligomeric styrene. The lines through the data show the predictions of the Zimm theory for η' and $2\eta''/\omega$ and the Zimm theory for $\Psi_1(\dot{\gamma})$ modified to account for finite extensibility, as discussed in Section 3.6.2.2.1. The dashed lines are the contributions of the individual Zimm relaxation modes to $2\eta''(\omega)/\omega$. (From McKinley 1996, private communication, with permission.)

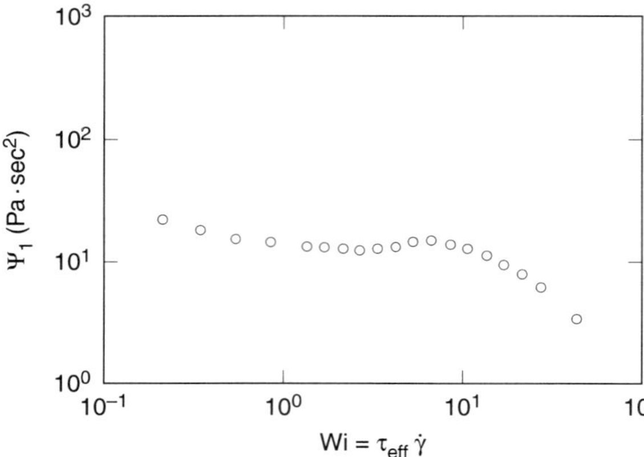

Figure 3.16 First normal stress coefficient Ψ_1 versus Weissenberg number, $\mathrm{Wi} \equiv \dot\gamma\,\tau_{\mathrm{eff}}$, for polystyrene ($M = 2 \times 10^7$) at a concentration of 1000 ppm in a binary solvent consisting of 28 wt% low-molecular-weight (48,000) polystyrene in dioctylphthalate. Here $\tau_{\mathrm{eff}} \equiv \Psi_{1,\mathrm{min}}/2\eta_{p,0}$; $\Psi_{1,\mathrm{min}}$ is Ψ_1 at its local minimum and $\eta_{p,0}$ is the polymer contribution to the zero-shear viscosity. (From Magda et al., 1993, with permission.)

Several causes of shear thinning in dilute solutions have been proposed. These include *finite extensibility*, nonequilibrium *hydrodynamic interaction*, *excluded volume*, and *internal viscosity* (Larson 1988). The most important of these is the finite extensibility of polymer molecules (Bird et al. 1987b); it is discussed in Section 3.6.2.2.1. For a review of the other causes of shear thinning, see Larson (1988).

3.6.2.1.2 EXTENSIONAL FLOW. In steady extensional flows, such as uniaxial extension, the single-relaxation-time Hookean dumbbell model and the multiple-relaxation-time Rouse and Zimm models predict that the steady-state extensional viscosity becomes infinite at a finite strain rate, $\dot\varepsilon$. With the dumbbell model, this occurs when the frictional drag force that stretches the dumbbell exceeds the contraction-producing force of the spring—that is, when the extension rate equals the critical value $\dot\varepsilon_c$:

$$\dot\varepsilon_c = \frac{1}{2\tau} \tag{3-54}$$

where τ is the relaxation time. For $\tau\dot\varepsilon < 1/2$, the spring force dominates and the dumbbell remains in a modestly deformed coiled state. For $\tau\dot\varepsilon > 1/2$, the drag force is more important, and the molecule undergoes a *coil-stretch transition* and becomes highly extended. Birefringence measurements on dilute solutions in extensional flows are consistent with this basic prediction (Keller and Odell 1985; Fuller and Leal 1980). Figure 3-2 shows the Trouton ratio of extensional viscosity $\bar\eta$ to zero-shear viscosity measured by stretching a filament of Boger fluid such that its length increases exponentially with time. Note in Fig. 3-2 that the steep rise in $\bar\eta$ is cut off at some large, but finite, value. This limiting value of $\bar\eta$ is apparently set by the finite extensibility of the polymer molecules (see Section 3.6.2.2.1).

In extensional flows at high Deborah numbers, $\dot\varepsilon\tau_1 > 1$, long polymer molecules can fracture (Keller and Odell 1985; Rabin 1987; Kausch 1985). If the strain is large, the breakage usually is near the middle of the molecule, indicating that fracture occurs after the molecule has undergone a coil-stretch transition (Keller and Odell 1985). For polystyrene of molecular weight 20 million in a low-viscosity solvent (dekalin), fracture occurs at extension rates above about 1000 sec^{-1}. In the "crossed-slot" device, the critical extension rate for fracture, $\dot{\varepsilon}_f$, scales with molecular weight as $\dot{\varepsilon}_f \propto M^{-2}$, which is a "Rouse"-like scaling that is stronger than the "Zimm"-like scaling for the coil-stretch transition, $\dot{\varepsilon}_c \propto M^{-1.5}$. This indicates that hydrodynamic drag is enhanced when the molecule is in the stretched state, as discussed shortly. Polymer chain fracture is an important practical problem in flows of dilute polymer solutions through syringes, in porous media, and in turbulent flows (Horn and Merrill 1984).

3.6.2.1.3 MIXED FLOW. Other flows with extensional components also have coil-stretch transitions. The smaller the extensional component is relative to the overall strain rate, the higher the overall strain rate at which the transition takes place (Giesekus 1962, 1966) A steady planar flow, for example, can be considered to be a mixture of a shearing and an extensional flow; in such a mixed flow, the velocity gradient tensor, $\nabla\mathbf{v}$, can be expressed as (Fuller and Leal 1980, 1981)

$$\nabla\mathbf{v} = \frac{1}{2}G \begin{pmatrix} 1+\alpha & 1-\alpha \\ -(1-\alpha) & -(1+\alpha) \end{pmatrix} \tag{3-55}$$

Here α is a parameter that determines the flow type; $\alpha = 1$ corresponds to planar extension (a strong flow), and $\alpha = 0$ corresponds to a simple shearing flow. Flows with intermediate values of α can be generated in a four-roll-mill device, as discussed in Section 1.3.3 and in Fuller and Leal (1980). It is found that the birefringence of a dilute solution of flexible polymers of a given molecular weight measured in such a device is a nearly universal function of $G\sqrt{\alpha}$ (see Fig. 3-17). $G\sqrt{\alpha}$ is the eigenvalue of the velocity gradient tensor and is equivalent to the *extensional component* of the flow. This implies that in a mixed flow, the

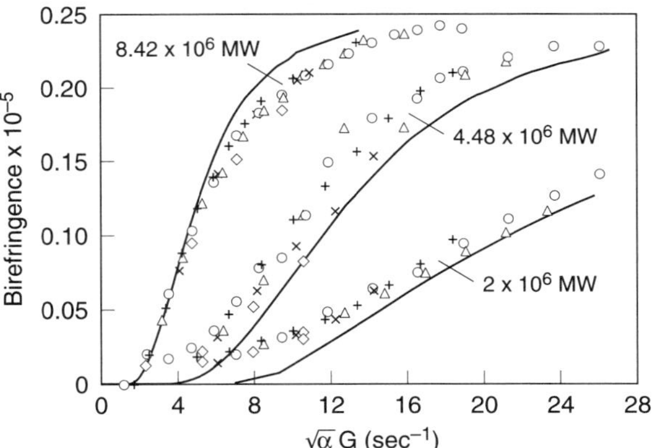

Figure 3.17   Birefringence as a function of the eigenvalue of the velocity gradient tensor, $\sqrt{\alpha}G$, for planar flows generated in a "four-roll mill," for dilute solutions of polystyrenes of three different molecular weights in polychlorinated biphenyl solvent. Here G is the strain rate and α the "flow type" parameter. For planar extension, $\alpha = 1$ and $G = \dot{\varepsilon}$ is the extension rate; for simple shear, $\alpha = 0$ and $G = \dot{\gamma}$ is the shear rate. The different symbols correspond to α values of 1.0 ($\bigcirc$), 0.8 ($\triangle$), 0.5 (+), and 0.25 (diamonds). The curves are theoretical predictions from the FENE dumbbell model, including conformation-dependent drag (discussed in Section 3.6.2.2.2). (From Fuller and Leal 1980, reprinted with permission from Steinkopff Publishers.)

stretching of the macromolecules is dominated by the extensional component of the flow; the influence of the shearing component is negligible unless the extensional component is almost entirely absent.

3.6.2.2 Simple Theories

3.6.2.2.1 FINITE EXTENSIBILITY. The rheological data for dilute solutions described above (in Section 3.6.2.1) indicate that the main limitation of the Hookean dumbbell (or the Oldroyd-B) model is that it assumes that the polymer molecules are infinitely extensible, which is objectionable for large molecular extensions. Fortunately, this defect of the elastic dumbbell model can be corrected simply by making the relation between the spring force, $\mathbf{F}^s$, and the molecular extension, $\mathbf{R}$, nonlinear, such that the force becomes very large as the molecular extension approaches the fully extended length L of the molecule. For example, for a *freely jointed chain* with many bonds (see Section 2.2.3.2), the force–extension relationship is an *inverse Langevin function*. The Langevin function is defined in Eq. (8-28); its inverse is not an analytic form, but it can be roughly approximated by the analytic *Warner spring law* (Warner 1972), also known as the "FENE" (finitely extensible, nonlinear-elastic) spring:

$$\mathbf{F}^s = \frac{2\beta^2 k_B T}{1 - (R/L)^2}\mathbf{R} \equiv H(R^2)\,\mathbf{R} \tag{3-56}$$

where $H \equiv 2\beta^2 k_B T/(1 - (R/L)^2)$ is the nonlinear spring coefficient. Also, $\beta^2 \equiv 3/(2N_K b_K^2)$, and $L = N_K b_K$. N_K is the number of links in the freely jointed chain, and b_K is the length of each link. Figure 3-18 compares the force law for the Warner spring with those for the inverse Langevin function and the linear Hookean spring. Significant departures ($\sim$10%) from linear behavior occur when the chain reaches a third of its full extension, and these deviations become large ($> 30\%$) when it exceeds half of its full extension. The inverse Langevin function and the Warner spring law are appropriate for most synthetic polymers whose flexibility can be approximated by the freely jointed chain model (see Section 2.2.3.2).

The freely jointed chain model is most appropriate for synthetic polymers, such as polyethylene and polystyrene. For other molecules, such as DNA and polypeptides, the molecular flexibility is better described by the *worm-like chain model* (described in Section 2.2.4), whose force law can be approximated by a simple expression due to Marko and Siggia (1995), namely,

$$\frac{F^s \lambda_p}{k_B T} = \frac{1}{4}\left(1 - \frac{R}{L}\right)^{-2} - \frac{1}{4} + \frac{R}{L} \tag{3-57}$$

where L is the contour length of the molecule, and λ_p is the persistence length (see Section 2.2.4). This force law has been confirmed by direct measurements on a single DNA molecule attached at one end to a surface and attached at the other end to a magnetic bead and then stretched under magnetic and hydrodynamic fields under a microscope (Smith et al. 1992). [The vector force $\mathbf{F}^s$ can be obtained from the scalar force F^s in Eq. (3-57) by multiplying it by a unit vector parallel to the molecule's end-to-end vector.] At high extensions, the Marko–Siggia expression approaches a limit somewhat similar to that of the Warner spring: $\mathbf{F}^s \to (k_B T/4\lambda_p L)(1 - R/L)^{-2}$.

When a nonlinear spring law is used in the dumbbell model, the Smoluchowski equation (3-28) is changed to

$$\frac{\partial}{\partial t}\psi = -\frac{\partial}{\partial \mathbf{R}} \cdot \left[\mathbf{R} \cdot \nabla \mathbf{v}\psi - \frac{2k_B T}{\zeta}\left(\frac{1}{k_B T}H(R^2)\mathbf{R}\psi + \frac{\partial \psi}{\partial \mathbf{R}} \right) \right] \tag{3-58}$$

Because $H(R^2)$ is not a constant, an analytic expression for the stress tensor cannot be obtained from Eq. (3-58) unless an approximation is made. A common approximation is to replace $H(R^2)$ by the *preaveraged* quantity, $H(\langle R^2 \rangle)$ (Peterlin 1961; Tanner 1985). In this approximation, H is taken to be independent of R^2, but dependent on the average of R^2. This means that Eq. (3-58) becomes a linear equation in $\mathbf{R}$ and thus can be solved rather easily. However, preaveraging can be dangerous if there are large fluctuations about the mean value of the preaveraged quantity. For steady-state flows, preaveraging seems to be an acceptable approximation (Tanner 1985). However, in some transient flows, the errors can be significant (Keunings 1997; Doyle et al. 1998). After making the preaveraging approximation, Eq. (3-58) can be multiplied by $\mathbf{RR}$ and integrated over configuration space in the usual way, yielding

$$\overset{\triangledown}{\mathbf{S}} + \frac{1}{\tau}\left(\frac{\mathbf{S}}{1 - (\mathrm{tr}\mathbf{S})/2L^2} - \beta^{-2}\boldsymbol{\delta} \right) = \mathbf{0} \tag{3-59a}$$

where $\mathbf{S} \equiv 2\langle \mathbf{RR} \rangle$, and $\tau = \zeta/8k_B T\beta^2$. Because the spring force law is not linear, the stress tensor is not proportional to $\langle \mathbf{RR} \rangle$, but from Eq. (3-11) is given by

$$\boldsymbol{\sigma} = \nu k_B T\beta^2 \left(1 - \frac{\mathrm{tr}\mathbf{S}}{2L^2} \right)^{-1} \mathbf{S} \tag{3-59b}$$

Since the nonlinearity in the spring law shows up mostly at high molecular extension, the predictions of the FENE dumbbell model are changed from those of a Hookean dumbbell at high shear rates in shearing flows, and in extensional flows when the extension rate $\dot{\varepsilon}$ exceeds the critical value $\dot{\varepsilon}_c$ for a coil-stretch transition. In shearing flow at high shear rates, the preaveraged FENE, or "FENE-P" dumbbell model, gives

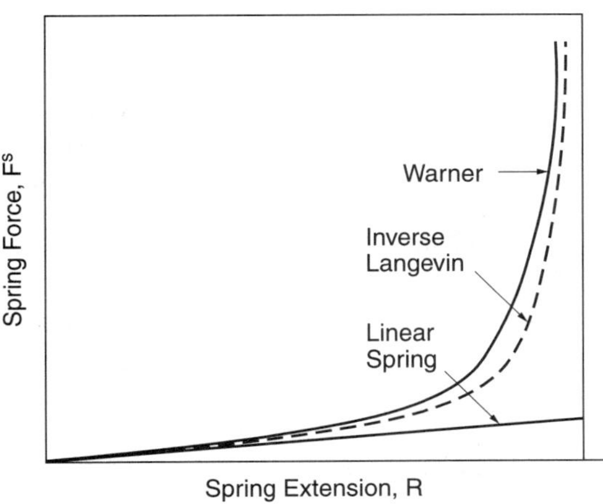

Figure 3.18 Elastic spring force versus molecular extension for the Warner spring, for the freely jointed chain (which is described by the inverse Langevin function), and for the linear spring. (From Tanner, copyright © 1985 by Oxford University Press, Inc. Used by permission of Oxford University Press, Inc.)

$$\frac{\eta_\infty^p}{\eta_0} = (\beta L)^{2/3}(\tau\dot\gamma)^{-2/3}, \qquad \frac{\Psi_{1,\infty}}{\Psi_{1,0}} = (\beta L)^{4/3}(\tau\dot\gamma)^{-4/3} \tag{3-60}$$

For this model, the second normal stress difference is zero at all shear rates. For the freely jointed chain, to which the FENE or FENE-P spring is an approximation, the polymer contribution to the shear viscosity at high shear rates is proportional to $\dot\gamma^{-1/2}$, rather than $\dot\gamma^{-2/3}$ (Doyle et al. 1997).

For an extensional flow, the polymer contribution to the normal stress difference at high $\dot\varepsilon$ is

$$\Delta\sigma_p \equiv \sigma_{11,p} - \sigma_{22,p} \sim \nu L F^s(R \rightarrow L) \approx \frac{1}{2}\nu L^2 \zeta\dot\varepsilon$$

because the spring constant approaches $H \rightarrow \dot\varepsilon\zeta/2$ at high $\dot\varepsilon$ for any finitely extensible spring. The friction coefficient is related to the relaxation time τ by $\zeta = 4H_0\tau$, where $H_0 \equiv 2k_BT\beta^2$ is the spring constant at small extensions. The polymer contribution to the elongational viscosity, $\Delta\sigma_p/\dot\varepsilon$, at high $\dot\varepsilon$ therefore approaches a constant:

$$\bar\eta_p \rightarrow \bar\eta_{p,\infty} = \frac{1}{2}\nu L^2\zeta = 2\nu L^2 H_0\tau = 2\nu k_BTB\tau = 2B\eta_{p,0} \tag{3-61}$$

where B is defined as $B \equiv 2\beta^2 L^2$, and $\eta_{p,0}$ is the polymer contribution to the zero-shear viscosity. Since $\beta^2 = 3/(2\langle R^2\rangle_0)$, the parameter B is three times the square of the ratio of the polymer's fully extended length to its root-mean-square end-to-end separation; that is, $B = 3L^2/\langle R^2\rangle_0$. The same asymptotic result is obtained with the Marko–Siggia force law. The fully extended length of a synthetic organic flexible polymer can be estimated as $L \approx 0.82n\ell$, where n is the number of carbon–carbon bonds in the backbone, and $\ell = 1.54\text{Å}$ is the length of a backbone bond (Flory 1969). The equilibrium mean-square end-to-end separation $\langle R^2\rangle_0$ in a theta solvent is given by $C_\infty n\ell^2$. For good solvents, $\langle R^2\rangle_0$

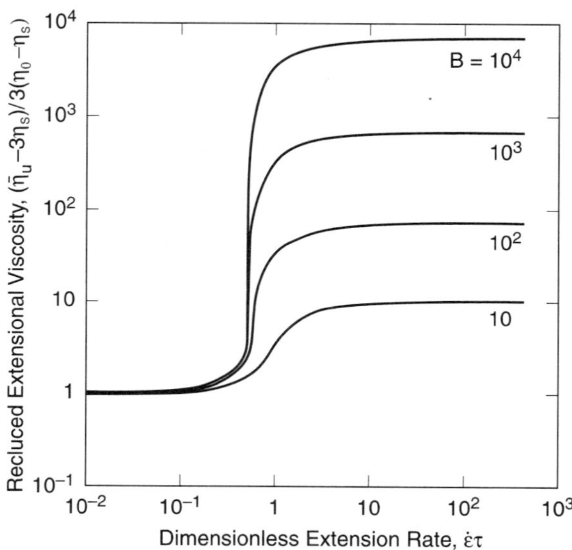

Figure 3.19 The polymer contribution to the steady-state uniaxial extensional viscosity $\bar\eta_u$ divided by the polymer contribution to the zero-shear viscosity $\eta_{p,0} = \eta_0 - \eta_s$ for the dumbbell model with a nonlinear "FENE" spring and various values of $B \equiv 2\beta L^2$. (From Bird et al. *Dynamics of Polymeric Liquids, Vol. 2*, Copyright © 1987. Reprinted by permission of John Wiley & Sons, Inc.)

is larger than this, and thus for a given molecular weight, $B \equiv 3L^2/\langle R^2 \rangle_0$ is smaller and the polymer is less "extensible." The reverse is true in solvents poorer than theta. Figure 3-19 shows the dependence of the steady-state uniaxial extensional viscosity $\overline{\eta}_u$ on $\dot{\varepsilon}$ for the Warner dumbbell for various values of B.

Comparisons of the predictions of the FENE dumbbell model with measurements of the extensional viscosity of "dilute" solutions have been fairly encouraging. Figure 3-2 compares the Trouton ratio predicted by a multimode FENE dumbbell model with experimental data for a Boger fluid. Good agreement is obtained if one uses a value of the parameter $B = 3L^2/\langle R^2 \rangle_0 = 40,000$, about a factor of two higher than the estimate from molecular parameters, $B = 20,000$. The discrepancy between the best-fit value of B and the estimate from molecular parameters might be caused by "conformation-dependent drag," discussed in Section 3.6.2.2.2 below.

Direct observations have recently been made of long ($\sim 20\,\mu$m) DNA molecules undergoing planar extensional flow in a cross-slot apparatus (Perkins et al. 1997). At steady state the degree of extension of the molecules as a function of extension rate agrees with the dumbbell model with a nonlinear worm-like chain spring law. When a molecule is suddenly exposed to an extensional flow with rate modestly above the critical rate for a "coil-stretch" transition, it often first forms a taut section in its middle; the extended portion grows by drawing portions of the chain out of the still-coiled portions attached at each end. This picture of chain unraveling is qualitatively similar to the so-called "yo-yo" model proposed by Ryskin (1987). However, other modes of stretching also occur, including the flattening of the molecule into a "folded" state (Acierno et al. 1974) that is much slower to unravel than the "yo-yo." Folds are predicted to become prevalent at high Deborah number (De $\gtrsim$ 10) (Larson 1990, 1998; Hinch 1994).

3.6.2.2.2 CONFORMATION-DEPENDENT DRAG COEFFICIENT. In the expressions (3-60) and (3-61) for the shear or extensional viscosities at high shear or extension rates, the relaxation time τ was taken to be independent of the average molecular conformation. The success of the Zimm model (Section 3.6.1.2) shows that the polymer relaxations in dilute solutions are dominated by hydrodynamic interactions between different segments on the chain, and one expects these interactions to change when the chain is extended in the flow. Thus, one might expect the effective drag coefficient, and hence the relaxation time, to increase significantly when the molecule is stretched out in a flow field. A simple model for such a change in hydrodynamic drag and relaxation is the *cylinder model* proposed by de Gennes (1974) and Hinch (1974). This model is really just an extension of the "hard sphere" model for the drag on a coiled-up polymer molecule. Recall from Section 3.6.1.1 that the hydrodynamic drag coefficient ζ_{coil} for drag on an unextended polymer coil is proportional to its radius of gyration, which, in turn, scales with molecular length L as L^ν, with $\nu = 0.5$ in a theta solvent. When the molecule is fully extended, the translational drag coefficient ζ_{rod} should be similar to that of a thin cylindrical rod of length L (Doi and Edwards 1986):

$$\zeta_{\text{rod}} = \frac{2\pi L \eta_s}{\ln(L/d)} = \frac{6.28 L \eta_s}{\ln(L/d)} \tag{3-62}$$

where d is the molecular diameter.

Since $\zeta_{\text{coil}} \propto L^\nu$ and $\zeta_{\text{rod}} \propto L$, the drag coefficient in the stretched state is very much larger than that in the coiled state *in the limit of high molecular weight*. This latter qualification is a very important one. Note, in particular, the logarithmic factor ln (L/d) dividing the right side of Eq. (3-62). From Eqns. (3-48) and (3-62), $\zeta_{rod}/\zeta_{coil} \approx \sqrt{n/C_\infty}/\ln(L/d)$ for a theta solvent, where n is the number of carbon bonds in the backbone. Because the molecular diameter is so small ($\sim$5–10 Å) compared to the length L, for a typical polymer such as polystyrene with molecular weight ranging from 10^5 to 10^7, L/d ranges from 250 to 25,,000, and thus $\ln(L/d) \approx 6$–10. Hence, ζ_{rod} is only a factor of about 2–14 times larger than ζ_{coil}, except for extraordinarily long molecules ($M > 10^7$). In agreement with this estimate, Doyle et al. (1998) find that the best fits of dumbbell models with conformation-dependent drag to extensional flow data for high-molecular-weight ($M \approx 2 \times 10^6$) polystyrene solutions are obtained when $\zeta_{rod}/\zeta_{coil} \approx 8$. Also, direct microscopic observations of the stretching of tethered, fluorescing DNA molecules in uniform flows give results consistent with the theoretical estimate, ζ_{rod}/ζ_{coil} around 2–3, for DNA aspect ratios L/d of 20,000–75,000 (Larson et al. 1997).

More rigorous treatments of conformation-dependent drag support these inferences. For the multiple beads-and-springs model, detailed theories of conformation-dependent hydrodynamic interactions in shear were developed by Fixman (1966), Öttinger (1985, 1986, 1987), Magda et al. (1988), Kishbaugh and McHugh (1990), and others. In these analyses, the nonequilibrium, flow-distorted distribution function is used to preaverage the Oseen tensors, and these preaveraged Oseen tensors, in turn, are used in the computation of the drag on each bead. The distribution function and the average Oseen tensors are determined *self-consistently*.

Figure 3-20 shows the first normal stress coefficient predicted by Kishbaugh and Mchugh for flexible polymer chains of various lengths. For $B \equiv 2\beta^2 L^2 = \infty$—that is, for an infinitely extensible, Hookean, molecule—the theory predicts *shear thickening* at high shear rates, because of the weakening of hydrodynamic interaction (and hence the increase in the effective drag coefficient) that occurs when the chain is greatly extended. Note, however, that this shear thickening occurs only at rather high dimensionless shear rates. Also, for chains of finite length, shear thickening is terminated at the highest shear rates by the onset of shear thinning produced by finite extensibility. Unless the macromolecule is unusually high in molecular weight, such that $B \gtrsim 3 \times 10^4$, the shear thinning produced by finite extensibility occurs at a shear rate low enough to suppress entirely the shear-thickening phenomenon. For polystyrene in a theta solvent, for example, $B = 3 \times 10^4$ corresponds to a molecular weight of around 10^7! Hence, dilute solutions of polystyrene will show shear thickening only for very high molecular weights. Consistent with this, the curve of Ψ_1 versus $\dot{\gamma}$ for a dilute polystyrene solution with $M = 2 \times 10^7$ shown in Fig. 3-16 has a weak shear thickening regime over the range of shear rates expected from the theoretical predictions in Fig. 3-20. (The predicted shear-thickening effect in Fig. 3-20 is highly exaggerated because of the expanded scale of the ordinate.) For polystyrene solutions of molecular weight lower than $\sim 10^7$, such as that for Fig. 3-15, this source of shear thickening can be neglected altogether.

Both the experimental shear-viscosity curve (Fig. 3-16) and the theoretical one (Fig. 3-20) also show a region of weak *shear thinning* at low shear rates. This rather minor phenomenon occurs, according to the theory, because the shear-induced changes in hydrodynamic drag that occur with increasing chain deformation are nonmonotonic. Hydrodynamic interaction can account for both shear thinning at low β^* and shear thickening at high β^* because the hydrodynamic interactions of chain segments that are close to each other are

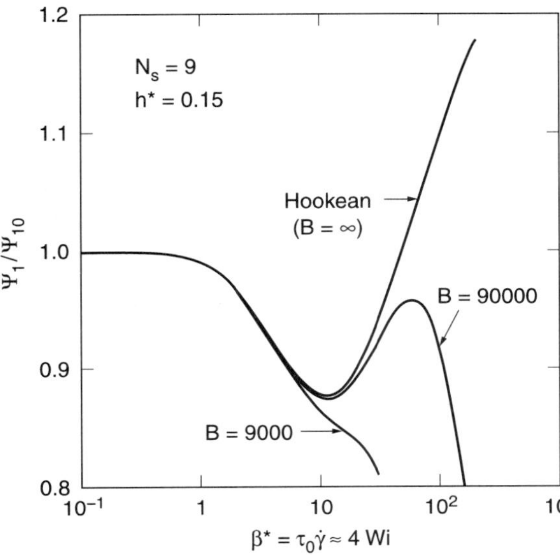

Figure 3.20 Dependence of the first normal stress coefficient on dimensionless shear rate β^*, as predicted by the bead-spring theory with conformation-dependent hydrodynamic interaction and finite extensibility with $N_s = 9$ springs, for various values of B. Here $\text{Wi} \equiv \dot{\gamma}\Psi_{1,0}/2\eta_{p,0}$. (Adapted from J. Non-Newt. Fluid Mech., 34:181, Kishbaugh and McHugh, (1990), with kind permission from Elsevier Science - NL, Sara Burgerhartstraat 25, 1055 KV Amsterdam, The Netherlands.)

qualitatively different from the interactions of segments that are on opposite ends of the chain (Larson 1988). The nonmonotonic effect of hydrodynamic interaction was first pointed out by Peterlin (1960).

In extensional flow, deformation-induced increases in hydrodynamic drag should steepen the coil-stretch transition and, if strong enough, produce hysteresis (de Gennes 1974). These effects can be captured qualitatively by the cylinder model described above. As in the case of shear, these effects are expected to be large in extensional flows only for molecules of high molecular weight (Larson et al. 1997). The interested reader can find further discussion of these issues in Larson (1988) and in articles by de Gennes (1974) and Hinch (1974).

The success of bead-spring models in describing the deformation of flexible polymer molecules is illustrated by some recent comparisons of the predictions of such models against very detailed data for DNA molecules in constant-velocity flows. Figure 3-21 shows the measured and predicted density of DNA mass as a function of position downstream of the point at which the chain is tethered to a small sphere held fixed in position by a laser optical trap (Perkins et al. 1995; Larson et al. 1997). The predictions were obtained without adjustable parameters by using the worm-like chain expression, Eq. (3-57), for the molecular elasticity, the low-shear-rate drag coefficient from diffusivity measurements, Eq. (3-48), and the high-shear-rate drag coefficient from Eq. (3-62). The excellent agreement between theory and experiment for the constant-velocity flow and in planar extensional flows (Larson 1998) indicates that the physics of macromolecular deformation in simple flow fields is well-described by the combination of Brownian motion, a nonlinear elastic spring law, and a weak dependence of the viscous drag coefficient on molecular extension. This inference is supported by recent comparisons of model predictions to extensional flow data for Boger fluids (Doyle et al. 1998). One experimental issue that is as yet unresolved, however, is the failure of light-scattering experiments to show much stretching of polymer

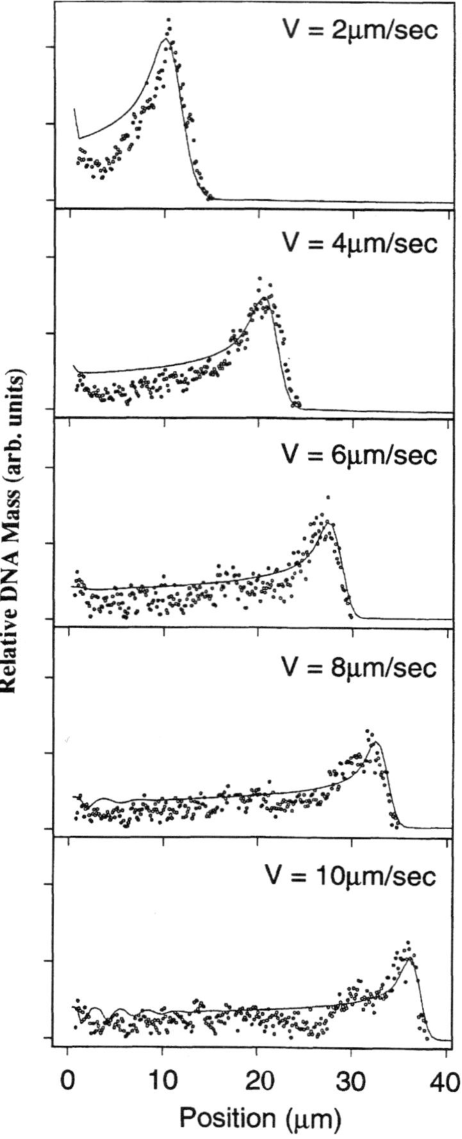

Figure 3.21 Distribution of bead mass as a function of position downstream of the tether point of a DNA molecule of length $L = 67.2 \, \mu$m for various velocities measured in experiments similar to those described in the caption to Fig. 3-1. The lines are the predictions of Monte Carlo molecular simulations using the elastic force from the "worm-like chain" model, Eq. (3-57), and conformation-dependent drag, as described in the text. The value of the parameter $\zeta_{\text{coil}}/k_B T = 4.8 \sec(\mu\text{m})^{-2}$ is obtained from the diffusivity measurements of Smith et al. (1995); $\zeta_{\text{rod}}/k_B T = 9.1 \sec(\mu\text{m})^{-2}$ is obtained from Eq. (3-62) for a fully stretched filament. (From Larson et al. 1997, reprinted with permission from the American Physical Society.)

chains, either in shearing flow (Cottrell et al. 1969; Link and Springer 1993; Lee et al. 1997) or even in strong extensional flows (Armstrong et al. 1980; Menasveta and Hoagland 1991).

• Problem 3.6 exercises your ability to compute nonlinear rheological properties of dilute solutions.

3.7 THE RHEOLOGY OF ENTANGLED POLYMERS

In nondilute polymer solutions and melts, the polymer coils interpenetrate each other enough that the molecular motions of one chain are greatly slowed by the interfering effects of other chains. These interferences are attributed to intermolecular *entanglements*.

Despite the complications produced by entanglements, melts and concentrated solutions are free of a couple of complications that exist in dilute solutions. First, in the melt, flexible polymer chains are "ideal"—that is, their configuration distribution is Gaussian. This is because the excluded-volume effect present when the chain is immersed in small-molecule solvent is *screened* by the surrounding chains (Flory 1953; de Gennes 1979). A second complication that appears in dilute solutions but not in melts is *hydrodynamic interaction*. In the melt, experiments show that hydrodynamic interaction is also screened out (Ferry 1980), so that the drag on one part of the chain does not influence the drag on a remote part of the same chain. The great complicating feature of melts is that the motion of each chain is affected by entanglements with the surrounding chains.

The effect of entanglements on the relaxation of polymer chains is illustrated in Fig. 3-22, which shows the storage modulus G' for a series of polystyrene melts of differing

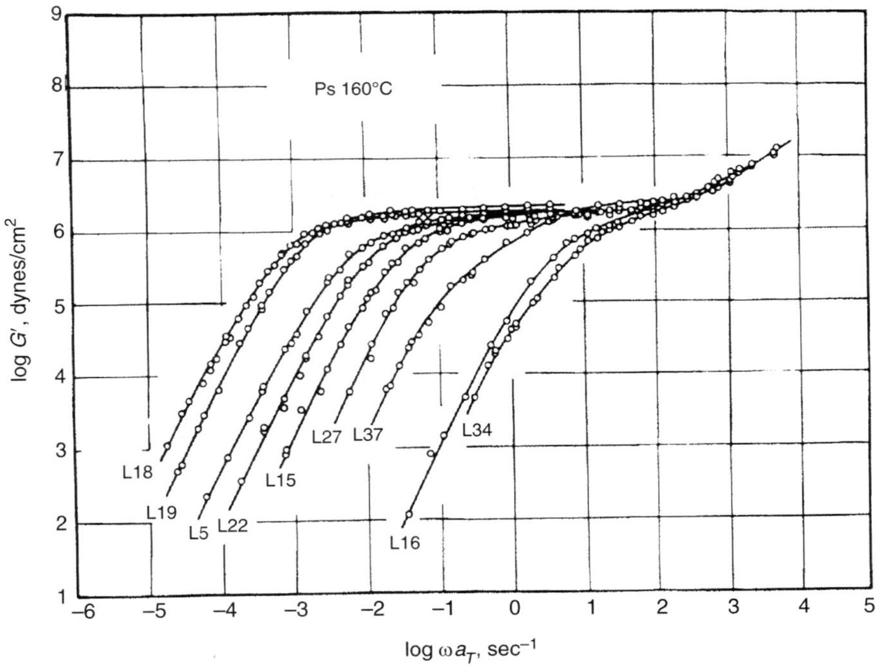

Figure 3.22   Storage modulus, G', as a function of frequency reduced to 160°C for nearly monodisperse polystyrenes of molecular weight ranging from 580,000 to 47,000, from left to right. (Reprinted with permission from Onogi et al., Macromolecules 3:109. Copyright 1970, American Chemical Society.)

molecular weight (Onogi et al. 1970). Notice that there is a *plateau* in G' whose width grows as the molecular weight increases. The plateau value of G' is the *entanglement plateau modulus*, G_N^0. The molecular weight at which the plateau first appears corresponds roughly to M_c, the molecular weight at which the zero-shear viscosity begins to rise as $M^{3.4}$ (see Fig. 3-3). The terminal, or longest, relaxation time τ_1 of these melts increases with molecular weight with the same power law, $\tau_1 \propto M^{3.4}$. In the frequency range of the plateau, the melt acts like a cross-linked elastic rubber, because G' is nearly constant. The plateau modulus G_N^0 can be related to the density of entanglements, or the density of effective cross-links, ν, by $G_N^0 = (4/5)\nu k_B T$. (The reason for the factor of 4/5 will become apparent in Section 3.7.4.3.) From ν and the bulk density ρ of the polymer, one can obtain the *molecular weight between entanglements, M_e*:

$$M_e = \frac{\rho N_A}{\nu} = \frac{4}{5} \frac{\rho N_A k_B T}{G_N^0} \tag{3-63}$$

It turns out that M_e is about a fifth to a half as large as M_c. The number of monomers between entanglements is denoted $N_e = M_e/M_0$, where M_0 is the molecular weight of a monomer.

The values of M_c and M_e vary from melt to melt; tabulations of these values can be found in Ferry (1980) and Fetters et al. (1994). [Differences in the M_e values between Ferry (1980) and Fetters et al. (1994) are due, in part, to a difference of a factor of 4/5 in the definition of M_e; here we follow the convention of Fetters et al. (1994).] Some of these M_e values are presented in Table 3-3; the chemical structures of many of these polymers can be found in Fig. 2-7. For highly flexible polymers, M_c usually corresponds to about 300–600 atoms in the backbone of the chain (Ferry 1980); for polystyrene, for example, $M_c \approx 38,000$ and $M_e \approx 13,300$. The magnitude of M_e can be estimated as the minimum molecular weight a chain would need so that the volume of space "pervaded" by the chain is twice that occupied by the chain itself, leaving enough room in the pervaded volume for just one other such chain. This argument suggests that "slender" polymer molecules, which pervade space without themselves taking up much volume, should have lower values of M_e than do "bulky" chains. This suggestion proves indeed to be the case; Fetters et al. (1994) have shown that values of M_e can accurately be predicted from the value of p, a "packing length" that is related to the bulkiness of the molecule (see the end of Section 2.2.3.4):

$$p \equiv \frac{M}{\langle R^2 \rangle_0 \rho N_A} \tag{3-64a}$$

Note that p is independent of molecular weight, since for Gaussian chains $M/\langle R^2 \rangle_0$ is a constant. At 140°C, the entanglement molecular weight M_e, plateau modulus G_N^0, and the "tube" diameter a (defined below in Section 3.7.1) are related to p by (Fetters et al. 1994)

$$M_e \approx [225.8 \text{ cm}^3 \text{ Å}^{-3} \text{ mol}^{-1}]p^3\rho \tag{3-64b}$$

$$G_N^0 \approx [12.16 \text{ MPa Å}^3]p^{-3} \tag{3-64c}$$

$$a \approx 19.36p \tag{3-64d}$$

M_e can also be related to K_θ (see Section 3.6.1.1) by

TABLE 3.3
Properties of Polymer Melts

	PE	PS	PDMS	PIB	PMMA	1,4PBd[a]	1,4-PI[b]
Properties at 140°C (from Fetters et al. 1994)							
$\rho\,(\mathrm{g\,cm^{-3}})$	0.784	0.969	0.895	0.849	1.13	0.826	0.830
C_∞	7.3	9.6	6.3	6.8	9.1	5.6	5.0
$p(\text{Å})$	1.6942	3.9480	4.0593	3.4309	3.4572	2.2946	3.2006
$G_N^0\,(\text{MPa})$	2.60	0.20	0.20	0.32	0.31	1.25	0.42
M_e	828	13,309	12,293	7288	10,013	1815	5429
$a(\text{calc.})(\text{Å})$	32.8	76.5	78.6	66.4	67.0	44.4	62.0
WLF temperature shift parameters (from Ferry 1980)							
$T_g\,(\text{K})$	c	373	150	205	381	205	200
$T_0\,(\text{K})$	c	373	303	298	381	263	248
$T_\infty\,(\text{K})$	c	323	81	101	301	149	146
c_1^0	c	12.7	1.90	8.61	34.0	5.97	8.86
$c_2^0\,(\text{K})$	c	50	222	200.4	80	123.2	101.6

PE, polyethylene; PS, polystyrene; PDMS, polydimethylsiloxane; PIB, polyisobutylene; PMMA, (atactic) polymethylmethacrylate; 1,4PBd, 1,4-polybutadiene; 1,4PI, 1,4-polyisoprene.

[a]WLF parameters are for "*cis–trans*–vinyl" polybutadiene from Ferry (1980).

[b]WLF parameters are for "Hevea rubber" from Ferry (1980).

[c]For polyethylene, the glass-transition temperature is far below the crystallization temperature; and time–temperature shifting satisfies an Arrhenius form, with activation energy $E_a = 6.5$ kcal/mol for high-density polyethylene.

$$M_e = \left(\frac{\Phi}{B' K_\theta \rho N_A}\right)^2 \tag{3-64e}$$

with $B' = 0.0516$. The above equations give accurate predictions for chains ranging from polyethylene, with $M_e = 830$, to poly(vinylcyclohexane), with $M_e = 39{,}000$. The coefficients in these equations are somewhat temperature-dependent. The above expressions are similar to those derived by Ronca (1983) and Lin (1987).

3.7.1 Reptation

An explanation for the slowdown of relaxation for melts with $M > M_c$ was given by de Gennes (1971). He considered a simpler problem, that of a single long polymer chain in a cross-linked rubber network, but the results of this analysis can qualitatively be applied to a chain moving in a mesh of other chains (see Fig. 3-23). Since the mesh cannot be crossed, the chain's lateral motion is limited, and the polymer must relax by sliding along its own contour like a snake. De Gennes called this motion *reptation*. The mesh of constraints confines the molecule laterally to a *tube-like* region (Edwards 1967) (see Fig. 3-24). The chain changes its conformation by sliding back and forth along the tube. Those portions of the molecule that escape from the ends of the tube are free to take on random orientations, and the portions of the tube that are vacated are forgotten. By moving back and forth, the chain gradually forgets more and more of the original tube ends; a "new" conformation diffuses from the ends of the chain inward (see Fig. 3-25). Because it is a diffusive process, the time required for the chain to vacate the original tube is proportional to the square of the *contour length* L_t of the tube divided by the diffusion coefficient of the snaking motion. The diffusion coefficient of the snaking motion is proportional to M^{-1}, while the square of the tube's contour length is proportional to M^2. Thus the *reptation time*, the time τ_d for disengagement from the tube, is proportional to $M^2/M^{-1} = M^3$. Hence, the longest relaxation time, $\tau_1 = \tau_d$, is predicted to be proportional to M^3, not too different from the measured scaling law, $\tau_1 \sim M^{3.4 \pm 0.1}$.

The diameter, a, of the tube corresponds to the entanglement spacing, M_e. That is, a strand of polymer having molecular weight M_e spans a random walk end-to-end distance a (Fig. 3-24). Thus, $\langle R^2 \rangle_0 = a^2 M/M_e$, and

$$a^2 = \frac{4}{5} \frac{\rho N_A k_B T \langle R^2 \rangle_0}{M G_N^0}$$

The tube itself is a random walk, each step of which has length a. This random walk is called the "primitive path" of the chain. The contour length of the tube, or the primitive path, is therefore $L_t = aM/M_e$. For polymers of high molecular weight, the tube's contour length is much less than the contour length of the chain (see Fig. 3-24). Thus, the chain meanders about the primitive path. Some values for the tube diameter a for typical polymer melts are presented in Table 3-3.

3.7.2 Nonreptative Relaxation Mechanisms

The reptation theory has been controversial. In large part, this is because experimental data and computer simulations usually show some deviations from the behavior expected for pure

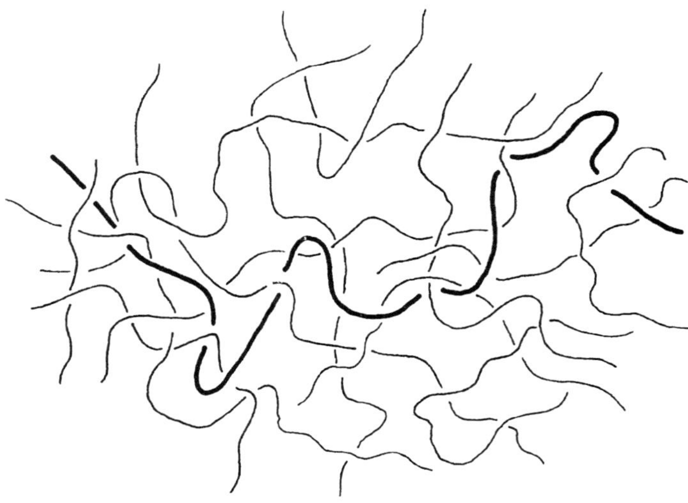

Figure 3.23   A polymer molecule entangled in a mesh of other polymer chains. (From Graessley 1982, reprinted with permission from Springer Verlag.)

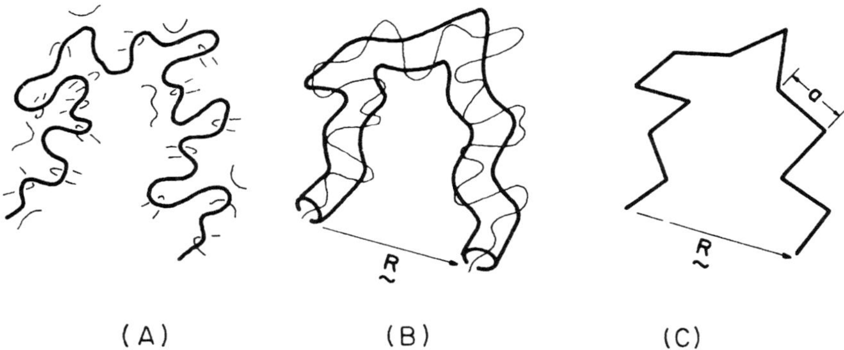

Figure 3.24   (**A**) A polymer molecule is entangled with neighboring molecules that (**B**) confine the given chain to a tube-like region. (**C**) The tube contour is roughly that of a random walk with step size equal to the tube diameter, *a*. This random walk is called the *primitive path*; its contour length is much less than the contour length of the chain itself. (From Graessley 1982, reprinted with permission from Springer Verlag.)

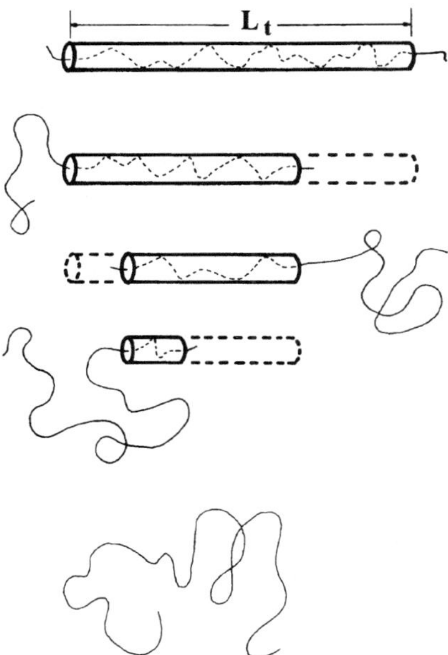

Figure 3.25   Reptation of a polymer molecule out of its tube, where, to aid visualization, the tube of Fig. 3-24 has been "straightened out." (From Graessley 1982, reprinted with permission from Springer Verlag.)

reptation. A prime example is the observed 3.4 power law for the viscosity (Kröger 1995), which differs somewhat from the predicted 3.0 power law (Doi 1983). It has gradually been realized that these deviations are due to relaxation processes other than reptation that are important in all but the most extreme cases of high molecular weight, low polydispersity, and slow flow. Thus, a satisfactory theory for flow and relaxation of entangled polymers needs to account for the nonreptative processes. The most important of these are *primitive-path fluctuations* and *constraint release*. In the highly nonlinear regime of strong flows, one must also account for *tube stretching*, discussed in Section 3.7.5.2.

3.7.2.1 Primitive-Path Fluctuations

This mechanism is most easily described for the case of an entangled polymer chain tethered at one end—for example, to a polymer branch point. In this situation, the entangled chain cannot slide back and forth as a whole, and hence it cannot reptate. de Gennes (1975) and Doi and Kuzuu (1980) proposed that the chain then relaxes by *primitive-path fluctuations*, sometimes called "breathing modes." In this mechanism, fluctuations draw the end of the chain in from the end of the tube (see Fig. 3-26). When the molecule re-expands, the end portion of the tube is forgotten, and the stress associated with the chain ends is lost. For complete relaxation to occur by this mechanism, the free end of the chain must diffuse to the tether point, and from there re-extend into a new tube. This diffusion process is entropically unfavorable, since it involves configurations that are ever more improbable as the size of the fluctuation increases [see Fig. 3-26 (bottom)]. Hence, the time for this process increases *exponentially* with the distance along the tube that the diffusion must occur. Using scaling arguments, Doi and Kuzuu (1980) predicted that the relaxation time $\tau(x)$ for a chain segment a fractional distance x from the tether point to relax is

$$\tau(x) = \tau_0(x) \, \exp\left[\frac{3}{2}\frac{M}{M_e}(1-x)^2\right] \tag{3-65}$$

where τ_0 is a time scale that is in principle dependent on x (Doi and Edwards 1986). This dependence is much weaker than that in the exponential; as a rough estimate, τ_0 is sometimes taken to be a constant, the Rouse relaxation time of the chain. A more accurate formula for $\tau_0(x)$ can be found in Milner and McLeish (1997). [Also, the factor of 3/2 in the exponential is changed to 15/8 if the entanglement spacing M_e is defined in Eq. (3-63) without the factor of 4/5.] According to Eq. (3-65), the segment of the chain at the tether point ($x = 0$) relaxes in a time longer than that of the free end ($x = 1$) by an exponential in the number of entanglements. Since the segments at intermediate positions

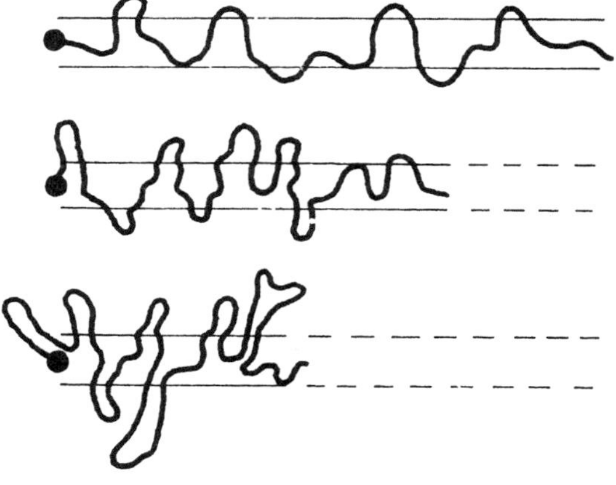

Figure 3.26 A fluctuation of the primitive-path length occurs when a chain randomly pulls its end away from the end of the tube. The probability of such a fluctuation decreases exponentially with the size of the fluctuation. (From Graessley 1982, reprinted with permission from Springer Verlag.)

have intermediate relaxation times, primitive-path fluctuations obviously create a very wide distribution of relaxation times. Also, since the zero-shear viscosity η_0 scales as the longest relaxation time, η_0 increases exponentially with the molecular weight, a prediction that has been confirmed experimentally in "star" branch polymers (Pearson and Helfand 1984).

If both ends of the molecule are free to move, and so the chain can reptate, segments in the interior of the chain will relax faster by reptation than by primitive-path fluctuations, and so reptation will control the longest relaxation time of the chain. However, because primitive-path fluctuations are so much faster for the chain ends than for the chain center, the chain ends will still relax by primitive-path fluctuations. Only for very high molecular weights $(M/M_e \gtrsim 100)$ are the contributions of fluctuations confined to small enough portions of the chain ends that these effects can be neglected.

3.7.2.2 Constraint Release

Another important relaxation process in entangled melts is *constraint release*, depicted in Fig. 3-27. When an end of a surrounding chain moves past a "test" chain, an entanglement constraint restricting the motion of the test chain is released, and a portion of the latter is freed to reorient (Graessley 1982; Montfort et al. 1986; Pearson 1987; Viovy et al. 1991). Constraint release can only be completely neglected for the case of an isolated chain

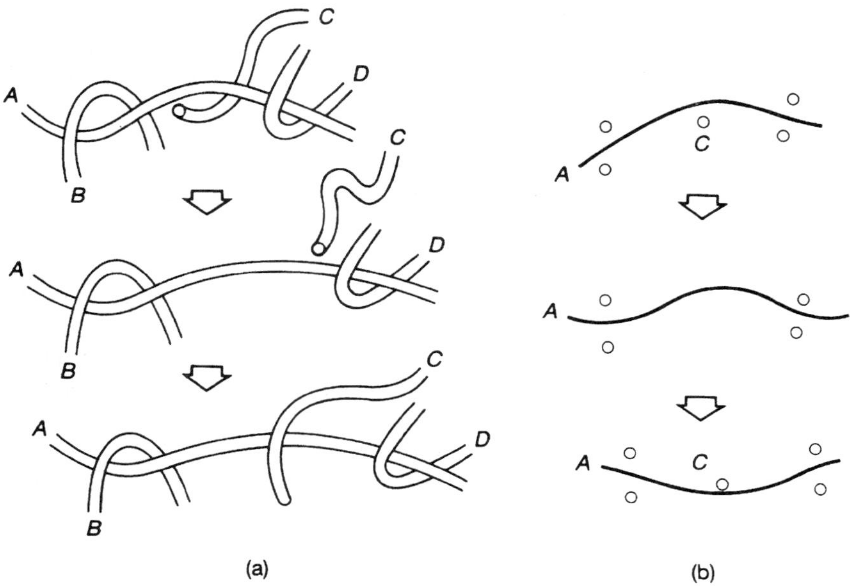

Figure 3.27   Depiction of the "constraint-release" mechanism of relaxation. In **(a)**, the topological constraint imposed on chain A by chain C is released, as the end of chain C crosses under chain A. Even if C eventually re-entangles with A, chain A has been given a chance to change its orientation, as illustrated in the two-dimensional depiction in **(b)** (From Doi and Edwards, copyright © 1986 by Oxford University Press, Inc. Used by permission of Oxford University Press, Inc.)

entangled with a matrix that is cross-linked, or whose molecular weight is higher than that of the test chain. Neglect of constraint release is marginally valid for monodisperse chains undergoing slow flows, but is grossly inaccurate for long chains entangled with shorter chains, since the shorter chains can release constraints by diffusing faster than the long chain can reptate.

3.7.3 Evidence for Reptation

Despite these complications, there are now numerous evidences that the tube model is basically correct. The signatory mark that the chain is trapped in a "tube" is that the chain ends relax first, and the center of the chain remains unrelaxed until relaxation is almost over. Evidence that this occurs has been obtained in experiments with chains whose ends are labeled, either chemically or isotopically (Ylitalo et al. 1990; Russell et al. 1993). These studies show that the rate of relaxation of the chain ends is distinctively faster than the middle of the chain, in quantitative agreement with reptation theory. The special role of chain ends is also shown indirectly in studies of the relaxation of "star" polymers. Stars are polymers in which several branches radiate from a single branch point. The arms of the star cannot reptate because they are anchored at the branch point (de Gennes 1975). Relaxation must thus occur by the slower process of primitive-path fluctuations, which is found to slow down exponentially with increasing arm molecular weight, in agreement with predictions (Pearson and Helfand 1984).

Perhaps the most graphic evidence of the validity of reptation-related ideas is to be found in the experiments of Chu and coworkers (Perkins et al. 1994a). They observed directly through a microscope the relaxation processes of very long (~ 16–$100\ \mu$m) fluorescently stained DNA molecules entangled in a sea of other, untagged—and therefore invisible— molecules (see Fig. 3-4). These experiments show convincing visual evidence that molecular motion is confined to a tube-like region. Molecular dynamics (MD) simulations also show chain motion that is highly anisotropic, suggesting that the diffusive motion of a long molecule is largely confined to a tube [see Fig. 3-28 (Kremer and Grest 1990)].

A variety of other experiments, measuring time-dependent pair correlation functions (Richter et al. 1992), diffusion coefficients (Tirrell 1984; Lodge et al. 1990), and viscoelastic relaxation phenomena, are more-or-less consistent with reptation theories. Complete quantitative agreement with simple reptation theories is rarely obtained, however, presumably because of additional mechanisms, such as "constraint release" and "primitive-path fluctuations," that occur along with pure reptation. Lodge et al. (1990) have published a comprehensive status report on the extent to which reptation theory is in agreement with experimental data on viscosity, diffusion, stress relaxation, gel electrophoresis, and other measurements. Since the main focus of this book is the prediction of rheological properties, we shall in the remainder of the chapter describe the predictions of "tube" theories for rheological properties, starting with the Doi–Edwards constitutive equation.

3.7.4 The Doi–Edwards Constitutive Equation

Doi and Edwards (1978a, 1979, 1986) developed a constitutive equation for entangled polymeric fluids that combines the linear viscoelastic response predicted by de Gennes

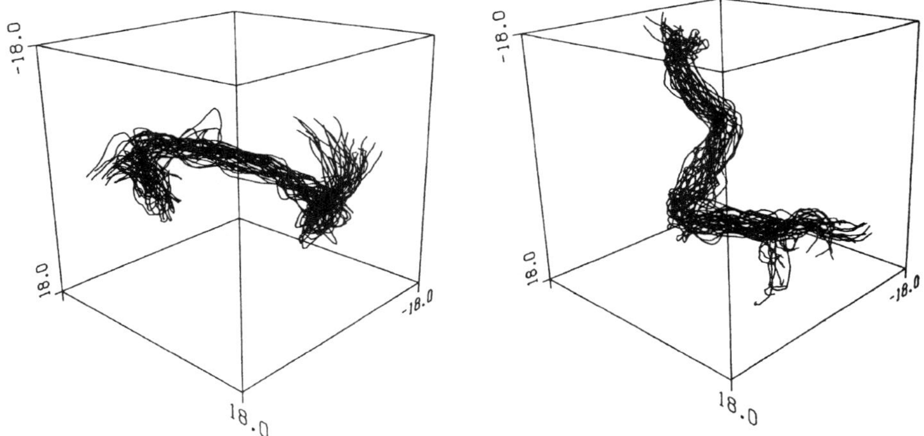

Figure 3.28 Each of the two images contains superimposed configurations of a chain at many different instants in time in a molecular-dynamics simulation of a melt of such chains in a box. Over the time scale simulated, each chain appears to be confined to a tube-like region of space, except at the chain ends. (From Kremer and Grest, reprinted with permission from J. Chem. Phys. 92:5057, Copyright 1990 American Institute of Physics.)

for a reptating chain with a nonlinear response to large deformations. Although the Doi–Edwards theory neglects primitive-path fluctuations and constraint release, and therefore is not quantitative, it is a good starting point for understanding the behavior of entangled melts and solutions.

3.7.4.1 Linear Relaxation Modulus

Consider a chain in a tube at time zero. Analysis of the reptation process (de Gennes 1971) shows that after a time t, only a fraction $P(t)$ of the original tube remains unvacated, namely

$$P(t) = \sum_{i \text{ odd}} \frac{8}{\pi^2 i^2} \exp\left[\frac{-i^2 t}{\tau_d}\right] \tag{3-66}$$

The linear relaxation modulus is $P(t)$ times G_N^0:

$$G(t) = \sum_{i \text{ odd}} G_i \exp[-t/\tau_i], \qquad G_i = 8G_N^0/\pi^2 i^2, \quad \tau_i = \tau_d/i^2 \tag{3-67}$$

The storage and loss moduli, G' and G'', are obtained from the relaxation spectrum in the usual way—that is, using $G' = \sum G_i[\omega^2\tau_i^2/(1 + \omega^2\tau_i^2)]$; $G'' = \sum G_i[\omega\tau_i/(1 + \omega^2\tau_i^2)]$. The longest relaxation mode of the relaxation modulus in Eq. (3-67) is the dominant one; it accounts for 96% of the zero-shear viscosity. Thus, the reptation model predicts that for a nearly monodisperse melt, the relaxation spectrum is dominated by a single relaxation time, $\tau_1 = \tau_d$. This is in reasonable accord with experimental data at low and moderate frequencies (see the dashed line in Fig. 3-29). As the frequency increases, however, there

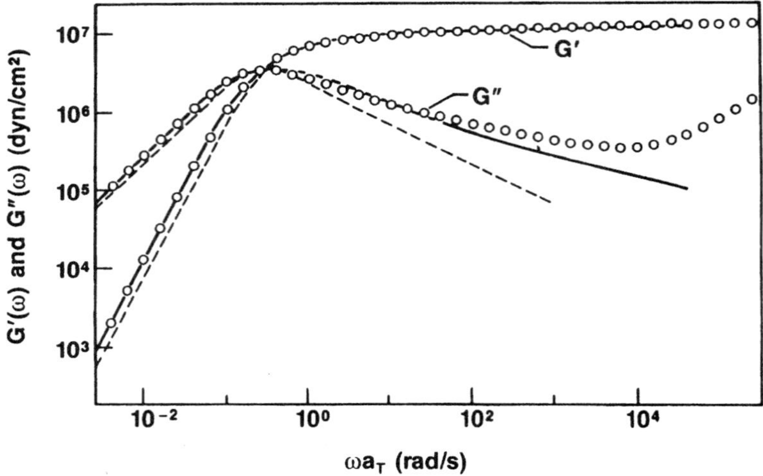

Figure 3.29   Linear moduli G' and G'' versus frequency shifted via time–temperature superposition to 27°C for a polybutadiene melt of molecular weight 360,000 and of low polydispersity. The dashed line is the prediction of reptation theory given by Eq. (3-67); the solid line includes effects of fluctuations in the length of the primitive path. (From Pearson 1987.)

is more deviation from the Doi–Edwards model, apparently because reptation is the only relaxation process considered. Any additional relaxation processes that might be present will tend to broaden the spectrum, thereby bringing $G'(\omega)$ and $G''(\omega)$ closer to the experimental curves. Inclusion of the process of tube fluctuation, for example, removes most of the discrepancy between experiment and theory (Pearson 1987; Milner and McLeish 1998) but does not account for the upturn in G'' at high ω (see the solid line in Fig. 3-29). This upturn is apparently caused by Rouse-like motions of parts of the molecule that lie within a primitive-path step (Milner and McLeish 1998). Even small levels of polydispersity broadens the relaxation spectrum (Lin 1984), and so does constraint release.

3.7.4.2 Nonlinear Modulus and Damping Function

Figure 3-30a shows the *nonlinear* modulus, $G(t, \gamma)$ after a series of "step" strains, γ, suddenly imposed on a concentrated polystyrene solution (Einaga et al. 1971). The nonlinear modulus is just the stress divided by the strain. Figure 3-30b shows that vertical shifting brings the relaxation moduli for various γ into coincidence at times greater than $\tau_r \sim 20$ sec. Thus, at times longer than τ_r, $G(t, \gamma)$ is factorable into time- and strain-dependent functions: $G(t, \gamma) = G(t)h(\gamma)$, where $G(t)$ is the linear modulus and $h(\gamma)$ is a strain-dependent *damping function*. The damping function, $h(\gamma)$, plotted in Fig. 3-31, shows considerable *strain softening*; that is, h decreases with increasing γ. The temporary network model of Green and Tobolsky (Section 3.4.4), on the other hand, has no strain softening since it predicts that $h(\gamma) = 1$ for all γ.

Doi and Edwards (1978a, 1979) explained this strain softening using an extension of the tube model of de Gennes. Suppose that the melt is subjected to a step strain and

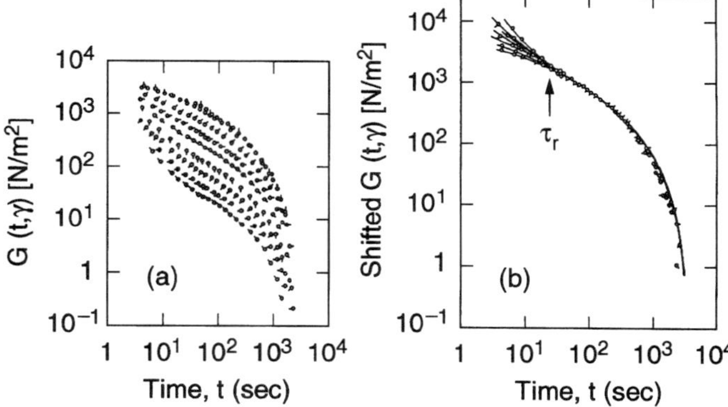

Figure 3.30 (a) The nonlinear shear relaxation modulus $G(t, \gamma)/\gamma$ as a function of time for various strain amplitudes for a 20% concentrated solution of polystyrene of molecular weight 1.8×10^6 in chlorinated diphenyl. Each curve corresponds to a different strain, ranging from 0.41 to 25.4, with the lower curves corresponding to the higher strains. (b) The curves are superimposed at times longer than $\tau_r = 20$ sec via vertical shifting by an amount $h(\gamma)$ plotted in Fig. 3-31. (From Einaga et al. 1971, with permission from the Society of Polymer Science, Japan.)

that the tube and the chain it contains are stretched affinely. Although the tube prevents lateral motion of the molecule, it does not stop the molecule from *retracting* along the tube contour. Since retraction does not violate the tube constraints, it occurs quickly compared to reptation. Specifically, retraction occurs in a time roughly equal to the Rouse time τ_r for the whole chain, which is smaller than τ_d by the ratio $\tau_r/\tau_d = (N_e/3N)$, where N/N_e is the number of entanglements per chain (Doi and Edwards 1986). (An even faster time scale than τ_r is the time for relaxation of the portion of a chain within a tube segment; this happens by Rouse-like processes in an *equilibration time* $\tau_e = a^4 \zeta_0/k_B T b^2$). The retraction of the chain brings the contour length of its primitive path back to its equilibrium value aM/M_e. The nonfactorable relaxation seen in Fig. 3-30 at times shorter than τ_r is presumed to be caused by incomplete retraction. At times longer than τ_r, retraction is complete; the remaining stress must relax by reptation, which occurs on a longer time scale. The strain softening in $h(\gamma)$ is therefore attributed to the retraction process.

Retraction moves a strand from one part of the tube to another; hence the strand's orientation is determined not by the orientation of the part of the tube it originally occupied, but by the orientation of the part of the tube into which it moves. To simplify the problem, however, Doi and Edwards invoked the *independent alignment approximation*, which assumes that after retraction each *strand is oriented independently of the others*, and the change in orientation produced by retraction is neglected.

Since the retraction process keeps the overall primitive path length constant, if one invokes the independent alignment approximation, each step in the primitive path returns to the same length, a, after retraction. The net effect of the deformation, therefore, is to *orient* strands *without stretching them*. Since rigid rods also respond to a deformation by rotating without stretching, the Doi–Edwards constitutive equation for melts is similar to that for the elastic stress for rigid-rod molecules [see Section 6.3.2.1 and Doi and Edwards (1978b)].

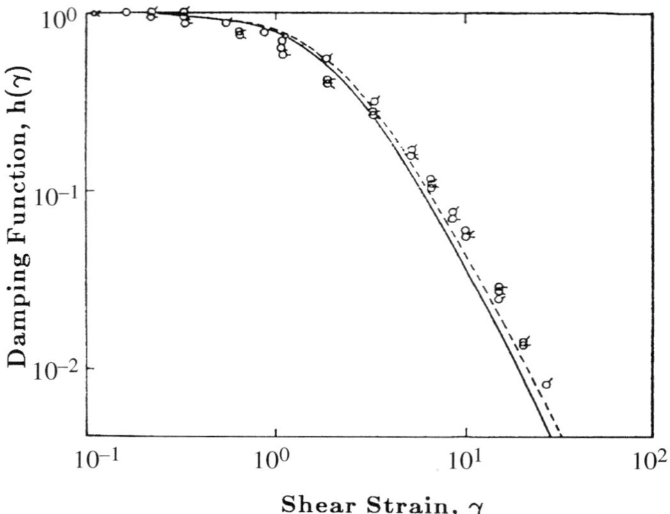

Figure 3.31 Damping function $h(\gamma)$ obtained by vertically shifting the time-dependent nonlinear moduli in Fig. 3-30a into superposition at long times. The data are from Fukuda et al. (1975). The solid and dashed lines are the prediction of the Doi–Edwards model, respectively, with and without the independent alignment approximation. (From Doi and Edwards 1978a, reproduced by permission of The Royal Society of Chemistry.)

3.7.4.3 Stress Tensor After a Step Strain

In the original Doi–Edwards model, retraction is assumed to occur infinitely fast. The stress tensor is then given by the "elastic," or Brownian, stress for rigid rods [see Eq. (6-36)]:

$$\boldsymbol{\sigma} = 3\nu k_B T \langle \mathbf{uu} \rangle = 3\nu k_B T a^{-2} \langle \mathbf{RR} \rangle \tag{3-68}$$

where $\mathbf{u}$ is the unit vector parallel to the end-to-end vector of a strand (or a step of the primitive path), a is the length of a step of the primitive path, and the brackets "$\langle\ \rangle$" denote an average over all strands. This expression for $\boldsymbol{\sigma}$ satisfies the stress-optic law.

Although the expression for the stress tensor in the Doi–Edwards model is the same as that of the temporary network model, except for the coefficient $3\nu k_B T a^{-2}$ [see Eq. (3-13)], the nonlinear response of a polymer strand to a step strain is different. In the temporary network model, $\mathbf{R} = \mathbf{R}' \cdot \mathbf{E}$, where $\mathbf{R}$ and $\mathbf{R}'$ are the strand's end-to-end vectors before and after the deformation, and $\mathbf{E}$ is the *inverse deformation gradient* tensor, defined in Eq. (1-12). In the Doi–Edwards theory, on the other hand, $\mathbf{R} = \mathbf{R}' \cdot \mathbf{E}/|\mathbf{u}' \cdot \mathbf{E}|$; thus, the strand's length $|\mathbf{R}| = |\mathbf{R}'|$ remains constant because of retraction. Hence, in the Doi–Edwards theory, after a step strain and after retraction, but before reptation occurs, the stress is given by

$$\boldsymbol{\sigma} = 3\nu k_B T \langle \mathbf{uu} \rangle = 3\nu k_B T \left\langle \frac{\mathbf{u}' \cdot \mathbf{E}\, \mathbf{u}' \cdot \mathbf{E}}{|\mathbf{u}' \cdot \mathbf{E}|^2} \right\rangle_0 = \frac{3}{5}\nu k_B T \mathbf{Q} \tag{3-69}$$

where "$\langle\cdot\rangle_0$" averages over an isotropic distribution of the unit vector $\mathbf{u}'$ so that $\psi_0(\mathbf{u}') = 1/4\pi$; thus

$$\langle\cdot\rangle_0 = \int \frac{d^2u}{4\pi} = \int_0^\pi \int_0^{2\pi} \cdot\, d\phi \sin\theta \, d\theta$$

where $\mathbf{u}' = (\sin\theta\cos\phi, \sin\theta\sin\phi, \cos\theta)$. In Eq. (3-69), $\mathbf{Q}$ is defined as

$$\mathbf{Q} \equiv 5\left\langle\frac{\mathbf{u}'\cdot\mathbf{E}\,\mathbf{u}'\cdot\mathbf{E}}{\left|\mathbf{u}'\cdot\mathbf{E}\right|^2}\right\rangle_0 \tag{3-70}$$

The factor of 5 is introduced into the definition of $\mathbf{Q}$ so that $\mathbf{Q}$ equals the linear viscoelastic strain, $\boldsymbol{\gamma}$, when the strain is small. $G_N^0 = {}^3\!/_5\nu k_B T$ is then the plateau modulus. When the independent alignment approximation is dropped, this is corrected to $G_N^0 = {}^4\!/_5\nu k_B T$.

In a step shear, Q_{12}, the shearing component of $\mathbf{Q}$ (where "1" is the flow direction and "2" is the gradient direction), can be expressed as $\gamma h(\gamma)$, where $h(\gamma)$ is the "damping function" defined earlier. Figure 3-31 shows that the predicted damping function agrees well with the experimental damping function for monodisperse polystyrene solutions. Figure 3-31 also shows that the damping function in a step shear is not much affected by the "independent alignment approximation." For other deformation histories, such as a "double-step" strain, however, larger errors are introduced by the independent alignment approximation (Doi and Edwards 1986). Damping functions can also be defined for other single-step-strain deformations, such as step biaxial extension and step uniaxial extension; for these, experimental data for nearly monodisperse melts are also in reasonably good agreement with the predictions of the Doi–Edwards theory (Urakawa et al. 1995).

3.7.4.4 Constitutive Equation

The stress that remains a time t after a step strain is the product of the stress immediately after the step, given by Eq. (3-69), multiplied by the fraction of tube length $P(t)$ vacated because of reptation, given by Eq. (3-66). If the strain is imposed gradually, so that reptation occurs during deformation, the stress is given by a *history integral*, analogous to the Lodge equation, (3-24). This history integral is the *Doi–Edwards constitutive equation*,

$$\sigma = \int_{-\infty}^t m(t - t')\mathbf{Q}(t', t)\, dt' \tag{3-71}$$

where the memory function $m(t - t')$ is given by

$$m(t - t') \equiv \frac{d}{dt'}G(t - t')$$

and $G(t - t')$ and $\mathbf{Q}(t', t)$ are given by Eqs. (3-67) and (3-70), respectively.

Equation (3-71) can be expressed in the form

$$\sigma = \int_{-\infty}^t m(t - t')[\phi_1(I_1, I_2)\mathbf{B}(t', t) + \phi_2(I_1, I_2)\mathbf{C}(t', t)]\, dt' \tag{3-72}$$

where $\mathbf{C}$ is the *Cauchy tensor*, which is the inverse of the Finger tensor, $\mathbf{C} \equiv \mathbf{B}^{-1}$; I_1 is the trace of the Finger tensor $\mathbf{B}$, and I_2 is the trace of its inverse $\mathbf{C}$. For the special case of a *separable K–BKZ equation* (Kaye 1962; Bernstein et al. 1963), the functions ϕ_1 and ϕ_2 are related to a *strain energy function*, $U(I_1, I_2)$, by

$$\phi_1 \equiv 2\frac{\partial U}{\partial I_1}, \qquad \phi_2 \equiv -2\frac{\partial U}{\partial I_2}$$

See Tanner (1985) or Larson (1988) for more details about the K-BKZ category of constitutive equations.

The Doi–Edwards equation is a special case of a separable K–BKZ equation, for which Currie (1980) found an accurate analytic approximation, namely

$$U(I_1,\ I_2) \approx \frac{5}{2}\ln\left(\frac{J-1}{7}\right) \tag{3-73}$$

with

$$J \equiv I_1 + 2(I_2 + 13/4)^{1/2} \tag{3-74}$$

From this potential function, one obtains

$$\mathbf{Q} \approx \left(\frac{5}{J-1}\right)\mathbf{B} - \left(\frac{5}{(J-1)(I_2+13/4)^{1/2}}\right)\mathbf{C} \tag{3-75}$$

A simpler and cruder approximation is (Larson 1984a)

$$U \sim \tfrac{5}{2}\ln(1 + \tfrac{1}{5}(I_1 - 3)) \tag{3-76a}$$

which gives

$$\mathbf{Q} = \left(\frac{1}{1 + \tfrac{1}{5}(I_1 - 3)}\right)\mathbf{B} \tag{3-76b}$$

This latter approximation shows that the strain dependence of the Doi–Edwards equation is "softer" than that of the temporary network model roughly by the factor $1 + (I_1 - 3)/5$.

There is also a *differential* approximation to the Doi–Edwards equation (Marrucci 1984; Larson 1984b):

$$\overset{\triangledown}{\sigma} + \frac{2}{3G}\mathbf{D} : \sigma\sigma + \frac{1}{\tau}(\sigma - G\delta) = \mathbf{0} \tag{3-77}$$

Equation (3-77) differs from the *upper-convected Maxwell equation*, Eq. (3-32), in that it includes the term $(2/3G)\mathbf{D} : \sigma\sigma$, which imparts strain softening and shear thinning to the behavior of the model.

3.7.5 Predictions of Reptation Theories

3.7.5.1 Steady-State Shear and Extension

The uniaxial extensional viscosity $\bar{\eta}(\dot{\varepsilon})$ and the viscometric functions $\eta(\dot{\gamma})$ and $\Psi_1(\dot{\gamma})$, predicted by the Doi–Edwards model for monodisperse melts, are shown in Fig. 3-32. The Doi–Edwards model predicts extreme thinning in these functions; the high-shear-rate asymptotes scale as $\bar{\eta} \propto \dot{\varepsilon}^{-1}$, $\eta \propto \dot{\gamma}^{-1.5}$, and $\Psi_1 \propto \dot{\gamma}^{-2}$. The second normal

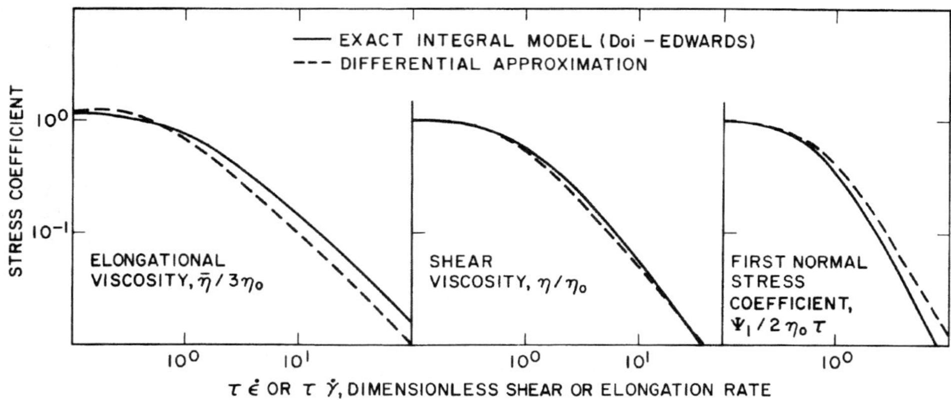

Figure 3.32   The predictions of the Doi–Edwards integral model for the normalized uniaxial extensional (or elongational) viscosity $\bar{\eta}$ and for the viscometric shear coefficients $\eta(\dot{\gamma})$ and $\Psi_1(\dot{\gamma})$. Also shown are the predictions of the differential model, Eq. (3-77). (From Larson, 1984b, with permission from the Journal of Rheology.)

stress coefficient, not shown in Fig. 3-32, scales as $\Psi_2 \propto -\dot{\gamma}^{-2.5}$ at high shear rates (Doi and Edwards 1979). The predicted shear thinning is so severe that as the shear rate increases, the shear stress $\eta\dot{\gamma}$ is predicted to pass through a maximum and then *decrease* with further increases in $\dot{\gamma}$ (see Fig. 3-33). Hence, at each shear stress there are at least *two* values of the shear rate. This, it is predicted, should lead to *material instabilities*—that is, apparent slip phenomena, such as "spurt," sometimes observed in flow through capillaries (see Section 3.7.5.3).

Experimentally, melts of low polydispersity that do not overtly "spurt," "slip," or succumb to other material instabilities will typically show steeply decreasing values of the viscosity and first normal stress coefficient in the shear-thinning region. Menezes and Graessley (1980) reported that $\eta \propto \dot{\gamma}^{-0.82}$ and $\Psi_1 \propto \dot{\gamma}^{-1.5}$ at large $\dot{\gamma}$. These dependencies become steeper as the polymer becomes more entangled, so that the power-law exponent for η approaches -1, implying that the shear stress is almost constant at high shear rate, as shown by the open triangles in Fig. 3-34. However, there is no direct evidence that η and Ψ_1 ever fall as steeply with shear rate as predicted by the Doi–Edwards theory. This is most likely

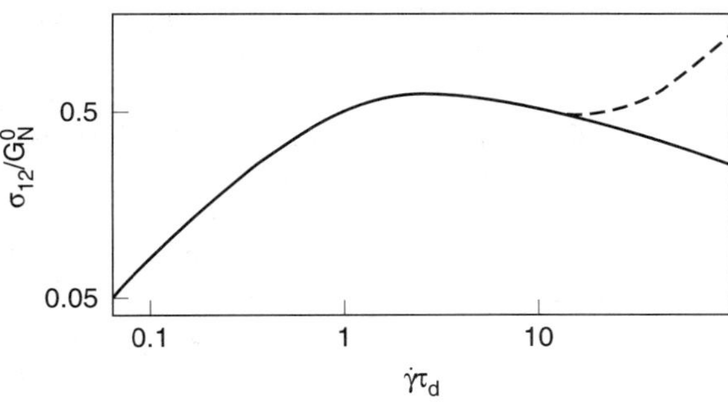

Figure 3.33   The solid curve is the dimensionless shear stress σ_{12}/G_N^0 versus dimensionless shear rate $\dot{\gamma}\tau_d$ predicted by the Doi–Edwards constitutive equation, Eq. (3-71). The dashed curve adds a speculated contribution to the stress from Rouse modes. (From Doi and Edwards 1979, reproduced by permission of The Royal Society of Chemistry.)

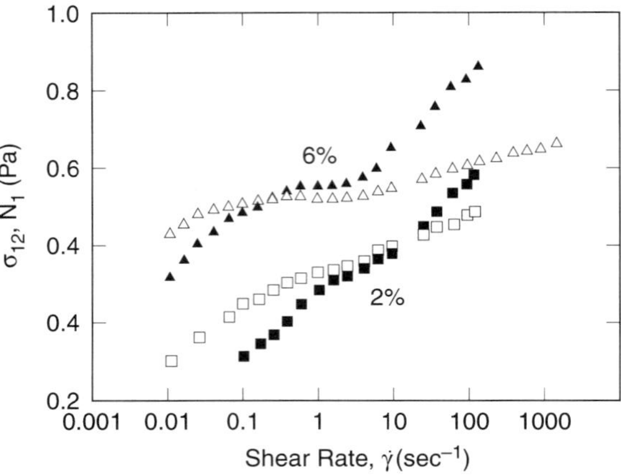

Figure 3.34 Shear stress (open symbols) and first normal stress difference (closed symbols) as functions of shear rate for two solutions of very-high-molecular-weight polymethylmethacrylate ($M = 23.8 \times 10^6$) in toluene at a concentration of 2 g/dL (squares) and 6 gm/dL (triangles). (From Bercea et al. 1993, with permission.)

because processes of relaxation other than reptation occur. In particular, Marrucci (1996) and Marrucci and Ianniruberto (1997) have pointed out that fast flows (De $\gg$ 1) convect away the polymer molecules constraining a given chain, and therefore destroy the tube surrounding that chain, faster than the chain itself can reptate out of the tube. Under these conditions, "convective constraint release," not reptation, is probably the dominant mechanism of relaxation. A new theory incorporating "convective constraint release" and "tube stretching" into the Doi–Edwards model predicts a nearly constant shear stress as a function of shear rate in the high shear rate regime [see Fig. 3-35 (Marrucci and Ianniruberto 1997; Larson et al. 1998)]. The predictions of this theory agree with many rheological measurements in both steady-state and transient shear flows (compare, for example, Fig. 3-35 with Fig. 3-34). The equations for this new theory are given and discussed in Problem 3.10.

The Doi—Edwards equation predicts that the ratio Ψ_2/Ψ_1 is $-2/7 = -0.29$ at low shear rates. This changes to $\Psi_2/\Psi_1 = -1/7 = -0.14$ when the "independent alignment approximation" is dropped (Osaki et al. 1981). With or without the independent alignment approximation, the ratio $-\Psi_2/\Psi_1$ is predicted to decrease towards zero as the shear rate increases. The prediction of Ψ_2/Ψ_1 for entangled solutions contrasts with that predicted for dilute solutions, for which Ψ_2/Ψ_1 is close to zero over the whole range of shear rates (see Fig. 3-36). These predictions for dilute and entangled solutions have been confirmed qualitatively in experiments of Magda et al. (1993) and Brown et al. (1995). Figure 3-37 shows Ψ_2/Ψ_1 versus an approximate Weissenberg number N_1/σ_{12} for a series of polystyrene solutions of high molecular weight, at concentrations ranging from dilute to concentrated. The data fall into two distinct groups, depending on the concentration: For the dilute solutions, $-\Psi_2/\Psi_1 \approx 0$, as predicted, while for the entangled solutions, $-\Psi_2/\Psi_1 \approx 0.2$ at low shear rates, decreasing toward zero as the shear rate increases. The measured ratios of $-\Psi_2/\Psi_1$ are in qualitative agreement with the predicted ones (compare Fig. 3-36 with Fig. 3-37). In concentrated solutions or melts of molecular weight low enough that the molecules are *unentangled,* one might expect the prediction of the Rouse theory to apply, namely $\Psi_2/\Psi_1 = 0$ for all shear rates. However, both experiments (Magda et al. 1993) and molecular dynamics simulations (Berker et al. 1992; Kröger et al. 1993) for unentangled solutions and melts give surprisingly high values of $-\Psi_{2,0}/\Psi_{1,0}$, around 0.15–0.45 or so, at low shear rates.

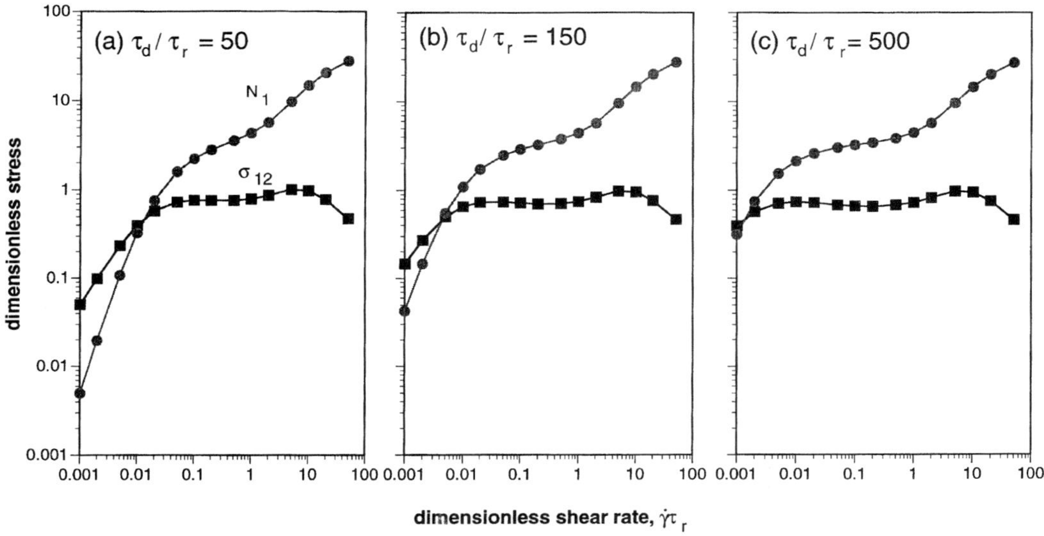

Figure 3.35   Steady-state values of the reduced shear stress σ_{12}/G_N^0 and first normal stress difference N_1/G_N^0 as functions of dimensionless shear rate $\dot{\gamma}\tau_r$ predicted by the equations of a constraint-release reptation theory (see Problem 3.10) for $\tau_d/\tau_r =$ **(a)** 50, **(b)** 150, and **(c)** 500, where τ_d is the reptation time and τ_r is the Rouse retraction time. See also Marracci and Ianniruberto (1997). (From Larson et al. 1998, with permission.)

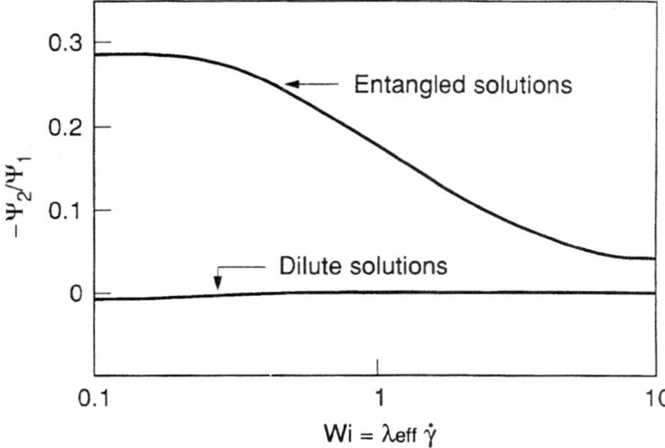

Figure 3.36   Negative ratio of the second to the first normal stress coefficients versus Weissenberg number predicted for entangled polymers by the Doi–Edwards theory, and for dilute solutions by the bead-spring theory with conformation-dependent hydrodynamic interaction. Here $\tau_{\text{eff}} \equiv \Psi_{1,0}/2\eta_{p,0}$. (Reprinted with permission from Magda et al., Macromolecules 26:1696. Copyright 1993, American Chemical Society.)

3.7.5.2 Stress Overshoots

The Doi–Edwards equation predicts an overshoot in shear stress as a function of time after inception of steady shearing, but no overshoot in the first normal stress difference (Doi and Edwards 1978a). Typical overshoots in these quantities for a polydisperse melt are shown in Fig. 1-10. For monodisperse melts, the Doi–Edwards model predicts that the shear-stress maximum should occur at a shear strain $\dot{\gamma}t = \gamma_p$, of about 2, roughly independently of

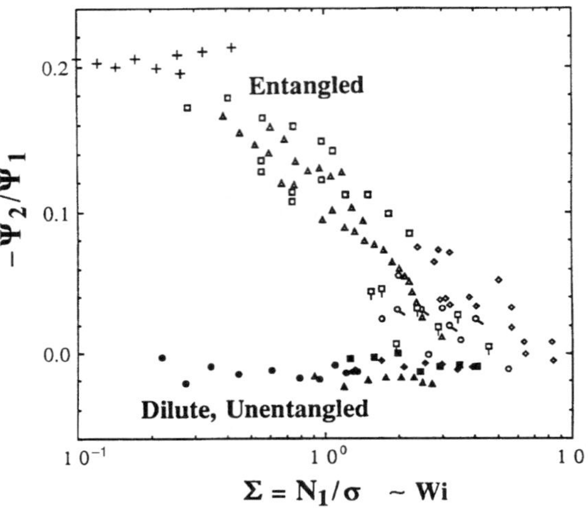

Figure 3.37 Negative ratio of the second to the first normal stress coefficients versus stress ratio N_1/σ for various dilute solutions (concentration ≤ 0.6 wt%; closed symbols) and entangled solutions (concentration ≥ 1 wt%; open symbols). (Reprinted with permission from Magda et al., Macromolecules 26:1696. Copyright 1993, American Chemical Society.)

the shear rate. For low strain rates, the experimental maximum is indeed at $\gamma_p \sim 2$; but for high strain rates it shifts to higher strains (Menezes and Graessley 1982). In addition, experiments show an overshoot in the first normal stress difference, which is *not* predicted by the Doi–Edwards equation; it occurs at a strain of about 5 at low shear rate. It has been observed that the strains at which these two overshoots occur increase with increasing shear rate when the shear rate exceeds the reciprocal of the *retraction time* τ_r (Menezes and Graessley 1982).

This behavior can be predicted if the Doi–Edwards theory is extended to allow for *tube stretching* (or incomplete retraction) (Marrucci and Grizzuti 1988; Pearson et al. 1991; Mead and Leal 1995; Larson et al. 1998). When shearing starts with $\dot{\gamma}\tau_r \gtrsim 1$, the polymer strands are stretched for a brief period, and the shear stress rises to a maximum. As the strands become highly oriented, their projected length in the direction of the shear gradient decreases, and the velocity difference between one end of the strand and the other begins to decrease. As a result, the shearing flow "loses its grip" on the molecules, and they are able to retract within their tubes. The first normal stress difference therefore overshoots and decreases towards its steady-state value. Equations that describe this tube-stretching process were presented in Pearson et al. (1991), and they are incorporated into the equations in Problem 3.10. They predict the shifting of the shear and normal stress maxima with shear rate.

3.7.5.3 Anomalous Rheology

As already noted, the measured nonlinear shear relaxation modulus, for linear molecules with little polydispersity, is in excellent agreement with the Doi–Edwards model at long times. However, for melts or concentrated solutions of very high molecular weight (e.g., $\phi M > 10^6$ for polystyrene, where ϕ is the polymer volume fraction), the measured damping function, $h(\gamma)$, is drastically lower than the Doi–Edwards prediction (Einaga et al. 1971; Vrentas and Graessley 1982; Larson et al. 1988; Morrison and Larson 1992). This anomalous result seems surprising at first, since one expects that the Doi–Edwards theory should be most accurate for polymers having the highest entanglement density and, hence, the highest molecular weight. However, it has recently been shown that by using small particle probes within the fluid, the anomalous modulus is a consequence of *slippage* of the polymeric material either at, or within a few microns of, the solid surfaces between which the fluid is sheared (Archer et al. 1995). This behavior is not necessarily inconsistent with the Doi–Edwards theory. The Doi–Edwards theory predicts that at a fixed time after a step strain the shear stress has a maximum at a strain $\gamma_p \approx 2$, because of the extreme strain softening of the nonlinear modulus. From principles of mechanical stability, the shear stress maximum is expected to lead to *strain localization*, which is a kind of material instability. It occurs because the material can reduce its stress level by spontaneously increasing the strain in some regions of the sample and decreasing it in others (Marrucci and Grizzuti 1983; Kolkka et al. 1991). In this way, the strain in each region of the sample evolves toward a value that is either far above γ_p or far below it, and all regions of the sample develop a low stress level. If the highly strained regions are at the rheometer walls, apparent slip will be the result. Such "slip" can only occur in samples whose molecular weight is high enough that the retraction time is much shorter than the reptation time, $\tau_r \ll \tau_d$, so that a stress–strain curve with a stress maximum adequately describes the material's rheology curve over a period of time long enough for strain localization to develop. Hence, anomalous "slip" produced by strain localization is expected (and observed) to be especially prominent for solutions and melts of high molecular weight (Marrucci and Grizzuti 1983).

A related phenomenon is predicted to occur in steady shearing flows (Doi and Edwards 1979). For such flows, the shear stress is predicted to exhibit a maximum as a function of shear rate (see Fig. 3-33). Thus, for each shear stress, there are two or more possible steady values of the shear rate; and a material instability, similar to slip, is again expected. Cates et al. (1993) have predicted that as a result of the material instability in a cone-and-plate or other simple shearing geometry, there should be a range of imposed shear rates over which the flow field becomes *stratified*; and two layers form, one with a high shear rate (on the dashed curve in Fig. 3-33) and the other with a low shear rate, but both having the same shear stress. As the imposed velocity V of the moving plate is increased, the shear rate in each of the two zones remains constant, but the thickness of the high-shear-rate layer increases at the expense of the low-shear-rate layer, so that the *average* shear rate is V/h, where h is the gap between the plates. Recent rheological data on very high molecular-weight ($M = 23.6 \times 10^6$) poly(methylmethacrylate) solutions (Bercea et al. 1993) are consistent with the theory of Cates et al., but direct confirmation of the two-layer flow has not yet been achieved in entangled polymer melts and solutions. Two-layer flow has been directly observed in flow of entangled "worm-like" micelles (Decruppe et al. 1995) (see Section 12.3.4.4). However, these solutions are complicated by the possibility that the two layers are induced by a flow-induced shift in the concentration at which a nematic phase appears.

It has also been suggested that a material instability, leading to stratified flow, is responsible for the so-called "spurt" phenomenon in which polymer melt flowing through a capillary suddenly increases its velocity by orders of magnitude when the pressure gradient crosses a critical threshold (McLeish and Ball 1986; McLeish 1987). However, theories for two-layer flows have generally ignored the role of "convective constraint release," which probably has a large effect on these phenomena.

A somewhat different, though related, mechanism of slip in polymers of high molecular weight has been postulated by Brochard and de Gennes (1992). In their picture, polymer chains adsorbed to the rheometer surfaces become highly stretched and then disentangle from chains in the bulk, leading to a disentangled layer near the surfaces which has a very low viscosity compared to the bulk. Large shear gradients accumulate in this low-viscosity zone, leading to apparent slip. In the Brochard–de Gennes theory, the density of chains absorbed to the surface controls the critical stress level at which the apparent slip velocity becomes very large. Evidence supporting the Brochard–de Gennes picture has recently been put forth by Leger and coworkers (Migler et al. 1993), who used (a) glass surfaces treated to reduce chain adsorption and (b) an optical bleaching and evanescent-wave interference technique to measure slip within 0.1 μm of the glass surfaces.

Theories for "material instabilities" or "slip" in highly entangled melts and solutions are still under active development.

• Problems 3.7 through 3.11 test your ability to work with reptation ideas and the Doi–Edwards equation.

3.7.6 Effects of Polydispersity and Branching

3.7.6.1 Polydisperse Melts

Most polymeric fluids that are of commercial importance are highly polydisperse (values of M_w/M_n of 2 or more); and some, such as low-density polyethylene, have long-chain branching. It is important for many applications that these effects be accounted for in the constitutive equation. Using *constraint-release* ideas, reasonably accurate predictions have been made of the linear modulus of linear *bidispersed* melts—that is, mixtures of two chemically identical polymers having distinctly different molecular weights (Rubinstein et al. 1987). Constraint release is important in such bidispersed melts, because the relaxation of a long chain is accelerated by release of the entanglements it has with the shorter, faster-relaxing chains around it.

For continuous, rather than bidispersed, distributions of molecular weight, systematic accounting of reptation and constraint-release processes for all different chain lengths in the mixture becomes an impractically complex problem. A much simpler way to account not only for bidisperse molecular weight distributions, but also for continuous ones, has been proposed, using a semiempirical scheme called "double reptation" (Tsenoglou 1987; des Cloizeaux 1988; Tuminello 1986). The double-reptation scheme allows accurate prediction of G' and G'' from a specified molecular weight distribution (Wasserman and Graessley 1992). The "double-reptation" formula for the relaxation modulus $G_{\text{blend}}(t)$ of a blend containing a continuous weight distribution $W(M)$ of components is

$$\left(\frac{G_{\text{blend}}(t)}{G_N^0}\right)^{1/2} = \int_{M_e}^{\infty} W(M) F^{1/2}(M, t) \, dM \tag{3-78}$$

where $F(M, t)$ is the relaxation function $G(t)/G_N^0$ of a monodisperse melt of molecular weight M, and G_N^0 is the plateau modulus. Thus, if $G(t)/G_N^0$ is measured (or predicted) for monodisperse components, $G_{\text{blend}}(t)$ can be predicted using Eq. (3-78). The double-reptation scheme can also be applied in reverse to infer a molecular weight distribution from measurements of the linear modulus (Mead 1994). Since the rheology of a melt is often easier to measure than its molecular weight distribution, this method of estimating molecular weight distributions from rheological data is a very useful tool.

The intuitive idea behind double reptation is that an entanglement between two chains is released if the ends of either of the two chains reptates past the entanglement point. When this idea is applied to a monodisperse polymer, the relaxation function $F(t)$ within double reptation is proportional to the *square* of $P(t)$ given by Eq. (3-66). While this postulate is inconsistent with the theory of reptation as proposed by de Gennes, the idea can be thought of as a simple way of combining ordinary reptation with constraint release, and thus it has a reasonable physical motivation (Milner 1996; Mead 1996). Empirically, squaring the function $P(t)$ enriches the relaxation-time spectrum with extra time constants, and it leads to better agreement with experimental G' and G'' curves at frequencies above the terminal region, even for monodisperse melts.

3.7.6.2 Star Molecules

Ideas based on the tube model can predict the rheological properties of entangled melts of polymer molecules with a branched architecture. The simplest branched structure is the "star" polymer. As mentioned above, entangled star polymers cannot relax by reptation, since one end of each arm is anchored to a cross-link point. The arm therefore relaxes by primitive-path fluctuations (see Section 3.7.2.1). Now the tube constraining a test molecule is defined by its entanglements with surrounding molecules. But these surrounding molecules are themselves relaxing by primitive-path fluctuations. Since the branch tips of the surrounding molecules relax quickly, the constraints they impose on the test chain also disappear rather quickly. As a result, many of the constraints confining the portion of the test chain near the branch point will have been released by the time that portion of the test chain is ready to relax by primitive-path fluctuations, and thus the tube confining the portion of the test chain near the branch point will be *widened* by the time the chain in it relaxes (Ball and McLeish 1989; McLeish 1995). This tube widening by constraint release is analogous to that produced by the addition of a small-molecule solvent, and it is therefore called "dynamic dilution" (Ball and McLeish 1989). Dynamic dilution leads to much faster relaxation of segments near the branch point than would otherwise occur. When primitive-path fluctuations are analyzed along with "dynamic dilution," the relaxation time of a tube segment a fraction x from the branch point changes from that of Eq. (3-65) to

$$\tau(x) = \tau_0 \exp\left[\frac{3M}{M_e}\left(\frac{(1-x)^2}{2} - \frac{(1-x)^3}{3}\right)\right] \tag{3-79}$$

The predictions of the storage and loss moduli that are obtained from these relaxation processes are in excellent agreement with experiment (see Fig. 3-38). The predicted exponential dependences of the longest relaxation time and zero-shear viscosity on the arm molecular weight are also well-confirmed experimentally (Pearson and Helfand 1984; Fetters et al. 1993), as is the insensitivity of these quantities to the number of arms at fixed arm molecular weight. With recent improvements in the theory (Milner and McLeish 1997), superb agreement with experiment is obtained without adjustable parameters.

The nonlinear damping function, $h(\gamma)$, measured in step shearing on star polymers follows the Doi–Edwards prediction, just as $h(\gamma)$ does for linear polymers. Although one end of each arm of the star is anchored, the other end is free to retract, leading to the same strain softening as in linear polymers (Pearson 1987). Molecules with strands that are anchored at two ends, such as molecules with the topology of an "H," a "comb," or a gel fractal, are expected to show nonlinear behavior that is very different from that of stars (McLeish 1988a; Bick and McLeish 1996).

3.7.6.3 Melts with Irregular Long-Chain Branching

Some polymer melts, such as commercial low-density polyethylene, are not only polydisperse, but also possess irregularly spaced long side branches. Here "long" means that the branches are longer than M_e, and hence are able to entangle with surrounding chains. Branches much shorter than this influence the friction coefficient but otherwise don't affect reptation. The effects of long side branches on rheological properties are profound, and are difficult to consider theoretically, especially when compounded by irregularity in side-branch length and spacing along the backbone. In low-density polyethylene, the long side branches can themselves have long branches, thus forming tree-like structures (McLeish 1988b, 1995). Such structures are also present in partially *cross-linked* polymers, discussed in Chapter 5. Polymer strands that terminate in a branching point at one end but are free at the other cannot reptate, but they can still undergo retraction. The contributions of such a polymer strand to the *nonlinear* properties of the melt [i.e., to the strain-energy function $U(I_1, I_2)$] are expected to be similar to those of a freely reptating chain. The contribution to the linear properties [i.e., $G'(\omega)$ and $G''(\omega)$] are, however, greatly affected by the presence

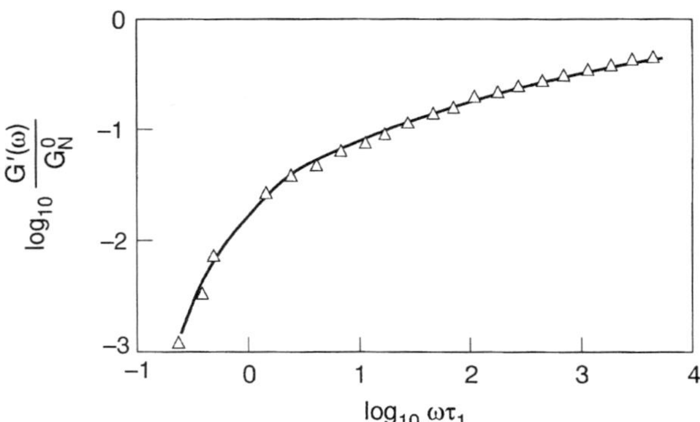

Figure 3.38   Reduced storage modulus G'/G_N^0 versus reduced frequency $\omega\tau_1$ for monodisperse star molecules, where G_N^0 is the plateau modulus for linear chains, and τ_1 is the terminal relaxation time of the star molecules. The symbols are experimental data for polyisoprene stars from Pearson and Helfand (1984), and the solid line is the prediction of the theory with primitive-path fluctuations and constraint release, with only τ_1 fitted (from Ball and McLeish 1989). (Reprinted with permission from Ball and McLeish, Macromolecules 22:1911 Copyright 1989, American Chemical Society.)

of a branch point at only one end of the chain, since one branch point is enough to suppress reptation so that relaxation must occur by primitive-path fluctuations.

Strands that terminate with a branch point at both of its ends can neither reptate nor completely retract. Relaxation of such strands presumably occurs by more complex, hierarchical processes discussed by McLeish (1988b). Here we simply note that the presence of branch points at both ends of a strand leads to *much more strain hardening* in extensional flows (Bishko et al. 1997; McLeish and Larson 1998). Low-density polyethylenes (LDPEs), which are highly branched, are well known for their extreme strain hardening behavior in extensional flows (Meissner 1972; Laun 1984) (see Fig. 3-39). The steady-state shear viscosity, as a function of shear rate, seems to be little affected by long-chain branching, however.

These characteristics of LDPEs, namely strain hardening in extension along with shear-thinning in shear, are highly desirable in some processing flows, such as film blowing. Film blowing is a process used to make thin polyethyelene sheets for garbage bags, grocery wrappings, and so on, in which melt emerging from an annular die is blown up like a bubble by hot air directed along the die axis. The shear thinning of LDPE permits extrusion rates in the annular film-blowing die to be high, while the extension hardening helps stabilize the bubble (Minoshima and White 1986). New *metallocene* catalyst systems allow synthesis of polyethylenes with better controlled long-chain branching characteristics (Colvin 1997, *Chemical Week* 1997). Thus, to take advantage of these new capabilities, it is important that the effect of long-chain branches on rheology and polymer processing behavior be well understood.

Another interesting complication of long-chain branched polyethylenes, such as LDPE, is that time–temperature superposition fails. At high frequencies, their activation energy E_a is similar to that of unbranched or short-chain branched polyethylenes, such as high-density polyethylene (HDPE); but at low frequencies, long-branched polyethylenes (PEs) have activation energies that depend on molecular weight and can be up to 15 kcal/mol, more than twice as high as that of HDPE (Raju et al. 1979; Carella et al. 1986). This important phenomenon is not yet completely understood.

- Problem 3.12 and Worked Example 3.13 illustrate the usefulness of the methods discussed here for calculating flow properties of entangled branched and polydisperse polymers.

3.7.6.4 Semiempirical Constitutive Equations

The development of *molecular* constitutive equations for commercial melts is still a challenging unsolved problem in polymer rheology. Nevertheless, it has been found that for many melts, especially those without long-chain branching, the rheological behavior can be described by empirical or semiempirical constitutive equations, such as the separable K–BKZ equation, Eq. (3-72), discussed in Section 3.7.4.4 (Larson 1988). To use the separable K–BKZ equation, the memory function $m(t)$ and the strain-energy function U, or its strain derivatives $\partial U/\partial I_1$ and $\partial U/\partial I_2$, must be obtained empirically from rheological data.

Molecular polydispersity has a large effect on the memory function $m(t)$ and has a weak or modest effect on the strain-energy function $U(I_1, I_2)$ or, equivalently, on the

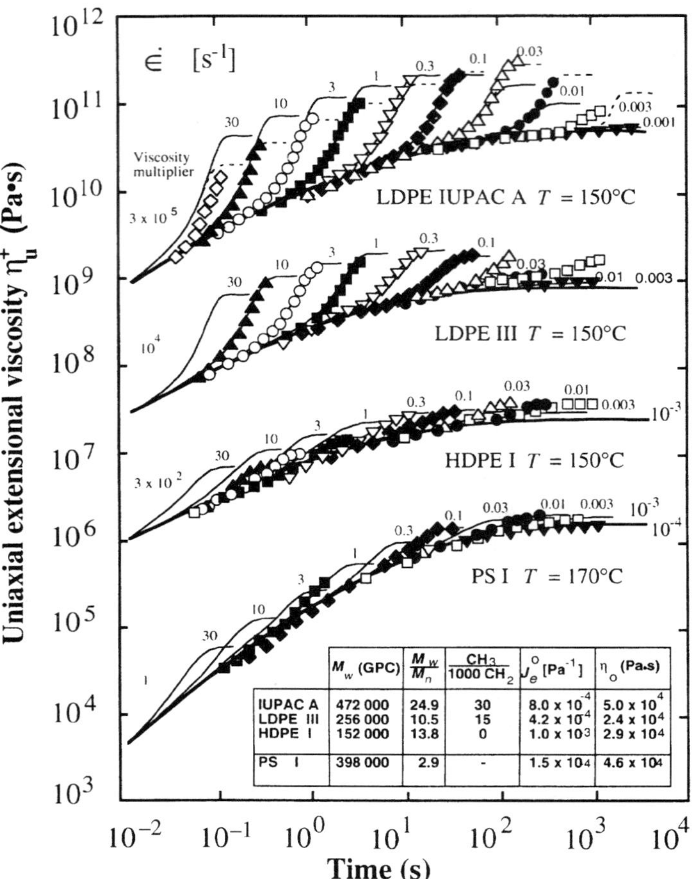

	M_w (GPC)	$\dfrac{M_w}{M_n}$	$\dfrac{CH_3}{1000\ CH_2}$	J_e^o [Pa^{-1}]	η_o (Pa·s)
IUPAC A	472 000	24.9	30	8.0×10^{-4}	5.0×10^4
LDPE III	256 000	10.5	15	4.2×10^{-4}	2.4×10^4
HDPE I	152 000	13.8	0	1.0×10^3	2.9×10^4
PS I	398 000	2.9	-	1.5×10^4	4.6×10^4

Figure 3.39 Uniaxial extensional viscosity $\overline{\eta}_u^+$ as a function of time following start-up of steady uniaxial extension at the extension rates $\dot{\varepsilon}$ indicated. Data are shown for an unbranched polystyrene (PS I), a high-density polyethylene with short, unentangled side branches (HDPE I), and two low-density polyethylenes (LDPE III and IUPAC A), with long side branches. (From Laun 1984, with permission from the Universidad Nacional Autónoma de México.)

damping function (Osaki 1993). Long-chain side branching, which is present in LDPE, strongly affects the strain-energy function, making such melts less strain-softening than melts composed of only linear chains, especially in uniaxial and planar extensional flows (Larson 1988; Osaki 1993; McLeish and Larson 1998). Empirically, the linear memory function $m(t - t')$ can be obtained from linear viscoelastic testing, such as small-amplitude oscillatory shearing. The function U determines the nonlinear viscoelastic properties of the material and must be obtained by large-strain experiments, such as a series of step-strain experiments. Various empirical expressions for the strain-energy function $U(I_1, I_2)$, or for ϕ_1 and ϕ_2, can be found in the literature (Wagner 1976; Wagner and Laun; 1978; Laun 1978; Larson 1988). Two of the most popular forms are

$$\phi_1 = f_1 e^{-n_1\sqrt{I-3}} + f_2 e^{-n_2\sqrt{I-3}}, \qquad \phi_2 = 0$$

$$\text{with } I \equiv \alpha I_1 + (1-\alpha)I_2 \quad \text{(Wagner et al. 1979)}$$

and

$$\phi_1 = \frac{1}{a(I_1-3) + b(I_2-3)}, \qquad \phi_2 = 0 \quad \text{(Papanastasiou et al. 1983)}$$

[These forms are not "true" K–BKZ kernels, since they are not derived from a potential function $U(I_1, I_2)$.] Fits to shear and extensional flow data for "IUPAC A," a well-studied LDPE, give $f_1 = 0.57$, $f_2 = 0.43$, $n_1 = 0.31$, $n_2 = 0.106$, and $\alpha = 0.032$ for the form proposed by Wagner et al. (1979), and $a = 0.0013$ and $b = 0.068$ for that of Papanastasiou et al. (1983).

While these functions have been adjusted to describe shear and uniaxial extensional flows, they seem to work poorly for *planar extension* of LDPE (Samurkas et al. 1989). Planar extensional flow represents a particularly difficult test for K–BKZ-type constitutive equations, since fits to shear data fix all the model parameters required for planar extension, and there is therefore no "wiggle" room left to obtain a fit to the latter. (This is because $I_1 = I_2$ in both shear and planar extension.) A recent non-K–BKZ *molecular* constitutive equation derived from reptation-related ideas shows improved qualitative agreement with planar extensional data (McLeish and Larson 1998).

The K–BKZ and other integral constitutive equations discussed above can be regarded as generalizations of the Lodge integral, Eq. (3-24). The upper-convected Maxwell (UCM) equation, which is the differential equivalent of the one-relaxation-time Lodge equation, can also be generalized to make possible more realistic predictions of nonlinear phenomena. Many differential constitutive equations of the Maxwell type have been proposed; most of them are of the following form:

$$\overset{\triangledown}{\sigma} + \frac{1}{\tau}\sigma + \mathbf{G}(\sigma, \mathbf{D}) + \mathbf{H}(\sigma) = 2G\mathbf{D} \qquad (3\text{-}80)$$

The stress tensor in this expression has been defined such that at equilibrium $\sigma = \mathbf{0}$, rather than $G\delta$. Some of the proposed possible forms for the functions $\mathbf{G}(\sigma, \mathbf{D})$ and $\mathbf{H}(\sigma)$ are

$$\mathbf{G} = \xi(\mathbf{D}\cdot\sigma + \sigma\cdot\mathbf{D}), \qquad \mathbf{H} = \mathbf{0} \quad \text{(Johnson and Segalman 1977)} \qquad (3\text{-}81\text{a})$$

$$\mathbf{G} = \mathbf{0}, \qquad \mathbf{H} = \alpha\frac{1}{\tau G}\sigma:\sigma \quad \text{(Giesekus 1966, 1982)} \qquad (3\text{-}81\text{b})$$

$$\mathbf{G} = \xi(\mathbf{D}\cdot\sigma + \sigma\cdot\mathbf{D}),$$

$$\mathbf{H} = \frac{1}{\tau}\left[\exp\left(\frac{\alpha}{G}\,\text{tr}\,\sigma\right) - 1\right]\sigma \quad \text{(Phan-Thien and Tanner 1977, 1978)} \qquad (3\text{-}81\text{c})$$

$$\mathbf{G} = \frac{2}{3}\frac{\alpha}{G}\mathbf{D}:\sigma(\sigma + G\delta), \qquad \mathbf{H} = \mathbf{0} \quad \text{(Larson 1984b)} \qquad (3\text{-}81\text{d})$$

$$\mathbf{G} = \alpha(2\mathbf{D}:\mathbf{D})^{1/2}\sigma, \qquad \mathbf{H} = \mathbf{0} \quad \text{(White and Metzner 1963)} \qquad (3\text{-}81\text{e})$$

The notation "tr" stands for the trace of the tensor. These expressions contain parameters, such as ξ and α, that must be obtained by fits to nonlinear rheological data. None of these equations fit time-dependent experimental data well unless a spectrum of relaxation modes is introduced in a way analogous to that described above for the UCM equation. That is, G, τ, and σ in Eqs. (3-80) and (3-81a)–(3-81e) must be subscripted with a mode index i; and the total stress must be given by a sum of the stresses from all modes, as in Eq. (3-37). Comparisons of the predictions of the various equations to rheological data for melts are presented elsewhere (Larson 1988; Bird et al. 1987a, 1987b; Tanner 1985).

> • Worked Example 3.14 and Problem 3.15 help you derive predictions from phenomeno-logical constitutive equations.

3.8 SUMMARY

The steady-state and transient behavior of solutions of long polymer molecules that are dilute enough that coils only rarely overlap can be described by beads-and-springs models. In steady flows, the shear thinning and strong extension hardening of dilute solutions are predicted by simple finitely extensible dumbbell models. Time-dependent phenomena can in principle be predicted by using more complex models with multiple beads and springs, and with hydrodynamic interactions accounted for using the Zimm theory.

If the molecules are rather short ($M < M_c \approx 3M_e$), they remain unentangled even in the melt. In that case, the low flow-rate properties of the melt seem to be described by the simple Rouse theory, whose rheological predictions are similar to the dumbbell model. At flow rates or frequencies that are high compared to the inverse of the longest relaxation time, the rheological behavior of such melts is influenced by "glassy" relaxation modes, similar to those influencing the behavior of small-molecule glass-forming liquids discussed in Chapter 4.

Melts and dense solutions of long polymer molecules ($M \gg M_e$) are entangled, and they can in general be described by versions of the de Gennes' reptation theory. Reptation theory assumes that long-range polymer motion occurs by a snake-like motion; thus, the gross molecular motion is confined to a tube that prohibits significant lateral motion. The escape from the tube is slow, leading at high molecular weights to a long plateau in the linear storage modulus, G'. Thus, for small deformations at frequencies in the plateau region, the material behaves almost elastically. If the sample is highly deformed, the tube is deformed, but a rapid retraction of the chain within it is permitted; this retraction accounts for the nonlinear strain softening in the shear modulus. Based on reptation and retraction, Doi and Edwards developed a constitutive model for concentrated solutions and melts. With some approximations, it reduces to a constitutive equation of K–BKZ class that requires only two parameters to define the linear viscoelastic response and none at all for the nonlinear response.

The Doi–Edwards model has been extended to allow processes of "primitive-path fluctuations," "constraint release," and "tube stretching." These extensions of the theory allow accurate prediction of many steady-state and time-dependent phenomena, including shear thinning, stress overshoots, and so on. Predictions of strain localization and slip at walls may in the future be possible using reptation ideas. The theory can be extended to account for the effects of simple branching and polydispersity. For complex commercial melts, however, it may be more convenient for some purposes to fit semiempirical constitutive equations, such as the K–BKZ equation or Maxwell-like differential equations, to experimental data.

REFERENCES

Acierno D, Titomanlio G, Marrucci G (1974). *J Polym Sci Polym Phys Ed* 12:2177.

Amelar S, Eastman CE, Morris RL, Smeltzly MA, Lodge TP, von Meerwall ED (1991). *Macromolecules* 24:3506.

Archer LA, Chen Y-L, Larson RG (1995). *J Rheol* 39:519.

Armstrong RC, Gupta SK, Basaran O (1980). *Polym Eng Sci* 20:466.

Baird DG, Dimitris IC (1995). *Polymer Processing: Principles and Design*, Butterworths, London.

Ball RC, McLeish TCB (1989). *Macromolecules* 22:1911.

Bercea M, Peiti C, Simionescu B, Navard P (1993). *Macromolecules* 26:7095.

Berker A, Chynoweth S, Klomp UC, Michopoulos Y (1992). *J Chem Soc Faraday Trans* 88:1719.

Bernstein B, Kearsley EA, Zapas LJ (1963). *Trans Soc Rheol* 7:391.

Berry GC, Fox TG (1968). *Adv Polym Sci* 5:261.

Bianchi U, Peterlin A (1968). *J Polym Sci A-2* 6:1011.

Bick DK, McLeish TCB (1996). *Phys Rev Lett* 76:2587.

Bird RB, Armstrong RC, Hassager O (1977). *Dynamics of Polymeric Liquids*, Vol 1, 1st ed, Wiley, New York.

Bird RB, Armstrong RC, Hassager O (1987a). *Dynamics of Polymeric Liquids*, Vol 1, 2nd ed, Wiley, New York.

Bird RB, Curtiss CF, Armstrong RC, Hassager O (1987b). *Dynamics of Polymeric Liquids*, Vol 2, 2nd ed, Wiley, New York.

Bishko G, McLeish TCB, Harlen O, Larson RG (1997). *Phys Rev Lett* 79:2352.

Boger DV, Binnington R (1977). *Trans Soc Rheol* 221:515.

Brochard F, de Gennes PG (1992). *Langmuir* 8:3033.

Brown EF, Burghardt WR, Kahvand H, Venerus DC (1995). *Rheol Acta* 34:221.

Byars JA, Öztekin A, Brown RA, McKinley GH (1994). *J Fluid Mech* 271:173.

Carella JM, Gotro JT, Graessley WW (1986). *Macromolecules* 19:659.

Cates ME, McLeish TCB, Marrucci G (1993). *Europhys Lett* 21:451.

Cathey CA, Fuller GG (1990). *J Non-Newt Fluid Mech* 34:63.

Chemical Week (1997). May 21:23.

Colvin R (1997). *Mod Plastics*, May:62.

Cottrell FR, Merrill EQ, Smith KA (1969). *J Polym Sci A-2* 7:1415.

Currie PK (1980). In *Rheology, Proceedings of the Eighth International Congress on Rheology*, Naples, Italy.

Decruppe JP, Cressely R, Makhloufi R, Cappelaere E (1995). *Colloid Polym Sci* 273:346.

de Gennes PG (1971). *J Chem Phys* 55:572.

de Gennes PG (1974). *J Chem Phys* 60:5030.

de Gennes PG (1975). *J Phys (Paris)* 36:1199.

de Gennes PG (1979). *Scaling Concepts in Polymer Physics*, Cornell University Press, Ithaca, New York.

Dealy JM, Wissbrun KF (1990). *Melt Rheology and Its Role in Plastics Processing*, Van Nostrand Reinhold, New York.

des Cloizeaux J (1988). *J Europhys Lett* 5:437; 6:475.

Doi M, Edwards SF (1978a). *J Chem Soc Faraday Trans II* 74:1789, 1802, 1818.

Doi M, Edwards SF (1978b). *J Chem Soc Faraday Trans II* 74:560, 918.

Doi M, Edwards SF (1979). *J Chem Soc Faraday Trans II* 75:38, 918.

Doi M, Edwards SF (1986). *The Theory of Polymer Dynamics*, Oxford University Press, New York.

Doi M, Kuzuu NY (1980). *J Polym Sci Polym Lett Ed* 18:775.

Doyle PS, Shaqfeh ESG, Gast AP (1997). *J Fluid Mech* 334:251.

Doyle PS, Shaqfeh ESG, McKinley GH, Spiegelberg SH (1998). *J Fluid Mech,* submitted.

Dünweg B, Kremer K (1991). *Phys Rev Lett* 66:2996.

Edwards SF (1967). *Proc Phys Soc* 92:9.

Einaga Y, Osaki K, Kurata M, Kimura S, Tamura M (1971). *Polym J* 2:550.

Ferry JD (1980). *Viscoelastic Properties of Polymers*, 3rd ed, Wiley, New York.

Fetters LJ, Kiss AD, Pearson DS, Quack GF, Vitus FJ (1993). *Macromolecules* 26:647.

Fetters LJ, Lohse DJ, Richter D, Witten TA, Zirkel A (1994). *Macromolecules* 27:4639.

Fixman M (1966). *J Chem Phys* 45:785; 793.

Flory PJ (1953). *Principles of Polymer Chemistry*, Cornell University Press, Ithaca, NY.

Flory PJ (1969). *Statistical Mechanics of Chain Molecules*, Wiley, New York, p 40.

Flory PJ, Rehner J (1943). *J Chem Phys* 11:512.

Fukuda M, Osaki K, Kurata M (1975). *J Polym Sci Polym Phys Ed* 13:1563.

Fuller GG (1995). *Optical Rheometry of Complex Fluids*, Oxford University Press, New York.

Fuller GG, Leal LG (1980). *Rheol Acta* 19:580.

Fuller GG, Leal LG (1981). *J Non-Newt Fluid Mech* 8:271.

Gao J, Weiner JH (1992). *Macromolecules* 25:3462.

Gao J, Weiner JH (1994). *Macromolecules* 27:1201.

Giesekus H (1962). *Rheol Acta* 2:50.

Giesekus H (1966). *Rheol Acta* 5:29.

Giesekus H (1982). *J Non-Newt Fluid Mech* 11:69.

Graessley WW (1982). *Adv Polym Sci* 47:68.

Green MS, Tobolsky AV (1946). *J Chem Phys* 14:80.

Hall CK, Helfand E (1982). *J Chem Phys* 77:3275.

Helfand E, Wasserman ZR, Weber TA (1980). *Macromolecules* 13:526.

Hinch EJ (1974). In *Proceedings of the Symposium on Polymer Lubrification,* Brest.

Hinch EJ (1994). *J Non-Newt Fluid Mech* 54:209.

Horn A, Merrill EW (1984). *Nature* 312:140.

Inoue T, Okamoto H, Osaki K (1991). *Macromolecules* 24:5670.

Inoue T, Hwang EJ, Osaki K (1992). *J Rheol* 36:1737.

James HM, Guth E (1943). *J Chem Phys* 11:455.

Janeschitz-Kriegl H (1983). *Polymer Melt Rheology and Flow Birefringence*, Springer-Verlag, New York.

Johnson RM, Schrag JL, Ferry JD (1970). *Polym J* 1:742.

Johnson MW Jr, Segalman D (1977). *J Non-Newt Fluid Mech* 2:225.

Kannan RM, Kornfield JA (1994). *J Rheol* 38:1127.

Kausch HH (1985). *Colloid Polym Sci* 263:306.

Kaye A (1962). College of Aeronautics, Cranford, UK, Note No 134.

Keentok M, Georgescu AG, Sherwood AA, Tanner RI (1980). *J Non-Newt Fluid Mech* 6:303.

Keller A, Odell JA (1985). *Colloid Polym Sci* 263:181.

Keunings R (1997). *J Non-Newt Fluid Mech* 68:85.

Kirkwood JG, Riseman J (1948). *J Chem Phys* 16:565.

Kishbaugh AJ, McHugh AJ (1990). *J Non-Newt Fluid Mech* 34:181.

Kolkka RW, Malkus DS, Rose TR (1991). *Rheol Acta* 30:430.

Kotaka T, Suzuki H, Inagaki H (1966). *J Chem Phys* 45:2770.

Kremer K, Grest GS (1990). *J Chem Phys* 92:5057.

Kröger M, Loose W, Hess S (1993). *J Rheol* 37:1057.

Kröger M (1995). *Rheology* 95(5):66.

Kuhn W (1934). *Kolloid-Z* 68:2.

Kuhn W, Grün F (1942). *Kolloid Z* 101:248.

Landry CJ (1985). PhD Thesis, University of Wisconsin.

Larson RG (1984a). In *Rheology, Proceedings of the Ninth International Congress on Rheology,* Acapulco, Mexico.

Larson RG (1984b). *J Rheol* 28:545.

Larson RG (1988). *Constitutive Equations for Polymer Melts and Solutions*, Butterworths, London.

Larson RG (1990). *Rheol Acta* 29:371.

Larson RG, Khan SA, Raju VR (1988). *J Rheol* 32:145.

Larson RG, Goyal S, Aloisio C (1996). *Rheol Acta* 35:252.

Larson RG, Mead DW, Doi M (1998). *Macromolecules*, in press.

Larson RG, Perkins TT, Smith DE, Chu S (1997). *Phys Rev* E 55:1794.

Larson RG (1998). *J Rheol,* submitted.

Laun HM (1978). *Rheol Acta* 17:1.

Laun HM (1984). In *Proceedings of the Ninth International Congress on Rheology*, Acapulco, Mexico.

Lee EC, Solomon MJ, Muller SJ (1997). *Macromolecules,* 30:7313.

Lin Y-H (1984). *J Rheol* 28:1.

Lin Y-H (1987). *Macromolecules* 20:3080.

Link A, Springer J (1993). *Macromolecules* 26:464.

Lodge AS (1956). *Trans Faraday Soc* 52:120.

Lodge AS (1968). *Rheol Acta* 7:379.

Lodge AS (1989). *J Rheol* 33:821.

Lodge AS, Wu Y (1972). University of Wisconsin Rheology Research Center Report No 19.

Lodge TP (1993). *J Phys Chem* 97:1480.

Lodge TP, Miller JW, Schrag JL (1982). *J Polym Sci: Polym Phys Ed* 20:1409.

Lodge TP, Rotstein NA, Prager S (1990). *Adv Chem Phys* 79:1.

Mackay ME, Boger DV (1987). *J Non-Newt Fluid Mech* 22:235.

Macosko CW (1994). *Rheology Principles, Measurements, and Applications*, VCH Publishers, New York.

Magda JJ, Larson RG, Mackay ME, (1988). *J Chem Phys* 89:2504.

Magda JJ, Lee C-S, Muller SJ, Larson RG (1993). *Macromolecules* 26:1696.

Man V (1984). PhD Thesis, University of Wisconsin.

Marko JF, Siggia ED (1995). *Macromolecules* 28:8759.

Marrucci G (1984). In *Advances in Transport Process*, Vol V, Mujamdar AS, Mashelkar RA (eds), Wiley, New York.

Marrucci G (1996). *J Non-Newt Fluid Mech* 62:279.

Marrucci G, Grizzuti N (1983). *J Rheol* 27:433.

Marrucci G, Grizzuti N (1988). *Gazz Chim Ital* 118:179.

Marrucci G, Ianniruberto G (1997). *Macromol Symp* 117:233.

Massa DJ, Schrag JL, Ferry JD (1971). *Macromolecules* 4:210.

McKinley GH (1996). Private communication.

McLeish TCB (1987). *J Polym Sci Polym Phys Ed* 25:223.

McLeish TCB (1988a). *Macromolecules* 21:1062.

McLeish TCB (1988b). *Europhys Lett* 6:511.

McLeish TCB (1995). *Physics World* March:32.

McLeish TCB, Ball RC (1986). *J Polym Sci Polym Phys Ed* 24:1735.

McLeish TCB, Larson RG (1998). *J Rheol,* 42:81.

Mead DW (1994). *J Rheol* 38:1797.

Mead DW (1996). *J Rheol* 40:633.

Mead DW, Leal LG (1995). *Rheol Acta* 34:339.

Meissner J (1972). *J Appl Polym Sci* 16:2877.

Menasveta MJ, Hoagland DA (1991). *Macromolecules* 24:3427.

Menezes EV, Graessley WW (1980). *Rheol Acta* 19:38.

Menezes EV, Graessley WW (1982). *J Polym Sci Polym Phys Ed* 20:1817.

Migler KB, Hervet H, Leger L (1993). *Phys Rev Lett* 70:287.

Milner ST (1996). *J Rheol* 40:303.

Milner ST, McLeish TCB (1997). *Macromolecules* 30:2159.

Milner ST, McLeish TCB (1998). *Phys Rev Lett,* submitted.

Minoshima W, White J (1986). *J Non-Newt Fluid Mech* 19:275.

Montfort JP, Marin G, Monge P (1986). *Macromolecules* 19:393.

Morawetz H (1979). *Science* 203:405.

Morris RL, Amelar S, Lodge TP (1988). *J Chem Phys* 89:6523.

Morrison FA, Larson RG (1992). *J Polm Sci Polym Phys Ed* 30:943.

Muller R, Froelich D (1985). *Polymer* 26:1477.

Muller SJ, Shaqfeh ESG, Larson RG (1993). *J Non-Newt Fluid Mech* 46:315.

Nguyêñ H and Boger DV (1979). *J Non-Newt Fluid Mech* 5:353.

Noda I, Yamada Y, Nagasawa M (1968). *J Phys Chem* 72:2890.

Onogi S, Masuda T, Kitagawa K (1970). *Macromolecules* 3:109.

Osaki K (1993). *Rheol Acta* 32:429.

Osaki K, Fukuda M, Kurata M (1975). *J Polym Sci Polym Phys Ed* 13:775.

Osaki K, Kimura S, Kurata M (1981). *J Polym Sci Polym Phys Ed* 19:517.

Osaki K, Okamoto H, Inoue T, Hwang E-J (1995). *Macromolecules* 28:3635.

Oseen CW (1910). *Arf Mat Astr Fys* 6(29):1.

Öttinger HC (1985). *J Chem Phys* 83:6535.

Öttinger HC (1986). *J Chem Phys* 84:4068.

Öttinger HC (1987). *J Chem Phys* 86:3731.

Papanastasiou AC, Scriven LE, Macosko CW (1983). *J Rheol* 27:387.

Pearson DS (1987). *Rubber Chem Technol,* Rubber Reviews.

Pearson DS, Helfand E (1984). *Macromolecules* 17:888.

Pearson DS, Kiss AD, Fetters LJ, Doi M (1989). *J Rheol* 33:517.

Pearson DS, Herbolzheimer E, Grizzuti N, Marrucci G (1991). *J Polym Sci B Polym Phys Ed* 29:1589.

Perkins TT, Smith DE, Chu S (1994a). *Science* 264:819.

Perkins TT, Smith DE, Chu S (1994b). *Science* 264:822.

Perkins TT, Smith DE, Larson RG, Chu S (1995). *Science* 268:83.

Perkins TT, Smith DE, Chu S (1997). *Science* 276:2016.

Peterlin A (1960). *J Chem Phys* 33:1799.

Peterlin A (1961). *Makromol Chem.* 44:338.

Phan-Thien N, Tanner RI (1977). *J Non-Newt Fluid Mech* 2:353.

Phan-Thien N, Tanner RI (1978). *J Rheol* 22:259.

Pierleoni C, Ryckaert J-P (1991). *Phys Rev Lett* 66:2992.

Plazek DJ (1965). *J Phys Chem.* 69:3480.

Rabin Y (1987). *J Chem Phys* 86:5215.

Raju VR, Rachapudy H, Graessley WW (1979). *J Polym Sci Polym Phys Ed* 17:1223.

Richter D, Butera R, Fetters LJ, Huang JS, Farago B, Ewen B (1992). *Macromolecules* 25:6156.

Ronca GJ (1983). *J Chem Phys* 79:1031.

Rouse PE Jr (1953). *J Chem Phys* 21:1272.

Rubinstein M, Helfand E, Pearson DS (1987). *Macromolecules* 20:822.

Russell, TP, Deline VR, Dozier WD, Felcher GP, Agrawal G, Wool RP, Mays JW (1993). *Nature* 365:235.

Ryskin G (1987). *J Fluid Mech* 178:423.

Sahouani H, Lodge TP (1992). *Macromolecules* 25:5632.

Sammler RL, Landry CJT, Woltman GR, Schrag JL (1990). *Macromolecules* 23:2388.

Samurkas T, Larson RG, Dealy JM (1989). *J Rheol* 33:559.

Smith SB, Finzi L, Bustamante C (1992). *Science* 258:1122.

Sridhar T, Gupta RK, Boger DV, Binnington R (1986). *J Non-Newt Fluid Mech* 21:115.

Talbott WH, Goddard JD (1979). *Rheol Acta* 18:507.

Tanner RI (1985). *Engineering Rheology*, Oxford University Press, New York.

Tirrell M (1984). *Rubber Chem Technol* 57:523.

Tirtaatmadja V, Sridhar T (1993). *J Rheol* 37:1081.

Treloar LRG (1943). *Trans Faraday Soc* 37:36; 241.

Treloar LRG (1975). *The Physics of Rubber Elasticity*, 3rd ed, Clarendon Press, Oxford.

Tsenoglou C (1987). *ACS Polym Preprints* 28:185.

Tuminello WH (1986). *Polym Eng Sci* 26:1339.

Urakawa O, Takahashi M, Masuda T, Golshan Ebrahimi N (1995). *Macromolecules* 28:7196.

Verhoef MRJ, van den Brule BHAA, Hulsen MA (1997). *J Non-Newt Fluid Mech,* submitted.

Viovy JL, Rubinstein M, Colby RH (1991). *Macromolecules* 24:3587.

Vrentas CM, Graessley WW (1982). *J Rheol* 26:359.

Wagner MH (1976). *Rheol Acta* 15:136.

Wagner MH, Laun HM (1978). *Rheol Acta* 17:138.

Wagner MH, Raible T, Meissner J (1979). *Rheol Acta* 18:427.

Wales JLS (1976). *The Application of Flow Birefringence to Rheological Studies of Polymer Melts,* Delft University Press, Delft, Holland.

Wall FT (1942). *J Chem Phys* 10:485.

Warner HR Jr (1972). *I&EC Fundam* 11:379.

Wasserman SH, Graessley WW (1992). *J Rheol* 36:543.

Weber TA, Helfand E (1983). *J Phys Chem* 87:2881.

White JL, Metzner AB (1963). *J Appl Polym Sci* 8:1367.

Williams ML, Landel RF, Ferry JD (1955). *J Am Chem Soc* 77:3701.

Yamamoto M (1956). *J Phys Soc Jpn* 11:413.

Yamamoto M (1957). *J Phys Soc Jpn* 12:1148.

Yamamoto M (1958). *J Phys Soc Jpn* 13:1200.

Ylitalo CM, Fuller GG, Abetz V, Stadler R, Pearson DS (1990). *Rheol Acta* 29:543.

Zawada JA, Fuller GG, Colby RH, Fetters LJ, Roovers J (1994). *Macromolecules* 27:6851.

Zhu W, Ediger MD (1995). *Macromolecules* 28:7549.

Zhu W, Ediger MD (1997). *Macromolecules* 30:1205.

Zimm BH (1956). *J Chem Phys* 24:269.

Zylka W, Öttinger HC (1989). *J Chem Phys* 90:474.

PROBLEMS AND WORKED EXAMPLES

Problem-3.1 (Worked Example)

From Eq. (3-24) for a "rubber-like liquid," assuming a single relaxation time τ and modulus G, calculate formulas for the extensional viscosity as a function of time after start-up of steady uniaxial extension at extension rate $\dot\varepsilon$.

ANSWER:

For a steady uniaxial extensional flow, the velocity gradient tensor is given by Eq. (1-9):

$$\nabla \mathbf{v} = \begin{pmatrix} \dot\varepsilon & 0 & 0 \\ 0 & -\dot\varepsilon/2 & 0 \\ 0 & 0 & -\dot\varepsilon/2 \end{pmatrix} \tag{A3-1}$$

The Finger deformation tensor for this flow is obtained from Eq. (1-19) with $\lambda_2 = \lambda_3 = \lambda_1^{-1/2}$. Hence, taking $\lambda \equiv \lambda_1$, we obtain

$$\mathbf{B} = \begin{pmatrix} \lambda^2 & 0 & 0 \\ 0 & \lambda^{-1} & 0 \\ 0 & 0 & \lambda^{-1} \end{pmatrix} \tag{A3-2}$$

where $\lambda = \lambda(t', t)$ is the stretch ratio in the direction of uniaxial stretch between times t' and t.

Suppose the sample is completely relaxed in a state of equilibrium until time 0, and then an extensional flow with extension rate $\dot\varepsilon$ begins at time 0. The stretch ratio history in this case is given by

$$\lambda(t', t) = \begin{cases} \exp[\dot\varepsilon(t - t')] & \text{for } t' > 0 \\ \exp(\dot\varepsilon t) & \text{for } t' \le 0 \end{cases} \tag{A3-3}$$

Note that for $t' < 0$, $\lambda(t', t) = \exp(\dot\varepsilon t)$ is independent of t', since the sample is not being stretched at times less than 0. From Eq. (A3-2), we obtain the components of the Finger tensor:

$$B_{11}(t', t) = \begin{cases} \exp[2\dot\varepsilon(t - t')] & \text{for } t' > 0 \\ \exp(2\dot\varepsilon t) & \text{for } t' \le 0 \end{cases} \tag{A3-4}$$

$$B_{22}(t', t) = B_{33}(t', t) = \begin{cases} \exp[-\dot\varepsilon(t - t')] & \text{for } t' > 0 \\ \exp(-\dot\varepsilon t) & \text{for } t' \le 0 \end{cases} \tag{A3-5}$$

We now wish to solve Eq. (3-24), the equation for a "rubber-like liquid":

$$\boldsymbol\sigma = \int_{-\infty}^{t} m(t - t')\mathbf{B}(t', t)\, dt' \tag{A3-6}$$

where we consider a single relaxation time so that $m(t - t') \equiv (G/\tau) \exp[(t' - t)/\tau]$. To compute the stress component σ_{11}, we insert Eq. (A3-4) into the "11" component of Eq. (A3-6); we must break the integral into two pieces:

$$\sigma_{11} = \int_{-\infty}^{0} \frac{G}{\tau} e^{(t' - t)/\tau} e^{2\dot\varepsilon t}\, dt' + \int_{0}^{t} \frac{G}{\tau} e^{(t' - t)/\tau} e^{2\dot\varepsilon(t - t')}\, dt' \tag{A3-7}$$

Likewise, the "22" component of Eq. (A3-6) yields

$$\sigma_{22} = \int_{-\infty}^{0} \frac{G}{\tau}\, e^{(t'-t)/\tau}\, e^{-\dot{\varepsilon}t}\, dt' + \int_{0}^{t} \frac{G}{\tau}\, e^{(t'-t)/\tau}\, e^{-\dot{\varepsilon}(t-t')}\, dt' \qquad \text{(A3-8)}$$

Carrying out these integrations gives

$$\sigma_{11} = Ge^{-(1-2\dot{\varepsilon}\tau)t/\tau} + \frac{G}{1-2\dot{\varepsilon}\tau}\,[1 - e^{-(1-2\dot{\varepsilon}\tau)t/\tau}] \qquad \text{(A3-9)}$$

$$\sigma_{22} = Ge^{-(1+\dot{\varepsilon}\tau)t/\tau} + \frac{G}{1+\dot{\varepsilon}\tau}\,[1 - e^{-(1+\dot{\varepsilon}\tau)t/\tau}] \qquad \text{(A3-10)}$$

Note that if $\dot{\varepsilon} > 1/(2\tau)$, σ_{11} diverges exponentially as time increases. If, however, $\dot{\varepsilon} < 1/(2\tau)$, a steady-state is reached in which

$$\sigma_{11} = \frac{G}{1-2\dot{\varepsilon}\tau}, \qquad \sigma_{22} = \frac{G}{1+\dot{\varepsilon}\tau} \qquad \text{(A3-11)}$$

The steady-state uniaxial extensional viscosity $\bar{\eta}_u$ is given by $(\sigma_{11} - \sigma_{22})/\dot{\varepsilon}$. Note that in the limit of small $\dot{\varepsilon}$, $\bar{\eta}_u$ approaches the Trouton limit, $\bar{\eta}_u \to 3G\tau$.

Problem 3.2 (Worked Example)

From Eq. (3-32) for a dilute solution of Hookean elastic dumbbells with relaxation time τ and modulus G, calculate polymer contributions to the extensional viscosity as a function of time after start-up of steady extension at extension rate $\dot{\varepsilon}$.

ANSWER:

Equation (3-32) is

$$\overset{\triangledown}{\boldsymbol{\sigma}^P} + \frac{1}{\tau}(\overset{\triangledown}{\boldsymbol{\sigma}^P} - G\boldsymbol{\delta}) = \mathbf{0} \qquad \text{(A3-12)}$$

where the "$\triangledown$" above $\boldsymbol{\sigma}^P$ is the upper-convected time derivative, defined in Eq. (3-33). Using this definition, we get Eq. (3-30):

$$\dot{\boldsymbol{\sigma}}^P - \boldsymbol{\nabla}\mathbf{v}^T \cdot \boldsymbol{\sigma}^P - \boldsymbol{\sigma}^P \cdot \boldsymbol{\nabla}\mathbf{v} + \frac{1}{\tau}(\boldsymbol{\sigma}^P - G\boldsymbol{\delta}) = \mathbf{0} \qquad \text{(A3-13)}$$

Note first that if the fluid is at a state of equilibrium with no flow, then the time derivative $\dot{\boldsymbol{\sigma}}^P$ is equal to zero, and the velocity gradient $\boldsymbol{\nabla}\mathbf{v}$ is also zero. This implies from the above equations that $\boldsymbol{\sigma}^P = G\boldsymbol{\delta}$. Hence $\sigma_{11}^P = \sigma_{22}^P = \sigma_{33}^P = G$ at equilibrium, and $\sigma_{ij}^P = 0$, for $i \neq j$. Thus, although the diagonal stress components are not zero at equilibrium, they are all equal to each other, and *the nondiagonal components are all equal to zero*. Hence, the stress tensor is *isotropic*, but nonzero at equilibrium. (If one redefines the stress tensor as $\boldsymbol{\Sigma}^P \equiv \boldsymbol{\sigma}^P - G\boldsymbol{\delta}$, then $\boldsymbol{\Sigma}^P = \mathbf{0}$ at equilibrium. The upper-convected Maxwell equation can then be rewritten in terms of $\boldsymbol{\Sigma}^P$.)

For a steady uniaxial extensional flow, we can assume by symmetry that the stress tensor contains only diagonal components. We can then evaluate the terms in Eq. (A3-13) containing the velocity gradient by using Eq. (A3-1):

$$(\boldsymbol{\nabla}\mathbf{v})^T \cdot \boldsymbol{\sigma}^P = \begin{pmatrix} \dot{\varepsilon} & 0 & 0 \\ 0 & -\dot{\varepsilon}/2 & 0 \\ 0 & 0 & -\dot{\varepsilon}/2 \end{pmatrix} \cdot \begin{pmatrix} \sigma_{11} & 0 & 0 \\ 0 & \sigma_{22} & 0 \\ 0 & 0 & \sigma_{33} \end{pmatrix}$$

$$
= \begin{pmatrix} \sigma_{11}^{P}\dot{\varepsilon} & 0 & 0 \\ 0 & \dfrac{-\sigma_{22}^{P}\dot{\varepsilon}}{2} & 0 \\ 0 & 0 & \dfrac{-\sigma_{33}^{P}\dot{\varepsilon}}{2} \end{pmatrix} \tag{A3-14}
$$

Since $\nabla \mathbf{v}$ is a diagonal tensor, and hence symmetric, the term $\boldsymbol{\sigma}^{P} \cdot \nabla \mathbf{v}$ equals $\nabla \mathbf{v}^{T} \cdot \boldsymbol{\sigma}^{P}$. Thus, for a steady uniaxial extensional flow, Eq. (A3-13) can be written as

$$
\begin{pmatrix} \dot{\sigma}_{11}^{P} & 0 & 0 \\ 0 & \dot{\sigma}_{22}^{P} & 0 \\ 0 & 0 & \dot{\sigma}_{33}^{P} \end{pmatrix} - \begin{pmatrix} 2\sigma_{11}^{P}\dot{\varepsilon} & 0 & 0 \\ 0 & -\sigma_{22}^{P}\dot{\varepsilon} & 0 \\ 0 & 0 & -\sigma_{33}^{P}\dot{\varepsilon} \end{pmatrix}
$$
$$
+ \frac{1}{\tau}\begin{pmatrix} \sigma_{11}^{P} - G & 0 & 0 \\ 0 & \sigma_{22}^{P} - G & 0 \\ 0 & 0 & \sigma_{33}^{P} - G \end{pmatrix} = \begin{pmatrix} 0 & 0 & 0 \\ 0 & 0 & 0 \\ 0 & 0 & 0 \end{pmatrix} \tag{A3-15}
$$

This tensor equation can be broken down into three scalar equations:

$$
\dot{\sigma}_{11}^{P} - 2\dot{\varepsilon}\sigma_{11}^{P} + \frac{1}{\tau}(\sigma_{11}^{P} - G) = 0 \tag{A3-16}
$$

$$
\dot{\sigma}_{22}^{P} + \dot{\varepsilon}\sigma_{22}^{P} + \frac{1}{\tau}(\sigma_{22}^{P} - G) = 0 \tag{A3-17}
$$

$$
\dot{\sigma}_{33}^{P} + \dot{\varepsilon}\sigma_{33}^{P} + \frac{1}{\tau}(\sigma_{33}^{P} - G) = 0 \tag{A3-18}
$$

These can be solved once an initial condition is specified. If the fluid is initially at equilibrium and we turn on the steady extensional flow at time 0, we have the initial conditions $\sigma_{11}^{P}(0) = \sigma_{22}^{P}(0) = \sigma_{33}^{P}(0) = G$. Then, rewriting Eq. (A3-16), we obtain

$$
e^{(2\dot{\varepsilon}-1/\tau)t}\frac{d}{dt}\left(e^{-(2\dot{\varepsilon}-1/\tau)t}\sigma_{11}^{P}\right) = \frac{G}{\tau} \tag{A3-19}
$$

Multiplying through by $\exp[-(2\dot{\varepsilon} - 1/\tau)t]$ and integrating gives

$$
e^{-(2\dot{\varepsilon}\tau-1)t/\tau}\sigma_{11}^{P}(t) - \sigma_{11}^{P}(0)
$$

$$
= \frac{G}{\tau}\int_{0}^{t} e^{-(2\dot{\varepsilon}\tau-1)t'/\tau}dt' = -\frac{G}{2\dot{\varepsilon}\tau - 1}[e^{-(2\dot{\varepsilon}\tau-1)t/\tau} - 1] \tag{A3-20}
$$

where $\sigma_{11}^{P}(0) = G$. We can then multiply through by $\exp[(2\dot{\varepsilon}\tau - 1)t/\tau]$ and rearrange to obtain

$$
\sigma_{11}^{P}(t) = Ge^{-(1-2\dot{\varepsilon}\tau)t/\tau} + \frac{G}{1 - 2\dot{\varepsilon}\tau}[1 - e^{-(1-2\dot{\varepsilon}\tau)t/\tau}] \tag{A3-21}
$$

Similarly, we obtain for the other two stress components:

$$
\sigma_{22}^{P}(t) = \sigma_{33}^{P}(t) = Ge^{-(1+\dot{\varepsilon}\tau)t/\tau} + \frac{G}{1 + \dot{\varepsilon}\tau}[1 - e^{-(1+\dot{\varepsilon}\tau)t/\tau}] \tag{A3-22}
$$

Note that these stresses are *identical* to those obtained with the integral equations, Eqs. (A3-9) and (A3-10).

We leave it as an exercise for the reader to derive the corresponding equations for the shear stress and normal stresses in start-up of steady shearing, using both the integral and differential equations.

Problem 3.3 From Eq. (3-4), estimate the "overlap concentration" in g/cm^3 for polystyrene of molecular weight 10^7 daltons in a theta solvent. Compare this to the value obtained from Eq. (3-8), where the intrinsic viscosity for a theta solvent is given by $[\eta]_0 = K_\theta M^{1/2}$, with K_θ given by Eq. (3-49).

Problem 3.4(a) (Worked Example) Suppose you want to estimate the viscosity of a very viscous colloidal gel, at 25°C. You pour the unaggregated liquid dispersion, or "sol," into a spherical mold, add a bit of acid, and voilà!, it gels in a minute or so. Now you carefully remove the gel from the mold; it is a 10-cm-diameter spherical ball. You place this ball on a flat surface in a humidity chamber at 25°C so that the gel does not dry out. After 1000 hours, you find that the gel has sagged so that its height has decreased to 90% of the original height. By using dimensional analysis and assuming that at the slow rate of sagging the gel is Newtonian, determine the scaling law showing how the time T to sag a given amount scales with sphere radius R, gel viscosity η, gel density ρ, and gravitational constant g. You do not need to derive an exact formula, only a law of proportionality.

ANSWER:

Consider the momentum and mass balance equations in the absence of inertia for an incompressible material:

$$-\nabla p + \nabla \cdot \boldsymbol{\sigma} = \rho g \mathbf{e}_z, \quad \nabla \cdot \mathbf{v} = 0 \qquad \text{(A3-23)}$$

where $\mathbf{e}_z$ is the downward-pointing unit vector. The gel is a very viscous, yet Newtonian, liquid at the conditions of the experiment. Hence, $\nabla \cdot \boldsymbol{\sigma} = \eta \nabla^2 \mathbf{v}$. The momentum balance equation then becomes

$$-\nabla p + \eta \nabla^2 \mathbf{v} = \rho g \mathbf{e}_z \qquad \text{(A3-24)}$$

Now, we perform a dimensional analysis. Let t be the time for the gel to sag 10%. Let all quantities superscripted by an asterisk be dimensionless, The velocity $\mathbf{v}$, gradient operator ∇, and pressure p can then be rescaled as

$$\mathbf{v} = \frac{R}{t}\mathbf{v}^*, \qquad \nabla = \frac{1}{R}\nabla^*, \qquad p = \frac{\eta}{t}p^* \qquad \text{(A3-25)}$$

Then we can rewrite Eq. (A3-24) as

$$-\left(\frac{\eta}{tR}\right)\nabla^* p^* + \left(\frac{\eta}{tR}\right)\nabla^{*2}\mathbf{v}^* = \rho g \mathbf{e}_z$$

Multiplying through by tR/η gives

$$-\nabla^* p^* + \nabla^{*2}\mathbf{v}^* = \frac{tR\rho g}{\eta}\mathbf{e}_z \qquad \text{(A3-26)}$$

The boundary conditions are zero stress at the gel surfaces that contact air (we neglect surface tension) and zero velocity at the surfaces that contact the flat solid surface. Thus all boundary conditions can be written so that the right sides are zero, and they rescale trivially. Likewise, the mass balance equation in Eq. (A3-23) rescales trivially.

Equation (A3-26) along with the mass balance equation and the boundary conditions form a linear problem which has a unique solution for each value of the coefficient $tR\rho g/\eta$. The same solution is therefore obtained if two or more constants in the coefficient $tR\rho g/\eta$ are varied in off-setting ways so that $tR\rho g/\eta$ remains constant. Hence, it follows that the sag time t must obey

$$t \propto \frac{\eta}{R\rho g} \tag{A3-27}$$

Problem 3.4(b) (Worked Example) Suppose you now mold a 1-cm-diameter spherical ball of 1,4-polyisoprene and place it on a flat surface at 25°C and find that the time for it to sag to 90% of its original height is 10 minutes. Now you place the polyisoprene in a rheometer at 25°C to measure its viscosity, but the viscosity is too high to measure accurately, so you raise the temperature to 100°C and measure a zero-shear viscosity of 10^5 P. Use this information and that in Problem 3.4(a) to determine the viscosity of the gel in Problem 3.4(a), given that the gel density is $\rho_{gel} = 3 \text{g/cm}^3$.

ANSWER:

From Table 3-3, the polyisoprene density is $\rho_{pi} = 0.830 \text{ g/cm}^3$, and the viscoelastic shift factor obeys the WLF relationship

$$\log_{10} a_T = -\frac{c_1^0(T - T_0)}{(T - T_\infty)} \tag{A3-28}$$

where for 1,4-polyisoprene $c_1^0 = 8.86$, $T_\infty = 146$ K, T_0 is the reference temperature (at which we take a_T to be unity), and T is the temperature at which the rheological properties are measured. We can choose T_0 to be the temperature of the sag experiments, T_0 to be 298 K, and T to be the temperature at which the viscosity of the polyisoprene is measured. According to Eq. (A3-28), we have

$$\log_{10} a_T = -\frac{8.86(373 - 298)}{373 - 146} = -2.93$$

Therefore, $a_T = 0.0012$. This means that the longest relaxation time τ of the polyisoprene at 100°C is 0.0012 times its value at 25°C. The viscosity η changes roughly in proportion to the relaxation time, if the small vertical shift factor is neglected. Thus,

$$\frac{\eta(25°C)}{\eta(100°C)} \approx \frac{\tau(25°C)}{\tau(100°C)} = \frac{1}{a_T} = 846$$

Because η for polyisoprene at 100°C is 10^5 P, we find that η at 25°C is 8.46×10^7 P.

Since the time for the gel to sag 10% at 25°C is 1000 hours, we can use the scaling law in Eq. (A3-27) to obtain

$$\text{(time for gel to sag 10\%)} = \left(\frac{R_{pi}}{R_{gel}}\right) \cdot \left(\frac{\rho_{pi}}{\rho_{gel}}\right) \cdot \left(\frac{\eta_{gel}}{\eta_{pi}}\right) \cdot \text{(time for pi to sag 10\%)}$$

where "pi" denotes "polyisoprene." Plugging in the information in the statements of Problems 3.4(a) and 3.4(b), we obtain

$$1000 \text{ hr} = \frac{1}{10} \cdot \frac{0.83}{3} \cdot \frac{\eta_{gel}}{8.46 \times 10^7 \text{ P}} \cdot \frac{1}{6} \text{ hr}$$

From this we find

$$\eta_{gel} \approx 2 \times 10^{13} \text{ P}$$

Problem 3.5 Using the "universal hydrodynamic constant" $\Phi = 2.5 \times 10^{21}$ dL cm^{-3} mol^{-1} (where "dl" is deciliters), calculate the intrinsic viscosity for polyethylene of molecular weight $M = 10^6$ daltons. Note that for polyethylene, $C_\infty = 7.3$.

Problem 3.6 Derive Eq. (3-60) for the dumbbell model with a Warner spring.

Problem 3.7 Consider an entangled melt of linear flexible polymer chains of molecular weight $M = 100{,}000$ and zero-shear viscosity $\eta_0 = 10^4$ P. If M is increased to 300,000, what would η_0 be?

Problem 3.8 Numerically solve Eq. (3-78), the differential approximation to the Doi–Edwards equation for entangled linear melts, in a steady-state shearing flow. Plot the dimensionless shear stress σ_{12}/G against Weissenberg number $Wi \equiv \dot\gamma\tau$ for Wi between 0.1 and 100.

Problem 3.9 Integrate the Doi–Edwards equation (3-71) using the Currie expression for the **Q** tensor, Eq. (3-75), for steady-state shearing, for $\dot\gamma\tau = 0.1$, 0.3, 1.0, 3.0, and 10.0, using only one relaxation time in the spectrum. Plot the values of dimensionless shear stress σ_{12}/G versus $\dot\gamma\tau$ on the same plot as in Problem 3.8. How close is the prediction of the approximate differential model to that of the "exact" integral model?

Problem 3.10 Suppose we approximate fast "Rouse" modes of entangled polymer molecules by a viscous stress, $\boldsymbol{\sigma}^R \approx 2\eta^R \mathbf{D}$, with $\eta^R = 0.01G\tau$, and let the total shear stress $\sigma_{12}^{\text{total}} = \sigma_{12} + \sigma_{12}^R$, where σ_{12} is the stress computed in Problem 3.8. (See, for example, Fig. 3-33.) By plotting $\sigma_{12}^{\text{total}}/G$ against $\dot\gamma\tau$, estimate the values of $(\dot\gamma\tau)_{\max}$ and $(\dot\gamma\tau)_{\min}$ at which $\sigma_{12}^{\text{total}}$ has a relative maximum and a relative minimum. What will happen in a rheometer if one attempts to impose a shear rate $\dot\gamma$ with $(\dot\gamma\tau)_{\max} < \dot\gamma\tau < (\dot\gamma\tau)_{\min}$?

Problem 3.11 A theory incorporating "convective constraint release" and "chain stretch" into the Doi–Edwards model gives the constitutive equations below (Larson et al. 1998):

$$\frac{1}{\tau} = \frac{1}{\lambda^2\tau_d} + e^{-(\lambda-1)}\left(\mathbf{D} : \mathbf{S} - \frac{\dot\lambda}{\lambda}\right) \tag{A3-29a}$$

$$\mathbf{S} = \int_{-\infty}^{t} \frac{1}{\tau(t')} \exp\left[-\int_{t'}^{t} \frac{dt''}{\tau(t'')}\right]\hat{\mathbf{Q}}\,dt' \tag{A3-29b}$$

$$\dot\lambda = \lambda\mathbf{D} : \mathbf{S} - \frac{1}{\tau_r}(\lambda-1) - \frac{1}{2}\left(\mathbf{D} : \mathbf{S} - \frac{\dot\lambda}{\lambda}\right)(\lambda-1) \tag{A3-29c}$$

$$\boldsymbol{\sigma} = 5G_N^0\lambda^2\mathbf{S} \tag{A3-29d}$$

In the above, λ is the "chain stretch," which is greater than unity when the flow is fast enough (i.e., $\dot\gamma\tau_r > 1$) that the retraction process is not complete, and the chain's "primitive path" therefore becomes stretched. This magnifies the stress, as shown by the multiplier λ^2 in the equation for the stress tensor $\boldsymbol{\sigma}$, Eq. (3-78d). The tensor $\hat{\mathbf{Q}}$ is defined as $\mathbf{Q}/5$, where $\mathbf{Q}$ is defined by Eq. (3-70). Convective constraint release is responsible for the last terms in Eqns. (A3-29a) and (A3-29c); these cause the orientation relaxation time τ to be shorter than the reptation time τ_d and reduce the chain stretch λ. Derive the predicted dependence of the dimensionless shear stress σ_{12}/G_N^0 and the first normal stress difference N_1/G_N^0 on the dimensionless shear rate $\dot\gamma\tau_r$ for $\tau_d/\tau_r = 50$ and compare your results with those plotted in Fig. 3-35.

Problem 3.12 Suppose you have monodispersed star-branched melts of some polymer with molecular weights 20,000, 40,000, and 100,000. You measure the longest relaxation times of the two lower-molecular-weight samples and find $\tau_1 = 1$ sec and 10 sec, respectively, for $M =$

20,000 and $M = 40,000$. Estimate τ_1 for $M = 100,000$. From these results and Eq. (3-79), estimate M_e, the entanglement molecular weight.

Problem 3.13(a) (Worked Example) You have a "binary blend" containing two different molecular weights, M_L and M_S, of the same polymer. Let the weight fraction of M_L be ϕ, where M_L corresponds to the high molecular weight. Approximate the linear relaxation moduli of the pure melts by $G_L(t) = G_0 \exp(-t/\tau_L)$ and $G_S(t) = G_0 \exp(-t/\tau_S)$. Derive an expression for $G(t)$ for the blend from "double reptation" theory.

ANSWER:

For a molecular weight distribution with only two components, the integral in Eq. (3-78) is replaced by a sum:

$$\left(\frac{G(t)}{G_0} \right)^{1/2} = \phi \exp\left(\frac{-t}{2\tau_L} \right) + (1 - \phi) \exp\left(\frac{-t}{2\tau_S} \right)$$

Thus,

$$G(t) = G_0 \left[\phi \exp\left(\frac{-t}{2\tau_L} \right) + (1 - \phi) \exp\left(\frac{-t}{2\tau_S} \right) \right]^2 \tag{A3-30}$$

Problem 3.13(b) (Worked Example) If $\tau_L \gg \tau_S$, then $G(t)$ of the blend has two plateaus, the second one corresponding to relaxation of the long molecules. The magnitude of the first plateau is obtained by taking the limit $t \ll \tau_S$, and the second plateau is obtained in the limit $t \gg \tau_S$, but $t < \tau_L$. Compute the modulus on the first and second plateaus. How does the second plateau modulus depend on ϕ?

ANSWER:

Putting $t \ll \tau_S$ into Eq. (A3-30), both exponentials become unity; thus

$$G(t) = G_0[\phi + (1 - \phi)]^2 = G_0$$

So the first plateau is G_0. For the second plateau, $t \gg \tau_S$, but $t < \tau_L$. Hence, the exponential $\exp(-t/\tau_S)$ is nearly zero. Thus, from Eq. (A3-30) we obtain

$$G(t) = G_0 \left[\phi \exp\left(\frac{-t}{2\tau_L} \right) \right]^2 = G_0 \phi^2 \exp\left(\frac{-t}{\tau_L} \right)$$

Thus, the modulus on the second plateau is $\phi^2 G_0$.

Problem 3.14 (Worked Example) Derive expressions for the shear viscosity and first and second normal stress coefficients in steady-state shearing of the Johnson–Segalman model, given by Eqs. (3-80) and (3-81a).

ANSWER:

From Eqs. (3-80) and (3-81a), the Johnson–Segalman model is

$$\overset{\triangledown}{\boldsymbol{\sigma}} + \frac{1}{\tau}\boldsymbol{\sigma} + \xi(\mathbf{D} \cdot \boldsymbol{\sigma} + \boldsymbol{\sigma} \cdot \mathbf{D}) = 2G\mathbf{D} \tag{A3-31}$$

At steady state, the upper-convected derivative, defined by Eq. (3-33), reduces to

$$\overset{\triangledown}{\boldsymbol{\sigma}} = -\nabla\mathbf{v}^T \cdot \boldsymbol{\sigma} - \boldsymbol{\sigma} \cdot \nabla\mathbf{v} \tag{A3-32}$$

For a simple shearing flow, we shall work in two dimensions, where "1" is the flow direction and "2" the gradient direction. The stress components σ_{i3} and σ_{3i} are zero, where $i = 1,\ 2,$ or 3. The velocity gradient tensor is given by Eq. (1-6):

$$\nabla\mathbf{v} = \dot\gamma \begin{pmatrix} 0 & 0 \\ 1 & 0 \end{pmatrix}, \qquad \nabla\mathbf{v}^T = \dot\gamma \begin{pmatrix} 0 & 1 \\ 0 & 0 \end{pmatrix} \tag{A3-33}$$

Then, from Eq. (A3-32), we obtain

$$\overset{\triangledown}{\boldsymbol{\sigma}} = -\dot\gamma \begin{pmatrix} 0 & 1 \\ 0 & 0 \end{pmatrix} \cdot \begin{pmatrix} \sigma_{11} & \sigma_{12} \\ \sigma_{12} & \sigma_{22} \end{pmatrix} - \dot\gamma \begin{pmatrix} \sigma_{11} & \sigma_{12} \\ \sigma_{12} & \sigma_{22} \end{pmatrix} \cdot \begin{pmatrix} 0 & 0 \\ 1 & 0 \end{pmatrix} = -\dot\gamma \begin{pmatrix} 2\sigma_{12} & \sigma_{22} \\ \sigma_{22} & 0 \end{pmatrix} \tag{A3-34}$$

We note that

$$\mathbf{D} = \frac{1}{2}\dot\gamma(\nabla\mathbf{v} + \nabla\mathbf{v}^T) = \frac{1}{2}\dot\gamma \begin{pmatrix} 0 & 1 \\ 1 & 0 \end{pmatrix}$$

Therefore,

$$\xi(\mathbf{D} \cdot \boldsymbol{\sigma} + \boldsymbol{\sigma} \cdot \mathbf{D}) = \frac{1}{2}\xi\dot\gamma \left\{ \begin{pmatrix} 0 & 1 \\ 1 & 0 \end{pmatrix} \cdot \begin{pmatrix} \sigma_{11} & \sigma_{12} \\ \sigma_{12} & \sigma_{22} \end{pmatrix} + \begin{pmatrix} \sigma_{11} & \sigma_{12} \\ \sigma_{12} & \sigma_{22} \end{pmatrix} \cdot \begin{pmatrix} 0 & 1 \\ 1 & 0 \end{pmatrix} \right\}$$

$$= \xi\dot\gamma \begin{pmatrix} \sigma_{12} & \tfrac{1}{2}(\sigma_{11} + \sigma_{22}) \\ \tfrac{1}{2}(\sigma_{11} + \sigma_{22}) & \sigma_{12} \end{pmatrix} \tag{A3-35}$$

Using Eqs. (A3-34) and (A3-35), we can express Eq. (A3-31) as a set of component equations:

$$\text{11 Equation:} \qquad -2\dot\gamma\sigma_{12} + \frac{1}{\tau}\sigma_{11} + \xi\dot\gamma\sigma_{12} = 0 \tag{A3-36}$$

$$\text{22 Equation:} \qquad 0 + \frac{1}{\tau}\sigma_{22} + \xi\dot\gamma\sigma_{12} = 0 \tag{A3-37}$$

$$\text{12 Equation:} \qquad -\dot\gamma\sigma_{22} + \frac{1}{\tau}\sigma_{12} + \frac{1}{2}\xi\dot\gamma(\sigma_{11} + \sigma_{22}) = G\dot\gamma \tag{A3-38}$$

Solving Eq. (A3-36) for σ_{12} gives

$$\sigma_{12} = \frac{\sigma_{11}}{(2 - \xi)\dot\gamma\tau} \tag{A3-39}$$

Likewise, we solve Eq. (A3-37), and use (A3-39) to replace σ_{12}:

$$\sigma_{22} = -\xi\dot\gamma\tau\sigma_{12} = -\xi\dot\gamma\tau \left(\frac{\sigma_{11}}{(2 - \xi)\dot\gamma\tau} \right) = -\frac{\xi}{2 - \xi}\sigma_{11} \tag{A3-40}$$

We now use Eq. (A3-39) and (A3-40) to replace σ_{12} and σ_{22} in Eq. (A3-38), giving

$$\frac{\xi\dot\gamma\sigma_{11}}{2 - \xi} + \frac{1}{2}\xi\dot\gamma \left(\sigma_{11} - \frac{\xi\sigma_{11}}{2 - \xi} \right) + \frac{1}{\tau}\frac{\sigma_{11}}{(2 - \xi)\dot\gamma\tau} = G\dot\gamma \tag{A3-41}$$

This equation can now be solved algebraically for σ_{11}, and the result can be used in Eq. (A3-39) and (A3-40) to obtain σ_{12} and σ_{22}. The results are

$$\sigma_{11} = \frac{(2 - \xi)G\dot\gamma^2\tau^2}{1 + (2\xi - \xi^2)\dot\gamma^2\tau^2} \tag{A3-42}$$

$$\sigma_{22} = \frac{-\xi G \dot{\gamma}^2 \tau^2}{1 + (2\xi - \xi^2)\dot{\gamma}^2\tau^2} \tag{A3-43}$$

$$\sigma_{12} = \frac{G\dot{\gamma}\tau}{1 + (2\xi - \xi^2)\dot{\gamma}^2\tau^2} \tag{A3-44}$$

We use the definitions of the shear viscosity η, first normal stress coefficient Ψ_1, and second normal stress coefficient Ψ_2 [from Eq. (1-24)] to obtain

$$\eta \equiv \frac{\sigma_{12}}{\dot{\gamma}} = \frac{G\tau}{1 + (2\xi - \xi^2)\dot{\gamma}^2\tau^2}$$

$$\Psi_1 \equiv \frac{\sigma_{11} - \sigma_{22}}{\dot{\gamma}^2} = \frac{2G\tau^2}{1 + (2\xi - \xi^2)\dot{\gamma}^2\tau^2}$$

$$\Psi_2 \equiv \frac{\sigma_{22} - \sigma_{33}}{\dot{\gamma}^2} = \frac{-\xi G\tau^2}{1 + (2\xi - \xi^2)\dot{\gamma}^2\tau^2}$$

To obtain this last result, remember that $\sigma_{33} = 0$. The interested reader should plot these functions and ponder how realistic they are.

Problem 3.15 Consider a K–BKZ integral equation in a shearing flow with the form shown below:

$$\boldsymbol{\sigma} = \int_{-\infty}^{t} m(t - t')\, h(\gamma)\, \mathbf{C}^{-1} dt' \tag{A3-45}$$

Suppose there is a single relaxation time, so that $m(t - t') = (G/\tau)\exp[-(t - t')]$, and suppose that $h(\gamma) = \exp(-\gamma)$. Calculate a formula for the steady-state shear viscosity as a function of shear rate. At high shear rate, what is the shear-thinning power-exponent p, with $\eta \propto \dot{\gamma}^{-p}$?

■ ───

Chapter **4**

GLASSY LIQUIDS

4.1 INTRODUCTION

When a liquid is cooled, the molecules composing it draw more closely together to maximize attractive interactions. If the shapes of the molecules are compatible with some regular packing, then the liquid will likely crystallize into an ordered solid via a first order transition at the freezing point. However, if the molecules are bulky and of irregular shape (see Fig. 4-1), or if the liquid is cooled too rapidly for the crystalline structure to form, then at low temperatures it *vitrifies* into a rigid phase that retains the disordered molecular arrangements of the liquid. This rigid disordered material is called a *glass*. Glasses are liquids whose molecules are so tightly packed, and hence are so sluggish, that they cannot relax to equilibrium even over periods of months or years. The arrangement of molecules in a glass is analogous to that of oddly shaped furniture and other household items packed tightly into a moving van. The tight packing is at the same time both irregular and carefully chosen (i.e., noncrystalline, yet of low entropy), and resists rearrangement under the jostling forces of the road (which are analogous to Brownian forces). Common examples of glasses include the inorganic silica glasses in ordinary window panes, the glassy polystyrene in disposable plastic cups, and the glassy polycarbonate in compact discs. Useful glasses often combine the optical properties of a simple liquid, including isotropy and clarity, with the rigidity of a hard solid. In addition, because they soften over a range of temperatures rather than at a single temperature as water does, molten glasses can be drawn into fibers or skillfully shaped by glass blowers into objects both useful and beautiful.

Because the *glass transition* of a liquid to a glassy state is not abrupt, but occurs over a range of a few degrees Celsius, the *glass transition temperature* is imperfectly defined. Although it is *not* a sharp thermodynamic melting transition, the glass transition can be recognized by a change in the slope of a plot of a thermodynamic property, such as specific volume, against temperature. Figure 4-2 is a schematic plot of some property P against temperature during a constant cooling-rate experiment. At a high temperature, such as at point A, the material is a liquid with a very fast rate of structural relaxation. As the sample is cooled, its structure at first has no trouble staying in equilibrium with the falling temperature, and the property P follows the equilibrium *liquidus* line A–B. (For supercooled liquids, this equilibrium is a metastable one.) With additional cooling, the sample densifies and becomes more sluggish; eventually, its structural relaxation cannot keep pace with the cooling regimen, and the sample falls out of equilibrium. This happens at point B for a fast rate of cooling, and at point D for a slow rate. After the sample has fallen out of equilibrium,

189

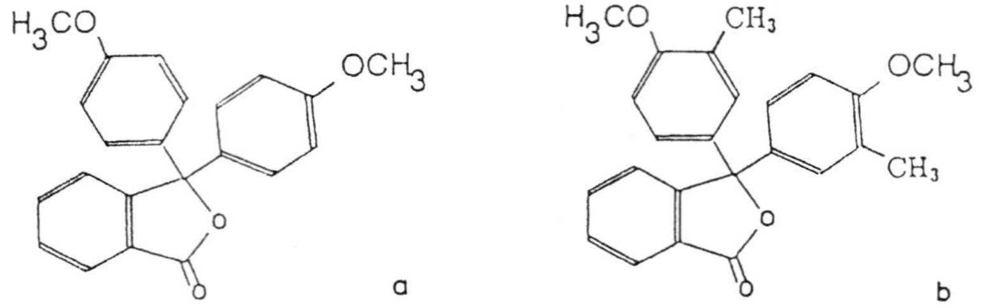

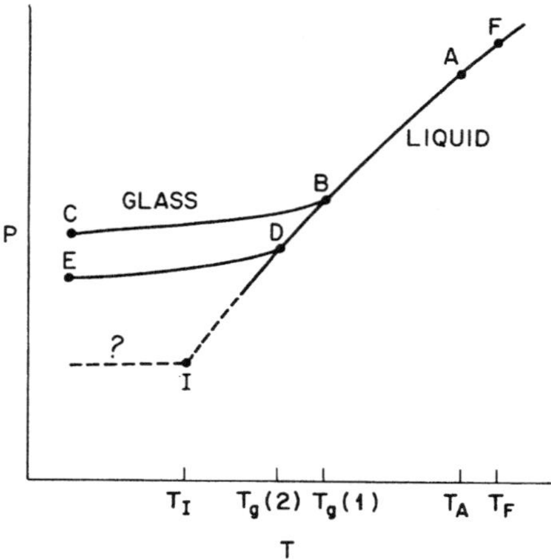

Figure 4.1 Chemical structures of two small-molecule glass formers: **(a)** phenolphthalein-dimethylether ($T_g = 294$ K) and **(b)** kresolphtalein-dimethylether ($T_g = 311$ K). (Reprinted from Physica, A201:318, Stickel et al. (1993), with kind permission from Elsevier Science - NL, Sara Burgerhartstraat 25, 1055 KV Amsterdam, The Netherlands.)

the dependence of property P on temperature T typically has a recognizably different slope, reflecting mainly vibrational degrees of freedom, as well as localized structural relaxations (so-called β relaxations) not "frozen in" at temperature B (or D). Thus the sample at point E has a denser structure and different property P than the material at point C, even though both samples were prepared from the same starting state A.

Hence, the apparent glass transition temperature is $T_g(1)$ for the fast cooling rate and $T_g(2)$ for the slow rate. To better define T_g, slow rates of cooling are often specified, around $1–100°$C/hr, although differential scanning calorimetry (DSC) uses faster cooling rates of order $10°$C/min. For each decade change in the cooling rate, the apparent glass transition temperature typically changes by only $2–3°$C; so for slow rates of cooling in the above range, the glass transition temperature is defined to within about $5°$C or so (Ferry 1980). The glass transition temperature is sometimes made more precise by defining it as the temperature at which the viscosity reaches some specified large value, say 10^{13} P, corresponding to a relaxation time of around 10^4 sec.

Figure 4.2 Generic dependence of property P on temperature during supercooling into the glassy state. The temperature T_F is the equilibrium crystallization temperature, and T_I is a hypothetical "ideal" glass-transition temperature obtained by extrapolating the liquid line to a point of zero configurational entropy. The two glass curves BC and DE are obtained at fast and slow cooling rates, respectively. (From Fredrickson 1988, with permission from the Annual Review of Physical Chemistry, Volume 39, © 1988, by Annual Review, Inc.)

The behavior discussed above is generic; glass transitions have been identified in materials as diverse as toluene, polystyrene, metallic alloys, silica glasses, and mixed salts (Ferry 1980; Angell 1985; Busch et al. 1995). Molecular dynamics simulations show that even atomic liquids, such as argon, would have a glass transition if they could be cooled fast enough to avoid crystallization (Fox and Andersen 1984). Although such computer "experiments" have not been replicated in the laboratory because the cooling rates required are too extreme, it has recently been shown that suspensions of spherical colloidal particles form a vitreous state at high packing density (Pusey and van Megen 1987). The transition to a glass in these suspensions is not brought about by changing the temperature, but by increasing the particle concentration—for example, by centrifugation.

The generality of the glass-transition phenomenon has excited interest in describing it theoretically. But before examining some of the theories, we first describe generic relaxation phenomena that occur in glasses and the phenomenology that has been successfully used to interpret these phenomena. Because of space limitations, many details about glassy behavior are omitted from this chapter. Additional material can be found in books by Scherer (1992), Matsuoka (1992), and Brawer (1985), as well as review articles by Fredrickson (1988), McKenna (1989, 1994), and Götze and Sjögren (1992).

4.2 PHENOMENOLOGY OF THE GLASS TRANSITION

As mentioned in Chapter 3, glassy relaxation processes are often associated with a fairly broad spectrum of relaxation times. A simple expression that describes this spectrum reasonably well over a wide range of time is the "stretched exponential," or *Kolrausch–Williams–Watts* (KWW) expression (Kolrausch 1847; Williams and Watts 1970; Shlesinger and Montroll 1984):

$$\frac{M(t)}{M(0)} = \exp[-(t/\tau)^{\beta}] \tag{4-1}$$

Here $M(t)$ is some linearly relaxing property, such as linear modulus, dielectric constant, or volumetric dilation, τ is a characteristic relaxation time, and $\beta \leq 1$ is an exponent that can vary with temperature. $\beta = 1$ implies monoexponential relaxation (i.e., a spectrum with a single relaxation time), while a smaller β corresponds to a wider spectrum. The best-fit exponent β has been found to depend somewhat on the linear property one is measuring (Götze and Sjögren 1992). For polystyrene, for example, $\beta = 0.34$ fits polarized light-scattering data, while $\beta = 0.40$ provides a better fit to depolarized scattering data (Lindsey et al. 1979). The time constant τ also seems to vary from one property to another. In particular, the process of *volumetric relaxation* following a small temperature jump is characterized by a KWW relaxation time τ_v that is typically a decade or so larger than that characteristic of stress relaxation (Scherer 1992; O'Connell et al. 1998).

The KWW expression is convenient, but not sacrosanct; other expressions can be used to approximate the spectrum. The Cole–Davidson spectrum is useful in the frequency domain, in particular for dielectric relaxation:

$$\varepsilon^* - \varepsilon_\infty = \frac{\Delta\varepsilon}{(1 + i\omega\tau)^{\beta}} \tag{4-2}$$

Here ε^* is the complex dielectric constant, ω the driving frequency of the electric field, and the material-dependent constants ε_∞, $\Delta\varepsilon$, and τ are the high-frequency dielectric constant, the relaxation strength, and the characteristic relaxation time, respectively. Figure 4-3 shows the dielectric loss spectrum at various temperatures for phenolphthalein-dimethylether (PDE), a simple organic glass-forming molecule drawn in Fig. 4-1. Figure 4-3 shows that the spectrum can be fit by a Cole–Davidson spectrum at temperatures ranging from 296 K to 461 K, except at high frequency. The KWW spectrum also tends to fail at high frequencies or short times (Matsuoka 1992; Dixon et al. 1990; O'Connell and McKenna 1997). Notice that as T decreases, the peak frequency ω_p shifts to lower frequencies, and the spectrum broadens, so that the Cole–Davidson β decreases from 0.9 to 0.6 (see Fig. 4-4).

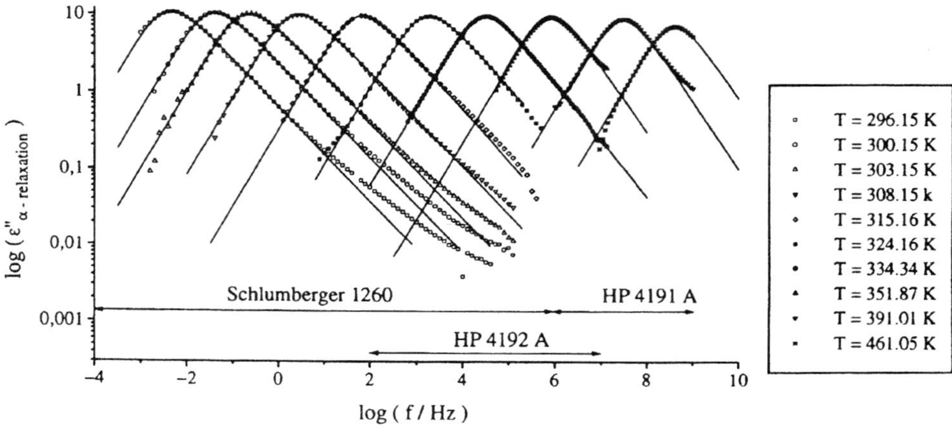

Figure 4.3   Frequency-dependence of the imaginary (loss) part of the dielectric relaxation function for PDE at different temperatures. The lines are fits by the Cole–Davidson function, Eq. (4-2), with $\omega = 2\pi f$ and temperature-dependent exponent given in Fig. 4-4. (Reprinted from Physica, A201:318, Stickel et al. (1993), with kind permission from Elsevier Science - NL, Sara Burgerhartstraat 25, 1055 KV Amsterdam, The Netherlands.)

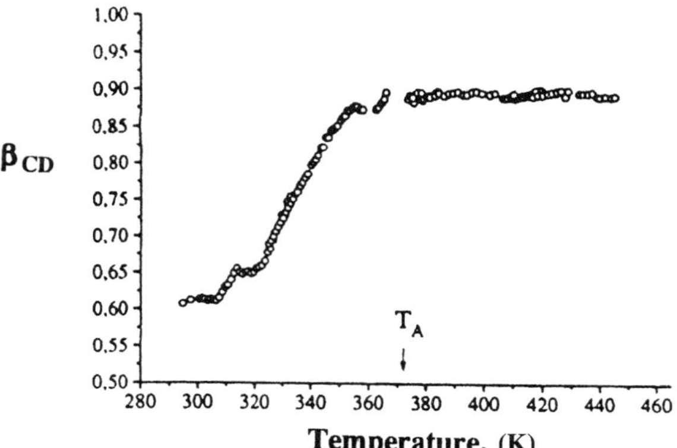

Figure 4.4 Temperature-dependence of the Cole–Davidson exponent β_{CD} for PDE. T_A is the temperature at which the temperature-dependence of the average relaxation time crosses over from Arrhenius to VFTH behavior. (Reprinted from Physica, A201:318, Stickel et al. (1993), with kind permission from Elsevier Science - NL, Sara Burgerhartstraat 25, 1055 KV Amsterdam, The Netherlands.)

For some simple small-molecule liquids, such as PDE, the exponent β in either the KWW or CD expressions is often not too far from unity at high temperature well above T_g, which implies that the relaxation at high temperatures is nearly monoexponential. For more complex molecules, such as polymers, there are Rouse-like modes that produce a broad spectrum of relaxation times even well above T_g, as discussed in Chapter 3. But even for polymers, there is a fast portion of the relaxation spectrum, corresponding to motions of small subunits of the polymer chain, whose behavior is similar in many respects to that of small-molecule liquids and can, for example, be approximated by the KWW equation (Matsuoka 1992). Not surprisingly, this "glassy" portion of the polymer relaxation spectrum is insensitive to the polymer molecular weight. For polymeric and nonpolymeric organic liquids, then, as the temperature approaches T_g, β_{KWW} usually decreases, typically down to values of around 0.3–0.5 at the glass transition.

The temperature-dependence of the average relaxation time τ is system-dependent. Since the high-frequency modulus G_∞ of most glasses is fairly temperature-insensitive, the temperature-dependence of the viscosity $\eta \approx G_\infty \tau$ is similar to that of the relaxation time. Figure 4-5 shows an "Arrhenius plot"—that is, a plot of the log of viscosity versus reciprocal reduced temperature T_g / T—for several glass formers (Angell 1985). Liquids like SiO_2 that form networks have a nearly Arrhenius dependence of viscosity on temperature:

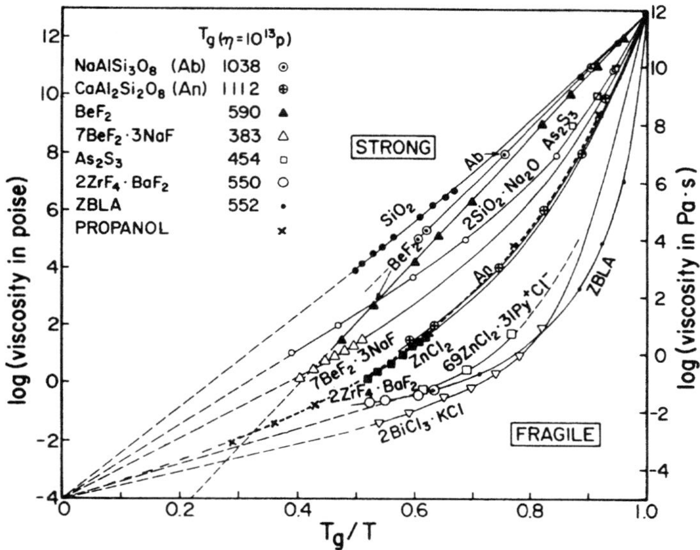

Figure 4.5 Viscosity versus inverse temperature for glass-forming liquids, showing behavior classified as "strong," typified by open tetrahedral networks, to "fragile," typical of ionic and molecular liquids. Here T_g is defined by the criterion that $\eta(T_g) = 10^{13}$ P. For most of the liquids, the viscosities seem to extrapolate to a common value of around 10^{-4} P at high temperatures, corresponding to a fundamental molecular vibrational frequency of around 10^{13} sec^{-1}. (Reprinted from J. Non-Cryst. Solids, 73:1, Angell (1985), with kind permission from Elsevier Science - NL, Sara Burgerhartstraat 25, 1055 KV Amsterdam, The Netherlands.)

$$\frac{\eta}{G_\infty} \sim \tau = \tau_\infty \exp\left(\frac{E_a}{RT}\right) \tag{4-3}$$

These are called "strong" glass formers; liquids for which the Arrhenius plot is highly curved are called "fragile" glass formers (Angell 1985). For fragile glass formers, the non-Arrhenius dependence of the viscosity or the characteristic relaxation time τ on temperature can often be represented by the *Vogel–Fulcher–Tammann–Hesse* (VFTH) equation (Vogel 1921; Fulcher 1925; Tammann and Hesse 1926):

$$\eta = \eta_0 \exp\left(\frac{A}{T - T_0}\right) \tag{4-4}$$

or

$$\frac{\eta}{G_\infty} \sim \tau = \tau_0 \exp\left(\frac{A}{T - T_0}\right) \tag{4-5}$$

where T_0 is typically around 30–50°C below T_g, and values of A are typically around 500–5000 K. The VFTH equation is equivalent to the WLF equation, (3-45b), as can be seen by a simple rearrangement of Eq. (4-5), where T_0 in Eq. (4-5) is the temperature T_∞ in Eq. (3-45b). Parameters of the WLF equation suitable for some common polymers can be found in Table 3-3. Equation (4-5) only applies when $T > T_g$; for $T < T_g$, the substance is out of equilibrium.

Equation (4-5) typically applies up to temperatures of $T_g + 50°C$ or so. For higher temperatures, an Arrhenius temperature-dependence often applies for small-molecule liquids, even if they are fragile glass formers. For example, Fig. 4-6 shows a plot of $1/\log_{10}(f_\infty/f_p)$ versus temperature for propylene carbonate, where $f_p \equiv 2\pi\omega_p$ is the peak frequency (in

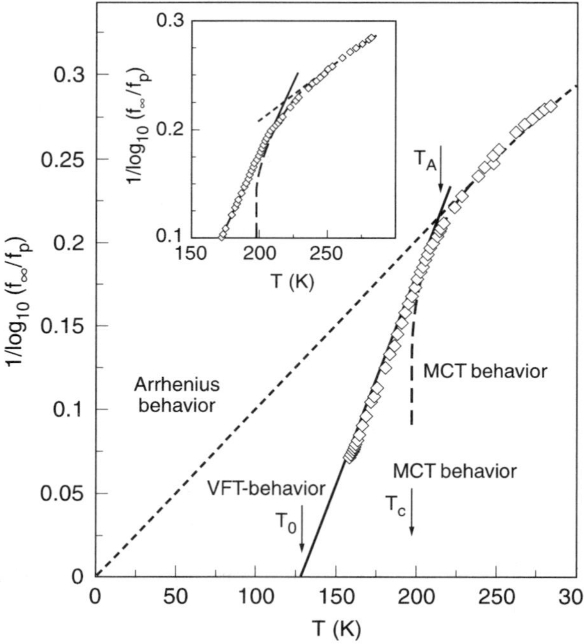

Figure 4.6 $1/\log_{10}(f_\infty/f_p)$ versus temperature for propylene carbonate. Here $\log_{10} f_\infty = 2\pi/\tau_\infty = 13.11$, where τ_∞ (in sec) is the high-frequency relaxation time in Eq. (4-3). T_A is the crossover temperature between Arrhenius and VFTH behavior. The prediction of the mode-coupling theory MCT (see Section 4.6) is also shown, where T_c is the critical temperature. (From Schönhals et al. 1993, with permission.)

hertz) of the dielectric loss curve, and $f_\infty \equiv 2\pi/\tau_\infty$. A straight line on this plot indicates either Arrhenius or VFTH behavior, depending on whether the intercept with the abscissa passes through a zero or a nonzero temperature. For temperatures above $T_A = 217$ K, the data extrapolate to $T = 0$ K, indicating Arrhenius behavior, while for lower temperatures, VFTH behavior is obtained. Figure 4-7 shows the Cole–Davidson β for propylene carbonate and two other organic glass-formers, glycerol and propylene glycol, as functions of temperature. For propylene carbonate, β_{CD} is close to unity for $T > T_A$, while for the other liquids, β_{CD} approaches unity only well above T_A, if at all.

Not all relaxation processes slow down drastically at the glass transition. In particular, there are usually localized relaxation modes whose relaxation times continue to follow an Arrhenius temperature-dependence even below T_A, where the modes governing the glassy response begin to slow dramatically. The modes that become "frozen in" at the glass transition are called α relaxation modes, and those that don't are called β modes (Johari 1970), where "β" here has nothing to do with the the exponent "β" used to describe the relaxation spectrum of the α modes. Figure 4-8 shows the dielectric spectrum for polyethyleneterephthalate at several temperatures. The high-frequency β modes follow an Arrhenius temperature-dependence even below the glass transition temperature, while the slower α modes slow down dramatically as T_g is approached (see Fig. 4-9). As the temperature increases, the characteristic relaxation times of the α and β modes approach

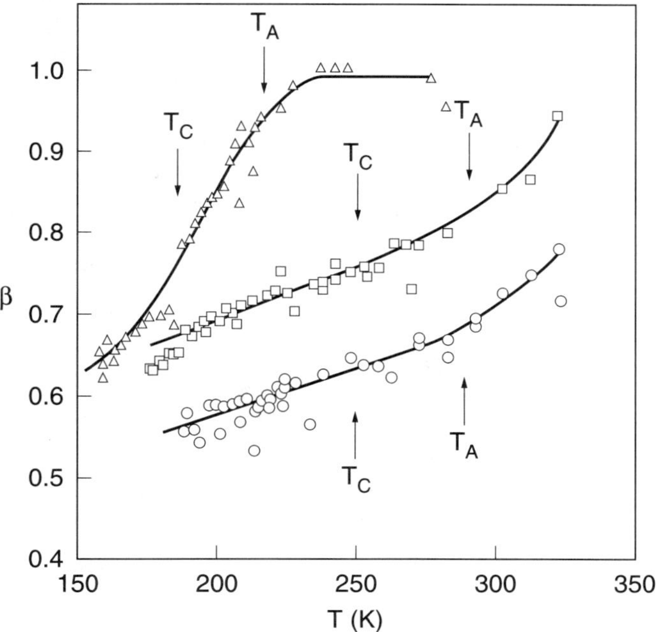

Figure 4.7 Cole–Davidson exponent β as a function of temperature for glycerol (circles), propylene glycol (squares), and propylene carbonate (triangles). T_A is the temperature at which the crossover from Arrhenius to VFTH behavior is observed, while T_c is the best fit to the critical temperature of the mode-coupling theory (see Section 4.6). (From Schönhals et al. 1993, reprinted with permission from the American Physical Society.)

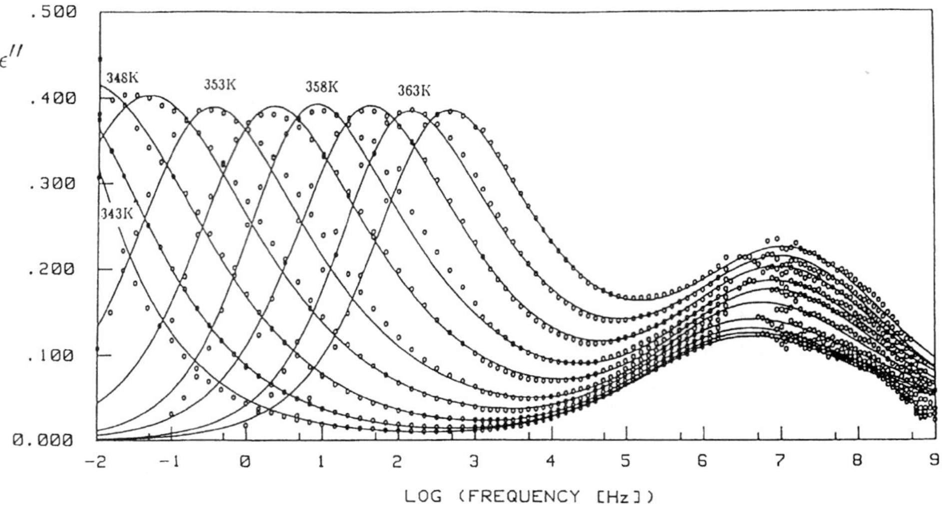

Figure 4.8 Imaginary part of the dielectric function for polyethyleneterephthalate at different temperatures, showing the low-frequency α peak and the high-frequency β peak. The lines are fits using two Havriliak–Negami (HN) functions, one for each peak. The HN function is a generalization of the Cole–Davidson function. (Reprinted from Physica, A201:106, Hofmann et al. (1993) with kind permission from Elsevier Science - NL, Sara Burgerhartstraat 25, 1055 KV Amsterdam, The Netherlands.)

each other and perhaps merge at high enough temperature at least for some glass formers (McCrum et al. 1967; Ferry 1980). For polymers, the β modes apparently correspond to motions of small pendant groups that are little affected when motions along the main chain "freeze" at T_g (McCrum et al. 1967). For some complex molecules, as the temperature is lowered far below T_g, additional, "γ" modes appear (Ferry 1980).

Thus, many aspects of the glass transition seem to be universal, but other aspects are not. Needless to say, it would be of great interest to understand theoretically both the "universal" features of the glass transition and the "nonuniversal" ones. Unfortunately, rigorous theories of the liquid state at densities approaching the glass transition are so intractable that drastic measures seem to be required to obtain results that can be compared to experiments. Thus, theories for the glass transition either are phenomenological, involve drastic mathematical approximations, or are simulations of "toy" or "analog" models which at best capture the spirit of glassy systems, but without physical realism. Below, we summarize some of the approaches taken.

4.3 FREE-VOLUME THEORIES

As discussed above, the extreme sluggishness of molecular rearrangements in glassy or near glassy materials is thought to derive from their dense, though disordered, packing. At high packing densities, the molecular configurations of an irregularly shaped molecule become

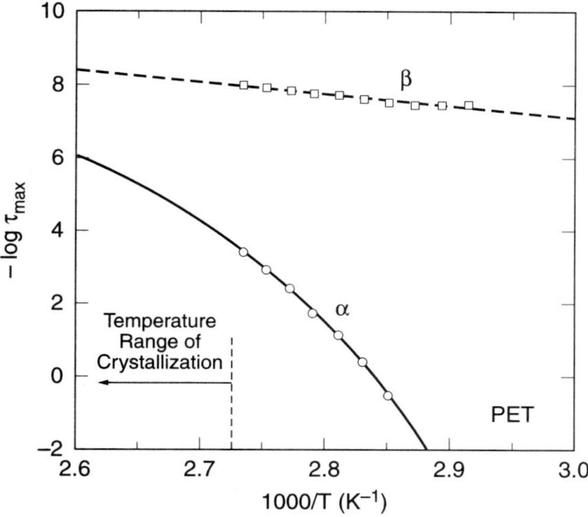

Figure 4.9 Negative logarithms of the dielectric relaxation times τ_{max}, the inverse peak frequencies in Fig. 4-8, for the α and β peaks versus $1/T$. (Reprinted from Physica, A201:106, Hofmann et al., (1993) with kind permission from Elsevier Science - NL, Sara Burgerhartstraat 25, 1055 KV Amsterdam, The Netherlands.)

so interlocked with those of its neighbors that the molecule must coordinate its internal rearrangements with those of its neighbors if it is to relax at all.

The packing density of a substance near T_g is illustrated in Fig. 4-10, where v is the specific volume, or volume per unit mass, $\alpha = d(\ln v)/dT \approx (1/v_g)dv/dT$ is the thermal expansion coefficient, and v_g is the specific volume at the glass transition. In the liquid state, $\alpha = \alpha_\ell$, while $\alpha = \alpha_g$ in the glass. Typically, $\alpha_\ell \approx 10^{-3}/K$ for organic liquids, and $\alpha_g/\alpha_\ell = 0.3$–$0.5$ (Ferry 1980).

As the liquid is cooled, its volume shrinks, presumably due to both a contraction of volume "occupied" by the molecules, v_o, and a contraction of the "free volume" v_f between molecules. The "occupied volume" is taken to include the volume within the van der Waals radii of the molecules, as well as contributions from atomic vibrations. The total volume v is just $v = v_o + v_f$. In the glassy state, the sample cannot relax its molecular configurations

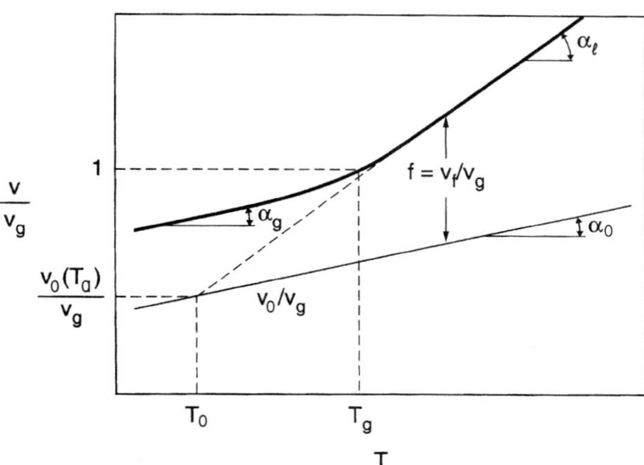

Figure 4.10 Graphical determination of free volume fraction f. The thick curved line is the total reduced volume v/v_g as a function of temperature during cooling. Extrapolating the linear high-temperature portion of this line to T_0 (the VFTH temperature at which the relaxation time becomes infinite) and drawing a thin line of slope $\alpha_0 = \alpha_g$ through this point, one obtains f at any temperature as the difference between the ordinates of the thick and thin lines.

in response to a change in temperature, so the free volume is "frozen" and it is assumed that the thermal expansion coefficient in the glassy state represents mainly the temperature-dependence of the occupied volume.

If the thermal expansion coefficient α_g of the glass can be identified with the thermal expansion coefficient of the occupied volume so that $\alpha_g = \alpha_0$, then the magnitude of the free volume can be estimated as shown in Fig. 4-10. Since T_0 is the temperature at which the relaxation time extrapolates to infinity [see Eq. (4-5)], it is assumed that the free volume is zero at this point. Hence, taking α_ℓ to be a constant, one can extrapolate v to its equilibrium value, $v_o(T_0)$, which is the occupied volume at T_0 (see Fig. 4-10). The occupied volume as a function of temperature can now be obtained from Fig. 4-10 by drawing a straight line with slope $\alpha_0 = \alpha_g$ through the point $(T_o, v_o(T_0))$. The difference between the measured volume v and the occupied volume v_o computed in this way is then taken to be the free volume v_f. The *fractional free volume* is defined as $f \equiv v_f/v_g$. At the glass transition, f defined this way typically has a value of 0.025 ± 0.005. The thermal expansion coefficient of the free volume is $\alpha_f = \alpha_\ell - \alpha_o \approx 0.6 \times 10^{-3}/°C$. Thus, $f = v_f/v_g$ increases from 0.025 to about 0.10 as T increases from T_g to $T_g + 120°C$.

The "free-volume" concept is by no means rigorous, but it can be the basis of a useful empirical method of analyzing relaxation processes near the glass transition. We should also remark that other ways of defining free volume have been postulated (Ferry 1980; Miller 1978), but the definition presented here is perhaps the most accepted.

Batchinski (1913), and more recently Doolittle and Doolittle (1957), developed an empirical relationship between the viscosity and the free volume, from which the relaxation time τ_v can be extracted:

$$\tau_v \propto \exp(B/f) \tag{4-6}$$

where B is a constant, found to be 0.9 ± 0.3 if one takes $\alpha_0 = \alpha_g$, as discussed earlier. Equation (4-6) has been rationalized by the theory of Cohen and Turnbull (1959, Turnbull and Cohen 1961), who assumed that a molecule diffuses by hopping into a void whose size must exceed a critical value. Noting that $f = v_f/v_g = \alpha_f(T - T_0)$, Eq. (4-6) gives

$$\tau_v \propto \exp\left(\frac{B}{\alpha_f(T - T_0)}\right) \tag{4-7}$$

which is equivalent to the WLF or VFTH equation, (4-5). The free-volume concept is very useful for describing the diffusion of penetrant molecules through a glassy material (Ferry 1980; Fujita 1961; Vrentas and Duda 1977). The loss of both free volume and orientational freedom that occur as the glassy state is approached affect the fluid entropy, and, not surprisingly, glassy relaxation behavior also correlates well with entropy, as discussed in the following section.

4.4 ENTROPY THEORIES

Although the glass transition is a kinetic phenomenon for any attainable cooling rate, it has been argued that a thermodynamic transition could, in principle, occur in the hypothetical limit of an infinitesimal cooling rate. This possibility follows from Kauzman's (1948) observation that the entropy of the equilibrium liquid, when extrapolated to low temperatures, eventually falls below that of any possible, or hypothetical, crystalline state. Figure 4-11, for example, shows the excess entropy $\Delta S \equiv S_{\text{liq}} - S_{\text{cryst}}$ of liquid *o*-terphenyl as a function of temperature, where S_{liq} and S_{cryst} are the entropies of the liquid and the crystalline states. This excess entropy ΔS can be obtained by measuring calorimetrically the heat capacities of the metastable liquid and the crystalline states over the same range of temperatures, as well as the entropy of fusion at the thermodynamic melting point, and using a thermodynamic identity to compute ΔS from these data. In Fig. 4-11 it appears that the excess entropy ΔS of the liquid extrapolates to zero at a temperature of around 210 K. Similar results are obtained for other liquids that have both a crystalline and a metastable liquid state over the same temperature range (Angell and Sichina 1976). The temperature T_2 at which the excess entropy of the liquid extrapolates to zero is often, but not always, close to T_0, the VFTH temperature at which the relaxation time becomes infinite (Angell and Rao 1972; Angell and Smith 1982). When T_2 differs significantly from T_0, it usually lies below T_0, as is the case for *o*-terphenyl (Greet and Turnbull 1967; Miller 1978).

To keep the liquid at metastable equilibrium while cooling it to $T_2 \approx T_0$, the cooling rate would have to be infinitely slow. It has been argued that in this hypothetical limit a thermodynamic transition of some kind, possibly second order, intervenes to prevent the excess entropy from becoming catastrophically negative. However, another possibility is that the true dependence of excess entropy on temperature deviates from the linear extrapolation to zero, and the excess entropy varies much more slowly with temperature near T_2 than it does at higher temperatures. This latter possibility is found in some simple models of the glass transition discussed below.

There is a third possibility. There could be a purely *dynamic transition* at or near T_0, in which *ergodicity* is broken, but all thermodynamic variables remain continuous, so that no thermodynamic transition occurs. This would mean that below a critical temperature in the vicinity of T_g, the system is kinetically prevented from exploring all microstates that are thermodynamically allowed, but gets permanently locked into a finite subset of these

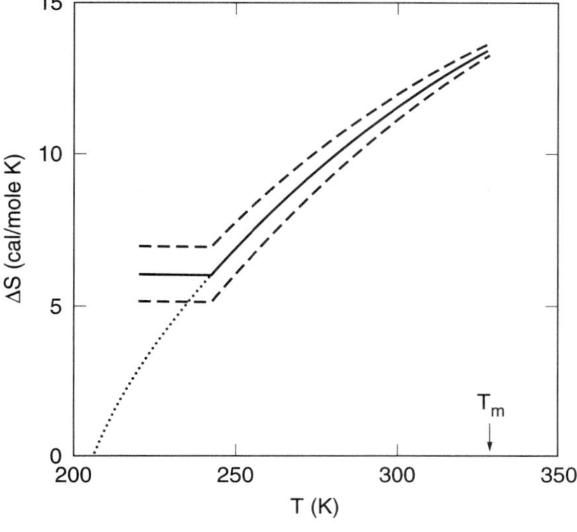

Figure 4.11   Configurational entropy ΔS measured calorimetrically for *o*-terphenyl. The dashed lines provide error bounds. T_m is the crystalline melting point. The horizontal portion is in the nonequilibrium glassy state. The dotted line extrapolates the curve to $\Delta S = 0$ at a finite temperature. (From Greet and Turnbull, reprinted with permission, from J. Chem. Phys. 47:2185, Copyright © 1967, American Institute of Physics.)

microstates. All three of these possibilities—a thermodynamic transition, no discontinuous transition, and a purely dynamic transition—have been suggested in different models of the glass transition. The Gibbs–DiMarzio theory (Section 4.4.1) predicts a second-order thermodynamic transition, the mode-coupling theory (Section 4.6) predicts a purely dynamic transition, and the Fredrickson–Andersen Ising model (Section 4.7.2) predicts no transition at all. As discussed in Section 4.6, the only convincing experimental evidence on this point is for colloidal glasses, where there appears to be a dynamic transition, in agreement with mode-coupling theory. The relevance of these findings for other glasses with orientational degrees of freedom, however, is speculative.

4.4.1 Gibbs–DiMarzio Theory

The connection between entropy and the glass transition was made concrete by Gibbs and DiMarzio, who developed a lattice model for glass-forming polymers (Gibbs 1956; Gibbs and DiMarzio 1958; DiMarzio and Gibbs 1958). While their original theory dealt only with equilibrium thermodynamics and not with glassy dynamics, the theory inspired the development of kinetic theories based on entropy. The Gibbs–DiMarzio theory uses Flory–Huggins ideas (Section 2.3.1.2) to derive the free energy of densely packed polymer chains on a lattice, with a fraction of vacant lattice sites. The polymer chains are assumed to have a preferred low-energy conformation, so that an energy $\Delta\varepsilon$ is assigned to each bond that is "flexed" out of its lowest energy state. Thus, at temperature T above the glass transition, the fraction f of flexed bonds is given by a Boltzmann weighting,

$$f = \frac{(z-2)\exp(-\Delta\varepsilon/k_BT)}{[1 + (z-2)\exp(-\Delta\varepsilon/k_BT)]} \tag{4-8}$$

where z is the coordination number of the lattice.

However, these chains must be packed together on the lattice. As each chain becomes more insistent on achieving a conformation close to its minimum-energy one, it has more trouble sharing the same space with neighboring chains, which are equally insistent on minimizing their conformational energy. If the minimum-energy conformations are straight, then the system can resolve the packing difficulty by breaking symmetry to form an aligned, nematic phase (see Section 2.2.2). However, if the lowest-energy configuration is crooked, then tight packing becomes impossible, and the system reaches a point of zero configurational entropy (i.e., negligibly few packing configurations) at finite temperature. The transition to the glass in this model is a second-order thermodynamic transition, and the temperature at which it occurs is the glass transition temperature; it can be computed analytically using the Flory–Huggins approximation. For a linear polymer chain with degree of polymerization x, the predicted glass transition temperature T_g is obtained from (Cowie and Toporowski 1968)

$$-\frac{x}{x-3}\left\{\frac{1}{1-v_0}\left(\ln v_0 + (1+v_0)\ln\left[\frac{(x+1)(1-v_0)}{2xv_0} + 1\right]\right) + \ln\frac{3(x+1)}{x}\right\}$$

$$= \frac{2\frac{\Delta\varepsilon}{k_BT_g}\exp\left(\frac{-\Delta\varepsilon}{k_bT_g}\right)}{1 + 2\exp\left(\frac{-\Delta\varepsilon}{k_BT_g}\right)} + \ln\left[1 + 2\exp\left(\frac{-\Delta\varepsilon}{2k_bT_g}\right)\right]$$

where a square lattice is assumed ($z = 4$) and v_0 is the volume fraction of vacancies at T_g; it can be obtained from the energy α required to create a vacancy.

To compare the predictions of the Gibbs–DiMarzio theory to experiments, one must specify the two parameters $\Delta\varepsilon$ and v_0; $\Delta\varepsilon$ is fixed by the value of T_g at infinite molecular weight. This leaves only v_0 (or α) adjustable. A typical value at the glass transition is v_0 = 0.025; the hole volume fraction rises with increasing temperature at constant pressure. Remarkably good fits are obtained of the above equation to plots of T_g versus molecular weight for polyvinylchloride and other polymers (Pezzin et al. 1970; McKenna 1989). Even more remarkably, the Gibbs–DiMarzio theory correctly predicts that T_g increases with increasing molecular weight for linear polymers, but decreases with increasing x for cyclic (or ring) polymers (DiMarzio and Guttman 1987; DiMarzio and Yang 1997). The theory also accurately predicts the effects of cross-linking on T_g (McKenna 1989). However, as we shall see in Section 4.7, computer simulations suggest that the equilibrium configurational entropy S_c does not reach zero near T_g (even at equilibrium); instead its dependence on temperature changes from a rapid decrease of S_c with decreasing T above T_g, to a slower decrease below it.

4.4.2 Adam–Gibbs Theory

Whether or not a thermodynamic transition lurks in the shadows of the dynamic glass transition, it is clear that the excess entropy of the slowly cooled equilibrium liquid becomes anomalously small for a noncrystalline substance when T is near T_g. This remarkable, low-entropy, yet disordered, state of matter evidently consists of molecular configurations that are so interlocked that only a fraction of them can fluctuate independently; thus the density of microscopic states is greatly reduced relative to the high-temperature liquid. The rich population of microstates that at high temperatures provides the "stepping stones" enabling the liquid to explore rapidly its configuration space are so denuded at temperatures near T_g that the remaining microstates are separated from each other by chasms that are jumped only infrequently. Hence, as T approaches T_g, the interlocking of molecular conformations produces both a rapid increase in the relaxation time τ_v and a decrease in the excess entropy ΔS.

The kinetic phenomenon—the increase in τ_v—has been related to the thermodynamic one—the decrease in ΔS—by the *Adam–Gibbs equation* (Adam and Gibbs 1965). This equation, originally inspired by the Gibbs–DiMarzio theory, can be simply derived as follows (Matsuoka 1992). Consider some possibly hypothetical high-temperature state in which each molecule can relax its conformations by overcoming internal barriers to conformational change, with no impedance from external barriers presented by the surrounding molecules. Let S_c^* be the *configurational* entropy per mole of that hypothetical state, and let S_c be the configurational entropy of the real liquid, in which intermolecular impediments exclude some conformations and slow the transitions among those that remain. Assume further that $S_c \approx \Delta S \equiv S_{\mathrm{liq}} - S_{\mathrm{cryst}}$. By statistical reasoning, $z \equiv S_c^*/S_c$ can be regarded as the average number of microscopic conformational states that are "locked" together, both thermodynamically and kinetically. Thus, near the glass transition, the liquid contains "domains" of z "conformers" where each conformer in the domain would be free to relax its configuration were it not locked to its surrounding $z - 1$ "conformers," such that the participation of z neighboring "conformers" is required to obtain the relaxation of a single conformer. In a polymeric liquid, a "conformer" can loosely be thought of as a carbon–carbon bond about which neighboring bonds can rotate, producing, for example, trans and gauche states.

Now, if E_a^* is the activation energy for relaxation of 1 mol of conformers, then zE_a^* is the barrier to relaxation in the liquid near T_g. Thus, the relaxation time τ_v is

$$\tau_v \propto \exp\left(\frac{zE_a^*}{RT}\right) = \exp\left(\frac{E_a^*}{R}\frac{S_c^*}{TS_c}\right) \tag{4-9}$$

and

$$\tau_v \propto \exp\left(\frac{B}{TS_c}\right) \tag{4-10}$$

with the constant $B = E_a^* S_c^*/R$, and $R = 1.98$ cal mol^{-1} K^{-1} the gas constant. Equation (4-10) is the *Adam–Gibbs equation* (Adam and Gibbs 1965), which links the relaxation time to the configurational entropy. This equation shows that the departure from an Arrhenius form $[\tau_v \propto \exp(E_a^*/RT)]$ as $T \to T_g$ is caused by the increasing cooperativity required for relaxation, because more and more microscopic states disappear at low temperature.

The excess entropy $\Delta S \approx S_c$ is related to the heat capacity difference ΔC_p between the liquid and the crystal by

$$\Delta S = \int_{T_0}^{T} \frac{\Delta C_p}{T'}\, dT' \tag{4-11}$$

where the entropy ΔS is assumed to go to zero at the temperature $T_2 \approx T_0$. The simplest reasonably accurate expression for the temperature-dependence of ΔC_p is

$$\Delta C_p = \frac{\text{constant}}{T} \tag{4-12}$$

Then Eq. (4-11) gives

$$\Delta S = \Delta C_{p,0}\left(1 - \frac{T_0}{T}\right) \tag{4-13}$$

where $\Delta C_{p,0}$ is the molar heat capacity difference at $T = T_0$. Taking $S_c \approx \Delta S$ and inserting Eq. (4-13) into Eq. (4-10) gives the VFTH equation (4-5) with $A = E_a^* S_c^*/R\Delta C_{p,0}$. Thus, both the Doolittle free-volume theory and the Adam–Gibbs entropy theory give the same (VFTH) temperature-dependence of the relaxation time of the equilibrium liquid.

The Adam–Gibbs equation (4-10) can be tested directly by using the calorimetrically measured entropy difference ΔS to compute the temperature-dependence of the relaxation time, with B then being a fitting parameter. This has been done, for example, with the data for o-terphenyl shown in Fig. 4-11, and the predicted temperature-dependence of the viscosity is found to be in qualitative, but not quantitative, agreement with the measured viscosity (see, for example, Fig 4-12). The main reason for the failure in Fig. 4-12 is that the temperature T_2 at which the entropy extrapolates to zero for o-terphenyl lies below the VFTH temperature T_0 required to fit the viscosity data; hence the predicted viscosity does not vary as rapidly with temperature as it should.

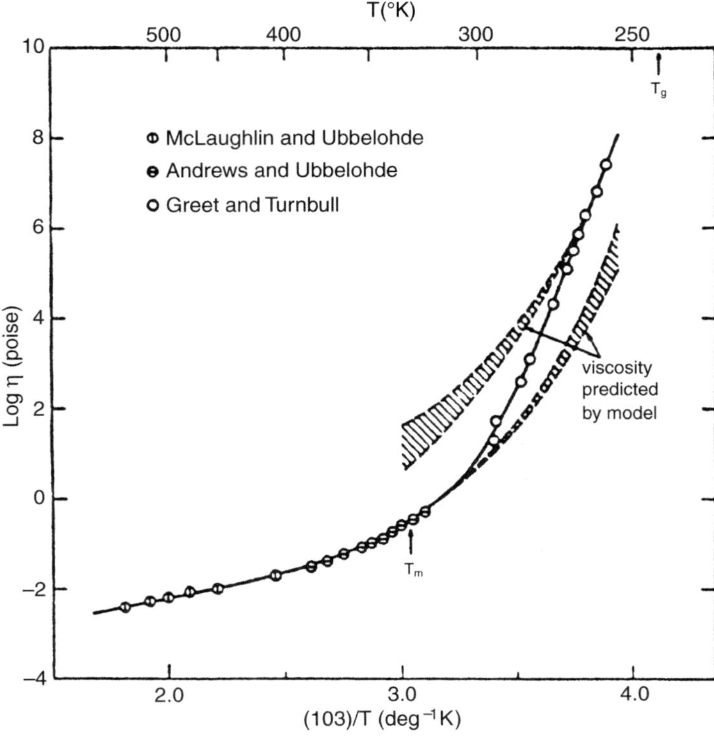

Figure 4.12   The points give the measured viscosity-temperature relationship for *o*-terphenyl, while the shaded regions are the viscosities predicted by the Adam–Gibbs equation (4-10) using $\Delta S(T)$ measured for *o*-terphenyl. The two shaded regions represent alternative fits of the Adam–Gibbs parameters, one fit to the high-temperature, and the other to the low-temperature, data. (From Greet and Turnbull, reprinted with permission, from J. Chem. Phys. 47:2185, Copyright © 1967, American Institute of Physics.)

4.4.3 Miller's Theory

The Adam–Gibbs theory can be "corrected" by assuming that a portion of the entropy ΔS measured calorimetrically does not couple to the glass transition and therefore remains finite at T_0; thus $S_c < \Delta S$ (Miller 1978). The "uncoupled" entropy $\Delta S - S_c$ is presumably associated with motions that are not quenched at the glass transition. If this portion of ΔS is assumed to be insensitive to temperature near the glass transition and is assumed to be of such a magnitude that it accounts for the difference between T_2 and T_0, then by subtracting it from ΔS the behavior closer to the Adam–Gibbs prediction can presumably be obtained.

Using these ideas, Miller (1978) has given a simple explanation for the validity of the VFTH equation using a rotational isomeric state (RIS) model for polymers, which also provides a molecular interpretation of its parameters, A and T_0. Miller assumes that only the configurational entropy associated with bond rotations should be included in the

Adam–Gibbs equation. Most flexible polymers have energetically favored trans states and disfavored gauche states, with one trans state for every two gauche states. If U is the energy difference per mole between the trans and gauche states, and all such states contribute independently to the energy, then the partition function for these conformations is

$$Q = 1 + 2\exp(-U/RT) \tag{4-14}$$

The configurational entropy per mole of bonds is given by

$$S_c = R\ln Q + \frac{U}{T}\frac{Q-1}{Q} \tag{4-15}$$

Thus, the Miller theory has trivial thermodynamics. Starting from the Gibbs–DiMarzio theory, one arrives at the Miller theory by setting $\Delta\varepsilon = U$ and $z - 2 = 2$ in Eq. (4-8) and by assuming that the conformation of one chain imposes no *thermodynamic* constraints which limit the possible conformations of other chains.

However, the low entropy obtained at low temperatures in the Miller theory is assumed to slow down the kinetics via the Adam–Gibbs equation. The configurational entropy, plotted in Fig. 4-13, has a linear portion extending from $U/RT = 0.75$ to 3.0; this can be fitted by

$$S_c = 2.5 - 0.595\frac{U}{RT} = \frac{2.5}{T}(T - T_0) \tag{4-16}$$

with

$$T_0 = U/4.2R \tag{4-17}$$

Inserting Eq. (4-16) into Eq. (4-10) gives the Vogel–Fulcher–Tammann–Hesse equation, where now the Vogel temperature $T_0 = U/4.2R$ can be computed from the energy U between trans and gauche states. The values of U obtained indirectly in this way, using

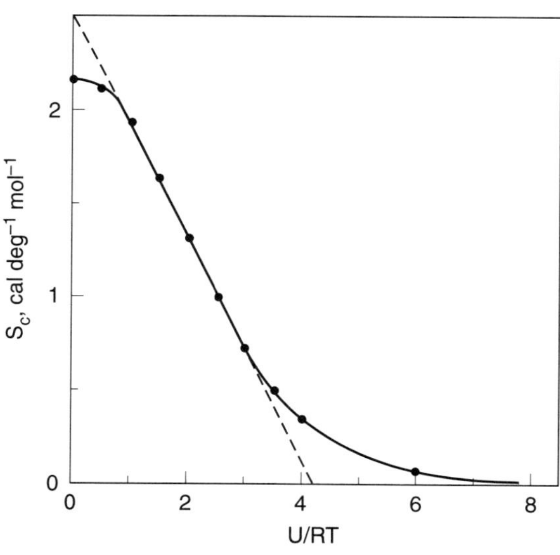

Figure 4.13 Configurational entropy computed from the RIS model for polymer molecules, discussed in the text. The points and the line are both from Eq. (4-15). (Reprinted with permission from Miller, Macromolecules 11:859. Copyright © 1978, American Chemical Society.)

best-fit values of T_0 from viscosity data, have in some cases been confirmed by more direct measurements, if those measurements are also interpreted using the simple RIS theory described above. For example, Saunders et al. (1968) found by photoelastic measurements that $U = 1.15$ kcal/mol for polyethylene, compared to $U = 1.35$ kcal/mol obtained from Eq. (4-17) using the accepted value of $T_0 = 160$ K. Likewise, Maxfield and Shepard (1973) found $U = 1.00$ kcal/mol for polydimethylsiloxane by polarized Raman scattering, while Eq. (4-17) gives $U = 1.13$ kcal/mol. In general, stiffer molecules are known to have higher values of T_0, which accords with this simple model (Miller 1978; Matsuoka 1992).

The VFTH parameter A can also be interpreted within this model. According to Eqs. (4-9) and (4-16), the parameter A in Eq. (4-5) is given by $A = S_c^* E_a^*/2.5\,R$. Then, taking for S_c^* the high-temperature entropy $S_c = R\ln(3)$, one obtains $A = 0.87 E_a^*/R$, where E_a^* is the energy *activation* barrier between trans and gauche states. (The barrier height is E_a^* for gauche to trans, and it is $E_a^* + U$ for trans to gauche.) Values of E_a^* computed from VFTH values of A are again in reasonable agreement with directly measured activation barriers (Miller 1978). Values of A range from 500 K to 5000 K, with most polymers having values between 1000 K and 2000 K, corresponding to E_a^* from 2 to 4 kcal/mol (Miller 1978; Matsuoka 1992).

According to the Miller argument, the VFTH equation should only work over the range of temperatures for which the entropy can be linearized (see Fig. 4-13). At both high and low temperatures, the entropy is more weakly dependent on temperature, and the Adam–Gibbs equation would then imply that the relaxation time or viscosity should revert towards Arrhenius behavior at both extremes in temperature. As discussed earlier, Arrhenius behavior is often found at high temperatures. Equilibrium behavior at low temperatures very close to T_0, on the other hand, cannot be observed since the glass transition intervenes. According to Fig. 4-13, linear behavior of S_c is predicted to prevail to reduced temperatures T/T_0 as low as 1.33, which is fairly close to the glass transition for many liquids. If one can maintain equilibrium below this temperature, reversions from VFTH behavior back toward Arrhenius behavior should be expected if Miller's argument is correct. And, indeed, the relaxation time frequently does revert from VFTH to Arrhenius behavior for temperatures near T_g [Angell and Sichina (1976) and references therein]. This suggests that the temperature T_0 indeed does *not* represent a thermodynamic or dynamic transition temperature, but is merely an artifact of an impermissible linear extrapolation to low temperatures of the temperature-dependence of the entropy.

As discussed below (in Section 4.7), Miller's simple rotational isomeric state argument not only gives an appealing explanation of the VFTH equation, but also can help explain the Adam–Gibbs equation itself, if one can assume that in the dense liquid state a bond can make a transition between a high-energy (gauche) and a low-energy (trans) state, or vice versa, only if the transition is *facilitated* by neighboring bonds on the same or a different chain that are in the high-energy state.

Although appealing, Miller's argument is oversimplified, and it is almost certainly incomplete or even incorrect for many glass-forming liquids. For polymers, the trans or gauche states of neighboring bonds along a polymer backbone are not independent of each other. When account is taken of the correlations, the energy difference U between trans and gauche states required to fit experimental measurements of average chain conformations is found to be considerably less than that extracted from Eq. (4-17). For polyethylene, for example, Flory (1969) found that, after accounting for these correlations, $U \approx 0.4$ kcal/mol, rather than the value obtained from Eq. (4-17), $U = 1.35$ kcal/mol. The smaller value of U is also more consistent with the observed dependence of the radius of gyration of polyethylene on temperature. For most polymers, the radius of gyration, or equivalently the Kuhn statistical segment length, changes by only about 10% or less for a 100°C change in temperature (Flory 1969). This implies a quite modest change in the relative proportion of trans to gauche bonds over this temperature range, and it is hard to see how such modest conformational changes could, by themselves, account for the enormous slowing down that occurs over this temperature range. Most likely the glass transition involves both (a) the densification of the liquid near the glass transition and (b) the stiffening of the polymer backbone.

4.5 NONLINEAR RELAXATION AND AGING

It was noted earlier that the specific volume of a glass depends on its thermal history. Figure 4-14 shows that glassy poly(vinyl acetate) is denser after it has been allowed to equilibrate for 100 hours at a given temperature than it is after equilibrating only a minute or so. The slow densification of a glass held at a fixed temperature is called physical "aging." Densification ceases only when thermodynamic equilibrium is reached. Figure 4-14 shows the time-dependence of volume during the aging of poly(vinyl acetate) at various temperatures near T_g. For $T \geq 30°C$, equilibrium is achieved within 100 hours; but for $T < 30°C$, densification is so slow that thousands of hours or more are needed for equilibrium to be attained. Aging and nonlinear relaxation are important in determining the long-term durability of mechanically stressed parts, such as seals, composites, and hard coatings. An example is an enamel glaze coated onto a metal ring. Changes in temperature produce viscoelastic stresses due to mismatched thermal expansion coefficients. Too great or too sudden a change in temperature can lead to cracking. Scherer (1992) gives numerous examples and worked problems of this kind.

If an equilibrated glass is subjected to only a few-degree change in temperature ΔT, the glass structure will not be far from equilibrium even immediately after the temperature change, and the process of relaxation toward equilibrium can be described by a linear spectrum of relaxation times. The shape of the relaxation spectrum governing densification

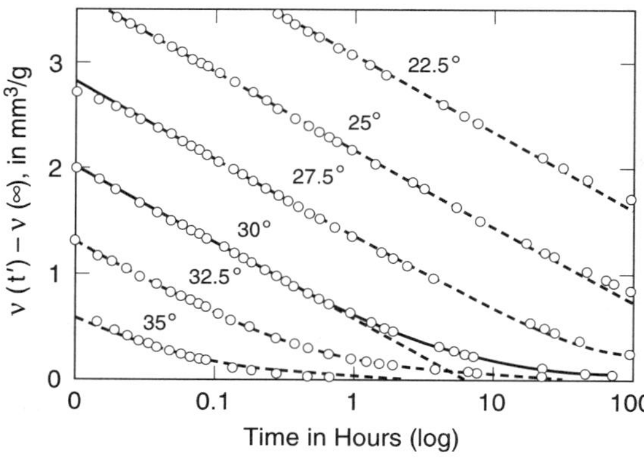

Figure 4.14 Volume contraction of poly(vinyl acetate) after cooling from well above T_g to the indicated temperature. The dashed curves are translations of the curve at 30°C along the abscissa. (From Kovacs, J. Polym. Sci. 30:131, Copyright © 1958, John Wiley & Sons, Inc.)

is similar to that controlling stress relaxation after a mechanical deformation. In particular, both spectra shift with temperature in accordance with the VFTH equation. However, as mentioned earlier, the characteristic relaxation time τ_v for the aging process is *not* the same as the time constant τ for shear relaxation; usually τ_v exceeds τ by a factor of 4–20 (Scherer 1992).

If the step in temperature is larger than a couple of degrees Celsius, then the aging process is not governed by a linear spectrum. Figure 4-15 shows the volume relaxation of poly(vinyl acetate) at a temperature of 35°C, after the sample had been equilibrated at initial temperatures T_i of 30°C and 40°C and then suddenly heated or cooled to $T_f = 35$°C. Note that for $\Delta T = \pm 5$°C, the volume relaxation is asymmetric; the relaxation following a jump up in temperature is very slow, and it eventually accelerates. The opposite occurs for a downward jump in temperature.

The asymmetry is easily explained by consideration of the effect of free volume on τ_v. Equilibration at a low temperature produces a low free volume; after an increase in temperature, the process of volume dilation is therefore sluggish until the free volume has expanded, and the rate of dilation then accelerates autocatalytically. The reverse occurs after a downward jump in temperature; the rate of volume shrinkage slows as the free volume collapses. Similar phenomena occur when the pressure, rather than the temperature, is increased or decreased (Ferry 1980).

If an equilibrium liquid is quenched to a temperature well below its glass transition, the aging process can continue for an enormous period of time, since it occurs by a self-delaying mechanism involving ever-decreasing free volume. Since the time to achieve equilibrium increases roughly tenfold with every 3°C decrease in temperature, one estimates that at 20°C below T_g, aging will continue for 100 years! As aging progresses, the average mechanical relaxation time τ increases enormously, but the distribution about the average does not change much. This was shown in a remarkable series of experiments by Struik (1976). Figure 4-16 shows Struik's tensile creep curves for polyvinylchloride quenched from 90°C to 20°C, well below the glass transition, and aged for periods of time up to 1000 days. The spectrum of mechanical relaxation times shifts with sample age, but at each aging time t_e, the spectrum shape is unchanged, as is shown by the collapse of the data onto a master curve, when shifted horizontally. The rate of shift of the characteristic mechanical relaxation time τ with sample age follows a simple law:

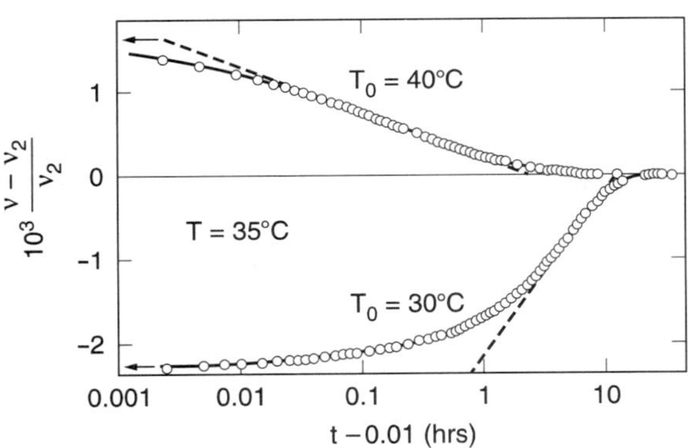

Figure 4.15 Isothermal volume changes of poly(vinyl acetate) at 35°C after sudden temperature changes from 30°C and from 40°C. Here v_2 is the final specific volume. (Kovacs 1964, reprinted with permission from Springer Verlag.)

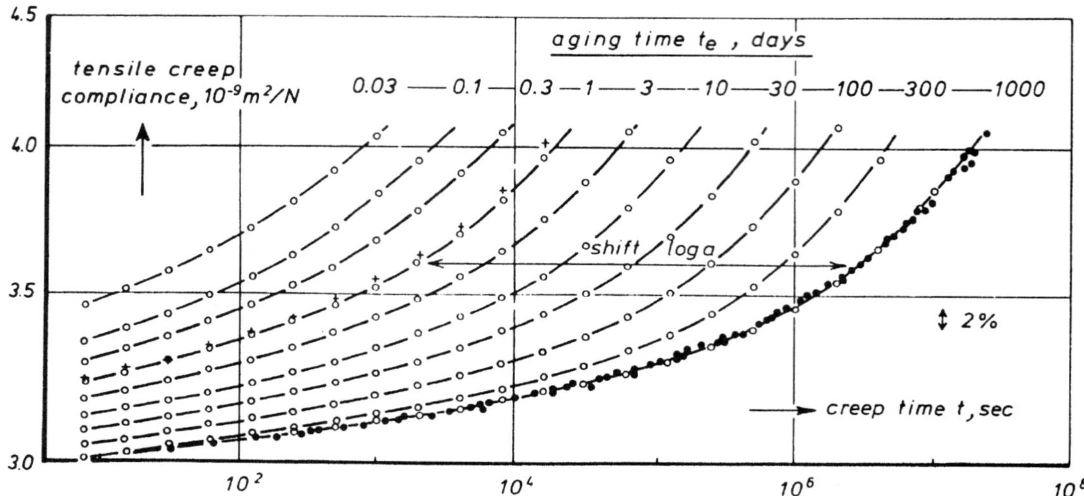

Figure 4.16 Creep compliance (strain per unit imposed tensile stress) versus time for glassy polyvinylchloride after aging for various times after a quench from equilibrium at 90°C to a glassy state at 20°C. The master curve with many symbols is the superposition of all the curves and is obtained by a horizontal shift. The pluses were obtained by reheating to 90°C after 1000 days of aging, and then quenching again to 20°C, followed by one day of aging. This result shows that the aging process is thermoreversible. (From Struik 1976, with permission from the New York Academy of Sciences.)

$$\frac{d \log \tau}{d \log t_e} = \mu \tag{4-18}$$

In Struik's experiments, $\mu = 1$, and thus the relaxation time τ is directly proportional to the sample age; that is, $\tau = Ct_e$, where the proportionality constant C is roughly 0.1. For other glasses, μ varies with temperature and strain (O'Connell and McKenna 1997). In addition, changes in the shape of the relaxation spectrum can occur on aging when β relaxations are important (McKenna 1994). In any event, for a deeply quenched glass, sample age is an extremely important parameter controlling the sample's relaxation characteristics. This is a significant consideration in the manufacture of products containing glassy materials.

In general, the prediction of the asymmetric volumetric relaxation processes in glassy materials requires that expressions such as (4-6) and (4-10) that describe the dependence of the characteristic relaxation time on free volume or entropy be extended so that they can describe the nonequilibrium state. Several different expressions for τ_v in the nonequilibrium glassy state have been proposed (Scherer 1992). One simple expression is obtained by applying the free volume expression (4-6) to the nonequilibrium state. Then

$$\tau_v = C \exp\left(\frac{B}{v}\right) = \tau_1 \exp\left(\frac{B}{f} - \frac{B}{f_1}\right) \tag{4-19}$$

where τ_1 is the relaxation time just after the jump in temperature, f_1 is the fractional free volume just after the jump, and C is a constant. Although the asymmetry of the experimental volumetric relaxation experiments is captured by this simple equation (Ferry 1980), it has a serious flaw; it predicts that τ_v depends on temperature only implicitly through the free volume fraction f. This implies that if the temperature of a glass is suddenly decreased from an initial temperature T_1 to some final temperature T_2, f will be unchanged initially, and the initial rate of relaxation of a glass will be independent of the temperature T_2 to which it is quenched. This prediction is demonstrably false.

Experiments show clearly that the rate of volumetric relaxation of a glass is controlled partly by temperature directly, and partly by the structure of the glass. The glass structure depends on temperature indirectly through the temperature-dependence of the structure. Hence, τ_v must be allowed to depend on both T and some structure-dependent variable. For convenience, the structural variable is often defined so that it has units of temperature, and it is called the *fictive temperature* T_f. The fictive temperature T_f of a nonequilibrium glass or liquid is the temperature at which a possibly hypothetical equilibrium liquid would have the same "structure" as that of the nonequilibrium glass or liquid. The "structure" of a glass can be defined in terms of enthalpy, viscosity, refractive index, specific volume, and so on. The fictive temperature for a particular glass structure depends, in general, on which of these variables is chosen. Figure 4-17 shows how T_f is defined when specific volume is the defining structural variable; analogous definitions apply to other possible structural variables. One extends a line with slope α_g from the point on the $v(T)$ curve corresponding to the nonequilibrium glass until it intersects a line with slope α_ℓ representing the equilibrium volume–temperature relationship for the liquid; the temperature at the intersection point is the fictive temperature.

As the glass or liquid ages and its volume decreases, the fictive temperature also decreases, finally reaching the actual temperature T if and when the glass or liquid reaches thermodynamic equilibrium. From Figs. 4-10 and 4-17, it can readily be deduced that T_f can be related to the fractional free volume by

$$f = \alpha_f(T_f - T_0) \tag{4-20}$$

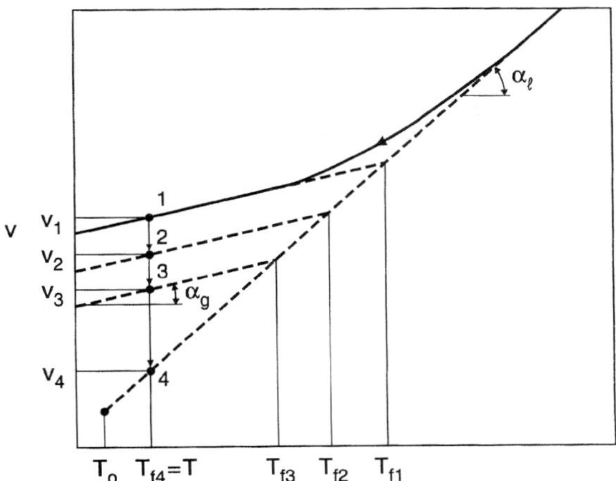

Figure 4.17 Graphical determination of the volume-based fictive temperature for a glass cooled to temperature T with specific volume v_1 and then aged so that its specific volume drops to v_2, v_3, and finally v_4. The fictive temperature for the glass at any point in its history is obtained by extending a line with slope α_g through the specific volume of the glass until it intersects the extrapolated liquidus line with slope α_ℓ. If aging continues until the specific volume lies on the liquidus line, then $T_f = T$, and aging stops.

From this, and the definitions of α_ℓ, α_g, and f, we obtain for the volume v at a time t after a step change in temperature from T_1 to T_2:

$$v(T_2, t) = v_{eq}(T_1) - \alpha_\ell(T_1 - T_2) + (\alpha_\ell - \alpha_g)\,(T_f(t) - T_2) \qquad (4\text{-}21)$$

It is assumed that the sample had been equilibrated to the volume $v_{eq}(T_1)$ at temperature T_1. Eq. (4-21) defines T_f in terms of directly measurable quantities. Note that with Eq. (4-21), T_f can be obtained not only for isothermal aging, but for any time–temperature thermal history.

We must now specify an appropriate expression for the dependence on T_f and T of the relaxation time τ_v. Gardon and Narayanaswamy (1970; Narayanaswamy 1971) proposed a simple generalization of the Arrhenius equation:

$$\tau_v = C \exp\left[\frac{x\,\Delta H}{RT} + \frac{(1-x)\Delta H}{RT_f}\right] \qquad (4\text{-}22)$$

with x being a constant that must be obtained by fitting the above expression to experimental data. In Narayanaswamy's equation (4-22), the usual Arrhenius activation enthalpy ΔH has been divided into two parts, one controlled directly by the temperature, and the other by the glass structure through the fictive temperature. Of course, at equilibrium we have $T_f = T$, and an ordinary Arrhenius expression, $\tau_v \propto \exp(\Delta H/RT)$, is recovered.

The Narayanaswamy expression fits volumetric relaxation data well over a range of temperatures for some glass formers (Rekhson et al. 1971; Mazurin 1977; Scherer 1992). But the equation has been criticized for its lack of a physical basis, as well as for its prediction of an Arrhenius temperature-dependence of the relaxation time at equilibrium. Furthermore, near the glass transition, the best-fit value of ΔH is much larger than the activation energy of the relevant molecular conformational transitions.

Other expressions for $\tau_v(T, T_f)$ have been proposed (Scherer 1992). The most satisfying of these, both conceptually and quantitatively, is a nonequilibrium version of the Adam–Gibbs equation:

$$\tau_v = C \exp\left[\frac{B}{T\,S_c(T_f)}\right] \qquad (4\text{-}23)$$

The configurational "entropy" S_c is a structural property and is therefore assumed to be a function of the fictive temperature T_f. The term "entropy" is used here loosely, since rigorously speaking, entropy is only defined at equilibrium. Thus, modifying Eq. (4-13) by making $S_c \approx \Delta S$ dependent on T_f rather than T, and inserting this into Eq. (4-23), we obtain

$$\tau_v = \tau_{v,0} \exp\left[\frac{A}{T(1 - T_0/T_f)}\right] \qquad (4\text{-}24)$$

At equilibrium, $T_f = T$, and the equilibrium version of the Adam–Gibbs theory, discussed earlier, is recovered. For a glass that is not able to relax, T_f is constant, and τ_v then follows an Arrhenius temperature-dependence, which accords well with experimental measurements on deeply quenched glasses.

Having proposed expressions for the characteristic relaxation time in terms of T and T_f, it is necessary to postulate an equation for the time evolution of the structural variable

T_f. The simplest possible equation assumes monoexponential relaxation after a step change in temperature; thus

$$\frac{dT_f}{dt} = \frac{T - T_f}{\tau_v} \tag{4-25}$$

Equation (4-25) is called *Tool's equation.*

Since glassy relaxation is definitely not monoexponential, Tool's equation is not adequate for quantitative predictions. However, Moynihan et al. (1976) and Kovacs et al. (1979) have developed a simple and successful way of incorporating a spectrum of relaxation times into nonlinear relaxation. First, consider a single step change in temperature ΔT from T_1 to $T_2 = T_1 + \Delta T$ at time $t = t_1$. Moynihan et al. (1976) proposed that the fictive temperature varies with time after this step according to

$$T_f(t) = T_1 + \Delta T[1 - m(t - t_1,\, t)] \tag{4-26}$$

where $m(t - t_1, t)$ is a nonlinear memory function, which can, for example, be obtained by generalizing the KWW equation:

$$m(t - t_1, t) = \exp\left[-\left(\int_{t_1}^{t} \frac{dt'}{\tau_v(t')} \right)^{\beta} \right] \tag{4-27}$$

For a small step in temperature, the fictive temperature T_f is never far from the actual temperature T; hence τ_v, as given by the Narayanaswamy or the Adam–Gibbs equations, doesn't vary much with time. Equation (4-27) then simplifies to the ordinary linear KWW equation, Eq. (4-1). For large ΔT, τ_v varies during the relaxation, and the asymmetry discussed earlier is predicted. Note, however, that β in Eq. (4-27) is assumed to be a constant; this is not strictly valid for large changes in temperature, but is usually acceptable even when ΔT is a few tens of degrees.

For two or more step changes in temperature, Moynihan et al. (1976) proposed that T_f should be given by a simple superposition principle; that is,

$$T_f(t) = T_1 + \sum_{j=1}^{N} \Delta T_j[1 - m(t - t_j,\, t)] \tag{4-28}$$

where the summation is over all temperature steps. Equation (4-28) can be generalized to an arbitrary temperature ramp by replacing the sum by a suitable integral; that is,

$$T_f = T_1 + \int_{t_1}^{t} dt' \frac{dT}{dt'} \left\{ 1 - \exp\left[-\left(\int_{t'}^{t} \frac{dt''}{\tau_v(t'')} \right)^{\beta} \right] \right\} \tag{4-29}$$

Using Eq. (4-28) or (4-29) and Narayanaswamy's equation (4-22) for τ_v, Moynihan et al. (1976) showed that many nonlinear relaxation phenomena in glassy materials can be predicted. Figure 4-18, for example shows a successful prediction of the overshoots in the growth rate of the fictive temperature during steady reheating of glassy As_2Se_3.

The above equations can also predict the celebrated "memory effect" of Kovacs (1964). The memory effect can be observed by quenching a glass former, such as B_2O_3, from equilibrium 583.2 K to a lower temperature, 498.7 K, and then aging the sample at this lower temperature long enough to bring its fictive temperature T_f down to some lower temperature,

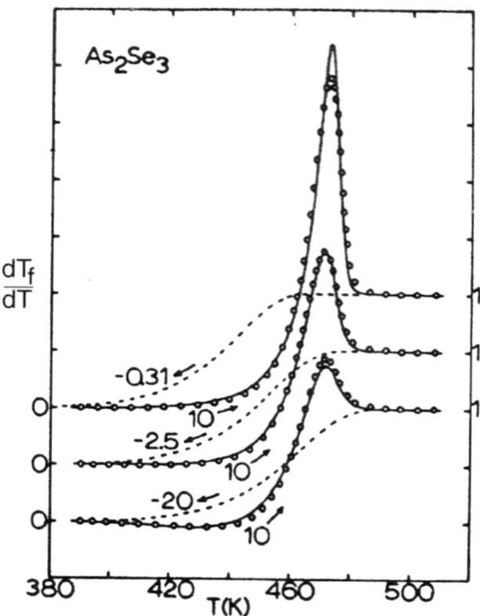

Figure 4.18   Time rate of change of fictive temperature (calculated from enthalpy) versus temperature during constant-rate heating of As$_2$ Se$_3$. The points are experimental data for heating at 10 K/min following constant-rate cooling at -20, -2.5, and -0.31 K/min. The dashed and solid lines are the calculated cooling and heating curves, respectively, using Eqs. (4-22) and (4-29), with $C = 7.6 \times 10^{-38}$ sec, $\Delta H = 81.8$ kcal/mol, $\beta = 0.67$, and $x = 0.49$. (From Moynihan et al. 1976, with permission from the New York Academy of Sciences.)

543.4 K. When the actual temperature T of the specimen is now suddenly raised to its fictive temperature, $T_f = 543.4$ K, one might expect that since $T = T_f$, the sample should be at equilibrium, and its volume and other properties would therefore not change following this second temperature jump. But Fig. 4-19 shows that the sample actually responds to such a thermal history by first increasing its fictive temperature (which corresponds to an expansion in its volume) and then decreasing its fictive temperature (decreasing its volume) back to the value it had immediately after the second temperature jump! A single relaxation-time model cannot predict this nonmonotonic behavior. This experiment thus shows that there are various levels of structure within a glassy material that respond on different time scales to a change in temperature. After the second temperature jump, the relaxation of the fast modes (which had come to equilibrium at the low temperature after the first temperature jump) tend to cause an increase in T_f, while the relaxation of the slow modes (which had remained equilibrated to the original high-temperature) tend to cause a decrease T_f. The result is the nonmonotonic response shown in Fig. 4-19, which is predicted by Moynihan's equations.

Although use of the Narayanaswamy equation is successful in the predictions of the data in Figs. 4-18 and 4-19, for polymeric liquids it fails badly and should be replaced by the Adam–Gibbs equation (Matsuoka 1992). For additional discussion of phenomenological theories of nonlinear relaxation, see Scherer (1992) and McKenna (1989, 1994).

4.6 MODE-COUPLING THEORY AND COLLOIDAL HARD-SPHERE GLASSES

A rigorous theory of the glass transition of molecular liquids is not yet available. Even the best theories are either partly phenomenological or highly approximate. However, one hydrodynamic theory, loosely derived from self-consistent *mode-coupling* arguments

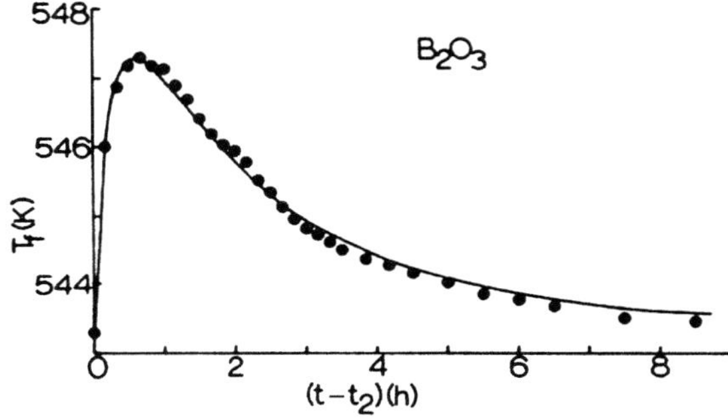

Figure 4.19 "Memory effect" in $B_2 O_3$. The fictive temperature (here calculated from the refractive index) is plotted versus time $t - t_2$ after a second temperature step. The first temperature step was from equilibrium at 583.2 K to $T_1 = 498.7$ K, and the second step at time t_2 was from T_1 to $T_2 = 543.4$ K. The solid curve is calculated from Eqs. (4-22), (4-27), and (4-28) using $C = 5.6 \times 10^{-34}$ sec, $\Delta H = 92$ kcal/mol, $\beta = 0.82$, and $x = 0.50$. In Kovacs' original experiments, the memory effect was monitored by careful measurement of an increase, followed by a decrease, in the sample's volume after the second temperature step. (From Moynihan et al. 1976, with permission from the New York Academy of Sciences.)

(Leutheusser 1984; Bentzelius et al. 1984), has stirred both excitement and controversy (Mezei et al. 1987; Richter et al. 1991; Petry et al. 1991; Schönhals et al. 1993; Mezei 1991; Kim and Mazenko 1992). While the details of the mode-coupling theory are beyond the scope of this chapter, the main idea is that at high fluid densities there is a nonlinear feedback mechanism by which fluctuations in the structure (or local density) of the fluid become arrested and cannot relax to equilibrium. The point at which this occurs is then a purely dynamic glass transition.

The dynamics predicted by the mode-coupling theory are controlled by the equilibrium liquid-state structure, and this structure is the principal input to the theory. It may seem surprising that the mode-coupling theory predicts a transition in which the structural relaxation time becomes infinite, even though the liquid structure itself undergoes little change at this transition. This behavior is explained qualitatively by imagining that each molecule (or atom)—which for the moment we take to be of spherical shape—is surrounded by a "cage" composed of the neighboring molecules (Götze and Sjögren 1992). At high specific volumes, the cage is rather open, and the molecule collides with it only a few times before it escapes the cage. As the liquid densifies, however, the confining effect of the cage becomes greater, and the molecule must collide many times with its neighbors before it finds an opening large enough to escape. Since each molecule is itself the constituent of its neighbors' cages, the trapping of one molecule decreases the likelihood that its neighbors will be able escape their cages. Hence, the mode-coupling glass transition is akin to a percolation phenomenon in that a gradual change in structure leads to a singularity in a transport property. Percolation phenomena are discussed in more detail in Chapter 5.

Also similar to percolation is the existence of a critical exponent; in the mode-coupling theory the longest relaxation time τ_α scales as

$$\tau_\alpha \propto (T - T_c)^{-\gamma} \tag{4-30}$$

where T_c is the critical temperature for formation of the glassy state. The transition is not a thermodynamic transition, and hence all thermodynamic variables can be linearized if one is close enough to the transition; this implies that the same scaling power law links τ_α to any other thermodynamic variable; for example, the scaling with density ρ is $\tau_\alpha \propto (\rho_c - \rho)^{-\gamma}$.

This and other predictions of the mode-coupling theory are derived from the mode-coupling equations for the time-dependent density correlation function $\psi(q, t)$, where q is the wavenumber. In the simplest version of the theory, one considers only a single wavenumber q, and the simplest mode-coupling equation is obtained:

$$\frac{\partial^2 \psi}{\partial t^2} + \nu \frac{\partial}{\partial t} \psi + \Omega_0^2 \psi + \Omega_0^2 \int_0^t dt' m(t - t') \frac{\partial \psi(t')}{\partial t'} = 0 \tag{4-31}$$

In Eq. (4-31), the first three terms describe a simple damped harmonic oscillator: the first term is due to molecular accelerations, the second is due to viscous drag, and the third is due to the restoring force. Ω_0 is the oscillator frequency, which is of order $10^{12}\ \text{sec}^{-1}$, and ν is a viscous damping coefficient. The crucial term producing the dynamic glass transition is, of course, the fourth term, which has the form of a memory integral, in which molecular motions produce a delayed response. The kernel $m(t - t')$ is determined self-consistently by the time-dependent structure. One simple choice relating $m(s)$ to the structure is

$$m(s) = \nu_2 \psi^2(s) \tag{4-32}$$

The coefficient ν_2 is taken to be a function of density or temperature, and the dynamic transition from a relaxing liquid to an arrested liquid or glass occurs when ν_2 is increased to 4. The simple theory with $m(s)$ given by Eq. (4-32) yields $\gamma = 1.76$ for the critical exponent of the α process described by Eq. (4-30).

Although Eq. (4-31) can be justified by high-frequency expansions, its general validity for low-frequency processes near the glass transition has not been established theoretically. Instead, the approach that has been taken is to examine the predictions of Eq. (4-31), as well as generalizations of it (which include multiple wavenumbers, for example), and to compare these predictions with experimental data for glass-forming systems. The general theory is so complex that detailed predictions are available at present only for liquids composed of spheres—that is, atomic or colloidal liquids. Since vitrification of atomic fluids such as argon would require imposition of impossibly high cooling rates, the most promising systems for quantitative tests of the mode-coupling theory are dense *suspensions* of purely repulsive spherical particles. The structure of these suspensions is determined by particle volume fraction instead of temperature. Direct visualization of the structure of glasses composed of 1050-nm-diameter silica spheres at volume fractions of 0.60–0.64 has recently been accomplished using confocal fluorescence microscopy by van Blaaderen and Wiltzius (1995). Since the relaxation time of such suspensions is nine or ten orders of magnitude longer than that of atomic liquids, the metastable disordered state remains free of crystallization long enough for the structural relaxation information to be studied by light scattering techniques.

Van Megen and coworkers (van Megen and Pusey 1991; van Megen and Underwood 1993) have made extensive light-scattering studies on dense suspensions of 340-nm-diameter poly(methylmethacrylate) spheres sterically stabilized by a chemically grafted layer of 12-hydroxystearic acid. These suspensions form a *colloidal glass* at particle volume fractions $\phi \geq 0.56$, which is above the volume fraction at which colloidal crystals first form. The equilibrium state for these highly concentrated suspensions is an ordered, macrocrystalline phase, but the suspensions cannot relax to this state at such high volume fractions. Figure 4-20 shows the time-dependent intermediate scattering function $f(q, t)$ at a wavenumber q near the main peak, $q = q_m$, in the static structure factor. These scattering measurements were made on samples that had been subjected to mechanical agitation to destroy any crystallinity. For samples with volume fraction ϕ less than 0.56, the scattering function relaxes toward zero, indicating fluid-like dynamics; while for $\phi = 0.565$, $f(q, t)$ decays only partially before its relaxation is arrested. The relaxation behavior of these colloidal glasses agrees with the predictions of mode-coupling theory in several respects. First, in the experiments on colloidal suspensions, the transition to a glass is observed at a concentration, ϕ_g, of around 0.56, not too far from the mode-coupling prediction, $\phi_c = 0.52$. Second, and more importantly, the experimental static structure factor shows no anomalous changes as ϕ increases through ϕ_g, which supports the prediction of mode-coupling theory that the transition to the glass is a purely dynamic one. Third, in the glassy state, the residual structure that remains unrelaxed at long times, $f(q, \infty)$, is in good agreement with the prediction of mode-coupling theory (see Fig. 4-21). Note in Fig. 4-21 that the portion of the structure that is frozen in the glassy state depends on both the wavenumber q and the particle concentration ϕ, in semiquantitative agreement with mode-coupling theory. According to van Megen and Underwood (1993), "the increase in $f(q, \infty)$ with particle concentration reflects the increasingly restricted particle motions, culminating in the effective cessation of motion at random close packing where $f(q, \infty) = 1$." The wavenumber dependence of $f(q, \infty)$ reflects the static structure of the underlying glass, as shown by the dashed curve in Fig. 4-21.

The relaxation that is not quenched at the glass transition, which is responsible for the partial relaxation in $f(q, t)$ even at ϕ above ϕ_g, is referred to as the β relaxation. It exists because of incompletely arrested structural relaxation. The longest relaxation time τ_β of this β process is predicted by the mode-coupling theory to become singular at the critical glass concentration, according to the power law:

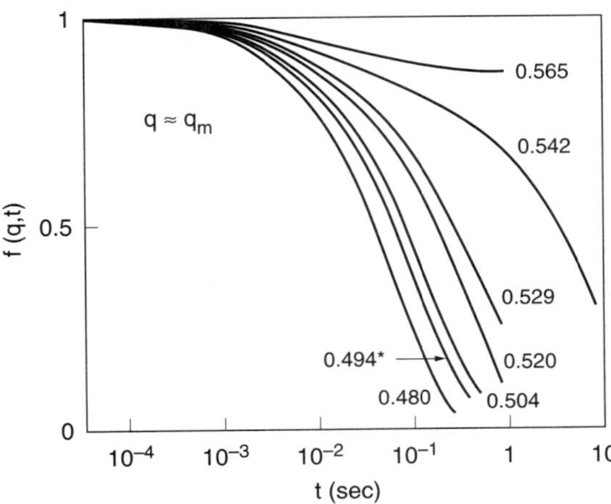

Figure 4.20 Normalized intermediate scattering function versus time at a wavenumber near the peak in the static scattering function, for suspensions of hard, noninteracting, spherical particles. The curves are labeled by the volume fraction. The curve for $\phi = 0.565$ is in the glassy state where the relaxation is arrested after a short time of relaxation. Above the concentration $\phi^* = 0.494$ the *equilibrium* structure would be colloidal crystalline. (From van Megen and Pusey 1991, reprinted with permission from the American Physical Society.)

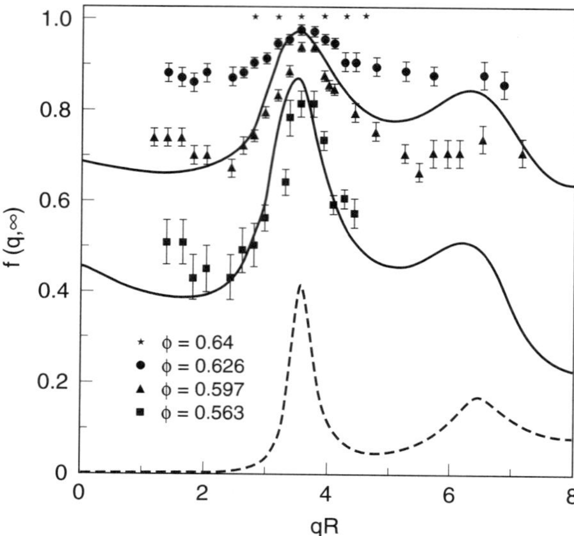

Figure 4.21   The long time limit of the scattering function versus dimensionless scattering vector qR, where R is the particle radius, for suspensions of hard spheres at the volume fractions ϕ given. The lower and upper solid lines are the predictions of the mode-coupling theory at $\phi - \phi_c = 0$ and 0.066, respectively. The dashed curve is the theoretical static structure fracture from the Percus–Yevick equation of state for hard spheres at $\phi = 0.563$, reduced in magnitude by a factor of 10. (From van Megen and Underwood 1993, reprinted with permission from the American Physical Society.)

$$\tau_\beta \propto |\phi - \phi_c|^{1/2a} \tag{4-33}$$

where $a = 0.301$ for hard-sphere fluids. According to Eq. (4-33), the β relaxation process speeds up as one moves away from the transition in either direction, either into the fluid state ($\phi < \phi_c$) or into the glass ($\phi > \phi_c$). The α process, on the other hand, is quenched in the glass, but in the fluid it scales as

$$\tau_\alpha \propto \left(\phi_c - \phi\right)^{-\gamma}, \qquad \phi < \phi_c \tag{4-34}$$

Thus, in the fluid state, there are two relaxation processes, the α and the β, with relaxation times that scale with proximity to the critical point with differing exponents, $-\gamma$ and $-1/2a$, respectively. For spherical particles, $\gamma = 2.58$ and $1/2a = 1.66$; thus the α process is predicted to slow more dramatically as the transition is approached than the β process. Figure 4-22 shows the relaxation times τ_α and τ_β extracted from the relaxation data of Fig. 4-20 for the colloidal fluids. The power laws given by Eqns. (4-33) and (4-34) fit these experimental concentration dependencies well, supporting the mode-coupling theory of this transition.

Other predictions of mode-coupling theory, along with detailed comparisons regarding the behavior of colloidal and other glasses, can be found in Götze and Sjögren (1992). By and large, the agreement between the predictions of mode-coupling theory and the behavior of colloidal glasses is impressive, raising hopes that this theory, or some generalization of it, could explain the behavior of small-molecule or polymeric glasses.

It appears, however, that the mode-coupling theory is not able to explain some of the most significant slow-relaxation processes of these more complex glass formers. In particular, it cannot explain the success of the Vogel–Fulcher–Tammann–Hesse (VFTH) equation for the temperature-dependence of the relaxation time near the glass transition. The mode-coupling theory predicts instead a power-law dependence of the longest relaxation

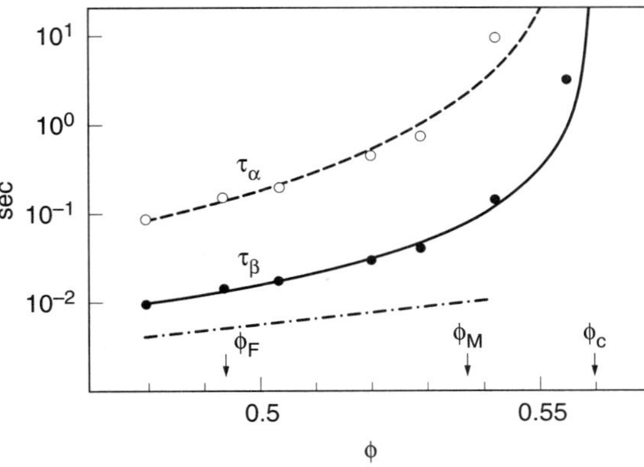

Figure 4.22 The two relaxation time scales τ_α and τ_β versus volume fraction ϕ of spheres in a dense suspension of spheres of radius 200 nm, extracted from the light-scattering data. The lines are fits of the mode-coupling predictions, Eqs. (4-33) and (4-34). ϕ_f and ϕ_M are the volume fractions at the freezing and melting transitions of the colloidal crystal, and ϕ_c is the volume fraction at the glass transition. (From Götze and Sjögren 1991, reprinted with permission from the American Physical Society.)

time (the α relaxation time) on temperature [see Eq. (4-30)]. A power law only fits relaxation-time or viscosity data over a limited range of temperatures not too close to the glass transition, while the VFTH equation fits the data much closer to the glass transition (see Figs. 4-6 and 4-23). Also, the relaxation time of the β transition in polymers and molecular glasses does not become singular at the glass transition, but retains an Arrhenius temperature-dependence through the glass transition.

Götze and Sjögren argue that mode-coupling theory *does* apply to *molecular* glass formers, but that the critical temperature T_c predicted by mode-coupling theory should not be identified with the phenomenological glass transition. Instead, they believe that for molecular glasses, the temperature T_c lies above the phenomenological glass transition, but below the transition temperature at which one recovers an Arrhenius temperature-dependence of the relaxation time; that is, $T_g < T_c < T_A$ (see Fig. 4-6). Evidence for the existence of a critical temperature in this range has been marshaled by Götze and Sjögren (1992; see also Kob and Andersen 1995); some of the most convincing is derived

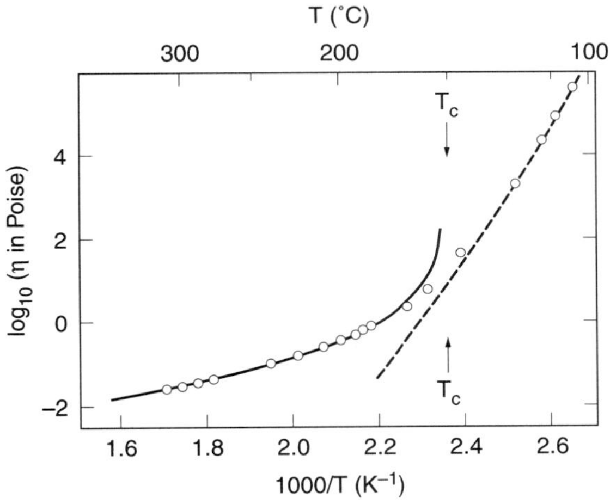

Figure 4.23 Viscosity of trinaphthylbenzene versus temperature fit by the VFTH equation (4-4) (dashed line) and by a power law with exponent 2.17 from mode-coupling theory (solid line). (From Götze and Sjögren, reprinted from Rep. Prog. Phys., 55:241, with permission from IOP Publishing Ltd.)

from neutron-echo experiments, which probe a time scale of $\sim 10^{-9}$ sec (Mezei et al. 1987; Richter et al. 1991). In recent molecular dynamics simulations of glass-forming NiZr (Teichler 1996), the decay of the intermediate scattering function on time scales of $\sim 10^{-9}$ sec is qualitatively nearly identical to that observed for nearly glass hard-sphere suspensions on time scales of seconds (compare Fig. 4-24 with Fig. 4-20). The vast difference in time scales is readily explained by the difference in diffusion times: The dynamics of molecular glasses are typically faster than colloidal suspensions by a factor of 10^9.

However, it is obvious from other experiments (Hofmann et al. 1993) that even if this apparent critical point in molecular glass-forming liquids is meaningful for high-frequency relaxation processes, the longest relaxation time does not really increase to infinity at temperature T_c. Götze and Sjögren argue that continued slow relaxation below T_c is caused by defect "hopping processes" that are not accounted for in simple mode-coupling theories (Lunkenheimer et al. 1996). These hopping processes might be more important in molecular glass formers than in colloidal glasses, because in the latter a sphere that has acquired a high momentum from thermal energy will expend much of that energy in moving through the viscous solvent and have less available to break out of its cage. This might help account for the mode-coupling theory's greater success in explaining the colloidal glass transition than that of molecular glasses. However, simple defect-hopping processes produce an Arrhenius temperature-dependence of the relaxation time, rather than the VFTH temperature-dependence, so this explanation cannot be accepted without qualification.

Thus, for molecular glass formers, mode-coupling theory might be helpful in understanding how the very fast molecular vibrations on the picosecond time scale begin to slow

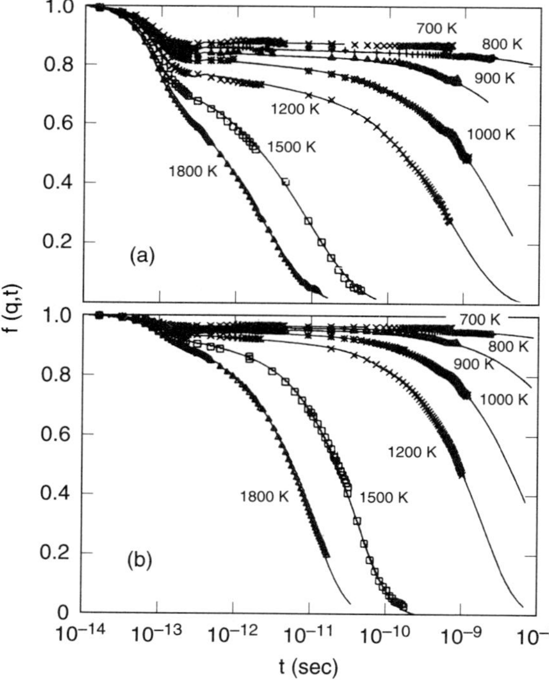

Figure 4.24 Normalized intermediate scattering function versus time for NiZr from molecular dynamics simulations at wavevectors $q =$ (a) 21.6 nm^{-1} and (b) 10.8 nm^{-1}. The lines are fits using a modified mode-coupling theory. (From Teichler 1996, reprinted with permission from the American Physical Society.)

to the nanosecond time scale as the temperature decreases and density increases; but *mode-coupling theory does not explain how molecular motions gradually slow to time scales of hours and days* as the temperature is reduced further still. The success of the VFTH equation in describing the temperature-dependence of the α relaxation time near the glass transition suggests that the slowing down of this relaxation time arises from the growing size of *cooperative domains*—that is, clusters of molecules or "conformers" that must move in concert or not at all. While there is as yet no realistic theoretical model of the liquid state that predicts the growth of such domains near the glass transition, there are "analog" models that capture the spirit of this idea.

4.7 SIMULATIONS OF ANALOG MODELS

So far, we have discussed phenomenological, free-volume, and entropic theories of the glass transition, as well as the molecular-hydrodynamic mode-coupling theory. None of these theories is completely satisfactory. A different approach involves the creation of "analog" or "toy" models that make no pretense to physical realism, but have the merits of simplicity, specificity, and computational tractability. The usefulness of these models arises from the generality of the glass transition phenomenon, which suggests that phenomena similar to those in molecular glasses can occur even in very different systems, if there is some means by which microscopic transitions become increasingly cooperative as the temperature is reduced. Analog models certainly do not supplant other theories, but do provide another outlook on the glass transition. Here, we limit our brief discussion to three such analog models, namely, the square tiling model of Stillinger and Weber (1986), the facilitated Ising model of Fredrickson and Andersen (1984, 1985), and the bond fluctuation model of Kremer, Binder, and coworkers (Carmesin and Kremer 1988; Okun et al. 1997).

4.7.1 Square Tiling Model

In the square tiling model, an underlying two-dimensional square lattice is tiled by squares of various sizes such that there are no gaps between squares (see Fig. 4-25). Periodic boundary conditions are used to remove edge effects. The energy of a given tiling is taken to be proportional to the total length of edges of the squares. Thus, as the temperature is lowered, the small squares would like to merge to form larger squares so that the total energy is lowered. The squares can then be thought of as cooperative "domains" of interlocked molecules or conformers. Kinetic rules are specified to mimic the processes by which such domains can grow or shrink. In one such set of rules, a square can fragment into smaller squares with some probability, and small squares can coalesce into larger squares, as long as these events do not produce any shapes other than squares. The probabilities for fragmenting and for coalescing satisfy "detailed balance"; that is, the ratio of the probability of a fragmentation event to its reverse coalescence event is given by $\exp(-\Delta E/k_B T)$, where ΔE is the energy difference between the fragmented and the coalesced states. As the temperature is reduced, the squares would like to coalesce to a greater and greater extent. At a critical temperature, there is predicted to be a first-order thermodynamic transition to a state in which all squares would like to coalesce into a single, large square. However, the kinetics of this transition are infinitely slow, since the formation of a large square requires a sequence of mergers of rightly configured smaller squares that becomes increasingly unlikely as the square to be formed becomes larger. Thus, the tiling becomes kinetically trapped in some nonequilibrium structure such as that shown in Fig. 4-25.

The square tiling model has some attractive features reminiscent of real glasses, such as "cooperativity," a relaxation spectrum that can be fit by the KWW equation, and a non-Arrhenius temperature-dependence of the longest relaxation time (Fredrickson 1988). However, the existence of an underlying first-order phase transition in real glasses is doubtful, and the characteristic relaxation time of the tiling model fails to satisfy the Adam–Gibbs equation.

4.7.2 Facilitated Ising Model

An analog model that can satisfy the Adam–Gibbs equation is the facilitated Ising model of Fredricksen and Andersen. Again, there is an underlying square lattice, or a cubic lattice if one wishes to consider a three-dimensional system. Each lattice site is occupied by a spin, which can be either up or down. It is assumed that the down state is energetically preferred, and in the simplest versions of the model the total energy is taken to be proportional to the number of down spins. Thus, the down spins can be thought of as low-energy configurations. As the temperature is lowered, the number of down spins increases. The number of down spins and all other thermodynamic properties, such as entropy, can be calculated analytically for this simple model. The partition function is simply $Q = 1 + \exp(-U/RT)$, differing from that of the rotational isomeric state model [Eq. (4-14)] by a trivial factor of 2 multiplying the exponential. In fact, the thermodynamics of the Ising model are almost equivalent to those of the rotational isomeric state (RIS) model described in Section 4.3,

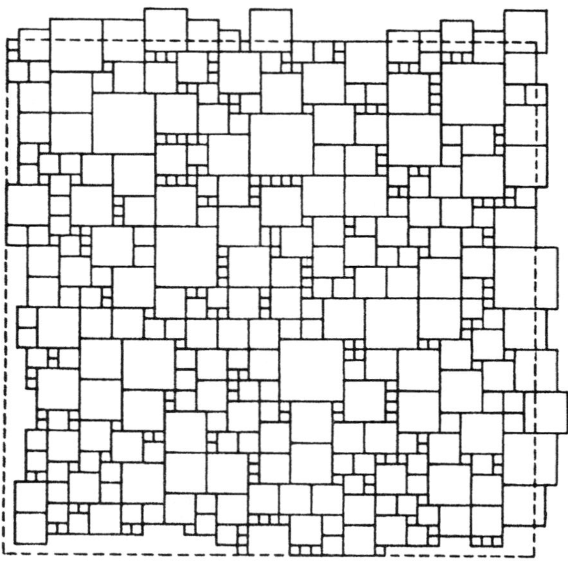

Figure 4.25 A "glassy" configuration of the square tiling model on a periodic 50 × 50 square lattice generated by a Monte Carlo simulation. (From Fredrickson 1988, with permission.)

if one regards the up and down spins to be "gauche" and "trans" bonds, respectively. The only difference is that in the RIS model there are two possible gauche states for every possible trans state, while in the Ising model there are equal numbers of possible up and down states.

Unlike the square tiling model, the facilitated Ising model has no underlying thermodynamic transition; the number of up states is a smoothly increasing function of temperature. Like the tiling model, "glassy" behavior is introduced by defining the transition kinetics such that cooperative behavior is required for the system to lower its energy. The cooperative behavior introduced into the facilitated Ising model is that a spin can flip from up to down state or vice versa only if it is "facilitated" by one or more neighboring spins that are in the up state. One might say that the high-energy up spin possesses the "free volume" needed to allow its neighboring spins to flip.

In the simplest version of the model, a spin can flip if *one* of its neighbors is an up spin. In this version of the model, at low temperatures, up spins are rare and isolated. An isolated up spin can diffuse by first facilitating the flipping up of a neighboring down spin, and then by flipping itself down, leaving an isolated up spin at a site adjacent to that occupied by the original up spin. The net result is that the original up spin has "hopped" to an adjacent site. The rate of this diffusive hopping motion has an Arrhenius temperature-dependence, and so this version of the model is not particularly relevant to the behavior of fragile glasses (but may be relevant to strong network-forming glasses).

If, however, *two* up spins are required to facilitate spin flipping on the square lattice, diffusion of a single up spin by the above mechanism is blocked, and *much more cooperation is required for relaxation to occur*; this cooperation becomes harder as the temperature is lowered and the number of up spins available to act as facilitators decreases. This behavior mimics that of real glasses, where at low temperature "conformers" must relax in concert. In fact, the linear and nonlinear relaxation behavior of the two-spin facilitated Ising model is in remarkable agreement with the observed relaxation behavior of real glasses. In particular, the relaxation of the single-spin time autocorrelation function is well fit by a KWW expression with an exponent β that decreases realistically with decreasing temperature. The average relaxation time $\bar{\tau}$ satisfies the Adam–Gibbs relationship amazingly well (see Fig. 4-26). On the three-dimensional cubic lattice, it turns out that the two-spin-facilitated model is not cooperative enough to produce Adam–Gibbs behavior; the facilitation of three spins seems to be required. It is interesting to note that the facilitated Ising model has no thermodynamic or even dynamic transition; the entropy and longest relaxation time depend on temperature nonlinearly but continuously down to temperatures of absolute zero, in such a way that the Adam–Gibbs relationship between them holds.

In nonlinear temperature-jump experiments, analogous to those discussed in Section 4.5, the two-spin facilitated Ising model also behaves very much like real systems, showing the characteristic asymmetry with respect to the direction of the temperature jump. Figure 4-27 shows that in a nonlinear temperature jump experiment, the two-spin model is well-described by the phenomenological equations (4-23), (4-26), and (4-27) that were developed for real glasses. In the nonequilibrium state, the entropy in Eq. (4-23) is taken from the equilibrium relationship between entropy and the number of up spins. The parameters of equations (4-23) and (4-27) were obtained by fitting the linear response of the two-spin facilitated Ising model, so that the agreement shown in Fig. 4-27 was obtained without the use of adjustable parameters.

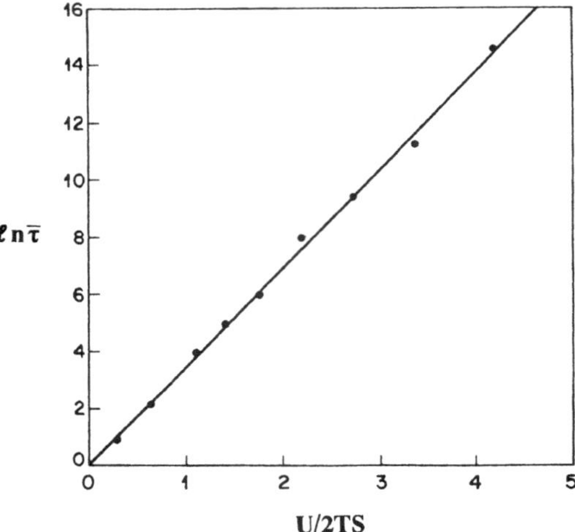

Figure 4.26 The log of the average relaxation time versus Adam–Gibbs exponent $U/2TS$ for the facilitated Ising model on the square lattice. The points are from Monte Carlo simulations and the straight line through the points is the prediction of the Adam–Gibbs theory. (From Fredrickson 1988, with permission from the Annual Review of Physical Chemistry, Volume 39, © 1988, by Annual Reviews, Inc.)

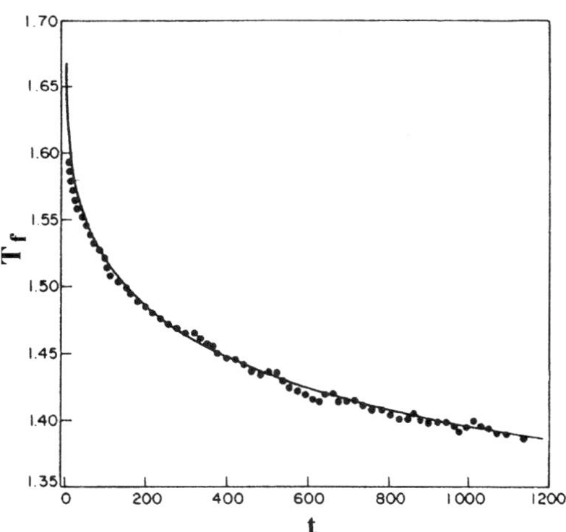

Figure 4.27 Nonlinear relaxation of the dimensionless fictive temperature following a step decrease in the dimensionless temperature $2T/U$ from 1.67 to 1.25 according to the Fredrickson–Andersen facilitated Ising model on a square lattice. The line was calculated from the phenomenological equations (4-23), (4-26), (4-27), with all adjustable parameters fixed by the linear relaxation behavior of the model. (From Fredrickson 1986, with permission from the Annual Review of Physical Chemistry, Volume 39, © 1988, by Annual Reviews, Inc.)

The encouraging agreement between the behavior of the facilitated Ising model and that of real glasses provides an incentive to look for some microscopic interpretation for the "spins" and their "up" and "down" states. One possible interpretation for polymeric glass formers is that the "up" spins are disfavored conformational states, such as gauche states, while the "down" spins are the favored trans states. The requirement that two adjacent "up" spins be available to "facilitate" the transition between states would then imply that conformational transitions can only occur if they are facilitated by neighboring bonds that are in a high-energy gauche state. It would be interesting to look for evidence of such facilitation in molecular dynamics simulations of realistic dense polymeric liquids. Another, more general interpretation of the "up" states is that they are micro-regions of low density that provide the free volume needed for neighboring states to relax. The usefulness of the facilitated Ising model has so far been limited by the lack of a definitive interpretation of the "up" and "down" spins, and of a precise mechanism of facilitation.

4.7.3 Bond Fluctuation Model

Recently, a coarse-grained polymer dynamic Monte Carlo simulation method, the "bond fluctuation model," has been applied to the problem of the glass transition in polymers (Okun et al. 1997). In this model, "monomers" residing on a cubic lattice each block eight sites of the lattice. The monomers are connected into chains by bonds that can fluctuate in length from 2 to $\sqrt{10}$ in units of lattice spacings. A monomer then hops from site to site, with the move being allowed if the target site is not blocked by another monomer and if the bond lengths after the hop all remain within the range 2 to $\sqrt{10}$. A glass transition is introduced into the model by assigning an energetic preference for long bonds. These long bonds use up more sites on the lattice than do shorter ones; as they become prevalent at lower temperatures, they block more moves. On a lattice crowded with chains, the rate of hopping is thereby slowed down.

The model exhibits the expected Rouse relaxation modes at high temperature. At lower temperature, the chain motions can still be decomposed into Rouse normal modes, but the normal modes no longer relax via single exponential decays. Instead, the decay of each mode is described by a stretched exponential, and the stretching increases (β decreases) as the mode number increases. In addition, the temperature-dependence of the relaxation rates is described by the VFTH expression. The emergence of these features of real glasses in such a simple model suggests that such features are insensitive to molecular details.

In addition, since it is a lattice model, the bond fluctuation model can be used to assess the validity of the Gibbs–DiMarzio theory described in Section 4.4.1. Baschnagel et al. (1997) show that the curve of entropy versus inverse temperatures (S_c versus $1/T$) does not cross zero when the temperature is lowered, but instead levels out at large $1/T$, in disagreement with the Gibbs–DiMarzio theory but qualitatively similar to the prediction of Miller's theory in Fig. 4-13.

4.8 RHEOLOGY OF GLASSY LIQUIDS

The rheological properties of glasses and glass-forming liquids is a vast topic. Since fluids are the focus of this book, we limit our discussion to a brief overview of the rheology of viscous, glassy liquids. For a description of the rheology of glassy solids, the reader should refer to Scherer (1992), Matsuoka (1992), McKenna (1994), Pesce and McKenna (1997), and references therein.

4.8.1 Linear Rheology

The rheological properties of glassy liquids are dominated by one or more very long relaxation times and a high modulus. The detailed linear viscoelastic response varies somewhat with the type of liquid. Some inorganic glassy liquids, such as zinc alkali phosphates, behave as nearly perfect single-relaxation-time Maxwell fluids (see Fig. 4-28). The complex viscosity $\eta^* \equiv G^*/i\omega$ in Fig. 4-28 follows the simple Maxwell formula

$$\eta^*(\omega) = \frac{G_0\tau}{1 + i\omega\tau} \tag{4-35}$$

Within a single-mode Rouse theory, the modulus G_0 can be related to the density ρ and the molecular weight M of the relaxing units by $G_0 = \nu k_B T = \rho RT/M$, where ν is the number density of relaxing units. The excellent fits in Fig. 4-28 are obtained by adjusting M and τ for each temperature; the best-fit values of M range from 463 to 859, while τ obeys a WLF temperature-dependence (Sammler et al. 1996). The fitted values of M are hard to interpret; Raman spectroscopy shows that such glasses have a wide range of species including orthophosphate tetrahedra, dimers, trimers, and longer chains. Why the motions of these multiple species do not produce a spectrum containing multiple relaxation processes, rather than only one, is not clear.

Bulky nonpolymeric molecules in the liquid state also show Maxwell-like linear rheology at low frequencies, but at higher frequencies have additional modes of relaxation, apparently because of internal flexing modes within individual molecules. This is illustrated in Fig. 4-29 by the tan δ measurements of small-molecule organic liquids hydroxypentamethyl flavan (HPF), glycerol sextol phthalate (GSP), and 2-phenyl-3-p-tolylindanone (PTI). For a Maxwell fluid, tan δ is inversely proportional to frequency, which would produce a straight line of slope -1 in Fig. 4-29. The small-molecule liquids HPF, GSP, and PTI show this behavior at low frequencies, but at high frequencies display a flattening of the curve, corresponding to the presence of additional high-frequency modes. The linear viscoelasticity of these liquids can, however, be fit by a rather simple empirical expression due to Barlow, Erginsav, and Lamb (1967; Ferry 1980). A derivation of the equation has been given based on a model that assumes diffusion of defect holes as the relaxation mechanism (Phillips et al. 1972). For comparison, Fig. 4-29 also shows tan δ for solid, glassy polymers at room temperature.

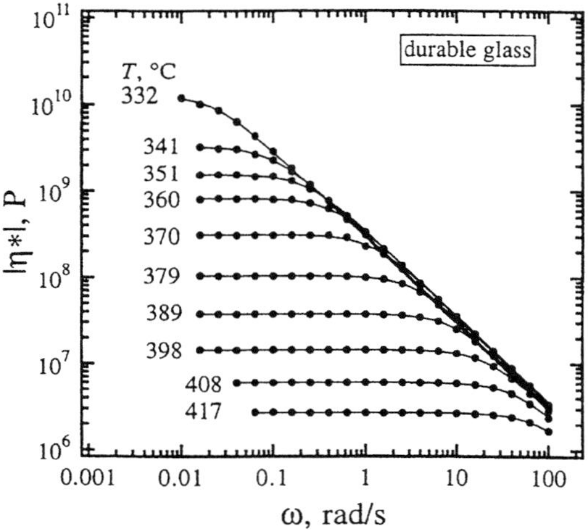

Figure 4.28 Complex viscosity η^* versus frequency at various temperatures for a durable zinc alkali phosphate glassy liquid containing P_2O_5, ZnO, Na_2O, K_2O, Al_2O_3, and SiO_2. The lines are fits of the Maxwell model, Eq. (4-35). (From Sammler et al. 1996, with permission from the Journal of Rheology.)

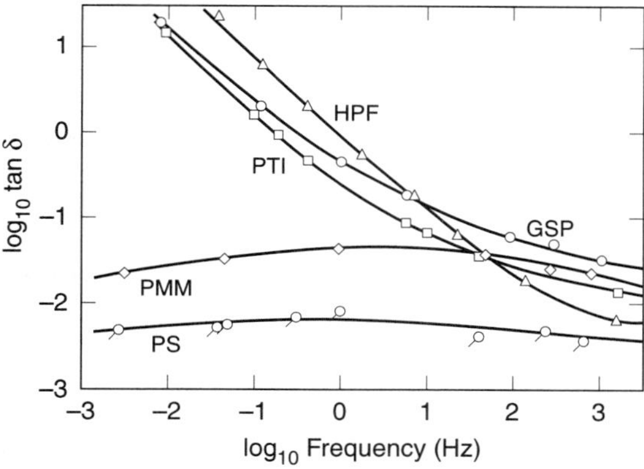

Figure 4.29 Loss tangent versus frequency for three low-molecular-weight organic glass-forming viscous liquids with viscosities around 10^9 P: hydroxypentamethyl flavan (HPF) at 16°C, glycerol sextol phthalate (GSP) at 26°C, and 2-phenyl-3-*p*-tolylindanone (PTI) at 20°C. Also shown are data for polymeric glasses polymethyl methacrylate (PMMA), and polystyrene (PS) at room temperature. (From Benbow and Wood 1958, with permission of the Royal Society of Chemistry.)

4.8.2 Nonlinear Rheology

At low shear stress, the steady-state shear viscosity of a glassy liquid is just the low-frequency dynamic viscosity $\eta'(\omega \to 0)$. At high shear stresses at ambient pressures, yield and fracture phenomena occur, as discussed in Section 1.5.3 (see Fig. 1-21). However, these processes can be suppressed by application of high pressures (up to 300 MPa), and the nonlinear rheology of glassy small-molecule liquids with viscosities around 10^2–10^5 Pa · s can then be studied in the highly nonlinear regime (Bair 1996). Under these conditions, shear thinning is observed, with viscosity–shear rate curves qualitatively similar to those of polymeric liquids. The normal stress difference appears to approach quadratic behavior at "low" shearing stresses, and it departs from this at higher stresses (see Fig. 4-30). For a simple one-relaxation-time Maxwell model, N_1 in the quadratic regime should be given by $N_1 = (2/\nu k_B T)\sigma^2$, where σ is the shear stress. The prefactor is in rough agreement with the data for the small molecule polyphenyl ether, as well as for a solution containing a short polymer of molecular weight 25,000 (see Fig. 4-30).

At low pressures, glassy liquids yield at high stresses, not only in shearing, but in other deformations as well (Matsuoka 1992). While yield, or *plastic deformation*, is a very complex phenomenon, it is generally agreed that it involves localization of strain, with shear banding being an extreme example. Figure 4-31 illustrates patterns of sheared local regions during creep predicted by a model of plastic deformation at (a) high temperatures where the highly sheared regions are uncorrelated spatially and (b) low temperatures where banding of highly sheared regions occurs. Further discussion of these topics is beyond the scope of this chapter; for a review of models of yielding, see Mott et al. (1993).

4.8.3 Thermorheological Behavior

To transform a polymer rapidly into a useful product, the molten polymer is often cooled while it flows, for example, into a mold or from a die. Modeling of these thermoshaping

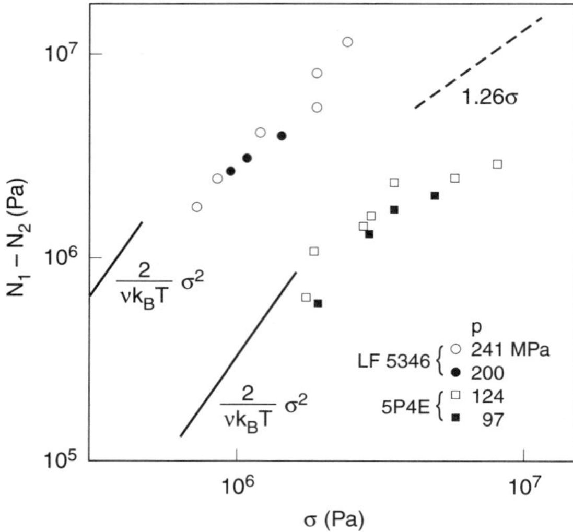

Figure 4.30 Plot of $N_1 - N_2$ versus shear stress σ for polyphenyl ether (5P4E) and for a 20% solution of polybutene with number average molecular weight 25,000 in mineral oil (LF 5346) at 20°C and high pressure (> 90 MPa), where the viscosities are in the range 10^2–10^5 Pa·s. For 5P4E, shear bands were observed at $\sigma = 20$ MPa; from their orientation, the Mohr–Coulomb theory suggests that $N_1 = 1.26\sigma$. The solid lines are the predictions of the single-mode Maxwell model with $N_2 = 0$. (From Bair 1996, reprinted with permission from Steinkopff Publishers.)

processes requires a constitutive equation that can describe the stress–strain behavior of the polymer over a range of temperature histories encompassing both the molten and solid (glassy or crystalline) states. General nonlinear constitutive equations of this kind can be extremely complex (Lustig et al. 1996), since they involve nonlinearity not only in the effects of thermal history, as described in Section 4.5, but also in the effects of strain history. However, a useful unifying concept is the "material time" or material "clock," which slows down as the material cools and densifies (Schapery 1969; Knauss and Emri 1987; Lustig et al. 1996). The material time t_T^* can be defined using a generalized shift factor a_T; for example,

$$t_T^* \equiv \int_0^t \frac{dt'}{a_T(T(t', v_f))}$$

Here the dependence of a_T on the free volume v_f allows nonequilibrium glassy behavior to be captured, so that the material time depends on the thermal history and not just the instantaneous temperature. Use of a material "clock" similar to this can be recognized in theories of nonlinear aging [see, for example, Eq. (4-29)]. The material time need not be based on free volume; material clocks based on strain (Zapas 1974) and stress (Bernstein and Shokooh 1980) have also been proposed. The theory of Zapas (1974) is of special interest since it extends the rather successful K–BKZ theory of polymer melts into the glassy regime. For a review of these and other material "clock" theories, see McKenna (1994, 1995).

If one is only interested in the steady-state shear stress, a much simpler approach is to fit the shear viscosity as a function of shear rate and temperature to an empirical function. Example functions can be found in Dealy and Wissbrun (1990). Such simplified empiricisms are often adequate for predicting processing flow rates and pressures and are helpful in engineering mold designs, for example. But they are useless for determining "molded-in" or "frozen-in" stresses resulting from incomplete viscoelastic relaxation during solidification.

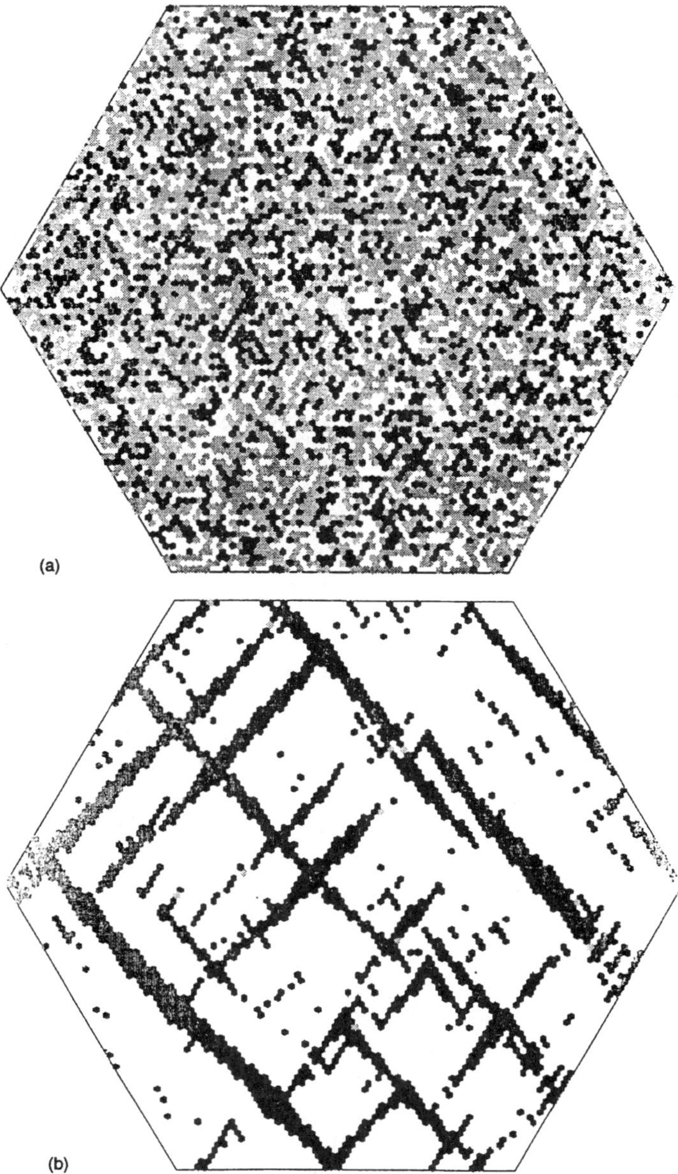

Figure 4.31   Regions of plastic strain (dark) predicted by a model for creep of a liquid at **(a)** high temperature above T_g and **(b)** low temperature in the glassy state. At low temperature, regions of high strain accumulate into shear bands. (From Argon et al. 1995, with permission from the Journal of Rheology.)

Such residual stresses lead to warping, anisotropic shrinkage, and other product flaws. Thus, adequate thermorheological modeling remains a major engineering problem.

4.9 SUMMARY

The transformation of a liquid into a glass on cooling is a common, yet mysterious process. From studies of suspensions of nonattracting hard spheres, it is apparent that an increase in packing density suffices to produce the glassy state. For bulky molecules, including polymers, the quenching of orientational relaxations also appears to be involved in the glass transition.

Simple phenomenological expressions, some involving free volume and entropy, can describe many glassy phenomena, including the temperature-dependence of the average relaxation time, the shape of the relaxation spectrum, and the nonlinear aging phenomenon. These predictions suffice in many cases for practical predictions of creep and deformation in parts containing glassy materials.

Yet no completely satisfactory microscopic theory of glassy behavior exists. At least for now, one must be content to draw theoretical understanding from a combination of several different theories, none of which is completely satisfactory. Theoretically, there are three basic ways that a glass transition can arise. The first is via an underlying thermodynamic transition in the equilibrium structure of a liquid, driven by molecular crowding, which so limits molecular conformations that the configurational entropy reaches zero at a finite temperature. This mechanism, with a second-order transition, is predicted by the Gibbs–DiMarzio lattice model of polymer glasses; a first-order transition is predicted by the square tiling model. The second possibility is a gentler version of the first. That is, the entropy S_c plunges toward zero, but veers away from zero when the temperature reaches the range where one would extrapolate S_c to zero. At lower temperatures, S_c decreases more slowly, reaching zero only at $T = 0$ K. The low-entropy state at low temperatures is glassy in that its dynamics are very slow. This behavior is predicted by the Miller rotational isomeric state model, the Fredrickson–Andersen facilitated spin model, and, apparently, the bond fluctuation model. The final possibility, which is predicted by the mode-coupling theory, is that there is no thermodynamic anomaly, but rather the formation of a rigid, percolated structure at a critical temperature which prevents the system from exploring its many thermodynamically allowed states.

Of these models, the mode-coupling theory has the clearest direct connection to liquid-state structure; it attempts to describe nonlinear density fluctuations in dense liquids. Although the mode-coupling theory shows promising agreement to the measured relaxation properties of densely packed suspensions, and perhaps also to high-frequency relaxations in molecular liquids, it does not seem able to describe the extremely slow relaxations that occur in molecular liquids near the glass transition. Semiempirical theories, such as that of Adam and Gibbs, and analogical or "toy" models, such as the facilitated Ising model of Fredrickson and Andersen, or the square tiling model of Stillinger and Weber, incorporate the "cooperative relaxation" that is thought to be essential to glassy behavior at long time scales. These models predict many of the phenomena observed in real glasses, but their connection to molecular structure and dynamics is tenuous at best.

Very recently, much progress has been made using powerful computer simulations, which can realistically model glass formation in simple liquids such as argon, or can model less realistically the formation of glassy characteristics in more complex liquids such as polymers. No doubt these methods will continue to enhance our understanding of glassy behavior.

REFERENCES

Adam G, Gibbs JH (1965). *J Chem Phys* 43:139.
Angell CA (1985). *J Non-Cryst Solids* 73:1.
Angell CA, Rao KJ (1972). *J Chem Phys* 57:470.
Angell CA, Sichina W (1976). *Ann NY Acad Sci* 279:53.
Angell CA, Smith DL (1982). *J Phys Chem* 86:3845.
Argon AS, Bulatov VV, Mott PH, Suter UW (1995). *J Rheol* 39:377.
Bair S (1996). *Rheol Acta* 35:13.
Barlow AJ, Erginsav A, Lamb J (1967). *Proc R Soc* A298:481.
Baschnagel J, Wolfgardt M, Paul W, Binder K (1997). *J Res Natl Inst Stand Technol* 102:159.
Batchinski AJ (1913). *Z Phys Chem* 84:644.
Benbow JJ, Wood DJC (1958). *Trans Faraday Soc* 54:1581.
Bentzelius U, Götze W, Sjölander (1984). *J Phys C* 17:5915.
Bernstein B, Shokooh A (1980). *J Rheol* 24:189.
Brawer S (1985). *Relaxation in Viscous Liquids and Glasses* American Ceramic Society, Columbus, OH.
Busch R, Schneider S, Peker A, Johnson WL (1995). *Appl Phys Lett* 67:1544.
Carmesin I, Kremer K (1988). *Macromolecules* 21:2819.
Cohen MH, Turnbull D (1959). *J Chem Phys* 31:1164.
Cowie JMG, Toporowski PM (1968). *Eur Polym J* 4:621.
Dealy JM, Wissbrun KF (1990). *Melt Rheology and Its Role in Plastics Processing*, van Nostrand Reinhold, New York.
DiMarzio EA, Gibbs JH (1958). *J Chem Phys* 28:807.
DiMarzio EA, Guttman CM (1987). *Macromolecules* 20:1405.
DiMarzio EA, Yang AJM (1997). *J Res Natl Inst Stand Technol* 102:135.
Dixon PK, Wu L, Nagel SR, Williams BD, Carini JP (1990). *Phys Rev Lett* 65:1108.
Doolittle AK, Doolittle DB (1957). *J Appl Phys* 28:901.
Ferry JD (1980). *Viscoelastic Properties of Polymers*, 3rd ed, Wiley, New York.
Flory PJ (1969). *Statistical Mechanics of Chain Molecules*, Carl Hanser Verlag, New York.
Fox JR, Andersen HC (1984) *J Phys Chem* 88:4019.
Fredrickson GH (1988). *Annu Rev Phys Chem* 39:149–180.
Fredrickson GH, Andersen HC (1984). *Phys Rev Lett* 53:1244.
Fredrickson GH, Andersen HC (1985). *J Chem Phys* 83:5822.
Fredrickson GH (1986). *Ann NY Acad Sci* 484:185.
Fujita H (1961). *Fortschr Hockpolym-Forsch* Bd 3:S1.
Fulcher GS (1925). *J Am Chem Soc* 8:339, 789.
Gardon R, Narayanaswamy OS (1970). *J Am Ceram Soc* 53:380.
Gibbs JH (1956). *J Chem Phys* 25:185.
Gibbs JH, DiMarzio EA (1958). *J Chem Phys* 28:373.
Götze W, Sjögren L (1991). *Phys Rev A* 43:5442.

Götze W, Sjögren L (1992). *Rep Prog Phys* 55:241.

Greet RJ, Turnbull D (1967). *J Chem Phys* 47:2185.

Hofmann A, Kremer F, Fischer EW (1993). *Physica A* 201:106.

Johari GP (1970). *J Chem Phys* 58:1766.

Kauzmann W (1948). *Chem Rev* 43:219.

Kim B, Mazenko G (1992). *Phys Rev A* 45:2393.

Knauss WG, Emri I (1987). *Polym Eng Sci* 27:86.

Kob W, Andersen HC (1995). *Phys Rev E* 51:4626.

Kolrausch F (1847). *Pogg Ann Phys* 12:393.

Kovacs AJ (1958). *J Polym Sci* 30:131.

Kovacs AJ (1964). *Adv Polym Sci* 3:394.

Kovacs AJ, Aklonis JJ, Hutchinson JM, Ramos AR (1979). *J Polym Sci Polym Phys Ed* 17:1097.

Leutheusser E (1984). *Phys Rev A* 29:2765.

Lindsey CP, Patterson GD, Stevens JR (1979). *J Polym Sci Polym Phys Ed* 17:1547.

Lunkenheimer P, Pimenov A, Dressel M, Goncharov YuG, Böhmer R, Loidl A (1996). *Phys Rev Lett* 77:318.

Lustig SR, Shay RM Jr, Caruthers JM (1996). *J Rheol* 40:69.

Matsuoka S (1992). *Relaxation Phenomena in Polymers*, Carl Hanser Verlag, New York.

Maxfield J, Shepard IW (1973). *Chem Phys Lett* 19:541.

Mazurin OV (1977). *J Non-Cryst Solids* 25:130.

McCrum NG, Read BE, Williams G (1967). *Anelastic and Dielectric Effects in Polymeric Solids*, Wiley, London.

McKenna G (1989). In *Comprehensive Polymer Science*, Vol 2: *Polymer Properties*, Booth C, Price C (eds), Pergamon, Oxford.

McKenna G (1994). *J Res Natl Inst Stand Technol* 99:169.

McKenna G (1995). *Mech Plast and Plast Composites* 68:309.

Mezei F, Knaak W, Farago B (1987). *Phys Rev Lett* 58:571.

Mezei F (1991). *Ber Bun Phys Chem* 95:1118.

Miller AA (1978). *Macromolecules* 11:859.

Mott PH, Argon AS, Suter UW (1993). *Philos Mag* 67:931.

Moynihan CT, Macedo PB, Montrose CJ, Gupta PK, DeBolt MA, Dill JF, Dom BE, Drake PW, Easteal AJ, Elterman PB, Moeller RP, Sasabe H, Wilder JA (1976). *Ann NY Acad Sci* 279:15.

Narayanaswamy OS (1971). *J Am Ceram Soc* 54:491.

O'Connell PA, McKenna GB (1997). *Polym Eng Sci* 37:1485.

O'Connell PA, Schultheisz CR, McKenna GB (1998). In *Physics of Glassy Polymers*, Hill A, Tant M (eds), ACS Books, Washington, DC, in press.

Okun K, Wolfgardt M, Baschnagel J, Binder K (1997). *Macromolecules* 30:3075.

Pesce J-J, McKenna GB (1997). *J Rheol* 41:929.

Petry W, Bartsch E, Fujara F, Kiebel M, Sillescu H, Farago B (1991). *Z Phys B Cond Mat* 83:175.

Pezzin G, Zilio-Grandi F, Sammartin P (1970). *Eur Polym J* 6:1053.

Phillips MC, Barlow AJ, Lamb J (1972). *Proc R Soc A* 329:193.

Pusey PN, van Megen W (1987). *Phys Rev Lett* 59:2083.

Rekhson SM, Bulaeva AV, Mazurin OV (1971). *Sov J Inorg Mater* 78:622.

Richter D, Zorn R, Frick B, Farago B (1991). *Ber Bun Phys Chem* 95:1111.

Sammler RL, Utaigbe JU, Lapham ML, Bradley NL, Monahan BC, Quinn CJ (1996). *J Rheol* 40:285.

Saunders DW, Lightfoot DR, Parsons DA (1968). *J Polym Sci A-2* 6:1183.

Schapery RA (1969). *Polym Eng Sci* 9:295.

Scherer GW (1992). *Relaxation in Glass and Composites*, Wiley, New York.

Shlesinger MF, Montroll EW (1984). *Proc Natl Acad Sci USA* 81:1280.

Schönhals A, Kremer F, Schlosser E (1991). *Phys Rev Lett* 67:999.
Schönhals A, Kremer F, Hofmann A, Fischer EW, Schlosser E (1993). *Phys Rev Lett* 70:3459.
Stickel F, Kremer F, Fischer EW (1993). *Physica A* 201:318.
Stillinger FH, Weber TA (1986). *Ann NY Acad Sci* 484:1.
Struik LCE (1976). *Ann NY Acad Sci* 279:78.
Tammann G, Hesse G (1926). *Z Anorg Allg Chem* 156:245.
Teichler H (1996). *Phys Rev Lett* 76:62.
Turnbull D, Cohen MH (1961). *J Chem Phys* 31:1164.
van Blaaderen A, Wiltzius P (1995). *Science* 270:1177.
van Megen W, Pusey PN (1991). *Phys Rev A* 43:5429.
van Megen W, Underwood SM (1993). *Phys Rev E* 47:248.
Vogel H (1921). *Phys Z* 22:645.
Vrentas JS, Duda J (1977). *J Poly Sci Polym Phys Ed* 15:441.
Vrentas JS, Duda JL, Hou AC (1986). *J Appl Polym Sci* 31:739.
Williams G, Watts DC (1970). *Trans Faraday Soc* 66:80.
Zapas LJ (1974). In *Deformation and Fracture of High Polymers*, Kausch HH, Hassell JA, Jaffee RI (eds), Plenum, New York.

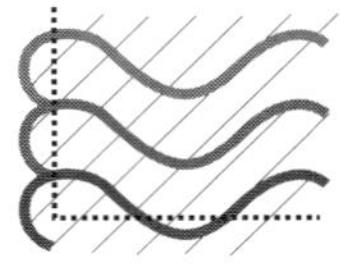

Chapter 5

Polymer Gels

5.1 INTRODUCTION

Gelation is the conversion of a liquid to a disordered solid by formation of a network of chemical or physical bonds between the molecules or particles composing the liquid. The liquid precursor is called the "sol," and the solid formed from it is the "gel." Gels can be as mundane as the epoxy glue used to mend a child's toy, or they can be as sublime as the jellies, meringues, and custards that delight the mavens of haute cuisine.

This chapter is devoted to the properties of *polymeric* gel-forming liquids. Particulate gels are discussed in Chapter 7. The structure of a polymeric gel is sketched in Fig. 5-1. Since this book is devoted to materials that are in some sense liquid, or at least liquefiable, we shall not say much about hard, irreversible, chemical gels such as cured epoxies or vulcanized rubber, but shall focus instead on chemical *pre-gels* and thermally reversible *physical gels*, both of which can be considered borderline fluids. This chapter is confined to a brief overview. Much more detail can be found in Winter and Mours (1997), and volume 101 of the *Faraday Discussions*.

Crucial to the formation of such gels is *branching* or *multifunctionality*. The functionality f of a molecule is the number of bonds it can form with other molecules; $f = 4$ in Fig. 5-1.

There are at least three generic types of *chemical* reaction that can produce such branching structures (de Gennes 1979). The first is a *condensation reaction*, whereby a molecule with three or more reactive groups, such as OH groups, reacts with a *cross-linker*. A second type of branching reaction is *addition* polymerization, whereby a double bond is opened by a free-radical reaction, creating additional bonds that link monomers together. This type of reaction will produce linear chains if there is only one double bond per monomer, but if there are two or more double bonds, branching can occur. The third way to create branching is to start with linear polymeric precursors, and *cross-link* or *vulcanize* them by introducing chemical links that bond them together. For an explanation of the chemistry of gelation, see Flory (1953).

Physical gelation occurs as a result of *intermolecular association*, leading to network formation (see Fig. 5-2). (Physical associations differ from chemical bonds in that the latter are covalent attachments between two atoms and are typically permanent at temperatures of interest here, while intermolecular associations are weak, reversible bonds or clusters produced by van der Waals forces, electrostatic attractions, or hydrogen bonding.) If physical associations are to produce gelation, rather than phase separation, it is crucial

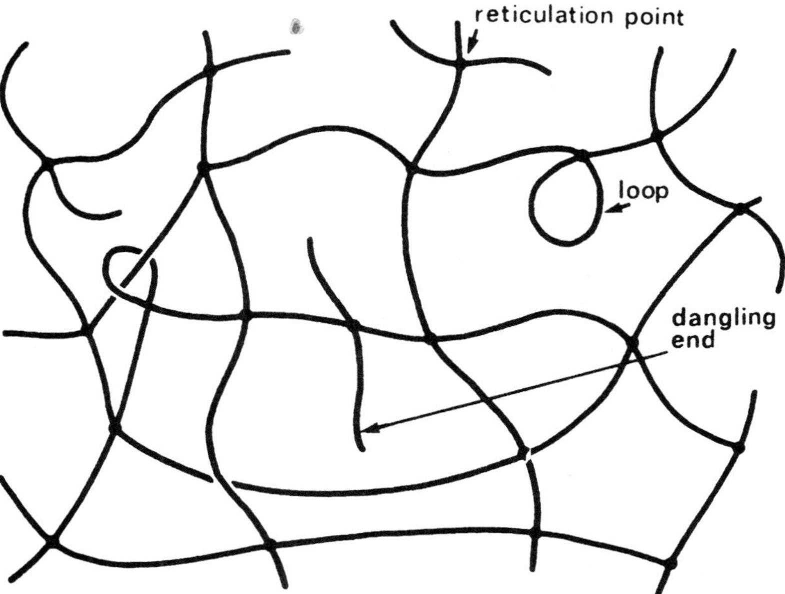

Figure 5.1 A typical polymer gel network. (Reprinted from Pierre-Gilles de Gennes, *Scaling Concepts in Polymer Physics.* Copyright © 1979 by Cornell University. Used by permission of the publisher, Cornell University Press.)

that the junctions between molecules that are formed by such associations do not grow too large. Thus, there must be some means of frustrating the growth of these associating domains, so that their size is limited. de Gennes identifies three types of interactions that can lead to physical gelation: (1) local *helical structures* whereby one molecule winds around another; (2) *microcrystallites*; and (3) *nodular domains*, in which the chain is chemically heterogeneous, and association only occurs at preferred sites along the chain. Examples of polymers that form nodular domains include water-soluble *associative thickeners*, which contain *hydrophobic* sites along an otherwise hydrophilic chain. At low concentrations, such thickeners greatly enhance the viscosity of water, and thus they are useful as additives to foods, shampoos, and other personal care products (see Fig. 1-4), or as mobility control agents in oil-field production. They form "flowable networks" that can, for example, be deposited in a capillary tube and used for electrophoretic separation of DNA (Menchen and Winnik 1994; Menchen et al. 1996). The reverse kind of associating polymer also exists—that is, hydrophobic molecules with hydrophilic sites, such as hydrogen-bonding or ionic groups. Peculiar rheological phenomena, such as "shear-induced gelation," have been ascribed to intermolecular "associations" for many years (Eliassaf et al. 1955; Lodge 1961; Peterlin and Turner 1965), but only recently has any detailed microscopic understanding been achieved. Further discussion of physical gelation is deferred to Section 5.4. More detail can be found in the book by Guenet (1992).

Because gels are disordered materials that are kinetically frozen, the method of preparation strongly influences the properties obtained. For example, a gel prepared in a "dry"

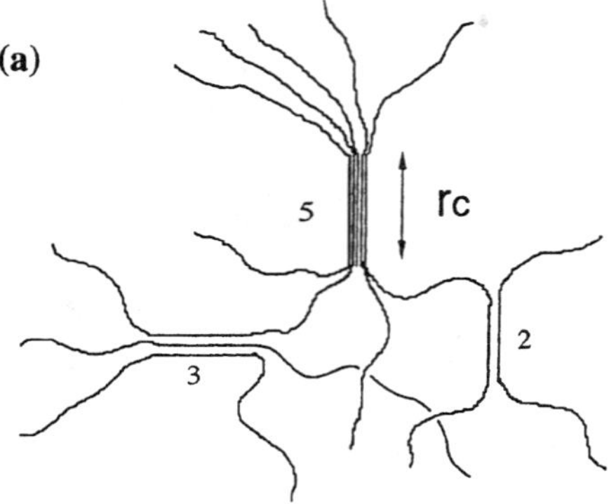

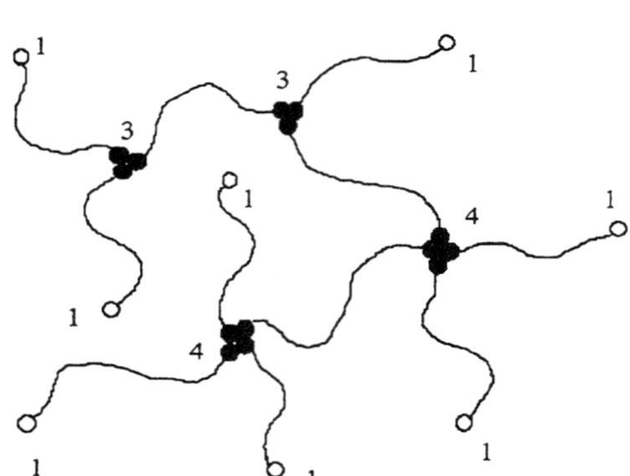

Figure 5.2 Illustrations of physical gels. In (a), the junctions are formed by microcrystallites, while in (b) the junctions are formed by the end groups of telechelic polymers. The functionalities of the various cross-link points are indicated by numbers beside the junctions. (Reprinted with permission from Tanaka and Stockmayer, Macromolecules 27:3943. Copyright 1994 American Chemical Society.)

state, with no solvent present, and then swollen by introduction of a solvent, will have a modulus that differs from that of a gel cross-linked with the solvent already present. Similar sensitivity to preparation conditions is found in physical gels. As a result, experiments on gels tend to be difficult to reproduce with precision.

Gels are indeed often prepared in the presence of a solvent, which is then removed to produce a solid with commercially valuable properties. If the solvent is removed by evaporation under normal conditions, the gel structure usually shrinks because of capillary forces acting on the liquid–air menisci. This produces a dense material with moderate or low porosity called a *xerogel*. On the other hand, if the solvent is removed by supercritical drying which prevents liquid–air menisci from forming, the product is an *aerogel*, which can have a solids volume fraction as low as 1% (Brinker and Scherer 1990).

Some polymeric gels with charged groups along their backbones can, when immersed in hydrophilic media, shrink or expand enormously in response to a change in temperature, pH, or electric field (Tanaka 1981; Osada and Ross-Murphy 1993). It has been proposed that such "intelligent gels," if they could be made to respond quickly enough to an electric field or temperature, might serve as "artificial muscles" (Osada and Ross-Murphy 1993; Hu et al. 1995b).

5.2 GELATION THEORIES

5.2.1 Percolation Theory

Stauffer (1976) and de Gennes (1976, 1979) have pointed out the connection between gelation and *bond percolation*. To illustrate, let each site (or lattice points) on the square lattice in Fig. 5-3 represent a polyfunctional molecular unit, and let each filled link represent a chemical bond between neighboring units. Chemical reaction then corresponds to the conversion of unfilled bonds to filled bonds. As one increases the fraction p of bonds that are filled, more and more units link together, producing *clusters* of bonds; and eventually, at the *percolation transition*, $p = p_c$ (which corresponds to the gel point), an infinite, lattice-spanning cluster appears. Generally speaking, *percolation* is the process of network formation by random filling of bonds (or sites) on a lattice, or by random

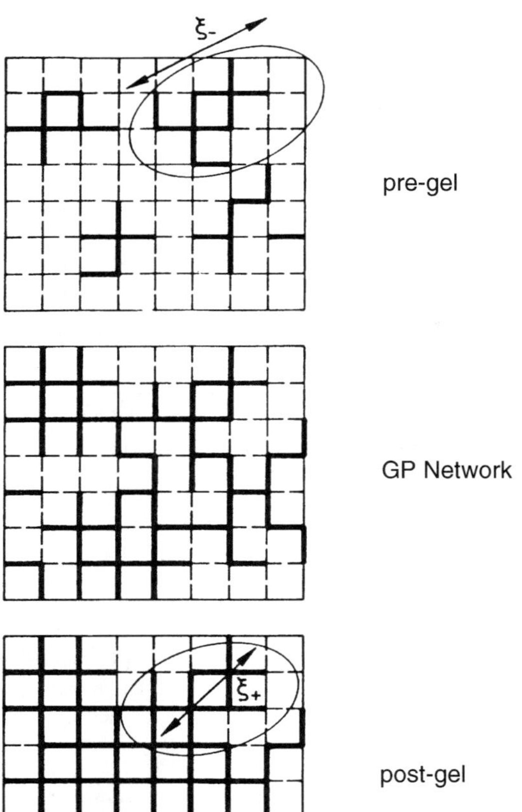

Figure 5.3 Typical configuration of closed bonds resulting from random filling of a square lattice. At small fractions p of filled bonds there are only isolated clusters whose correlation length is ξ_-; when p exceeds the percolation threshold p_c, a sample-spanning cluster appears. The correlation length ξ is infinite at p_c and is finite both above and below it. (From Hess et al. 1988), (reprinted with permission from Hess et al., Macromolecules 21:2536. Copyright 1988 American Chemical Society.)

filling of regions of space (Broadbent and Hammersley 1957; Kirkpatrick 1973). For bond percolation on a square, $p_c = 0.5$. On other 2-D lattices, one finds empirically that $p_c \approx 2/z$, where z is the lattice coordination number; on 3-D lattices, $p_c \approx 1.5/z$ (Shante and Kirkpatrick 1971; Brinker and Scherer 1990). Although in a gelling system there are molecular diffusive motions and other complications, percolation theory nevertheless provides useful predictions, especially for properties near the gel point.

5.2.2 Flory–Stockmayer Theory

An earlier way of viewing gelation is due to Flory (1941, 1942) and Stockmayer (1943). In this *classical* theory, one also considers the buildup of large clusters by random bonding, but loops or cycles are ignored. Thus, the bonding process is effectively *tree-like*, as depicted in Fig. 5-4. Each new branch of the tree has as much freedom to grow new branches as its predecessor, without restrictions due to excluded volume or cycle formation. Because of the absence of closed cycles, the statistical properties of tree-like clusters can be computed analytically (Fisher and Essam 1961; Stinchcombe 1974; Larson and Davis 1982; Straley 1977, 1982), which makes the Flory–Stockmayer model a very convenient one that captures the essence of the gelation process.

The gel point in the classical theory is

$$p_c = \frac{1}{f-1}$$

where f is the coordination number of the tree—that is, the number of bonds that can form at each site of the network (Flory 1953). If the gel is formed by reacting precursor molecules (A) with a chemical cross-linkers (B), then the gel point, measured as a fraction $p_{c,A}$ of A's reaction sites, depends on the functionalities (f_A and f_B) of both A and B as

$$p_{c,A} = \frac{1}{\sqrt{(f_A - 1)(f_B - 1)/r}}$$

where r is the "stoichiometric ratio" of B to A reactive sites:

$$r \equiv \frac{f_B n_B}{f_A n_A}$$

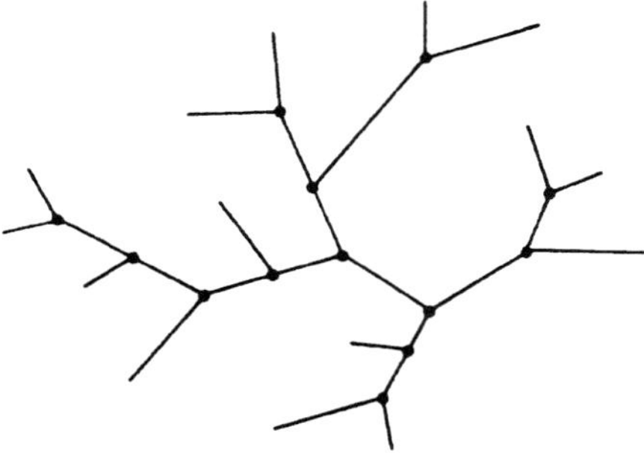

Figure 5.4 Illustration of a tree-like gel cluster. (Reprinted from Pierre-Gilles de Gennes, *Scaling Concepts in Polymer Physics*. Copyright © 1979 by Cornell University. Used by permission of the publisher, Cornell University Press.)

where n_A and n_B are the number of moles of A and B in the reactive mixture. Despite the limitations of the classical theory (e.g., the neglect of loops), the above formula for p_c seems to be reasonably accurate (Vallés and Macosko 1979; Venkataraman et al. 1989).

In the classical theory, however, the neglect of loops significantly affects the size distribution and other properties of the clusters as one approaches the gel point. Some of the "critical exponents" that describe these properties in the classical theory and in percolation theory near $p = p_c$ are compiled in Table 5-1 (Martin and Adolf 1991).

In Table 5-1, $\varepsilon \equiv |p - p_c|$, $N(m)$ is the number of clusters containing m bonds, R is the radius of a cluster of molecular weight M, and M_z and and M_w are the z-averaged and weight-averaged molecular weights of the clusters, namely,

$$M_z = \frac{\sum m^3 N(m)}{\sum m^2 N(m)}, \qquad M_w = \frac{\sum m^2 N(m)}{\sum m N(m)} \tag{5-1}$$

When $p > p_c$, one can define $P(p)$ to be the fraction of bonds belonging to the infinite cluster. The percolation predictions of the modulus G, the longest relaxation time τ, and the viscosity η depend on whether one uses the Rouse–Zimm (R–Z) theory, or the analogy to an electrical network (EN). The exponent for the modulus G is predicted to be greater than either of these (i.e., around 3.7) if bond-bending dominates (Arbabi and Sahimi 1988). Further details about these exponents can be found in Chapter 5 of Brinker and Scherer (1990), as well as in Martin and Adolf (1991).

5.2.3 Fractals and Self-Similarity

The power-law scaling of the cluster properties shown in Table 5-1 arises from their *fractal* or *self-similar* character. Self-similarity implies that the huge clusters formed near the gel point look the same at any magnification, as long as elementary units making up the cluster are too small to see. Furthermore, the cluster size distribution at one value of $\varepsilon(\varepsilon_1)$ is the

TABLE 5-1
Scaling Exponents for Classical and Percolation Theories of Gelation

Exponent	Relation	Classical	3-D Percolation	Experimental
λ	$N(m) \sim m^{-\lambda}$	5/2	2.20	2.18–2.3
σ	$M_z \sim \varepsilon^{-1/\sigma}$	1/2	0.45	—
γ	$M_w \sim \varepsilon^{-\gamma}$	1	1.76	1.0–2.7
ν	$R_z \sim \varepsilon^{-\nu}$	1	0.89	—
D_f	$R^{D_f} \sim M$	4	2.5	1.98
β	$P \sim \varepsilon^{\beta}$	1	0.39	—

			R–Z	EN	
t	$G \sim \varepsilon^t$	3	2.7	1.94	1.9–3.5
ζ	$\tau \sim \varepsilon^{-\zeta}$	3	4.0–2.7	2.6	3.9
$k = \zeta - t$	$\eta \sim \varepsilon^{-k}$	0	0–1.35	0.75	0.75–1.5

same as at a smaller value of $\varepsilon(\varepsilon_2)$, if one uniformly magnifies in size all clusters formed at ε_1. The exponent D_f in Table 5-1 is called the *fractal dimension* of the cluster; it is the exponent relating the linear size to the mass. For any dense three-dimensional ($D = 3$) object, this exponent is $D_f = D = 3$; clusters with $D_f < D$ are ramified, open structures.

5.3 RHEOLOGY OF CHEMICAL GELS AND NEAR-GELS

When a precursor liquid, composed of either small molecules or polymers, is cross-linked to form a gel, the rheological properties change from those of a viscous liquid to those of an elastic solid. Thus, at the gel point, the viscosity of the liquid diverges to infinity, and the low-frequency modulus G_0 rises from zero, as shown schematically in Fig. 5-5. The modulus of the fully cured elastic solid can be estimated as (Wall 1943; Treloar 1975)

$$G_0 = \nu kT \tag{5-2}$$

where ν is the number of "elastically effective" network strands per unit volume. Equation (5-2) assumes that the cross-link points or junctions of the network move *affinely*, or in proportion to, the macroscopic strain. This is only expected to occur when the functionality of the network is high. For low functionality, the junctions are liable to move nonaffinely to produce a lower overall stress. If the junctions and the chains can move nonaffinely without interfering with each other (i.e., they are so-called phantom chains), then ν in Eq. (5-2) should be replaced by $\nu - \mu$, where μ is the number of junctions per unit volume (James and Guth 1953; Ferry 1980). Erman and Flory (1983) have developed equations for the more realistic case of "constrained junction fluctuations." Additional prefactors

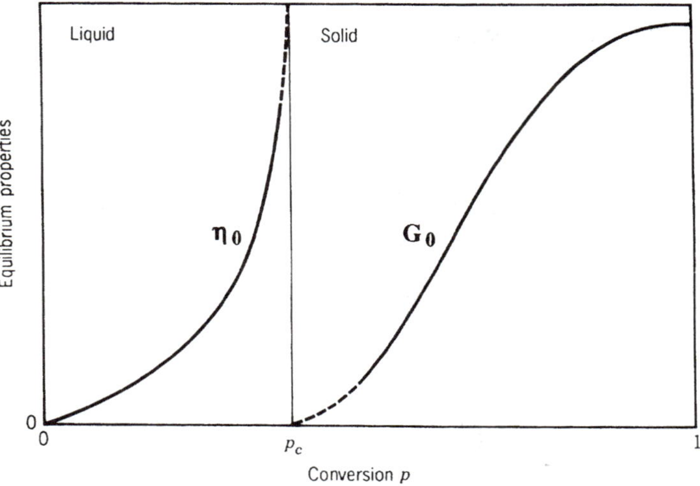

Figure 5.5   Illustration of the dependence of zero-shear viscosity η_0 and equilibrium modulus G_0 on conversion p for a cross-linking system. (From Winter, *Encyclopedia of Polymer Science and Engineering*, Copyright © 1989. Reprinted by permission of John Wiley & Sons, Inc.)

can be introduced into Eq. (5-2) due to the presence of "trapped entanglements" and other considerations (Ferry 1980). In addition, "dangling ends" that are not "elastically effective" must be excluded from ν. Near the gel point, such "ineffective" bonds are common, and the gel modulus typically follows a power law $G \sim |p - p_c|^t$ as indicated in Table 5-1.

At large deformations, outside the linear regime, the stress tensor for a polymer gel, according to the classical affine-motion rubber-elasticity theory (Section 3.4.2), is

$$\sigma = G_0 \mathbf{B} \tag{5-3}$$

where $\mathbf{B}$ is the *Finger tensor* defined in Eq. (1-16).

Equation (5-3) does not describe real gels very well. The empirical *Mooney–Rivlin* expression (Mooney 1940; Rivlin 1948; Treloar 1975) does better:

$$\sigma = 2C_1 \mathbf{B} + 2C_2 \mathbf{C} \tag{5-4}$$

where $\mathbf{C} \equiv \mathbf{B}^{-1}$ is the Cauchy tensor, the inverse of the Finger tensor, and C_1 and C_2 are empirical constants tabulated for various polymer gels by Horkay and McKenna (1996).

Further discussion of models of the elasticity of gels is beyond the scope of the present work; the interested reader can find a thorough description of the elasticity and viscoelasticity of polymer chemical gels in Ferry (1980), Treloar (1975), and Flory (1953).

More relevant to this book on complex *fluids* are the properties of the partially gelled liquids formed on the way toward complete gelation. The rheology of partially cured materials has been studied in detail by Winter, Chambon, and coworkers (Chambon et al. 1986; Winter and Chambon 1986; Winter et al. 1988; Scanlan and Winter 1991; Izuka et al. 1992; Richtering et al. 1992). Such partially cured or lightly cross-linked materials not only are scientifically interesting, but also are technologically important, for example as *adhesives*. Their rheology is intermediate between fluid and solid, making them sticky or *tacky* (Winter 1989; Zosel 1991).

Figure 5-6 shows the storage and loss modulus, at a fixed frequency, for poly(dimethylsiloxane) cross-linked with a tetrasilane cross-linker, as a function of reaction time. At short times after the start of cross-linking, the material is a liquid with $G'' \gg G'$; but as the reaction continues, the storage modulus rises from close to zero toward a long-time asymptote of around 10^5 Pa. At the point marked t_c, the storage and loss moduli cross each other, marking a transition from liquid-like to solid-like behavior. These measurements were made after quenching the reaction at various times after the start of the reaction. Quenching can be avoided for photocurable samples cured in the rheometer with transparent fixtures [see Chiou et al. (1996)].

Figure 5-7 shows the frequency dependences of the storage and loss moduli at various times during the reaction, from 6 minutes before t_c to 6 minutes after it. Note that at t_c (labeled "Gel Point" in Fig. 5-7), G' and G'' follow *power laws* over the entire frequency range! For times less than this (labeled -2 and -6 in Fig. 5-7), the curves slope downward at low frequencies, which is indicative of fluid-like behavior, while at times after the "gel point" (labeled $+2$ and $+6$), G' flattens at low frequency—a characteristic of solid-like behavior. Thus, the intermediate state with a power-law frequency dependence over the whole frequency range is the transitional state between liquid-like and solid-like behavior, and therefore it defines the gel point. This rheologically determined gel point coincides with the conventional value, namely the maximum degree of cure at which

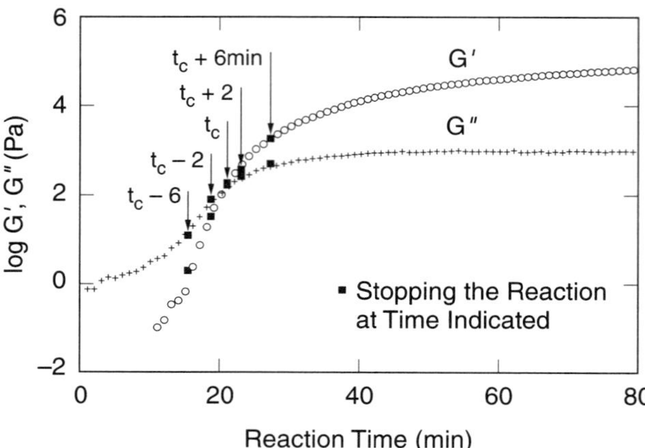

Figure 5.6 Time-dependence of G' (○) and G'' (+) during cross-linking reaction of subentangled poly(dimethylsiloxane) at balanced stoichiometry with a tetrasilane crosslinker. The gel point is marked as t_c. (From Winter and Chambon 1986, with permission from the Journal of Rheology.)

the partially cross-linked system still dissolves completely in a good solvent (Winter et al. 1988).

The relaxation modulus of the transitional state at the gel point is therefore described by a simple power law:

$$G(t) = St^{-n} \tag{5-5}$$

where S is called the *strength* of the gel. The exponent n in Eq. (5-5) is 0.5 for the data in Figs. 5-6 and 5-7. It has been found to vary over the wide range 0.19–0.92 for chemically cross-linking systems (Scanlan and Winter 1991), with even lower values of n in some physically gelling systems (Richtering et al. 1992). By the Kramers–Kroenig relationship, Eq. (5-5) implies that

$$G'(\omega) = \frac{G''(\omega)}{\tan(n\pi/2)} = \Gamma(1-n) \cos\left(\frac{n\pi}{2}\right) S\omega^n \tag{5-6}$$

where $\Gamma(\ )$ is the gamma function. For $n < 0.5$ we have $G' > G''$, while for $n > 0.5$ we have $G' < G''$. Figure 5-8 shows the variation of n and S with the molecular weight of polycaprolactone precursors (Izuka et al. 1992). As the molecular weight M_n crosses the entanglement threshold (see Section 3.1), which is around $M_n \approx 6600$, the exponent n drops from near unity to much lower values. Evidently, entanglements among the precursor polymer molecules make the critical gel more elastic, giving it a lower value of n. The exponent n is also affected by the stoichiometric ratio of cross-linker to precursor. Defining r as the molar ratio of cross-linker reactive end groups to precursor reactive end groups, $r = 1$ corresponds to a "balanced" stoichiometry, in which, at complete reaction, there is no excess of either precursor or cross-linker molecules. Figure 5-9b shows that n decreases as the stoichiometric ratio increases towards unity. Thus, the critical gel is more "fluid like," or dissipative, when there is an excess of precursor end groups (low r).

With increasing r, Fig. 5-9a shows that the decrease in n is accompanied by an increase in the parameter S in Eq. (5-5). At least for some of these critical gels, S can be estimated from the low-frequency modulus G_0 of the fully cured gel and the viscosity η_{sol} of the unreacted sol, or prepolymer, as (Scanlan and Winter 1991)

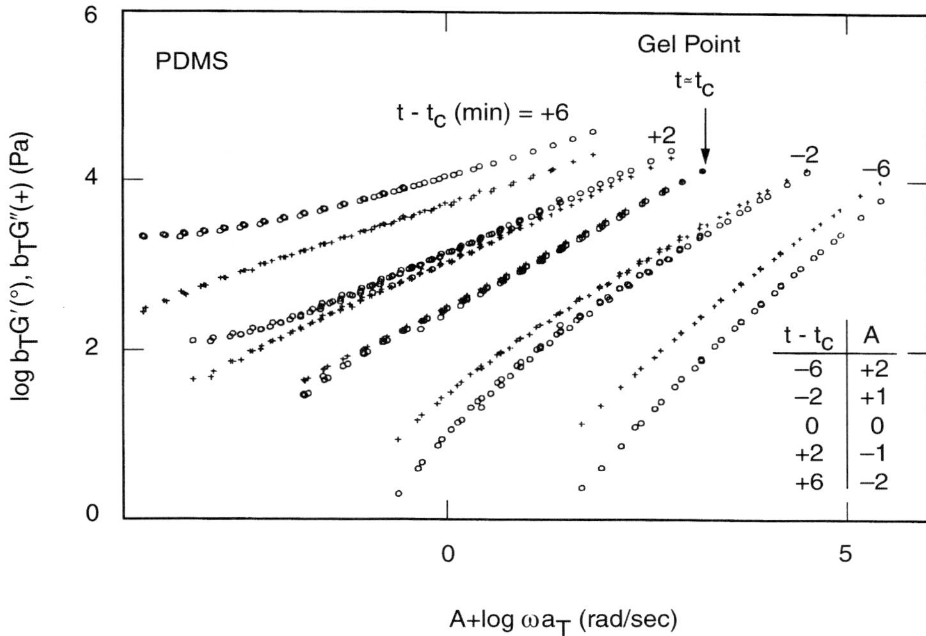

Figure 5.7 Frequency dependences of the storage ($\bigcirc$) and loss ($+$) moduli for poly(dimethylsiloxane) (PDMS) samples whose reactions were quenched at the times indicated (see Fig. 5-6). The data are time–temperature-shifted to the reference temperature T_{ref} of 34°C, and they are shifted additionally by an amount A on the logarithmic axis to keep the curves from overlapping. The vertical shift factors b_T are given by $\rho(T_{\mathrm{ref}})T_{\mathrm{ref}}/(\rho(T)T)$, where ρ is the density. (From Winter and Chambon 1986, with permission from the Journal of Rheology.)

$$S \approx G_0 \left(\frac{\eta_{\mathrm{sol}}}{G_0} \right)^n \tag{5-7}$$

The power laws for viscoelastic spectra near the gel point presumably arise from the fractal scaling properties of gel clusters. Adolf and Martin (1990) have attempted to derive a value for the scaling exponent n from the universal scaling properties of percolation fractal aggregates near the gel point. Using Rouse theory for the dependence of the relaxation time on cluster molecular weight, they obtain $n = D/(2 + D_f) = 2/3$, where $D_f = 2.5$ is the fractal dimensionality of the clusters (see Table 5-1), and $D = 3$ is the dimensionality of space. The theoretical value of $n = 2/3$ agrees with only a small subset of the data published by Winter et al. One system for which $n \approx 2/3$ is observed is a partially cross-linked epoxy system studied by Adolf and Martin (1990) (see Fig. 5-10). For this system, data for samples at various degrees of cure, from the gel point to nearly full cure, can be superposed using *time-cure superposition,* a remarkable law by which the data are shifted horizontally and vertically using the power-law scaling $\tau \propto |p - p_c|^{-3.9}$ and $G_{\mathrm{char}} \propto |p - p_c|^{2.8}$ (see Fig. 5-10). These exponents, -3.9 and 2.8, are close to the values, -4 and $8/3 = 2.67$, derived

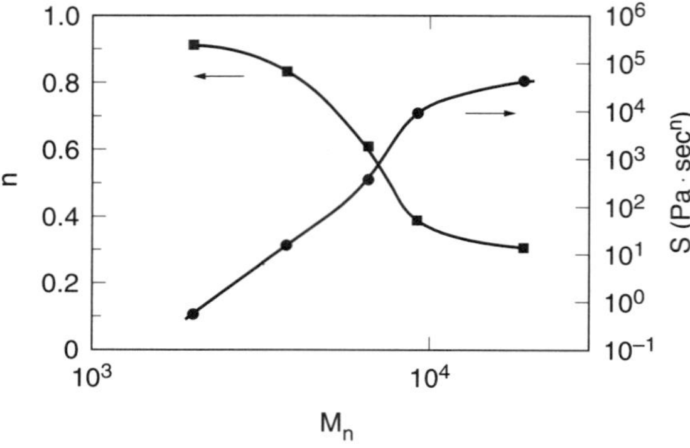

Figure 5.8 Relaxation exponent n and relaxation strength S as functions of the number-average molecular weight of precursor polycaprolactone molecules. (Reprinted with permission from Izuka et al., Macromolecules 25:2422. Copyright 1992 American Chemical Society.)

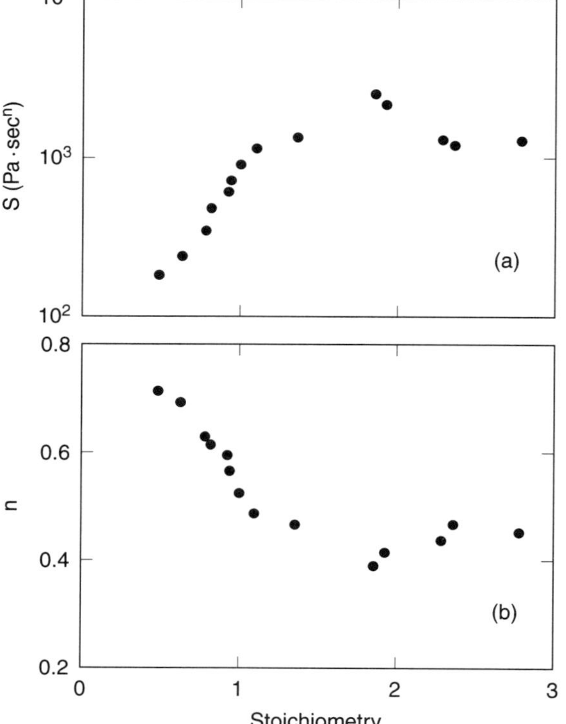

Figure 5.9 **(a)** Relaxation strength S and **(b)** relaxation exponent n as functions of stoichiometric ratio r (moles of cross-linker ends/moles of prepolymer ends) for poly(dimethylsiloxane) with prepolymer number-averaged degree of polymerization equal to 142. (Reprinted with permission from Scanlan and Winter, Macromolecules 24:47. Copyright 1991 American Chemical Society.)

from "Rouse" theory for gel clusters (Martin and Adolf 1991). Also, the power law for the frequency dependence at the gel point, $G' \propto G'' \propto \omega^{0.72}$, is close to that predicted by the Rouse theory for fractal clusters, $n = 2/3$. An analogous time-cure superposition works for extents of cure less than the gel point, except that then the low-frequency data show terminal (i.e., fluid-like) behavior. The viscosity diverges near the gel point as $\eta_0 \propto |p_c - p|^{1.1}$, where the exponent 1.1 is close to the predicted value 4/3.

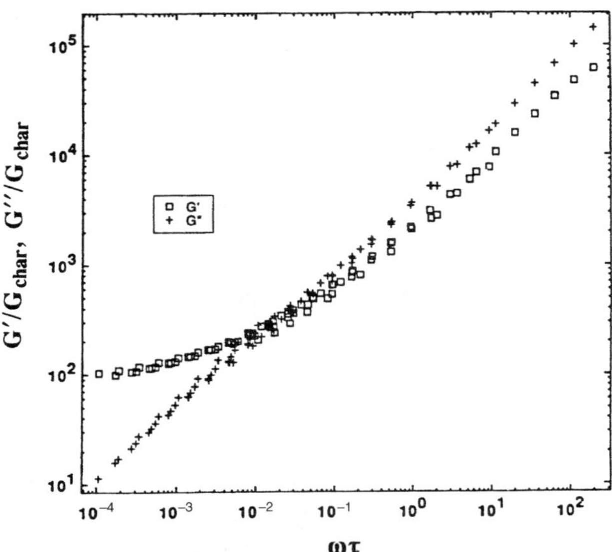

Figure 5.10 Time-cure superposition for a partially cross-linked epoxy obtained by multiplying the frequency ω by $\tau \propto \varepsilon^{-3.9}$ and dividing the modulii G' and G'' by $G_{\text{char}} \propto \varepsilon^{2.8}$, where $\varepsilon \equiv p - p_c$. In the power-law frequency regime, $G' \propto G'' \propto \omega^{0.72}$. (From Martin and Adolf 1991, with permission from the Annual Review of Physical Chemistry, Volume 42, © 1991, by Annual Reviews, Inc.)

However, as remarked earlier, for many gelling systems, particularly those with relatively large precursor molecules, the exponent n can be much less than the theoretical value $n = 2/3$. Winter and Mours (1996) provide a thorough summary of these and other rheological studies of chemical gels.

5.4 RHEOLOGY OF PHYSICAL GELS

We now turn our attention from chemical to physical gels. As mentioned in the Introduction to this chapter, the junctions in physical gels can consist of *locally helical structures, microcrystallites*, or *nodular domains*.

Gelation by formation of *helical structures* is still mysterious. Helix formation has been implicated in the gelation of many polymers, including poly(methyl methacrylate) in toluene, bromobenzene, and o-xylene (Berghmans et al. 1994; Fazel et al. 1994; Spevacek and Schneider 1974, 1987), isotactic polystyrene in carbon disulfide, *cis*-decalin, *trans*-decalin, and 1-chlorodecane (François et al. 1988; Guenet and McKenna 1988), agarose in water/dimethylsulfoxide (Rochas et al. 1994), and polypeptides in water (Reid et al. 1974; Michon et al. 1993). While a generic "two-step" process for the formation of such gels has been proposed (Berghmans et al. 1994), the precise structure of the intermolecular associations seems to be uncertain. It is clear, however, that tacticity has a strong influence on gel formation; this is consistent with the postulated helix formation. However, even atactic polystyrene can form a gel in many solvents (Tan et al. 1983); thus, even short syndiotactic sequences in atactic polystyrene can induce physical gelation (François et al. 1988). Gelation is often very solvent specific, beyond what can be attributed to generic "solvent quality" (François et al. 1988; Guenet and McKenna 1988). This suggests that solvent–polymer complexes form in at least some cases. For poly(vinyl chloride) in dioctyl phthalate, diethyl oxalate, esthers, and ketones (Alfrey et al. 1949; Mijangos et al. 1993; Lopez et al. 1994), the formation of "sheet-like structures" (presumably analogous to microcrystals) has been postulated as a cause of physical gelation. Molecules with some *rigidity* seem especially prone to physical gelation.

Polymers that form *nodular domains* have specific groups, or "stickers," attached to them that physically bond with each other, thereby producing physical networks. One class of such associating polymers consists of water-soluble polymers containing hydrophobic groups that huddle together to shield themselves from their aqueous environment. These polymers readily form gel-like networks that greatly enhance the solution viscosity at low concentrations (0.5–5.0 wt%), and thus they are used as "thickeners" in paints, paper coatings, and other such products (Yekta et al. 1995; Karunasena and Glass 1989). Conversely, one can have hydrophobic polymers to which hydrophilic groups are attached; in the melt state the hydrophilic groups associate. For *ionomers*, such as polystyrene sulfonate or sulfonated ethylene-propylene-diene, aggregation occurs via dipole–dipole interactions among ionic groups (Hollday 1983; Eisenberg 1980). Aggregation can also be produced by groups that *hydrogen bond* with each other (Longworth and Morawetz 1958; Stadler and de Lucca Freitas 1986). The enthalpy of formation of a hydrogen bond is on the order of 3–6 kcal/mol, or around 5–10 $k_B T$ per hydrogen bond at room temperature (Pimentel and McClellan 1960). Thus, hydrogen bonds are by no means permanent; but with many such bonds along its backbone, the diffusion of a polymer chain will be drastically slowed down. An example of a hydrogen-bond-forming group is urazole, which, when attached to a polybutadiene chain, leads to the formation of hydrogen-bonded urazole dimers (see Fig. 5-11). Hydrogen-bonding polymers are used as "anti-misting" additives to prevent fuel from ruptured tanks (such as those on damaged airplanes) from atomizing into small, and therefore highly inflammable, droplets (Ballard et al. 1988).

Since attractive interactions between polymer molecules can promote both gelation and phase separation, one might expect phase separation and gelation to occur under similar conditions. Indeed, Fig. 5-12 shows the occurrence of both phase separation and gelation for atactic polystyrene in carbon disulfide. Note that gelation occurs in both the one- and two-phase regions of the phase diagram. If a first-order solid–liquid phase separation occurs by *spinodal decomposition*, the network that forms may be rigid and unable to coarsen, thereby leaving a kinetically trapped gel-like phase (Prasad et al. 1993). That phase separation has occurred may be deduced, however, from sample cloudiness, or from its tendency to undergo *syneresis*, which is the slow exudation of solvent from the gel mass. Even when a network forms without any tendency for bulk phase separation, the formation of infinitely large clusters at the gel point leads to a thermodynamic singularity, which in the model of Tanaka

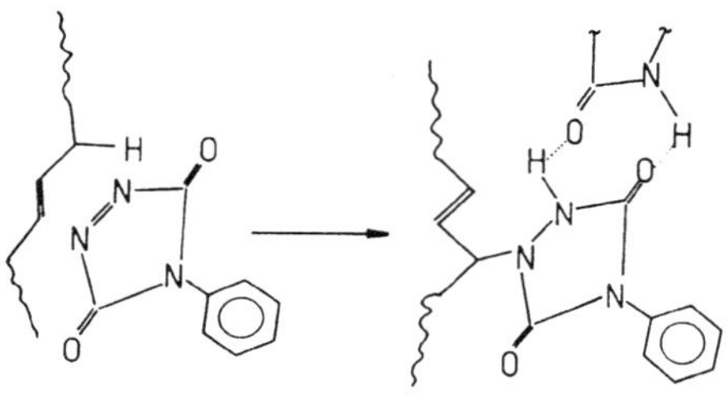

Figure 5.11 A polybutadiene chain reacts with a 4-phenyl-1,2,4-triazoline-3,5-dione group. This group, once attached to the polybutadiene chain, has a hydrogen and an oxygen, both of which can form hydrogen bonds (shown by dashed lines) with the same group on a different chain. (From Stadler and de Lucca Freitas 1986, reprinted with permission from Steinkopff Publishers.)

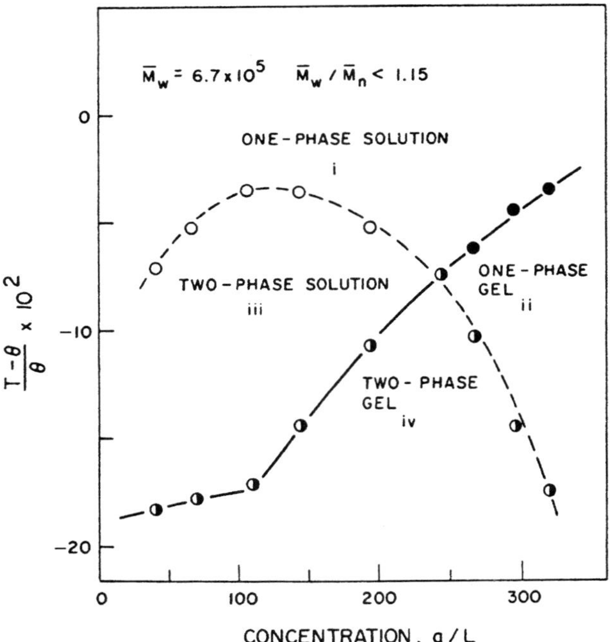

Figure 5.12 Phase diagram of atactic polystyrene in nitropropane. The theta temperature is $\Theta = 200$ K. (Reprinted with permission from Tan et al., Macromolecules 16:28. Copyright 1983 American Chemical Society.)

and Stockmayer (1994) is either second or third order, depending on whether the primary chains making up the physical gel are polydisperse or monodisperse. The phase diagrams predicted by the Tanaka–Stockmayer model for physical gelation are qualitatively similar to those observed experimentally (compare Fig. 5-13 with Fig. 5-12).

The subtle relationship between gelation and phase separation is well illustrated by solutions of poly(γ-benzyl-L-glutamate) (PBLG) in solvents such as dimethylformamide (DMF), benzyl alcohol, or toluene. PBLG molecules are stiff, and they form chiral nematic phases when sufficiently concentrated in solution (see Section 2.2.2.1). Figure 5-14a shows the experimental phase diagram for PBLG in DMF, determined by Miller and coworkers using polarimetry and nuclear magnetic resonance (NMR) measurements (Wee and Miller 1971; Miller et al. 1974; see also Tipton and Russo 1996). At high temperatures, there is an isotropic-to-liquid crystal phase transition at polymer volume fractions of around 0.08–0.15, depending on the temperature, with a *narrow* biphasic gap. Both phases at high temperature are fluid, and if the overall composition is in the two-phase "chimney," they macroscopically separate from each other over time. At lower temperatures, the biphasic gap is huge, and compositions within this wide region form viscous "gels" even when the polymer concentration is as low as 1%! The observed phase diagram is in excellent qualitative agreement with the diagram predicted by Flory (1956) (see Fig. 5-14b). In this theory of Flory, the athermal free energy of rod-like molecules in a solvent is supplemented by a solvent–polymer interaction term, proportional to a parameter χ (see Sections 2.3.1 and 13.2.1). Agreement between theory and experiment is obtained by setting $\chi = -3.51 + 1035/T$, which has a form typical for polymers (see Sections 2.3.1 and

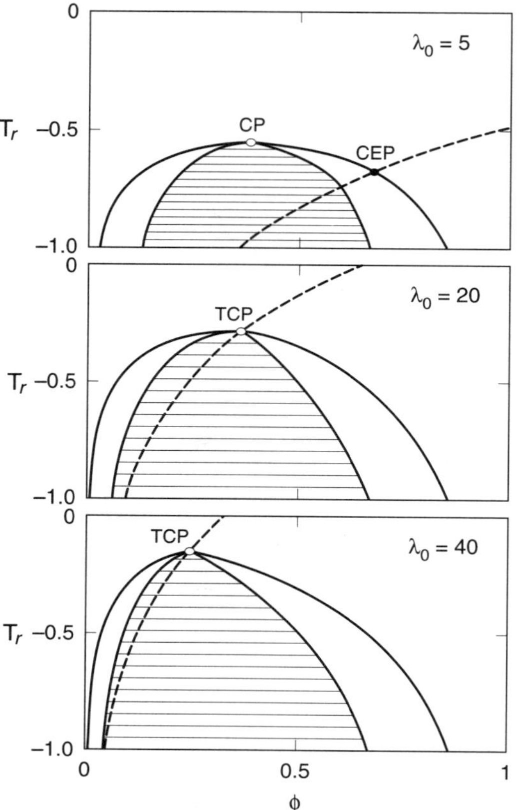

Figure 5.13 Predicted phase diagrams for physical gels made from low-molecular-weight molecules with junctions of unrestricted functionality; ϕ is the total volume fraction of polymer, and T_r is here the reduced distance from the theta temperature, $T_r \equiv 1 - \Theta/T$. The parameter λ_0 controls the equilibrium constant among aggregates of various sizes. The outer solid lines are binodals, the inner solid lines are spinodals, and the dashed lines are gelation transitions. CP is a critical solution point, CEP is a critical end point, and TCP is a tricritical point. (Reprinted with permission from Tanaka and Stockmayer, Macromolecules 27:3943. Copyright 1994 American Chemical Society.)

13.2.1). The agreement between theory and experiment supports the conclusion drawn from NMR measurements that the lower region is indeed a two-phase zone, despite the fact that it does not macroscopically phase separate. The implication is that the highly concentrated ordered phase that forms at low temperature is too rigid to separate macroscopically from the solvent, thereby forming a rigid network that pervades the solvent. The resulting material has the rheological properties of a gel (Shukla and Muthukumar 1988). Analogous "gels" form when "waxy crude oil" is cooled, leading to crystallization of the long parafinic components, which then form a rigid network percolating through the oil, imparting to it a yield stress (Wardhaugh and Boger 1991). Phase-separated "gels" might also occur in flocculated suspensions of rigid spherical particles, to be described in Chapter 7.

In addition to the interesting connections they have to phase-separated systems, physical gels are also similar in some sense to *glasses* (de Gennes 1979; Shukla and Muthukumar 1988). Glasses are disordered solids formed by the progressive freezing of some of the liquid degrees of freedom as the temperature is lowered, resulting in a liquid structure that is too slow to relax on human time scales (see Chapter 4). Physical gelation involves a quenching of mobility due to the formation of a network of bonds that relaxes slowly, if at all. One might suppose that physical gelation can be distinguished from glass formation

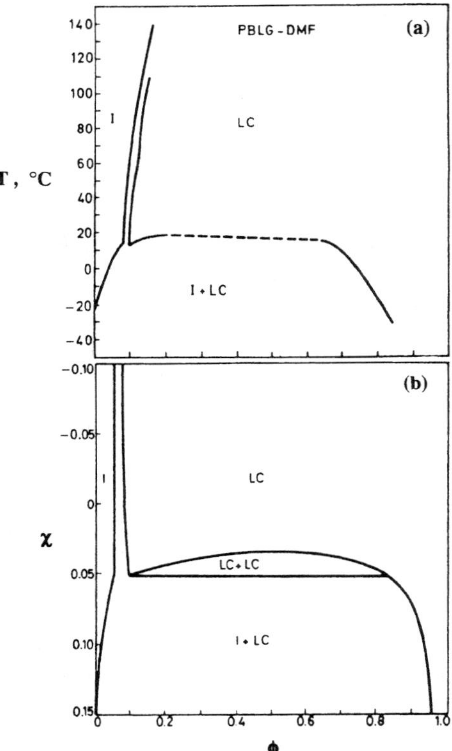

Figure 5.14 (a) Temperature–volume fraction phase diagram for PBLG ($M_w = 310,000$) in DMF, where I denotes an isotropic phase, LC denotes a chiral nematic liquid-crystalline phase, and I + LC is a "gel" that is presumed to be two coexisting phases that are unable to separate macroscopically. (b) The χ-volume fraction phase diagram predicted by the Flory lattice theory for rigid rods of axial ratio (length/diameter) = 150. (From Miller et al. 1974, with permission.)

by the existence in the former of a network; however, so-called strong glasses are also believed to be network formers (see Section 4.2). Thus, conceptually, there is no clear-cut distinction between glasses and gels. It might be helpful to regard gelation and vitrification (or glass formation) as two ends of a continuum. At one extreme, the formation of a network of irreversible chemical bonds can be called *strong gelation* (de Gennes 1979), while the gradual, reversible slowing down of molecular motion due to changes in molecular packing or "free volume" can be called "fragile vitrification," in accord with Angell's classification (see Section 4.2). The intermediate case, in which a network of physical bonds forms whose strength is perhaps 5–30 times $k_B T$, could be called either a "weak gel" or a "strong glass." By convention, a "weak gel" is distinguished from a "strong glass" by the presence in the former of a network of polymer molecules in a solvent, and in the latter of a network of small molecules or ions. Consequently, a glass is typically a hard substance with a higher modulus than a gel.

The rheology of associating polymers can be very complex: They may be solids that fracture under flow, or, conversely, they may be highly fluid at rest and form gels only under flow! The type of behavior depends strongly on the deployment of the stickers along the chain, as well as on molecular weight, concentration, and the method of solution preparation (Pedley et al. 1989). Reproducibility is often a major problem with such materials; for example, samples can "age," or change slowly over time, especially when ionic groups are present which might absorb moisture from the atmosphere (Witten 1988). In addition, the presence of associating groups gives the chains an effective attraction for each other that can lead to phase separation, even from a solvent that would be considered "good" in the absence of associations (Witten 1988).

5.4.1 Telechelic Polymers

However, the behavior of one class of associating polymers, the *telechelic polymers*, seems to be reasonably well understood, thanks to recent experimental and theoretical work. Telechelic polymers are linear chains containing associating "sticker" groups only on the chain ends (Jerome et al. 1985). Under steady shear flow, solutions of telechelic polymers typically show a viscosity increase with increasing shear rate (shear thickening), followed by a viscosity decrease (shear thinning) at higher rates.

Examples of telechelic polymers include hydrophobically modified ethoxylated urethanes (HEURs) with hydrophopic end caps consisting of aliphatic alcohols, alkylphenols (Emmons and Stevens 1978; Lundberg et al. 1991), or fluorocarbons (Amis et al. 1996). Figure 5-15 shows a typical structure. The alkane-containing end groups clump together to form nodular "micelles" containing several end groups in aqueous solutions; this substantially enhances the solution viscosity. Further viscosity enhancement, even at a low concentration of HEUR, is achieved by addition of a surfactant. Paints formulated using these components have been found to spatter less when rolled onto surfaces (Lundberg et al. 1991). For a review of the literature on surfactant interactions with associative thickeners, see Winnik and Yekta (1997) and Winnik and Regismond (1996).

The micelles in telechelic polymers differ from micelles formed by typical small-molecule surfactants in that the water-loving "head" groups of telechelic chains are long polymer chains, while in small molecules the head groups are small ionic or hydrophilic nonionic groups. In addition, the telechelic chains have two hydrophobic "tail" groups, one on each end of the hydrophilic chain. In Section 12.3.1, it will be shown that the number of hydrophobic units N_{agg} contained in a micelle is related to the volume v of each hydrophobe and the area a of the micelle surface required to accommodate each hydrophobe within the micelle. The area a is controlled by the bulkiness of the hydrophilic part of the molecule. The formula for N for a spherical micelle is given by [see Eq. (12-9)] $N_{agg} = 36\pi v^2/a^3$. For an alkane hydrophobe, $v \approx 27n_c$ Å^3, where n_c is the number of carbon atoms in the alkane chain [see Eq. (12-1)]. Thus, for $n_c = 16$, we find the relationship $N_{agg} \approx 20/a^3$, where a is in nm^2. Because the "head" groups of the telechelic

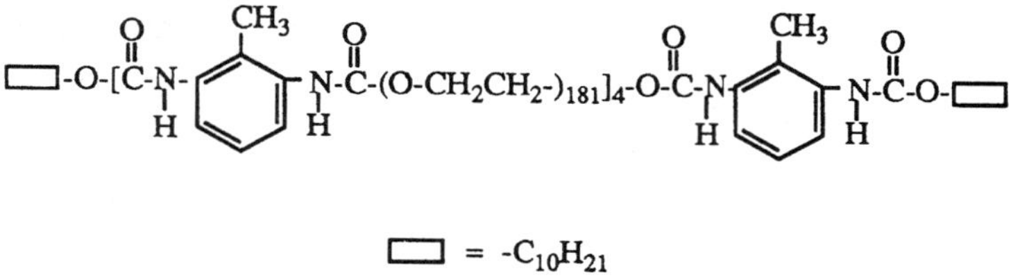

$$\square\text{-O-}[\overset{\overset{\text{O}}{\|}}{\text{C}}\text{-N-}\underset{\underset{\text{H}}{|}}{}\overset{\text{CH}_3}{\bigcirc}\underset{\underset{\text{H}}{|}}{\text{N}}\text{-}\overset{\overset{\text{O}}{\|}}{\text{C}}\text{-(O-CH}_2\text{CH}_2\text{-)}_{181}]_4\text{-O-}\overset{\overset{\text{O}}{\|}}{\text{C}}\text{-N}\underset{\underset{\text{H}}{|}}{}\overset{\text{CH}_3}{\bigcirc}\underset{\underset{\text{H}}{|}}{\text{N}}\text{-}\overset{\overset{\text{O}}{\|}}{\text{C}}\text{-O-}\square$$

$$\square = \text{-C}_{10}\text{H}_{21}$$

Figure 5.15 Structure of a HEUR polymer. (From Lundberg et al. 1991, with permission from the Journal of Rheology.)

"surfactants" are long, we expect each chain to occupy a large patch of the micelle surface, compared to that occupied by a typical small-molecule surfactant. As a consequence, the aggregation number of telechelic "micelles" should be significantly smaller than that of a small-molecule micelle with a comparably sized hydrophobe. Using fluorescence decay studies, Yekta et al. (1995) have deduced a micelle aggregation number of $N_{agg} \approx 18$–28 hydrophobes per micelle, while modeling of rheological data suggests a smaller value, $N_{agg} \approx 7$ (Annable et al. 1993). These aggregation numbers are much smaller (by a factor of 5–10) than ordinary surfactant micelles with tails of comparable length (see Section 12.3 and Fig. 12-6). Telechelic polymers are also analogous to the *triblock copolymers* discussed in Chapter 13. The difference is in the shortness of the aliphatic "tail" group compared to the block size of typical triblocks. A telechelic polymer is therefore a cross between a surfactant and a block copolymer; it contains two surfactant-sized hydrophobic groups attached to a polymer-sized hydrophilic one.

Because the telechelic polymer has two "tails" or stickers, separated by a long hydrophilic chain, the micelles formed by telechelics are expected to contain "loops" and "bridges," as depicted in Fig. 5-16. Loops are expected to predominate at concentrations

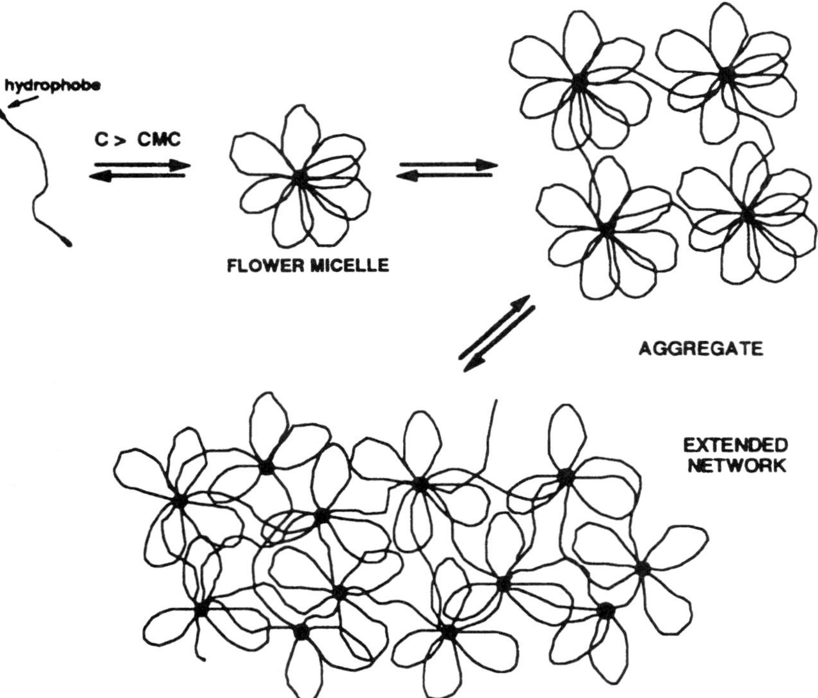

Figure 5.16   Model for associations of telechelic polymers as a function of increasing concentration. For strong associations, isolated "flower" micelles form just above the critical micelle concentration (CMC), which is often around 2 to 10 ppm (Winnick and Yekta 1997). At higher concentrations, the flowers are expected to be connected by "bridges." (From Winnik and Yekta 1997, with permission from Current Chemistry Ltd.) © 1997 Current Opinion in Colloid + Interface Science.

too low for the hydrophilic chains to bridge between adjacent micelles. These isolated, loop-dominated micelles are called "flowers." Semenov et al. (1995) have predicted that there is a concentration range where a phase dense in flower micelles separates from a phase lean in them. At very high concentrations (>20 wt%), x-ray scattering and other evidence points to the formation of an ordered cubic array of bridged micelles (Abrahmsen-Alami et al. 1996). As discussed in Chapter 12, similar ordered micellar phases occur in aqueous solutions of ordinary surfactants.

The various types of association that a single telechelic molecule can experience are depicted in Fig. 5-17a. At high enough concentration, one expects networks to form (see Fig. 5-17b). If the molecular weight of the telechelic polymer is low enough, or its concentration in solution is high enough, that the chains do not entangle, then relaxation of a network junction occurs rapidly whenever a sticker group manages to release itself

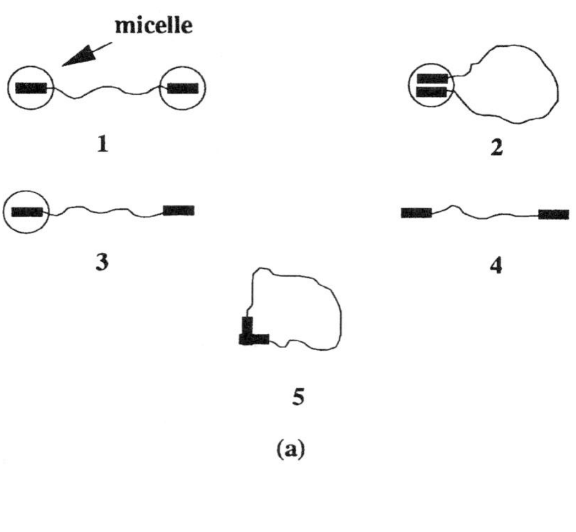

Figure 5.17 (**a**) Illustration of types of chain association in telechelic polymers. (**b**) Chain architectures that can form in solution; micelles that have a network functionality greater than two are shown in black. (From Annable et al. 1993, with permission from the Journal of Rheology.)

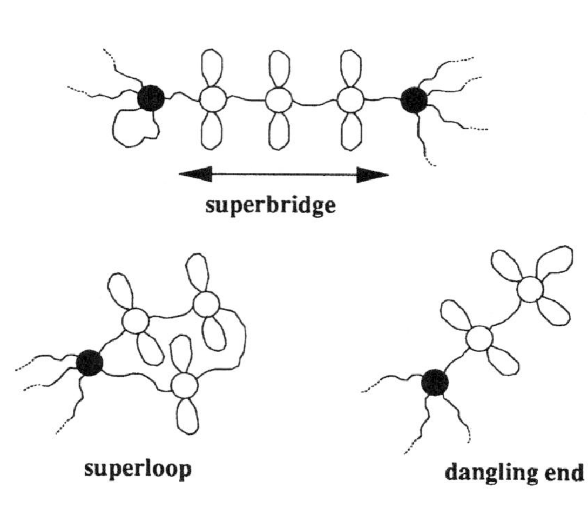

from one of the micelles. The rheological properties of a such a network are therefore especially simple, as has been shown by Jenkins et al. (1991) and Annable et al. (1993). In particular, for the telechelic polymers studied by Annable et al., the relaxation of the network structure is described by a single-relaxation-time "Maxwell model" (see Fig. 5-18). Such perfect single-relaxation-time behavior is rare. Other known cases of such behavior are for solutions of wormy micelles (discussed in Section 12.3.4), some inorganic glasses (Section 4.8.1), and some dense emulsions (Section 9.3.4). The relaxation time τ of the network is the time constant τ_{diss} for dissociation of a sticker from a micelle, which can be related to the activation barrier energy $\Delta\mu$ for dissociation by

$$\tau_{\mathrm{diss}} = \Omega_0^{-1} e^{\Delta\mu/k_B T} \tag{5-8}$$

where $\Delta\mu$ is the free energy of micellization per sticker, and Ω_0 is a fundamental vibrational frequency, $\Omega_0^{-1} \sim 10^{-10}$ sec (Tanaka and Edwards 1992b). (This free energy difference is roughly equal to the chemical potential difference $\mu_N^0 - \mu_1^0$ for micellization discussed in Section 12.3.1.) For an alkane chain, $\Delta\mu$ increases by roughly $1.5k_B T$ per CH_2 unit. Roughly consistent with this, Annable et al. (1993) found that the relaxation time τ and the zero-shear viscosity η_0 of a typical telechelic HEUR solution increase exponentially with the number of CH_2 units in the sticker, with an increment of around $0.9k_B T$ in $\Delta\mu$ per methylene unit for stickers containing 12–22 CH_2 units.

When $\Delta\mu/k_B T \gg 1$, there will be few free stickers. According to the classical theory for gels, the modulus of a telechelic gel should be simply given by Eq. (5-2), $G_0 = \nu k_B T$,

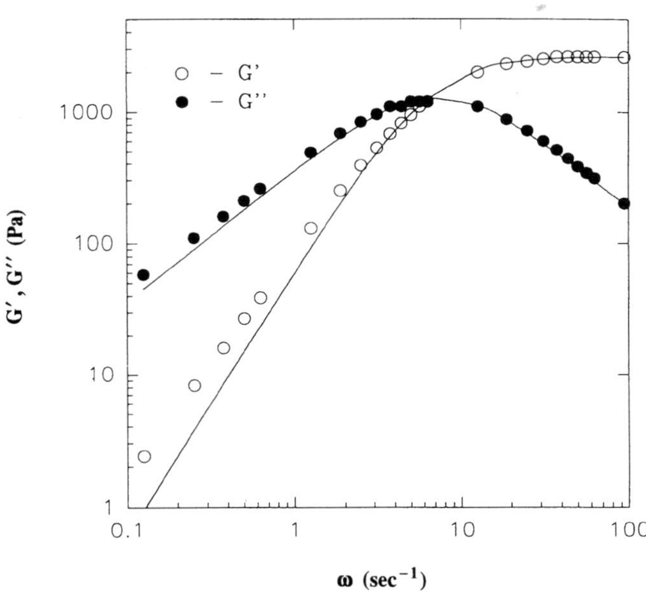

Figure 5.18   Storage and loss moduli for a 7% w/v HEUR associative thickener ($M_w = 33{,}100$; $M_w/M_n = 1.47$) end-capped with hexadecanol at 25°C. The lines are a fit to a one-mode Maxwell model. (From Annable et al. 1993, with permission from the Journal of Rheology.)

where ν is the number of elastically active chains per unit volume. The zero-shear viscosity is then just $\eta_0 = G_0 \tau$. If all chains are elastically "active," the modulus will be proportional to polymer concentration ν. However, a chain is "active" and contributes to the modulus only if each of its stickers is in a different micelle, as depicted in structure 1 in Fig. 5-17(a). Structure 2 and 5 in Fig. 5-17(a) depict loops which are inactive. If the solution is dilute enough that the chains are, on average, separated from each other by a distance roughly as great or greater than the chain's radius of gyration, most chains will have to stretch in order to link separate micelles; and thus the probability of loops will be high, so that micelles are mostly unbridged flowers (see Fig. 5-16). Since the radius of gyration is proportional to the square root of the chain's length, this implies that the probability that a chain is active increases towards unity as the product $c\sqrt{M}$ increases, where c is the concentration of polymer and M is its molecular weight. Figure 5-19a shows that the ratio $G_0/\nu k_B T$ does indeed increase with c, as expected by this argument. The relaxation time also increases with c, as shown in Fig. 5-19b. Annable et al. (1993) argued that τ decreases at low c because the loop formation produces "superbridges" in which $n = 2$ or more chains string

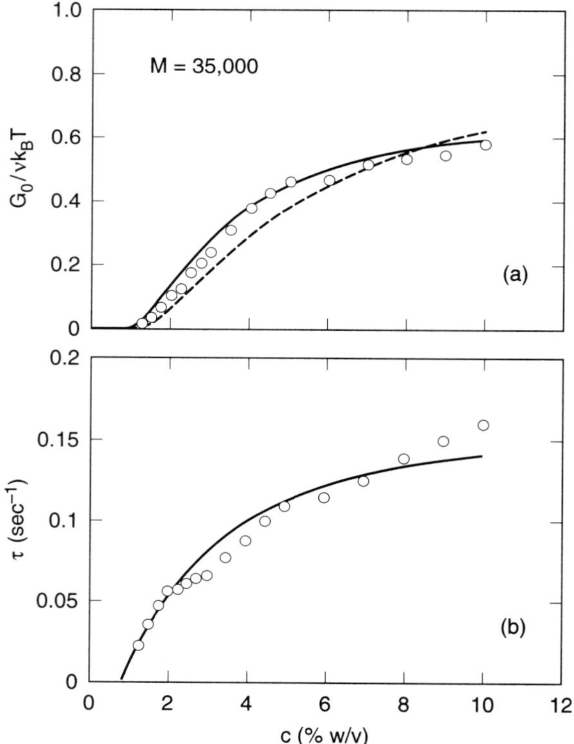

Figure 5.19   (a) The reduced modulus $G_0/\nu k_B T$, and (b) the relaxation time τ, as functions of concentration of the polymer described in the caption to Fig. 5-18. The solid and dashed lines are theoretical predictions assuming, respectively, 70% and 100% for the end-cap efficiencies. (From Annable et al. 1993, with permission from the Journal of Rheology.)

together like paper dolls (see Fig. 5-17b). Since a superbridge is broken when a sticker on any one of the n chains composing it is released from its micelle, the relaxation time is faster by a factor of n than that of a simple bridge. The lines in Fig. 5-19 are predictions of a simple model developed by Annable et al. (1993), which gives good agreement with the measurements. Annable et al. (1993) also confirmed the prediction that τ is a function of the combined variable $c\sqrt{M}$.

A plot of viscosity versus shear rate for a model HUER polymer is shown in Fig. 5-20, and compared to the dynamic viscosity versus frequency. Note that the Cox–Merz rule (see Section 1.3.1.5) fails in that at the frequency ($\omega \approx 1 \ \mathrm{sec}^{-1}$) where the dynamic viscosity begins to decrease, the steady shear viscosity begins to *increase* with increasing shear rate, followed by shear thinning at a somewhat higher shear rate. Shear thickening, followed by shear thinning, is often observed in associating polymers (Witten 1988; Marrucci et al. 1993; Hu et al. 1995a; Ketz 1993). The shear thinning can readily be attributed to shear-induced breakup of the gel structure. For stickers whose association energy is significantly greater than $k_B T$, one would expect the gel network to break under shear only when the chains are nearly fully extended. The onset of shear thinning then should occur at a shear rate $\dot{\gamma}_c$ of around $N_K^{1/2}/\tau$, where N_K is the number of "Kuhn steps" in the telechelic polymer, and the relaxation time τ can be estimated by the inverse of the frequency at which the dynamic viscosity begins to shear thin (see Fig. 5-20). This seems to agree with experimental observations (Marrucci et al. 1993).

Figure 5-21 is a plot of the shear viscosity of a HEUR solution, along with superposed illustrations of the structural changes that are believed to occur in the shear-thickening and shear-thinning regions. As explained by Marrucci et al. (1993), weak shear thickening, similar to that shown in Figs. 5-20 and 5-21, can be accounted for by the non-Hookean elastic behavior of network strands that are stretched to more than half their full extension (see Section 3.6.2.2.1). Shear thickening would be expected at shear rates just below those at which shear thinning occurs. Since highly stretched strands pull out of their micelles, only a weak shear-thickening effect can be accounted for by non-Hookean elasticity (Marrucci et al. 1993). Fluorescence studies show that the degree of association of the sticker groups

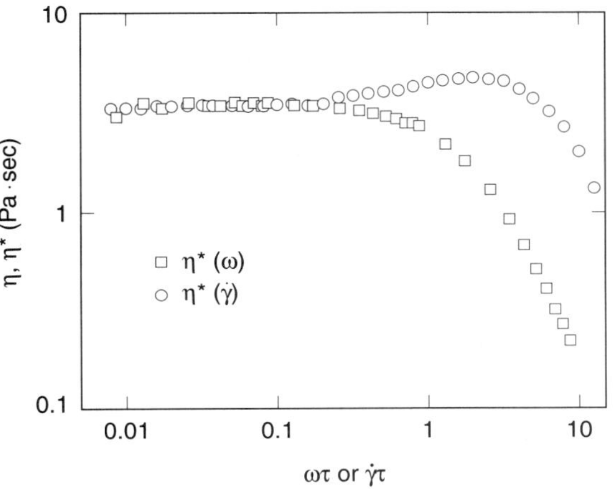

Figure 5.20 Steady-state viscosity $\eta(\dot{\gamma})$ and dynamic complex viscosity $\eta^*(\omega)$ as functions of reduced shear rate ($\dot{\gamma}\tau$) or frequency ($\omega\tau$), for a 1.5% w/v solution of the associative thickener described in the caption to Fig. 5-18. (From Annable et al. 1993, with permission from the Journal of Rheology.)

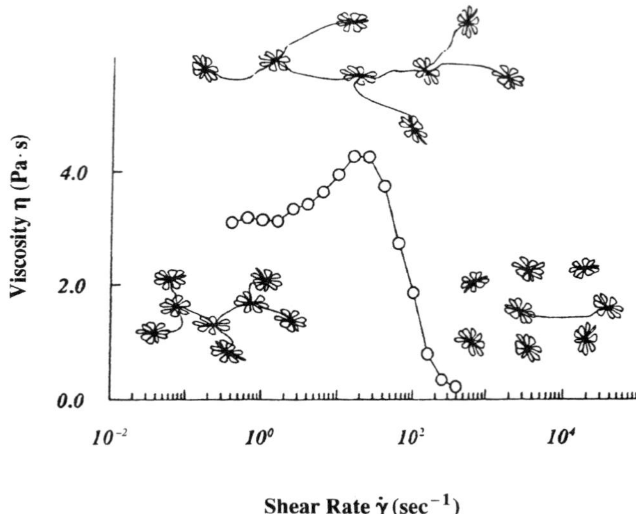

Figure 5.21   Viscosity versus shear rate for 1.0 wt% HEUR ($M_n = 51,000\ M_w/M_n = 1.7$) telechelic polymers with hexadecanol end caps at 22°C. The illustrations show the structural transitions that are thought to occur as the shear rate is increased. First, the bridging chains are stretched, producing shear thickening. Then, many bridging chains are pulled out at one end from the micelles to which they were attached, and shear thinning occurs. (Reprinted with permission from Yekta et al., Macromolecules 28:956. Copyright 1995 American Chemical Society.)

does not change over the shear-rate range depicted in Fig. 5-21 (Yekta et al. 1995). This would seem to imply that the decrease in bridging that occurs in the shear-thinning region is accompanied by an increase in loop formation, so that free ends are avoided. Thus, at high shear rates, "flowers" may predominate.

Some associating polymers show very strong shear thickening, with the viscosity increasing by more than an order of magnitude over a narrow range of shear rates. Massive shear thickening of this kind seems to be common in polymers with many stickers distributed along each chain. We discuss the behavior of such polymers in the next section. We end this section by noting that Tanaka and Edwards (1992a, 1992b) and Marrucci et al. (1993) have developed promising temporary-network kinetic models for telechelic polymers, by applying ideas originally formulated by Green and Tobolsky (1946) and Yamamoto (1956, 1958) (see Section 3.4.2).

5.4.2 Entangled "Sticky" Chains

Telechelic polymers have stickers only on their ends, and they are often of small enough molecular weight to be unentangled. There are, of course, many other ways of deploying stickers on a polymer. There can be several, or many, stickers, arranged either regularly or randomly along the chain. The stickers can be attached directly to the polymer backbone, or they can be offset by a nonsticky "spacer" (Winnik and Yekta 1997). Clever balancing of various constituents can lead to unusual solution properties with important technological applications. An example is "hydrophobic alkali-swellable emulsion" (HASE) polymers that contain carboxylic and acrylate ester groups in a composition balanced so that the polymer collapses into an insoluble ball at low pH, but at pH > 6 it expands and dissolves (Jenkins et al. 1996; Winnik and Yekta 1997).

A simpler case to consider theoretically is that of many associating sites more or less regularly spaced along the contour of a chain that is long enough and concentrated enough to be entangled with other chains. An example is the melt of polybutadiene with randomly attached urazole groups studied by Stadler and de Lucca Freitas (1986, 1989). Each urazole group is apparently capable of forming two hydrogen bonds with another such group. Figure 5-22 shows G' as a function of reduced frequency for polybutadiene with various mole percentages of attached urazole groups. The added urazole groups dramatically slow down the relaxation, and change the shape of the curve, such that transition to true terminal behavior (for which $G' \propto \omega^2$) becomes more gradual. This change in the shape of G' versus ω at low frequency is reminiscent of that produced by molecular-weight polydispersity.

The frequency-dependent loss modulus for such samples often has two peaks. One of the peaks corresponds to the longest relaxation time of the molecule; this peak shifts to lower frequency (longer relaxation time) as the number of urazole groups per chain increases. The second peak occurs at a frequency of around 2×10^4 sec^{-1} at 0°C and is independent of the number of urazole groups. This frequency appears to correspond to the inverse lifetime of an association between two urazole groups. The presence of a time constant that is much longer than the association lifetime makes this many-sticker system differ markedly from the telechelic chains discussed in Section 5.4.1. For unentangled telechelics, the relaxation

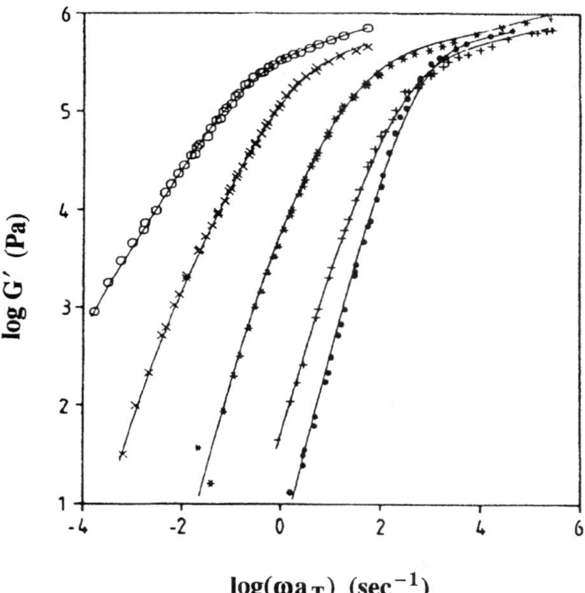

Figure 5.22 Master curves of the storage modulus at a reduced temperature of 0°C for polybutadiene ($M_n = 26{,}000$) which has been modified by attachment of 4-phenyl-1,2,4-triazoline-3,5-dione groups, as illustrated in Fig. 5-11. The degree of modification is $x = 0$ (●), 0.5 (+), 2(∗), 5(×), and 7.5(○), where $x = 7.5$ corresponds to 36 functional groups per chain. (Reprinted with permission from de Lucca Freitas and Stadler, Macromolecules 20:2478. Copyright 1987 American Chemical Society.)

time constant of the gel is either equal to or less than the time a sticker spends in an association.

This difference presumably exists because, for multisticker chains, the dissociation of one sticker from another does not permit relaxation of the entire chain, since the chain is anchored in place by many other stickers, and because it is confined by entanglements with other chains to a "tube-like" region (see Section 3.7). The other stickers and entanglements prevent the chain from diffusing very far before any newly released sticker is captured by a new association (see Fig. 5-23). Thus, one might expect that the chain can only relax its conformations during those exponentially rare moments when all stickers are released. However, Ballard et al. (1988) pointed out that a chain with many stickers can move like a *centipede*: at any one time only a few of the centipede's legs are moving freely, but since the animal is somewhat flexible and each leg eventually gets a turn to move, the whole animal can slowly creep forward. Likewise, even if only a small fraction of its stickers are free to move at any one instant, the polymer molecule can alter its shape and center-of-mass position slightly to accommodate the movement of a few stickers. Over time the whole chain slowly moves back and forth in its tube, like a drunken centipede in a maze, and slowly relaxes its configuration, even though at no time are all the stickers released.

Leibler et al. (1991) have developed a model for this process, which they call "sticky reptation." For long chains with many stickers, the self-diffusion coefficient of a sticky reptating chain turns out to be

$$D_{\text{self}} \approx \frac{a^2}{2\tau_{\text{diss}}S^2} \left(1 - \frac{9}{p} + \frac{12}{p^2} \right) \tag{5-9}$$

where a is the reptation "tube diameter" (see Section 3.7), S is the number of stickers per chain, p is the average fraction of stickers that are associated at a given time, and τ_{diss} is the lifetime of the association. Apart from the factor involving p, Eq. (5-9) is analogous to the corresponding formula for ordinary reptation, $D_{\text{self}} = (a^2/\tau_e)(N_e/N)^2$, with S in Eq. (5-9) playing the role of the number of entanglements per chain, N/N_e, and τ_{diss} in Eq. (5-9) playing the role of the equilibration time of a chain segment between entanglement points, τ_e (see Section 3.7.4.2). Here, as elsewhere, N is the number of monomers per macromolecule and N_e is the number of monomers in an entanglement spacing.

For a chain moving by reptation or by "sticky reptation," one expects the reptation time, which scales roughly as the 3.5 power of N, to be related to the diffusion coefficient (which scales as N^{-2}), by

$$\tau \approx \left(\frac{N}{N_e} \right)^{1.5} \frac{a^2}{D_{\text{self}}} = \left(\frac{N}{N_e} \right)^{1.5} \frac{2S^2 \tau_{\text{diss}}}{1 - 9/p + 12/p^2} \tag{5-10}$$

The predictions of Eq. (5-10) are in good agreement with measured τ values for urazole-modified polybutadienes (Leibler et al. 1991).

The plateau modulus is given by the usual formula for entangled polymers, $G_0^N \approx \nu k_B T$, where ν is the number of entanglement strands per unit volume of melt; that is, $\nu = N \nu_m / N_e$, where ν_m is the number of molecules per unit volume, $\nu_m = \rho N_A / M$, N_A is Avogadro's number, ρ is the melt density, and M is the chain's molecular weight. The zero-shear viscosity is estimated to be simply $\eta_0 \approx G_0^N \tau$, as usual.

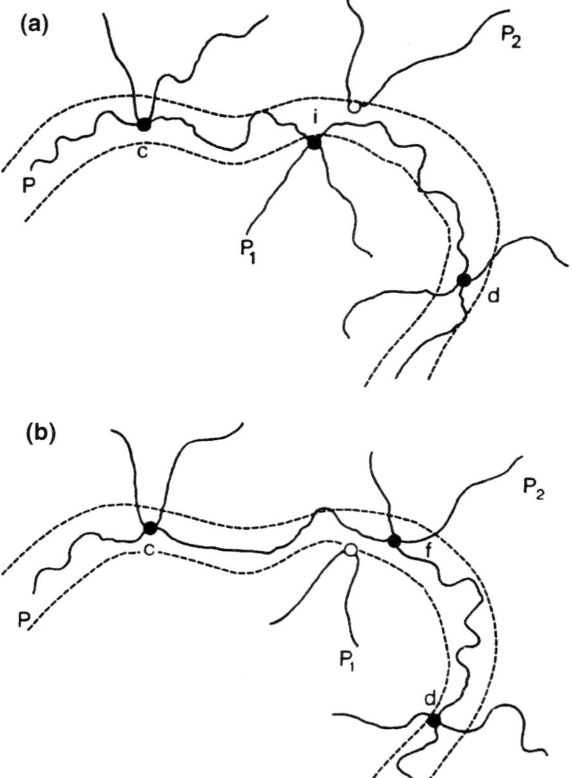

Figure 5.23 "Sticky reptation": In **(a)** the chain P is cross-linked to chain P_1 at point i, but in **(b)**, it has released this cross-link and attached itself to chain P_2 at point f. (Reprinted with permission from Leibler et al., Macromolecules 24:4701. Copyright 1991 American Chemical Society.)

Shear thickening in polymers with multiple stickers is thought to be caused by a shear-induced change in the balance between intramolecular and intermolecular associations (Witten and Cohen 1985). According to this idea, at low shear rates, many of the associations are *intramolecular* and therefore contribute little or nothing to the viscosity. Shearing flow stretches the molecules, and thus it makes intermolecular associations more probable. The result is an increase in viscosity. The increase in viscosity causes the chain to stretch even more, and this promotes even more interchain associations. The result can be a runaway increase in the viscosity, or *shear-induced gel formation.*

While this mechanism for shear thickening is plausible, it has not yet been confirmed by direct probes of the association behavior of the chains. Pedley et al. (1989) found that shear thickening in such systems is not accompanied by any measurable change in average extension of the chains. This could imply that only a small fraction of chains participate in the shear-induced thickening phenomenon, while the rest remain balled up in self-aggregated clusters (Marrucci et al. 1993). Witten (1988) has argued that chains with associations strong enough to produce dramatic shear-thickening effects are likely to be prone to phase separation. This may explain the poor reproducibility and sensitivity to sample preparation frequently experienced with these solutions.

Severe shear thickening is most likely to occur for multisticker, entangled chains. For such chains, relaxation after a sticker is released is slower than reassociation, so that chains reassociate while still in a stretched state, and very high viscosities can then build up. In short unentangled telechelic polymers, on the other hand, chain relaxation is expected to be fast enough that chains are unstretched when they reassociate, and the shear thickening is then modest.

5.5 SUMMARY

Chemical gels, and perhaps physical gels also, show power-law frequency-dependences of the linear viscoelastic moduli G' and G'' at the transition from sol to gel, and thus the spectrum can be characterized completely by a power law exponent n and a relaxation strength S. The constants n and S vary systematically with molecular weight of the prepolymer and with the ratio of prepolymer to cross-linker.

The rheological properties of physical gels, which have associating groups along their backbones or on their ends, are, on the whole, not yet well-understood, in part because of their sensitivity to preparation and poor reproducibility. However, much progress has recently been made toward understanding the rheology of telechelic polymers, which have associating groups or "stickers" only on their ends. Telechelic polymers seem to be describable by a temporary network model in which the relaxation is dominated by the rate of release of stickers from the micelles to which they are associated. Molecules with many stickers along their backbone have rheological properties that depend on the number of such sticker groups as well as the sticker release rate. It might be possible to model the rheology of long molecules with many stickers by the "sticky reptation" model of Leibler et al. Under steady shear, telechelics and other associating polymers usually show shear thickening, followed at higher shear rate by shear thinning. The shear thinning is probably caused by stress-induced breakdown of the network. A weak shear-thickening phenomenon can by explained by non-Gaussian chain statistics, but massive shear-thickening or shear-induced gelation seems to imply that intermolecular associations can be enhanced by shearing flow.

REFERENCES

Abrahmsen-Alami S, Alami E, François F (1996). *J Colloid Interface Sci* 179:20.
Adolf D, Martin JE (1990). *Macromolecules* 23:3700.
Alfrey T, Wiederhorn N, Stein RS, Tobolsky AV (1949). *Ind Eng Chem* 41:701.
Amis EJ, Hu N, Seery TAP, Hogen-Esch TE, Yassini M, Hwang F (1996). In *Hydrophilic Polymers, Performance with Environmental Acceptability*, Glass JE (ed), American Chemical Society, Washington, DC.
Annable T, Buscall R, Ettelaie R, Whittlestone D (1993). *J Rheol* 37:695.
Arbabi S, Sahimi M (1988). *Phys Rev B* 38:7173.
Ballard MJ, Buscall R, Waite FA (1988). *Polymer* 29:1287.
Berghmans M, Thijs S Cornette M, Berghmans H, De Schryver FC, Moldenaers P, Mewis J (1994). *Macromolecules* 27:7669.
Brinker CJ, Scherer GW (1990). *Sol–Gel Science: The Physics and Chemistry of Sol–Gel Processing,* Academic Press, New York.
Broadbent SR, Hammersley JM (1957). *Proc Camb Philos Soc* 53:629.

Chambon F, Petrovic ZS, MacKnight, Winter HH (1986). *Macromolecules* 19:2146.

Chiou B-S, English RJ, Khan SA (1996). *Macromolecules* 29:5368.

de Gennes PG (1976). *J Phys (Paris) Lett* 37L:61.

de Gennes PG (1979). *Scaling Concepts in Polymer Physics*, Cornell University Press, Ithaca, NY.

de Lucca Freitas L, Stadler R (1987). *Macromolecules* 20:2478.

Eisenberg A (ed) (1980). *Ions in Polymers*, Advances in Chemistry Series 187, American Chemical Society, Washington, DC.

Eliassaf J, Silberberg A, Katchalsky A (1955). *Nature* 176:1119.

Emmons WD, Stevens TS (1978). US Patent 4,079,028.

Erman B, Flory PJ (1983). *Macromolecules* 16:1600.

Fazel N, Brulet A, Guenet J-M (1994). *Macromolecules* 27:3836.

Ferry JD (1980). *Viscoelastic Properties of Polymers*, 3rd ed, Wiley, New York.

Fisher ME, Essam JW (1961). *J Math Phys* 2:609.

Flory PJ (1941). *J Am Chem Soc* 63:3083.

Flory PJ (1942). *J Phys Chem* 46:132.

Flory PJ (1953). *Principles of Polymer Chemistry*, Cornell University Press, Ithaca, NY.

Flory PJ (1956). *Proc R Soc (Lond)* A234:73.

François J, Gan J, Sarazin D, Guenet J-M (1988). *Polymer* 29:898.

Green MS, Tobolsky AV (1946). *J Chem Phys* 14:80.

Guenet J-M (1992). *Thermoreversible Gelation of Polymers and Biopolymers*, Academic Press, London.

Guenet J-M, McKenna GB (1988). *Macromolecules* 21:1752.

Hess W, Vilgis TA, Winter HH (1988). *Macromolecules* 21:2536.

Hollday L (ed) (1983). *Developments in Ionic Polymers*, Applied Science, London.

Horkay F, McKenna GB (1996). In *Physical Properties of Polymers Handbook*, Mark JE (ed), AIP Press, New York.

Hu Y, Wang SQ, Jamieson AM (1995a). *Macromolecules* 28:1847.

Hu Z, Zhang X, Li Y (1995b). *Science* 269:525.

Izuka A, Winter HH, Hashimoto T (1992). *Macromolecules* 25:2422.

James HM, Guth E (1953). *J Chem Phys* 21:1039.

Jenkins RC, Silebi CA, El-Asser MS (1991). *ACS Symp Ser* 462:222–233.

Jenkins RC, DeLong LM, Bassett DR (1996). In *Hydrophilic Polymers* Glass JE (ed), ACS Advances in Chemistry Series 248, American Chemical Society, Washington, DC.

Jerome R, Broze G, Teyssie P (1985). In *Microdomains in Polymer Solutions*, Dubin P (ed), Plenum Press, New York.

Karunasena A, Glass JE (1989). *Prog Org Coatings* 17:301.

Ketz RJ Jr (1993). PhD Thesis, Princeton University.

Kirkpatrick S (1973). *Rev Mod Phys* 45:574.

Larson RG, Davis HT (1982). *J Phys C Solid State Phys* 15:2327.

Leibler L, Rubinstein M, Colby RH (1991). *Macromolecules* 24:4701.

Lodge A (1961). *Polymer* 2:195.

Longworth R, Morawetz H (1958). *J Polym Sci* 29:307.

Lopez D, Dahmani M, Mijangos C, Brûlet A, Guenet J-M (1994). *Macromolecules* 27:7415.

Lundberg DJ, Glass JE, Eley RR (1991). *J Rheol* 35:1255.

Marrucci G, Bhargava S, Cooper SL (1993). *Macromolecules* 26:6483.

Martin JE, Adolf D (1991). *Annu Rev Phys Chem* 42:311.

Menchen S, Winnik MA (1994). US Patent 5,290,418.

Menchen S, Johnson B, Winnik MA, Xu B (1996). *Electrophoresis* 17:1451.

Michon C, Cuvelier G, Launay B (1993). *Rheol Acta* 32:94.

Mijangos C, López D, Muñoz ME, Santamaría A (1993). *Macromolecules* 26:5693.

Miller WG, Wu CC, Wee EL, Santee GL, Rai JH, Goebel KG (1974). *Pure Appl Chem* 38:37.

Mooney M (1940). *J Appl Phys* 11:582.

Osada Y, Ross-Murphy SB (1993). *Sci Am*, May: 82.

Pedley AM, Higgins JS, Peiffer DG, Rennie AR, Staples E (1989). *Polymer Comm* 30:162.

Peterlin A, Turner DT (1965). *J Polym Sci Polym Lett* 3:517.

Pimentel GC, McClellan AC (1960). *The Hydrogen Bond*, WH Freeman, San Francisco.

Prasad A, Marand H, Mandelkern L (1993). *J Polym Sci Polym Phys Ed* 32:1819.

Reid DS, Bryce TA, Clark AH, Rees DA (1974). *Faraday Discuss* 57:230.

Richtering HW, Gagnon KD, Lenz RW, Fuller RC, Winter HH (1992). *Macromolecules* 25:2429.

Rivlin RS (1948). *Philos Trans R Soc* A241:379.

Rochas C, Brulet A, Guenet J-M (1994). *Macromolecules* 27:3830.

Scanlan JC, Winter HH (1991). *Macromolecules* 24:47.

Semenov AN, Joanny J-F, Khokhlov AR (1995). *Macromolecules* 28:1066.

Shante VKS, Kirkpatrick S (1971). *Adv Phys* 20:325.

Shukla P, Muthukumar M (1988). *Polym Eng Sci* 28:1304.

Spevacek J, Schneider B (1974). *Makromolek Chem* 175:2939.

Spevacek J, Schneider B (1987). *Colloid Interface Sci* 27:81.

Stadler R, de Lucca Freitas L (1986). *Colloid Polym Sci* 264:773.

Stadler R, de Lucca Freitas L (1989). *Makromol Chem Macro Symp* 26:451.

Stauffer D (1976). *J Chem Soc Faraday Trans II* 72:1354.

Stinchcombe RB (1974). *J Phys C Solid State Phys* 7:179.

Stockmayer WH (1943). *J Chem Phys* 11:45.

Straley JP (1977). *J Phys C Solid State Phys* 10:3009.

Straley JP (1982). *J Phys C Solid State Phys* 15:2333.

Tan H-M, Moet A, Hiltner A, Baer E (1983). *Macromolecules* 16:28.

Tanaka T (1981). *Sci Amer* 244:124.

Tanaka F, Edwards SF (1992a). *Macromolecules* 25:7003.

Tanaka F, Edwards SF (1992b). *J Non-Newt Fluid Mech* 43:247,273,289.

Tanaka F, Stockmayer WH (1994). *Macromolecules* 27:3943.

Tipton DL, Russo PS (1996). *Macromolecules* 29:7402.

Treloar LRG (1975). *The Physics of Rubber Elasticity*, 3rd ed, Clarendon Press, Oxford.

Vallés EM, Macosko CW (1979). *Macromolecules* 12:521.

Venkataraman SK, Coyne L, Chambon F, Gottlieb M, Winter HH, (1989). *Polymer* 30:2222.

Wall FT (1943). *J Chem Phys* 11:527.

Wardhaugh LT, Boger DV (1991). *J Rheol* 35:1121.

Wee EL, Miller WG (1971). *J Phys Chem* 75:1446.

Winnik MA, Regismond STA (1996). *Colloid Surf A* 118:1.

Winnik MA, Yekta A (1997). *Curr Opin Colloid Interface Sci* 2:424.

Winter HH (1989). In *Encyclopedia of Polymer Science and Engineering*, Supplement Volume, Second ed, Wiley, New York.

Winter HH, Chambon (1986). *J Rheol* 30:367.

Winter HH, Morganelli P, Chambon F (1988). *Macromolecules* 21:532.

Winter HH, Mours M (1997). *Adv Polym Sci* 134:165.

Witten TA Jr, Cohen MH (1985). *Macromolecules* 18:1915.

Witten TA (1988). *J Phys (France)* 49:1055.

Yamamoto M (1956). *J Phys Soc Jp* 11:413.

Yamamoto M (1958). *J Phys Soc Jp* 13:1200.

Yekta A, Xu B, Duhamel J, Adiwidjaja H, Winnik MA (1995). *Macromolecules* 28:956.

Zosel A (1991). *J Adhesion* 34:201.

SUSPENSIONS

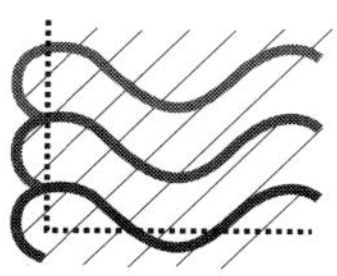

Chapter **6**

PARTICULATE SUSPENSIONS

6.1 INTRODUCTION

Suspensions or "dispersions" of particles in a liquid medium are ubiquitous. Blood, paint, ink, and cement are examples that hint at the diversity and technological importance of suspensions. Suspensions include drilling muds, foodstuffs, pharmaceuticals, ointments and cremes, and abrasive cleansers and are precursors of many manufactured goods, such as composites and ceramics. Control of the structure and flow properties of such suspensions is often vital to the commercial success of the product or of its manufacture. For example, in consumer products, such as toothpaste, the rheology of the suspension can often determine consumer satisfaction. In *ceramic processing*, dense suspensions are sometimes molded (Lange 1989) and then dried and sintered or fired into optical components, porcelin insulators, turbine blades, fuel cells, and bricks (Rice 1990; Simon 1993). Crucial to the success of the processing is the ability to transform a liquid, moldable suspension into a solid-like one that retains its shape when removed from the mold. These examples could be multiplied many times over.

Suspensions of small particles are often called "colloids," a term derived from the Greek word κολλα for "glue." This name was coined by Thomas Graham (1805–1869), who defined colloids as substances that could not diffuse through a membrane. Already in the seventeenth century, however, alchemists had produced stable suspensions, or *sols*, of inorganic particles, such as gold. By the nineteenth century, rubber was produced from the milky sap, or *latex*, of certain plants and trees; this sap is really a stable suspension of small balls of rubber that are insoluble in the suspending solvent. The term "latex" is now used for any stable suspension of polymeric particles. In the 1950s, suspensions of highly monodisperse polymeric particles first became available, and the study of these has greatly aided in the development of quantitative theories for the dynamics and flow properties of colloidal dispersions.

In this chapter, we review the rheological properties of suspensions of solids in a liquid medium, under conditions in which the particles do not clump together—that is, do not *gel*. Gelled colloidal suspension are discussed in Chapter 7, while *emulsions* and *foams*—where the suspended phase is another liquid or a gas—are discussed in Chapter 9. Even within these limits, the scope of this chapter is extensive, and there is only room for major topics. Additional detail can be found in Russel et al. (1989), Hiemenz and Rajagopalan (1997), Kim and Karrila (1991), and Chapter 10 of Macosko (1994). For reviews of the most recent work, see Brady (1996) and Mellema (1997).

In any multiphase liquid, stability is a paramount concern. Thermodynamics drives clumping of dispersed components, and this is sometimes enhanced by flow. However, tricks for stabilizing suspensions are as old as the inks of Egypt. Electrostatic and steric stabilization are the most common. By matching the dielectric properties, some particle–fluid combinations can be found that are inherently stable. A more detailed discussion of suspension stabilization is deferred to Chapter 7.

Spheroidal particles larger than about 1 μm or so tend to settle under gravity unless the particle density matches that of the suspending medium, or the suspending medium is very viscous. Small particles are maintained in suspension by Brownian motion, but this agitation also promotes particle collisions, which often leads to aggregation, followed by gelation or by gravitational settling of the particle clumps. To avoid sedimentation, the ratio of gravitational to Brownian forces, $a^4 \Delta \rho g / k_B T$, must be less than unity, where a is the radius of the particle or clump of particles, $\Delta \rho$ is the density difference between the particle and the liquid medium, and $g = 980$ cm/sec^2 is the gravitational constant. For $\Delta \rho \sim 1$ g/cm^3, particles or clumps bigger than 1 μm or so in radius tend to settle.

Even when sedimentation is avoided, suspensions of particles of size larger than a micron or so show other complicating effects when flow is present. Migration of a spherical particle across a streamline in a *prolonged* shearing flow occurs when the jump in the Bernoulli stress across the particle $\rho_s V \dot{\gamma} a = \rho_s h \dot{\gamma}^2 a$ is comparable to the Brownian stress $k_B T / a^3$, where ρ_s is the fluid density and h is the gap in the shearing device. Thus, migration occurs when $\rho_s h \dot{\gamma}^2 a^4 / k_B T$ exceeds 0.01–0.1 (Ho and Leal 1974). When $h = 1$ mm, 1-μm particles will show significant migration when the shear rate exceeds about 10 sec^{-1}. Particle migration can also be induced by stress or viscosity gradients (Leighton and Acrivos 1987; Phillips et al. 1992). Complications due to particle inertia can occur for $\gtrsim$ 10-μm particles in a medium of low viscosity $\eta_s \approx 1$ cP, because then the *particle Reynolds number* Re$_p \equiv \rho_s \dot{\gamma} a^2 / \eta_s$ exceeds 0.1 at a shear rate of 10^3 sec^{-1}. Thus, to avoid particle-inertia effects or inhomogeneities in particle concentration due to sedimentation or flow-induced migration, the particle size should be kept below about 1 μm.

Another possible complication of flows of dense particle suspensions is *wall slip*. This can be minimized by using roughened shearing surfaces (Persello et al. 1994).

Assuming that all such difficulties are avoided or accounted for, one can measure the rheological properties of suspensions using the techniques described in Chapter 1. As an example of the complex rheology of such suspensions, Fig. 6-1 shows the shear viscosity as a function of shear stress for 250-nm polystyrene ethylacrylate (St/EA) spheres in water at various particle volume fractions ϕ. Note that at low ϕ, the suspension viscosity is nearly constant, and is only slightly greater than that of the solvent, while at high volume fraction, $\phi \gtrsim 0.40$, the viscosity shows pronounced *shear thinning* at low shear stress and can show *shear thickening* at high stress. In the following sections, data such as these are explored more fully.

6.2 HARD, AND SLIGHTLY DEFORMABLE, SPHERES

The simplest suspensions are composed of so-called hard spheres in which the only interactions between particles are rigid repulsions that occur when particles come into contact. Even suspensions as simple as these can show rather complex rheological phenomena.

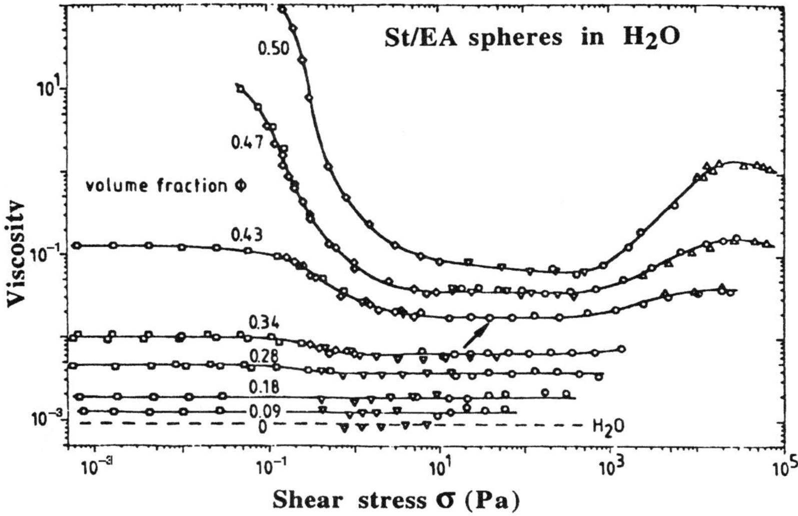

Figure 6.1 Viscosity versus shear stress for aqueous suspensions of charged poly(styrene-ethylacrylate) copolymer spheres of diameter $d = 250$ nm at various volume fractions ϕ and $T = 25°C$. The NaOH concentration was varied with ϕ over the range 0.022–0.230 M to keep the pH in the range 6.2–6.5. (From Laun 1984a, reprinted with permission from Hühig & Wepf Verlag.)

6.2.1 Zero-Shear Viscosity

At very low volume fractions ($\phi \lesssim 0.03$), the shear viscosity η of a suspension of hard spheres can be predicted by the simple formula

$$\eta = \eta_s(1 + 2.5\phi) \tag{6-1}$$

which Einstein (1906, 1911) calculated from the viscous dissipation produced by the flow around a single sphere. Thus, Einstein's formula, Eq. (6-1), is only valid when the suspension is dilute enough that the flow field around one sphere is not appreciably influenced by the presence of neighboring spheres. When two spheres are close enough that the drag on one of them is influenced by a second nearby sphere, the spheres are said to experience *hydrodynamic interactions.* Such interactions between two spheres leads to a contribution to η that is proportional to ϕ^2, three-body hydrodynamic interactions produce a term proportional to ϕ^3, and so on. The effect of two-body interactions on η was computed by Batchelor (1977); when combined with Einstein's result, his calculation gives

$$\eta_r \equiv \frac{\eta}{\eta_s} = 1 + 2.5\phi + 6.2\phi^2 \tag{6-2}$$

Figure 6-2 compares this formula with data of Saunders for polystyrene latices and shows that Batchelor's formula holds good for $\phi \lesssim 0.10$.

The expansion, Eq. (6-2), was extended to higher order in ϕ by a simple, though approximate, effective medium argument of Arrhenius (1917). According to this argument, we suppose that we increase by $d\phi$ the particle concentration in a suspension of viscosity

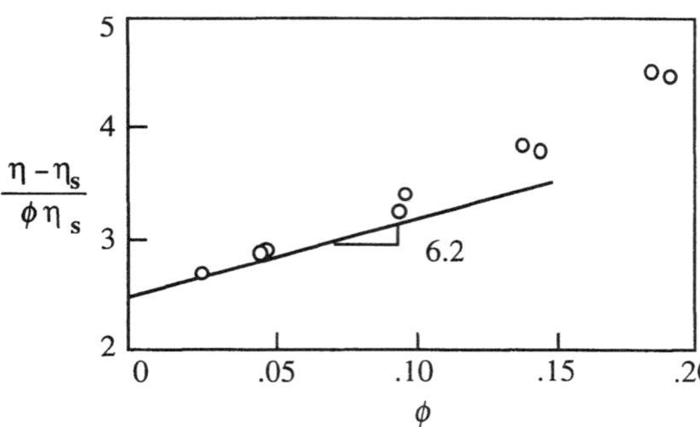

Figure 6.2 Reduced viscosity of suspensions of sterically stabilized polystyrene "hard spheres" of diameter $d = 420$ and 870 nm, as a function of volume fraction ϕ, replotted from data of Saunders (1961), compared to the prediction of Eq. (6-2). (From Macosko, *Rheology Principles, Measurements, and Applications,* Copyright © 1994. Reprinted by permission from John Wiley & Sons.)

$\eta(\phi)$ by adding a few new particles. If we treat the suspension into which we add these particles as a *homogeneous* viscous medium of viscosity $\eta(\phi)$, so that the increment in viscosity $d\eta$ resulting from the addition of the new particles is given by Einstein's formula, then

$$d\eta = 2.5\eta(\phi)\,d\phi \tag{6-3}$$

Integrating this, we obtain

$$\eta = \eta_s \exp(5\phi/2) \tag{6-4}$$

A similar argument can be made for arbitrarily shaped particles, yielding

$$\eta = \eta_s \exp([\eta]\phi) \tag{6-5}$$

Here $[\eta]$ is the *intrinsic viscosity*, which for suspensions is the dilute limit of the viscosity increment per unit particle volume fraction, divided by the solvent viscosity. Thus, it is a dimensionless quantity defined as

$$[\eta] = \lim_{\phi \to 0} \frac{\eta - \eta_s}{\phi \eta_s} \tag{6-6}$$

For spheres, $[\eta] = 2.5$. An exponential dependence on $[\eta]\phi$ has been found for the relaxation time of flexible polymer molecules in the crossover regime from dilute to concentrated solutions (Amelar et al. 1991). [For polymers, the intrinsic viscosity is a dimensional quantity; see Eq. (3-5).]

For suspensions, Eq. (6-5) fails at high ϕ because of correlations in the positions of the particles due to crowding. A simple "fix" to account for this divergence is to replace the viscosity increment $d\eta$ by (Ball and Richmond 1980)

$$d\eta = \frac{[\eta]\eta_s\,d\phi}{(1 - \phi/\phi_m)} \tag{6-7}$$

so that $d\eta$ diverges as ϕ approaches the *maximum-packing* volume fraction ϕ_m. For hard spheres, $\phi_m \approx 0.63\text{–}0.64$ (Onoda and Liniger 1990). Integrating Eq. (6-7) gives

$$\eta = \eta_s\left(1 - \frac{\phi}{\phi_m}\right)^{-[\eta]\phi_m} \tag{6-8}$$

Equation (6-8) is the *Krieger–Dougherty equation* (Krieger and Dougherty 1959), which is a general empirical expression for suspensions of particles of spherical or other shape. The parameters of this equation, $[\eta]$ and ϕ_m, have been tabulated by Barnes et al. (1989) for a variety of particles, from spheres to glass fibers. Generally, as the particle aspect ratio increases, $[\eta]$ increases and ϕ_m decreases, while the product $[\eta]\phi_m$ usually remains in the range 1.4–3. Note that the viscosity is very sensitive to volume fraction ϕ at large ϕ; small errors in the determination of ϕ can therefore lead to large errors in the value of η estimated from Eq. (6-8) (Meeker et al. 1997). Also, at high ϕ, viscosities become sensitive to small variations in particle properties such as surface roughness, size and shape polydispersity, and so on. Other useful empirical expressions for the concentration and shear-rate dependences of the shear viscosity can be found in Barnes et al. (1989) and in Dealy and Wissbrun (1990).

For particles that have steric-stabilization layers grafted onto them, the "hard-sphere" radius and the volume fraction must be adjusted to account for the layer thickness, Δ. When the volume fraction is not near close packing, the grafted layer can be considered a "hard" coating of thickness Δ, and the volume fraction ϕ can be corrected using the simple formula

$$\phi = \phi_0 \left(1 + \frac{\Delta}{a}\right)^3 \tag{6-9}$$

where ϕ_0 is the volume fraction of "bare" uncoated particles, and a is the particle radius. Alternatively, the dilute solution intrinsic viscosity $[\eta]$ can be measured, and the effective volume fraction can be extracted by using the Einstein formula. Typically, the layer thickness obtained this way agrees well with direct estimates (Mewis et al. 1989).

For high loadings of particles, the particle-size distribution has a strong effect on viscosity. In particular, when particles of two differing sizes are mixed, the viscosity can be much lower than it is for suspensions containing the same volume fraction of monosized particles. This is shown in Fig. 6-3, which plots the relative viscosity against the fraction f of large particles in a bidisperse suspension of particles of size ratio 5:1 for various total particle volume fractions. Notice that for $\phi \geq 0.60$, the viscosity drops by more than a decade as f increases from 0 to around 0.6. The viscosity minimum at $f \approx 0.6$ is known as the "Farris effect"; it is apparently a consequence of the packing of small particles into the interstices between the large ones. As a result, higher volume fractions of bidispersed particles than of monosized ones can be packed into a suspension. This fact is of much practical importance in the formulation of heavily loaded suspensions, such as plastic molding compounds, where especially high particle loading allows the thermal expansion coefficient to be more closely matched to that of silicon devices encapsulated within the compound. (A close match in thermal expansion coefficient helps prevent cracking and debonding during temperature cycling of packaged devices.)

6.2.2 Shear Thinning

Figure 6-1 shows that for $\phi \gtrsim 0.3$, the viscosity becomes sensitive to shear rate (or, equivalently, to shear stress). This shear-rate-dependence of the viscosity occurs when the shear rate is high enough to disturb from equilibrium the distribution of interparticle spacings. The rate at which the particle equilibrium is regained is controlled by the particle diffusivity, given in dilute solutions by

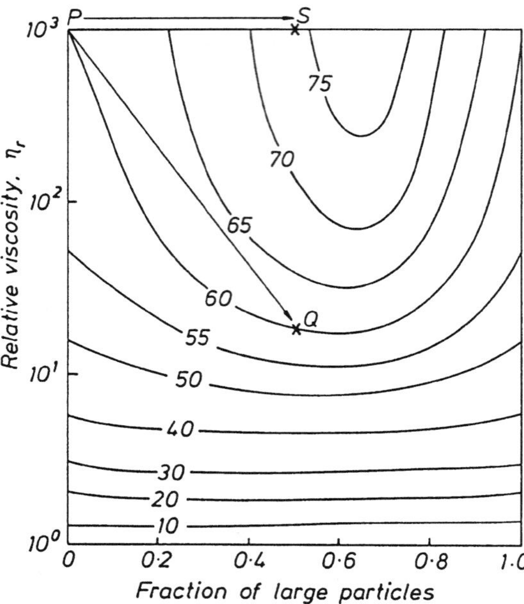

Figure 6.3 Relative viscosity as a function of the fraction of large spheres in a bimodal distribution of particle sizes with a 5:1 ratio of diameters, at various total volume percentages of particles. The arrow $P \rightarrow Q$ illustrates the 50-fold reduction in viscosity that occurs when monosized particles in a 60 vol% suspension are replaced by a 50–50 mixture of large and small spheres. The arrow $P \rightarrow S$ shows that if monosized spheres are replaced by a bimodal size distribution, the concentration of spheres can be increased from 60% to 75% without increasing the viscosity. (Reprinted from Barnes et al., *An Introduction to Rheology* (1989), with kind permission from Elsevier Science - NL, Sara Burgerhartstraat 25, 1055 KV Amsterdam, The Netherlands.)

$$D_0 = \frac{k_B T}{6\pi \eta_s a} \tag{6-10}$$

The time t_D for a particle to diffuse a distance equal to its radius a is therefore

$$t_D \approx \frac{a^2}{D_0} = \frac{6\pi \eta_s a^3}{k_B T} \tag{6-11}$$

One can then define a dimensionless shear rate, or *Peclet number*, as

$$\mathrm{Pe} \equiv \frac{\eta_s \dot{\gamma} a^3}{k_B T} \propto \dot{\gamma} t_D \tag{6-12}$$

[In the literature, Pe is sometimes defined as $\mathrm{Pe} \equiv \dot{\gamma} t_D = 6\pi \eta_s a^3 \dot{\gamma}/k_B T$, which is 6π times the value given by Eq. (6-12). Here, the definition in Eq. (6-12) will be used.]

Krieger (1972) has argued that the shear-rate-dependent suspension viscosity $\eta(\phi, \dot{\gamma})$ is more appropriate than the solvent viscosity η_s to use in defining a dimensionless shear rate. Since $\eta(\phi, \dot{\gamma})\dot{\gamma}$ is the shear stress σ, Krieger's suggested dimensionless group is really a reduced shear stress:

$$\sigma_r \equiv \frac{\sigma a^3}{k_B T} \tag{6-13}$$

Thus, the relative shear viscosity η/η_s of a concentrated suspension of monosized spherical particles should be a universal function of two dimensionless quantities: ϕ and σ_r (or Pe). Figure 6-4 shows that for three different particle radii: 85, 141, and 310 nm, the relative viscosity $\eta_r \equiv \eta/\eta_s$, is indeed a nearly universal function of σ_r for fixed ϕ. The dependence of η_r on σ_r can be fit by a simple expression

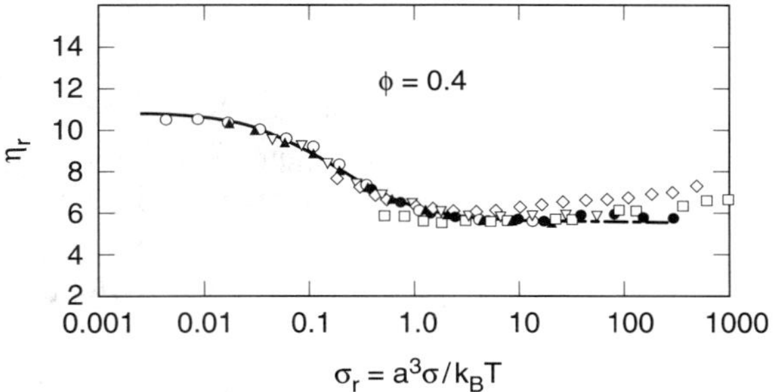

Figure 6.4 Relative viscosity $\eta_r \equiv \eta/\eta_s$ verses reduced shear stress for latices of poly(methylmethacrylate) spheres of radii 85, 141, 204, and 310 nm, sterically stabilized by adsorbed triblock copolymer, poly(dimethylsiloxane)–polystyrene–poly(dimethylsiloxane), in silicone fluids of viscosity 7.98 cP and 44.1 cP at 30°C. The line is the fit from Eq. (6-14a). (From Choi and Krieger 1986, reprinted with permission from Academic Press.)

$$\eta_r = \eta_{r\infty} + \left(\frac{\eta_{r0} - \eta_{r\infty}}{1 + b\sigma_r}\right) \tag{6-14a}$$

where η_{r0} is the low-shear-rate viscosity, $\eta_{r\infty}$ is the high-shear-rate viscosity, and b is a fitting parameter, with $b = 5.70$ for the data in Fig. 6-4. Note in Fig. 6-4 that there is a small deviation from the "plateau" value $\eta_{r\infty}$ at the highest shear stress, because of the onset of shear thickening.

As discussed earlier, the ϕ-dependence of the zero-shear viscosity η_{r0} obeys the Krieger–Dougherty equation (6-8). It turns out that $\eta_{r\infty}$ also obeys the Krieger–Dougherty equation, but with a larger value of ϕ_m. Figure 6-5 shows experimental ϕ-dependences of η_{r0} and $\eta_{r\infty}$ and fits by the Krieger–Dougherty equation. For the polystyrene spheres in

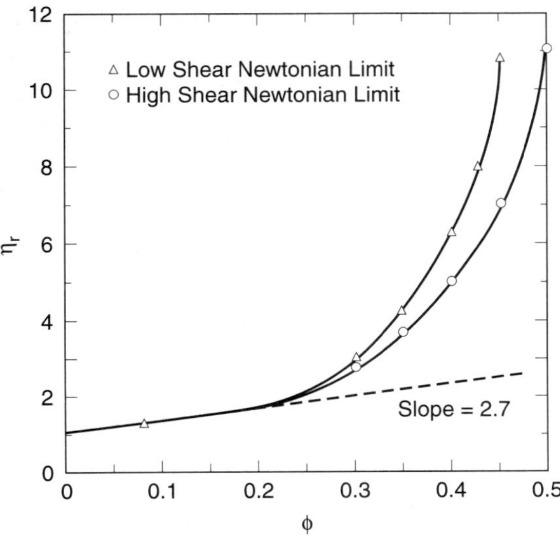

Figure 6.5 Experimental ϕ dependences of $\eta_{r0}(\triangle)$ and $\eta_{r\infty}$ (○) for "hard-sphere" polystyrene spheres of diameter $d = 150$–430 nm in benzyl alcohol or metacresol. The lines are fits to the Krieger–Dougherty equation (6-8) with $\phi_{m0} = 0.57$, $\phi_{m\infty} = 0.68$, and $[\eta] = 2.67$. (From Papir and Krieger 1970, reprinted with permission from Academic Press.)

these studies, the high-shear-rate plateau viscosity is fit using $\phi_m = \phi_{m\infty} = 0.68$, while at low shear rates we have $\phi_m = \phi_{m0} = 0.57$. For concentrations between the limiting values for high and low shear rates (i.e., for $\phi_{m0} < \phi < \phi_{m\infty}$), the suspension shows an apparent *yield stress* σ_y below which there is no flow, but above which the viscosity decreases toward η_∞. A thorough comparison of data for η_{r0} and $\eta_{r\infty}$ for nominally hard-sphere suspensions can be found in Phan et al. (1996).

Equation (6-14a) can be rewritten in an equivalent form,

$$\frac{\eta - \eta_\infty}{\eta_0 - \eta_\infty} = \frac{1}{1 + \sigma/\sigma_c} \tag{6-14b}$$

where $\eta_\infty = \eta_s \eta_{r\infty}$; $\eta_0 = \eta_s \eta_{r0}$, and $\sigma_c = k_B T/(a^3 b)$ is a critical shear stress for shear thinning. A typical dependence of σ_c on volume fraction is shown in Fig. 6-6. Note that at high particle concentrations, the critical stress for shear thinning plunges toward zero.

The flow curves for fluids with a yield stress are often fit by the constitutive equation of a "Bingham plastic":

$$\sigma = \sigma_y + \eta_{\rm pl}\dot{\gamma} \tag{6-15a}$$

where $\eta_{\rm pl}$ is the "plastic viscosity." However, a better fit is usually obtained with the so-called Casson equation (Dealy and Wissbrun 1990)

$$\sigma^{1/2} = \sigma_y^{1/2} + C\dot{\gamma}^{1/2} \tag{6-15b}$$

Fits of these expressions to experimental data produce values of σ_y that often differ from the "true" dynamic or static yield stress (see Sections 8.2.2.2 and 8.2.2.3 for a more general discussion of yield stresses).

Mewis et al. (1989) have shown that the limiting concentrations ϕ_{m0} and $\phi_{m\infty}$ are sensitive to the deformability of stabilization layers that are grafted or adsorbed onto the particles. By studying polymethyl methacrylate (PMMA) particles of radius 42–610 nm with grafted poly(12-hydroxystearic acid) layers of thickness 8–10 nm, they were able to obtain ratios of particle radius to layer thickness in the range 5–61 and thereby vary the effective particle softness, or "squishiness." For the largest, effectively hardest particles

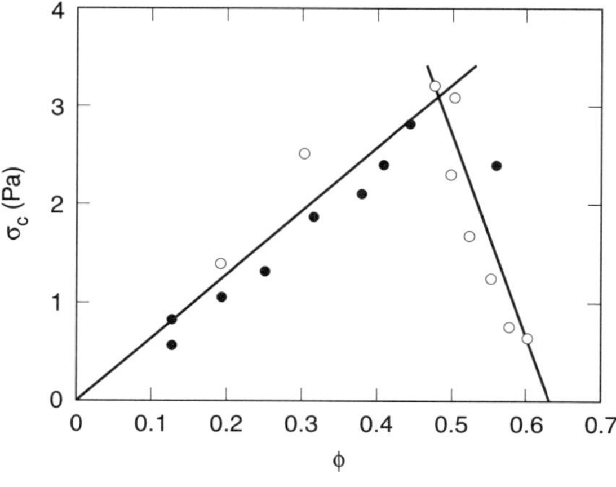

Figure 6.6 Dependence of critical reduced stress σ_c for shear thinning on volume fraction ϕ of sterically stabilized "hard" silica spheres with $a = 78$ nm in cyclohexane. The lines are guides to the eye. (From de Kruif et al., reprinted with permission from J. Chem. Phys. 83:4714, Copyright 1985 American Institute of Physics.)

($a = 610$ nm), they found $\phi_{m0} = 0.49$ and $\phi_{m\infty} = 0.62$, while for effectively softer particles with $a = 42$ nm, they found that $\phi_{m0} = 0.72$, $\phi_{m\infty} = 0.96$. These latter values are above the maximum packing density for hard spheres ($\phi_m = 0.63$), but the uncorrected volume fraction ϕ_0 of the bare PMMA particles remains below ϕ_m. Thus, the high value of the corrected volume fraction ϕ is achieved by deformation of the grafted layers, which allows the particles to squeeze into spaces too tight for hard particles of the same size.

The tendency of particles to deform increases with increasing volume fraction. At effective volume fractions below about $\phi \approx 0.4$, suspensions of "squishy" spheres have viscosities similar to those of harder spheres (see Fig. 6-7). But at higher ϕ, there are substantial differences between the two, and layer deformability becomes important. Thus, for soft spheres at high ϕ, the dependence of relative viscosity on shear stress no longer obeys Eq. (6-14). Mewis et al. (1989) found that an equation of Cross (1965), which contains an extra parameter, works better:

$$\eta_r = \eta_{r\infty} + \frac{\eta_{r0} - \eta_{r\infty}}{1 + (b\sigma_r)^m} \tag{6-16}$$

with the best-fit value of m becoming larger than unity (as large as 3.5) when ϕ exceeds ≈ 0.5.

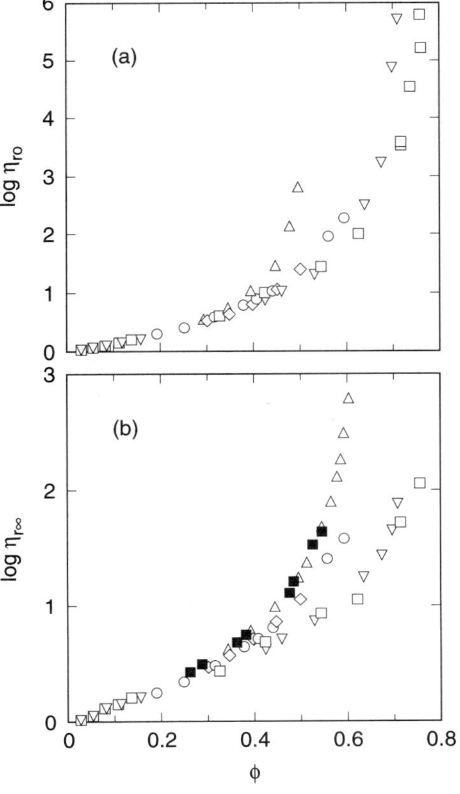

Figure 6.7 (a) Low-shear-rate relative viscosity η_{r0} and (b) high-shear-rate relative viscosity $\eta_{r\infty}$ versus particle volume fraction ϕ for sterically stabilized PMMA spheres with $a = 42$ nm PMMA in decalin (∇), $a = 42$ nm in exsol ($\square$), $a = 237$ nm in decalin ($\triangle$), and $a = 610$ nm in decalin (■). The small spheres are effectively softer, and thus show smaller viscosities than the large spheres at the same volume fraction. Also shown are data from Papir and Krieger (1970) ($\diamond$) and from de Kruif et al. (1985) ($\bigcirc$). (From Mewis et al. 1989, with permission from the American Institute of Chemical Engineers.)

The mechanism of shear thinning in hard-sphere suspensions has been elucidated by computer simulations. Using "Stokesian dynamics" computer simulations of hard-sphere suspensions, Brady and coworkers (Bossis and Brady 1989; Phung et al. 1996; Brady 1996) have shown that the Brownian contribution to the shear stress disappears at high shear stress, leaving only the hydrodynamic contribution (see Fig. 6-8). The disappearance of the Brownian contribution leads to a viscosity reduction of about a factor of two for $\phi = 0.45$ in three dimensions, in good agreement with experiments (compare Fig. 6-8 with Fig. 6-4). Simulations of Visscher and Heyes (1994) and Phung et al. (1996) at $\phi = 0.30$ and $\phi = 0.45$ show the formation of lines of particles, or "strings," parallel to the flow direction. These occur at shear rates high enough to be in the shear-thinned state, but disappear at still higher shear rates where shear thickening occurs (Phung et al. 1996). In the plane normal to the flow direction, these strings register into a regular structure with hexagonal order (see also Laun et al. 1992). Light-scattering studies on "hard-sphere" suspensions are consistent with "string" formation (Ackerson and Pusey 1988). The structure of a sheared hard-sphere suspension appears to be similar to that found in simulations of small-molecule liquids at very high shear rates (Hanley et al. 1983); these also show shear thinning.

More severe shear thinning would be expected in suspensions that form three-dimensionally ordered phases at rest. Ideal hard spheres begin to form an ordered close-packed phase (FCC and/or HCP) at $\phi = \phi_1 = 0.494$; charged electrostatically stabilized spheres can form ordered phases at lower volume fractions, as will be discussed in Section 6.4. When ordered phases are forced to flow, one typically observes a yield stress above which the three-dimensionally ordered structure transforms into a layered structure that permits continuous deformation. The viscosity then drops dramatically. Even below ϕ_1, partial ordering can be induced by shearing at modest rates or shear strains. Ackerson and Pusey (1988), for example, have shown by light-scattering experiments that a disordered "hard-sphere" suspension with $\phi = 0.48$ can be induced to form FCC order under oscillatory shearing with strains of around unity. At higher shear strains, the ordered structure is partially broken down. Thus, even below ϕ_1, one might expect significant shear-induced ordering or disordering.

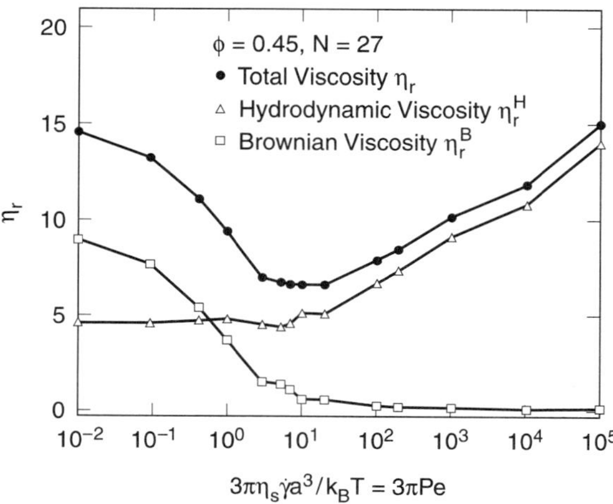

Figure 6.8   Stokesian dynamics simulation results for the steady shear viscosity at $\phi = 0.45$; also shown are the separate contributions of the Brownian and hydrodynamic stresses. (From Phung et al. 1996, with peremission from Cambridge University Press.)

6.2.3 Shear Thickening

As shown in Fig. 6-1, at high particle volume fractions, $\phi \gtrsim 0.40$, shear thickening can occur at high shear stresses, above the range of shear stresses where shear thinning occurs. Shear thickening has been called *dilatancy*, because it was thought to be caused by a tendency of the interparticle spacing to increase under shear. Reynolds (1885) showed the existence of such volume dilation by pouring sand into a rubber balloon along with a small excess of water over that necessary to fill the interstices between grains. A glass tube attached to the balloon showed that the amount of excess water decreased when the sand pack was deformed, thus showing that deformation increases the space between grains. In a more recent experiment, Onoda and Liniger (1990) showed that at particle volume fractions in excess of the "random loose-packed" limit of $\phi = 0.55$, shearing flow causes dilation of a pack of 250-μm spheres. Volume dilation is important in granular flows (Goddard and Bashir 1990), and it occurs because each grain needs more room in the flowing state than it does at rest (Bagnold 1954). A simple manifestation of this phenomenon occurs when one leaves dry footprints while walking along a wet beach: the deformed sand dilates, and it therefore sucks free water into its interstices. Metzner and Whitlock (1958) showed, however, that there is no simple connection between volume dilation and shear thickening, and so the distinctiveness of two phenomena should be kept clearly in mind. Thus, to avoid confusion, it is best that the term "dilatancy" *not* be used as a synonym for "shear thickening."

In Brownian suspensions, as ϕ increases, the slope of the viscosity–shear rate curve in the shear-thickening regime typically increases, and for electrostatically stabilized suspensions at high-volume fractions, it can even become a discontinuous jump. At shear rates above the shear-thickening regime, there is typically a second shear-thinning regime (see Fig. 6-1).

Dimensional analysis implies that for a given value of ϕ, all *monodisperse hard-sphere* suspensions ought to show an onset of shear thickening at a universal value of the Peclet number Pe, or reduced stress σ_r. Thus, the critical shear rate $\dot{\gamma}_c$ for shear thickening ought to be proportional to the inverse cube of the particle radius, $\dot{\gamma}_c \propto a^{-3}$. A compilation of data for a wide range of sterically or electrostatically stabilized suspensions of spheres with diameters ranging from 0.03 to 100 μm show a scaling law that deviates somewhat from this, namely $\dot{\gamma}_c \propto a^{-2}$ (Barnes 1989). The systematic deviation from the expected scaling law probably arises from deviations from a hard-sphere repulsive potential; since "softness" caused by electrostatic charges or deformability of the steric stabilizing layer is generally more important for small particles than for large ones, a deviation toward a weaker power law should be expected.

It has been suggested that particles with "soft" or long-ranged repulsive interparticle potentials, such as sterically stabilized "plastisol" particles (Willey and Macosko 1982), and especially electrostatically stabilized particles (Hoffman 1972), might be the most shear thickening. Such a tendency might be explained by the greater tendency of such suspensions to form layered structures under modest shear, and hence the greater potential they have for shear thickening once a shear rate is reached at which the layering partially decomposes. Some shear thickening is expected even for pure hard spheres, however. In the "Stokesian dynamics" simulations of Brady and coworkers (Phung et al. 1996), a factor-of-two increase in viscosity is observed as the Peclet number $\mathrm{Pe} \equiv \eta_s \dot{\gamma} a^3 / k_B T$ is increased from around 10 to 10^4 (see Fig. 6-8). In the simulations, the increase in viscosity is associated with the formation of *clusters* containing particles driven by shear into close proximity (Brady 1996). The deformation of such clusters under shear presumably produces large lubrication stresses in the thin films separating closely spaced particles. Experimental evidence for cluster formation in the shear thickening regime is provided in the experiments of D'Haene et al. (1993). They found that there is a long-time tail in the shear stress relaxation of a suspension after cessation of shearing from the shear thickened state, if the shear thickening is discontinuous in shear rate. Suspensions with gradual shear thickening show no such tail. A long-time tail in the stress–relaxation curve for a hard-sphere suspension implies the existence of particle clusters that rearrange very slowly after cessation of shearing. While shear thickening always seems to involve cluster formation, the details of how such clusters are produced might well differ depending on the nature of the repulsive potential between particles. Therefore, caution should be exercised when comparing shear-thickening phenomena in different types of suspensions. Shear thickening in electrostatically stabilized suspensions will be discussed in Section 6.4.4.

6.2.4 Suspension Modulus

The linear viscoelastic properties of hard-sphere suspensions have been measured by de Kruif, Mellema, and coworkers (van der Werff et al. 1989) and by Shikata and Pearson (1994). Figure 6-9 shows G' and $G'' - \eta'_\infty \omega a_T$ measured as a function of reduced frequency $a_T \omega \tau_w$. A shift factor a_T for the solvent is used to collapse data gathered at different temperatures onto a single curve. (The use of shift factors is explained in Section 3.4.2.) Viscous properties dominate elastic ones in these suspensions; still, even hard-sphere suspensions are slightly elastic; that is, G' is nonzero. The weak elasticity of these suspensions is produced by Brownian motion, which tends to restore to equilibrium the shear-distorted particle configurations. This elasticity should also produce small normal stress differences, as we will see in Section 6.2.5.

The shape of these $G'(\omega)$ and $G''(\omega)$ curves look similar for all concentrations in the range $0.30 \lesssim \phi \lesssim 0.55$; however, the curves shift toward lower frequencies and higher moduli as ϕ increases. The shift toward lower frequencies results from the increase in the

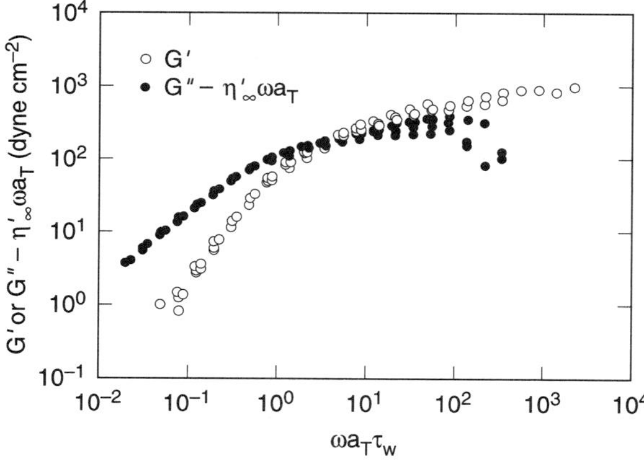

Figure 6.9 Dynamic moduli G' and $G'' - \eta'_\infty \omega a_T$ as functions of reduced frequency $\omega a_T \tau_w$ for a concentrated suspension, $\phi = 0.48$, of $a = 60$ nm silica particles in a mixture of ethylene glycol and glycerol chosen to match the refractive index of the particles. The reference temperature at which the shift factor $a_T = 1$ is $-10°C$. $\tau_w \approx 0.5\tau_p$ is the longest relaxation time. (From Shikata and Pearson, reprinted from J. Rheol. 38:601, Copyright 1994, with permission from the Journal of Rheology.)

suspension relaxation time with increasing ϕ. Spheres that are close together cannot move with respect to each other rapidly without generating large lubrication forces; hence closely packed spheres are slow to relax their interparticle spacings. The characteristic particle relaxation time $\tau_p(\phi)$ in a suspension is expected to be inversely proportional to the particle diffusivity. Since the particles need only to move short distances to restore the equilibrium structure, the *short-time* diffusivity $D_s(\phi)$ is the relevant one. Hence, we expect

$$\tau_p(\phi) = \frac{a^2}{6D_s(\phi)} \tag{6-17}$$

where the factor of 6 has been included for convenience in what follows.

At high frequencies, the viscoelastic behavior of suspensions is primarily dissipative, as the particles are forced to move through the solvent much faster than they can relax by Brownian motion. The high-frequency behavior is characterized by a constant high-frequency viscosity $\eta'_\infty(\phi) = \lim_{\omega\to\infty} G''/\omega$, which has been subtracted from the data plotted in Fig. 6-9. Extending the Stokes–Einstein law to the high-frequency regime, we expect

$$\eta'_\infty(\phi) = \frac{k_B T}{6\pi D_s(\phi)a}, \quad \text{or} \quad \frac{\eta'_\infty(\phi)}{\eta_s} = \frac{D_0}{D_s(\phi)} \tag{6-18}$$

where $D_0 = D_s^0$ is the diffusivity of an isolated particle in dilute solution. This prediction (6-18) has been nicely confirmed by Shikata and Pearson (1994) (see Fig. 6-10). As the particle concentration increases, the short-time diffusivity decreases and η'_∞ increases, because the movement of a particle with respect to its neighbors requires the flow of solvent through ever tighter interparticle gaps. The dependence of η'_∞ on ϕ is in agreement with a

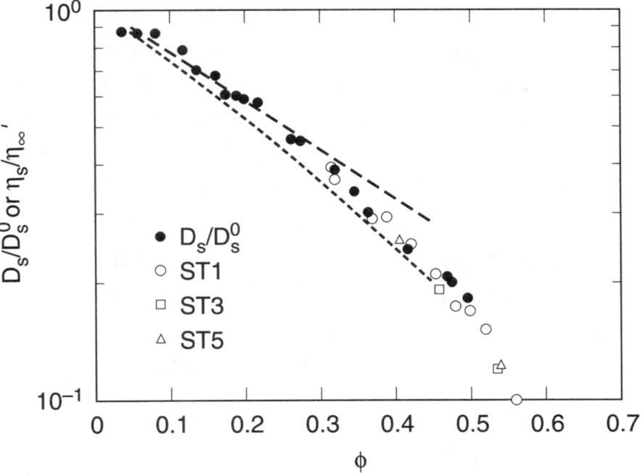

Figure 6.10   The short-time diffusion coefficient reduced by that of the zero-concentration limit, $D_s(\phi)/D_0$, as a function of particle concentration ϕ, taken from van Megan and Underwood (1990), compared to the reciprocal of the reduced high-frequency viscosity η_s/η'_∞. The long-dashed and short-dashed lines are, respectively, the predictions of Beenakker and Mazur (1984) and Beenakker (1984). ST1, ST3, and ST5 are silica particles of radius 60, 125, and 225 nm, respectively, in mixtures of ethylene glycol and glycerol. (From Shikata and Pearson, reprinted from J. Rheol. 38:601, Copyright 1994, with permission from the Journal of Rheology.)

theory of Beenakker (1984; see Shikata and Pearson 1994). One can express $\tau_p(\phi)$ in terms of $\eta'_\infty(\phi)$, by using Eqs. (6-17) and (6-18):

$$\tau_p(\phi) = \frac{\pi \eta'_\infty(\phi) a^3}{k_B T} \tag{6-19}$$

The entire relaxation spectrum of these suspensions shifts to longer time scales with increasing ϕ, roughly in proportion to $\tau_p(\phi)$. The width of the spectrum, measured by $\langle \tau^2 \rangle / \langle \tau \rangle^2$, where $\langle \cdot \rangle$ is an average over the spectrum, is around 4.

Note in the data of Shikata and Pearson (Fig. 6-9) that the storage modulus appears to approach a constant, G'_∞, at high frequencies. Theories for the high-frequency modulus of a suspension show that G'_∞ is highly sensitive to the interaction potential $W(r)$ between particles in close contact. We define $\overline{W}(r') \equiv W(r)$, where $r' \equiv r/d$ is the separation between centers of mass of two particles made dimensionless by the particle diameter d. Thus, $r' = 1$ at particle–particle contact. At very high frequencies, Zwanzig and Mountain (1965) assumed that the particles are convected much faster than they diffuse, and thus they move affinely. Affine displacements of particles from their equilibrium relative positions would then produce a stress whose magnitude is controlled by the shape of the interparticle potential $\overline{W}(r')$ and the probability $g(r')$ for finding a given separation r'. [$g(r')$ is called the *radial distribution function* (see Section 2.2.1).] The following classic result is thereby computed (Zwanzig and Mountain 1965; Evans and Lips 1990; Lionberger and Russel 1994):

$$\frac{a^3 G'_\infty}{k_B T} = \frac{3\phi}{4\pi} + \frac{3\phi^2}{5\pi} \int_0^\infty g(s) \frac{d}{ds} \left[s^4 \frac{d}{ds} \left(\frac{\overline{W}(s)}{k_B T} \right) \right] ds \tag{6-20}$$

where the osmotic term $3\phi/4\pi = \nu a^3$, with ν the number of particles per unit volume, is usually negligible. Equation (6-20) involves two derivatives of $\overline{W}$, because the first derivative gives a force, and a second derivative gives the coefficient of the linear relationship between force and deflection, which is proportional to the modulus. For hard spheres, $\overline{W}(r')$ is a step function, and Eq. (6-20) predicts an infinite modulus. This result is a consequence of the assumed affine motion, which forces nearly contacting hard particles into overlap with each other. Of course, in reality, the lubrication layers between the hard particles will become highly stressed and will deflect the particle motion away from the affine limit. Lionberger and Russel (1994) have accounted for these hydrodynamic effects by using a lubrication approximation and have predicted a bounded high-frequency modulus G'_∞. An approximate expression for this is (Shikata and Pearson 1994)

$$\frac{a^3 G'_\infty}{k_B T} \approx 0.78 \phi^2 \frac{\eta'_\infty}{\eta_s} g(1, \phi) \tag{6-21}$$

where $g(1, \phi) = g(r', \phi)$ at $r' = 1$—that is, at particle–particle contact. It can be approximated by (Shikata and Pearson 1994)

$$g(1, \phi) = \frac{1 - \frac{1}{2}\phi}{(1 - \phi)^3}, \qquad \phi < 0.5$$

$$\tag{6-22}$$

$$g(1, \phi) = \frac{1.2 \phi_m}{\phi_m - \phi}, \qquad \phi > 0.5$$

where $\phi_m = 0.63$–0.64 is the volume fraction at random close packing. The predictions of Eq. (6-21), and numerically calculated values of G'_∞, are seen in Fig. 6-11 to be in excellent agreement with the data of Shikata and Pearson.

Suspensions of *sterically stabilized* particles, studied by van der Werff et al. (1989), seem to show no high-frequency plateau in G'_∞, but rather $G'_\infty \propto \omega^{1/2}$ at large ω. This behavior has been predicted by Brady (1993) using a "free draining" approximation for the hydrodynamic interactions, which yields

$$\frac{d^3 G'_\infty}{k_B T} = \frac{24\phi^2}{5\pi} \left(\frac{\omega d^2}{D_s(\phi)} \right)^{1/2} g(1, \phi) \tag{6-23}$$

Lionberger and Russel (1994) suggested that the stabilizing layers on the spheres of van der Werff et al. produce different lubrication forces than those between bare particles, such as those of Shikata and Pearson, and that this accounts for the differences between the high-frequency moduli of these two systems. Thus, it would appear, perhaps not surprisingly, that the high-frequency behavior of concentrated suspensions is sensitive to the details of interactions between spheres in near contact.

Brady has noted that the high-frequency viscosity η'_∞ consists only of the hydrodynamic contribution, since at high frequencies Brownian motion hasn't sufficient time to contribute to the viscosity. The hydrodynamic contribution should be the same at any frequency, since it is an instantaneous stress. Thus, the zero-shear—or, equivalently, zero-frequency—viscosity is given by

$$\eta_0 = \eta'_\infty + \eta_0^b \tag{6-24}$$

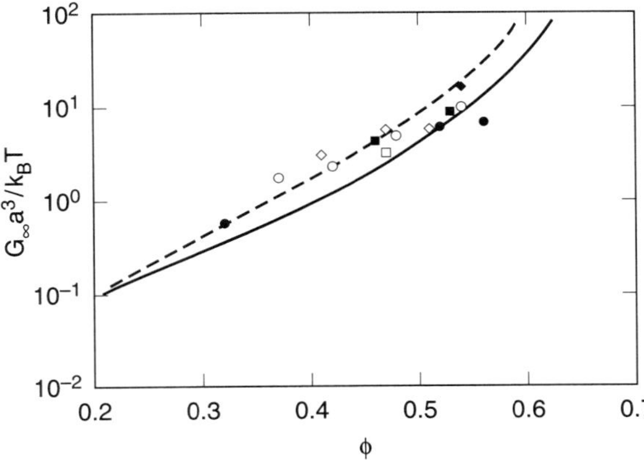

Figure 6.11   The reduced high-frequency modulus $G'_\infty a^3 / k_B T$ as a function of particle concentration ϕ for particles ST1 ($\bigcirc$), ST2 ($\square$), and ST3 ($\diamond$) described in the caption to Fig. 6-10. The filled symbols are estimates based on the behavior in the high-frequency limit. The solid line is a numerical prediction from the theory of Lionberger and Russel (1994), while the dashed line is their approximate result, given by Eq. (6-21). (From Shikata and Pearson, reprinted from J. Rheol. 38:601, Copyright 1994, with permission from the Journal of Rheology.)

where η_0^b is the zero-frequency Brownian contribution, which Brady computed using the "free-draining" approximation:

$$\eta_0^b = \frac{12}{5}\eta_s\phi^2\frac{g(1,\phi)D_0}{D_s(\phi)} \tag{6-25}$$

Figure 6-12a shows that Brady's expression is in excellent agreement with the data of Shikata and Pearson. Similar agreement is obtained with other data sets for "hard spheres" (Brady 1993). At high ϕ, the viscosity is predicted by this theory to approach $\eta_{r0} \rightarrow 1.3(1-\phi/\phi_m)^{-2}$, while the relaxation time should be inversely proportional to $D_s(\phi)$, which follows $D_s(\phi) \propto (1-\phi/\phi_m)$. For a recent discussion of these theories, see Lionberger and Russel (1997).

6.2.5 Normal Stress Differences

Experimental measurements of the normal stress differences N_1 and N_2 for suspensions are rare, especially for hard spheres. According to a recent theory by Brady and Vicic (1995), in the dilute regime the low shear-rate values of the normal stress differences are $N_{1,0}/\eta_0\dot{\gamma} = 0.8996\pi\phi^2\text{Pe}$ and $N_{2,0}/\eta_0\dot{\gamma} = -0.7886\pi\phi^2\text{Pe}$. These values are quite small at concentrations and shear rates where the theory might apply ($\phi \lesssim 0.15$; Pe $\lesssim 0.1$). The

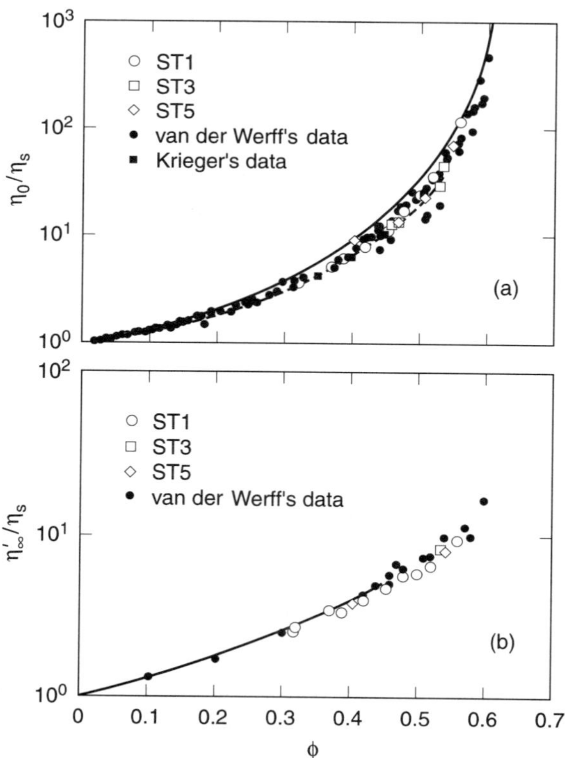

Figure 6.12 (a) The dependence on particle concentration of the relative zero-shear viscosity η_0/η_s for suspensions of various particle sizes described in the caption to Fig. 6-10. Also plotted are the data of Krieger (1972) and van der Werff et al. (1989). The solid line is a prediction of Brady (1993), given by Eqns. (6-24)–(6-25), and the dashed line is that of Lionberger and Russel (1994). (b) The concentration dependence of the relative high-frequency viscosity η'_∞/η_s; data of van der Werff et al. (1989) are also plotted. The solid line is from the theory of Beenaker (1984). (From Shikata and Pearson 1994, with permission from the Journal of Rheology.)

small values of $N_1/\eta\dot{\gamma}$ and $N_2/\eta\dot{\gamma}$ reflect the lack of orientability of spherical particles, and hence their weak elasticity. At higher ϕ, theory and simulations give larger values of the normal stress differences; these predictions have not yet been confirmed by corresponding experiments. In the shear-thickening region, unusual *negative N_1* and *positive N_2* values are reported, as will be discussed in Section 6.4.4.

6.2.6 Other Nonlinear Properties

Nonlinear transient rheological properties of hard-sphere suspensions at high volume fractions ($\phi \approx 0.53$) were measured by Watanabe et al. (1996). Experimental tests included (a) stress growth after start-up of steady shearing and (b) relaxation of stress after cessation of steady and step shearing. The step-strain data showed approximate "time-strain factorability" similar to that seen in entangled polymer melts and solutions (see Section 3.7.4.2). From these data, a nonlinear "damping function" could be defined, and the nonlinear time-dependent data could then be predicted using a factorized "K-BKZ equation" (see Section 3.7.4.4). The K–BKZ description proved adequate only in the shear-thinning regime; it failed to describe the shear-thickening regime.

> • Problems and Worked Examples 6.1 through 6.5, at the end of this chapter, will sharpen your skills in obtaining simple, practical estimations of the viscosity, modulus, and relaxation time of hard-sphere suspensions.

6.3 NONSPHERICAL PARTICLES

When suspended particles are nonspherical, they can be oriented by a flow field, producing much stronger elastic effects (including large normal stress differences) than with a suspension of spherical particles at a similar volume fraction. The simplest non-spherical shapes are *axisymmetric*; that is, they have an axis of rotational symmetry. Simple examples are rods, disks, and prolate or oblate spheroids (see Fig. 6-13). A "spheroid" is an axisymmetric ellipsoid—that is, an ellipsoid in which two of the principal axes are of equal length. Elongated particles, such as fibers, are often added to liquids, either to bulk up their viscosity or to add strength to the solids that are made from them. Examples of the latter include fiber-reinforced concrete and fiber-reinforced molded plastic parts. The starting point for understanding the behavior of such suspensions is to consider the response of a single nonspherical particle in a dilute suspension, where the particles do not interact.

6.3.1 Dilute Suspensions

6.3.1.1 Jeffery Orbits

If **u** is a unit vector parallel to the axis of symmetry of a spheroidal particle, then in the absence of Brownian motion (Pe $\rightarrow \infty$) and of interparticle interactions, the time rate of change of **u** in a flow field is

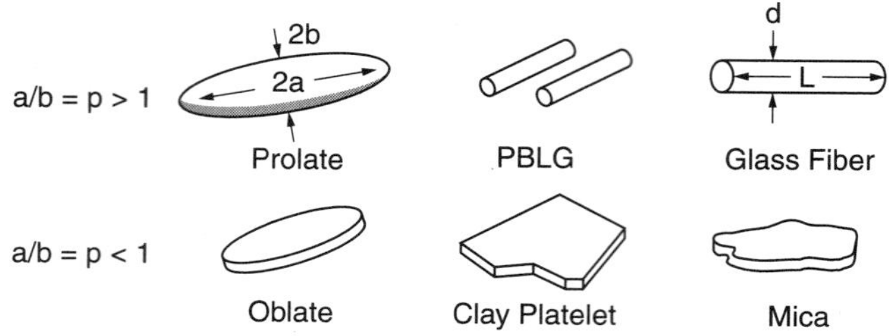

Figure 6.13 Prolate and oblate shapes of various particles. (Adapted from Macosko, *Rheology Principles, Measurements, and Applications*, Copyright © 1994. Reprinted by permission from John Wiley & Sons.)

$$\dot{\mathbf{u}} = \mathbf{u} \cdot \boldsymbol{\omega} + \left(\frac{p^2 - 1}{p^2 + 1}\right) (\mathbf{u} \cdot \mathbf{D} - \mathbf{uuu} : \mathbf{D}) \tag{6-26}$$

where p is the ratio of the length of the particle along its axis of symmetry to its length perpendicular to this axis. $\mathbf{D}$ and $\boldsymbol{\omega}$ are the strain rate and vorticity tensors, defined in Eqs. (1-27) and (1-28). For prolate spheroids, p is greater than unity, while for oblate ones, $p < 1$. Equation (6-26) also applies to other axisymmetric particles, but p must be replaced by an effective aspect ratio p_{eff}. For cylindrical rods, $p_{\text{eff}} \approx 0.7(L/d)$, where L/d is the true aspect ratio, the ratio of length (L) to diameter (d) of the cylinder (Trevelyn and Mason 1951). In a shearing flow with the particle long axis in the plane of deformation (the plane parallel to both the flow and the gradient directions), this expression integrates to

$$\tan \theta = p \tan \left(\frac{\dot{\gamma} t}{p + 1/p}\right) + \tan \theta_0 \tag{6-27}$$

where θ is the angle of the axis of symmetry measured in the clockwise direction from the flow direction, and θ_0 is the angle at time 0.

Note that θ is *time-periodic*; a non-Brownian axisymmetric particle rotates indefinitely in a shearing flow. This rotation is called a *Jeffery orbit* (Jeffery 1922). The period P required for a rotation of π in a Jeffery orbit is

$$P = \frac{\pi}{\dot{\gamma}} \left(p + \frac{1}{p}\right) \tag{6-68}$$

Thus, $P \to \infty$ for either $p \to \infty$, which is the limit of an infinitely thin prolate ellipsoid, or $p \to 0$, which is the limit of an infinitely flat oblate one. In these two limits, the particle rotates until a long axis is parallel to the flow direction, and then rotation slows to a halt. For large, but finite, aspect ratios, the particle rotates slowly when its long axis is nearly parallel to the flow direction, and rapidly otherwise. The Jeffery orbits of rod-like and disk-like particles have been observed directly (Anczurowski and Mason 1967a, 1967b) and indirectly by optical dichroism (Frattini and Fuller 1986).

6.3.1.2 Rotary Diffusivities

Brownian motion and interparticle interactions can produce deviations from this Jeffery orbit. When particle rotations are disturbed by Brownian motion, the orbits become stochastic and must be described probabilistically. We therefore let $\psi(\mathbf{u})$ be the probability that the axis of symmetry of an axisymmetric particle is parallel to the unit vector $\mathbf{u}$. ψ obeys a *Smoluchowski equation*,

$$\frac{\partial \psi}{\partial t} = -\frac{\partial}{\partial \mathbf{u}} \cdot [\dot{\mathbf{u}}\psi] + D_{r0}\frac{\partial^2 \psi}{\partial u^2} \tag{6-29}$$

where $\dot{\mathbf{u}}$ is given by Eq. (6-26), $\partial/\partial\mathbf{u}$ is the gradient on the unit sphere, $\partial/\partial\mathbf{u} = [(\partial/\partial\theta, (1/\sin\theta)(\partial/\partial\phi)]$, where θ and ϕ are the polar and azimuthal angles, respectively, and D_{r0} is the dilute-solution *rotary diffusivity* of the particle—that is, the rate at which a particle reorients by Brownian motion. For a thorough discussion of Eq. (6-29), including the use of spherical coordinates, see Bird et al. (1987).

For a particle of nearly spherical shape and diameter d we have

$$D_{r0} = \frac{k_B T}{\pi \eta_s d^3} \tag{6-30}$$

while for a circular disk-like particle of diameter d we obtain

$$D_{r0} = \frac{3k_B T}{4\eta_s d^3} \tag{6-31}$$

and for a spheroid of aspect ratio p we have

$$D_{r0} = \frac{3k_B T(\ln(2p) - 0.5)}{\pi \eta_s L^3} \tag{6-32a}$$

For rods of length L and diameter d (Doi and Edwards 1986; Kirkwood and Auer 1951), D_{r0} is estimated by a formula similar to Eq. (6-32a):

$$D_{r0} = \frac{3k_B T(\ln(L/d) - 0.8)}{\pi \eta_s L^3} \tag{6-32b}$$

Equation (6-32b) is in good agreement with D_{r0} measured in dilute solutions of PBLG (Ookubo et al. 1976; Warren et al. 1973). A slightly more rigorous equation for D_{r0} is given below, in Eqs. (6-37) and (6-38). Rotary diffusivities for spheroids and other shapes are given by Brenner (1974).

The crossover from Brownian to non-Brownian behavior in a flowing suspension is controlled by a *rotational Peclet number*,

$$\text{Pe} \equiv \frac{\dot{\gamma}}{D_r} \tag{6-33}$$

The term "Peclet number" is common in the suspension literature, while the corresponding quantity is usually called the "Deborah number" or "Weissenberg number" in the polymer literature. From Eqs. (6-30) through (6-33) we find, in general, for a solvent of viscosity ~ 1 cP, that $D_{r0} \sim 1/b^3$, where $b(= d$ or $L)$ is the particle's longest dimension in units of μm, and D_{r0} is in sec^{-1}. Since typical shear rates are in the range $10^{-3} \lesssim \dot{\gamma} \lesssim 10^3$ sec^{-1}, we find that for particles whose longest dimension exceeds $\sim 10\,\mu$m, we have Pe $\gg 1$, and such particles can therefore be considered non-Brownian. Thus, if the aspect ratio exceeds 10, and the thinnest dimension of a plate- or rod-like particle is greater than a micron or so, then $b > 10\,\mu$m, and the particle is usually non-Brownian.

6.3.1.3 Stresses for Suspensions of Spheroids

For spheroidal particles—that is, axisymmetric ellipsoids—the equation for the stress tensor contains an elastic term from Brownian motion:

$$\sigma^e = 3\left(\frac{p^2 - 1}{p^2 + 1}\right) \nu k_B T \langle \mathbf{uu} \rangle \tag{6-34}$$

The viscous stress has been given by Hinch and Leal (1973) as

$$\sigma^v = 2\eta_s \phi \left\{ A \langle \mathbf{uuuu} \rangle : \mathbf{D} + B[\langle \mathbf{uu} \rangle \cdot \mathbf{D} + \mathbf{D} \cdot \langle \mathbf{uu} \rangle] + C\mathbf{D}] \right\} \tag{6-35}$$

where the coefficients A, B, and C depend on the particle aspect ratio p. Values for them are given in Hinch and Leal (1972). The viscous stress arises from drag produced by the solvent as it flows past the ellipsoid. It goes to zero immediately when flow ceases, while the elastic, or Brownian, term relaxes gradually, as flow-induced orientation disappears by Brownian motion. The total stress σ is the sum of the elastic and viscous contributions, as well as that from the Newtonian solvent, $\sigma^s = 2\eta_s \mathbf{D}$; thus $\sigma = \sigma^e + \sigma^v + \sigma^s$.

Brenner (1974) has presented numerical results for the suspension stresses in various flows. Figure 6-14 plots the intrinsic viscosity [defined in Eq. (6-6)] for oblate and prolate spheroids of various aspect ratios as functions of the Peclet number. Note that as the aspect ratio of the spheroid increases, the zero-shear viscosity increases, and the suspension shows more shear thinning. The suspension also becomes more elastic when the aspect ratio (p for prolate or $1/p$ for oblate spheroids) is large; see Fig. 6-15, which plots $N_3 \equiv N_1 - N_2$ versus Pe for prolate spheroids of various aspect ratios p. Typically, N_2 is roughly an order of magnitude less than N_1, so this plot of N_3 mainly reflects the behavior of N_1.

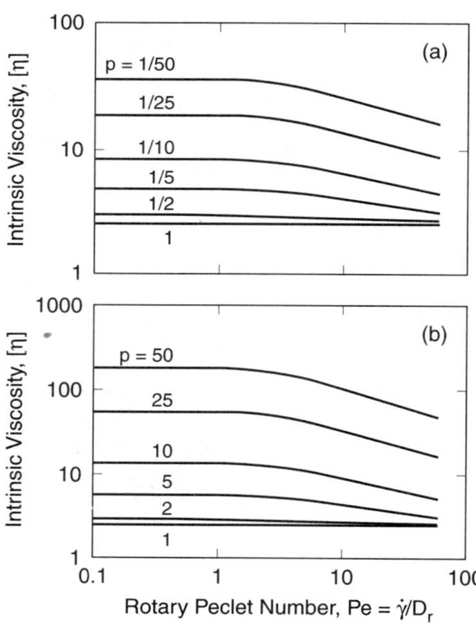

Figure 6.14   Intrinsic viscosity versus Peclet number for dilute suspensions of spheroidal particles of (a) oblate shape and (b) prolate shape. (From Macosko 1994, adapted from Brenner 1974, with permission from Pergamon Press.)

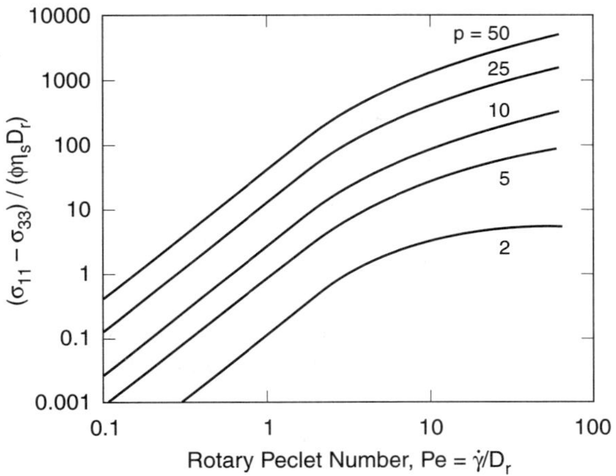

Figure 6.15 Reduced third normal stress difference, $(\sigma_{11}-\sigma_{33})/(\phi\eta_s D_r)$, as a function of Peclet number for prolate spheroids of various aspect ratio. (From Macosko 1994, adapted from Brenner 1974, with permission from Pergamon Press.)

The quadratic dependence of N_3 on $\dot{\gamma}$ gives way to a constant magnitude of N_3 at high Pe. The magnitude of $N_3/(\phi\eta_s D_{r0})$ increases with aspect ratio p roughly as p^2. Table 6-1 summarizes the behavior of suspensions of spheroidal particles predicted by Hinch and Leal (1972; see also Brenner 1974) for shearing flow at low and high Peclet number. In this table, ε is a small ($\ll 1$) deviation of p from unity; $\varepsilon \equiv p - 1$. The low Peclet-number values are denoted by the subscript 0, and the high Peclet-number values are denoted by the subscript ∞. The subscript p on η denotes the particle contribution to the viscosity; that

TABLE 6-1
Viscometric Coefficients for Dilute Suspensions of Rigid Spheroids

	Low Shear Rates		
Aspect Ratio p	$\eta_{p0}/\phi\eta_s$	$\Psi_{1,0}D_{r0}/(\phi\eta_s)$	$\psi_{2,0}/\psi_{1,0}$
$p \to \infty$	$\dfrac{4p^2}{15\ln(p)}$	$\dfrac{p^2}{15\ln(p)}$	$\dfrac{-1}{7}$
$p = 1 + \varepsilon$	$2.5 + 0.777\,\varepsilon^2$	$\dfrac{1}{5}\varepsilon^2$	$\dfrac{-1}{7}$
$p \to 0$	$\dfrac{32}{15\pi}\dfrac{1}{p}$	$\dfrac{4}{15\pi}\dfrac{1}{p}$	$\dfrac{-2}{7}$

	High Shear Rates		
Aspect Ratio p	$\eta_{p\infty}/\eta_{p0}$	$\psi_{1,\infty}/\psi_{1,0}$	$\psi_{2,\infty}/\psi_{2,0}$
$p \to \infty$	$\dfrac{1.18}{p}$	$\dfrac{15}{4}p^2\mathrm{Pe}^{-2}$	$\dfrac{105}{4}\mathrm{Pe}^{-2}$
$p = 1 + \varepsilon$	$1 - 0.24\varepsilon^2$	$36\mathrm{Pe}^{-2}$	$36\mathrm{Pe}^{-2}$
$p \to 0$	$4.61p$	$\dfrac{25}{4}\dfrac{1}{p^2}\mathrm{Pe}^{-2}$	$\dfrac{35}{8}\dfrac{1}{p^2}\mathrm{Pe}^{-2}$

is, $\eta_p \equiv \eta - \eta_s$. The high Peclet-number behavior of the normal stress coefficients is only reached at Pe values high enough that $\psi_{1,\infty}/\psi_{1,0}$ and $\psi_{2,\infty}/\psi_{2,0}$ are much less than unity.

6.3.1.4 Stresses for Suspensions of High-Aspect-Ratio Particles and Molecules

Brownian rod-like objects of high aspect ratio are usually molecules, not colloidal particles. As exception is tobacco mosaic virus (TMV), which is a Brownian particle of length 300 nm and diameter 18 nm (Caspar 1963). For completeness, we shall discuss the theory of Brownian rod-like particles in this chapter, with the understanding that the theory for such "particles" is actually more relevant to long stiff molecules than to rod-like fibers. The behavior of non-Brownian fiber suspensions is covered in Section 6.3.2.2.

The stress tensor for solutions or suspensions of cylindrical rods in solution is

$$\boldsymbol{\sigma} = 2\eta_s \mathbf{D} + \nu\zeta_{\text{str}} \langle \mathbf{uuuu} \rangle : \mathbf{D} + 3\nu k_B T \left\{ \langle \mathbf{uu} \rangle - \frac{1}{3}\boldsymbol{\delta} \right\} \tag{6-36}$$

The terms on the right side are the solvent, viscous, and elastic terms, $\boldsymbol{\sigma}^s$, $\boldsymbol{\sigma}^v$, and $\boldsymbol{\sigma}^e$, in that order. The coefficient ζ_{str} is a viscous drag coefficient, which for long, slender fibers is given by (Batchelor 1970)

$$\zeta_{\text{str}} = \frac{\pi \eta_s L^3}{6 \ln(2L/d)} f(\varepsilon) = \frac{k_B T}{2D_{r0}} \tag{6-37}$$

with $\varepsilon \equiv [\ln(2L/d)]^{-1}$, and

$$f(\varepsilon) = \frac{1 + 0.64\varepsilon}{1 - 1.5\varepsilon} + 1.659\varepsilon^2 \tag{6-38}$$

for long cylinders, and

$$f(\varepsilon) = \frac{1}{1 - 1.5\varepsilon} \tag{6-39}$$

for long, rigid spheroids. According to Eq. (6-38), $f(\varepsilon)$ slowly decreases from 2.3 to 1.4 as L/d increases from 15 to 1000. Equations (6-37) and (6-38) for cylinders are slightly more rigorous than Eq. (6-32b).

From Eqs. (6-26), (6-29), and (6-36)–(6-38), the stresses for arbitrary flows of dilute suspensions of rods can be computed. At low shear rates,

$$\eta_0 = \eta_s + \frac{2\nu k_B T}{15 D_{r0}} \tag{6-40}$$

In an oscillatory shearing flow for such a suspension,

$$G' = \frac{3}{5}\nu k_B T \frac{\tau^2\omega^2}{1 + \omega^2\tau^2}, \quad G'' - \omega\eta_s = \frac{3}{5}\nu k_B T \left(\frac{\tau\omega}{1 + \omega^2\tau^2} + \frac{1}{3}\tau\omega \right) \tag{6-41}$$

where τ, the rotational relaxation time, is

$$\tau = \frac{1}{6 D_{r0}} \tag{6-42}$$

The contribution $\tau\omega/3$ to G'' in Eq. (6-41) comes from the viscous stress contribution $\boldsymbol{\sigma}^v$. This contribution changes somewhat if the long particles are modeled not as cylindrical rods, but as long spheroids, in which case the term $\tau\omega/3$ should be replaced by $2\tau\omega/5$. Another model sometimes used for high-aspect-ratio particles is the rigid "dumbbell" model, in which all the viscous drag is lumped onto spheres placed at each end of a rigid connector. For the rigid-dumbbell model, the appropriate viscous term in the absence of hydrodynamic interaction between the two spheres is $2\tau\omega/3$. For each of these latter two cases, the elastic stress contribution remains the same, although the formulas for the relaxation time τ are somewhat different. Figure 6-16 shows that the predictions of G' and G'' for long rigid spheroids ($p \rightarrow \infty$) compare well to data for a dilute suspension of TMV. Figure 6-17 shows that the shear viscosity versus shear rate $[\eta]/[\eta]_0$ for a dilute solution of rod-like PBLG molecules can be fit by the shear viscosity curve for rigid dumbbells. At high $\dot{\gamma}$, $\eta_p \approx 0.68\nu k_B T \tau (\dot{\gamma}\tau)^{-2/3}$ and $\psi_1 \approx 1.20\nu k_B T \tau^2 (\dot{\gamma}\tau)^{-4/3}$ for a suspension of rigid dumbbells (Bird et al. 1987). ψ_2/ψ_1 at low shear rates is zero for the rigid dumbbell-model without hydrodynamic interactions, because the negative contribution to ψ_2 from $\boldsymbol{\sigma}^e$ is canceled by a positive contribution from $\boldsymbol{\sigma}^v$. Bird et al. (1987) discuss the rigid-dumbbell model thoroughly.

6.3.2 Semidilute Solutions and Suspensions of Rods

6.3.2.1 Brownian Rods

A suspension of long rods with aspect ratio of 50 or more can only be considered "dilute" if its concentration is very low, less than 1% by volume. The reason is that diluteness requires that the rods be able to rotate freely without being impeded by neighboring rods (Fig. 6-18a), and the volume that a single long rod can sweep out by rotation about its center of mass must be large, around L^3. Thus, rod–rod interactions should be expected when the number concentration of rods, ν, reaches a value proportional to L^{-3}. Experimentally (Mori et al. 1982), found that the transition occurs at more than 30 times this estimate, apparently because a rod can easily "dodge" several other rods that invade its sphere of rotation. Thus

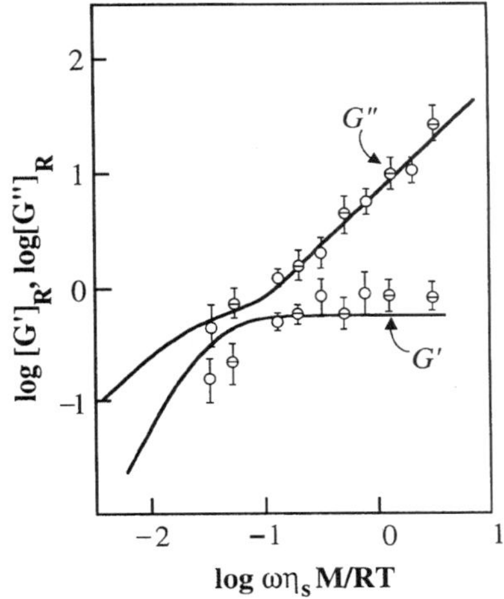

Figure 6.16   Frequency-dependence of reduced intrinsic moduli $[G']_R \equiv (5/3)\lim_{\nu\to 0} G'/\nu k_B T$; $[G'']_R \equiv (5/3)\lim_{\nu\to 0}(G'' - \omega\eta_s)/\nu k_B T$ for dilute suspensions of tobacco mosaic virus (TMV). The concentration ν is the number of particles per unit volume of solution, and is given by $\nu = c N_A/M$, where c is mass of TMV per unit volume, N_A is Avogadro's number, and M is the mass of a TMV particle. The lines are the predictions for a dilute suspension of long prolate spheroids. (From Nemoto et al., Biopolymers, 14:407, Copyright © 1975. Reprinted by permission of John Wiley & Sons, Inc.)

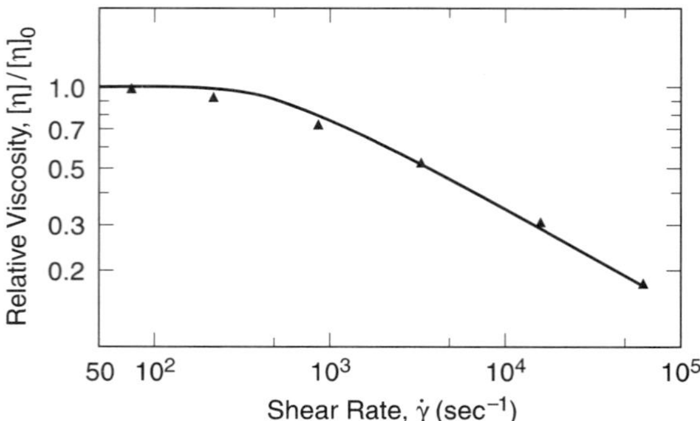

Figure 6.17   Normalized intrinsic viscosity $[\eta]/[\eta]_0$ for a dilute solution of poly(γ-benzyl-L-glutamate) (PBLG); $M_w = 208{,}000$) in *m*-cresol. The line is a calculation for the rigid-dumbbell model, with the relaxation time $\tau = 1/6D_{r0}$ adjusted to the value 10^{-3} sec to obtain a fit. The stress tensor for a suspension of rigid dumbbells is given by Eq. (6-36) with ζ_{str} replaced by $k_B T/D_{r0}$. (From Bird et al. 1987; data from Yang 1958, *Dynamics of Polymeric Liquids, Vol. 2*, Copyright © 1987. Reprinted by permission of John Wiley & Sons, Inc.)

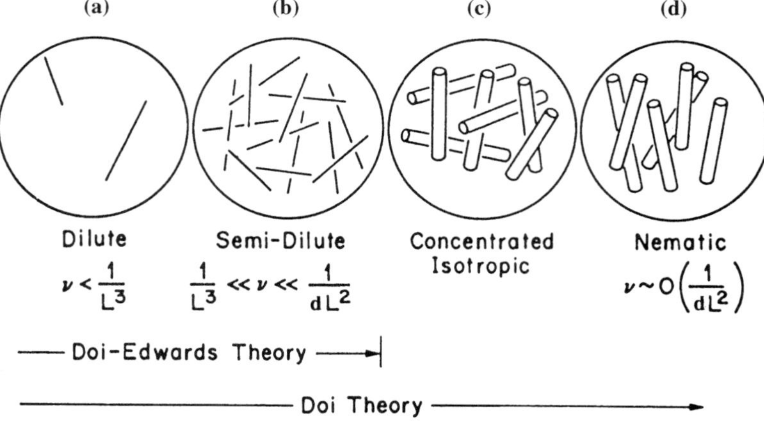

Figure 6.18   Concentration regimes for rodlike molecules. **(a)** Dilute; **(b)** semidilute; **(c)** isotropic concentrated; and **(d)** nematic. (From Doi and Edwards, copyright © 1986 by Oxford University Press, Inc. Used by permission of Oxford University Press, Inc.)

$\nu_0 L^3 \approx 30$ marks the transition from "dilute" to "semidilute" behavior. Since the volume fraction ϕ of rods is given by

$$\phi = \pi d^2 L \nu/4 \tag{6-43}$$

the crossover from dilute to semidilute behavior occurs when $\phi_0 \approx 30\pi d^2/4L^2 = 24(d/L)^2$. For an aspect ratio L/d of 50, this corresponds to a volume percentage of 1%, while for $L/d = 100$ the crossover occurs at only 0.24% of the dispersed phase. In practical applications, semidilute suspensions are much more common than dilute ones.

At concentrations somewhat above ϕ_0, the solution is said to be "semidilute," not "concentrated," because the static properties of the solution are almost unchanged from that of a dilute solution, even though molecular rotations are greatly impeded. Thus, at rod concentrations somewhat greater than ϕ_0, a new rod can be randomly inserted into the solution with negligible probability of it intersecting other rods (see Fig. 6-18b). Since there is little correlation between orientations of neighboring rods, rods can then be thought of as lines with zero volume. The semidilute regime is distinguished from concentrated isotropic regime (Fig. 6-18c), in that in the latter the rods begin to have difficulty packing isotropically because of excluded-volume interactions. The onset of the concentrated isotropic regime is around $\nu_0' \approx 1/dL^2$ or $\phi_0' \approx \pi d/4L$.

Eventually, at still higher concentrations (Fig. 6-18d), excluded-volume interactions lead to the formation of a nematic liquid-crystalline state, discussed in Section 2.2.2. According to Onsager's theory, the maximum concentration at which the solution remains isotropic is $\phi_1 = 3.3d/L$ (see Section 2.2.2.1). For $d/L = 100$, this gives $\phi_1 = 0.033$, a low concentration. However, the rod must be very stiff for this result to apply; if the persistence length λ_p is less than the rod length, one has instead $\phi_1 \propto d/\lambda_p$. At even higher concentrations, $\phi \gtrsim 0.35$, there can be transitions to positionally ordered liquid crystalline states, such as hexagonal or smectic liquid crystals to be described in Section 11.5.5.

Doi and Edwards have developed a "cage" model to predict the rotary diffusivity of rods in the semi-dilute regime. The "cage" in which a rod is confined by surrounding rods is analogous to the "tube" that forces entangled flexible polymer molecules to diffuse primarily along their own contour (see Section 3.7.1). If the cage is modeled as a cylinder surrounding the rod into which other rods do not penetrate, its radius a is given by $a \sim 1/\nu L^2$ (Doi and Edwards 1986). The crucial difference between rods and flexible polymers is that the rod must remain almost completely oriented as long as any part of it lies in the tube, whereas any portion of the flexible molecule that escapes the tube becomes disoriented. When a rod leaves its tube, it can only change its orientation by a small angle, $\varepsilon \approx a/L, \approx 1/\nu L^3$ (Doi and Edwards 1986). If τ_D is the time required to escape the cage, the time required for the rod to become totally disoriented by this mechanism is approximately $\tau_D/\varepsilon^2 \approx \tau_D L^2/a^2 \approx \tau_D (\nu L^3)^2$. The reciprocal of this is the rotational diffusion coefficient; since $\tau_D \approx 1/D_{r0}$, this works out to be

$$D_r = \beta D_{r0}(\nu L^3)^{-2} \propto \phi^{-2} L^{-7} \ln(L/d) \tag{6-44}$$

The dimensionless constant β turns out to be large (around 1350) for perfectly rigid rods (Teraoka and Hayakawa 1989; Bitsanis et al. 1990). The transitional concentration ν_0 from dilute to semidilute occurs when D_r departs from its dilute solution value D_{r0}; according to Eq. (6-44), this occurs when $\nu_0 L^3 \approx \sqrt{\beta} \approx 30$, or so, as noted above. Measurements (Mori et al. 1982) of dynamic electric birefringence in solutions of rodlike poly(γ-benzyl-L-glutamate) (PBLG) are in agreement with Eq. (6-44).

Under flow, the orientation distribution of the rods becomes *anisotropic*, and the cage diameter increases such that $a \propto 1/\langle \sin(\theta) \rangle$, where θ is the angle of orientation of a cage rod with respect to the trapped test rod (Doi and Edwards 1986). If $\mathbf{u}$ is the orientation of the test rod, then

$$\langle \sin\theta \rangle = \frac{4}{\pi} \int |\mathbf{u} \times \mathbf{u}'| \psi(\mathbf{u}')\, du'^2 \tag{6-45}$$

Here $|\mathbf{u} \times \mathbf{u}'|$ is the absolute value of the sine of the angle between $\mathbf{u}$ and $\mathbf{u}'$. Then, defining $\hat{D}_r$ as the *orientation-dependent* rotary diffusivity, we have $\hat{D}_r \propto a^2 \propto 1/\langle \sin\theta \rangle^2$, or

$$\hat{D}_r(\mathbf{u}) = D_r \left[\frac{4}{\pi} \int |\mathbf{u} \times \mathbf{u}'| \psi(\mathbf{u}')\, du'^2 \right]^{-2} \tag{6-46}$$

As the rods become more oriented, $\hat{D}_r(\mathbf{u})$ becomes more direction dependent, and its average value increases.

Except for the dependence of $\hat{D}_r$ on $\mathbf{u}$, the Smoluchowski equation for rigid rods in a flow field is the same as that for dilute rigid rods; that is,

$$\frac{\partial \psi}{\partial t} + \frac{\partial}{\partial \mathbf{u}} \cdot \left[(\mathbf{u} \cdot \nabla\mathbf{v} - \mathbf{u}\mathbf{u}\mathbf{u} : \mathbf{D})\psi - \hat{D}_r \frac{\partial \psi}{\partial \mathbf{u}} \right] = 0 \tag{6-47}$$

This equation can be simplified by replacing $\hat{D}_r$ with an orientation-independent average diffusivity, $\overline{D}_r$. Doi and Edwards suggested using an average of Eq. (6-46):

$$\hat{D}_r \to \overline{D}_r \equiv D_r \left[\frac{4}{\pi} \int \int \psi(\mathbf{u})\psi(\mathbf{u}')|\mathbf{u} \times \mathbf{u}'|\, du^2\, du'^2 \right]^{-2} \tag{6-48}$$

The *preaveraged* rotary diffusivity $\overline{D}_r$ can be pulled outside the derivative $(\partial/\partial\mathbf{u})$ in Eq. (6-47).

The stress tensor for a semidilute solution of rods is given by Eq. (6-36), the formula for dilute solutions. However, if in a thought experiment one holds the shear rate fixed at a low value while increasing the concentration of rods from dilute to semidilute, the Brownian contribution to the stress will greatly increase, since the rotary diffusivity decreases according to (6-44). The viscous stress contribution, however, only increases in proportion to ν. Thus, as Doi and Edwards (1986) argued, the ratio of viscous to Brownian stresses decreases as $(\nu L^3)^{-2}$ as the concentration increases in the semidilute regime. Hence, in the semidilute regime at low shear rates the viscous contribution should become negligible compared to the Brownian term. However, at high shear rates, where the rods are highly oriented, the rotary diffusivity in the semidilute regime begins to increase toward its dilute solution value, due to the cage-dilation effects described by Eq. (6-46) or (6-48). Thus as the shear rate increases, the Brownian contribution doesn't rise as rapidly with shear rate as the viscous term, and at high Peclet number the viscous term can become dominant, even for semidilute or concentrated solutions of rods. This is especially true of the shear stress, since the viscous contribution to it rises roughly linearly with shear rate, while the Brownian contribution actually decreases with shear rate at high Pe, because of flow-induced orientation.

Doi and Edwards (1978) and Kuzuu and Doi (1980) have solved the Smoluchowski equation (6-47)–(6-48) for simple shearing and elongational flows, and they obtained predictions of rheological behavior that are similar to those of the reptation theory for concentrated flexible polymers discussed in Section 3.7.5.1. Figure 6-19, for example, shows the shear-rate-dependence of the shear viscosity and first and second normal stress coefficients predicted by the Doi–Edwards theory for semidilute rods; these results are similar to those predicted by the Doi–Edwards theory for entangled flexible molecules. At

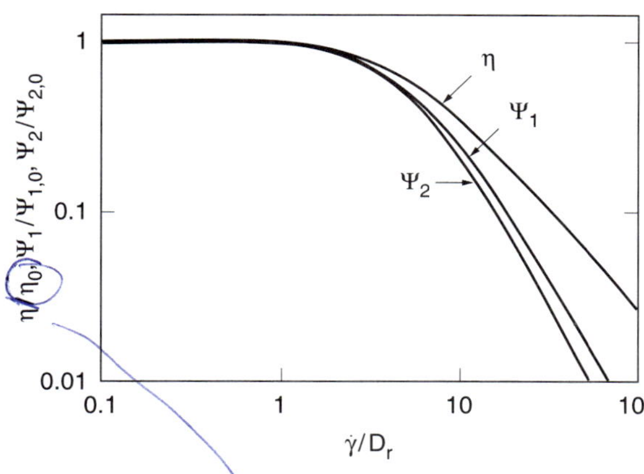

Figure 6.19 Normalized shear viscosity η, first normal stress coefficient ψ_1, and second normal stress coefficient ψ_2, as functions of Peclet number for the Doi–Edwards model of semi-dilute solutions of rods, with the viscous contribution neglected. (From Kuzuu and Doi 1980, with permisssion from the Society of Polymer Science, Japan.)

low shear rates, the viscosity and first and second normal stress coefficients are (Kuzuu and Doi 1980)

$$\eta_0 = \frac{\nu k_B T}{10 D_r}; \quad \Psi_{1,0} = \frac{\nu k_B T}{30 D_r^2}; \quad \frac{\Psi_{2,0}}{\Psi_{1,0}} = -\frac{2}{7} \tag{6-49}$$

Comparisons of the Doi–Edwards (DE) theory for semidilute solutions of rods with experiments are complicated by the lack of ideality of most rod-like molecules. The behavior is sensitive to molecular flexibility and polydispersity, which are difficult to avoid in such solutions. Because the rotary diffusivity of a rod has such a strong dependence on the molecular weight M of the rod, scaling as $\phi^{-2} M^{-7}$ in semidilute solution, the effects of even small levels of polydispersity are large. However, there is an extension of the DE theory to polydisperse solutions of rods by Marrucci and Grizzuti (1983). According to this extended theory, a long rod in a matrix of shorter rods can reorient more quickly as a result of dissolution of the cage produced by motion of the short rods than it can by sliding many times out of a fixed cage. Chow et al. (1985b) found that the predictions of Marrucci and Grizzuti's generalization of the Doi–Edwards theory are in good agreement with steady-state and transient birefringence measurements on semidilute solutions of slightly flexible collagen molecules in start-up of steady shearing. The predicted scaling, $\tau \propto \nu^2$, of relaxation time with concentration was observed to hold; however, the parameter β required to fit theory to experiment was $\beta \approx 10^4$, larger by an order of magnitude than the theoretical value $\beta = 1350$. Presumably this discrepancy arises from the flexibility of collagen, whose persistence length, 170 nm, is somewhat less than the average molecular length $L \approx 300$ nm (Chow et al. 1985a; Nestler et al. 1983).

For semiflexible solutions of poly(γ-benzyl-L-glutamate) in the semidilute and concentrated isotropic concentration regimes, qualitative agreement is found with the Doi–Edwards or Marrucci–Grizzuti theories. Figure 6-20a, for example, shows the shear viscosity of PBLG solutions as a function of shear rate at various concentrations (Mead and Larson 1990). Notice that as the concentration enters the semidilute regime at a concentration of around 0.31% (by mass), the zero-shear viscosity begins to rise very rapidly with concentration, in qualitative agreement with the semidilute theory, which predicts $\eta_0 \propto \phi^3$ (see Fig. 6-20b). At concentrations of 5–10%, which is just below the transition to the nematic

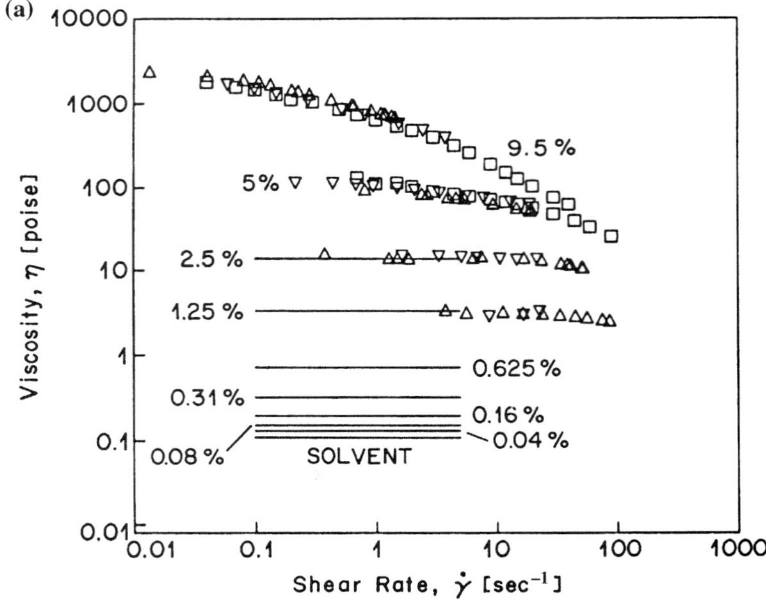

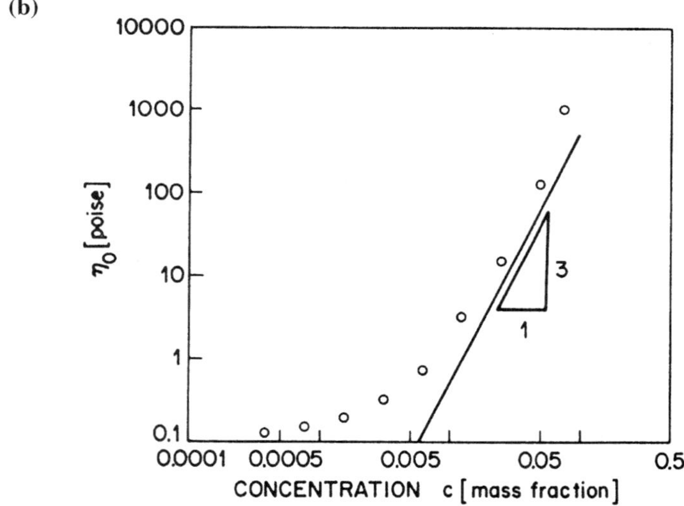

Figure 6.20 (a) Viscosity of solution of poly(γ-benzyl-L-glutamate) (PBLG; molecular weight 231,000), in m-cresol as a function of shear rate at 29°C for various mass percentages ranging from dilute ($\lesssim 0.31\%$) to semidilute (0.31–2.5%) to concentrated isotropic (5–9.5%). The different symbols and lines refer to data taken on different instruments. (b) Zero-shear viscosity as a function of concentration. (reprinted with permission from Mead and Larson, Macromolecules 23:2524. Copyright 1990 American Chemical Society.)

state, the viscosity rises more rapidly with concentration than predicted by the semidilute theory (Fig. 6-20b), evidently because excluded-volume effects, such as a so-called "rod-jamming" effect, become severe. In addition, at high concentration, pronounced shear thinning occurs at modest shear rates, in qualitative agreement with theory (Larson and Mead 1991). Similar phenomena were observed earlier in semidilute and concentrated isotropic solutions of poly(1,4-phenylene-2,6-benzobisthiazole) (PBZT) in methane sulfonic acid (Einaga et al. 1985).

Quantitative agreement between theory and the rheological behavior of PBLG solutions is not obtained, even when polydispersity is accounted for by the Marrucci–Grizzuti theory. In particular, Larson and Mead (1991) found that the Marrucci–Grizzuti theory fails to predict quantitatively the frequency dependence of G' and G'' for these PBLG solutions; the experimental spectrum of relaxation times is not as wide as predicted. The rheology of polydisperse solutions of rod-like polymers is therefore still not completely understood. Further discussion of the dynamics and rheology of semidilute solutions of Brownian rods can be found in Doi and Edwards (1986) or in Larson (1988). For a recent comprehensive light scattering study of isotropic solutions of rod-like PBG molecules, see Bu et al. (1994).

6.3.2.2 Non-Brownian Fiber Suspensions

As mentioned earlier, suspensions of particulate rods or *fibers* are almost always non-Brownian. Such fiber suspensions are important precursors to composite materials that use fiber inclusions as mechanical reinforcement agents or as modifiers of thermal, electrical, or dielectrical properties. A common example is that of glass-fiber-reinforced composites, in which the matrix is a thermoplastic or a thermosetting polymer (Darlington et al. 1977). Fiber suspensions are also important in the pulp and paper industry. These materials are often molded, cast, or coated in the liquid suspension state, and the flow properties of the suspension are therefore relevant to the final composite properties. Especially important is the distribution of fiber orientations, which controls transport properties in the composite. There have been many experimental and theoretical studies of the flow properties of fibrous suspensions, which have been reviewed by Ganani and Powell (1985) and by Zirnsak et al. (1994).

Fibers are composed of glass, carbon, nylon, or many other materials; they are typically millimeters long, with aspect ratios (L/d) ranging from 6 up to several hundred. Their flexibility is usually unspecified, and the effect of flexibility on flow properties has only recently begun to be studied (Yamamoto and Matsuoka 1995). A wide variety of suspending media are used; to suppress settling, either the medium must be very viscous, or its density must match that of the fibers. Fibers are often as large, or even larger than, rheometer gaps; hence the rheology and orientation of fibers might be affected by interactions with solid boundaries.

Since Brownian motion can be neglected, the stress tensor for a suspension of rigid fibers can be obtained from Eq. (6-36) simply by dropping the Brownian term:

$$\sigma = 2\eta_s \mathbf{D} + \nu \zeta_{str} \langle \mathbf{uuuu} \rangle : \mathbf{D} \tag{6-50}$$

In a dilute suspension of cylindrical fibers, the coefficient ζ_{str} is given by Eqs. (6-37) and (6-38). As discussed in Section 6.3.2.1, suspensions of high-aspect-ratio rods are not dilute unless their volume fraction is very low, $\phi \lesssim (L/d)^{-2}$. More typically, concentrations are in the semidilute range, or higher. (For non-Brownian rods, the crossover from dilute to semidilute concentrations occurs when the rods begin to influence each other hydrodynamically; this is not necessarily precisely the same concentration at which the Brownian motion of one rod is affected by its neighbors.)

For semidilute suspensions of long rods, Batchelor (1971) developed a theory of "hydrodynamic screening," from which he estimated

$$\zeta_{\text{str}} = \frac{\pi \eta_s L^3}{3 \ln(\pi/\phi)} \tag{6-51}$$

A semirigorous "multiple-scattering" theory of Shaqfeh and Fredrickson (1990) that accounts for multiparticle hydrodynamic interactions for slender bodies has verified Batchelor's theory and has given a slightly improved formula for ζ_{str}:

$$\zeta_{\text{str}} = \frac{\pi \eta_s L^3}{3[\ln(1/\phi) + \ln(\ln(1/\phi)) + A]} \tag{6-52}$$

where $A = -0.66$ for suspensions in which the rods are isotropically oriented and $A = 0.16$ when they are aligned in a single direction. These expressions differ only modestly from those for a dilute suspension. For example, for $\phi = 0.01, L/d = 100, \zeta_{\text{str}}$ from Eq. (6-52) differs from that of Eqs. (6-37) and (6-38) by only a few percent. Hence, the dramatic viscosity increase at the transition from dilute to semidilute behavior in *Brownian* suspensions is *not* matched by any similar effect in non-Brownian suspensions.

In fact, the fiber contribution to the shear viscosity of a fiber suspension at steady state is modest, at most. The reason is that, without Brownian motion, the fibers quickly rotate in a shear flow until they come to the flow direction; in this orientation they contribute little to the viscosity. Of course, the finite aspect ratio of a fiber causes it to occasionally "flip" through an angle of π in its Jeffery orbit, during which it dissipates energy and contributes more substantially to the viscosity. The contribution of these rotations to the shear viscosity is proportional to the ensemble- or time-averaged quantity $\langle u_x^2 u_y^2 \rangle$, where u_x is the component of fiber orientation in the flow direction and u_y is the component in the shear gradient direction. Figure 6-21 shows $\langle u_x^2 u_y^2 \rangle$ as a function of νL^3 for rods of aspect ratio L/d varying from 10 to 31.9, obtained from a computer simulation, from the Leal–Hinch calculations for spheroids, and from the experiments of Stover et al. (1992). Note that $\langle u_x^2 u_y^2 \rangle$ is small, around 0.03 for $L/d = 10$, decreasing to only 0.01 for $L/d = 31.9$. The full angular distribution of fiber orientations in a Couette shearing flow determined by Stover et al. (1992) is shown in Fig. 6-22. The distribution is sharply peaked and approximately (but not exactly—see below) symmetric about the flow direction.

Even very weak hydrodynamic interactions among fibers in the semidilute regime are enough to disturb the Jeffery orbits of individual fibers enough that a steady-state distribution of fiber orientations is obtained after long shearing times; without such interactions, no steady state would ever be reached, since the initial fiber orientation would just be revisited over and over with periodicity equal to that of the Jeffery orbit.

From Eqs. (6-50) and (6-52), in shearing flow, we obtain for the shear viscosity

$$\frac{\eta - \eta_s}{\eta_s} = \nu \zeta_{\text{str}} \langle u_x^2 u_y^2 \rangle$$

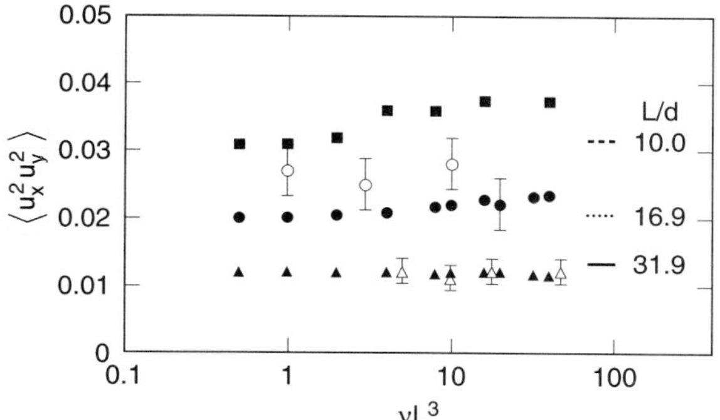

Figure 6.21 Steady-state values of $\langle u_x^2 u_y^2 \rangle$ for non-Brownian rod-like particles as a function of reduced concentration νL^3, obtained by simulation for $p \equiv L/d = 10$(■), 16.9 (●), and 31.9 (▲) and from the experimental data of Stover et al. (1992) for $L/d = 16.9$ (○) and 31.9 (△). Dotted lines are the theoretical predictions of Leal and Hinch (1971) for spheroids with equivalent rod aspect ratio $L/d = p/0.7 = 10$, 16.9, and 31.9. (Reprinted from J Non-Newt Fluid Mech 54:405, Yamane et al. (1994), with kind permission from Elsevier Science - NL, Sara Burgerhartstraat 25, 1055 KV Amsterdam, The Netherlands.)

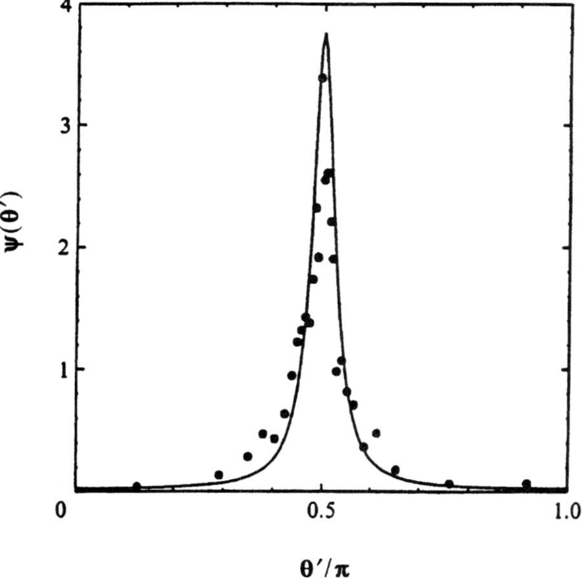

Figure 6.22 Orientation distribution function $\psi(\theta')$ in a shearing flow of fibers of cellulose acetate propionate of diameter $d = 95\ \mu$m, $L/d = 16.9$, at a dimensionless concentration νL^3 of 20 in an index-matched mixture of glycerine and polyethylene glycol. The flow direction corresponds to $\theta' = 0.5\pi$ radians. The line is the prediction of the Jeffery theory, Eq. (6-27), with $p_{\mathrm{eff}} = 0.7(L/d)$. (From Stover et al. 1992, with permission from Cambridge University Press.)

$$= \frac{\pi \nu L^3 \langle u_x^2 u_y^2 \rangle}{3[\ln(1/\phi) + \ln(\ln(1/\phi)) + A]} = \frac{4\phi p^2 \langle u_x^2 u_y^2 \rangle}{3[\ln(1/\phi) + \ln(\ln(1/\phi)) + A]} \tag{6-53}$$

For $L/d = p \approx 30$, both ϕ and $\langle u_x^2 u_y^2 \rangle$ are of order 0.01, and Eq. (6-53) predicts that the fibers will make only a small or modest contribution to the shear viscosity. This remains true for semidilute suspensions even if the aspect ratio $p = L/d$ is made large, since from Fig. 6-21, $\langle u_x^2 u_y^2 \rangle$ decreases roughly as $1/p$ with increasing p, and the maximum value of ϕ at which the suspension is semidilute also decreases as $1/p$. The predicted modest contribution of the fibers to the viscosity is confirmed in Fig. 6-23, which shows the predictions of $\eta_r \equiv \eta/\eta_s$ from the simulations of Yamane et al. (1994) and compares them to the viscosity measurements of Bibbo (1987). The experiments show values of η_r no higher than 3, while the maximum value of η_r in the simulations is only around 1.5, even when $\nu L^3 = 40$. The modest particle contributions to the viscosity of semidilute solutions of non-Brownian rods are in stark contrast to the enormous viscosity increases for Brownian rods (compare Figs. 6-23 and 6-20b).

Since at steady state the angular distribution of fiber orientations is predicted to be symmetric about the flow direction in a shearing flow, Eq. (6-50) implies that the normal stresses (e.g., $\sigma_{11} \propto \langle u_x^3 u_y \rangle$) will be identically zero. However, nonzero positive values of N_1 have frequently been reported for fiber suspensions (Zirnsak et al. 1994). Figure 6-24 shows N_1, normalized as discussed below, as a function of shear rate for various suspensions of high fiber aspect ratio. These normal stress differences are *linear* in the shear rate and can be quite large, as high as 0.4 times the shear stress, which is dominated by the contribution of the solvent medium, $\sigma \approx \eta_s \dot\gamma$. In Fig. 6-24, the N_1 data are normalized to test a theory of Carter (1967), who proposed that fiber–fiber collisions occur, yielding a prediction $N_1 \propto \eta_s \dot\gamma \phi p^{3/2}/(\ln(2p) - 1.8)$.

One possible explanation for nonzero normal stress differences is that the orientation distribution function is not completely symmetric about the flow direction. Indeed, Stover et al. (1992) found that for an aspect ratio of $p = 31.9$, there appears to be a small, but systematic, offset in the average orientation angle of around $-1°$ to $-2°$, where the negative sign implies that the offset is counterclockwise from the flow direction. Stover et al. state that hydrodynamic interactions between fibers would not produce such an asymmetry, although presumably fiber–fiber collisions could. Folgar and Tucker (1984) have proposed a phenomenological theory in which fiber–fiber interactions are accounted for by an *effective* rotary diffusive process which mimics in an average sense the role of fiber–fiber interactions. Thus, in their treatment, the rotary diffusivity $\hat{D}_r$ in the Smoluchowski equation (6-47) is not set to zero, but is made linear in the shear rate,

$$\hat{D}_r = D_r = C\dot\gamma \tag{6-54}$$

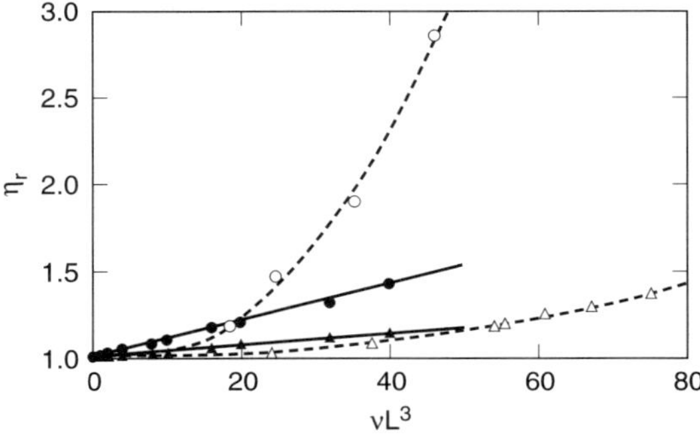

Figure 6.23 The relative viscosity at steady state of suspensions of non-Brownian rod-like particles versus dimensionless concentration νL^3. Simulations for $p \equiv L/d = 16.9$ (●), 31.9 (▲); Bibbo's (1987) experimental results for $L/d = 16.9$ (○), and 31.9 (△). (Reprinted from J Non-Newt Fluid Mech 54:405, Yamane et al. (1994), with kind permission from Elsevier Science - NL, Sara Burgerhartstraat 25, 1055 KV Amsterdam, The Netherlands.)

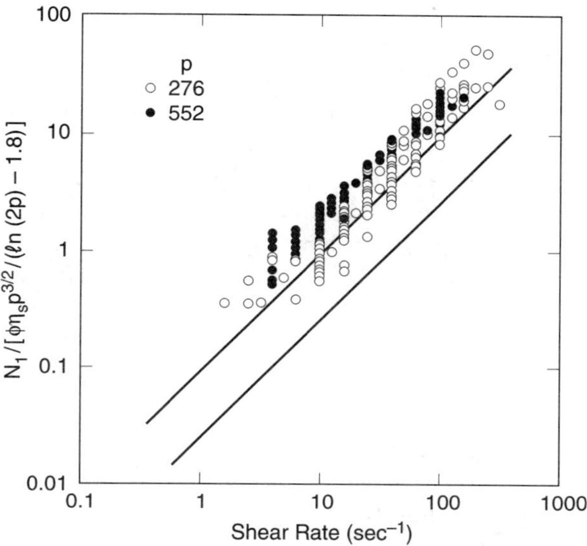

Figure 6.24   The normalized first normal stress difference $N_1/[\phi\eta_s p^3/(\ln(2p) - 1.8)]$ versus shear rate for glass fibers of aspect ratios $p = 276$ and 552 in Newtonian solvents, namely, glucose wheat syrup and epoxy resin. For $p = 276$, the volume fraction ϕ is in the range 0.0002–0.0011; and for $p = 552$, $\phi = 0.00045$–0.00095. The solid lines are the upper and lower boundaries of data from Carter (1967), Kitano and Kataoka (1981), and Goto et al. (1986). (Reprinted from J Non-Newt Fluid Mech 54:153, Zirnsak et al. (1994), with kind permission from Elsevier Science - NL, Sara Burgerhartstraat 25, 1055 KV Amsterdam, The Netherlands.)

The value C is called the "Folgar–Tucker constant." This effective diffusivity causes back-diffusion of rods against the vorticity in a shear flow and leads to an asymmetry about the flow direction in the fiber orientation distribution. Folgar and Tucker obtained values of $C \approx 0.003$–0.016 by fitting the orientation distribution functions measured in fiber suspensions with fiber aspect ratios of 16 and 83 to the predictions of (6-47) and (6-54) for infinitely thin rods. For infinitely thin rods, the orientation distribution would be infinitely sharp in the absence of diffusion; hence Folgar and Tucker attribute all of the dispersion in the measured distribution function to the "effective diffusivity" term. Folgar and Tucker's theoretical orientational distributions are, however, significantly more asymmetric than the data, suggesting that much of the spread in the orientation distribution in the experiments is due to factors other than "effective diffusivity," factors such as the finite aspect ratio of the particles. Indeed Yamane et al. (1994) have extracted values of C from simulations of fiber suspensions under shearing flow, and for $p = 10$ and 16.9 they have obtained values of C in the range $\approx 10^{-8}$–10^{-4}, much smaller than that assumed by Folgar and Tucker.

Rather rigorous analyses of hydrodynamic interactions among cylindrical among fibers using slender-body theory (Koch 1995) leads to the prediction that the effective diffusivity should be *anisotropic* and dependent on flow type. The diffusivity in Eq. (6-47) should then be replaced by a tensor dotted into $\partial\psi/\partial\mathbf{u}$. A reasonably accurate estimate of this diffusion tensor is

$$\hat{\mathbf{D}}_r \approx \frac{\nu L^3}{\dot{\gamma}\ln^2 p}\left(\lambda_1 \delta\mathbf{D} \,:\, \langle \mathbf{uuuu}\rangle \,:\, \mathbf{D} + \lambda_2 \mathbf{D} \,:\, \langle \mathbf{uuuuuu}\rangle \,:\, \mathbf{D}\right)$$

where $\lambda_1 = 3.16 \times 10^{-3}$ and $\lambda_2 = 1.13 \times 10^{-1}$. For shear with average rod orientation in the flow direction, the Folgar–Tucker constant that one obtains from this more rigorous theory is around $C \approx 6 \times 10^{-3}\nu L^3/p\ln^2 p$. For $\nu L^3 = 10$, $p = 20$, this gives $C \approx 3.3 \times 10^{-4}$, a decade or more higher than predicted by the simulations of Yamane et al.

From Table 6-1 and Eq. (6-54), we find that the first normal stress difference for long, slender dilute spheroids in the limit of high Peclet number is predicted to be

$$N_1 = \frac{p^4}{4\ln(p)}D_r\phi\eta_s = \frac{p^4}{4\ln(p)}C\dot{\gamma}\phi\eta_s \tag{6-55}$$

which can be rewritten as

$$\frac{N_1}{\phi\eta_s p^{3/2}\ln(p)} = \frac{p^{5/2}C}{4}\dot{\gamma} \tag{6-56}$$

[For semidilute suspensions, one expects π/ϕ to replace p in the logarithmic term in Eq. (6-55), but otherwise the expression for N_1 should be similar to that for dilute suspensions.] Hence, a plot of $N_1/(\phi\eta_s p^{3/2}\ln(p))$ versus $\dot{\gamma}$ will be "universal" only if C scales as $p^{-5/2}$ and is independent of ϕ; this is consistent with neither slender-body theory nor the simulations of Yamane et al. Hence, the "effective diffusivity" of rigid rods does not seem able to account for the behavior of the measured values of N_1 in fiber suspensions. However, other possible sources may contribute to the first normal stress difference in these suspensions. For example, according to recent simulations, fiber flexibility produces a positive first normal stress difference (Yamamoto and Matsuoka 1995). Other possible sources of nonzero N_1 include interactions of long fibers with rheometer walls, or streamline curvature.

In transient shear flows starting from an *isotropic* distribution of fiber orientations, considerably higher viscosities will be initially observed, until the fibers become oriented. In Bibbo's experiments, η_r for isotropically oriented fibers is around 3.5 for $\nu L^3 = 75$. These viscosities can also be predicted reasonably well by semidilute theory and by simulations (Mackaplow and Shaqfeh 1996). Figure 6-25 shows the shear stress as a function of strain for a polyamide 6 melt with 30% by weight glass fibers of various aspect ratios, where the fibers were initially oriented in the flow-gradient direction. Notice the occurrence of a stress overshoot (presumably due to polymer viscoelasticity), followed by a decrease in viscosity, as the fibers are reoriented into the flow direction.

While the fiber contribution to the steady-state stress tensor at steady-state is modest for shearing flow, its contribution to the stress in *extensional flow* is large at steady state. In a uniaxial extensional flow, the fibers orient in such a way that the viscous dissipation is maximized. Large values of the extensional viscosity are the result; from Batchelor's (1971) theory the uniaxial extensional viscosity is

$$\bar{\eta} = 3\eta_s + \nu\zeta_{\text{str}} = 3\eta_s\left[1 + \frac{4\phi p^2}{9\ln(\pi/\phi)}\right] \tag{6-57}$$

Figure 6-26 shows that large stresses are produced in extensional flow of fiber suspensions; $\bar{\eta}/\eta_s$ can be as large as 75 (Mewis and Metzner 1974), in agreement with the theory.

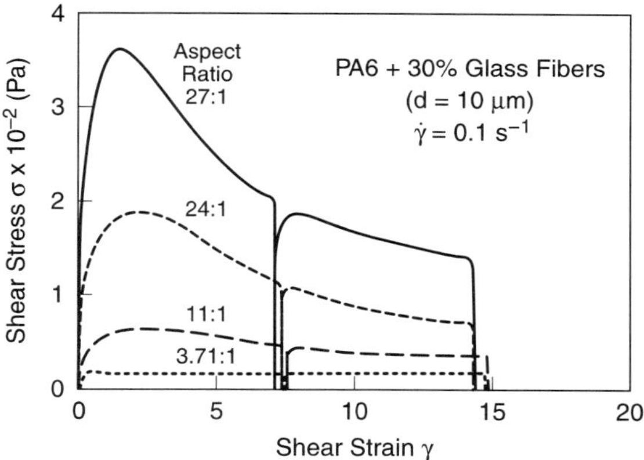

Figure 6.25   Shear stress as a function of shear strain for a polyamide 6 filled with 30% by weight glass fibers of different aspect ratios, where the fibers were initially oriented in the velocity-gradient direction. The shear was interrupted after about 7 strain units, and the stress was allowed to relax. The shear was then restarted; the shear stress quickly returned to the value it had before cessation of shearing, showing that fiber orientation only changes during shearing. (From Laun 1984b, reprinted with permission from Steinkopff Publishers.)

- Problem 6.6 and Worked Examples 6.7 and 6.8 show by practical example how to estimate the rheological properties of suspensions of elongated particles, both in the dilute and semidilute concentration regimes.

6.4 ELECTRICALLY CHARGED PARTICLES

The "hard-sphere" or "soft-sphere" particles discussed in Section 6.2 are typically coated with an organic layer that provides a steric barrier to prevent flocculation. The other common method of stabilizing particle suspensions is through electrostatic charging of the particle surfaces. The charges lead to long-range repulsions that can keep the particles far enough apart that they are not drawn together by short-range van der Waals forces. Because electrostatic interactions are long-ranged, strong electrostatic effects are possible even at minute particle volume fractions, as low as $\phi \approx 0.001$. Much higher concentrations than this are used in the preparation of suspensions used as ceramic precursors, where surface charging helps prevent particle clumping, which can lead to inhomogeneities and defects in the product (Ulrich 1990).

Two spherical particles whose centers are separated by a distance r repel each other with an interaction potential $W(r)$ of [see Section 2.4.4; Eq. (2-59) with $D = r - 2a$]:

$$W(r) = 2\pi \varepsilon \varepsilon_0 a \psi_s^2 \ln[1 + \exp(-\kappa(r - 2a))] \tag{6-58}$$

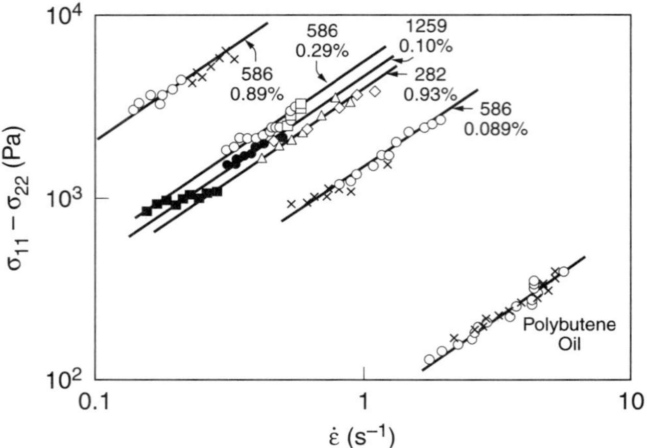

Figure 6.26 Extensional stress $\sigma_{11} - \sigma_{22}$ versus extension rate for glass fibers of aspect ratios 282, 586, and 1259 suspended in Newtonian polybutene oil (viscosity 283 P) at the volume percents shown. The straight lines have slopes of unity; the data are in excellent agreement with the equation of Batchelor (1971), Eq. (6-57). [From Macosko 1994 (adapted from Mewis and Metzner 1974), with permission from Cambridge University Press.]

where ψ_s is the *surface potential* of the particles, $\varepsilon_0 = 8.8 \times 10^{-12} \mathrm{C}^2\mathrm{J}^{-1}\mathrm{m}^{-1}$ is the permittivity of space, ε is the dielectric constant, a is the particle radius, and κ^{-1} is the Debye screening length. Equation (6-58) is valid for constant surface potential when $\kappa D \lesssim 2$ (see Fig. 4-14 of Russel et al. 1989). For widely separated particles, so that $\kappa D \gtrsim 2$, Eq. (6-58) should be replaced by Eq. (2-58), which can be written as

$$\frac{W(r)}{k_B T} = \frac{\alpha \exp(-\kappa r)}{\kappa r} \tag{6-59}$$

where

$$\alpha \equiv \frac{4\pi \varepsilon \varepsilon_0 \psi_s^2 a^2 \kappa \exp(2a\kappa)}{k_B T} \tag{6-60}$$

For charged particles, the net charge contained by the solvent must balance that carried by the particles. Suppose, in addition to these counterions, there is a concentration n_b of a symmetric electrolyte, where n_b is the number of ion pairs per unit volume *of suspension*; the concentration of ion pairs in the liquid phase is $n_1 = n_b/(1 - \phi)$. The inverse square Debye length from both contributions then works out to be (Russel et al. 1989)

$$\kappa^2 = \frac{e^2}{\varepsilon \varepsilon_0 k_B T} \frac{2z^2 n_b + \dfrac{3|\sigma|z\phi}{ae}}{1 - \phi} \tag{6-61}$$

where $|\sigma|$ is the absolute value of the surface charge per unit area of particle surface (note that elsewhere in this chapter "σ" is the shear stress), z is the charge valence of the electrolyte (e.g., $z = 1$ for NaCl), and $e = 1.6 \times 10^{-19}$ C is the charge of an electron.

Since they keep particles apart, *the surface charges increase the effective particle diameter*. The effective particle diameter d_{eff} can be estimated from the condition

$$W(r = d_{\text{eff}}) \approx k_B T \tag{6-62}$$

This condition, when applied to Eq. (6-59) for the low-salt case, gives

$$d_{\text{eff}} \approx \frac{1}{\kappa} \ln\{\alpha/\ln[\alpha/\ln(\alpha/\ldots)]\} \tag{6-63}$$

where the infinite concatenations of logarithms can in practice be truncated after the third one.

The effective volume fraction of particles is therefore $\phi_{\text{eff}} = \phi(d_{\text{eff}}/2a)^3$. Since a hard-sphere suspension forms a macrocrystalline lattice at a volume fraction of around $\phi \approx 0.55$, charged spheres will do so when

$$\phi_{\text{eff}} \equiv \phi\left(\frac{d_{\text{eff}}}{2a}\right)^3 \tag{6-64}$$

is greater than around 0.55. At low ionic strength, $d_{\text{eff}} \gg 2a$; thus charged colloids can form iridescent, macrocrystalline phases at very low particle volume fractions. Figure 6-27 shows the phase diagram for some charged polystyrene latices, compared to the predicted phase diagram using the theory described above. At very low ionic strength, the ordered phases that form at $\phi \leq 0.01$ are body-centered cubic (BCC), while face-centered cubic (FCC) phases form at higher ϕ. The transition from BCC to FCC is expected, since FCC phases are produced by "hard" (or short-ranged) interparticle potentials, while BCC phases are produced by "soft" or (long-ranged) ones.

6.4.1 Viscosity

The effective enlargement of the diameter of charged spheres leads to enhancement of the low-shear-rate viscosity. According to Russel (1978; Russel et al. 1989), the zero-shear viscosity out to second order in ϕ of a disordered suspension of charged spheres is

$$\frac{\eta}{\eta_s} = 1 + 2.5\phi + \left[2.5 + \frac{3}{40}\left(\frac{d_{\text{eff}}}{a}\right)^5\right]\phi^2 \tag{6-65}$$

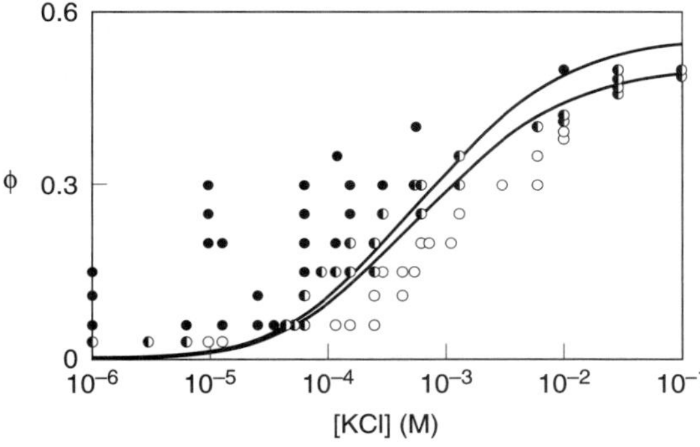

Figure 6.27 Phase diagram for order–disorder transition for charged spheres in an electrolyte solution of KCl. Data of Hachisu, Kobayashi, and Kose (1973) for polystyrene latices with particle radius $a = 85$ nm; samples are disordered (○), two-phase (◑), or ordered (●). The lines are the predictions of the phase boundaries from a perturbation theory for $a = 100$ nm and $4\pi a^2 \sigma = 5000e$. (From Russel et al. 1989, with permission from Cambridge University Press.)

where d_{eff} is given by Eq. (6-63). This prediction for the ϕ^2 term is compared to experimental results for small charged spheres in Fig. 6-28. Note the dramatic enhancement of this second-order contribution to the viscosity due to surface charges.

For effective volume fractions ϕ_{eff} above 0.10 or so, one would not expect Eq. (6-65) to be accurate. However, by extending the Krieger–Dougherty equation, one predicts that

$$\frac{\eta}{\eta_s} = \left(1 - \frac{\phi_{\text{eff}}}{\phi_m}\right)^{-(5/2)\phi_m} \tag{6-66}$$

Figure 6-29 shows that this simple model gives a good fit to the concentration-dependence of the shear viscosity data for polystyrene latices in 5×10^{-4} M NaCl.

As the shear rate increases, the hydrodynamic contributions to the viscosity increase, and the electrostatic ones do not keep pace. Thus, the electrostatic contribution to the viscosity shear thins away until the viscosity is close to that expected for uncharged particles (see Fig. 6-30). This can be accounted for by letting d_{eff} depend on the shear rate such that d_{eff} is the interparticle separation at which hydrodynamic and electrostatic forces balance (Blatchford et al. 1969; Russel et al. 1989). Thus as the shear rate goes up, the particles approach each other more closely and d_{eff} goes down. Specifically, one includes the viscous energy $\eta \dot{\gamma} d_{\text{eff}}^3/8$ along with the interparticle and Brownian energies in the balance equation (6-62):

$$W(d_{\text{eff}}) - k_B T - \frac{\sigma d_{\text{eff}}^3}{8K} = 0 \tag{6-67a}$$

where the shear stress is $\sigma = \eta \dot{\gamma}$, and the viscosity η is given by the modified Krieger–Dougherty equation (6-66) (Buscall 1991). K is an adjustable constant taken to be 0.10. This equation is adequate for estimating d_{eff}, except at low stresses, where a more refined equation should be used:

$$d_{\text{eff}} = d + \int_d^\infty \left[1 - \exp\left(\frac{W(r)}{k_B T + \sigma(r/2)^3/K}\right)\right] dr \tag{6-67b}$$

Figure 6-30 shows that this simple approach can successfully fit shear stress data for

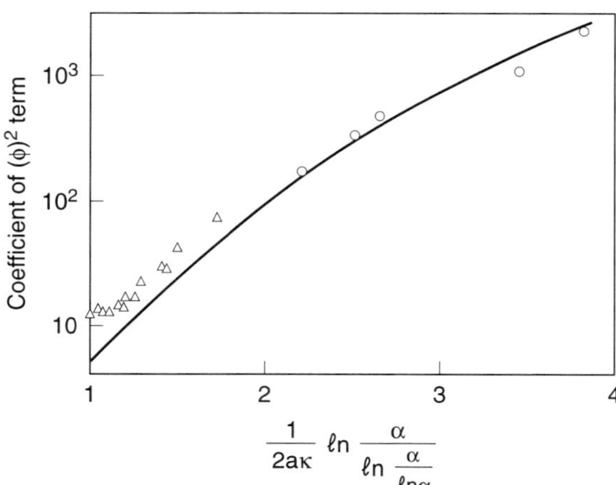

Figure 6.28   The order ϕ^2 coefficient in Eq. (6-65) for the low-shear-rate viscosity of a suspension of charged spheres of polystyrene ($\bigcirc$) with $a = 25$–45 nm (Stone-Masui and Watillon 1968), and of bovine serum albumin ($\triangle$) with $1 = 3.6$ nm (Tanford and Buzzell 1956). The line is the prediction of Eqs. (6-65) and (6-63). (From Russel et al. 1989, with permission from Cambridge University Press.)

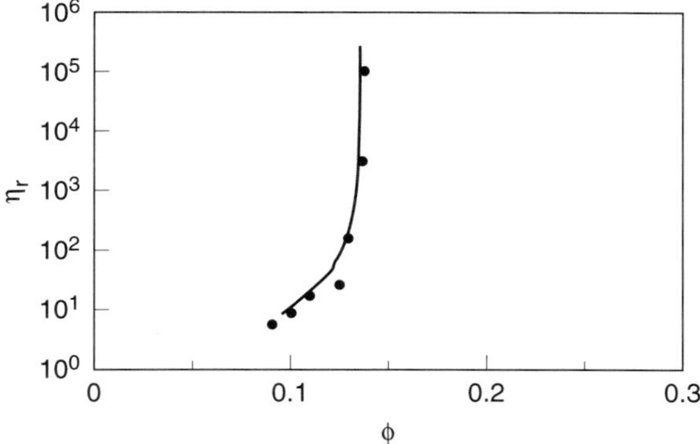

Figure 6.29 Zero-shear relative viscosity versus particle volume fraction for aqueous suspensions of charged polystyrene spheres ($a = 34$ nm) in 5×10^{-4} M NaCl ($\bullet$) (Buscall et al. 1982a). The line is calculated by using Eq. (6-66) for the viscosity, with ϕ_{eff} given by Eq. (6-64), and d_{eff} by Eq. (6-67a) or (6-67b). The potential $W(R)$ is given by Eq. (6-58) or (6-59) with κ given by Eq. (6-61); the constant K is 0.10, and $|\psi_s|$ is in the range 50–90 mV. (From Buscall 1991, reproduced with permission of the Royal Society of Chemistry.)

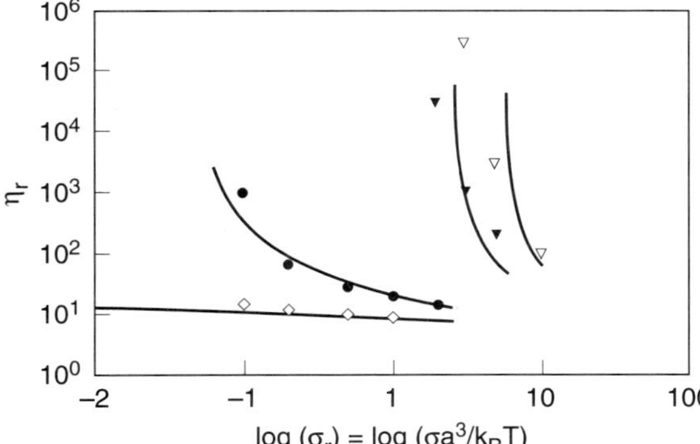

Figure 6.30 Relative viscosity versus reduced shear stress for aqueous suspensions of charged polystyrene spheres ($a = 110$ nm) at a concentration of $\phi = 0.40$ at HCl concentrations of 0 (∇), 1.88×10^{-4} ($\blacktriangledown$), 1.88×10^{-3} ($\bullet$), and 0.0188 ($\diamond$) (from Krieger and Eguiluz 1976). The solid lines are calculated using the same equations as in Fig. 6-29. (From Buscall 1991, reproduced with permission of the Royal Society of Chemistry.)

polystyrene latices in electrolytes of various ionic strengths. An alternative approach for calculating viscosities of charged latices, one that uses the Eyring transition rate theory, has been proposed by Ogawa et al. (1997).

6.4.2 Yield Stress and Modulus

Note in Fig. 6-30 that as the HCl concentration decreases to 1.88×10^{-3} M, the viscosity appears to diverge below a dimensionless stress of around 0.1, which is the dimensionless *yield stress*. The yield stress appears when the particle repulsions are strong enough to induce macrocrystallization. As the ionic strength decreases further, the repulsions between spheres become stronger, and the yield stress becomes larger. The magnitude of the yield stress σ_y

can be estimated from Eq. (6-67a) by noting that at $\sigma = \sigma_y$ the effective particle diameter is just high enough for effective "interparticle contact"; that is, the effective volume fraction ϕ_{eff} equals ϕ_m, the volume fraction at which the effective particles touch. This occurs when d_{eff} is equal to r_m, the average interparticle separation, that is,

$$r_m = 2a \left(\frac{\phi_m}{\phi} \right)^{1/3} \tag{6-68}$$

where ϕ_m is the volume fraction at which particles touch. Thus, setting the shear stress σ in Eq. (6-67a) equal to the yield stress σ_y, we obtain

$$\sigma_y \approx K \left(\frac{W(r_m) - k_B T}{(r_m/2)^3} \right) \tag{6-69}$$

In the ordered state, the suspension can have a substantial modulus (see Fig. 6-31). Buscall et al. (1982b, 1991) have derived an expression for the high-frequency modulus G'_∞ from the interparticle potential by calculating the work required for particles to move affinely with the flow. The result of Buscall et al. follows from the Zwanzig–Mountain expression, Eq. (6-20), where $g(r')$ can be replaced by a delta function since the suspension is crystalline. Neglecting the osmotic term, the result is (Evans and Lips 1990; Buscall 1991)

$$G'_\infty = \frac{\phi_m N}{5\pi r_m} \left\{ \frac{4}{r_m} \left(\frac{dW}{dr} \right)_{r=r_m} + \left(\frac{d^2 W}{dr^2} \right)_{r=r_m} \right\} \approx \frac{\phi_m N}{5\pi r_m} \kappa^2 W(r_m) \tag{6-70}$$

where N is the number of nearest neighbors per particle. For an FCC array, $\phi_m = 0.74$ and $N = 12$; while for a BCC array, $\phi_m = 0.68$ and $N = 8$. The expression on the far right

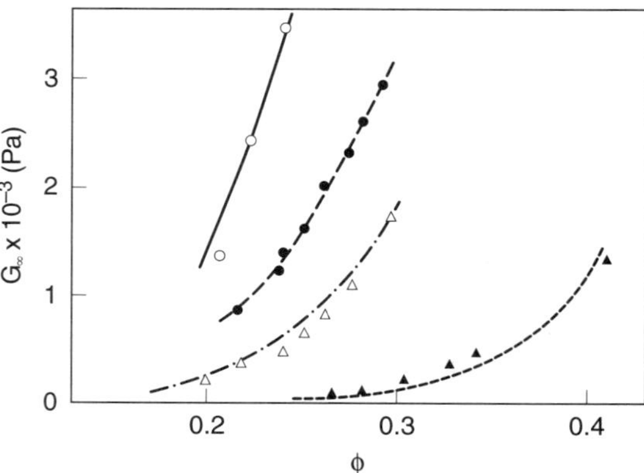

Figure 6.31   High-frequency modulus G_∞ versus particle volume fraction ϕ for charged polystyrene latices of radius $a = 26.3$ nm ($\bigcirc$), 34.3 nm ($\bullet$), 39.2 nm ($\triangle$), and 98.3 nm ($\blacktriangle$) in aqueous solutions with 5×10^{-4} M NaCl. The lines are the predictions of the theory of Buscall et al. (1982b) for an FCC lattice with $N = 12$ and $\phi_m = 0.74$, and a best-fit value of $|\psi_s|$ increasing from 50 to 89 mV as the particle radius increases from 26.3 to 98.3 nm. These predictions differ from Eq. (6-70) only in that a prefactor (3/32) in Buscall et al. (1982b) was corrected to $1/(5\pi)$ in Buscall (1991). (From Buscall et al. 1982b, reproduced with permission of the Royal Society of Chemistry.)

of Eq. (6-70), which neglects the first-derivative term and assumes an exponential form for the potential, is due to Buscall et al. (1991). Equation (6-70) also neglects electrostatic screening produced by the counterions for the surface charges; these become important at low electrolyte strength and high particle concentrations. For inclusion of these effects, see Wagner (1993) and van der Vorst et al. (1995). $W(r)$ can be obtained from Eq. (6-58) or Eq. (6-59). Buscall et al. (1982b) found that for polystyrene latices in 5×10^{-4} mol/dm^3 NaCl, the variation of modulus with particle volume fraction accords well with Eq. (6-70). Figure 6-31 shows that for particle sizes ranging from 26.3 to 98.3 nm, good agreement between predicted and measured high-frequency modulus (measured by shear-wave propagation at 185–225 Hz) could be obtained by assuming that the surface potential depends somewhat on particle size. Equation (6-70) applies to both ordered and disordered samples, since at high frequencies only local structure matters and no time is allowed for slow relaxations that occur in disordered samples.

From Eqs. (6-69) and (6-70), the ratio of the yield stress to the modulus works out to be (Buscall 1991)

$$\frac{\sigma_y}{G'_\infty} = \frac{40\pi K}{\phi_m N (\kappa r_m)^2} \left(1 - \frac{k_B T}{W(r_m)} \right) \tag{6-71}$$

From this, σ_y / G_∞ is typically ≈ 0.01, in rough agreement with experiments (Buscall et al. 1982a; Chen and Zukoski 1990; Fagan and Zukoski 1997). However, Eq. (6-71) predicts a dependence of σ_y / G_∞ on ionic strength which seems not to be observed (Buscall 1991). For ordered colloidal suspensions with $\phi = 0.4$–0.6, the ratio σ_y / G_0 of the yield stress to the zero-frequency modulus has been reported to be 0.035 ± 0.005 (Chen et al. 1994), in excellent agreement with the predictions from computer simulations (Stevens et al. 1991).

Theoretical expressions similar to those presented in this section have recently been successfully used to predict the modulus and viscosity of suspensions of charged rod-like particles (Solomon and Boger 1998).

6.4.3 Flow Mechanisms

When a colloidal crystal is forced to flow by the imposition of a steady shearing rate or of a constant stress in excess of the yield stress, the three-dimensionally ordered macrolattice restructures to accommodate continuous deformation. In principle, deformation could concentrate in fracture or slip layers, with the rest of the lattice retaining its static crystalline structure. However, based on light- and neutron-scattering experiments, Ackerson and coworkers (Ackerson and Clark 1984; Ackerson et al. 1986) have suggested that colloidal crystals find more complex ways of accommodating shearing deformations. While the details depend on the symmetry of the colloidal crystal (recall that at low particle concentrations, $\phi \lesssim 0.01$, a BCC phase is formed, while at higher ϕ the structure is typically FCC), for either symmetry, there are similarities in the behavior under shear. At low shear rates, the macrocrystal orients so that the direction of closest packing of the spheres is in the flow direction ($\mathbf{v}$), while the planes containing the densest packings are parallel to the shearing surfaces. In this orientation, the spacing between spheres in the gradient direction is maximized, and spheres can then most readily move over one another. For BCC packing, this means that the 111 crystal direction is parallel to $\mathbf{v}$, while 110 planes are perpendicular to the gradient direction (∇). For FCC symmetry, the closest packing direction is 110, which is therefore parallel to **v**, while the slipping planes parallel to the rheometer walls are 111 planes. In this orientation, particles in the FCC slipping planes are in a two-dimensional *hexagonal* arrangement, which can be stacked in two different ways, called *twins*. Hopping between these twinned structures apparently occurs during the slipping layer motion. The self-selection of slipping planes, slip direction, and twinned structures is apparently similar to that observed in block copolymers with ordered spherical domains (Koppi et al. 1994) and in sheared metals (Hirth and Lothe 1982), both of which show BCC symmetry. For a more detailed comparison between the flow of BCC colloidal crystals and that of BCC-ordered diblock copolymer domains [see Koppi et al. (1994)].

In addition to the slipping of layers over one another in the flow direction, a zigzag motion of one slipping plane with respect to its neighbor is inferred from the scattering data (see Fig. 6-32). This zigzag motion carries each sphere through a path of least resistance in the repulsive potential field created by the surrounding spheres, but distorts the three dimensional symmetry of the crystal, creating a "strained-crystal" microstructure (Chen et al. 1994). The zigzag motions have also been seen in molecular dynamics simulations of sheared colloidal crystals (Stevens et al. 1991). As the shear rate increases, the zigzag paths straighten out to reduce the particle path length, and the lateral offset (in the $\mathbf{v} \times \nabla$ direction) of one layer relative to the next shifts so that the straightened particle paths pass halfway between the rows of particles in the adjacent layers (see Fig. 6-32c). Thus, the lattice distorts to allow easy *sliding* of layers over one another with a minimum of interparticle interference. In the "sliding-layer configuration," the spheres can pack more closely within sliding planes to allow more space between these planes (Tomita and van de Ven 1984). Further increases in the shear rate break down the layered structure, leaving only a tendency for particles to be aligned in "strings" along the flow direction (Ackerson and Clark 1984).

The above features of a sheared colloidal crystal appear to be similar in both BCC and FCC structures. However, there are differences in details, and perhaps even within a given symmetry the flow behavior might vary with particle concentration or charge density. For example, Chen et al. (1994) have shown that between the "strained crystal" and "sliding-layer" microstructures there can be a polycrystalline structure, the formation of which produces a discontinuous drop in shear stress (see Fig. 6-33). Ackerson and coworkers gave a detailed description of the fascinating shear-induced microstructures of these systems (Ackerson and Clark 1984; Ackerson et al. 1986; Chen et al. 1992, 1994).

Despite significant differences in detail from one system to the next, Fagan and Zukoski (1997) found that a crude steady-state "master curve" can be obtained by plotting σ/G versus $\dot{\gamma}\eta_s/G$, where the modulus G is frequency-insensitive for concentrated charged-sphere suspensions. Such a "master curve" may be a useful engineering tool in applications of these suspensions.

6.4.4 Shear Thickening and Normal Stress Differences

For very low particle concentrations, the shearing of colloidal crystals produces only a weak stress above that of the solvent. However, in more concentrated suspensions, the shear viscosity and normal stress differences have been found to have quite unusual behavior, which can, in part, be explained by (a) the formation of sliding layers and (b)

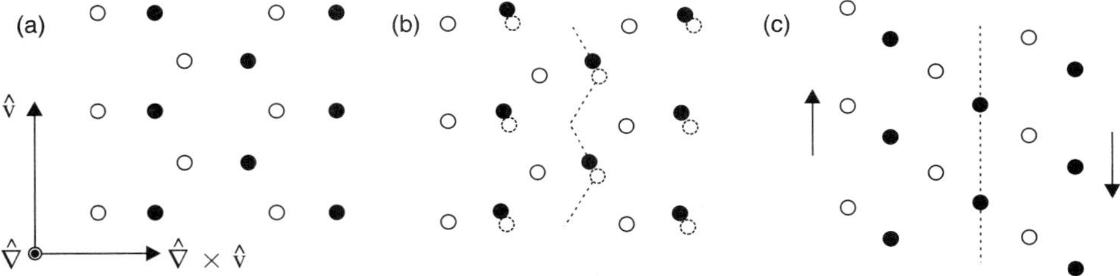

Figure 6.32 **(a)** An FCC structure oriented with 110 direction parallel to flow ($\hat{v}$ direction) and 111 planes parallel to the shearing surfaces. The open circles represent one 111 layer of spheres, and the closed circles represent an adjacent layer. **(b)** For low rates of shear, the layer represented by closed circles slips over the layer represented by open circles, with the closed circles moving in a zigzag fashion from one FCC registration site to another in the layer plane. The dashed circles show the equilibrium positions of the closed circles before deformation. **(c)** For higher rates of shear, the particles move on straight lines centered between the rows created by the particles in the adjacent layer; this is called the "sliding-layer configuration," because each layer moves rigidly in the flow direction with no in-plane particle motion. Notice, however, that the positions of the particles within the layer are distorted (sheared) away from their equilibrium positions, as indicated by the arrows. (From Ackerson et al., reprinted with permission from J. Chem. Phys. 84:2344, Copyright 1986 American Institute of Physics.)

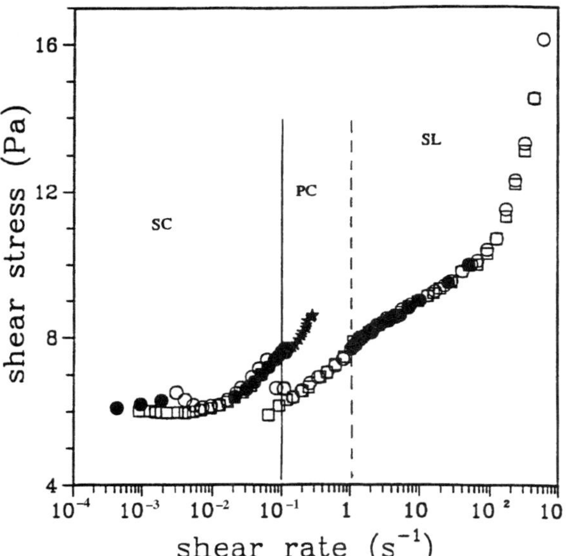

Figure 6.33 Shear stress as a function of shear rate for $a = 73$ nm charged polystyrene spheres at a volume fraction of $\phi = 0.33$ in 10^{-3} M KCl. The various symbols stand for measurements made under different conditions, namely, increasing shear rate ($\bigcirc$), decreasing shear rate ($\square$), constant stress ($\bullet$), or metastable shear rates ($*$). SC denotes the "strained-crystal" configuration shown in Fig. 6-32b, SL is the "sliding-layer" configuration in Fig. 6-32c, and PC is an intermediate "polycrystalline configuration." (From Chen et al. 1994, with permission from the Journal of Rheology.)

the breakdown of these layers as the shear rate increases. Figure 6-34 shows the shear viscosity versus shear rate for charged styrene-ethylacrylate copolymer spheres ($d = 165$ nm) in glycol at concentrations of 0.355 (A3G), 0.434 (A4G), and 0.523 (A5G). At the lowest particle concentration, $\phi = 0.335$, only shear-thinning behavior is observed;

for $\phi = 0.434$, however, shear thinning gives way at high shear rate to a steep, but continuous, *shear-thickening* phenomenon at a critical rate of shear. For $\phi = 0.523$, the shear thickening becomes discontinuous in the shear rate. Sudden shear thickening can cause serious industrial problems when, for example, it occurs during pumping or coating of a dense colloidal suspension.

The shear-thinning phenomenon in Fig. 6-34 can readily be attributed to the slipping of layers past each other, as described above. In fact, the slope of the viscosity–shear rate curve in Fig. 6-34 for the two more concentrated samples is around -1 at low shear rates, implying that a yield stress must be exceeded to induce the layers to move over one another (see also Chen et al. 1994). Figure 6-35 shows the scattering pattern in the plane normal to the shear gradient for the suspension with $\phi = 0.434$ at various shear rates. In the yielding region ($\dot{\gamma} \lesssim 1 \text{ sec}^{-1}$), where the viscosity curve has a -1 slope, the spheres are ordered hexagonally within each layer, as described by Ackerson and coworkers and by Hoffman (1972) (see above). Thus, it is logical to infer that at higher shear rates the loss of shear thinning and the onset of shear thickening is caused by a breakdown of the sliding-layer flow.

Indeed, Hoffman has presented evidence from light scattering and rheology that for his charged colloids the shear thickening phenomenon is associated with a shear-induced breakup of the layered structure. In his experiments, suspensions of highly monodisperse spheres (ratio of weight to number-average diameter $= 1.01$) of polyvinyl chloride (PVC)

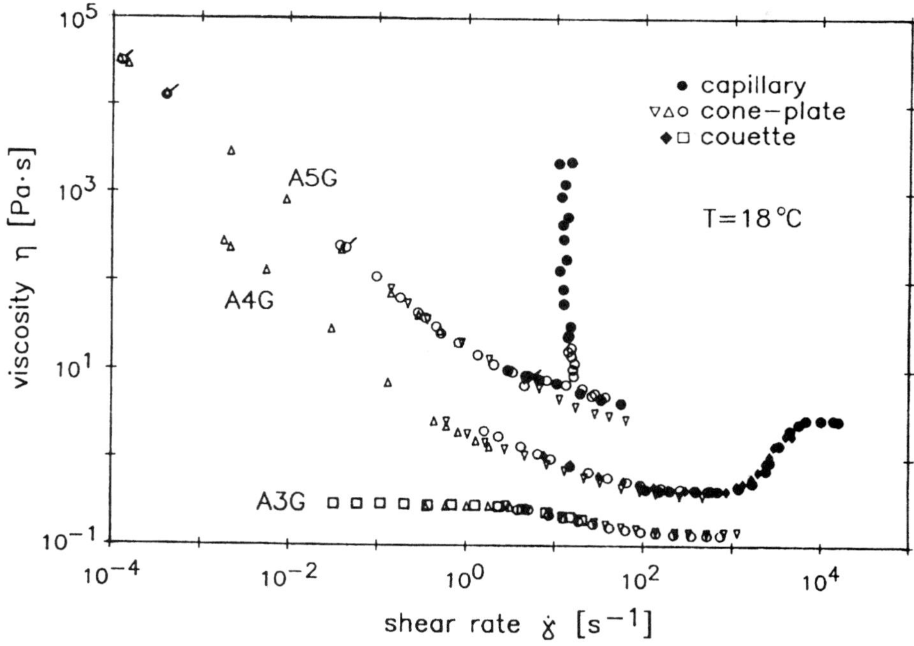

Figure 6.34   Viscosity as a function of shear rate for $a = 82.5$ nm charged particles of styrene-ethylacrylate in glycol at volume fractions ϕ of 0.523 (A5G), 0.434 (A4G), and 0.355 (A3G). (From Laun et al. 1992, with permission from the Journal of Rheology.)

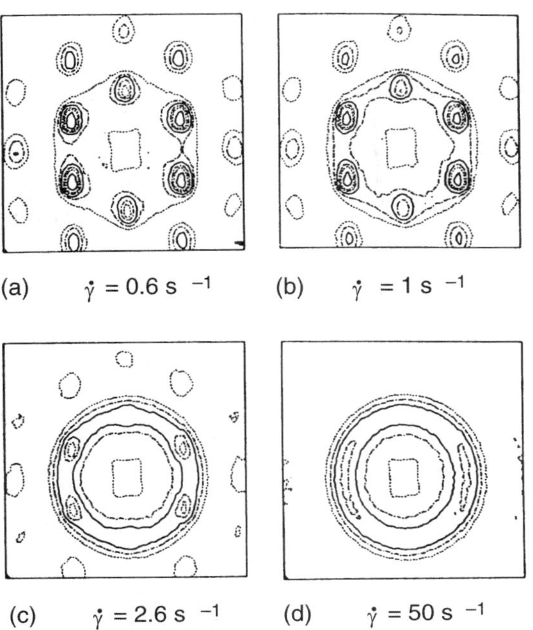

(a) $\dot\gamma = 0.6\ \text{s}^{-1}$ (b) $\dot\gamma = 1\ \text{s}^{-1}$

(c) $\dot\gamma = 2.6\ \text{s}^{-1}$ (d) $\dot\gamma = 50\ \text{s}^{-1}$

Figure 6.35 Neutron scattering patterns with the neutron beam directed in the *gradient direction* for various shear rates $\dot\gamma$ of suspension A4G described in the caption to Fig. 6-34. The horizontal direction is the flow direction and the vertical direction is the vorticity direction. (From Laun et al. 1992, with permission from the Journal of Rheology.)

and (poly)styrene-acrylonitrile (PSAN) of sizes 0.416–0.775 μm in dioctyl phthalate (DOP) show a discontinuous jump in viscosity at a critical shear rate $\dot\gamma_c$ which decreases with increasing volume fraction ϕ (see Fig. 6-36). The volume fraction at which $\dot\gamma_c \rightarrow 0$ corresponds to the concentration $\phi = \pi/(3\sqrt{3}) \approx 0.60$ at which a plane of hexagonally close-packed spheres touches an adjacent layer of such spheres at some point during its sliding motion. This suggests that in these experiments, shear thickening is caused by *layer–layer interactions,* an inference supported by Hoffman's light-scattering experiments, which show a decrease in the intensity of the hexagonal pattern at $\dot\gamma = \dot\gamma_c$. Hoffman (1974) developed a particle-double model from which the critical condition for layer breakdown can be estimated. In the model, a pair of particles within a layer rotates slightly out of the layer due to fluctuations. At low shear rates, electrostatic forces between layers drives the particle pair to rotate back into the layer plane, but at high shear rates, viscous torques overcome electrostatic ones, and rotation of the particle pair continues, thereby disrupting neighboring layers. This mechanism is supported by recent Stokesian dynamics simulations and a simplified linear stability analysis (Dratler et al. 1997).

In studies of Laun et al. (1992), however, the breakup of the hexagonal pattern begins at shear rates lower than the onset of shear thickening (compare Figs. 6-34 and 6-35). Furthermore, the hexagonal layered structures formed by charged suspensions at lower volume fractions ($\phi < 0.5$) also break up at high shear rates, but without any abrupt shear thickening (Chow and Zukoski 1995). Thus, breakup of a hexagonally ordered layered structure appears to be an insufficient condition for shear thickening. More recent evidence from simulations (Bossis and Brady 1989; Phung and Brady 1992; Boersma et al. 1995), optical dichroism (Bender and Wagner 1995), and transient nonlinear rheology (Watanabe

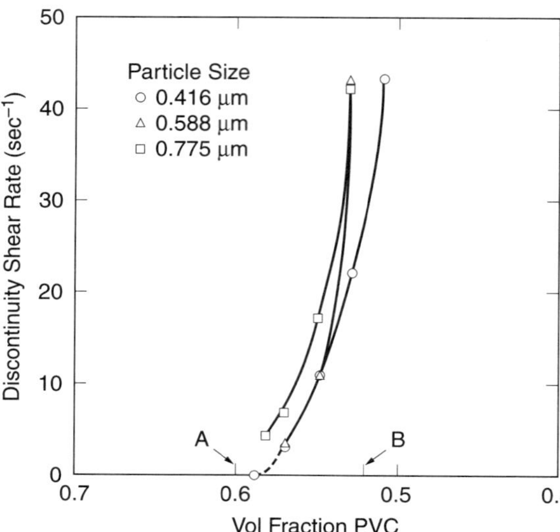

Figure 6.36 Shear rate at which the shear viscosity suddenly jumps discontinuously upward as a function of the volume fraction of poly(vinyl chloride) (PVC) spheres of various diameters in dioctyl phthalate. Point *A* denotes the volume fraction at which two-dimensional hexagonal close-packed layers of spheres first touch. Point *B* is the volume fraction for three-dimensional simple cubic packing. (From Hoffman, reprinted from Trans. Soc. Rheol. 16:155, Copyright 1972 American Institute of Physics.)

et al. 1996) suggest that the stresses in shear-thickening suspensions are predominately hydrodynamic, or viscous, in origin and are due to *particle* clustering, which produces effectively elongated aggregates that dissipate more than do nonaggregated spheres. This conclusion is supported by recent computer simulations of Melrose and coworkers (Ball and Melrose 1995). Intriguing experimental results of Chow and Zukoski (1995) on dense ($\phi = 0.5$), charged spheres in very thin rheometer gaps ($\lesssim = 15\mu$m), show that the critical shear rate for shear thickening decreases with decreasing gap. Shear thickening in these suspensions is therefore associated with clusters that grow to a size comparable to that of the gap. These results suggest that shear thickening requires not only that sliding layers be broken down by shear, but that the fragments of these layers must rotate and collide with each other to form structures whose average dimensions in the *flow-gradient direction* are large. Such structures can "jam" the flow, leading to abrupt shear thickening. If the particle concentration is not high enough, layer breakdown does not lead to jamming, and there is no abrupt shear thickening.

Laun (1994) has measured the first and second normal stress differences N_1 and N_2 of suspensions at steady state in the shear-thickened state. He found, remarkably, that N_1 is *negative* and N_2 is *positive*, which are opposite the usual signs for these quantities! He also found the following relationship between N_1, N_2, and the shear stress σ:

$$2N_2 \approx -N_1 \approx |\sigma| \tag{6-72}$$

(In the low-shear-rate shear-thinning regime, N_1 has its more usual positive sign.) The only other materials known to have steady-state negative values of N_1 are (a) polymeric liquid crystals (LCPs) composed of rod-like molecules (Section 11.3.3) and (b) electrorheological suspensions which form chain-like aggregates under an electric field (Section 8.2.2). The similarity between the normal-stress behavior of rod-like molecules and that of shear-thickening suspensions supports the notion that in the shear-thickening region the relevant flow units are no longer spheres, but are instead rod-like or disk-like particle aggregates.

Intuitively, one would expect that at the high particle concentrations required to enter the shear-thickening regime, small differences in particle-size distribution, or particle shape, might produce significant changes in flow properties; thus shear-thickening phenomena might vary substantially from one system to the next, as seems to be the case in the experiments of Hoffman vis à vis those of Laun et al. Slip is also a significant factor in the shearing flow of these suspensions (Laun et al. 1991); this could lead to some geometry-dependence of the observed flow properties and could account for differences among experiments. For further discussion of shear thickening in dense colloidal suspensions, see Barnes (1989).

> • Worked Example 6.9 shows how to make practical use of the theory for the rheology of charged-particle suspensions.

6.5 PARTICLES IN VISCOELASTIC LIQUIDS: "FILLED MELTS"

Solid particles are frequently added to polymeric materials to impart to them thermal, dielectric, or mechanical properties of the solid, while retaining the processability of the molten polymer. For example, potentially damaging thermal expansion stresses on a silicon or metal component induced by the surrounding plastic "packaging" can be minimized if particle filler is added to the packaging to more closely match the thermal expansion coefficient of the plastic compound to that of the electronic part. Particles fillers are also added to plastic as a cost-saver, so that a given volume of compound requires less of the expensive polymer. The rheological properties of such "filled melts" of course affect their processing characteristics, and therefore they need to be understood.

The rheological properties of filled melts are governed by the diverse properties of polymeric melts, discussed in Chapter 3, compounded by the equally diverse and complex properties of suspensions, discussed in this chapter. In addition, the adhesive characteristics of the junction between polymer and particle can affect the rheology of the filled melt. Thus, the range of possible rheological phenomena in filled melts is immense. Because of limited space, only a few aspects can be touched on here; the interested reader is directed to a review article by Khan and Prud'homme (1987), and references therein.

6.5.1 Spherical Particles

In limiting cases, the properties of filled melts are dominated either by the polymer or by the filler. Thus, for small loadings ($\phi < 0.05$) of nonaggregating spheroidal particles in a polymer melt, the flow properties of the compound are, for the most part, not much different from those of the melt. Or, if the particle loading is high, the properties of the filled melt can in some cases qualitatively resemble those of dense suspensions. Figure 6-37, for example, shows the effect of various loadings of carbon black on the shear viscosity of a polystyrene melt. At low volume fraction ($\phi \approx 0.05$), the carbon black has only a small effect; the shear thinning is essentially that of the polystyrene. However, at high volume fraction ($\phi \gtrsim 0.2$),

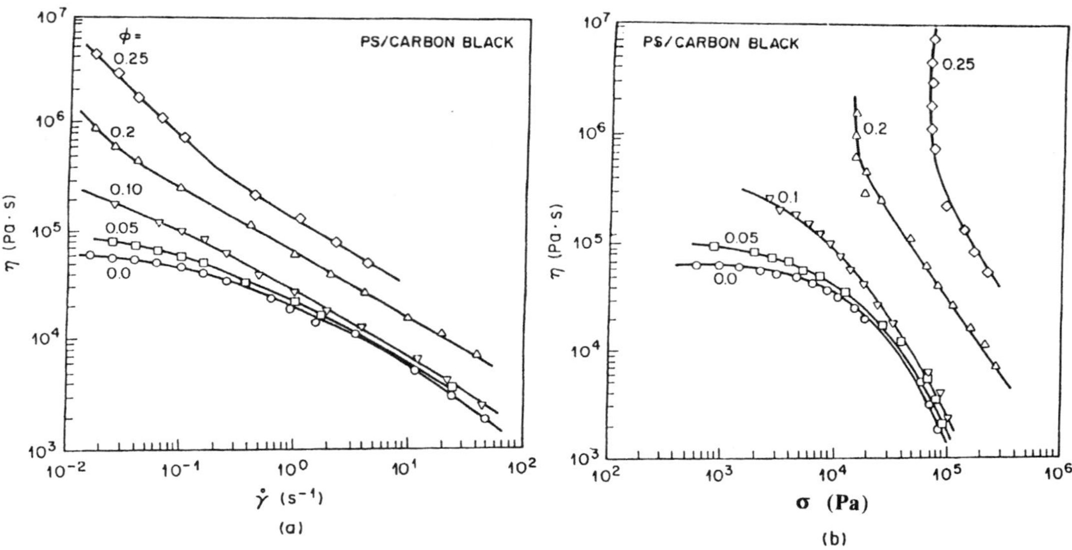

Figure 6.37 Viscosity as a function of (**a**) shear rate or (**b**) shear stress of a polystyrene melt ($M_w = 214{,}000$) filled with carbon black (surface area $= 124$ m²/g) at various volume fractions ϕ at 170°C. (From Lobe and White 1979, reprinted with permission from the Society of Plastics Engineers.)

a yield stress develops (Fig. 6-37b), similar to that of dense suspensions of particles with attractive interactions, to be discussed in Section 7.3.1.

In addition to familiar phenomena such as these, new phenomena arise because of the *synergism* between the properties of the polymer and those of the particle filler. One such synergistic effect is enhanced shear thinning, which occurs because the shear rate experienced by the polymer confined between two particles can be much larger than the overall imposed shear rate (Khan and Prud'homme 1987). Another general observation is that the filled melt is often *effectively less elastic* than the polymer alone, evidently because the filler enhances the viscosity more than it does the first normal stress difference N_1 (Han 1981; Han et al. 1981). Thus, Fig. 6-38 shows that at *fixed shear stress* the first normal stress difference N_1 for polypropylene *decreases* upon addition of CaCO$_3$ particles.

Another synergistic effect is the strong sensitivity of the rheology of a filled melt to the surface treatment of the particles. Figure 6-38 shows that when a titanate "coupling agent" is added to promote adhesion between the polymer and the CaCO$_3$ particles, the viscosity decreases. Drawing from our knowledge of the effect of surface treatment on the rheology of particles suspended in small-molecule liquids, we infer that the "coupling agent" increases polymer–particle attractive interactions relative to particle–particle attractions, thereby hindering particle aggregation; this lowers the viscosity and prevents the development of a yield stress (see Fig. 6-38). However, surface treatments can have other effects. A particle surface coating (such as a fluorosilane) can cause the viscosity of the filled melt to *decrease*, evidently by promoting slip between the particles and the matrix (Inn and Wang 1995). The high viscosities of polymer melts, combined with the enhanced shear rates in small gaps between particles, easily produces stresses high enough ($\gtrsim 0.1$ MPa) that slip

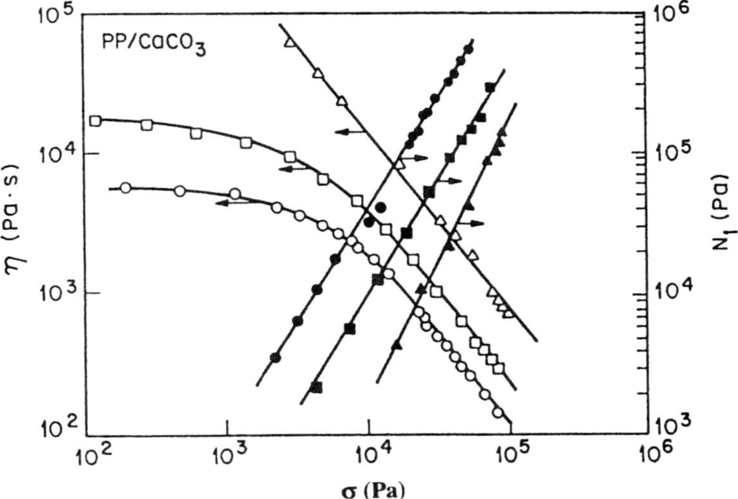

Figure 6.38   Viscosity (open symbols) and first normal stress difference (closed symbols) as a function of shear stress for neat polypropylene melt ($\bigcirc$, $\bullet$), the same melt filled with 50% by weight $CaCO_3$ particles of size 2.5 μ, ($\triangle$, $\blacktriangle$), and the filled melt with a titanate "coupling agent" ($\square$, $\blacksquare$). (From Han et al. 1981, reprinted with permission from the Society of Plastics Engineers.)

becomes likely. Also, even very small slip velocities V_s at particle surfaces will create observable effects in filled melts, since the slip extrapolation length $b \equiv V_s/\dot{\gamma}$ need only be of order the particle radius a, and not the rheometer gap h, for large effects to occur (Inn and Wang 1995).

In a filled melt at low particle loadings ($\phi \lesssim 0.05$), even particles that interact strongly with each other agglomerate very slowly, because the high viscosity of the polymer matrix impedes particle movement. In a melt with viscosity $\eta \geq 100$ P, the diffusion time constant of 1-μm particles is $\sigma \sim \eta a^3/k_B T \gtrsim 1$ hr. Thus, after gel-like particle structures in such a melt are disrupted by a shearing flow, they take a long time to re-form. Hence, flow induces changes in fluid structure that are erased only after hours of quiescence. This phenomenon, called *thixotropy*, is really a kind of viscoelasticity, but with a very long relaxation time. Thixotropy is displayed when the shear viscosity measured by progressively *increasing* the shear rate differs from that measured when one progressively decreases it. Figure 6-39 shows the dependence on shear rate of the viscosity of polydimethylsiloxane (PDMS) melt with 3% by weight added fumed silica particles, measured under increasing and under decreasing shear rate. Each arrow in Fig. 6-39 marks the highest shear rate in a ramp of increasing $\dot{\gamma}$, after which the shear viscosity was measured along a path of decreasing shear rate. The viscosities measured along the paths of decreasing shear rate are lower than they are on the branch of increasing $\dot{\gamma}$, and on each decreasing branch the viscosity is controlled by the highest shear rate accessed along the upper branch before the direction of the shear-rate ramp was changed. The viscosities on the lower branches are thus only *apparent steady-state* viscosities, which reflect a long-lived metastable state induced by the highest shear rate in the sample's recent past. If one allows the sample to relax for the requisite time period (many hours), steady-state viscosities on the upper branch are recovered. More highly concentrated suspensions of fumed silica (up to 10 wt%) in polypropylene glycol can show pronounced shear thickening in steady and oscillatory flow (Raghavan and Khan 1997). This is believed to be caused by shear-induced clustering of fumed silica aggregates; the low volume fraction at which this occurs may be related to the open, ramified shape of the fumed silica aggregates.

The structure of suspensions of fumed silica in PDMS has been explored in recent studies by Macosko and coworkers (Aranguren et al. 1992). Fumed silica consists of primary particles of order 10 nm in diameter that are fused into aggregates of about $1\,\mu$m in diameter, which can be broken down to a smaller size by vigorous mixing in a viscous solvent. The aggregates are believed to be fractal-like with fractal dimension $D_f \approx 1.7$ (Raghavan and Khan 1997). These aggregates, when dispersed in PDMS, form large *agglomerates*. The agglomerates are believed to be held together by *bridging* interactions, in which a single PDMS polymer molecule hydrogen bonds to silanol groups on two different silica aggregates. This inference is supported by experiments showing that the low-frequency modulus of the filled melt decreases dramatically when the particles are treated to remove surface silanol groups (see Fig. 6-40). These bridging interactions can lead to some very peculiar rheological properties. For example, the elastic modulus G' of suspensions of fused silica in PDMS *decreases* as the molecular weight of the PDMS *increases* from 17,000 to 88,000. At molecular weights higher than 88,000, however, the modulus increases with increasing molecular weight in the usual way (Aranguren et al. 1992).

6.5.2 Rod-Like Particles

In a weakly elastic polymeric fluid (Weissenberg number $\ll 1$), experiments (Saffman 1956; Karnis and Mason 1966; Iso et al. 1996a) and analysis (Leal 1975) show that the Jeffery orbit of a single isolated fiber is modified by a spiraling drift of the fiber axis towards the *vorticity direction*, which is the stable orientation in this case. This drift can be accounted

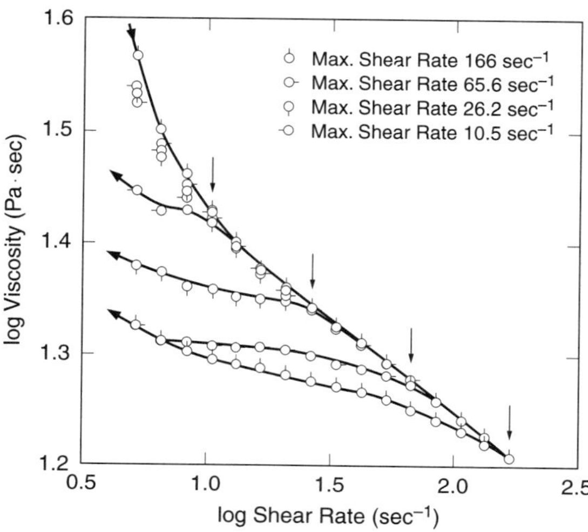

Figure 6.39 Hysteresis loops of viscosity versus shear rate of a 3% by weight suspension of fumed silica (surface area $= 325$ m²/g) in poly(dimethylsiloxane), (PDMS; molecular weight $= 67,000$, $\eta_s \approx 125$ P) at 30°C. In each run, the shear rate was first increased up to a maximum shear rate $\dot{\gamma}_{max}$ located at the arrow, and then decreased. After a rest of 23 hours, another run was made, with a different $\dot{\gamma}_{max}$, thus producing the series of curves shown. (Reprinted from J Non-Newt Fluid Mech 17:45, Ziegelbaur and Caruthers (1985), with kind permission from Elsevier Science - NL, Sara Burgerhartstraat 25, 1055 KV Amsterdam, The Netherlands.)

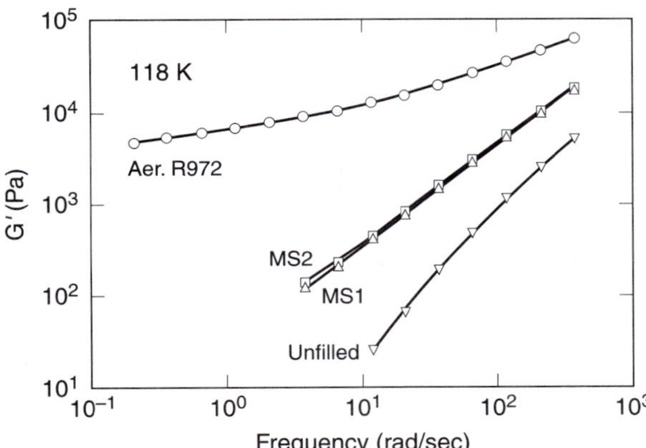

Figure 6.40 Storage modulus versus frequency for suspension of fumed silica (Aerosil R972; surface area $= 108$ m²/g; aggregate size $\approx$ 200 nm) at a concentration of $\phi = 0.085$ in PDMS ($M_W = 118,000$). The modulus for the unfilled PDMS is also shown, as is the modulus for the suspension treated in two different ways (MS1, MS2) to remove the active silanol groups from the silica particle surface. (From Aranguren et al. 1992, with permission from the Journal of Rheology.)

for by a modification of Jeffery's equation, (6-26). For nondilute fiber concentrations, however, fiber–fiber interactions produce an effective diffusivity (such as discussed in Section 6.3.2.2) that at dimensionless concentrations $\nu L^3 \gtrsim 15$ leads to fiber orientations in the flow direction, with a distribution similar to that in Newtonian fluids (Iso et al. 1996a). In highly elastic liquids, even isolated fibers orient in the flow direction (Iso et al. 1996b). For intermediate elasticities and fiber concentrations, broad distributions of fiber orientations between the flow and vorticity directions are observed (Iso et al. 1996b). Fibers in a polymer melt can also be slowly aligned by oscillatory shearing outside the linear regime (Kim and Song 1997), a phenomenon resembling that seen in block copolymers (see Section 13.3.1.2).

There is much more to tell about the rheology of filled melts, but space limitations preclude further discussion here. The interested reader is directed to the articles by Khan and Prud'homme (1987), Metzner (1985), amd White (1982), the book by Han (1981), and references therein. Viscoelastic theories for filled melts, especially for rubbers containing carbon black, can be found in Montes and White (1993), Witten et al. (1993), and references therein.

• Problem 6.10 tests your understanding of flow-induced particle alignment in injection-molding.

6.6 SUMMARY

Stable particle suspensions exhibit an extraordinarily broad range of rheological behavior, which depends on particle concentration, size, and shape, as well as on the presence and type of stabilizing surface layers or surface charges, and possible viscoelastic properties of the suspending fluid. Some of the properties of suspensions of spheres are now reasonably well understood, such as (a) the concentration-dependence of the zero-shear viscosity of "hard-sphere" suspensions and (b) the effects of deformability of the steric-stabilization layers on the particles. In addition, qualitative understanding and quantitative empirical equations are available for the shear-stress dependence of the viscosity in such suspensions. However, detailed understanding of shear thinning and shear thickening at high concentrations in such suspensions is incomplete; in the future, computer simulations should help clarify these issues. Although such suspensions can be very viscous, their elasticity is typically low, as shown by predictions and measurements of the normal stress differences and storage moduli.

The viscous and elastic properties of orientable particles, especially of long, rod-like particles, are sensitive to particle orientation. Rods that are small enough to be Brownian are usually stiff molecules; true particles or fibers are typically many microns long, and hence non-Brownian. The steady-state viscosity of a suspension of Brownian rods is very shear-rate- and concentration-dependent, much more so than non-Brownian fiber suspensions. The existence of significant normal stress differences in non-Brownian fiber suspensions is not yet well understood.

Suspensions of charged spheroidal particles readily form (a) ordered macrocrystals at rest and (b) ordered sliding-layer structures under shear. These suspensions show yield stresses or severe shear thinning at modest shear stress, followed by continuous or discontinuous shear thickening at high shear stress. In the shear-thickening regime, these suspensions (and uncharged ones) can show negative first normal stress differences. Qualitative explanations of some of these phenomena have been suggested, and simple theories are available to predict viscosities and linear moduli. However, much work is still required, especially to explain the shear-thickening behavior of these suspensions.

Filled polymer melts show diverse behavior that can include rheology typical of unfilled melts, rheology typical of dense suspensions, and novel thixotropic behavior and sensitivity to particle surface treatment.

REFERENCES

Ackerson BJ, Clark NA (1984). *Phys Rev A* 30:906.

Ackerson BJ, Pusey PN (1988). *Phys Rev Lett* 61:1033.

Ackerson BJ, Hayter JB, Clark NA, Cotter L (1986). *J Chem Phys* 84:2344.

Amelar S, Eastman CE, Morris RL, Smeltzly MA, Lodge TP, von Meerwall ED (1991). *Macromolecules* 24:3505.

Anczurowski E, Mason SG (1967a). *J Colloid Interface Sci* 23:522.

Anczurowski E, Mason SG (1967b). *J Colloid Interface Sci* 23:533.

Aranguren MI, Mora E, DeGroot JV Jr, Macosko CW (1992). *J Rheol* 36:1165.

Arrhenius Z (1917). *Biochem J* 11:112.

Bagnold RA (1954). *Proc R Soc Lond A* 225:49.

Ball R, Richmond P (1980). *J Phys Chem Liquids* 9:99.

Ball RC, Melrose JF (1995). *Adv Colloid Interface Sci* 59:19.

Barnes HA (1989). *J Rheol* 33:329.

Barnes HA, Hutton JF, Walters K (1989). *An Introduction to Rheology*, Elsevier, New York.

Batchelor GK (1970). *J Fluid Mech* 41:545.

Batchelor GK (1971). *J Fluid Mech* 46:813.

Beenakker CWJ (1984). *Physica A* 128:48.

Beenakker CWJ, Mazur P (1984). *Physica A* 126:349.

Bender JW, Wagner NJ (1995). *J Colloid Interface Sci* 172:171.

Bibbo MA (1987). Ph.D. Thesis, Department of Chemical Engineering, Massachusetts Institute of Technology, Cambridge, MA.

Bird RB, Curtiss CF, Armstrong RC, Hassager O (1987). *Dynamics of Polymeric Liquids,* Vol. 2: *Kinetic Theory,* Wiley, New York.

Bitsanis I, Davis HT, Tirrell M (1990). *Macromolecules* 23:1157.

Blatchford J, Chan FS, Goring DAI (1969). *J Phys Chem* 73:1062.

Boersma WH, Laven J, Stein HN (1995). *J Rheol* 39:841.

Bossis G, Brady JF (1987). *J Chem Phys* 87:5437.

Bossis G, Brady JF (1989). *J Chem Phys* 91:1866.

Brady JF (1993). *J Chem Phys* 99:567.

Brady JF (1996). *Curr Opin Colloid Interface Sci* 1:472.

Brady JF, Vicic M (1995). *J Rheol* 39:545.

Brenner H (1974). *Int J Multiphase Flow* 1:195.

Bu Z, Russo PS, Tipton DL, Negulescu II (1994). *Macromolecules* 27:6871.

Buscall R (1991). *J Chem Soc Faraday Trans* 87:1365.

Buscall R, Goodwin JW, Hawkins MW, Ottewill RH (1982a). *J Chem Soc Faraday Trans I* 78:2873.

Buscall R, Goodwin JW, Hawkins MW, Ottewill RH (1982b). *J Chem Soc Faraday Trans I* 78:2889.

Carter LF (1967). PhD. Thesis, University of Michigan, Ann Arbor, Michigan.

Caspar DLD (1963). *Adv Protein Chem* 18:37.

Chen LB, Zukoski CF (1990). *J Chem Soc Faraday Trans* 86:2629.

Chen LB, Zukoski CF, Ackerson BJ, Hanley HJM, Straty GC, Barker J, Glinka CJ (1992). *Phys Rev Lett* 69:688.

Chen LB, Ackerson BJ, Zukoski CF (1994). *J Rheol* 38:193.

Choi GN, Krieger IM (1986). *J Colloid Interface Sci* 113:101.

Chow AW, Fuller GG, Wallace DG, Madri JA (1985a). *Macromolecules* 18:793.

Chow AW, Fuller GG, Wallace DG, Madri JA (1985b). *Macromolecules* 18:805.

Chow MK, Zukoski CF (1995). *J Rheol* 39:15.

Cross MM (1965). *J Colloid Sci* 20:417.

Darlington MW Gladwell PK, Smith GR (1977). *Polymer* 18:1269.

de Kruif CG, van Iersel EMF, Vrij A, Russel WB (1985). *J Chem Phys* 83:4717 .

Dealy JM, Wissbrun KF (1990). *Melt Rheology and Its Role in Plastics Processing*, Van Nostrand Rheinhold, New York.

D'Haene PD, Mewis J, Fuller GG (1993). *J Colloid Interface Sci* 156:350.

Doi M, Edwards SF (1978). *J Chem Soc Faraday II* 74:560; 74:918.

Doi M, Edwards SF (1986). *The Theory of Polymer Dynamics*, Oxford Press, New York.

Dratler DI, Schowalter WR, Hoffman RL (1997). Presentation at the 68th Annual Meeting of the Society of Rheology, Galveston, TX.

Einaga Y, Berry GC, Chu SG (1985). *Polym J* 17:239.

Einstein A (1906). *Ann Phys* 19:289.

Einstein A (1911). *Ann Phys* 34:591.

Evans ID, Lips A (1990). *J Chem Soc Faraday Trans* 86:3413.

Fagan ME, Zukoski CF (1997). *J Rheol* 41:373.

Folgar F, Tucker CL (1984). *J Reinforced Plast Composites* 3:98.

Frattini PL, Fuller GG (1986). *J Fluid Mech* 168:119.

Ganani L, Powell RL (1985). *J Compos Mater* 19:194.

Goddard JD, Bashir YM (1990). In *Recent Developments in Structured Continua*, Vol 2, DeKee D, Kaloni N (eds), Pitman Research Notes in Applied Mathematics, No. 229, Wiley, New York.

Goto S, Nagazono H, Kato H (1986). *Rheol Acta* 25:119.

Hachisu S, Kobayashi Y, Kose A (1973). *J Colloid Interface Sci* 42:342.

Han CD van den Weghe T, Shete P, Haw JR (1981). *Polym Eng Sci* 21:196.

Han CD (1981). *Multiphase Flow in Polymer Processing*, Academic Press, New York.

Hanley HJM, Rainwater J, Clark NA, Ackerson BJ (1983). *J Chem Phys* 79:4448.

Hiemenz PC, Rajagopalan R (1997). *Principles of Colloid and Surface Chemistry*, 3rd ed, Marcel Dekker, New York.

Hinch EJ, Leal LG (1972). *J Fluid Mech* 52:683.

Hinch EJ, Leal LG (1973). *J Fluid Mech* 57:753.

Hirth JP, Lothe, J (1982). *Theory of Dislocations*, Wiley, New York.

Ho BP, Leal LG (1974). *J Fluid Mech* 65:365.

Hoffman RL (1972). *Trans Soc Rheol* 16:155.

Hoffman RL (1974). *J Colloid Interface Sci* 46:491.

Inn YW, Wang SQ (1995). *Langmuir* 11:1589.

Iso Y, Koch DL, Cohen C (1996a). *J Non-Newt Fluid Mech* 62:115.

Iso Y, Koch DL, Cohen C (1996b). *J Non-Newt Fluid Mech* 62:135.

Jeffery GB (1922). *Proc R Soc Lond A* 102:161.

Karnis A, Mason SG (1966). *Trans Soc Rheol* 10:571.

Khan SA, Prud'homme RK (1987). *Rev Chem Eng* 4:205.

Kim JK, Song JH (1997). *J Rheol* 41:1061.

Kim S, Karrila SJ (1991). *Microhydrodynamics: Principles and Selected Applications*, Butterworth-Heinemann, Boston.

Kirkwood JG, Auer PL (1951). *J Chem Phys* 19:281.

Kitano T, Kataoka T (1981). *Rheol Acta* 20:390.

Koch DL (1995). *Phys Fluids* 7:2086.

Koppi KA, Tirrell M, Bates FS Almdal K, Mortensen K (1994). *J Rheol* 38:999.

Krieger IM (1972). *Adv Colloid Interface Sci* 3:111.

Krieger IM, Dougherty TJ (1959). *Trans Soc Rheol* 3:137.

Krieger IM, Eguiluz M (1976). *Trans Soc Rheol* 20:29.

Kuzuu NY, Doi M (1980). *Polym J* 12:883.

Lange FF (1989). *J Am Ceram Soc* 72:3.

Larson RG (1988). *Constitutive Equations for Polymer Melts and Solutions,* Butterworths, Boston.

Larson RG, Mead DW (1991). *J Polym Sci Polym Phys Ed* 29:1271.

Laun HM (1984a). *Angew Makromol Chem* 123/124:335.

Laun HM (1984b). *Colloid Polym Sci* 262:257.

Laun HM (1994). *J Non-Newt Fluid Mech* 54:87.

Laun HM, Bung R, Schmidt F (1991). *J Rheol* 35:999.

Laun HM, Bung R, Hess S, Loose W, Hess O, Hahn K, Hädicke, Hingmann R, Schmidt F, Lindner P (1992). *J Rheol* 36:743.

Leal LG (1975). *J Fluid Mech* 69:305.

Leal LG, Hinch EJ, (1971). *J Fluid Mech* 46:685.

Leighton D, Acrivos A (1987). *J Fluid Mech* 181:415.

Lionberger RA, Russel WB (1994). *J Rheol* 38:1885.

Lionberger RA, Russel WB (1997). *J Rheol* 41:399.

Lobe VM, White JL (1979). *Polym Eng Sci* 19:617.

Mackaplow MB, Shaqfeh ERG (1996). *J Fluid Mech* 329:155.

Macosko CW (1994). *Rheology Principles, Measurements, and Applications*, VCH Publishers, New York.

Marrucci G, Grizzuti N (1983). *J Polym Sci Polym Lett Ed* 21:83.

Mead DW, Larson (1990). *Macromolecules* 23:2524.

Meeker SP, Poon WCK, Pusey PN (1997). *Phys Rev E* 55:5718.

Mellema J (1997). *Curr Opin Colloid Interface Sci* 2:411.

Metzner AB (1985). *J Rheol* 29:739.

Metzner AB, Whitlock M (1958). *Trans Soc Rheol* 2:239.

Mewis J, Metzner AB (1974). *J Fluid Mech* 62:249.

Mewis J, Frith WJ, Strivens TA, Russel WB (1989). *AIChE J* 35:415.

Montes S, White JL (1993). *J Non-Newt Fluid Mech* 49:277.

Mori Y, Ookubo N, Hayakawa R, Wada Y (1982). *J Polym Sci* 20:2111.

Nemoto N, Schrag JL, Ferry JD, Fulton RW (1975). *Biopolymers* 14:407.

Nestler FHM, Hvidt S, Ferry JD, Veis A (1983). *Biopolymers* 22:1747.

Ogawa A, Yamada H, Matsuda S, Okajima K, Doi M (1997). *J Rheol* 41:769.

Onoda GY, Liniger ER (1990). *Phys Rev Lett* 64:2727.

Ookubo N, Komatsubara M, Nakajima H, Wada Y (1976). *Biopolymers* 15:929.

Papir YS, Krieger IM (1970). *J Colloid Interface Sci* 34:126.

Persello J, Magnin A, Chang J, Piau JM, Cabane B (1994). *J Rheol* 38:1845.

Phan S-E, Russel WB, Cheng Z, Zhu J, Chaikin PM, Dansmair JH, Ottewill RH (1996). *Phys Rev E* 54:6633.

Phillips RJ, Armstrong RC, Brown RA, Graham AL, Abbott JR (1992). *Phys Fluids A* 4:30.

Phung T, Brady JF (1992). *Am Inst Phys Conf Proc* 256:391.

Phung T, Brady JF, Bossis G (1996). *J Fluid Mech* 313:181.

Raghavan SR, Khan SA (1997). *J Colloid Interface Sci* 185:57.

Reynolds O (1885). *Philos Mag* 8:20.

Rice RW (1990). *AIChE J* 36:481.

Russel WB (1978). *J Fluid Mech* 85:209.

Russel WB, Saville DA, Schowalter WR (1989). *Colloidal Dispersions,* Cambridge University Press, Cambridge.

Saffman PG (1956). *J Fluid Mech* 1:540.

Saunders FL (1961). *J Colloid Sci* 16:13.

Shaqfeh ERG, Fredrickson GH (1990). *Phys Fluid A* 2:7.

Shikata T, Pearson DS (1994). *J Rheol* 38:601.

Simon C (1993). In *Stabilization of Aqueous Powder Suspensions in the Processing of Ceramic Materials,* Dobais B (ed), Marcel Dekker, New York.

Solomon MJ, Boger DV (1998). *J Rheol,* 42:929.

Stevens MJ, Robbins MO, Belak JF (1991). *Phys Rev Lett* 66:3004.

Stone-Masui J, Watillon A (1968). *J Colloid Interface Sci* 28:187.

Stover CA, Koch DL, Cohen C (1992). *J Fluid Mech* 238:277.

Tanford CF, Buzzell JG (1956). *J Phys Chem* 60:225.

Teraoka I, Hayakawa R (1989). *J Chem Phys* 91:2643.

Tomita M, van de Ven TGM (1984). *J Colloid Interface Sci* 99:374.

Trevelyan BJ, Mason SG (1951). *J Colloid Sci* 6:354.

Ulrich DR (1990). *Chem Eng News* 68:28.

van der Vorst B, van den Ende D, Mellema J (1995). *J Rheol* 39:1183.

van der Werff JC, de Kruif CG, Blom C, Mellema J (1989). *Phys Rev A* 39:795.

van Megen W, Underwood SM (1990). *Langmuir* 6:35.

Visscher PB, Heyes DM (1994). *J Chem Phys* 101:6096.

Wagner NJ (1993). *J Colloid Interace Sci* 161:169.

Warren TC, Schrag JL, Ferry JD (1973). *Biopolymers* 12:1905.

Watanabe H, Yao M-L, Yamagishi A, Osaki K, Shikata T, Niwa H, Morishima Y (1996). *Rheol Acta* 35:433.

White JL (1982). *Plast Compd* Jan/Feb:47.

Willey SJ, Macosko CW (1982). *J Rheol* 26:557.

Witten TA, Rubinstein M, Colby RH (1993). *J Phys II (France)* 3:367.

Yamamoto S, Matsuoka T (1995). *J Chem Phys* 102:2254.
Yamane Y, Kaneda Y, Doi M (1994). *J Non-Newt Fluid Mech* 54:405.
Yang JT (1958). *J Am Chem Soc* 80:1783.
Ziegelbaur RS, Caruthers JM (1985). *J Non-Newt Fluid Mech* 17:45.
Zirnsak MA, Hur DU, Boger DV (1994). *J Non-Newt Fluid Mech* 54:153.
Zwanzig R, Mountain RD (1965). *J Chem Phys* 43:4464.

PROBLEMS AND WORKED EXAMPLES

Problem 6.1 (Worked Example) Estimate the zero-shear viscosity of a suspension of hard spheres 100 nm in diameter at a volume fraction of $\phi = 0.35$ in a solvent of viscosity 1 cP.

ANSWER:

From the Krieger–Dougherty equation, (6-8), we obtain

$$\eta = \eta_s \left(1 - \frac{\phi}{\phi_m} \right)^{-2.5\phi_m} = (1\,\text{cP}) \left(1 - \frac{0.35}{0.57} \right)^{-1.6} = 3.88\,\text{cP} \qquad \text{(A6-1)}$$

Problem 6.2 Estimate the diffusion coefficient for a hard sphere of radius 100 nm in water at room temperature.

Problem 6.3(a) (Worked Example) At what critical shear rate $\dot{\gamma}_c$ do you expect the onset of shear thinning in a 40% (by volume) suspension of hard spheres of radius 1 μm in water at room temperature? (*Hint*: use the data of Fig. 6-4, plus scaling principles.)

ANSWER:

From Fig. 6-4, shear thinning begins at a reduced shear stress σ_r of around 0.01, where the relative viscosity η_r is around 10. By definition of η_r, we have

$$\sigma = \eta_r \eta_s \dot{\gamma} = 10 \eta_s \dot{\gamma} \qquad \text{(A6-2)}$$

Therefore, by definition of σ_r and by using $k_B T \approx 4 \times 10^{-14}$ erg at room temperature, we get

$$\sigma_r \equiv \frac{a^3 \sigma}{k_B T} = \frac{a^3 \cdot 10 \cdot \eta_s \dot{\gamma}}{4 \times 10^{-14}} \qquad \text{(A6-3)}$$

Since $\sigma_r \approx 0.01$ at the onset of shear thinning, the viscosity of water is 0.01 P, and the particle radius a is 10^{-4} cm, we find from Eq. (A6-3) that $\dot{\gamma} = 0.004\,\text{sec}^{-1}$ at the onset of shear thinning.

Problem 6.3(b) (Worked Example) If the particle radius in part (a) is reduced from $a_\text{old} = 1$ μm to $a_\text{new} = 10$ nm, what would this shear rate be?

ANSWER:

Shear thinning sets in at a critical value of the Peclet number, $\text{Pe} = \text{Pe}_c = \eta_s \dot{\gamma}_c a^3 / k_B T$. Hence, the critical shear rate $\dot{\gamma}_c$ must be inversely proportional to the cube of the particle radius. Thus, using the answer in part (a), we obtain

$$\dot{\gamma}_{c,\text{new}} = \left(\frac{a_{\text{new}}}{a_{\text{old}}}\right)^{-3} \cdot \dot{\gamma}_{c,\text{old}} = 100^3 \cdot 0.004 \ \text{sec}^{-1} = 4,000 \ \text{sec}^{-1}$$

Problem 6.4(a) Estimate the high-frequency modulus G_∞ of a suspension of hard spheres of radius 1 μm at a concentration ϕ of 0.4, at room temperature in a solvent with viscosity 1 P.

Problem 6.4(b) What would G_∞ be for a similar suspension of 10-nm-radius particles?

Problem 6.5(a) and (b) Estimate the particle diffusivity $\tau_p(\phi)$ for the two suspensions in 6.4(a) and 6.4(b).

Problem 6.6(a) How long does it take for a non-Brownian cylindrical particle of aspect ratio $L/d = 10$ suspended in a viscous liquid to rotate through an angle of π radians in a shearing flow at a shear rate $\dot{\gamma} = 1 \ \text{sec}^{-1}$?

Problem 6.6(b) How long would it take if the aspect ratio L/d were 100?

Problem 6.7(a) (Worked Example) Estimate the first normal stress difference N_1 for a suspension of long, thin particles (approximated as spheroids) with $p = 100$ and $L = 0.1\mu$m, if the solvent viscosity is 1 P, the shear rate $\dot{\gamma}$ is 100 sec^{-1}, and the particle concentration is $\phi = 0.001$, which is in the dilute regime.

ANSWER:

From Eq. (6-32a), we obtain the rotary diffusivity of prolate spheroids:

$$D_{r0} = \frac{3k_B T\,(\ln(2p) - 0.5)}{\pi \eta_s L^3}$$

$$= \frac{3 \times 4 \times 10^{-14}\,(\ln\,200 - 0.5)}{3.14 \times 1 \times (10^{-5})^3} = 180 \ \text{sec}^{-1} \qquad \text{(A6-4)}$$

Then the rotary Peclet number is

$$\text{Pe} \equiv \frac{\dot{\gamma}}{D_{r0}} = \frac{100}{180} \approx 0.5$$

According to Fig. 6-15, this Peclet number is low enough to be in the low Peclet-number limit, where the normal stress differences are quadratic in the shear rate, and hence the first normal stress coefficient is a constant. This constant can be obtained from the low-shear-rate portion of Table 6-1. Since p is large ($p \to \infty$), this table gives

$$\Psi_1 \to \Psi_{1,0} = \frac{\phi \eta_s}{D_{r0}} \frac{p^2}{15\ln(p)} \qquad \text{(A6-5)}$$

$$= \frac{10^{-3} \times 1}{180} \times \frac{10^4}{69} = 0.8 \times 10^{-3} \ \text{dyn sec}^2/\text{cm}^2 \qquad \text{(A6-6)}$$

Then

$$N_1 = \dot{\gamma}^2 \Psi_1 = 8 \ \text{dyn/cm}^2 \qquad (\text{at } \dot{\gamma} = 100 \ \text{sec}^{-1}) \qquad \text{(A6-7)}$$

This is a small value, but it is measurable with a sensitive rheometer. However, N_1 is much smaller than the shear stress, which is at least as large as the solvent contribution, $\sigma_{xy} \approx \eta_s \dot{\gamma} = 100$ dyn/cm^2.

Problem 6.7(b) (Worked Example) What would N_1 be at $\dot{\gamma} = 1$ sec^{-1}?

ANSWER:

At a lower shear rate, Ψ_1 will remain in the low-shear-rate regime, so Ψ_1 will still be obtained from Eq. (A6-6). Thus,

$$N_1 = \dot{\gamma}^2 \Psi_1 = 8 \times 10^{-4} \text{dyn/cm}^2 \qquad (\text{at } \dot{\gamma} = 1 \text{ sec}^{-1}) \qquad \text{(A6-8)}$$

This value of N_1 is too small to be measured with available rheometers, and it is three orders of magnitude smaller than the shear stress.

Problem 6.7(c) (Worked Example) Suppose the particle volume fraction is increased to $\phi = 0.01$, and the shear rate is held at $\dot{\gamma} = 1$ sec^{-1}. What will N_1 be? (You may take the particles to be cylindrical, rather than spheroidal, if you wish, with $L/d = 100$).

ANSWER:

We need to check what concentration regime we are in. According to the beginning of Section 6.3.2.1, crossover from dilute to semidilute for rods of aspect ratio $p = 100$ occurs when

$$\phi \geq 30\pi d^2/4L^2 \approx 24p^{-2} = 24 \times 10^{-4} = 0.0024 \qquad \text{(A6-9)}$$

Thus, $\phi = 0.01$ is in the semidilute concentration regime. (It is actually at the lower end of the concentrated isotropic regime, which begins at a concentration $\phi = \pi d/4L = 0.008$. However, even in the concentrated isotropic regime, the viscosity versus concentration behavior does not change much from that of the semidilute until near the transition to the nematic regime at $\phi = 3.3d/L = 0.033$.)
In the semidilute regime, Eq. (6-44) applies; thus

$$D_r = 1350 D_{r0} (\nu L^3)^{-2} \qquad \text{(A6-10)}$$

Since $L = 0.1\mu$m and the aspect ratio is 100, the rod diameter is $d = 10^{-3}\mu$m. Therefore, from Eq. (6-43), the number of rods per unit volume, ν, is

$$\nu = \frac{4\phi}{\pi d^2 L} = \frac{0.04}{3.14(10^{-7})^2\, 10^{-5}} \text{ cm}^{-3} = 0.013 \times 10^{19}\text{cm}^{-3} \qquad \text{(A6-11)}$$

Thus,

$$\nu L^3 = 0.013 \times 10^{19} \cdot (10^{-5})^3 = 1.3 \times 10^2$$

and

$$(\nu L^3)^2 = 1.7 \times 10^4$$

Then, from Eq. (A6-10) and using Eq. (A6-4) for D_{r0}, we obtain D_r:

$$D_r = \frac{1350 \times 180}{1.7 \times 10^4} = 14 \text{ sec}^{-1} \qquad \text{(A6-12)}$$

Thus, at $\dot{\gamma} = 1$ sec^{-1}, the rotary Peclet number is

$$\text{Pe} = \frac{\dot{\gamma}}{D_r} = \frac{1}{14} = 0.07$$

From Fig. 6-19, we see that this value of $\dot{\gamma}/D_r$ is low enough to be in the "zero-shear" regime. Hence, we can use Eq. (6-49) to obtain $\Psi_{1,0}$ in the semidilute regime:

$$\Psi_{1,0} = \frac{\nu k_B T}{30 D_r^2} \tag{A6-13}$$

Inserting for ν from Eq. (A6-11) and D_r from Eq. (A6-12) gives

$$\Psi_{1,0} = \frac{0.013 \times 10^{19} \times 4 \times 10^{-14}}{30 \times (14)^2} = 0.9 \text{ dyn sec}^2/\text{cm}^2$$

Hence,

$$N_1 = \Psi_1 \dot{\gamma}^2 = 0.9 \text{ dyn/cm}^2 \tag{A6-14}$$

Comparing this with the result for $\phi = 0.001$ at the same shear rate, Eq. (A6-8), we see that a 10-fold increase in ϕ, from the dilute to the semidilute regime, has produced a 1000-fold increase in N_1. The rate of increase is even steeper with respect to increases in the length of the rods. As an exercise, show that within the semidilute regime, at fixed ϕ, $\Psi_{1,0}$ is proportional to L raised to the power 14!

Problem 6.8 (Worked Example) Estimate the steady-state uniaxial viscosity of a suspension of 0.1% by volume of rod-like particles $L = 6\,\mu$m long and $d = 10$ nm in diameter in a Newtonian oil of viscosity 100 P at an extension rate of 1 sec^{-1}.

ANSWER:

First, we need to determine if the rods are Brownian or non-Brownian under the above conditions. So, we compute the rotational diffusivity under dilute conditions using Eq. (6-32b) for rods of aspect ratio $p = L/d = 600$:

$$D_{r0} = \frac{3k_B T\,(\ln(L/d) - 0.8)}{\pi \eta_s L^3} = \frac{3 \times 4 \times 10^{-14}(\ln(600) - 0.8)}{3.14 \times 100 \times (6 \times 10^{-3})^3} \tag{A6-15}$$

which gives $D_{r0} = 9.9 \times 10^{-9}$ sec^{-1}. The rotational Peclet number is $\text{Pe} \equiv \dot{\gamma}/D_{r0} = 10^8$. This is large enough to put us well into the non-Brownian regime. This conclusion will not be altered if the particle concentration is not dilute, since this would only serve to make Pe even larger.

We now use Batchelor's formula, Eq. (6-57), for the uniaxial extensional viscosity of non-Brownian dispersions of rods:

$$\bar{\eta}_u = 3\eta_s \left[1 + \frac{4\phi p^2}{9\,\ln(\pi/\phi)} \right] = 6.2 \times 10^3 \text{ P}$$

Since the steady-state shear viscosity is only slightly affected by the rods, the ratio of the extensional to the shear viscosity is of order 50 or so.

Problem 6.9 (Worked Example) You need to know if silica particles 20 μm in diameter with density 2 g/cm^3 are likely to settle in an aqueous dispersion containing 10^{-5} M NaCl and 10% by volume of small, well-dispersed, 200-nm-diameter polystyrene spheres, each with a charge of $+1000e$ on its surface. The dielectric constant is 50.

ANSWER:

Since the density difference between the silica particles and water is $\Delta\rho$ is 2 g/cm^3 $-$ 1 g/cm^3 = 1 g/cm^3, and the radius of the large particles is $a_p = 10\,\mu$m, we find (in cgs units) that $a_p^4\Delta\rho g/k_B T \approx (10^{-3})^4 \times 980/4 \times 10^{-14} = 2.5 \times 10^4$. Thus the silica particles are large enough to be non-Brownian, and they will settle unless the dispersion of small particles has a yield stress high enough to support their weight. The dispersion will have a yield stress if the particles are repulsive enough to form a colloidal crystal. Our strategy, then, is to compute the "effective volume fraction" of particles using Eq. (6-64), to see if it exceeds that required for a colloidal crystal, namely, $\phi_{\text{eff}} \approx 0.6$. To determine if this is the case, we first compute the surface charge σ on each particle in units of C/m^2:

$$\sigma = \frac{1000e}{4\pi a^2} = \frac{1000 \times 1.6 \times 10^{-19}}{4 \times 3.14 \times (10^{-7})^2} = 0.13 \times 10^{-2}\,\text{C/m}^2 \tag{A6-16}$$

where we have used the particle radius $a = 100$ nm $= 10^{-7}$ m, and we are from now on working in mks units. Next, we need the Debye length, κ^{-1}. We must remember that the 10^{-5} M salt NaCl is augmented by the counterions needed to neutralize the surface charges on the particles. Thus, we use Eq. (6-61):

$$\kappa^2 = \frac{e^2}{\varepsilon\varepsilon_0 k_B T}\,\frac{2z^2 n_b + \dfrac{3|\sigma|z\phi}{ae}}{1-\phi} \tag{A6-17}$$

where n_b is the number of ion pairs in the added electrolyte. Thus, multiplying the molarity by Avogadro's number, n_b is $10^{-5} \times 6 \times 10^{23}$ liter^{-1}. Converting from inverse liters liter^{-1} to m^{-3} gives $n_b = 6 \times 10^{21}$ m^{-3}. We then obtain

$$2z^2 n_b + \frac{3|\sigma|z\phi}{ae} = 12 \times 10^{21} + \frac{3 \times 0.13 \times 10^{-2} \times 10^{-1}}{10^{-7} \times 1.6 \times 10^{-19}}$$

$$= 1.2 \times 10^{22} + 2.4 \times 10^{22} = 3.6 \times 10^{22}$$

Inserting this into Eq. (A6-17) gives

$$\kappa^2 = \frac{(1.6 \times 10^{-19})^2}{50 \times 8.8 \times 10^{-12} \times 4 \times 10^{-21}} \times 3.6 \times 10^{22}\,\frac{1}{1-0.1} \approx 6 \times 10^{14}\,\text{m}^{-2}$$

Hence

$$\kappa = 2.5 \times 10^7\,\text{m}^{-1} \tag{A6-18}$$

In the limit of weak electrostatic interactions, we can compute ψ_s from Eq. (2-51):

$$\psi_s = \frac{\sigma}{\varepsilon\varepsilon_0\kappa} = \frac{0.13 \times 10^{-2}}{50 \times 8.8 \times 10^{-12} \times 2.5 \times 10^7} = 0.12\,\text{J/C} = 120\,\text{mV} \tag{A6-19}$$

Now, we check our assumption that electrostatic forces are weak; we compute $ez\psi_s/4k_B T = 1.2$. This is just about unity, so our assumption of a weak electrostatic interaction is barely acceptable. Although we will complete the calculation under the assumption of a "weak" electrostatic force, the result should be checked by comparison against a numerical solution for the surface potential. We next compute α from Eq. (6-60):

$$\alpha \equiv 4\pi\varepsilon\varepsilon_0\psi_s^2 a^2\kappa\,\exp(2a\kappa)/k_B T = 5000\exp(5) = 7.4 \times 10^5 \tag{A6-20}$$

We now use Eq. (6-63) to get d_{eff}:

$$d_{\text{eff}} \approx \frac{1}{\kappa} \ln\{\alpha / \ln[\alpha / \ln(\alpha / \ldots)]\} = 11/\kappa = 4.4 \times 10^{-7}\,\text{m} \tag{A6-21}$$

Equation (A6-21) is valid when $d_{\text{eff}} > 2a$, which holds in this case and justifies using the electrostatic potential in Eq. (6-59).

At last, we use Eq. (6-64) to obtain

$$\phi_{\text{eff}} \equiv \phi \left(\frac{d_{\text{eff}}}{2a}\right)^3 = 10.6\phi = 1.06$$

This hypothetical "effective" concentration is certainly higher than that required for colloidal crystallization, so we should expect the suspension to have a yield stress. To estimate the magnitude of the yield stress σ_y, we use Eq. (6-69):

$$\sigma_y \approx K \left(\frac{W(r_m) - k_B T}{(r_m/2)^3}\right) \tag{A6-22}$$

$W(r_m)$ is obtained from Eq. (6-59),

$$\frac{W(r_m)}{k_B T} = \frac{\alpha \, \exp(-\kappa r_m)}{\kappa r_m} \tag{A6-23}$$

with r_m given by Eq. (6-68), where we take ϕ_m from close packing in an FCC lattice:

$$r_m = 2a \left(\frac{\phi_m}{\phi}\right)^{1/3} = 2 \times 10^{-7}\,\text{m} \left(\frac{0.74}{0.1}\right)^{1/3} = 3.9 \times 10^{-7}\,\text{m} \tag{A6-24}$$

Putting this into Eq. (A6-23) gives

$$\frac{W(r_m)}{k_B T} = 4.7, \qquad \text{so } W(r_m) = 19 \times 10^{-21}\,\text{J}$$

Inserting this and Eq. (A6-24) into Eq. (A6-22), along with taking $K = 0.1$, gives

$$\sigma_y = 0.1 \left(\frac{19 \times 10^{-21} - 4 \times 10^{-21}}{(1.95 \times 10^{-7})^3}\right) = 0.2\,\text{Pa} \tag{A6-25}$$

By dimensional analysis, we estimate the order of magnitude of the stress σ_g needed to support the weight of the $a_p = 10\text{-}\mu\text{m-radius}$ silica spheres to be

$$\sigma_g \sim \frac{4}{3}\pi g(\rho_p - \rho_s)a_{\text{silica}} \approx 0.4\,\text{Pa}$$

where we have been careful to express all quantities (including densities) in mks units. The above estimate of the stress needed to support a large particle is twice the estimated yield stress of the suspension in Eq. (A6-25). Given the approximations involved in the calculations, σ_y and σ_g are too close to each other to draw a final conclusion, So, in this case one must do the experiment!

Problem 6.10 To make automobile panels, you inject a polymer melt filled with $10\text{-}\mu\text{m-long}$ fibers into a mold with a thin gap. The melt flows along the long direction x of the mold; the thin direction is y. In what direction do you expect the fibers to orient?

Chapter 7

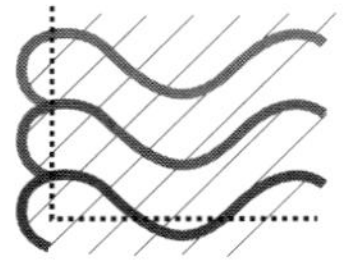

PARTICULATE GELS

7.1 INTRODUCTION

Ceramic coatings, films, and parts, such as fibers and catalyst supports, are often manufactured using the "sol–gel" process, in which a liquid suspension, or "sol," of colloidal particles, such as ZrO_2, SiO_2, or Al_2O_3, is "gelled," or flocculated, into a quasi-solid mass by addition of a chemical agent (Brinker and Scherer 1990). Solvent is then removed from the gel by drying or extraction, any organics are burned out, and the gel is sintered at high temperature to form a useful glass or ceramic coating or part. Sometimes the particles are themselves grown in solution from monomeric species such as aqueous tetraethoxysilane (TEOS) by polymerization reactions; these particles then flocculate into a gel under appropriate pH and salinity conditions (Brinker and Scherer 1990).

For an initially stable sol composed of colloidal particles, the gelling agent, which is usually a pH modifier, an electrolyte, or a polymer, produces gelation by reducing repulsive particle–particle interactions, so that attractive van der Waals forces can draw particles into near contact. If the particle concentration is high enough (around a few volume percent or higher), a sample-spanning network of such contacts forms, producing a solid-like gel phase.

Given the universality of attractive van der Waals forces, the gelation phenomenon is perhaps less in need of explanation than is the existence of a stable dispersed particulate sol phase under appropriate conditions. Clumping of micron- and submicron-sized particles is in fact the norm; preparation of stable sols, on the other hand, requires special techniques, of which there are several. One is simply to match the index of refraction of the particles to that of the suspending medium, so that the Hamaker constant, which determines the magnitude of the van der Waals interactions, is small. Such a dispersed phase cannot be conveniently gelled, however, except perhaps by changing the temperature enough to increase the van der Waals interactions among the particles.

A more useful way to produce a sol phase is to use particles whose surfaces in solution are charged, resulting in *electrostatic stabilization* of the sol. Many oxide particles, such as SiO_2 or TiO_2, contain hydroxyl (–OH) groups at their surfaces that can hydrolyze in aqueous media to form negatively charged $-O^-$ groups; these can stabilize the suspension (Israelachvili 1991; Adamson and Gast 1997). Addition of an acid or an acid former tends to neutralize these groups, producing gelation. Alternatively, salt can be added, which at high enough concentration, often around 0.1 M, collapses the diffuse electrostatic double layer so that particles can approach closely enough to be drawn into near contact by van der Waals forces.

Yet another way to stabilize colloidal dispersions is to *graft* or *adsorb* surfactant, hydrocarbon, or polymer chains to the surfaces of particles, producing a steric barrier to flocculation (Russel et al. 1989). If the chains are long enough and interact favorably with the solvent, the particles are kept at arm's length from each other and cannot flocculate. When desired, flocculation can then be induced by changing the temperature so that the solvent is repelled from the adsorbed chains, and the chains then "stick" to each other, thereby binding together the particles to which they are attached.

As discussed below, a dispersion that has been somehow stabilized can also be made to flocculate by adding to the suspension a *nonadsorbing* polymer, which induces *depletion flocculation* (Asakura and Oosawa 1954, 1958; Vrij 1976; Fleer and Scheutjens 1982; Li-in-on et al. 1975; Vincent et al. 1988; Liang et al. 1994).

Long polymer molecules that strongly adsorb to the particle surfaces can also induce flocculation by *bridging* the gap between neighboring particles (Russel et al. 1989; Otsubo 1993). The rheological properties of mixtures of particles and adsorbing polymers in a solvent bear a resemblance to those of polymeric physical gels (see Section 5.4), wherein the particles play the role of cross-linkers, binding different polymer molecules together. When the concentration ratio of particles to polymers is not too extreme, these suspensions form bridging-flocculated gels. An example is silica particles and polyethylene oxide polymer in water, which gels and exhibits shear-thickening transitions (Cabane et al. 1997), analogous to the behavior of polymeric physical gels.

7.2 PARTICLE INTERACTIONS IN SUSPENSIONS

7.2.1 Interparticle Potentials

In principle, it should be possible to predict the critical conditions for flocculation and the rheological properties of the flocculated gel from the effective interparticle potential $W(D)$, where D is the distance or gap separating the surfaces of neighboring particles. Typically, $W(D)$ is estimated by adding together contributions from hard-core steric interactions, van der Waals interactions, electrostatic interactions, and possibly other interactions such as those from absorbed, grafted, or dissolved chains, or from thin (so-called Stern) layers of adsorbed ions, along with their hydration shells (Israelachvili 1991; Russel et al. 1989; Adamson and Gast 1997).

For spherical particles, some of these interactions can be described by simple formulas. The hard-core steric potential is simply

$$W_{\text{steric}} = \begin{cases} \infty & D < 0 \\ 0, & D \geq 0 \end{cases} \tag{7-1}$$

The van der Waals potential is [Russel et al. 1989; see Eq. (2-36)]

$$W_{\text{vdw}} = -\frac{A_H}{12} \left\{ \frac{4a^2}{4aD + D^2} + \left(\frac{2a}{2a + D} \right)^2 + 2\ln\left[1 - \left(\frac{2a}{2a + D} \right)^2 \right] \right\} \tag{7-2}$$

where A_H is the Hamaker constant and a is the particle radius. For small separations D, this reduces to simply

$$W_{\text{vdw}} \approx -\frac{a A_H}{12 D} \qquad \text{for } D \ll a \tag{7-3}$$

The electrostatic potential, in the limit that the the gap between particles is small and the surface potential ψ_s is constant, is [Eq. (2-59)]

$$W_e = 2\pi \varepsilon_0 \varepsilon a \psi_s^2 \ln[1 + \exp(-\kappa D)] \tag{7-4a}$$

while in the limit of constant surface charge [Eq. (2-60)] we obtain

$$W_e = 2\pi \varepsilon_0 \varepsilon a \psi_s^2 \ln\left[\frac{1}{1 - \exp(-\kappa D)}\right] \tag{7-4b}$$

Here $\varepsilon_0 = 8.8 \times 10^{-12} \text{ C}^2 \text{ J}^{-1} \text{ m}^{-1}$ is the permittivity of space, ε is the dielectric constant of the medium, ψ_s is the electrostatic potential at the particle surfaces, and κ^{-1} is the Debye screening length, $\kappa \equiv (\sum_i n_{\infty i} e^2 z_i^2 / \varepsilon \varepsilon_0 k_B T)^{1/2}$. For simple 1:1 electrolytes such as NaCl at room temperature, κ^{-1} is given by (Israelachvili 1991)

$$\kappa^{-1} = \frac{0.304}{\sqrt{[\text{NaCl}]}} \qquad \text{nanometers} \tag{7-5}$$

where [NaCl] is the molarity of the salt. For particle separations greater than a Debye length κ^{-1}, there is little difference between W_e for the constant-potential boundary condition and that for the constant-surface-charge boundary condition.

The surface potential ψ_s is often equated with the *zeta potential* ζ, which is obtained by measuring the rate of migration of particles under an electric field. The zeta potential is the potential at the particle's *shear surface*, which can be displaced from the true particle surface when there is a Stern layer of tightly bound ions. The pH at which the zeta potential is zero is known as the *isoelectric point*; in general it can be different from the *point of zero charge*, which is the pH required to neutralize the surface of the particle.

The potentials (7-1), (7-2), and (7-4a), when combined, form the basis of the celebrated DLVO (Derjaguin and Landau, 1941; Verwey and Overbeek, 1948) theory of colloid stability. This theory is useful in predicting the conditions of surface potential, ionic strength, and so on, under which flocculation will occur. But the theory has important limitations, in part because it only considers van der Waals, electrostatic, and hard-core interactions.

The theory can be extended to include additional interactions. For example, if non-adsorbing polymer is present in solution, exclusion of this polymer from regions where the particles are closer together than the radius of gyration of the polymer molecules produces a potential W_{depl} that is roughly the osmotic pressure Π times the volume of layers from which polymer is depleted. Thus (Patel and Russel 1987; Russel et al. 1989)

$$W_{\text{depl}} = \begin{cases} -\frac{4\pi}{3}(a+\Delta_d)^3 \left[1 - \frac{3(2a+D)}{4(a+\Delta_d)} + \frac{(2a+D)^3}{16(a+\Delta_d)^3}\right] \Pi, & 0 \le D < 3\Delta_d \\ 0, & D > 3\Delta_d \end{cases} \tag{7-6}$$

where Δ_d is the depletion-layer thickness. In dilute solutions, Δ_d is roughly the polymer radius of gyration R_g. In more concentrated semidilute solutions, where polymer molecules overlap, Δ_d is smaller than this.

The osmotic pressure Π is given by

$$\Pi \approx \nu^{(R)} k_B T \left(1 + \frac{A_2 M_n^2 \nu^{(R)}}{N_A} \right) \tag{7-7}$$

where $\nu^{(R)}$ is the number of polymer molecules per unit volume of "free solution"—that is, solution not occupied by colloidal particles or their depletion regions (Ilett et al. 1995). $\nu^{(R)}$ is also the number of coils per unit volume of a particle-free polymer solution in osmotic equilibrium with the sample. $\nu^{(R)}$ is related to ν, the number of polymer molecules per unit total volume of colloidal solution by $\nu = f\nu^{(R)}$, where f is the fraction of solution that is "free." It can be obtained from (Lekkerkerker et al. 1992; Ilett et al. 1995):

$$f = (1 - \phi) \exp[-A\gamma - B\gamma^2 - C\gamma^3]$$

with $\gamma \equiv \phi/(1-\phi)$, $A \equiv 3\xi + 3\xi^2 + \xi^3$, $B \equiv 4.5\xi^2 + \xi^3$, and $C \equiv 3\xi^3$, with $\xi \equiv \Delta_d/a$. M_n in Eq. (7-7) is the number-averaged polymer molecular weight, and A_2 is the second virial coefficient; for a theta solvent, $A_2 = 0$, while in a good solvent, $A_2 > 0$. The better the solvent, the stronger the depletion flocculation.

When the particles are in contact, W_{depl} is at its minimum value, which is

$$(W_{\text{depl}})_{min} = -\frac{4\pi \Delta_d^3}{3} \left(1 + \frac{3a}{2\Delta_d} \right) \Pi \tag{7-8}$$

Recently, direct experimental measurements of the depletion force using the surface forces apparatus were reported by Kuhl et al. (1996); the magnitude of the force was found to be in agreement with theoretical expectations.

Figure 7-1 gives examples of some of the above interaction potentials. The total interaction potential is usually assumed to be just the sum the individual contributions:

$$W = W_{\text{steric}} + W_{\text{vdw}} + W_e + W_{\text{depl}} + W_{\text{hyd}} \tag{7-9}$$

where W_{hyd} is the "hydration layer" potential to be discussed shortly. Although the individual potentials are typically monotonic functions of particle separation D, the sum of all contributions has one or more local minima. Since the magnitude of the van der Waals interaction grows more steeply with decreasing separation than any other interactions except very short range steric interactions, there is a deep (in theory, infinitely deep) attractive "well" at separations near zero. This well is called the *primary minimum*. There may also be a much shallower *secondary minimum* at larger separations (see Fig. 7-1).

In principle, spherical particles that come close enough together should always fall into an infinitely deep van der Waals primary minimum. In practice, as particle separations shrink to a couple of nanometers or less, there are strong steric forces that can prevent closer approach. In aqueous systems, steric forces are generated by adsorbed ions and hydration layers—that is, water that is hydrogen-bonded with ions adsorbed onto the particle surface. The effective potential, W_{hyd}, produced by these ions and hydration layers is sensitive to the type of absorbed ions that are present on the surface. W_{hyd} has been represented as an exponentially decaying function of the particle–particle separation (Israelachvili 1991):

$$W_{\text{hyd}} = W_0 \exp(-D/2\Delta_{\text{hyd}}) \tag{7-10}$$

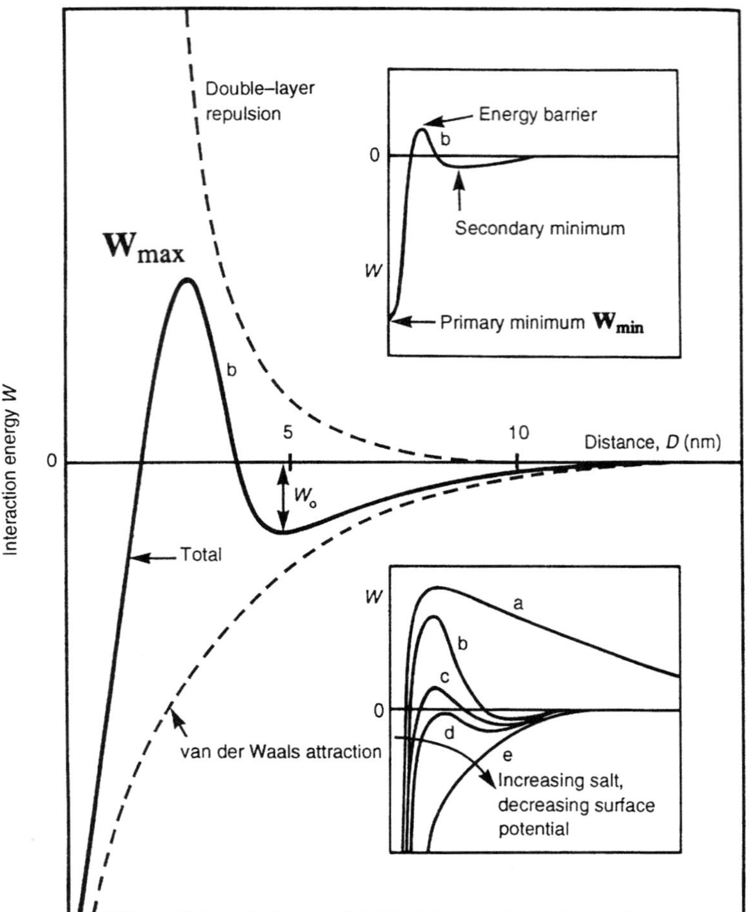

Figure 7.1 Schematic diagram of interaction potential versus separation distance D for van der Waals and electrostatic "double-layer" interactions. The lower inset shows the collapse of the repulsive barrier as the electrolyte concentration is increased or the surface potential is decreased. At a separation distance of zero, there is an infinitely steep hard-core repulsive (or positive) interaction. (From Israelachvili 1991, reprinted with permission from Academic Press.)

where W_0 is typically around 3–30 mJm^{-2} and the range of the interaction is small, so that $\Delta_{\mathrm{hyd}} \approx 1$ nm. Because of its short range, the "hydration" interaction is sometimes incorporated into the total potential W simply by slightly enlarging the effective hard-core radius of the particles used to compute W_{steric} from Eq. (7-1), while using just the "bare" radius a in the formula for the van der Waals interaction. In this way, the infinite van der Waals well is cut off, and it is replaced by a well of finite depth.

Once flocculation has occurred, the properties of the flocculated dispersion depend strongly on the details of the interaction at separations near the primary minimum, including the depth of this primary minimum. This depth is often estimated simply by evaluating the potential W at the separation D_0 given by twice the thickness Δ_{hyd} of the assumed hydration or Stern layer. Since Δ_{hyd} is very small, on the order of a nanometer or so, the electrostatic interaction is sensitive to the choice of a constant potential or a constant surface-charge boundary condition (see Fig. 2-14). The electrostatic potential W_e obtained using the constant potential boundary condition, Eq. (7-4a), is bounded at small separation distances, while W_e from the constant charge boundary condition, Eq. (7-4b), is singular as $D \to 0$, although it is less singular than the van der Waals potential. Experimental data from the surface-forces apparatus (Israelachvili 1991) indicate that the electrostatic contribution lies between the two limits, but is closer to that for a constant surface charge. Rheological measurements also seem to support the constant-surface-charge condition (Friend and Hunter 1971). Thus, considering only van der Waals interactions and electrostatic interactions with a constant surface charge, one obtains for the depth of the primary minimum

$$W_{\min} \approx -\frac{a A_H}{12 D_0} + C \zeta^2 a \tag{7-11a}$$

where

$$C \equiv 2\pi \varepsilon_0 \varepsilon \ln \left[\frac{1}{1 - \exp(-\kappa D_0)} \right] \tag{7-11b}$$

and we have replaced the surface potential ψ_s with the more easily measurable zeta potential ζ.

If D_0 is small, the van der Waals interaction usually dominates the electrostatic one, and the primary attractive minimum is then very deep, so that $-W_{\min}/k_B T \gg 1$. Particles that fall into this minimum will therefore "stick" to other particles, and at thermodynamic equilibrium all particles will be clumped together into a single mass. Note, however, that the *depth of the well can be reduced by increasing the magnitude of the zeta potential ζ*.

7.2.2 Electrostatic Stabilization

The stabilization produced by electrostatic forces is often a *kinetic* effect. As one increases the surface potential ψ_s, or the Debye length κ^{-1}, the local maximum in W becomes increasingly positive (see inset in Fig. 7-1). When $W_{\max}$ exceeds many $k_B T$, particles that are initially separated from each other rarely acquire enough kinetic energy to surmount the high repulsive potential barrier $W_{\max}$ and fall into the deep van der Waals primary minimum. [When D_0 and ζ are unusually large, the van der Waals minimum disappears altogether (Frens and Overbeek 1972; Israelachvili 1991).] The probability that a particle–particle collision will be energetic enough to overcome the barrier is proportional to $\exp[-W_{\max}/k_B T]$. Since the time scale for a particle to diffuse a distance a equal to its own radius is $\sim a^2/D_0$, where D_0 is here the diffusivity of the particle, the time t_D required for particles to diffuse into their attractive minima scales as (Russel et al. 1989)

$$t_D \approx \frac{a^2}{D_0} \exp \left(\frac{W_{\max}}{k_B T} \right) = \frac{6\pi \eta_s a^3}{k_B T} \exp \left(\frac{W_{\max}}{k_B T} \right) \tag{7-12}$$

where we have used the Stokes–Einstein value for the diffusivity, $D_0 = k_B T/6\pi \eta_s a$, and η_s is the viscosity of the solvent medium. Because of the exponential dependence of t_D on

$W_{\max}$, even small 100-nm-diameter particles in a low-viscosity solvent take months or even years to aggregate if the potential barrier $W_{\max}/k_B T$ exceeds 25 or so.

Thus, although thermodynamic stability is not attained, surface charges can generate *kinetic stability* of essentially indefinite duration, provided that the sol is stored in conditions that keep $W_{\max}/k_B T$ high, and provided that the sol is not subjected to strong flows that might impart to the particles enough kinetic energy to induce flocculation (Russel et al. 1989).

7.2.3 Flocculation and Gelation

If desired, flocculation of the stabilized sol can be deliberately induced. This can be done, for example, by adding electrolyte to the suspension or by changing the surface charge by altering the pH. A condition for rapid flocculation can be derived (Russel et al. 1989) by finding the electrolyte concentration at which the potential barrier is eliminated, so that $W_{\max} = 0$. Using Eq. (7-3) for the van der Waals interactions, using Eq. (2-57) for the electrostatic interactions, and applying the condition $W' = 0$ at $W = 0$ gives for a symmetric electrolyte the critical bulk electrolyte concentration (number of cations per unit volume) n_{crit}:

$$n_{\text{crit}} = \frac{49.6}{z^6 \ell_b^3} \left[\frac{k_B T}{A_H} \right]^2 \tanh^4 \left(\frac{ez\psi_s}{4k_B T} \right) \tag{7-13}$$

where

$$\ell_b \equiv \frac{e^2}{4\pi \varepsilon \varepsilon_0 k_B T} \tag{7-14}$$

is the *Bjerrum length*. At room temperature, $\ell_b \approx 58/\varepsilon$ nm.

Note from Eq. (7-13) that when $ez\psi_s/4k_B T > 1$, there is a strong dependence of the critical electrolyte concentration on charge valency, namely $n_{\text{crit}} \propto z^{-6}$. Thus, multivalent ions are predicted to be very effective at inducing flocculation; this is called the Schulze–Hardy rule (Russel 1989; Israelachvili 1991).

For a symmetric electrolyte and a weak surface potential, using Eq. (2-51) for the relationship between the surface potential ψ_s and the surface charge density σ (Coulombs per unit area) and using Eq. (2-48) for κ (with $n_1 = n_2 = n_{\text{crit}}$), we obtain

$$n_{\text{crit}} = \frac{0.36 \ell_b^{-1} z^{-2} \sigma^{4/3}}{(\varepsilon \varepsilon_0 A_H)^{2/3}} \tag{7-15}$$

This result implies that for the typical values $A_H \approx 10^{-20}$ J and $\varepsilon \approx 50$, a suspension of particles with a surface charge of 0.1 charges/nm^2 will be rapidly flocculated by a 0.1 M solution of univalent electrolyte (see Worked Example 7.4 at the end of this chapter).

As flocculation continues, particle pairs, or doublets, become triplets, and so on, so that flocs containing many particles appear. This growth process has been studied by light scattering in dilute or modestly concentrated suspensions, with particle volume fractions ϕ less than 0.10. These studies (Dimon et al 1986; Schaefer et al. 1984; Aubert and Cannell 1986) show that the radius a_k of a floc containing k particles scales as a power law with k; that is,

$$\frac{a_k}{a} \approx k^{1/D_f}, \qquad k \approx \left(\frac{a_k}{a}\right)^{D_f} \tag{7-16}$$

If the flocs had consisted of densely packed particles, then the exponent D_f would equal 3, the dimensionality of space. Instead, experiments with aggregated neutral gold particles show $D_f \approx 1.7$–1.8; and for slightly charged silica particles, $D_f \approx 2.0$–2.2. A power-law scaling with $D_f < 3$ implies that the flocs are open, ramified structures, or *fractals* (see Fig. 7-2). In general, a fractal is a self-similar object which looks the same at different magnifications. The smaller the dimension D_f, the more open and porous the fractal floc. Theoretical models show that $D_f = 1.75$ is consistent with a "fast flocculation" process called "diffusion-limited aggregation" (Jullien et al. 1984; Russel 1989), in which flocs grow predominantly by fusing with other flocs as soon as they come into contact with them.

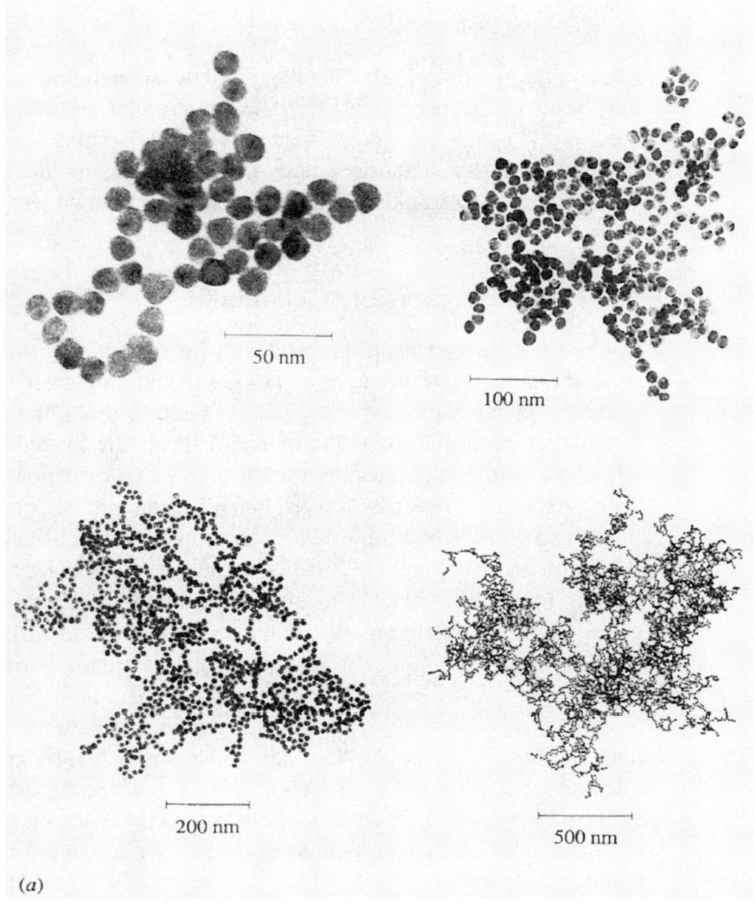

Figure 7.2   Transmission electron micrographs of flocculated gold sols with particle radii $a = 7.2 \pm 0.8$ nm. (From Weitz and Huang 1984, with permission.)

If flocs can interpenetrate to some extent after contacting, the floc becomes denser and D_f increases to $D_f \approx 2.0$–2.2, which corresponds to "slow flocculation," or "reaction-limited aggregation."

As a floc grows, its porosity increases. The volume fraction of solid in a floc containing k particles decreases with increasing k as

$$\phi_k \sim \left(\frac{a_k}{a}\right)^{D_f-3} \approx k^{(D_f-3)/D_f} \tag{7-17}$$

Hence, as the flocs grow larger, the volume fraction of solution they encompass increases. Once the volume fraction permeated by the flocs fills the entire solution, the flocs connect together to form a percolating (sample-spanning) network (Feng and Sen 1984). This is *gelation*, and it occurs when the floc size k reaches a value k^* such that ϕ_{k^*}, the volume fraction of particles in the floc, equals ϕ, the volume fraction of particles in solution. This condition gives

$$k^* \sim \phi^{-D_f/(3-D_f)} \tag{7-18}$$

From this equation, we find that there will be isolated flocs containing 100 or more particles only when ϕ is small, $\phi < 0.1$. Since the particles are usually denser than the suspending medium, isolated large flocs, formed when ϕ is small, can sediment under gravity. For larger ϕ, however, gelation usually occurs, and sedimentation is then avoided, unless the gel structure is so fragile that it collapses under its own weight.

7.2.4 Thermoreversible Gelation

Chemically induced gelation, produced for example by addition of an electrolyte as described above, often produces a hard gel that cannot easily be restored to a fluid state. To study the gelation process in more detail and to probe the rheology of gels as a function of the strength of interparticle bonds, it is desirable to study *thermoreversible* gelation, in which the gel transition can be traversed in either direction as many times as desired with a single sample, merely by lowering and raising the temperature.

Since particles within a few angstroms of contact bind together strongly by van der Waals forces, weak gels only form if the particle surfaces are somehow prevented from coming into near contact. One successful method for accomplishing this is to adsorb or graft polymeric or oligomeric chains onto the particle surfaces (Russel et al. 1989). The required thickness of the adsorbed layer can be estimated from the van der Waals interaction potential $W_{vdw} \approx -aA_H/12D$. This potential will readily be overcome by Brownian motion if $-W_{vdw} < k_BT$. Thus, the spacing D between particles must be kept larger than $D_{crit} \approx aA_H/12k_BT$. This can be accomplished by adsorbing or grafting a layer of thickness $\Delta_{graft} = \frac{1}{2}D_{crit} = aA_H/24k_BT$ onto each particle's surface. Since a typical value of A_H/k_BT is 2.5, we predict that we shall need a layer whose thickness is about $a/10$, about 10% of the particle's radius. For lower values of A_H/k_BT, layers even thinner than this will suffice.

If these particles are suspended in a solvent that is marginal for the grafted chains, then at high temperature, above the "theta" point for the chains in the solvent (see Section 2.3.1.2), the chains repel each other, and the particles remain dispersed. If, however, the temperature

T is reduced below the theta temperature θ, attractive particle–particle interactions occur, eventually producing flocculation at low enough T. Thus, in such a dispersion, the strength of the particle–particle interactions can be controlled by adjusting the temperature, and a dispersion that has been flocculated at a low temperature can be redispersed merely by increasing T.

One system with these desirable features is a suspension of small ($a \approx 50$ nm) silica particles onto which octadecyl chains have been densely grafted (Stöber et al. 1968; van Helden et al. 1981; Woutersen and de Kruif 1991). The octadecyl chains have a theta point near room temperature in various solvents, including benzene, dodecane, and hexadecane.

The properties of the above system at modest particle concentrations are relatively simple to model, because the grafted octadecyl layer is thin compared to the particle radius and because the particle–particle interactions are weak enough that the properties of the dispersion are not sensitive to the detailed shape of the particle–particle interaction potential. These considerations have motivated the use of a simple "square-well potential" as a model of the particle–particle interactions (Woutersen and de Kruif 1991) (see Fig. 7-3). This potential consists of an infinite repulsion at particle–particle contact (where $D = 0$), bounded by an attractive well of width Δ and depth ε. There are no interactions at particle–particle gaps greater than Δ. Near the theta point, the well depth ε depends on temperature as follows (Flory and Krigbaum 1950):

$$\varepsilon = \alpha \left(\frac{\theta}{T} - 1 \right) k_B T \qquad \text{for } T < \theta \tag{7-19}$$

where α is a constant.

The strength of the particle–particle interactions produced by this potential depend on both the depth ε and width Δ of the potential. However, for narrow wells, $\Delta/a \ll 1$, the particle–particle interactions are controlled only by the combination parameter (Baxter 1968)

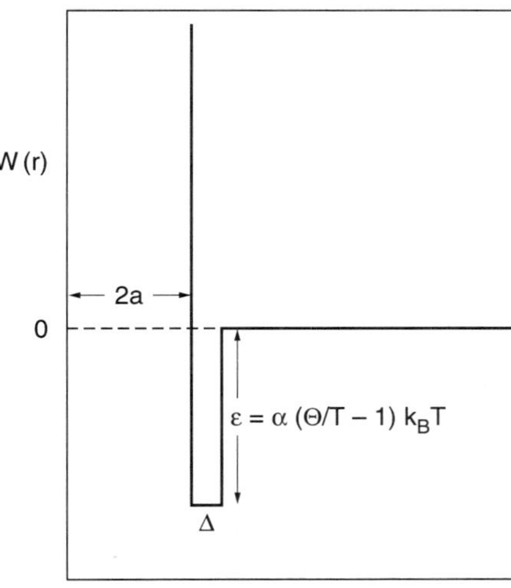

Figure 7.3   The square well potential where $r = 2a + D$ is the separation of the particles' centers of mass. (From Woutersen and de Kruif, reprinted with permission from J. Chem. Phys. 94:5739, Copyright © 1991, American Institute of Physics.)

$$\tau_B = \frac{2a + \Delta}{12\Delta} \exp\left(\frac{-\varepsilon}{k_B T}\right) \tag{7-20}$$

In the limit $\Delta/a \to 0$ at fixed τ_B, the structure and rheology of the suspension depend only on τ_B and not on Δ and ε separately. This limit is Baxter's "adhesive hard-sphere," or "sticky hard-sphere," model (Baxter 1968). The parameter τ_B of the model is a monotonically increasing function of temperature, and thus it can be thought of as a rescaled temperature. The predictions of the adhesive hard-sphere model are in reasonable agreement with light-scattering data for the weakly flocculated silica particles with grafted octadecyl chains at low and modest volume fractions of particles. However, at high particle volume fractions, neutron scattering shows substantial deviations between the measured structure factor and the theoretical one (Woutersen et al. 1993).

7.2.5 Gelation and Phase Separation

For the adhesive hard-sphere model, the theoretical phase diagram in the τ_B–ϕ plane has been partially calculated (Watts et al. 1971; Barboy 1974; Grant and Russel 1993). According to this model, there is a critical point $\tau_{B,c} = 0.0976$ below which the suspension is predicted to *phase separate* into a phase dilute in particles and one concentrated in them (see Fig. 7-4). The particle concentration at the critical point of this phase transition is $\phi_c = 0.1213$. This phase transition is analogous to the gas–liquid transition of ordinary

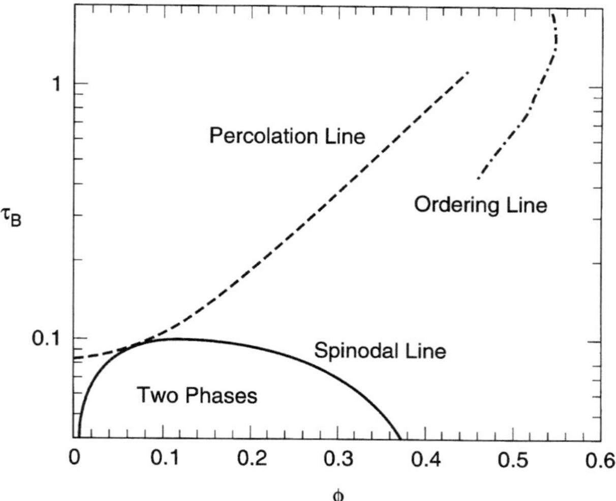

Figure 7.4   Phase diagram for adhesive hard spheres as a function of Baxter temperature τ_B. The solid line is the spinodal line for liquid–liquid phase separation (the dense liquid phase is probably metastable), the dot–dashed line is the "freezing" line for appearance of an ordered packing of spheres, and the dashed line is the percolation transition. (Adapted from Grant and Russel 1993, reprinted with permission from the American Physical Society.)

small-molecule fluids. The phase dilute in colloidal particles is analogous to the "gas," and the concentrated phase corresponds to "liquid" phase of small molecules. (The phase separation of a concentrated from a dilute colloidal fluid phase is known as *coacervation*, if it is induced by addition of nonadsorbing polymer.)

At high particle concentrations, the adhesive hard-sphere model also predicts a transition from the dense disordered "liquid" phase to a macrocrystalline solid-like packing of spheres, analogous to a freezing transition of ordinary small molecules (see Fig. 7-4). (The freezing line in Fig. 7-4 terminates at a relatively high value of τ_B because of computational difficulties in extending it to lower τ_B.) The phase behavior of adhesive hard spheres is in some respects similar to that of ordinary small molecules because both systems are driven by a balance of van der Waals attraction, hard-core repulsion, and thermal agitation.

There are, however, important differences between the phase behavior of "sticky" spheres and that of small molecules, which arise from differences in the relative ranges of the attractive potentials. These differences have been explored in a wonderful set of calculations and experiments by Gast et al. (1983) and Pusey and coworkers (Ilett et al. 1995) for suspensions of spheres that are made to attract each other by the polymer-depletion mechanism. In such systems, the range of the attractive potential relative to the sphere size can be varied by controlling the ratio $\xi \equiv \Delta_d/a$ of the polymer depletion-layer thickness to the sphere radius. For $\xi \to 0$ the potential is short-ranged, like that of sticky hard spheres, while for $\xi \approx 1$ the potential is long-ranged, like that of ordinary molecular liquids. Figure 7-5 shows the phase diagrams computed for the depletion potential, Eq. (7-6), with $\xi = 0.08, 0.33,$ and 0.57. These predicted diagrams were confirmed by experiments on PMMA

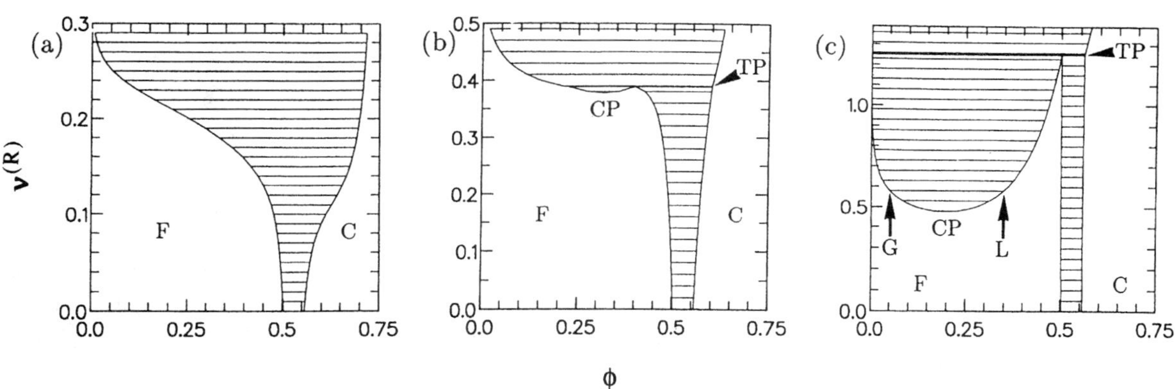

Figure 7.5   Theoretical phase diagrams for mixtures of polymers and colloids for size ratios $\Delta_d/a =$ **(a)** 0.08, **(b)** 0.33, and **(c)** 0.57 of depletion layer thickness (or polymer radius of gyration) to colloidal sphere radius. The ordinate $\nu^{(R)}$ is the number of polymer molecules per unit volume of "free solution." In **(a)**, the range of the attractive depletion interaction is short, and when the potential becomes strong at high $\nu^{(R)}$, the coexistence region of colloidal crystal and disordered fluid becomes broad. For a long-ranged potential in **(c)**, a fluid–fluid coexistence region emerges with a dilute colloidal fluid, or "gas" coexisting with a concentrated colloidal fluid, or "liquid." CP denotes the fluid–fluid critical point, and TP is the triple point, which is a line on these diagrams. (From Ilett et al. 1995, reprinted with permission from the American Physical Society.)

spheres stabilized with a thin layer of grafted chains, in decaline, with added long-chain polystyrene. The ordinate in Fig. 7-5 is the polymer concentration in a hypothetical polymer phase in osmotic equilibrium with the colloidal phase, as discussed in Section 7.2.1; it is closely related to the chemical potential of the polymer. For the long-ranged potential in Fig. 7-5c, the phase diagram is analogous to that of ordinary molecular substance such as water, where $\nu^{(R)}$ is analogous to an inverse temperature, since it controls the strength of the attractive interaction. Thus, if the ordinate of Fig. 7-5c is inverted, the phase diagram is similar to that of a molecular liquid, with its gas, liquid, and crystalline phases.

As the range of the potential is narrowed (Fig. 7-5a), the gas–liquid coexistence region disappears, and at modest $\nu^{(R)}$ the coexisting macrocrystalline solid phase appears at much lower volume fractions of spheres. Comparing Fig. 7-5a with Fig. 7-4 (and inverting the ordinate of Fig. 7-5a, since increasing $\nu^{(R)}$ corresponds to decreasing τ_B), we conclude that in the limit of a very short range interaction, that of adhesive spheres, the ordering line shown in Fig. 7-4, if calculations could be performed to extend it to lower values of τ_B, would curve toward much lower volume fractions, and the fluid–fluid coexistence region in Fig. 7-4 is almost certainly a metastable one. The *equilibrium* phase diagram of sticky hard spheres is thus probably similar to Fig. 7-5a, with the ordinate inverted. Although the fluid–fluid coexistence region in Fig. 7-4 is probably not *thermodynamically* stable, it may play an important role in the *kinetics* of phase separation and gelation of these systems.

The phase diagram of the adhesive hard-sphere model suggests analogies between the *gelation* and *phase separation*. For example, when a dilute suspension is flocculated, large isolated flocs form, which settle under gravity to form a dense "phase" that is separated macroscopically from the remaining dilute suspension. An analogous process occurs in molecular fluids, such as dilute polymer solutions from which solid polymer is precipitated by addition of a nonsolvent. In the latter, the precipitated polymer particles are usually crystalline, while obvious crystalline order is usually absent from particulate flocs. Nevertheless, dense-phase crystalline order can also be produced in particulate flocs, if the particles are nearly monodisperse and if the repulsive barrier is reduced gradually, giving the attracting particles time to develop crystalline structure before they lose the ability to rearrange (Brinker and Scherer 1990). Even rather rigid gels formed by primary van der Waals attractions in dense suspensions manage to contract slowly in volume over a period of hours and days, by exuding solvent; this process is called *syneresis* (Brinker and Scherer 1990).

These examples suggest that at least in some cases of gelation there is a thermodynamic driving force for separation of a dense particulate phase from a more dilute one. But exactly what is the relationship between gelation and phase separation? Figures 7-4 and 7-5 suggest that for low and modest particle volume fractions not far from the critical concentration, an increase in the interparticle forces (by lowering τ_B or raising $\nu^{(R)}$) ought to produce macroscopic phase separation. This prediction is supported by experiments of Rouw et al. (1989), who temperature-quenched initially stable dispersions of 24-nm octadecyl-grafted silica particles ($\phi = 0.13$) in benzene, and then monitored structural changes by light scattering. Several aspects of the observed structural changes were analogous to the process of *spinodal decomposition* in simple liquids, to be described in Section 9.2.1. In particular, soon after the quench, there was an early growth regime characterized by a rapidly growing scattering peak at a constant scattering angle. Later, the peak shifted to smaller angles, corresponding to coarsening of a well-developed pattern of inhomogeneities in particle concentration.

While these aspects of gelation are similar to those of classical spinodal decomposition, the power-law exponents governing the late-stage coarsening in the particulate system differ greatly from the classical values for spinodal decomposition. This is perhaps not surprising, since late-stage coarsening in spinodal decomposition of liquid–liquid mixtures involves interfacial-tension-driven compaction of continuum fluid labyrinths, while aggregated or flocced particles, on the other hand, lack continuum interfaces and may be too rigid to form compact structures. Thus, the fluid-like spinodal pattern of inhomogeneities that appears at short times after quenching a modestly concentrated particle dispersion evidently develops into a rigid fractal-like structure at later times.

When rigid fractal-like structures link up to span the medium, a gel is formed. Hence, the process of gelation has been compared to *percolation* (see Section 5.2.1). From the sticky hard-sphere model, the percolation line on the τ_B–ϕ phase diagram can be calculated (see Fig. 7-4). At the percolation transition, the theory gives for the Baxter parameter

$$\tau_{B,\text{perc}} = \frac{19\phi^2 - 2\phi + 1}{12(1 - \phi)^2}$$

Below the percolation line, there is predicted to be a sample-spanning cluster of contacting spheres. Woutersen et al. (1994) found that the gel point for 47-nm octadecyl-grafted silica spheres in benzene is in reasonable agreement with the predicted percolation transition. However, Grant and Russel (1993) found that the gelation line is below the percolation line for a similar suspension in hexadecane. According to Fig. 7-4, when a dispersion with particle concentrations well above the critical point $\phi_c \approx 0.12$ is cooled, the percolation transition is encountered well before the liquid–gas phase boundary. Hence, if gelation does correspond to the percolation line, Fig. 7-4 implies that *gelation can occur without phase separation or syneresis,* if τ_B is not too small—that is, if the gel is not too strongly flocculated. We should note, however, that since the position of the freezing line for adhesive spheres at low τ_B is not known, there remains the possibility that gelation of adhesive spheres corresponds not to percolation, but to an arrested fluid–solid transition. The relationship between gelation and phase separation is no mere academic issue; phase separations produce inhomogeneities in density and structure that are deleterious to a gel body's strength, rigidity, and resistance to cracking (Brinker and Scherer 1990).

7.3 RHEOLOGY OF PARTICULATE GELS

Rheological measurements on concentrated, strongly flocculated gels are hampered by the following difficulties (Buscall et al. 1986, 1993; Rueb and Zukoski 1997):

1. Poor reproducibility
2. Sensitivity to gel preparation
3. Sensitivity to shear history
4. Extremely limited range of linear viscoelastic response
5. Slip

The first three or four of these experimental difficulties arise because of the nonequilibrium structure of strongly flocculated gels. Particles bound strongly together in their primary van der Waals potential minima are unable to rearrange within laboratory time scales; hence the structures cannot relax to achieve thermodynamic equilibrium. Therefore, the gel structure depends on preparation history, including any deformation experienced by the gel prior to the rheological measurement.

These problems can be avoided in weakly flocculated gels, which attain thermodynamic equilibrium on experimental time scales. Since an understanding of weakly flocculated gels can form the basis for study of strongly flocculated gels, we shall consider the former first.

7.3.1 The Rheology of Weakly Flocculated Gels

Figure 7-6 shows the viscosity of thermoreversible dispersions (discussed in Section 7.2.4) of $a \approx 50$-nm silica particles onto which octadecyl chains have been densely grafted in benzene at particle volume fractions $\phi = 0.088$–0.133, as a function of temperature T (Woutersen and de Kruif 1991). For $T \geq \theta = 316$ K, the viscosity relative to that of the solvent, $\eta_r \equiv \eta/\eta_s$, is independent of temperature, and its dependence on volume fraction ϕ is exactly as expected for hard spherical particles without attractive interactions. As T is lowered below θ, however, the viscosity rises rapidly, because of the onset of attractive interactions.

The effect of these attractive interactions is even more dramatic at higher particle concentrations. Figure 7-7 shows the relative viscosity at $\phi = 0.47$, as well as the longest relaxation time for $\phi = 0.40$ and 0.47, at low shear rates. Note that there is a large increase in η_r as the temperature is lowered. Figure 7-8 shows the shear-rate dependence of η_r for $\phi = 0.367$ at several temperatures. The bland, Newtonian behavior at high T gives way to strong shear thinning when $T \leq \theta - 13$ K $= 303$ K. The complex viscosity η^* scales roughly as $\omega^{-1/2}$ at high frequency ω (Woutersen et al. 1994).

The expression for the zero-shear viscosity for the adhesive hard-sphere model at modest particle volume fraction is (Woutersen and de Kruif 1991)

$$\eta_{r,0} = 1 + 2.5\phi + \left(6.2 + \frac{2.1}{\tau_B}\right)\phi^2 \tag{7-21}$$

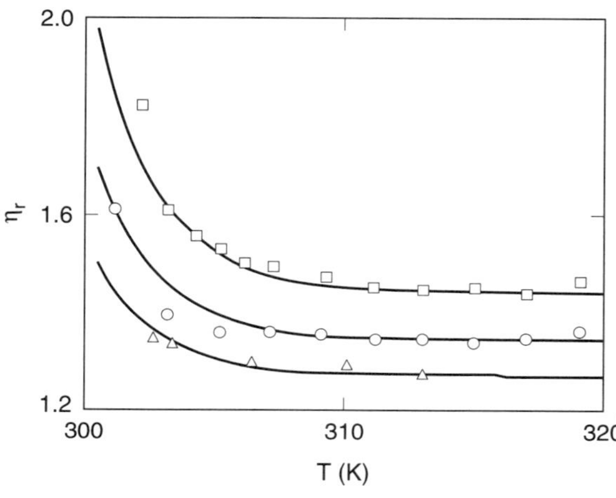

Figure 7.6 Relative viscosity as a function of temperature T for dispersions in benzene of octadecyl-grafted silica spheres with radii $a = 47$ nm at volume fractions ϕ of 0.133 ($\square$), 0.106 ($\bigcirc$), and 0.088 ($\triangle$). The lines are the predictions of Eq. (7-21), with square well parameters of $\Delta = 0.3$ nm, $\alpha = 117$ and $\theta = 316$ K. (From Woutersen and de Kruif, reprinted with permission from J. Chem. Phys. 94:5739, Copyright © 1991, American Institute of Physics.)

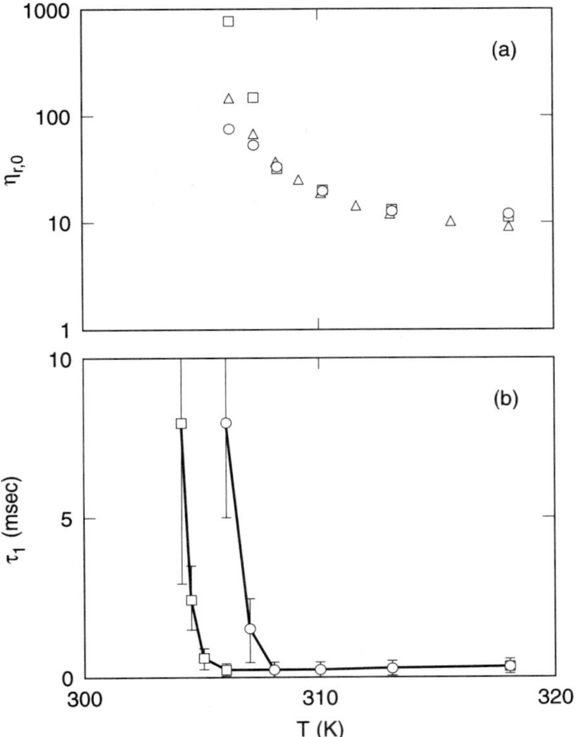

Figure 7.7 **(a)** Relative viscosity as a function of temperature T for a dispersion described in Fig. 7-6 at a volume fraction ϕ of 0.47, extrapolated to low shear rates ($\bigcirc$, $\triangle$), and at low shear frequencies ($\square$), from steady shearing and oscillatory shearing, respectively. **(b)** Longest relaxation time τ_1 as a function of temperature for $\phi = 0.40$ ($\square$) and 0.47($\bigcirc$). (From Woutersen et al., reprinted with permission from J. Chem. Phys. 101:542, Copyright © 1994, American Institute of Physics.)

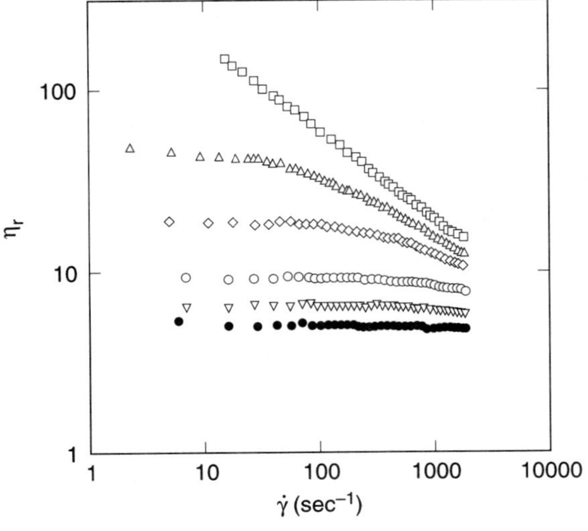

Figure 7.8 Shear-rate dependence of the relative viscosity for a dispersion similar to that described in Fig. 7-6 at a volume fraction ϕ of 0.367 and at temperatures (from bottom to top) of 317.28 K, 308.13 K, 306.20 K, 304.17 K, 303.16 K, and 302.16 K. (From Woutersen and de Kruif, reprinted with permission from J. Chem. Phys. 94:5739, Copyright © 1991, American Institute of Physics.)

The lines in Fig. 7-6 show that this expression fits the temperature-dependence of the viscosity for $\phi \lesssim 0.15$. Combining this with Eq. (7-20), we find that the attractive interactions between particles contribute a term proportional to $\phi^2 \exp(\varepsilon/k_B T)$ to the relative viscosity. Thus, *the viscosity increases exponentially with the depth of the potential well.*

A similar, and even more dramatic, viscosity enhancement was observed by Buscall et al. (1993) for dispersions of 157-nm acrylate particles in "white spirit" (a mixture of high-boiling hydrocarbons). These particles were stabilized by an adsorbed polymer layer, and then they were depletion-flocculated by addition of a nonadsorbing polyisobutylene polymer. Figure 7-9 shows curves of the relative viscosity versus shear stress for several concentrations of polymer at a particle volume fraction of $\phi = 0.40$. Note that a polymer concentration of 0.1% by weight is too low to produce flocculation, and the viscosity is only modestly elevated from that of the solvent. For weight percentages of 0.4–1.0%, however, there is a 3–6 decade increase in the zero-shear viscosity!

Figure 7-10 shows the zero-shear viscosity $\eta_{r,0}$ plotted semilogarithmically against polymer concentration. The lines through the data are fits to an exponential dependence of $\eta_{r,0}$ on the depth of the potential well $-W_{\min}$, that is,

$$\eta_{r,0} \propto \exp\left(\frac{-\alpha W_{\min}}{k_B T}\right) \tag{7-22}$$

(Keep in mind that $W_{\min}$ is negative.) The potential well depth was calculated using Eq. (7-8) for depletion flocculation. At the highest polymer concentrations, the dimensionless well depth reaches $-W_{\min}/k_B T \approx 18$. (At polymer concentrations high enough that polymer coils begin to overlap, which is the beginning of the semidilute regime, the depletion layer thickness begins to decrease with increasing polymer concentration, which

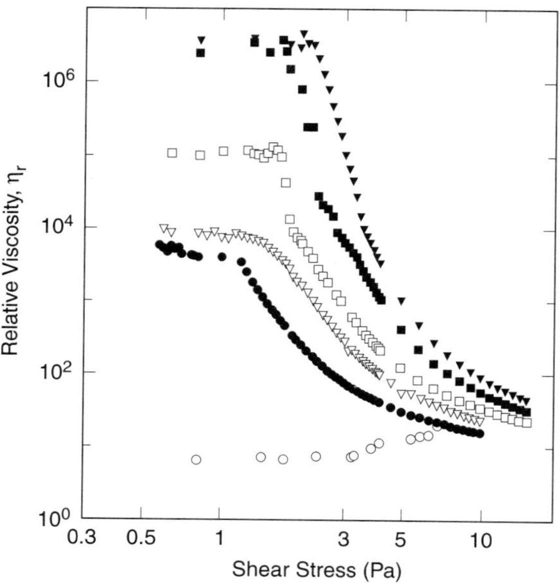

Figure 7.9 Shear-stress dependence of the relative viscosity for dispersions in "white spirit" of acrylic copolymer particles of radius $a = 157$ nm at a volume fraction of $\phi = 0.40$ for differing concentrations of nonadsorbing polyisobutylene polymer of molecular weight 411,000. The particles had been stabilized by addition of a comb–graft copolymer of PMMA backbone (which adsorbed to the particles) with non-adsorbing poly(12-hydroxystearic acid) teeth. The concentrations (in weight per unit volume) of polyisobutylene are 1.0% (▼), 0.85%(■), 0.6%(□), 0.5%(∇), 0.4%(●), and 0.1%(○). (From Buscall et al. 1993, with permission from the Journal of Rheology.)

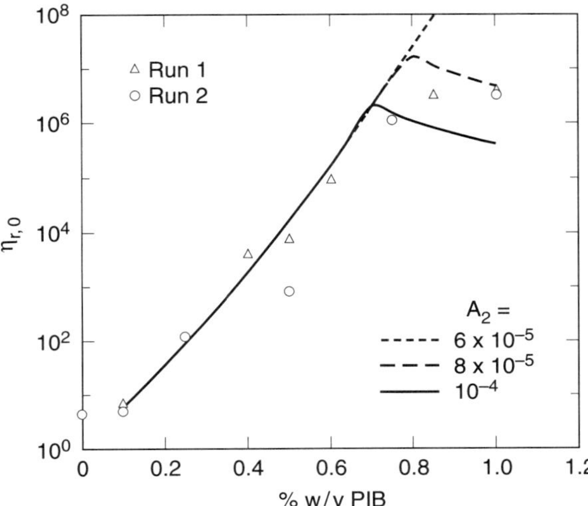

Figure 7.10 The effect of polyisobutylene (PIB) concentration on the zero-shear viscosity of the suspensions described in Fig. 7-9. The lines were calculated assuming $\eta_{r,0}(0) = K\,\exp(-\alpha W_{\min}/k_B T)$, with values of the second virial coefficient A_2 of 6×10^{-5}, 8×10^{-5}, and 10^{-4}. (From Buscall et al. 1993, with permission from the Journal of Rheology.)

is apparently responsible for the turnover of $\eta_{r,0}$ at the highest polymer concentration in Fig. 7-10.)

The increased zero-shear viscosity of more strongly flocculated gels is a consequence of an increase in the *relaxation time* τ required for particles to rearrange their positions (see Fig. 7-7). The relaxation time τ is estimated to be

$$\tau \approx \frac{a^2}{D_0}\,\exp\left(\frac{-\alpha W_{\min}}{k_B T}\right) = \frac{6\pi\eta_s a^3}{k_B T}\,\exp\left(\frac{-\alpha W_{\min}}{k_B T}\right) \tag{7-23}$$

where D_0 is the particle free diffusivity. Hence, as noted earlier, the larger $-W_{\min}/k_B T$ is and the more strongly a gel is flocculated, the longer it takes to reach thermodynamic equilibrium. In a strongly flocculated alkoxide silica gel formed by polymerization of tetrethyl orthosilicate (TEOS), a slow creep test showed the relaxation time to be a day or longer (Scherer et al 1988) and showed the low-shear-rate viscosity to be as high as 10^{12} Pa!

Note in Fig. 7-9 that as the gel becomes more strongly flocculated, not only does the low-shear-rate plateau viscosity becomes larger, but the drop in viscosity in the shear thinning region becomes steeper (Fedotova et al. 1967). The onset of shear thinning occurs at a critical shear stress of about 3 Pa. An even more sudden decrease in viscosity at a stress of about 10 Pa is shown in Fig. 7-11 for 2.5% flocculated silica particles in methyl laurate. When the plateau viscosity becomes very high, it is only accessed at very low shear rates, $\dot{\gamma} \sim 10^{-5}$–$10^{-3}$ sec^{-1}. Since experiments at shear rates less than 10^{-3} sec^{-1} require hours or days to carry out, plateaus that exist only at shear rates less than $\sim 10^{-3}$ sec^{-1} will usually not be observed in rheological experiments, unless great patience is exercised. Instead, at the lowest experimental shear rates, a steep decrease in viscosity with increasing shear stress will be seen. Figure 7-12 is an example of such viscosity curves (Patel and Russel 1987), obtained for depletion-flocculated dispersions of surfactant-stabilized polystyrene particles ($\phi = 0.20, 0.30$; $a = 115$ nm). Behavior of this sort—a sudden decrease in viscosity above

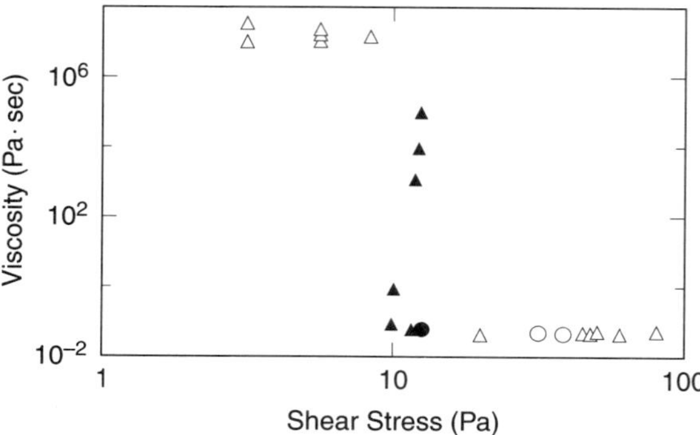

Figure 7.11 Viscosity versus shear stress for flocculated 2.5% silica particles in methyl laurate. The open triangles are from a stress-controlled instrument, while the other symbols are from a shear-rate-controlled one. (Reprinted from Coll. Surf., 69:15, Van der Aerschot and Mewis (1992), with kind permission from Elsevier Science - NL, Sara Burgerhartstraat 25, 1055 KV Amsterdam, The Netherlands.)

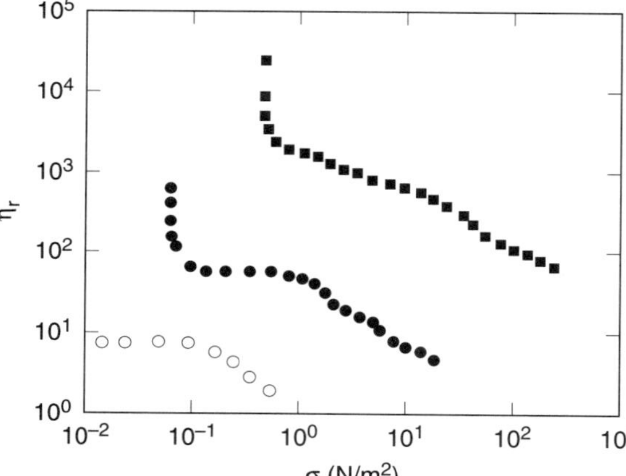

Figure 7.12 Shear-stress dependence of the relative viscosity for dispersions in water of charged polystyrene particles of radius $a = 115$ nm with nonadsorbing Dextran T-500 polymer (synthesized from glucose) added as a depletion flocculant. The polymer molecular weight is 298,000, and its radius of gyration R_g is 15.8 nm. Volume fractions and polymer concentrations are: $\phi = 0.3$, $C_p = 2.5$ wt% (■), $\phi = 0.2$, $C_p = 2.5$ wt% (●), and $\phi = 0.2$, $C_p = 0.5$ wt% (○). (From Patel and Russel 1987, with permission from the Journal of Rheology.)

a critical shear stress—is often referred to as *yield*, and the critical stress is called the *yield stress* σ_y.

There has been debate in the literature regarding the propriety of referring to a critical stress as a yield stress when there is a possibility of a Newtonian flow region at suitably low shear stress (Barnes and Walters 1985). However, from a practical point of view, the sudden onset of a measurable rate of shear above a critical shear stress, and no measurable rate of shear below it, is a useful experimental criterion for defining the yield phenomenon. Still, the possible presence of Newtonian creep at very low rates of strain should not be forgotten, since this such behavior can affect the drying and sagging behavior of gel bodies.

7.3.2 The Rheology of Strongly Flocculated Gels

Let us now consider strongly flocculated dispersions at particle concentrations high enough to produce rigid gel networks. The term "strong flucculation" is ambiguous, but for practical purposes we take it to refer to gels in which the attractive potential minimum is large, $-U_{\min}/k_B T \geq 20$. In such cases, once particle–particle contacts are formed, they are released by thermal agitation so infrequently that particle rearrangements are strongly suppressed. Thus, the time for equilibration of the structure, given by a relaxation time τ, is too long to occur within the experimental time frame, which is usually no more than several hours. The "Newtonian" zero-shear viscosity is only attained at shear rates $\dot{\gamma} \lesssim \tau^{-1}$, and these rates are too low to be accessed. Thus, strongly flocculated gels are characterized by a yield stress, rather than a zero-shear viscosity. Other rheological quantities that are important for strongly flocculated gels include compactive strength, linear (and nonlinear) elastic moduli at high and low frequencies, and the shear-rate-dependent viscosity. The dependencies of these on particle size, particle concentration, and particle–particle interactions are of obvious importance for the processing of colloidal gels. One would like to develop theoretical understanding of such dependencies, or even quantitative models, if possible.

Rheological measurements are difficult on strongly gelled colloids, and data often do no reproduce well. Multiple runs often must be averaged together to reduce data scatter. Strongly flocculated suspensions are by definition not at equilibrium, and so their properties are sensitive to preparation technique and deformation history. In addition, in large-deformation or continuous shearing, slipping of the gel against the surfaces of the rheometer tools is always a danger, although using roughened tools (Buscall et al. 1993) or vane-type rheometer fixtures (Dzuy and Boger 1985; Leong et al. 1993) seems to be an effective countermeasure.

Despite these difficulties, several experimental studies have established trends that seem to be generic for the rheological properties of these materials. In particular, Buscall et al. (1986, 1987, 1988) have reported extensively on suspensions of well-characterized silica and polystyrene spheres coagulated by addition of electrolyte. Figure 7-13 shows measured values of the compactive strength P_y and yield stress σ_y as functions of particle volume fraction ϕ for $a = 245$-nm polystyrene particles coagulated at 0.1 M $BaCl_2$. Analogous results were obtained with $a = 13$-nm silica particles. The compactive strength P_y is a compressive yield stress above which compression of the gel occurs in a centrifuge. A similar volume-fraction dependence was found for the high-frequency modulus G_∞, obtained by measuring the velocity of propagation of a shear wave; the characteristic frequency of this wave is high (around $\sim 10^3 \ \sec^{-1}$), and its amplitude is small (around 3×10^{-4}). Just above $\phi_g \approx 0.05$, the minimum particle concentration for gelation, one expects the gel modulus to depend on ϕ as $G_\infty \propto (\phi - \phi_g)^n$ (de Gennes 1976; Feng and Sen 1984; Stauffer 1985). However, over most of the range of ϕ, both G_∞ and P_y increase with ϕ roughly as (Buscall et al. 1987; Rueb and Zukoski 1997)

$$P_y \propto G_\infty \propto \phi^\mu \tag{7-24}$$

The power-law exponent μ is around 4.0–5.0. A similar result is obtained for the yield stress in shear σ_y, except the exponent is smaller, $\sigma_y \propto \phi^{3.0}$, and σ_y is a couple of decades smaller in magnitude than P_y. Chen and Russel (1991) found a similar power law for flocculated octadecyl-coated silica particles. Such power laws have been derived from theories that model the gel as a network of interconnected fractal clusters (Buscall et al. 1988; Shih et al. 1990; Potanin et al. 1995).

The influence of particle size on the yield stresses P_y and σ_y have also been measured; Buscall et al. (1988) found the power-law dependencies

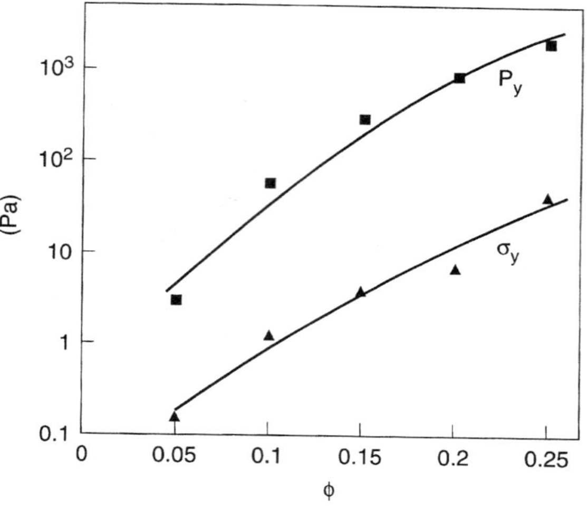

Figure 7.13 Compressional, P_y, and shear, σ_y, yield stresses versus particle volume fraction ϕ for dispersions in water of charged polystyrene particles of radius $a = 245$ nm coagulated by addition of $BaCl_2$. (From Buscall et al. 1987). (reprinted from J Non-Newt Fluid Mech 24:183, Buscall et al. (1987), with kind permission from Elsevier Science - NL, Sara Burgerhartstraat 25, 1055 KV Amsterdam, The Netherlands.)

$$P_y \sim a^{-2.3}, \qquad \sigma_y \sim a^{-2} \tag{7-25}$$

Data for σ_y versus a, corroborating the above scaling, are reported in Fig. 7-14.

The low-frequency modulus G_0 for flocculated polystyrene particles, measured by creep tests at small stresses and strains as low as 10^{-4}, was found to be nearly identical to the high-frequency modulus G_∞ (Buscall et al. 1987). Thus, these strongly flocculated systems are highly elastic at small strains, with little relaxation over a wide range of frequencies. Figure 7-15 confirms this interpretation, showing for a strongly flocculated dispersion of silica particles that there is little frequency dependence of the elastic storage modulus G' over a wide range of frequencies, and that the loss modulus G'' is much smaller than G'.

The modulus of strongly flocculated gels tends to be highly strain-dependent, with linear behavior confined to very low strain amplitudes (Buscall et al. 1988). Figure 7-16 shows the low-frequency modulus in creep versus normalized strain amplitude for strongly flocculated polystyrene particles with volume fractions ϕ between 0.1 and 0.25. In

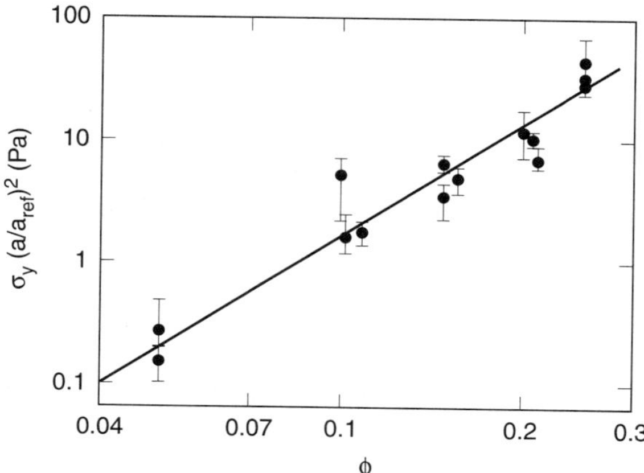

Figure 7.14 Shear yield stress σ_y versus particle volume fraction ϕ for charged polystyrene particles aggregated in water by addition of $BaCl_2$. σ_y is normalized by the square of the particle radius to account for the scaling $\sigma_y \propto a^{-2}$. Data for particles of radii 245 nm, 480 nm, and 1700 nm are included, with $a_{ref} = 245$ nm. The line has a slope of 3.0. (From Buscall et al. 1988, reproduced by permission of The Royal Society of Chemistry.)

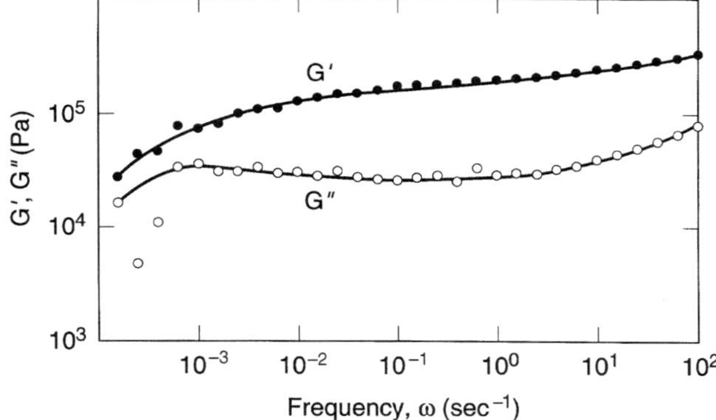

Figure 7.15 Storage and loss moduli G' and G'' as functions of frequency in small-amplitude oscillatory straining at a strain amplitude of 0.5% for a silica particulate dispersion ($\phi \approx 0.25$, $a \approx 25$ nm) gelled by reduction of pH.

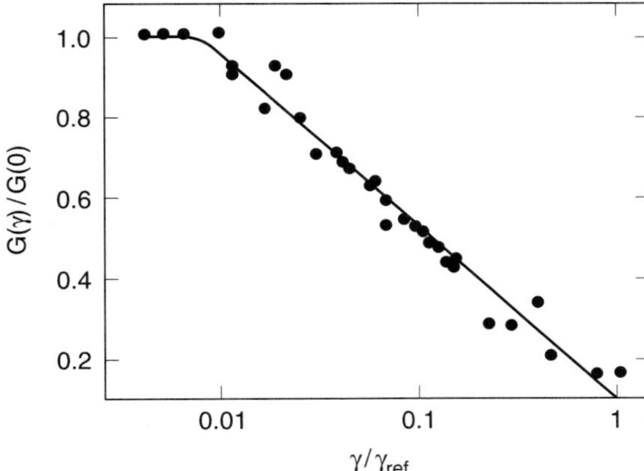

Figure 7.16 Low-frequency shear modulus $G(\gamma)$ divided by the small-amplitude modulus $G(0)$, versus normalized strain $\gamma/\gamma_{\text{ref}}$ for charged polystyrene particles of radii $a = 245$ nm and 1300 nm and volume fractions between 0.1 and 0.25 aggregated in water by addition of $BaCl_2$. γ_{ref} ranges from 0.015 to 0.073. (From Buscall et al. 1988, reproduced by permission of The Royal Society of Chemistry.)

Fig. 7-16, γ is normalized by a reference strain γ_r to bring the data for different values of ϕ into superposition; γ_r varies from 0.073 at $\phi = 0.1$ to 0.015 at $\phi = 0.25$. Thus, according to Fig. 7-16, for the most concentrated gels, the modulus is strain-independent only when $\gamma \leq \gamma_c = 0.0005$, or 0.05%! Above this tiny strain, the modulus shows strain softening. The strain γ_y required for yield is, however, much higher than this, around 3% or so. Another surprise is that if a strain of 1% or so is imposed on the sample, and then the stress is suddenly removed, a large percentage, around 80%, of the imposed strain is recovered, even though the imposed strain is high enough for the modulus to be highly nonlinear.

Weakly flocculated gels are less strain sensitive than the strongly flocculated gels discussed above. Figure 7-17 shows the elastic modulus G versus strain γ for two gels made from weakly flocculated octadecyl-coated silica spheres (Chen and Russel 1991). For the more weakly flocculated of the two, obtained by lowering the temperature to 29°C, nonlinearity occurs above a critical strain γ_c of around 10%, while for a somewhat more

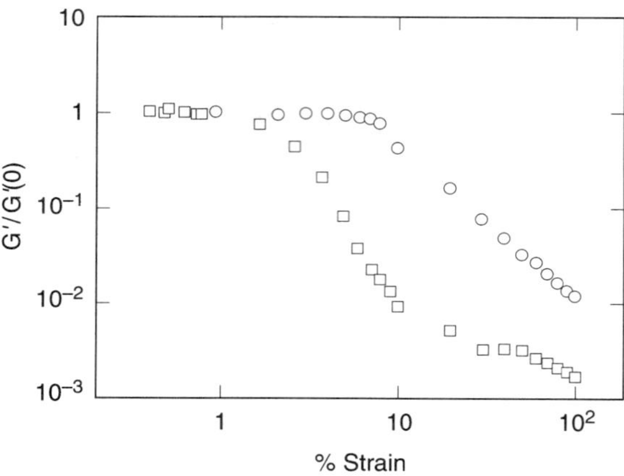

Figure 7.17 Shear elastic modulus G' divided by the small-strain modulus $G'(0)$ versus strain γ for dispersions in hexadecane of octadecyl-grafted silica spheres with radius $a = 56$ nm and concentration $\phi = 0.182$. The temperatures are 29°C (○), and 20°C (□). (From Chen and Russel 1991, reprinted with permission from Academic Press.)

strongly flocculated gel at 20°C, γ_c is only 1%. As we saw, for the strongly flocculated gels of Buscall and coworkers, γ_c is around 0.05%. Thus, as the flocculation becomes stronger, the strain sensitivity increases. In addition, *strongly flocculated gels are likely to be more brittle than weakly flocculated ones*, in that although they are harder than weakly flocculated gels, they cannot be deformed as much without fracturing (Velamakanni et al. 1990). This finding is of considerable importance for the processing of gel bodies.

Thus, the mechanical properties of a gel depend not only on particle radius a and concentration ϕ, but also on the flocculation strength. For gels in aqueous media, the mechanical properties therefore depend on the charge on the particle surfaces, as well as on the type of ions that might be bound to them. Figure 7-18a from Leong et al. (1993) shows the yield stress for suspensions of zirconia particles ($a = 150$ nm) with particle volume fractions ranging from 0.12 to 0.24, flocculated by addition of a strong acid (nitric acid, HNO_3) or a strong base (potassium nitrate, KOH). These pH adjusters change the surface charge of the particles, presumably through surface binding of H^+ or OH^-. The surface charge of the particles was probed by measurements of electroacoustic mobility of the particles in diluted suspensions (see Fig. 7-18b). The electroacoustic-mobility data were converted to an approximate surface potential, or zeta potential ζ (see below). (Although the extrapolation of these electroacoustic data for diluted suspensions to a ζ potential for concentrated suspensions is not entirely proper, the electroacoustic technique is a simple and useful means of obtaining a qualitative estimate of the surface potential, at least near the isoelectric point.) Note in Fig. 7-18 that the maximum yield stress occurs at the pH (~ 7) for which the dynamic mobility, and hence the surface charge, is zero.

Figure 7-19 makes this point even more dramatically, showing that the yield stress σ_y decreases linearly with the square of the zeta potential, for both positive and negative ζ. *Thus σ_y is maximum at $\zeta = 0$;* that is, at pH $= 7$ for the zirconia particles. A similar finding was reported much earlier by Hunter and Nicol (1968) and by Friend and Hunter (1971). This linear dependence of σ_y on ζ^2 can be predicted directly from the linear dependence on ζ^2 of the primary minimum ($-W_{min}$) of the interparticle potential [see Eq. (7-11a)].

The pH at which ζ is zero can be adjusted by adding surface-binding anions, such as phosphate, citrate, or sulfate, to the suspension (Leong et al. 1993). Figure 7-20 shows

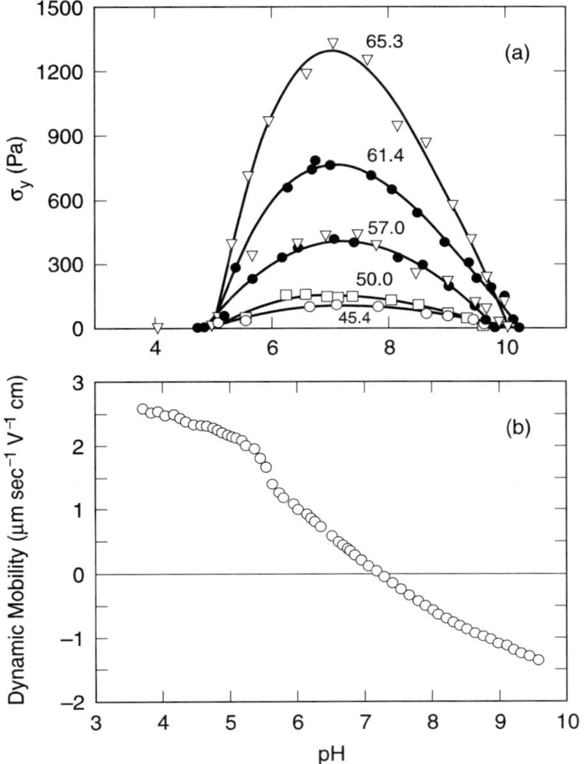

Figure 7.18 (a) Shear yield stress σ_y versus pH for dispersions in water of "blocky"-shaped ZrO_2 particles with radius $a = 150$ nm and volume fractions $\phi = 0.124, 0.145, 0.184, 0.213,$ and 0.242. These correspond to the mass percentages 45.4%, 50.0%, 57.0%, 61.4%, and 65.3%. The pH was adjusted using HNO_3 and KOH. (b) The dynamic mobility of particles in diluted suspensions as a function of pH. (From Leong et al. 1993, reproduced by permission of The Royal Society of Chemistry.)

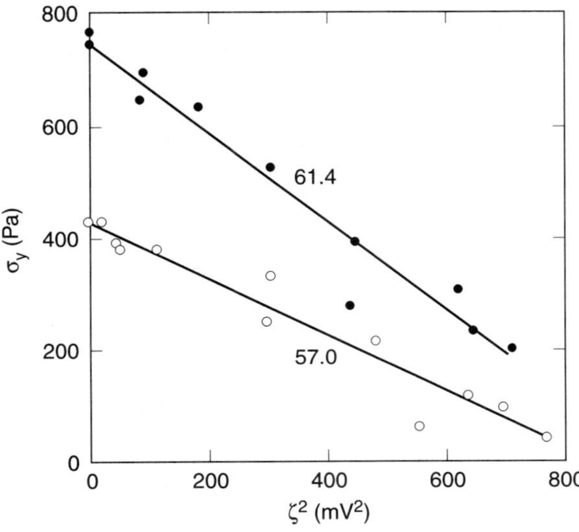

Figure 7.19 Shear yield stress versus square of the zeta potential ζ^2 for the dispersions described in Fig. 7-18 at particle volume fractions $\phi = 0.184$ and 0.213, or mass percentages of 57.0% and 61.4%. The zeta potential was obtained at low ϕ from the dynamic mobility. (From Leong et al. 1993, reproduced by permission of The Royal Society of Chemistry.)

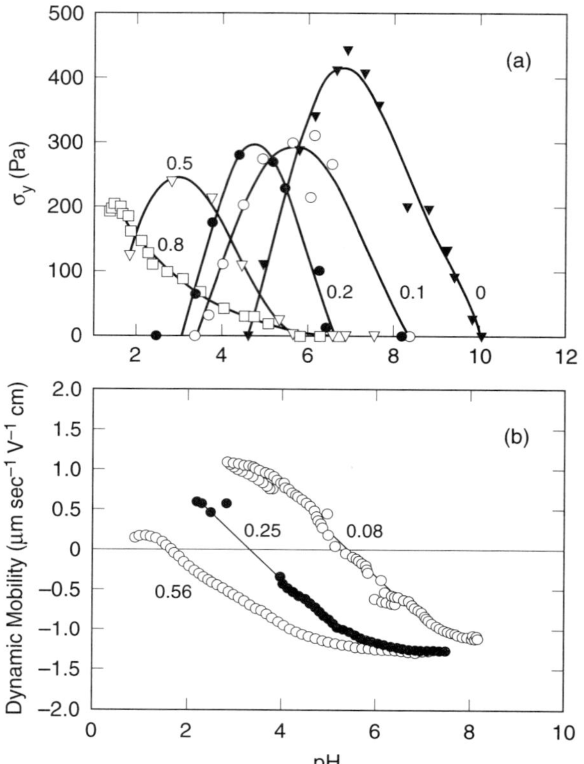

Figure 7.20 (a) Shear yield stress σ_y versus pH for dispersions described in Fig. 7-18 at a volume fraction $\phi = 0.184$, with varying amounts of added phosphate shown as a percentage on a dry weight basis. (b) The dynamic mobility of diluted ($\phi = 0.025$) suspensions as a function of pH with varying amounts of added phosphate. (From Leong et al. 1993, reproduced by permission of The Royal Society of Chemistry.)

the yield stress versus pH and the dynamic mobility versus pH for a series of zirconia suspensions containing varying amounts of phosphate ion. As they are added to the suspension, phosphate anions bind to the particle surfaces, rendering it more negatively charged; to neutralize this charge, the pH must be lowered to supply H^+ ions. Thus, the pH at which $\zeta = 0$ is less when phosphate is present in solution than when it is absent (see Fig. 7-20b). Note in Fig. 7-20a that the pH at which σ_y is maximized is correspondingly reduced! The pH at which σ_y is maximum corresponds well with the pH at which $\zeta = 0$ for various concentrations of phosphate (see Fig. 7-21). A similar result holds for other anions, although different ions differ in their ability to bind to the surface and shift the isoelectric point (i.e.p.) (see Fig. 7-22).

Thus, *the yield stress σ_y is maximized at the isoelectric point*. Note, however, in Fig. 7-20 that when the i.e.p. is shifted by the binding of ions to the particle surfaces, the maximum yield stress is reduced. As noted in Section 7.2.1, the binding of ions to the particle surfaces is likely to increase the thickness of the hydration layers on the particles that keep the particle surfaces from coming closer than a few nanometers from each other. Leong et al. (1993) have found a correlation between the size of the adsorbed anion and the magnitude of the decrease in the yield stress. However, the size of the *hydrated* ion, rather than the size of the ion itself, should, in principle, control the closest approach of the particles (Israelachvili 1991).

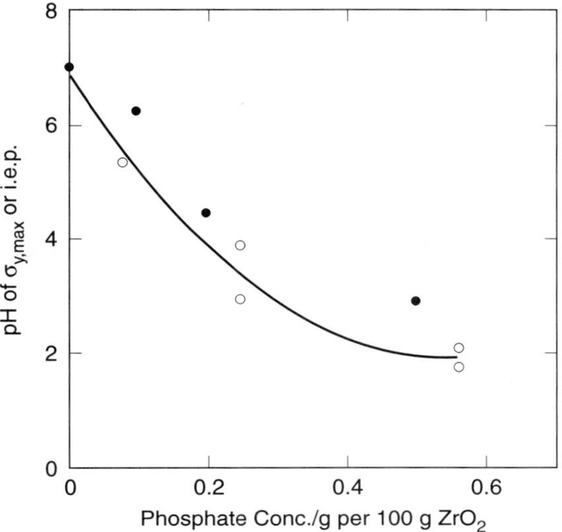

Figure 7.21 pH at which the yield stress σ_y is maximum (●) and pH at the isoelectric point (○) measured electroacoustically, versus concentration of phosphate, for suspensions described in Fig. 7-20. (From Leong et al. 1993, reproduced by permission of The Royal Society of Chemistry.)

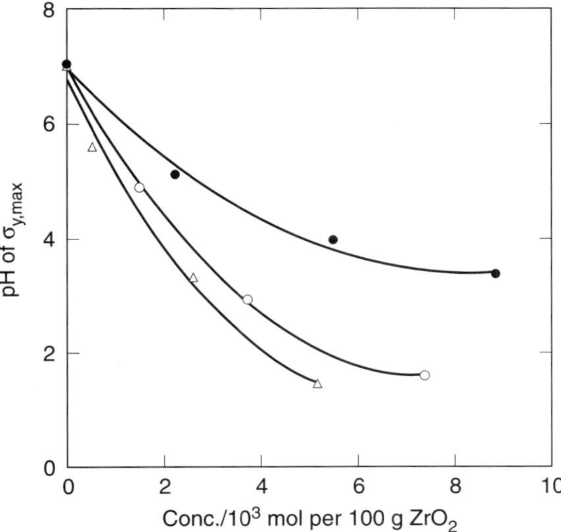

Figure 7.22 pH at which the yield stress σ_y is maximum versus concentration of lactate (●), malate (○), and citrate (△) for suspensions described in Fig. 7-18, with $\phi = 0.184$. (From Leong et al. 1993, reproduced by permission of The Royal Society of Chemistry.)

Analogous results have been reported by Velamakanni et al. (1990) for the shear-thinning viscosity of alumina-particle suspensions ($a \approx 100$ nm, $\phi = 0.20$). Gels produced by raising the electrolyte (NH_4 Cl) concentration at a pH away from the i.e.p. have lower viscosities than those produced at the i.e.p. Velamakanni et al. refer to the former as "coagulated" gels and the latter as "flocculated" gels; they argue that "coagulated" gels are more easily sedimented or filter-pressed to high density than are "flocculated" gels. High-density gels are desirable because they are more easily dried and sintered.

7.3.3 Theory

7.3.3.1 Yield Stress and Elastic Modulus

We have seen that the rheological properties of weakly flocculated gels can be predicted at least qualitatively using reasonable particle–particle interaction potentials derived from van der Waals and polymer depletion forces. Can a similar approach succeed in predicting the mechanical properties of strongly flocculated gels?

Developing accurate theories for strongly flocculated gels is challenging, since the structures of such gels are not at thermodynamic equilibrium. At best, one might assume that such gels are in a state of *static equilibrium* in which the forces acting on each particle are in balance. Since the interaction potential between particles in a strongly flocculated gel has a minimum $W_{min} = W(D_0)$ that is deep compared to $k_B T$, gaps between neighboring particle surfaces in such gels will presumably almost always be close to D_0, unless the gel is subjected to a mechanical strain. Therefore, the shape of the potential $W(D)$ near D_0 is important in determining the gel's mechanical properties.

Sensitivity to the shape of $W(D)$ differentiates weakly from strongly interacting particles. For the former, the precise shape of the potential is not important; we saw in Section 7.2.4 that even a simple square-well potential is an adequate approximation. But insensitivity to the shape of the potential can only be expected when the particles are only weakly bound by that potential, so that rapid, thermally driven changes in particle–particle separation average out the details of the shape of the potential. For strongly flocculated gels, the particle–particle separations remain trapped near the minimum in the potential well, and the shape of the well near this minimum matters much more.

Nevertheless, if one assumes a static, rather than a thermodynamic, equilibrium, one can attempt to estimate the dependence of the yield stress σ_y and the modulus G on the shape and depth of the interparticle potential. Imagine that a gel is subjected to a shear strain γ that *homogeneously* displaces particles from their positions of static equilibrium. Pairs of particles are pulled apart by this strain, and separations between particle centers of mass should increase roughly by an amount γr_0, where $r_0 \equiv 2a + D_0$ is the separation between centers of mass in the absence of strain. Hence, the imposition of a strain γ increases the gap between particle surfaces from D_0 to

$$D \approx D_0 + \gamma(2a + D_0) \tag{7-26}$$

(Some particles will be tend to be pushed together by a homogeneous strain, but these can't move much closer together, because they encounter hard steric repulsions.) If γ is small, this increased separation of particles is small *relative to the initial separation of centers of mass*. But the increase is much larger *relative to the initial gap D_0*. Thus, the ratio of the gap between particles *after* the strain to that *before* the strain is $(\gamma r_0 + D_0)/D_0 = [\gamma(2a + D_0) + D_0]/D_0 \approx 2\gamma a/D_0$. Since the ratio $2a/D_0$ is usually large ($\gtrsim 100$), even a strain of only 1% multiplies the gap between neighboring particles by a factor of two or more! From this, we can see why strongly flocculated gels, with particle–particle gaps as low as 1 nm, are so strain-sensitive (Buscall et al. 1987).

A force $F = -W'(D)$ with $D = \gamma(2a + D_0) + D_0$ is produced by this increased separation between the particles, where W' is the derivative of W with respect to D. This force would restore the original interparticle spacing if the shearing stress were removed.

The macroscopic stress σ is this force times the number of interparticle bonds that cross a unit area of the sample; this latter factor should scale as ϕ^2/a^2 (Russel et al. 1989). As long as the local applied force increases with increased strain, σ increases with increasing strain, and the gel maintains its mechanical stability. But once the strain reaches the point that the slope W' of the potential is a maximum (see Fig. 7-23), any further strain produces a decreasing force, and the interparticle structure breaks apart. This corresponds to the point of *yield*. Thus, the yield strain γ_y is given by the condition that the second derivative W'' of $W(D)$ is zero; that is, $W''(D_y) = 0$, where $D_y = 2\gamma_y a + (\gamma_y + 1)D_0$ is the value of D for which $W'' = 0$. Very roughly, we might expect that W' is a maximum ($W'' = 0$) when separation $D = D_y$ is on the order of twice D_0, the value of D at static equilibrium. This would imply that the yield strain γ_y is roughly $D_0/2a$; hence, for particles 100 nm in radius, $\gamma_y \approx 0.005$ (0.5%), or less, not too far from experimental observations.

The yield stress (i.e., the stress at the yield point) is proportional to $F_{\max} = W'_{\max} = W'(D_y)$ times the number of interparticle bonds that cross a unit area of the sample, ϕ^2/a^2; thus

$$\sigma_y \sim \frac{\phi^2}{a^2} W'(D_y) \tag{7-27}$$

We can estimate W' at the yield point to be roughly $-W_{\min}/D_0$. Hence, from Eqs. (7-11a) and (7-11b) we obtain

$$\sigma_y \sim \frac{\phi^2}{a} \left(\frac{A_H}{12 D_0^2} - \frac{C}{D_0}\zeta^2 \right) \tag{7-28}$$

where $C \equiv 2\pi\varepsilon_0\varepsilon \ \ln[(1 - \exp(-\kappa D_0))^{-1}]$. For a specific example, let us take 0.01 M 1:1 electrolyte so that $\kappa \approx 0.3$ nm^{-1}. For zirconia particles in water, $A_H \approx 6 \times 10^{-20}$ J (Leong et al. 1993). Taking $D_0 \approx 2$ nm, and $a \approx 100$ nm, we obtain from Eq. (7-28)

$$\sigma_y[\text{Pa}] \sim \phi^2(1.2 \times 10^4 - 14\zeta^2[(mV)^2]) \tag{7-29}$$

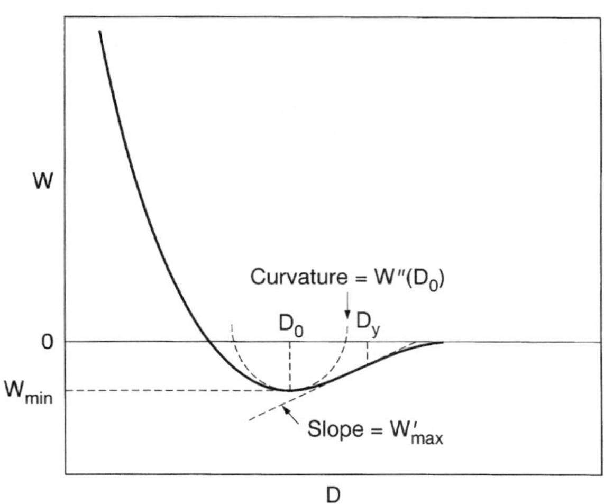

Figure 7.23   Schematic of pair potential $W(D)$, showing location of maximum slope, which should control the yield stress and yield strain, and the curvature of the potential at its minimum, which in theory controls the low-strain modulus. (Adapted from Russel et al. 1989, with permission.)

For a particle volume fraction of $\phi = 0.2$, this predicts a yield stress of 500 Pa at the i.e.p., and it also predicts that this yield stress is reduced to zero when ζ is increased to about 30 mV. These results are within about a factor of two of the experimentally measured yield stresses (see Fig. 7-19 for example). However, the scaling of σ_y with particle size and concentration ($\sigma_y \propto \phi^3 a^{-2}$) is stronger than predicted by Eq. (7-28), and so agreement is not as good at other volume fractions.

The linear modulus G can be estimated by an analogous argument (Russel et al. 1989). G is defined as the stress divided the strain γ, where γ is small enough that G is independent of γ. As we argued above, the stress σ is given roughly by $(\phi^2/a^2)W'(D)$, with D given by Eq. (7-26). If the quantity σ/γ is to be independent of strain, then γ must be small enough that W' can be linearized in D; that is, $W'(D) \approx W''(D_0)(D - D_0) \approx W''(D_0)2a\gamma$. Hence $G \sim (\phi^2/a^2)W'/\gamma$ is

$$G \sim \frac{2\phi^2}{a} W''(D_0) \tag{7-30}$$

Thus, the linear modulus is controlled by the *curvature* of the particle–particle potential W at its minimum. Of course, this local curvature is extremely sensitive to the details of the particle–particle interactions at close separations (Goodwin et al.1986), and thus the modulus will also depend strongly on these details. Nevertheless, if we estimate $W''(D_0) \approx -W_{\min}(D_0)/D_0^2$, then Eq. (7-30), combined with Eqs. (7-11a) and (7-11b), gives

$$G \sim \frac{2\phi^2}{D_0^2} \left(\frac{A_H}{12D_0} - C\zeta^2 \right) \tag{7-31}$$

Or, for the example considered above, we have

$$G \; [\text{kPa}] \sim \phi^2(1.2 \times 10^3 - 1.4\zeta^2 \; [(mV)^2])$$

For $\phi = 0.2$ and $\zeta = 0$, this gives $G \sim 50$ kPa. The above arguments imply that $G/\sigma_y \approx 2a/D_0$; that is, the modulus is a couple of orders of magnitude larger than the yield stress. This is consistent with the data of Buscall et al. (1987). Note in Eq. (7-31) that the modulus is predicted to be *independent* of the particle size a, a prediction consistent with the data of Chen and Russel (1991) and Goodwin et al. (1986); see also Russel et al. (1989). However, again, the dependence on ϕ in Eq. (7-30) is considerably weaker than that found experimentally (Buscall et al. 1988).

The steeper dependences of σ_y and G on ϕ found in the experiments can be understood if one notes that Eqs. (7-28) and (7-30) assume that every interaction between neighboring particles contributes to the modulus or yield stress. However, as ϕ decreases, because of the fractal character of the gel structure, an ever smaller fraction of these contacts is likely to support the stress applied to the gel, while the remaining are "dead-end," or "ineffective," contacts. The growing fraction of "ineffective" contacts as ϕ decreases implies that G and σ_y should depend more strongly on ϕ than ϕ^2, and that Eqs. (7-28) and (7-30) will be most accurate at large ϕ. And, indeed, the concentration dependences of the yield stress data of Buscall and coworkers do show a tendency to bend over toward a lower power law at large ϕ (see Fig. 7-13).

7.3.3.2 Shear Viscosity at High Shear Rates

At high shear rates, when the gel network is broken down, the dominant viscoelastic contribution comes from flocs that break apart and reform rapidly. For such dispersions, at modest particle volume fractions, a typical relationship between steady-state shear stress σ and shear rate $\dot{\gamma}$ is shown in Fig. 7-24 (Friend and Hunter 1971). Note that at the highest shear rates, the $\dot{\gamma}$–σ relationship appears to be linear, but the extrapolation of this linear relationship to zero shear rate intersects the stress axis at a positive value, σ_B, rather than zero. This intercept is called the "Bingham" yield stress, derived from the Bingham equation for shear stress (Friend and Hunter 1971):

$$\sigma = \eta_{\text{pl}}\dot{\gamma} + \sigma_B \tag{7-32}$$

where η_{pl}, the "plastic viscosity," is the slope of the linear relationship between σ and $\dot{\gamma}$. As Fig. 7-24 shows, the Bingham yield stress σ_B differs from the "true" yield stress σ_y; σ_y is measured at small shearing strains by finding the minimum stress required to induce flow.

Theories for the steady-shear viscosity are complex. They involve assumptions about the dependences on shear rate of floc size, shape, and floc–floc interactions. The simplest case one might consider is the limit of very high shear rates and not-too-high particle concentrations. In this limit, σ_B is assumed to arise from the work that must be done to break apart *particle pairs* (Friend and Hunter 1971). These particle pairs re-form again because of shear-induced collisions. At steady state, the rate at which particle pairs are produced by the collisions must equal the rate at which shearing pulls these pairs apart. This rate, per unit volume of suspension, can be estimated from simple kinetic theory as $3\phi^2\dot{\gamma}/\pi^2a^3$. Each time a particle pair is pulled apart, the energy expended is $-W_{\text{min}}$, given by Eq. (7-11), which assumes that the particles must be separated from their primary attractive minima. Thus, the rate at which energy is dissipated by this process is $(-W_{\text{min}})3\phi^2\dot{\gamma}/\pi^2a^3$. This dissipation rate is then set to $\sigma_B\dot{\gamma}$, so that (Friend and Hunter 1971)

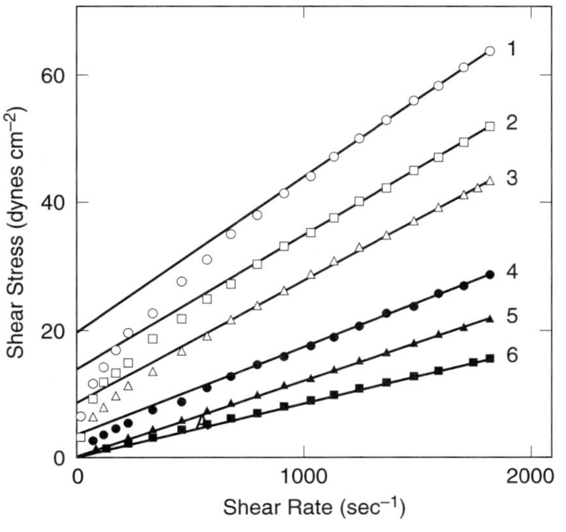

Figure 7.24 Shear stress versus shear rate for dispersions in water of polymethylmethacrylate particles with radius $a = 220$ nm and volume fraction $\phi = 0.070$ at ionic strength 0.02 g ions/liter of NaCl and zeta potentials of $+14.6$ (1), $+21.6$ (2), $+29.3$ (3), $+35.9$ (4), $+55$ (5), and supernatant (6). (From Friend and Hunter 1971, reprinted with permission from Academic Press.)

$$\sigma_B = \frac{3\phi^2}{\pi^2 a^3}(-W_{\min}) = \frac{3\phi^2}{\pi^2 a^3}\left[\frac{a A_H}{24 D_0} - C\zeta^2 a\right] \tag{7-33}$$

The second equality is obtained by using Eq. (7-11a) for $W_{\min}$. Equation (7-33) predicts that σ_B should scale with particle volume fraction as $\sigma_B \propto \phi^2$, and with zeta potential as $\sigma_B \approx \sigma_{B,0} - k\zeta^2$, where $\sigma_{B,0}$ is the Bingham yield stress at zero zeta potential, and k is a constant. These scalings have been observed in several experiments (Firth 1976).

Despite these successes, this "particle-pair" model is untenable at the shear rates typical of most experiments, because, as pointed out by Firth and Hunter (1976a), the maximum hydrodynamic force available to pull two spheres apart is orders of magnitude too small to pull the spheres out of their deep potential well. This maximum hydrodynamic force is given by

$$F_H = 6.12\pi \eta_s a^2 \dot{\gamma} \sim 10^{-13} \text{ N} \tag{7-34}$$

for $\dot{\gamma} = 10^3 \text{ sec}^{-1}$, while the van der Waals force F_{vdw} holding the spheres together is obtained by differentiating Eq. (7-3) with respect to the separation D. At the i.e.p., we have

$$F_{\text{vdw}} \approx \frac{a A_H}{12 D^2} \sim 10^{-10} \text{ N} \tag{7-35}$$

The above estimates of these forces are based on the following typical parameter values: $D = 1$ nm, $a = 100$ nm, $\eta_s = 10^{-3}$ Pa·s, $A_H = 10^{-20}$ J, and $\dot{\gamma} = 10^3 \text{ sec}^{-1}$. Of course, if D is taken to be much larger, say 20 nm or so, then the van der Waals force would be weak enough that the particles could be pulled apart by the hydrodynamic force. But if somehow D were as large as this, then Eq. (7-33) would predict a value for the Bingham yield stress that is orders of magnitude smaller than the measured values. Also, if the sheared suspension consists mainly of sphere singlets and doublets, the plastic viscosity η_{pl} would not be much larger than that for a suspension of noninteracting hard spheres; but in fact, η_{pl} is much larger than this. Furthermore, both direct visualization and scattering measurements show that even under flow, the flocs are bigger than doublets (Reich and Vold 1957; Firth 1976; Rueb and Zukoski 1997). Using a Coulter counter, Hunter and Frayne (1980) report that the floc radius R scales with shear rate roughly as $R \sim \dot{\gamma}^{-0.4}$.

Thus, it seems that only at much higher shear rates, $\dot{\gamma} \sim 10^5 \text{ sec}^{-1}$ or so, will the stresses be dominated by doublets and single particles. Starting with Firth and Hunter (1976a), several models of the sheared dispersions have been developed that attempt to account for stresses generated by shear-induced distortion and breakage of larger flocs. A general framework seems to be provided by the "elastic floc model" of Firth and Hunter (1976a, 1976b; see also Hunter 1992). According to this model, most of the stress is generated by tension in particle–particle bonds within large flocs. Large values of the plastic viscosity are accounted for by assuming that individual flocs can be treated as "hard spheres" whose volume exceeds that occupied by individual particles in the floc, because of the flocs' open structure. A more recent theory by Potanin et al. (1995) differentiates between "soft" and "rigid" bonds between particles, with only the latter contributing to the stress. This model has been used to interpret the rate at which the elastic modulus of a gel rebuilds after being broken down in a strong shearing flow (Rueb and Zukoski 1997). Such predictions are potentially important in designing suspensions as thixotropic agents for commercial applications, such as encapsulants of microelectronic devices. However, to date, the detailed experimental probes of floc structure necessary to validate models of this kind, and to provide measurements of the parameters that go into them, are still lacking.

> • Problems and Worked Examples 7.1 through 7.4 test your practical working knowledge of colloidal gels.

7.4 SUMMARY

Many experimental data are now available on the rheology of particulate gels, including the effects of particle concentration, size, and strength of interaction. For weakly flocculated gels, reasonable theories exist that allow for at least qualitative, and even quantitative, prediction of the zero-shear viscosity. These theories predict that the relaxation time τ and the zero-shear viscosity η_0 increase with the depth $-W_{\min}$ of the attractive potential well as $\tau \sim \eta_0 \sim \exp(-W_{\min}/k_B T)$. For strongly flocculated gels with $-W_{\min}/k_B T \gtrsim 20$, τ becomes too long for equilibration of the gel structure to occur within reasonable experimental times. Also, for strongly flocculated gels, the zero-shear viscosity is so large, and is accessed at such low shear rates, that these gels are characterized by a yield stress, rather than a zero-shear viscosity.

Rheological data on strongly flocculated gels are hard to reproduce and are sensitive to sample preparation, pre-shearing, and experimental protocol. Nevertheless, clear trends are evident in the experimental dependences of yield stress σ_y and modulus G of strongly flocculated gels on particle volume fraction ϕ, radius a, and zeta potential ζ. The magnitudes of σ_y and G can be estimated theoretically based on an estimate of the gap D_0 between particles; D_0 appears to be sensitive to the presence and type of ions adsorbed onto the particle surfaces. The predicted dependence of σ_y on ζ is in good agreement with experiment, while experiments show a steeper dependence on ϕ than is predicted by the simplest theory. In general, as the net interparticle attractions are made stronger (for example, by reducing electrostatic repulsions), the particles become more tightly bonded to each other, and the elastic modulus and yield stress increase. Also, strongly flocculated gels yield or weaken at smaller strains, and hence are more brittle than more weakly flocculated ones. Since optimization of gel properties usually requires that the modulus be as high as possible, while the brittleness be minimized, understanding and control of the rheology of gels is of considerable importance in the processing of colloidal gel bodies.

REFERENCES

Adamson AW, Gast AP (1997). *Physical Chemistry of Surfaces*, 6th ed., Wiley, New York.
Asakura S, Oosawa F (1954). *J Chem Phys* 22:1255.
Asakura S, Oosawa F (1958). *J Polym Sci* 33:183.
Aubert C, Cannell DS (1986). *Phys Rev Lett* 56:738.
Barboy B (1974). *J Chem Phys* 61:3194.

Barnes HA, Walters K (1985). *Rheol Acta* 24:323.

Baxter RJ (1968). *J Chem Phys* 49:2770.

Brinker CJ, Scherer GW (1990). *Sol–Gel Science. The Physics and Chemistry of Sol-Gel Processing.* Academic Press, New York.

Buscall R, Mills PDA, Yates GE (1986). *Colloid Surf* 18:341.

Buscall R, McGowan IJ, Mills PDA, Stewart RF, Sutton D, White LR, Yates GE (1987). *J Non-Newt Fluid Mech* 24:183.

Buscall R, Mills PDA, Goodwin JW, Lawson DW (1988). *J Chem Soc Faraday Trans* 84:4249.

Buscall R, McGowan IJ, Morton-Jones AJ (1993). *J Rheol* 37:621.

Cabane B, Wong K, Lindner P, Lafuma F (1997). *J Rheol* 41:531.

Chen M, Russel WB (1991). *J Colloid Interface Sci* 141:564.

de Gennes PG (1976). *J Phys (Paris) Lett* 37L:61.

Derjaguin BV, Landau L (1941). *Acta Physiocochim URSS* 10:25.

Dimon P, Sinha SK, Weitz DA, Safinya CR, Smith GS, Varady WA, Lindsay HM (1986). *Phys Rev Lett* 57:595.

Dzuy NQ, Boger DV (1985). *J Rheol* 29:335.

Fedotova VA, Zhodzhaeva K, Rehbinder PA (1967). *Dokl Akad Nauk SSR* 177:155.

Feng S, Sen P (1984). *Phys Rev Lett* 52:216.

Firth BA (1976). *J Colloid Interface Sci* 57:257.

Firth BA, Hunter RJ (1976a). *J Colloid Interface Sci* 57:248.

Firth BA, Hunter RJ (1976b). *J Colloid Interface Sci* 57:266.

Fleer GJ, Scheutjens JMHM (1982). *Adv Colloid Interface Sci* 16:341.

Flory PJ, Krigbaum WR (1950). *J Chem Phys* 18:1086.

Frens G, Overbeek JThG (1972). *J Colloid Interface Sci* 38:376.

Friend JP, Hunter RJ (1971). *J Colloid Interface Sci* 37:548.

Gast AP, Hall CK, Russel WB (1983). *J Colloid Interface Sci* 96:251.

Goodwin JW, Hughes RW, Partridge SJ, Zukoski CF (1986). *J Chem Phys* 85:559.

Grant MC, Russel WB (1993). *Phys Rev E* 47:2606.

Hunter RJ (1992). *Foundations of Colloid Science*, Vols I and II, Oxford University Press, New York.

Hunter RJ, Frayne J (1980). *J Colloid Interface Sci* 76:107.

Hunter RJ, Nicol SK (1968). *J Colloid Interface Sci* 28:250.

Ilett SM, Orrock A, Poon WCK, Pusey PN (1995). *Phys Rev E* 51:1344.

Israelachvili J (1991). *Intermolecular and Surface Forces,* 2nd ed, Academic Press, London.

Jullien RL, Kolb M, Botet R (1984). *J Phys (Paris) Lett* 45:L977.

Kuhl T, Guo Y, Alderfer JL, Berman AD, Leckband D, Israelachvili J, Hui SW (1996). *Langmuir* 12:3003.

Lekkerkerker HNW, Poon WCK, Pusey PN, Stroobants A, Warren PB (1992). *Europhys Lett* 20:559.

Leong YK, Scales PJ, Healy TW, Boger DV, Buscall R (1993). *J Chem Soc Faraday Trans* 89:2473.

Liang W, Tadros ThF, Luckham PF (1994). *Langmuir* 10:441.

Li-in-on FK, Vincent B, Waite FA (1975). *ACS Symp Ser* 9:165.

Otsubo Y (1993). *J Rheol* 37:799.

Patel PD, Russel WB (1987). *J Rheol* 31:599.

Potanin AA, de Rooij R, van den Ende D, Mellema J (1995). *J Chem Phys* 102:5845.

Reich I, Vold RD (1957). *J Phys Chem* 63:1497.

Rouw PW, de Kruif CG (1989). *Phys Rev A* 39:5399.

Rouw PW, Woutersen ATJM, Ackerson BJ, de Kruif CG (1989). *Physica A* 156:876.

Rueb CJ, Zukoski CF (1997). *J Rheol* 41:197.

Russel WB, Saville DA, Schowalter WR (1989). *Colloidal Dispersions* Cambridge University Press, Cambridge.
Schaefer DW, Martin JE, Wiltzius P, Cannell DS (1984). *Phys Rev Lett* 52:2371.
Scherer GW, Pardenek SA, Swiatek RM (1988). *J Non-Cryst Solids* 107:14.
Shih W-H, Shih WY, Kim S-I, Liu J, Aksay IA (1990). *Phys Rev A* 42:4722.
Stauffer D (1985). *Introduction to Percolation Theory*, Taylor & Francis, London.
Stöber W, Fink A, Bohn E (1968). *J Colloid Interface Sci* 26:155.
Van der Aerschot E, Mewis J (1992). *Colloid Surface* 69:15.
van Helden AK, Jansen JW, Vrij A. (1981). *J Colloid Interface Sci* 81:354.
Velamakanni BV, Chang JC, Lange FF, Pearson DD (1990). *Langmuir* 6:1323.
Verwey EJW, Overbeek JT (1948). *Theory of Stability of Lyophobic Colloids* Elsevier, Amsterdam.
Vincent BJ, Edwards J, Emmett S, Croot R (1988). *Colloid Surf* 31:267.
Vrij A (1976). *Pure Appl Chem* 48:471.
Watts RO, Henderson D, Baxter RJ (1971). *Adv Chem Phys* 21:421.
Weitz DA, Huang JS (1984). In *Kinetics of Aggregation and Gellation,* Family P, Landau DP (eds), Elsevier, New York.
Woutersen ATJM, de Kruif CG (1991). *J Chem Phys* 94:5739.
Woutersen ATJM, May RP, de Kruif CG (1993). *J Rheol* 37:71.
Woutersen ATJM, Mellama J, Blom C, de Kruif CG (1994). *J Chem Phys* 101:542.

■ ____

PROBLEMS AND WORKED EXAMPLES

Problem 7.1 Name two ways of inducing flocculation in a suspension of electrostatically stabilized spheres.

Problem 7.2 Consider a suspension of charged spheres. Suppose the suspension is a soft solid (i.e., with a yield stress) when no electrolyte has been added. After adding 0.01 M NaCl, the suspension becomes a runny liquid. After adding an additional 0.1 M NaCl, it is a solid again! Can you give an explanation?

Problem 7.3(a) (Worked Example) Consider a flocculated suspension, 20% by volume, of silica particles of 100-nm radius in 0.1 M KNO$_3$, where $A_H = 10^{-20}$ J, $\varepsilon = 50$, and the particles are held 2 nm apart by adsorbed ions. Estimate the yield stress at the i.e.p.

ANSWER:

From Eq. (7-28), we obtain

$$\sigma_y \sim \frac{\phi^2}{a} \left(\frac{A_H}{12 D_0^2} - \frac{C \zeta^2}{D_0} \right) \tag{A7-1}$$

At the isoelectric point $\zeta = 0$. Hence in mks units, Eq. (A7-1) gives

$$\sigma_y \sim \frac{(0.2)^2}{10^{-7}\text{m}} \left(\frac{10^{-20} \text{ J}}{12(2 \times 10^{-9})^2 \text{m}^2} \right) = 83 \text{ J/m}^3 \text{ or Pa}$$

Problem 7.3(b) (Worked Example) At what ζ potential (in mV) do you expect the colloidal gel to lose its yield stress?

ANSWER:

From Eq. (A7-1), the yield stress will disappear when

$$\zeta^2 = \frac{A_H}{12CD_0} \tag{A7-2}$$

Now, from Eq. (7-11b), we obtain

$$C = 2\pi\varepsilon\varepsilon_0 \ln\left[\frac{1}{1 - \exp(-\kappa D_0)}\right] \tag{A7-3}$$

In 0.1 M KNO_3, Eq. (7-5) gives the following for the Debye length of any 1:1 electrolyte:

$$\kappa^{-1} = \frac{0.3}{\sqrt{0.1}} \approx 1 \text{ nm}$$

Therefore $\kappa D_0 = 2$. Hence, $\ln[(1 - \exp(-\kappa D_0))^{-1}] = \ln[(1 - \exp(-2))^{-1}] = 0.15$. Also,

$$2\pi\varepsilon\varepsilon_0 = 6.28 \times 4.4 \times 10^{-10} \text{ C}^2 \text{ J}^{-1} \text{ m}^{-1} = 27.6 \times 10^{-10} \text{ C}^2 \text{ J}^{-1} \text{ m}^{-1}$$

Therefore, C in Eq. (A7-3) is 4×10^{-10} C^2 J^{-1} m^{-1}. Then, from Eq. (A7-2), we obtain

$$\zeta^2 = \frac{10^{-20} \text{ J}}{12 \times 4 \times 10^{-10} \times 2 \times 10^{-9} \text{ C}^2 \text{ J}^{-1}} = 10^{-3} \text{ J}^2/\text{C}^2$$

which implies that

$$\zeta = 0.03 \text{ J/C} = 30 \text{ mV} \tag{A7-4}$$

Problem 7.3(c) (Worked Example) The electrostatic contribution in Eq. (A7-1) is valid in the limit of a weak electrostatic force, so that the Debye–Hückel theory applies. Show that this limit is applicable under the conditions described in part (a).

ANSWER:

According to Section 2.2, just before Eq. (2-49), the Debye–Hückel theory applies when $ez\psi_s/4k_BT < 1$. Since $\zeta = \psi_s$ and the charge valence z is unity 1, we have the following from Eq. (A7-4):

$$\frac{ez\psi_s}{4k_BT} = \frac{1.6 \times 10^{-19} \times 0.03 \text{ J}}{4 \times 4 \times 10^{-21} \text{ J}} = 0.3 < 1$$

Thus, the electrostatic forces are weak enough for Eq. (A7-1) to be valid.

Problem 7.4 (Worked Example) Consider a suspension of silica particles in water for which the Hamaker constant is 10^{-20} J and the dielectric constant is $\varepsilon = 50$. If the surface charge is 0.1 charges/nm^2, calculate how high the molarity of NaCl must be to induce flocculation. Remember, each surface charge is that of an electron, $e = 1.6 \times 10^{-19}$ C, the permittivity of space is $\varepsilon_0 = 8.8 \times 10^{-12}$ C^2 J^{-1} m^{-1}, and the Bjerrum length is $\ell_b = 58/\varepsilon$ nm. Assume a weak surface potential.

ANSWER:

According to Eq. (7-15), flocculation is induced when the number of cations per unit volume reaches the value

$$n_{crit} = \frac{0.36\ell_b^{-1}z^{-2}\sigma^{4/3}}{(\varepsilon\varepsilon_0 A_H)^{2/3}} \tag{A7-5}$$

where $\ell_b = 58/50 = 1.16$ nm $= 1.6 \times 10^{-9}$ m. For NaCl, the valence z is unity. Converting the charge density into coulombs per square meter, we obtain

$$\sigma = 0.1 \times 10^{18} \text{ charges/m}^2 = 0.1 \times 10^{18} \times 1.6 \times 10^{-19} \text{C/m}^2 = 1.6 \times 10^{-2}\text{C/m}^2$$

Hence,

$$\sigma^{4/3} = 4 \times 10^{-3} \text{ C}^{4/3}\text{m}^{-8/3} \tag{A7-6}$$

Now

$$\varepsilon\varepsilon_0 A_H = 50 \times 8.8 \times 10^{-12} \times 10^{-20} \text{ C}^2\text{m}^{-1} = 4.4 \times 10^{-30} \text{ C}^2\text{m}^{-1}$$

Then

$$(\varepsilon\varepsilon_0 A_H)^{2/3} = 2.7 \times 10^{-20} \text{ C}^{4/3} \text{ m}^{-2/3} \tag{A7-7}$$

From Eqs. (A7-6) and (A7-7), we find

$$\frac{\sigma^{4/3}}{(\varepsilon\varepsilon_0 A_H)^{2/3}} = 1.5 \times 10^{17} \text{ m}^{-2}$$

Inserting this into Eq. (A7-5) gives

$$n_{crit} = \frac{0.36}{1.16 \times 10^{-9}\text{m}} 1.5 \times 10^{17}\text{m}^{-2} \approx 0.5 \times 10^{20}\text{cm}^{-3} = 0.5 \times 10^{23} \text{ liter}^{-1}$$

This can be converted to molarity by dividing by Avogadro's number, 6×10^{23}, giving 0.08 M for the critical salt concentration to induce flocculation.

■ ____

Chapter **8**

ELECTRO- AND MAGNETORESPONSIVE SUSPENSIONS

8.1 INTRODUCTION

Over the last 50 years, suspensions have been made that are unusually responsive to electric or magnetic fields. These include *electrorheological fluids*, *magnetorheological fluids*, and *ferrofluids*. Electrorheological (ER) fluids are liquids that solidify, or become very viscous, under an electric field. Analogously, magnetorheological fluids become much more viscous under a magnetic field. The viscosity of a ferrofluid, on the other hand, remains small when a magnetic field is applied; however, a drop or blob of ferrofluid acts like a "liquid magnet" that is drawn into, and held in place by, a magnetic field. One can well imagine the kinds of uses to which such fluids might be put in the design of motors, clutches, and so on. An intermediate case is that of suspensions of larger ferromagnetic particles than those used for ferrofluids; such suspensions (which spontaneously gel) are used as coatings for magnetic audio- and video-recording tapes.

In this chapter, electrorheological fluids are covered in some detail. Magnetorheological fluids have been less studied, and their behavior is analogous to ER fluids; hence, they are discussed more briefly (in Section 8.3). There is a large literature on ferrofluids; however, the rheological properties of ferrofluids are not profoundly different from those of other suspensions, and so for the sake of brevity, only a few highlights of these fluids are discussed in Section 8.4. Much more complete reviews of ER fluids can be found in Gast and Zukoski (1989) and in Parthasarathy and Klingenberg (1996). For a thorough review of the physics of ferrofluids, the reader should turn to a book by Rosensweig (1985).

8.2 ELECTRORHEOLOGICAL FLUIDS

ER fluids are typically are suspensions of 1- to 100-μm particles of cornstarch, silica, calcium titanate, or other semiconductors at volume fractions of 0.05–0.50 in a hydrophobic liquid, such as mineral oil or corn oil. For an electric field E of $\sim$ 50–5000 V/mm, the particles form chains that span the gap between the field-generating electrodes (see

Fig. 8-1). If one attempts to slide one electrode relative to the other the particle chains resist with a force that increases roughly as E^2 (Winslow 1949; Klingenberg and Zukoski 1990). Electrorheological fluids were discovered and patented by W. M. Winslow (1947, 1949) some 50 years ago. Since then, Winslow and others have dreamed of widespread applications of these fluids in fast clutches, actively controlled shock absorbers, variable-flow pumps, and robotic activators.

Although early ER fluids, such as a mixture of corn starch in corn oil, clearly demonstrated the electrorheological effect (and were even edible!), commercialization was hobbled by problems of thermal breakdown, particle settling, irreversible clumping, and abrasiveness. The effectiveness of many early fluids depended on the presence of moisture in the fluid, which tended to evaporate at the high temperatures produced by viscous heating. Fortunately, however, more recent theoretical and experimental work has shown that moisture is not essential to the ER effect, and working dry ER fluids have been formulated (Block et al. 1990; Filisko and Radzilowski 1990). An interesting recent development has been the polymerization of the ER-fluid matrix after suspension of the particles (Bohon and Kraus 1997). The resulting rubbery solid is an "artificial muscle" that changes shape rapidly upon application of an electric field! Electrorheological effects in immiscible polymer blends are also now under investigation (Tajiri et al. 1997).

The simplest electrorheological mechanism is the chaining of particles brought about by their polarization in an electric field (Winslow 1949). In principle, such field-induced chaining should occur whenever there is a mismatch between the bulk dielectric constant of the particles, ε_p, and that of the suspending medium ε_s. Usually $\varepsilon_r \equiv \varepsilon_p/\varepsilon_s > 1$. In addition to true *bulk* polarizability, particles are *effectively* polarizable if charged species can adsorb and migrate along particle surfaces, or if there are polarizable electric double layers around the particles (Block et al. 1990). These effects may in some cases swamp any effects of bulk polarization (Zukoski 1993). It is also noteworthy that a particle with low dielectric constant that has been coated with a thin highly polarizable layer has an effective polarizability almost as large as a particle composed entirely of the highly polarizable material (Gast and Zukoski 1989). Thus, the addition to an ER fluid of a small quantity of a highly polarizable liquid, such as water, can have a large effect on its ER response. No matter

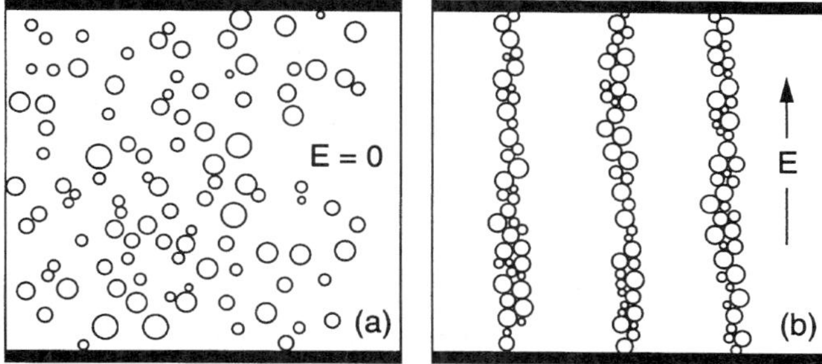

Figure 8.1 **(a, b)** Schematic illustration of the chain-forming effect of an electric field **E** on an ER suspension. (Reprinted with permission from Kerr, Copyright © 1990, American Association for the Advancement of Science.)

how it is acquired, an effective polarization causes particles to interact electrostatically with each other.

Figure 8-2 illustrates the interactions between particles that are polarized by a vertically oriented electric field. Particle pairs for which the line separating their centers of mass is parallel to the electric field are attracted to each other, while those for which the alignment is perpendicular to the field are mutually repelled. Letting θ be the angle between the field and the line joining the centers of two particles, one finds that the particles attract each other when $\theta < 55°$; otherwise they repel. This mechanism of particle interaction is the basis of the *polarization model* of ER fluids, which predicts rapid particle chain formation when an electric field is turned on, in agreement with direct visual observations (Klingenberg and Zukoski 1990; Whittle 1990; Melrose 1991). If the field is maintained, the chains tend to consolidate slowly into thicker columns. Columns of monodisperse particles can eventually organize into a dense phase with body-centered tetragonal symmetry (Chen et al. 1992; Tao 1993; Halsey and Martin 1993).

The shear stress σ in a sheared ER fluid is often simply represented by the Bingham model (Uejima 1972; Klingenberg and Zukoski 1990),

$$\sigma = \sigma_y + \eta_{pl}\dot{\gamma} \tag{8-1}$$

where η_{pl} is the *plastic viscosity*, $\dot{\gamma}$ the shear rate, and σ_y is the *yield stress*, which typically scales as E^2, while η_{pl} is nearly independent of the field (Gast and Zukoski 1989). Figure 8-3 shows examples of shear-stress data for an ER fluid; the Bingham law fits the data reasonably well. As defined by Eq. (8-1), σ_y is the *dynamic* yield stress. The *static* yield stress is defined as the minimum shear stress required to induce continuous shearing flow in an initially static sample. In general, these two definitions of the yield stress are not equivalent (Bonnecaze and Brady 1992b), although they are closely related.

Since most potential applications of ER fluids depend on a fast electrically driven transformation of the ER liquid to a near-solid of appreciable rigidity, the most important rheological characteristics of an ER fluid are that it achieve a high magnitude of σ_y, and that it do so rapidly after the voltage has been applied. Although σ_y can be made larger by

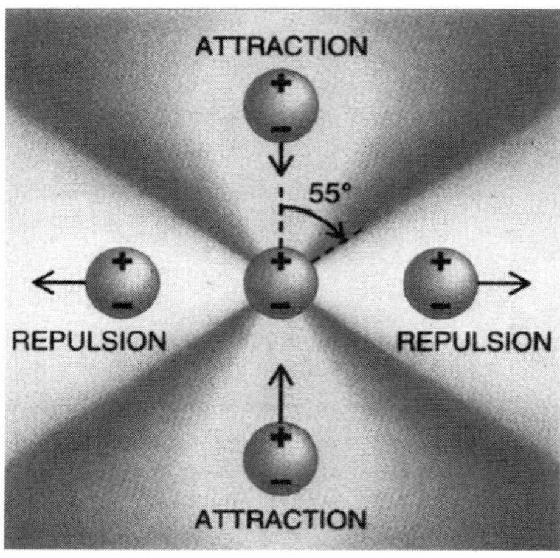

Figure 8.2 Interactions between particles polarized by an electric field. Particles whose separation is perpendicular to the field repel each other, while those whose separation is parallel to the field attract. (From Halsey and Martin 1993, Copyright © Jared Schneidman Design, used with permission.)

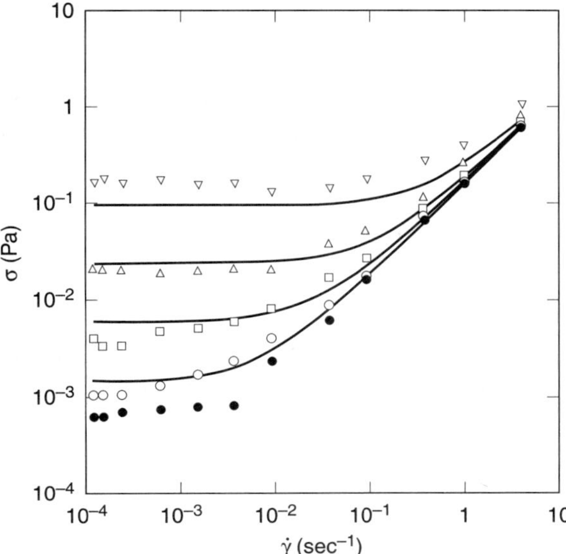

Figure 8.3   Shear stress as a function of shear rate for an ER fluid (symbols) at electric field strengths E of 0, 50, 100, 200, and 400 V/mm (from bottom to top curve), compared to the predictions of the Bingham equation with yield stress proportional to E^2. (From Zukoski 1993.)

increasing the electric field E, electrorheological fluids are inevitably slightly conductive, and too high a field can dissipate so much electrical power that the fluid overheats. Thus, one would like to design the ER fluid so that for a given limited field strength, say 1000 V/mm, the yield stress is as large as possible. A "good" ER fluid nowadays attains a yield stress of around 1 kPa at 1000 V/mm.

The yield stress can in principle be predicted from the polarization model. Rigorous calculation of the movement of one sphere in a flowing ER fluid under an electric field requires computation of the dielectric and hydrodynamic forces on that sphere. But these forces depend on the location and movement of all surrounding spheres, which are themselves responding to similar forces. Thus, one must solve a many-body problem, and this requires computer simulation.

8.2.1 The Simple Polarization Model

The forces one must include in such a simulation include *electrostatic*, *hydrodynamic*, and *steric* forces. For small particles, *Brownian* forces might also be present, but since these break up particle structures, it is desirable to use particles big enough ($> 1\mu$m) to suppress Brownian motion. Ordinarily, particle inertia can be neglected. Simulations can be greatly simplified by making drastic approximations, including the *point-dipole* approximation, and the *Stokes'-drag* approximation. Both of these approximations are only really valid for widely separated particles.

8.2.1.1 Electrostatic Forces

In the point-dipole approximation, one assumes that the dipole moment of a particle is not affected by the surrounding particles, and that the ratio $\varepsilon \equiv \varepsilon_p/\varepsilon_s$ is near unity. The dipole moment u of an isolated sphere in a field E is $u = \beta a^3 E$, where

$$\beta \equiv \frac{\varepsilon_r - 1}{\varepsilon_r + 2} \tag{8-2}$$

is the effective polarizability of the particle, and a is the particle radius. In the point-dipole approximation, the dipole moments of all particles of a given size are therefore equal and are all aligned parallel to the field. Let θ_{ij} be the angle between the field and the line joining the centers of mass of particles i and j. In the point-dipole approximation, the interaction potential between two particles is

$$W_{e,ij} = -\frac{4\pi \varepsilon_0 \varepsilon_s u^2}{r_{ij}^3} (3\cos^2 \theta_{ij} - 1) \tag{8-3}$$

where r_{ij} is the center-of-mass separation between the two particles, $r_{ij} \equiv |\mathbf{r}_i - \mathbf{r}_j|$, and ε_0 is the permittivity of space, $\varepsilon_0 = 8.85 \times 10^{-12} \; \mathrm{C^2 \, J^{-1} m^{-1}}$.

The force $\mathbf{F}_{ij}^e = -\partial W_{e,ij}/\partial \mathbf{r}_i$ on particle i produced by particle j is then

$$\mathbf{F}_{ij}^e = \frac{12\pi \varepsilon_0 \varepsilon_s u^2}{r_{ij}^4} [(3\cos^2 \theta_{ij} - 1)\mathbf{e}_r^{ij} + \sin 2\theta_{ij} \mathbf{e}_\theta^{ij}]$$

$$= 12\pi \varepsilon_0 \varepsilon_s \beta^2 a^2 E^2 \left(\frac{a}{r_{ij}^4}\right)^4 [(3\cos^2 \theta_{ij} - 1)\mathbf{e}_r^{ij} + \sin 2\theta_{ij} \mathbf{e}_\theta^{ij}] \tag{8-4}$$

where, again, $u \equiv \beta a^3 E$, $\mathbf{r}_j - \mathbf{r}_i = r_{ij}\mathbf{e}_r^{ij}$, and $\mathbf{e}_r^{ij}$ is a unit vector oriented parallel to the line joining the centers of the two spheres, while $\mathbf{e}_\theta^{ij}$ is a unit vector perpendicular to $\mathbf{e}_r^{ij}$ and in the plane containing the field direction (see Fig. 8-4). Under the point-dipole approximation, the electrostatic forces are pairwise additive, so that the total force $\mathbf{F}_i^e$ acting on particle i is given by

$$\mathbf{F}_i^e = \sum_{j \neq i} \mathbf{F}_{ij}^e \tag{8-5}$$

Anderson (1992) and Davis (1992) have noted that because all ER fluids have some electrical conductivity, then for direct current (dc) or slow alternating current (ac) electric fields, charges will migrate to the particles, eventually completely screening out the dipoles within the particles. If the electrical conductivity of the medium, Σ_s, differs from that of the particles, Σ_p, then there will still be an effective particle polarizability β, but its magnitude

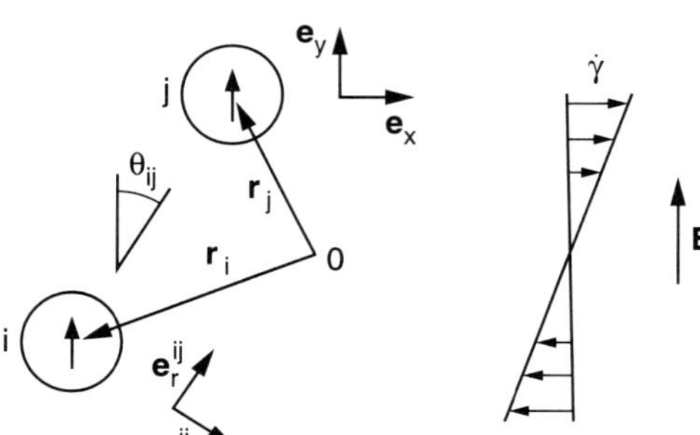

Figure 8.4   Coordinate system for a pair of particles in an electric field undergoing a shearing deformation. (From See and Doi 1992, with permission from the Journal of Rheology.)

will be determined not by the polarizabilities of the particles and the medium [as per Eq. (8-2)], but by the conductivities; that is,

$$\beta = \frac{\sum_p - \sum_s}{\sum_p + 2\sum_s} \tag{8-2a}$$

The effective polarizability β should still be given by Eq. (8-2) when an ac field is applied whose frequency ω is fast compared to the inverse of the Maxwell–Wagner time t_{MW} required for the mobile charges to screen the particle dipoles. When $t_{MW}\omega \ll 1$, however, β should be given instead by Eq. (8-2a). The Maxwell–Wagner time is given by (Parthasarathy and Klingenberg 1996)

$$t_{MW} = \varepsilon_0 \frac{\varepsilon_p + 2\varepsilon_s}{\sum_p + 2\sum_s}$$

For typical conductivities, $\sum_p \sim 10^{-9}\text{–}10^{-5}\Omega^{-1}m^{-1}$, t_{MW} is in the range 1 μsec to 10 msec. The important effect of even small levels of electrical conductivity was not appreciated until 1992; hence in earlier work, the simple polarization model with β given by Eq. (8-2) was often applied even in the case of a dc electric field.

8.2.1.2 Hydrodynamic Forces

The simplest form for the hydrodynamic "drag" force $\mathbf{F}^d$, namely Stokes' drag, is, like the point-dipole approximation, strictly valid only when the particles are widely spaced. In the Stokes'-drag approximation, for a shearing flow at shear rate $\dot{\gamma}$, we have

$$\mathbf{F}_i^d = -6\pi\eta_s a \left(\frac{d}{dt}\mathbf{r}_i - \dot{\gamma}\, y_i\mathbf{e}_x \right) \tag{8-6}$$

where y_i is the coordinate of $\mathbf{r}_i$ parallel to the electric field, with $y_i = 0$ on one electrode and $y_i = L$ on the other. The flow direction is x.

8.2.1.3 Steric Forces

Finally, a steric repulsive force $\mathbf{F}_i^{\text{steric}} = \sum_{j\neq i} \mathbf{F}_{ij}^{\text{steric}}$ must be included to keep particles from overlapping. This steric force would not be required if the hydrodynamic forces were treated more realistically, because as particles approach each other closely, strong lubrication forces are produced by the solvent that must be squeezed from between the particles (Bonnecaze and Brady 1992a). However, these lubrication forces are omitted from one-particle Stokes' hydrodynamics, and a steric force must be introduced.

8.2.1.4 Brownian Forces

Brownian forces can disrupt the particle chains that produce a yield stress, but can be made negligible if the electrostatic forces holding particle structures together are strong enough. From Eq. (8-3), we can compute the characteristic depth of the potential well holding together a pair of particles. For particles in near contact, $r_{ij} \approx 2a$, the electrostatic interaction is the strongest, and we define the ratio of electrostatic to thermal energy, W_e/k_BT, at this minimum in the potential by the parameter λ:

$$\lambda \equiv \frac{-W_{e,\min}}{2k_BT} = \frac{\pi\varepsilon_0\varepsilon_s\beta^2 E^2 a^3}{2k_BT} \tag{8-7}$$

For typical values of the parameters, and $E = 100$ mV/m, Eq. (8-7) gives $\lambda \sim 10a^3$, with a measured in microns. Thus, when a exceeds about 10 μm, $\lambda \approx 10^4$. Since the escape probability of a particle decreases exponentially with $\lambda = -W_e/2k_BT$, the rate of structural change by Brownian motion should be negligible for particles 10 μm or bigger. In fact, one might naively expect Brownian motion to be negligible at even smaller values of λ; nevertheless, both simulations and experiments indicate that Brownian motion remains important even for values of λ well in excess of 10^3 (Lemaire et al. 1995; Halsey et al. 1992; Parthasarathy and Klingenberg 1996). Thus, although at $\lambda = 10^3$ particles remain strongly bound together in chains, they can still experience localized Brownian-induced rearrangements.

8.2.1.5 The Particle Evolution Equation

Neglecting Brownian and inertial forces, a force balance on particle i, namely $\mathbf{0} = \mathbf{F}_i = \mathbf{F}_i^d + \mathbf{F}_i^e + \mathbf{F}_i^{\text{steric}}$, gives an evolution equation for the particle position:

$$\frac{d}{dt}\mathbf{r_i} - \dot{\gamma}y_i\mathbf{e}_x = \frac{1}{6\pi\eta_s a}\sum_{j\neq i}(\mathbf{F}_{ij}^e + \mathbf{F}_{ij}^{\text{steric}}) \tag{8-8}$$

This is a set of *evolution equations* for the particle positions. It can be integrated numerically by computer to obtain the history dependence of the particle positions under a given imposed flow and electric field. Usually periodic boundary conditions are imposed in the x and z directions, which are orthogonal to the electric field. Special periodic boundary conditions that permit shearing flow can also be imposed in the y direction (See and Doi 1992). Alternatively, the electrodes can be simulated explicitly, in which case the electrostatic forces between particles and the electrodes need to be included in Eq. (8-8); this can be done using the method of images (Klingenberg et al. 1991a).

If one makes the above equations dimensionless, a couple of scaling parameters immediately drop out, namely a time scale

$$t_s \equiv \frac{\eta_s}{2\varepsilon_0\varepsilon_s\beta^2 E^2} \tag{8-9}$$

and a dimensionless ratio of hydrodynamic to polarization forces, or *Mason number*:

$$\mathrm{Ma} \equiv \frac{\eta_s\dot{\gamma}}{2\varepsilon_0\varepsilon_s\beta^2 E^2} = \dot{\gamma}t_s \tag{8-10}$$

The Mason number is related to the parameter λ defined in Eq. (8-7) and the *Peclet number* Pe, defined in Chapter 6 [Eq. (6-12)] by $\mathrm{Ma}\,\lambda = (\pi/4)$ Pe. For the typical values $\eta_s \approx 0.02$Pa $\cdot$ s, $E \approx 1000$ V/mm, $\varepsilon_r \approx 3$ (and hence $\beta = 0.4$), and $\varepsilon_s \approx 3$, we obtain $t_s \approx 1$ msec, Ma $\approx 10^{-4}$ for $\dot{\gamma} = 0.1$, and Ma ≈ 1 for $\dot{\gamma} = 1000$. The time scale t_s is very roughly the time required for a particle to move a distance equal to its own diameter, under a polarization interaction with a nearby particle. Thus, the Mason number is near unity when the shearing is rapid enough to pull particle pairs apart as fast as they can form. Even at Mason numbers much smaller than this, chains of particles can be broken, although particle pairs cannot be. The time for particles to form a gap-spanning chain is much longer than t_s, as discussed below.

8.2.1.6 The Stress Tensor

The stress tensor $\boldsymbol{\sigma}$ for an electrorheological fluid can be obtained from the Kirkwood formula [see Eq. (1-42)],

$$\boldsymbol{\sigma} = \frac{1}{V} \sum_{j \neq i} \mathbf{r}_{ij} \, \mathbf{F}^e_{ij} \tag{8-11}$$

where V is the volume of the system. [The sign change between Eq. (1-42) and (8-11) arises from the definition of $\mathbf{F}^{ij}$ as a force on particle i rather than on j.] This expression neglects hydrodynamic-stress contributions from flow of solvent around particles; one expects these contributions to be negligible at low shear rates. However, at high shear rates they should dominate. "Stokesian dynamics" simulations by Bonnecaze and Brady (1992a) that include hydrodynamic as well as electrostatic stresses show that the hydrodynamic stress becomes dominant at Ma $\gtrsim 10^{-1}$, while it is negligible when Ma $\lesssim 10^{-4}$. Hence the yield stress σ_y is controlled by electrostatic stresses, while the plastic viscosity η_{pl} is governed mainly by hydrodynamic stresses.

So far, components of the stress tensor other than the shear stress have excited little interest, presumably because they are thought to be of marginal relevance to the design of ER fluids. However, it has been noted that the first normal stress difference in ER fluids is *negative* (Jordan et al. 1992), which is not a surprising finding, since shearing of particle chains that span the gap produces a reactive force that tends to pull together the shearing surfaces. Also, the full stress tensor for ER fluids is asymmetric (Kraynik 1993), because of the external couple from the electric field acting on particle chains.

8.2.2 Electrorheological Phenomena

8.2.2.1 Aggregation Kinetics

In the absence of flow, the aggregation kinetics of interacting dipolar particles has been studied by See and Doi (1991) (see Fig. 8-5). The time-dependence of the aggregation process seems to be reasonably well described by a hierarchical model, in which singlets form doublets after a time t_1, doublets form quadruplets after a time t_2, and so on. By

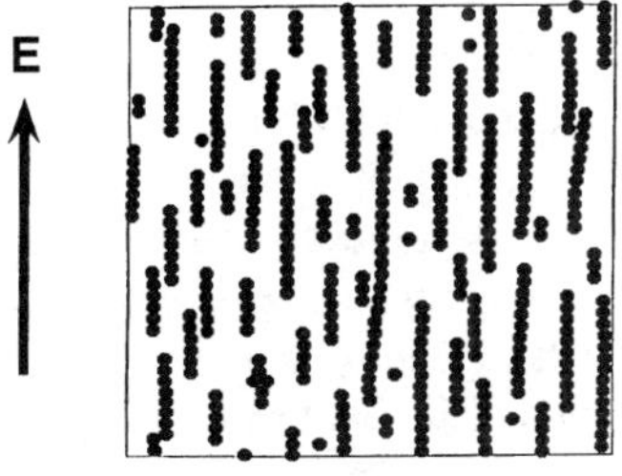

Figure 8.5   Particle chaining soon after application of an electric field in a two-dimensional simulation using the simple polarization model. (From See and Doi 1991, J Phys Soc Japan 60:2778, reprinted with permission.)

estimating both the electrostatic force and the hydrodynamic drag acting on elongated multiplets, these times can be calculated; one finds that after time t the average cluster size is $s_{\mathrm{ave}} \propto (t\varepsilon_0\varepsilon_s\beta^2\mathrm{E}^2/\eta_s)^{D/(D+3)}$, where D is the dimensionality of the system (See and Doi 1991). This prediction agrees with both simulations in two dimensions and experiments on three-dimensional systems (Vorob'eva and Vlodavets 1974). The dependence on concentration and time is predicted to be $s_{\mathrm{ave}} \propto \phi^{3/(D+3)} t^{D/(D+3)} (= \phi^{0.5} t^{0.5}$ in three dimensions and $\phi^{0.6} t^{0.4}$ in two dimensions). This prediction is in fair agreement with two-dimensional simulations, which show $s_{ave} \propto \phi^{0.85} t^{0.39}$ (See and Doi 1991). If the hierarchical model were to be valid up to the time t_p at which a sample-spanning "percolating" cluster forms, in three dimensions one would predict that $t_p \propto L^2$, where L is the gap. Simulations within the point-dipole approximation, however, indicate that $t_p \propto L^1$ (Toor 1993). In any event, we expect the ratio of particle diameter to gap size should influence the rheological time scales of ER fluids, a prediction that seems not to have been tested experimentally. Other simulations show $t_p \propto \phi^{-3.2}$ (Klingenberg et al. 1993), in reasonable agreement with experiments (Zukoski 1993; Parthasarathy and Klingenberg 1996). Experiments also show that t_p scales with E^{-2} as expected (Ginder and Elie 1992). Although t_p is much longer than the time for particle pairs to form, Marshall et al. (1989) have reported that t_p can still be less than 0.1 sec, a result favorable for fast switching in ER devices. Chains that are ruptured by shear presumably reform much faster than this if the shear is stopped while the electric field is maintained.

After the particle chains have formed, they slowly coarsen into columns (Martin et al. 1998). It may seem surprising that this should occur, since a pair of particles repel each other if the line connecting their centers is orthogonal to the field. But when two chains interact, each particle in one chain interacts with all particles in the other chain, and most of these interactions are attractive. Halsey (1992) has argued that Brownian motion produces the fluctuations needed for the coarsening to occur, and that the column width should grow with time t as $t^{5/9}$. Light-scattering experiments (Martin et al. 1992) indicate a somewhat weaker power law with an exponent of 0.4.

8.2.2.2 Static Yield Stress

Once chains or columns of particles span the gap between electrodes, a finite stress, the *static yield stress* σ_y, is required to break the chains so flow can occur. As deformation is applied to an ER fluid under a field, the chains tilt and stretch; an idealized picture of this is given in Fig. 8-6. At a critical tilt angle θ_m, the force exerted by the chains reaches a maximum, and continued shearing therefore produces an instability that breaks the chains. If the chain is assumed to be a single particle wide, to deform affinely, and to have negligible interactions with other chains, then the tilt angle θ_c of the chains at which this instability occurs in the point-dipole limit is 21° (Klingenberg and Zukoski 1990). This is considerably less than the angle at which the particles no longer attract each other—that is, around 55°. The tilt angle, 21°, corresponds to a yield strain $\gamma_c \approx \tan\theta_c \approx 0.4$. More rigorous calculations of this point of instability show that γ_c depends on particle concentration and dielectric ratio ε_r (Bonnecaze and Brady 1992b). Tilting and breaking of chains have been observed experimentally; the maximum tilt angle was observed to be around 28°, not far from the predicted value (Klingenberg and Zukoski 1990). From the dipole forces between particles

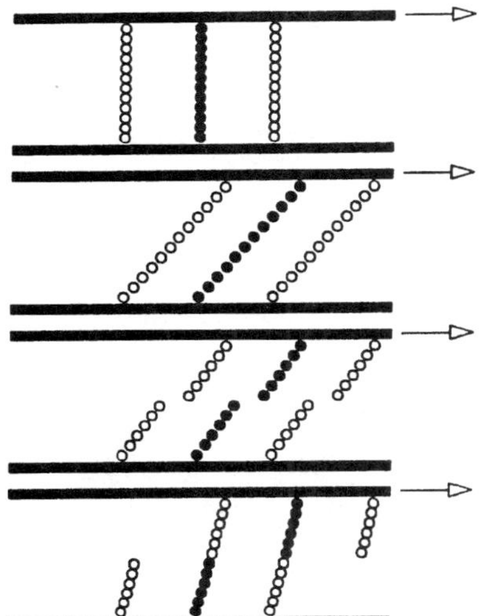

Figure 8.6 Schematic illustration of particle chaining, tilting, breaking, and reforming during shearing of an ER fluid under an electric field. The affine motion of the spheres in the second image is naive; instabilities lead to more complicated particle rearrangements. (From Klingenberg and Zukoski 1990, reprinted with permission from Klingenberg and Zukoski, Langmuir 6:15. Copyright © 1990, American Chemical Society.)

in the chain, the yield stress σ_y has been estimated by Klingenberg and Zukoski (1990) for gaps much greater than the particle diameter to be

$$\sigma_y \sim 18\phi\varepsilon_0\varepsilon_s\beta^2 E^2 f_m \tag{8-12}$$

where f_m is a dimensionless force equal to 0.057 in the point-dipole approximation. For $\phi = 0.3$, $\varepsilon_s \approx 3$, $\varepsilon_r \approx 3$, and E = 1000 V/mm, this gives a yield stress of around 1 Pa, which is lower by at least a factor of 2–50 than the experimental values (Klingenberg and Zukoski 1990; Klingenberg et al. 1991b). However, for $\varepsilon_r > 1$, the polarization forces are much larger than predicted by the theory because of multipolar and multibody effects that are beyond the scope of the point-dipole approximation, and one must use a larger value of f_m in Eq. (8-12).

One can go beyond the point-dipole approximation by generalizing the expression (8-4) for the electrostatic force to include *multipole* effects:

$$\mathbf{F}^e_{ij} = 12\pi\varepsilon_0\varepsilon_0\beta^2 a^2 \mathbf{E}^2 \left(\frac{a}{r^4_{ij}}\right)^4 [(2f_{\parallel}\cos^2\theta_{ij} - f_{\perp}\sin^2\theta_{ij})\mathbf{e}^{ij}_r + f_{\Gamma}\sin 2\theta_{ij}\mathbf{e}^{ij}_{\theta}] \tag{8-13}$$

In the point-dipole limit, each f_k equals unity, while when multipole effects are dominant, $f_{\parallel} \to \infty$ for particles in near contact. For conducting spheres ($\beta = 1$), if $\overline{D} \equiv r_{ij}/(2a) - 1$ is the dimensionless gap between two particles, then (Arp and Mason 1977; Klingenberg et al. 1991b)

$$f_{\parallel} \to \left(\frac{2\pi^4}{27}\right)\frac{1}{\overline{D}\ln^2(\overline{D}/6.344)} \tag{8-14}$$

Equation (8-13) is only rigorous for two spheres in a dielectric medium. Hence, it does not account for *multibody* effects. Bonnecaze and Brady (1992a) have included multipole and multibody effects accurately by using a moment expansion to compute a capacitance matrix for the entire set of particle configurations at each time step. Such simulations are expensive, however. Klingenberg et al. (1991b) accounted for multipole and multibody effects in an ad hoc fashion, by computing two factors—one for each effect—and multiplying these factors by the yield stress computed from the point-dipole limit. While this procedure is crude, it does indicate that the true yield stress of a typical ER fluid is likely to be at least 10 times higher than that calculated in the point-dipole limit. With this approach, Klingenberg et al. (1991b) have estimated the highest yield stress likely in ER fluids for large (100 μm) conductive particles ($\varepsilon_r \to \infty$), whose surfaces are kept a small distance ($\sim$ 1 nm) apart to prevent conduction; for such a suspension, σ_y will be around 1.2 kPa at a field strength of E $=$ 1 kV/mm.

In practice, highly conductive particles do not make good ER fluids, in part because of excessive conduction in the resulting suspension. An acceptable conductivity of the particles seems to be around $10^{-5}\Omega^{-1}m^{-1}$ (Block et al. 1990), at least for particles of the polymer poly(acenequinone) in chlorinated oil. These ER fluids show a yield stress of about 1 kPa at E $=$ 1 kV/mm, roughly equal to the predicted "maximum possible" value for this electric field. As expected, even higher yield stresses exceeding 10 kPa are possible when E exceeds 3 kV/mm.

8.2.2.3 Steady Shearing Flow

Yield stresses of ER fluids are usually measured by lowering the shear rate until the shear stress approaches a constant that can be identified as the *dynamic* yield stress. The mechanism producing this dynamic yield stress has been explored in simulations and analysis of Bonnecaze and Brady (1992a, 1992b). They found that sample-spanning chains in an ER fluid that are initially perpendicular to the shearing surfaces tend to deform until they exceed the critical strain γ_c; they then break near the middle and rapidly reconnect with fragments from adjacent chains (Klingenberg and Zukoski 1990; Bonnecaze and Brady 1992a). The deformation is largely elastic, so the energy dissipation occurs mostly during the rapid process of breakage and reformation of the chains (Bonnecaze and Brady 1992b). At low shear rates, Ma $\lesssim 10^{-4}$, the shearing process approaches a limit in which there are long periods of slow, quasi-static straining of particle chains (in which little energy is dissipated), punctuated by rapid, irreversible particle rearrangements, including chain breakage and reformation. The rapid rearrangements occur when the particle configuration reaches a point of instability.

Once the point of instability is reached, particle rearrangements are driven by energy ΔE stored up during the period of slow quasi-state straining. For slow enough flows, ΔE is independent of the shear rate. Thus, the rate of rapid particle rearrangement, and hence the time Δt that it takes, are independent of strain rate in the limit Ma $\to$ 0. The time between rearrangements is given by $\gamma_c/\dot{\gamma}$. The *average* rate of energy dissipation for the fluid is the rate of energy dissipation during rapid rearrangement, $\Delta E/\Delta t$, divided by the time $\gamma_c/\dot{\gamma}$ between these rearrangements; that is, $(\Delta E/\Delta t)/(\gamma_c/\dot{\gamma}) \propto \dot{\gamma}$. Thus, the average rate of energy dissipation $\dot{W}_{\text{diss}}$ is proportional to the shear rate $\dot{\gamma}$. Since $\dot{W}_{\text{diss}}$ is the product $\sigma\dot{\gamma}$ of the shear stress and the shear rate, the stress σ must approach a constant, the dynamic yield stress σ_y, at small Ma. This result contrasts with the Newtonian result in which $\dot{W}_{\text{diss}} \propto \dot{\gamma}^2$, and $\sigma \propto \dot{\gamma}$. From more detailed versions of this argument, one can estimate the dynamic yield stress from the energy lost per unit volume in the sudden rearrangements (Bonnecaze and Brady 1992b). In general, a dynamic yield stress will be present in any material that can "store energy on a slow time scale and then rapidly release and viscously dissipate that energy" on a much faster shear-rate-independent time scale (Bonnecaze and Brady 1992b). A qualitatively similar mechanism, with different microscopic details, seems also to account for the dynamic yield stress of foams (see Section 9.5).

Klingenberg et al. (1991a) find in their simulations a dependence of the dynamic yield stress on particle volume fraction ϕ that is in qualitative agreement with experiment (see Fig. 8-7). Note that in both experiments and simulation, σ_y is roughly linear in ϕ for $\phi \lesssim 0.30$. For higher ϕ, σ_y levels off, or even has a maximum, evidently because chains are more efficient at momentum transfer when they are a single particle wide than when they are clumped into columns, as they tend to be when ϕ is large (Klingenberg et al. 1991b). Note, also, however, that the magnitude of σ_y from the simulations is much lower, by a factor or 50 or so, than the experimental values. At least part of this discrepancy can readily be attributed to the point-dipole approximation, which is only accurate when particles are widely spaced and when $\varepsilon_r \rightarrow 1$, conditions that aren't met. The magnitude of the errors made in the point-dipole approximation can be appreciated when one considers the limit of conducting particles, $\varepsilon_r \rightarrow \infty$. In this limit, $\beta = (\varepsilon_r - 1)/(\varepsilon_r + 2) \rightarrow 1$, and the interactions within the point-dipole approximation remain bounded; the true dipolar interaction becomes singular,

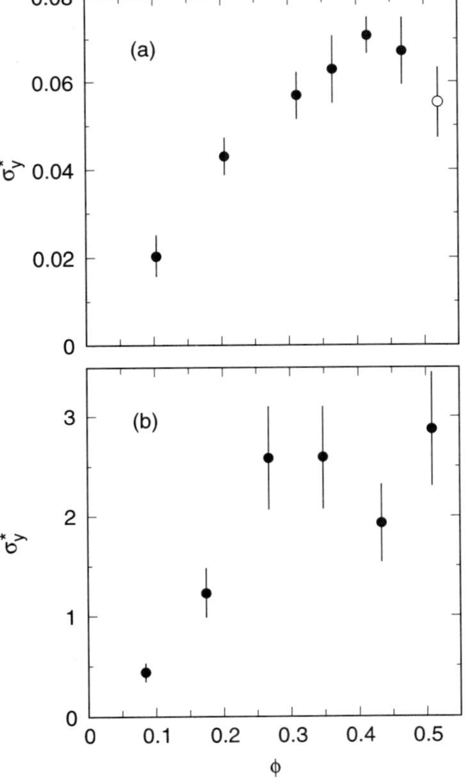

Figure 8.7 Dimensionless dynamic yield stress σ_y^* as a function of volume fraction ϕ of (**a**) particles in a three-dimensional simulation using the simple polarization model. (**b**) Hollow silica spheres in corn oil. The stress is made dimensionless by dividing by $3\pi \varepsilon_0 \varepsilon_s \beta^2 E^2 / 16$. (From Klingenberg et al., reprinted with permission from J. Chem. Phys. 94:6160, Copyright © 1991, American Institute of Physics.)

however, as the particles approach each other. The force is only kept finite by particle surface roughness, solvation forces, or other factors that keep particle surfaces a few nanometers apart on average. Thus, as ε_r increases, these "near-contact" interactions, rather than point-dipole interactions, dominate the yield stress (Klingenberg et al. 1991b).

When Brownian forces are not negligible—that is, when λ defined in Eq. (8-7) is less than around 10^4—an alternative picture of the steady-flow behavior of ER fluids, suggested by Halsey et al. (1992; see also Halsey 1992), seems to hold. In this picture, the strings of particles are broken down by shearing into ellipsoidal "droplets" of condensed particles. The major axis of each droplet assumes an angle relative to the field direction that is dictated by the balance of hydrodynamic and electrostatic forces. The higher the shear rate, the greater this angle becomes, and the smaller the size of the droplet. Scaling arguments give an angle proportional to the one-third power of the Mason number, $\mathrm{Ma}^{1/3}$, and a droplet size inversely proportional to Ma. The resulting shear-thinning viscosity η is predicted to scale as $\mathrm{Ma}^{-2/3}$, which disagrees with that of the Bingham model, Eq. (8-1), the latter predicting $\eta \propto \mathrm{Ma}^{-1}$. Experimentally, the power-law exponent p (in $\eta \propto \mathrm{Ma}^{-p}$) has been found to vary with λ from 0.68, close to the Halsey prediction for $\lambda = 280$, to $p = 0.93$, closer to the Bingham limit for $\lambda = 1700$ (Halsey et al. 1992; Martin et al. 1994).

At high shear rates (or high Mason numbers), hydrodynamic contributions to the stress tensor become important. By modifying their "Stokesian dynamics" method to include multipole and multibody effects, Bonnecaze and Brady (1992a) were able to combine accurate descriptions of both hydrodynamic and electrostatic forces, permitting, in principle, accurate simulations even for dense suspensions with ε_r far from unity and for a wide range of shear rates. The results of these simulations in two dimensions were compared with the three-dimensional steady-shear experiments of Marshall et al. (1989) by invoking the "two-thirds rule" in which the areal fraction ϕ_A occupied by particles in the simulations is converted to a volume fraction by multiplying ϕ_A by 2/3 (see Fig. 8-8). The viscosity appears to follow the formula $\eta/\eta_\infty = \mathrm{Ma}^*/\mathrm{Ma} + 1$, where η_∞ is the high-shear-rate viscosity, and Ma^* is a parameter that experimentally is proportional to ϕ. This expression is equivalent to the Bingham formula, Eq. (8-1). Three-dimensional simulations of Melrose and Heyes (1993), however, are better fit by a Casson equation [see Eq. (6-15b)]. In either case, there is a dynamic yield stress, the existence of which is expected when the suspension is non-Brownian. For particles smaller than a micron, there should instead be a low-shear-rate plateau viscosity η_0.

Figure 8-8 also shows that for different shear rates $\dot{\gamma}$ and field strengths E, the relative viscosity $\eta_r \equiv \eta/\eta_s$ in the experiments depends only on the Mason number Ma $\equiv \eta_s \dot{\gamma}/(2\varepsilon_0\varepsilon_s\beta^2\mathrm{E}^2)$, and not $\dot{\gamma}$ and E separately. This scaling is predicted by the evolution equation (8-8), and it is an important confirmation of the polarization model of ER fluids.

8.2.2.4 Oscillatory Electric Field

The effect of ac, rather than dc, fields on the steady shear viscosity has also been investigated. An ac field might be desirable as a means of periodically reversing, and thereby minimizing, electrically induced migration or chemical reactions. High-frequency ac fields are also required if one is to avoid a crossover to a conduction-dominated regime in which the effective polarizability is controlled by conductivities rather than polarizabilities of the particles and medium (see Section 8.2.1.1). Klass and Martinek (1967) found that at a fixed

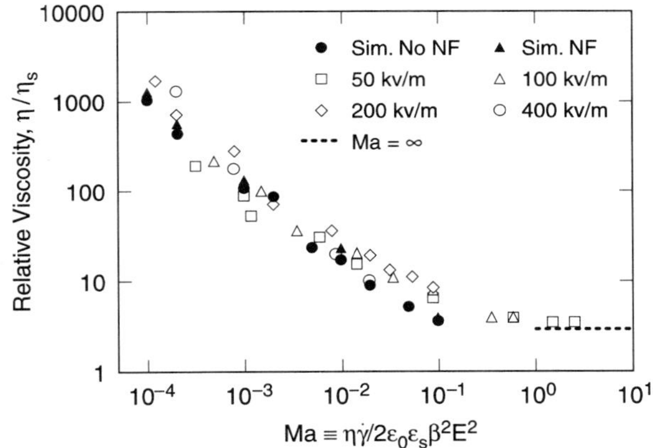

Figure 8.8 Relative viscosity η/η_s versus Mason number for an ER fluid consisting of hydrated lithium poly(methacrylate) particles in a chlorinated hydrocarbon studied by Marshall et al. (1989) with $\phi = 0.23$ at various field strengths, compared to predictions of two-dimensional "Stokesian dynamics" simulations (closed symbols) with and without "near-field" (NF) interactions at areal fraction $= 0.4$. Since β in the above was taken from the polarization model with Eq. (8-2), while the experiments were carried out under dc fields for which the effective polarization should be controlled by conductivities [Eq. (8-2a)], the quantitative agreement between simulations and experiment is presumably fortuitous [see Parthasarathy and Klingenberg (1996)]. (From Bonnecaze and Brady, reprinted with permission from J. Chem. Phys. 96:2183, Copyright © 1992, American Institute of Physics.)

shear rate of 753 sec^{-1}, the viscosity of a silica-sphere suspension shows a gradual decrease with increasing frequency ω of the electrical field (see Fig. 8-9). This effect can readily be attributed to the gradual decrease with increased frequency of the real part of the particles' dielectric constant ε_p (See and Doi 1992).

However, superimposed on this gradual decrease in η they found a sharp, pronounced minimum in η at an electric-field frequency of around 200 Hz (again see Fig. 8-9). A similar

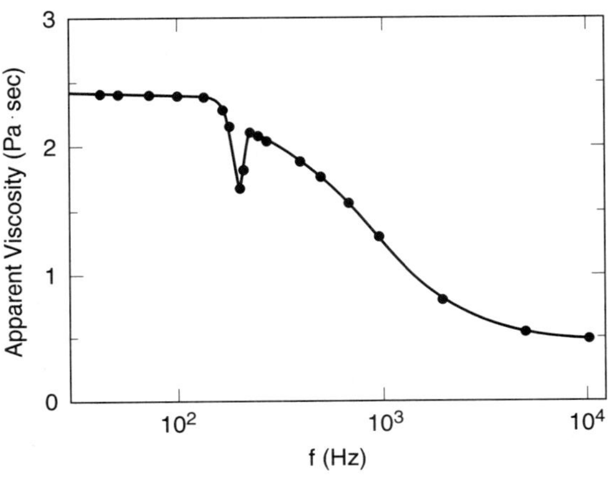

Figure 8.9 Apparent viscosity versus oscillation frequency f of the electric field for a dispersion of silica particles in naphthenic fluid at a field strength of 930 V/mm and a shear rate of 753 sec^{-1}. (From Klass and Martinek, reprinted with permission from J. Appl. Phys. 38:67, Copyright © 1967, American Institute of Physics.)

minimum in η shows up in some computer simulations of See and Doi (1992), based on the Stokes' law/point-dipole description of ER fluids in Eq. (8-8). An analysis by See and Doi shows that this minimum occurs at the ratio $\omega/\dot{\gamma} \approx 1.6$, for a wide range of shear rates. The minimum evidently occurs because the particle chains are rapidly broken apart by shear during the low-voltage part of the cycle, but not so rapidly rebuilt during the high-field part of the cycle. As a result, near $\omega/\dot{\gamma} \approx 1$, the rigid, fully formed particle chains only exist during a small fraction of the cycle.

Although this mechanism is consistent with available data, a full test of the theory is still lacking. Also, some ER fluids don't seem to show a minimum in viscosity over the range of frequencies tested. Frequency dependencies can also occur because of a crossover to a conduction-dominated regime at low frequencies, and this was not included in the See–Doi calculations.

8.2.2.5 Dynamic Moduli

Dynamic oscillatory experiments have been performed on ER fluids, both with dc and ac applied fields, to probe in more detail the relaxation processes (Jordan et al. 1992; Parthasarathy et al. 1994; Parthasarathy and Klingenberg 1996). These experiments show that at small strains and high fields, G' is roughly independent of frequency and is larger than G'' by more than a decade, as would be expected for an elastic solid. If the ER structure is modeled as chains of spheres one sphere in width, then in the point-dipole limit, McLeish et al. (1991) find the storage modulus to be $G' = 3\phi\varepsilon_0\varepsilon_s\beta^2E^2$, which is smaller than the measured values, probably because of the point-dipole approximation. If the field is reduced, G' drops and becomes frequency-dependent (see Fig. 8-10a). Parthasarathy et al. (1994) have shown that curves of G' versus frequency ω at various electric field strengths can be collapsed onto a single "master" curve when plotted in dimensionless form as shown in Fig. 8-10b. This method of rescaling can be justified by computer simulation and by small-strain analysis (Klingenberg 1993).

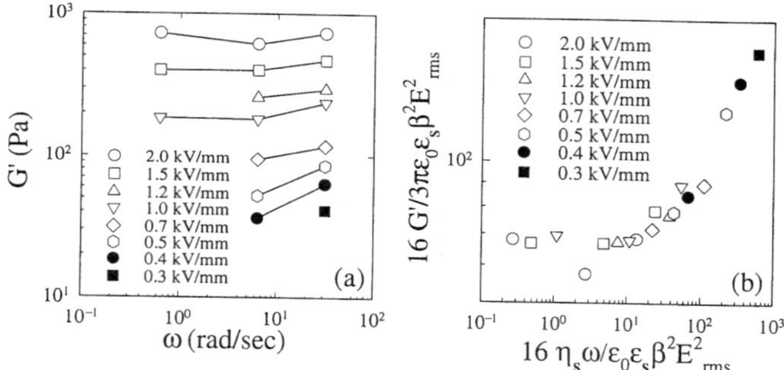

Figure 8.10 (a) G' versus frequency ω for an ER fluid composed of 20 wt% alumina particles in poly(dimethylsiloxane) at an ac field frequency of 500 Hz. (b) "Master" plot superposing the data of (a). E_{rms}^2 is the root-mean-square field strength at 500 Hz. (From Parthasarathy et al. 1994, with kind permission from Elsevier Science–NL, Sara Burgerhartstraat 25, 1066 KV Amsterdam, The Netherlands.).

Experimentally, while G'' is small, it is still much larger than the viscous contribution to G'' from the solvent. Hence, relaxation processes do occur in ER fluids, even when they have been "solidified" by a strong electric field. Proposed sources of relaxation in ER fluids include (a) relaxation of "free" chains of particles that are attached to only one or no electrodes (McLeish et al. 1991) and (b) competition between hydrodynamic and electrostatic forces on particles in columns (Klingenberg 1993).

In addition, the moduli are very sensitive to strain (Jordan et al. 1992; Parthasarathy and Klingenberg 1996). Figure 8-11, for example, shows that at high fields, G' is a function of strain even at strain amplitudes lower than 0.1%! In point-dipole simulations, Parthasarathy and Klingenberg (1995) found that this nonlinearity is produced at small strain amplitudes by microrearrangements of particles. It seems that particle configurations are rendered unstable when particles are even slightly displaced by shear. At low shear frequencies, these rearrangements theoretically show up as (a) small discontinuities in the stress waveform and (b) nonlinearities in the moduli.

8.2.3 Other Electrorheological Effects

The bulk polarization model, despite its successes, is incomplete. If bulk polarization were the whole story, then the ER effect would not change if the electrodes that produce the dc field were painted with an insulating layer. But, in fact, the ER effect disappears when the electrodes are insulated, because mobile ions build up at the electrodes and screen the field (Zukoski 1993). Even when the electrodes are conducting, mobile charges can play an important role in the ER effect, beyond making the fluid somewhat conducting, with a nonohmic conductivity of 10^{-9}–$10^{-5}\Omega^{-1}m^{-1}$ (Winslow 1949; Klingenberg and Zukoski 1990; Block et al. 1990; Foulc et al. 1994). These charges can collect at particle surfaces, making them much more polarized than would otherwise be the case (Gast and Zukoski 1989). Small amounts of water, for example, could easily carry such charges to particle surfaces, a phenomenon on which the effectiveness of early ER fluids apparently depended. An alternative, or supplemental, role of water is to form bridges that hold particles together by surface tension forces (Stangroom 1991; Deinega and Vinogradov 1984; See et al.

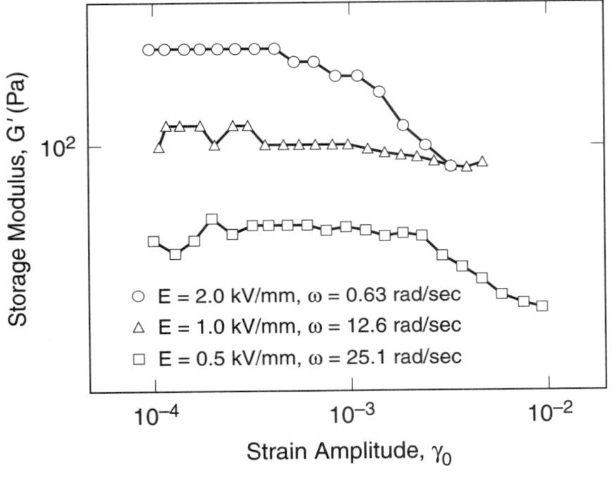

Figure 8.11 G' as a function of shear strain amplitude γ_0 for a 20 wt% ER suspension of silica–alumina particles in silicone oil with 2 wt% surfactant at various electric fields, with field frequency 500 Hz, strength E, and shear oscillation frequency ω. (From Parthasarathy and Klingenberg 1996, reprinted with permission from Steinkopff Publishers.)

1993). There is also the possibility that the particles develop a net charge that reverses when the particle reaches an electrode bearing the opposite charge. The result will be a circulating motion of the particles that can dissipate energy (Deinega and Vinogradov 1984; Parthasarathy and Klingenberg 1996).

Klingenberg and Zukoski (1990) have observed inhomogeneities in dilute ($\phi = 0.02$) ER fluids under continuous shearing. They found two distinct layers: (1) a nondeforming "solid" zone dense in particles and (2) a flowing zone more dilute in particles. As the shear rate increased, the fluid region grew at the expense of the solid region. Jordan et al. (1992) noted that the following occurs during shearing of an ER fluid: While thin strands of spheres tended to break in the middle, thick columns tended to break at the electrode. These observations suggest that surface effects may be important in determining the yield stress of ER fluids. The appropriateness of the sticking boundary conditions used at the electrodes (Klingenberg et al. 1991a) can be questioned; however, pure slip at the boundaries would presumably destroy the yield-stress phenomenon.

The effects of particle-size polydispersity have been largely unexamined so far. Non-spherical non-Brownian particles have also been unexplored, as have deformable particles, even though one might imagine that elongated particles might show an even more dramatic increase in viscosity when they orient and line up under an electric field than do spherical particles. However, Yang and Shine (1992) have studied the electrorheological properties of long semirigid Brownian *molecules*, namely poly(n-hexyl isocyanate) in p-xylene. Although these are *nematic* liquid crystalline solutions and not suspensions, they show a large (35-fold) increase in viscosity when an electric field is applied, because the field orients the molecules in a direction for which the relevant Miesowicz viscosity is high (see Section 11.5.1). These solutions do not show a yield stress, however, nor would they be expected to, since the molecules are strongly affected by Brownian motion, which would prevent the formation of any solid-like microstructure. As discussed earlier, the effect of Brownian motion is negligible in typical ER fluids when the particle radius a exceeds about 10 μm, and it is important when $a < 1\mu$m. Thus, as the particle size increases, there is a rather sharp crossover from "liquid-like" to "solid-like" ER behavior.

8.3 MAGNETORHEOLOGICAL FLUIDS

The magnetic analogs of electrorheological fluids are called *magnetorheological fluids*. These fluids contain particles that are *paramagnetic*, that is, they become magnetized on application of a magnetic field. The most interesting case is one in which the particles are *superparamagnetic*; that is, they are aggregates containing many ferromagnetic single-domain inclusions with no long-range magnetic order. An example of a superparamagnetic material is polystyrene studded with small (10 nm) inclusions of iron oxide (Fermigier and Gast 1992; Promislow et al. 1995). Micron-size particles of this composite dispersed in a solvent make an ideal magnetorheological (MR) fluid.

MR fluids were first developed in the late 1940s by Rabinow (1948). Like ER fluids, applications have been slow to develop; however, MR fluids has recently been used commercially in exercise equipment as a means for controlling resistance. Perhaps the most important advantage an MR fluid can offer over present-day ER fluids is a higher yield stress, around 100 kPa, more than an order of magnitude larger than is attainable with the best ER fluids (Parziale and Tilton 1950; Weiss and Duclos 1994; Ginder et al. 1996). A disadvantage is that the MR effect requires a magnetic field, which requires current and hence draws power; but in practice, ER fluids also consume power, since at high fields there is inevitably some current leakage. A more serious disadvantage of MR fluids may be that the best particles contain iron, making them heavy and prone to settle.

As model systems for scientific study, MR fluids are superior to ER fluids, because complications due to charging and conductivity in ER fluids have no counterpart in MR fluids, since magnetic "monopoles," the analog of electric charges, are unknown in nature. Thus, a magnetic analog of the simple polarization model described in Section 8.2.1 for ER fluids should be even more appropriate for MR fluids.

To develop the magnetic-polarization model, note that an isolated magnetizable particle of radius a acquires a magnetic dipole $\mathbf{m}$ in the presence of a magnetic field $\mathbf{H}$:

$$\mathbf{m} = \frac{4}{3}\pi a^3 \mu_0 \chi \mathbf{H} \tag{8-15}$$

where χ is the *magnetic susceptibility* of the particulate phase. Here $\mu_0 = 4\pi \times 10^{-7}$ Tm/A is the permeability of free space, where "T" is the tesla, where 1 tesla $=$ 1 N/A-m. (An equivalent unit is 1 henry/meter $=$ 1 tesla-meter/ampere.) The units of the field H are A/m, and χ is dimensionless. Two such magnetically polarized particles (i and j) each with dipole moment m then interact with energy

$$W_{m,ij} = -\frac{m^2}{4\pi \mu_0 r_{ij}^3}(3\cos^2\theta_{ij} - 1) \tag{8-16}$$

Equation (8-16) is obviously similar to the corresponding equation (8-3) for interacting electric dipoles. The magnitude of the magnetic interaction energy relative to thermal energy is $-W_{m,\min}/k_B T$:

$$\lambda \equiv \frac{-W_{m,\min}}{2k_B T} = \frac{\pi \mu_0 \mu_0 \chi^2 \, H^2 a^3}{18 k_B T} \tag{8-17}$$

Deriving a pairwise additive magnetic force from Eq. (8-16) and adding it to the drag forces and steric forces, a magnetic analog of the polarization model for ER fluids can be derived (Fermigier and Gast 1992). The aggregation kinetics of such a model in the absence of flow have been studied and compared to experiments on MR fluids with low and modest values of λ ($\sim$ 8–34) by Promislow et al. (1995).

At low field strengths, the yield stress σ_y of an MR fluid can be computed from the polarization model by an analog of Eq. (8-12), which then predicts $\sigma_y \propto H^2$. However, this prediction neglects the important, and unique, role played by magnetic-field *saturation*, which occurs near particle–particle contacts, because of the concentration of magnetic field lines in these regions. When this *local* saturation of magnetization at its maximum value M_s is accounted for, one finds for the yield stress σ_y and elastic modulus G_0 the predictions (Ginder et al. 1996):

$$\sigma_y = \sqrt{6}\phi\mu_0 M_s^{1/2}H^{3/2}, \qquad G_0 = 3\phi\mu_0 M_s H \qquad \text{(intermediate field strengths)}$$

The prediction of a 3/2 power law for the field-dependence has been confirmed experimentally (Ginder et al. 1996). At still higher applied fields, the magnetization saturates throughout the particles, and the yield stress and modulus should become independent of applied field:

$$\sigma_y^{\text{sat}} = 0.086\phi\mu_0 M_s^2, \qquad G_0^{\text{sat}} = 0.3\phi\mu_0 M_s^2 \qquad \text{(high field strengths)}$$

For iron-based MR fluids, $\mu_0 M_s \approx 2.0$ T, and one predicts $\sigma_y^{\text{sat}} \approx 80$ kPa for $\phi = 0.3$. These high yield stresses make MR fluids potentially attractive for applications requiring fast conversion from a liquid-like to a near solid-like response.

8.4 FERROFLUIDS

Ferrofluids are suspension of small, ~ 10-nm particles, each of which contains a single permanent ferromagnetic domain. Thus, each particle is a permanent magnet, which, in the absence of an external magnetic field, rotates randomly under the action of Brownian forces, which are strong because of the particle's small size. Hence, typical ferrofluids differ from magnetorheological fluids in two ways: (a) the magnetic field within each particle is permanent, and (b) the particles are much smaller than a micron in diameter. Indeed, since ferrofluid particles are permanent dipoles, if they were as large as a micron they would cluster in the absence of a magnetic field the way particles in a magnetorheological fluid do only when the field is present. Suspensions of micron-sized ferromagnetic particles (of γ-Fe_2O_3 or CrO_2) are actually of technological importance in the manufacture of magnetic recording films for audio and video tapes (di Rico 1963), but these particles agglomerate and form gel-like networks (Yang et al. 1986; Scholten and Felius 1990). Thus, the fluidity of a ferrofluid depends on the presence of strong Brownian forces to break up clusters.

Of course, particles of such small size tend to flocculate because of van der Waals forces, as discussed in Chapter 7, unless the particle surfaces are suitably treated. The surfaces of a ferrofluid are typically coated with surfactant, as depicted in Fig. 8-12. Starting from a magnetic powder, a ferrofluid can be made by *wet-grinding* for long periods of time (~ 1000 hours) in the presence of a surfactant concentrated enough to produce a monolayer on the 10-nm particles that are eventually produced. This, and other methods of preparing small surfactant-coated particles, such as arrested chemical precipitation, are discussed in Rosenzweig (1985). The carrier fluid for ferrofluid particles can be either aqueous or organic.

There are many existing and potential uses for ferrofluids. Most personal computer disk drives use ferrofluid as a rotary shaft seal to prevent entry by contaminants (Rosenzweig 1996). Ferrofluids are also used as heat conductors for the coils of high-power speakers; and there are potential applications as magnetic tracers and drug carriers, high-speed printing ink, signal enhancers in medical magnetic-resonance imaging, and so on (Rosenzweig 1985).

A sensitive ferrofluid requires particles that are highly magnetized. Magnetite (Fe_3O_4) has a *magnetization* M_d of around 4.5×10^5 A/m, where A denotes amperes of current. The magnetic dipole moment $\mathbf{m}$ of a particle of volume V is then $\mathbf{m} = \mu_0 \mathbf{M}_d V$. $\mathbf{M}_d$ is the vector magnetization—that is, M_d multiplied by a unit vector directed along the magnetization direction. The product $\mu_0 M_d$ is then around 0.56 T for magnetite, which is equivalent to 5.6 kG in cgs units. (The quantity $4\pi\mu_0$ is, by definition, unity in cgs units.) The units of the magnetic field H are the same as those of M_d; thus $\mu_0 H$ is measured in

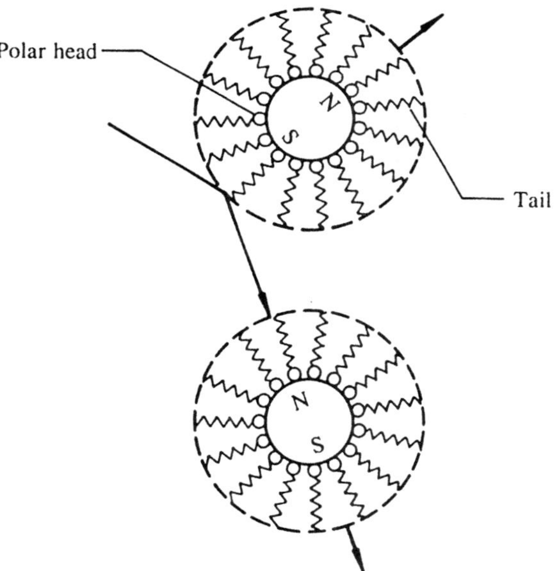

Figure 8.12 Illustration of permanently magnetized, single-domain, ferrofluid particles coated with a surfactant. (From Rosenzweig 1985, with permission.)

tesla or gauss. The product $\mu_0 H M_d$ has units of energy density—that is, joules/m^3 in mks units and ergs/cm^3 in cgs.

8.4.1 Magnetic Dipole–Dipole Interactions

As alluded to earlier, even in the absence of a field, large particles spontaneously agglomerate because of interactions of the permanent magnetic dipoles. This interaction energy between particles i and j given by

$$W_{m,ij} = -\frac{\mathbf{m}_i \cdot \mathbf{m}_j}{4\pi\mu_0 r_{ij}^3}[3(\hat{\mathbf{m}}_i \cdot \hat{\mathbf{r}}_{ij})(\hat{\mathbf{m}}_j \cdot \hat{\mathbf{r}}_{ij}) - \hat{\mathbf{m}}_i \cdot \hat{\mathbf{m}}_j] \tag{8-18}$$

where the hatted quantities are unit vectors in the directions of the magnetic dipoles $\mathbf{m}_i$ and $\mathbf{m}_j$. As before, $\mathbf{r}_{ij} \equiv \mathbf{r}_j - \mathbf{r}_i$ is a unit vector parallel to the vector separating the centers of mass of the particles i and j. Equation (8-18) generalizes Eq. (8-16), which applies when the two dipoles are equal and point in the same direction (which is appropriate for magnetorheological fluids because their magnetic dipoles are induced by, and therefore parallel to, the applied field). In ferrofluids, the dipoles exist without a field and rotate randomly by Brownian motion. If the dipoles $\mathbf{m}_i$ and $\mathbf{m}_j$ in a ferrofluid are equal in magnitude, the strongest attractive interaction (minimum in W_m) occurs when the dipoles of the two particles are parallel to each other and to the line connecting their centers (i.e., $\hat{\mathbf{m}}_i = \hat{\mathbf{m}}_j = \pm\hat{\mathbf{r}}_{ij}$) and the particles are in contact (i.e., $r_{ij} = d = 2a$). In this case, if the dipole moments of the two particles are equal, the attractive interaction energy divided by $k_B T$ is

$$\frac{W_{m,\min}}{k_B T} = -\frac{2m^2}{4\pi\mu_0 d^3 k_B T} = -\frac{\pi\mu_0 M_d^2 d^3}{72} \tag{8-19}$$

from which we obtain λ, the *coupling coefficient*,

$$\lambda \equiv \frac{-W_{m,\min}}{2k_BT} = \frac{\mu_0 M_d^2 \pi d^3}{144 k_B T} \tag{8-20}$$

We have used $m = \mu_0 M_d \pi d^3/6$ as the dipole strength. This definition of λ for magnetic clustering is the analog of the definitions in Eqs. (8-7) and (8-17) for clustering of polarizable particles in ER and MR fluids. In a ferrofluid, the threshold particle size at which the particles begin to aggregate spontaneously is therefore defined by the condition $2\lambda = -W_{m,\min}/k_BT \approx 1$. This corresponds to

$$d \approx \left(\frac{72 k_B T}{\pi \mu_0 M_d} \right)^{1/3} \qquad \text{(aggregation threshold)} \tag{8-21}$$

For magnetite particles ($M_d = 4.5 \times 10^5$ A/m), this gives a particle size of about $d = 7$ nm, at which aggregation becomes important.

If λ is only somewhat greater than unity, aggregates will form which do not percolate across the sample. The average size of these aggregates has been estimated by de Gennes and Pincus (1970) and Jordan (1973) for low concentrations of particles. They find that in the absence of a field the average aggregate contains n_0 particles, while in the presence of a strong field there are n_∞ particles per cluster, with

$$n_0 = \frac{1}{1 - \frac{2}{3}(\phi/\lambda^3)e^{2\lambda}}, \qquad n_\infty = \frac{1}{1 - \frac{2}{3}(\phi/\lambda^2)e^{2\lambda}} \tag{8-22}$$

Each expression ceases to be valid when λ is large enough that the denominator becomes negative. As expected, the clusters are larger when the field is on. Also, the clusters are collinear with the field when it is on; when there is no field, the strings of particles "wander" and are analogous to "wormlike" polymer molecules. Figure 8-13 shows clustering in the absence and the presence of a field, as simulated by a Monte Carlo method. Recent, more sophisticated theory and simulations of aggregation can be found in Satoh and Kamiyama (1995).

8.4.2 Dipole Orientations in an Applied Magnetic Field

If now a magnetic field H is applied to a suspension of spherical particles of diameter d, the orientation-dependent potential energy W_H of a particle in this field is expressed as

$$W_H = -\mathbf{m} \cdot \mathbf{H} = -\mu_0 M_d \cdot \mathbf{H} \frac{\pi}{6} d^3 = -\alpha k_B T \cos\theta \tag{8-23}$$

Here θ is the angle of orientation of the magnetic dipole with respect to the field, and we define

$$\alpha \equiv \frac{m\mathrm{H}}{k_B T} = \frac{\mu_0 M_d \mathrm{H} \pi d^3}{6 k_B T} \tag{8-24}$$

We find that the magnetic energy equals the thermal energy $k_B T$ when $\alpha \approx 1$, or

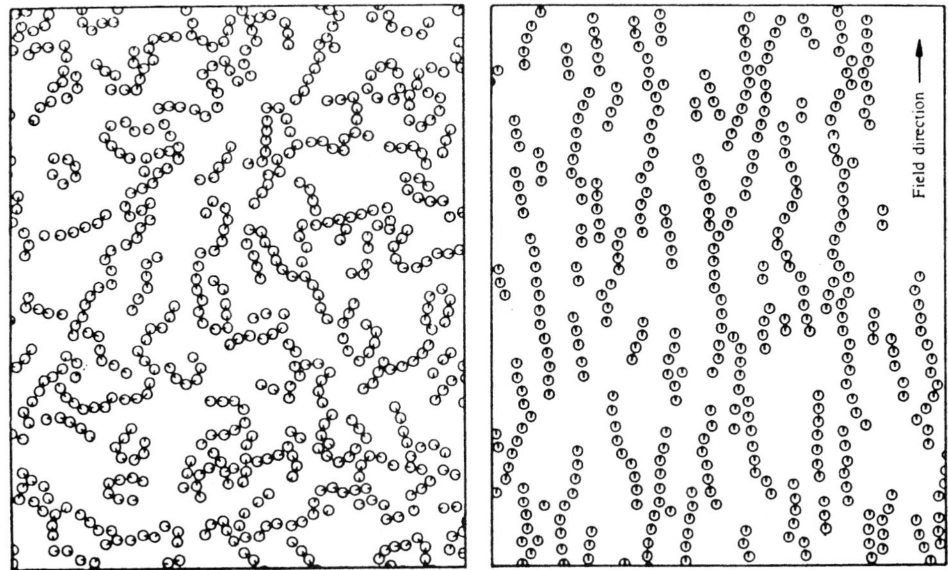

Figure 8.13 Ferrofluid particles in a Monte Carlo simulation showing clustering **(a)** in the absence of a magnetic field and **(b)** in the presence of a field. (From Chantrell et al., reprinted with permission from J. Appl. Phys. 53:2742, Copyright © 1982, American Institute of Physics.)

$$d \approx \left(\frac{6k_B T}{\pi \mu_0 \mathrm{M}_d \mathrm{H}} \right)^{1/3} \qquad \text{(orientation, segregation threshold)} \qquad (8\text{-}25)$$

Thus, in a field $\mu_0\mathrm{H}$ of 0.1 T, with $\mathrm{M}_d = 4.5 \times 10^5$ A/m, at room temperature, we obtain from Eq. (8-25) that particles of around diameter 10 nm and above will orient their magnetic dipoles with the field. Furthermore, if only part of the ferrofluid is exposed to the magnetic field, particles larger than 10 nm will be drawn into the field and will concentrate where the field is strongest. Thus, in practical applications of ferrofluids, *the particles must be kept small to avoid field-induced particle segregation.*

If the ferrofluid is dilute, so that particle–particle interactions can be neglected, then the *distribution of dipole orientations* $\psi(\theta)$ in a field H is given by the usual Boltzmann equation:

$$\psi(\theta) = C \, \exp\left(\frac{-W_H}{k_B T} \right) \qquad (8\text{-}26)$$

where C is a normalization constant. The average particle dipole moment $\overline{m}$ along the field direction is then given by $\langle m \cos \theta \rangle$, or

$$\overline{m} = \frac{\int_0^\pi m \cos\theta \, \exp(\alpha \cos\theta) \sin\theta \, d\theta}{\int_0^\pi \exp(\alpha \cos\theta) \sin\theta \, d\theta} \qquad (8\text{-}27)$$

The integration can be carried out analytically to give

$$\frac{\overline{m}}{m} = L(\alpha) \equiv \coth \alpha - \frac{1}{\alpha} \tag{8-28}$$

where $\coth \alpha \equiv (e^{\alpha} + e^{-\alpha})/(e^{\alpha} - e^{-\alpha})$ is the hyperbolic cotangent, and $L(\alpha)$ is called the *Langevin function*. (Its inverse appears in the expression for the elastic force of an extended freely jointed chain model of a macromolecule; see Section 3.6.2.2.1.) The magnetization M of the ferrofluid is given by

$$\mathrm{M} = \frac{1}{\mu_0} \nu \overline{m} = \phi \mathrm{M}_d \left(\frac{\overline{m}}{m} \right) \tag{8-29}$$

where ν is the number of particles per unit volume and ϕ is the volume fraction of particles. Thus, for a strong field in which $\overline{m} \to m$, the magnetization of the fluid approaches a saturation value which is ϕ times the magnetization of the dispersed solid particles; that is,

$$\mathrm{M} \to \phi \mathrm{M}_d \qquad (\alpha \gg 1) \tag{8-30}$$

For a weak field H, $L(\alpha) \to \alpha/3$, and the magnetization is linear in H. The low-field *susceptibility* $\chi_i \equiv \mathrm{M/H}$ is then given by

$$\chi_i = 8\phi\lambda = \frac{\pi}{18}\phi\mu_0 \frac{\mathrm{M}_d^2 d^3}{k_B T} \qquad (\alpha \ll 1) \tag{8-31}$$

If M/H is not small, the magnetic moments of the particles interact, and a better approximation is (Shliomis 1974; Rosenzweig 1985):

$$\frac{\chi_i (2\chi_i + 3)}{\chi_i + 1} = 24\phi\lambda \tag{8-32}$$

The particles of a ferrofluid rotate very rapidly in response to an imposed field or after the field is turned off. The rotary Brownian diffusivity of spherical particles is given by Eq. (6-30), $D_{r0} = k_B T / \pi \eta_s d^3$, which yields $\sim 10^6$ sec^{-1}, for particles of size 10 nm in a solvent of viscosity 4 cP. Thus, the relaxation time constant is of order 1 μsec. The time constant might be even smaller than this in cases for which the particle switches its magnetization not by rotating but by flipping the domain orientation direction within the particle. The time constant for this process, which is called the Néel mechanism, is exponential in the volume of the particle and is estimated to decrease enormously from 1 sec to 10^{-18} sec as the particle size decreases from 12.5 nm to 8 nm in one particular case (Rosenzweig 1985).

8.4.3 Viscosity of a Ferrofluid

The viscosity of a ferrofluid is qualitatively similar in some respects to that of dilute to moderately concentrated suspensions discussed in Chapter 6. Useful ferrofluids often have viscosities that are a few times that of the solvent. In a strong magnetic field, the shear viscosity of a ferrofluid is higher than that in the absence of a field, if the field is in the plane of deformation, since the particle rotation is suppressed by the field and the dissipation is therefore enhanced (McTague 1969; Rosenzweig 1985). The viscosity enhancement becomes relatively more important as the flow becomes more vorticity dominated. In the limit of solid-body rotation, a ferrofluid still has a finite viscosity (i.e., it dissipates energy), while an ordinary fluid has none. There is no viscosity enhancement, however, if the flow is extensional, or is a shear flow with the field in the vorticity direction. Ferrofluids can also be somewhat shear thinning, evidently because of shear-induced breakup of particle clusters (Rosenzweig 1985). At very high frequencies, $\omega \sim 1/D_{r0} \sim 10^6 \text{ sec}^{-1}$, the viscosity of a ferrofluid should be frequency-dependent (Saluena and Rubi 1995).

A remarkable recent discovery is that if the magnetic field applied to a ferrofluid is oscillated in the kilohertz frequency range, the viscosity of the ferrofluid measured in flow through a tube can be reduced to a value lower than that in the absence of a field (Rosenzweig 1996)! This "negative-viscosity" phenomenon occurs because of a symmetry-breaking coupling between rotation produced by the field and that produced by flow vorticity. As a result, on average the field assists the particle rotation in the direction of the flow vorticity; some of the field's energy is thereby transformed into kinetic energy in the fluid, and the apparent fluid viscosity is reduced.

8.4.4 Ferrofluid Phenomena

Ferrofluids spontaneously flow toward regions of strong magnetic fields and then stubbornly remain there, even in the presence of other forces. Quantitative predictions of how this occurs can be made using the appropriate ferrofluid constitutive equation, the equation of motion, and the boundary conditions for ferrofluids. These are presented and discussed in detail in Rosenzweig (1985). The equation of motion is (Rosenzweig 1985)

$$\rho\left(\frac{\partial}{\partial t}\mathbf{v} + \mathbf{v}\cdot\nabla\mathbf{v}\right) = -\nabla p^* + \mu_0\mathbf{M}\cdot\nabla\mathbf{H} + \eta\nabla^2\mathbf{v} + \rho\mathbf{g} \qquad (8\text{-}33)$$

where $\mathbf{g}$ is the gravitational acceleration vector, and p^* is a pressure modified to include isotropic stresses produced by the coupling of the magnetic field to the fluid.

To give the reader a flavor of the kinds of phenomena that can be predicted, we consider the simple case of a pool of ferrofluid, a portion of which is subjected to a weak magnetic field, at static equilibrium (see Fig. 8-14a). Thus, we drop the time derivative and the flow terms from Eq. (8-33), yielding

$$p^* + \rho g h - \mu_0\overline{M}H = \text{const} \qquad (8\text{-}34)$$

where

$$\overline{M} \equiv \frac{1}{H}\int_0^H M d\mathrm{H}' \qquad (8\text{-}35)$$

When we choose points 1 and 2 in Fig. 8-14 that lie in the ferrofluid at the interface with air (1, out of the field; 2, in the field), Eq. (8-34) yields

$$p_1^* + \rho g h_1 = p_2^* + \rho g h_2 - \mu_0\overline{M}H \qquad (8\text{-}36)$$

If the magnetic field at point 2 is normal to the interface and the field strength does not jump when the interface is crossed, and if the interface curvature is negligible (so that the capillary pressure term can be dropped), then $p_1^* = p_2^* = p_0$, the pressure in the air. The difference $\Delta h \equiv h_2 - h_1$ in the height of the interface is therefore

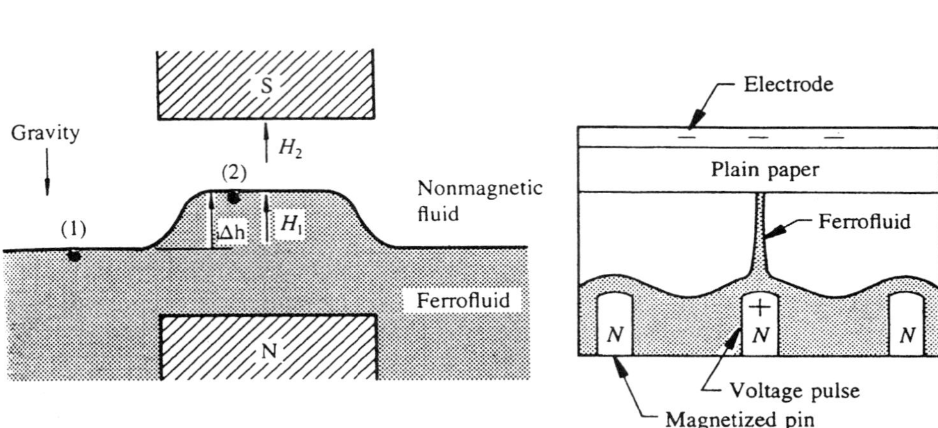

Figure 8.14   **(a)** When a magnetic field is imposed across a portion of a pool of ferrofluid, the free surface of the ferrofluid in the field rises relative to that outside of the field. **(b)** The use of a ferrofluid ink for printing. (From Rosenzweig, Copyright © 1985, by Oxford University Press, Inc. Used by permission of Oxford University Press.)

$$\Delta h = \frac{\mu_0 \overline{M} H}{\rho g} \tag{8-37}$$

For a strong enough field, the magnetic fluid fills the gap between the poles of the magnet.

By analyzing the equations of motion for ferrofluids, a variety of unusual phenomena observed in ferrofluids can be predicted. These include the *normal-field instability*, in which a field applied normal to a layer of ferrofluid lying underneath an ordinary fluid produces proturberances of the interface between the two fluids (Rosenzweig 1985). If the field is applied in a direction parallel to the interface, disturbances to the interface are suppressed. Such disturbances can be generated via a Kelvin–Helmholtz instability if the two layers are flowing. Other instabilities unique to ferrofluids can occur, such as the *labryinthine instability* produced when a ferrofluid and an immiscible nonmagnetic fluid are confined between two glass plates (a Hele–Shaw cell) with a magnetic field applied normal to the plates. A thorough discussion of these and other phenomena can be found in Rosenzweig (1985).

Ferrofluids have recently been combined with other complex fluids to produce fluids with very interesting behavior. When ionic ferrofluids are doped into nematic liquid crystals (Brochard and de Gennes 1970), the resulting nematic liquid orients under a magnetic field 10^3 times lower than that required in the absence of the dopant (Chen and Amer 1983; Bacri and Figueiredo Neto 1994). Another interesting complex fluid is a *ferrofluid emulsion*—that is, micron-sized droplets of ferrofluid in a continuous medium (Bibette 1993). A ferrofluid emulsion has two distinct levels of microstructure: The magnetic particles are around 10 nm in size, which is 100 times smaller than the droplet size; and the droplets are much smaller than macroscopic dimensions. The droplets are stabilized by surfactants and can be made to cluster by a magnetic field. Careful measurements of droplet separation as a function of magnetic field make possible the inference of surface forces in the thin layers separating the droplets.

8.5 SUMMARY

The last few years have seen impressive advances in our understanding of the rheology of electrorheological (ER) and magnetorheological (MR) fluids. Using a simple polarization model, combined with computational integration of the evolution equations for particle positions, the mechanism of particle chaining has been established, and the yield stress can now be estimated, at least for some ER fluids, from the field strength, particle concentration, and the polarizability of the particles relative to that of the solvent. At high shear rates, the dimensionless Mason number seems to correlate the electric-field and shear-rate dependences of the viscosity, in some cases at least. One should not, however, lose sight of the many challenges that remain, especially for ER fluids. The effect of small quantities of water on the ER effect is still controversial. The time scales of rheological response of an ER fluid are still not fully understood. In addition, the appropriate boundary conditions to apply at the electrodes are not known with certainty. Present understanding of electrorheological mechanisms should help in commercial application of ER technologies; however, practical problems associated with viscous and ohmic heating, particle settling, and abrasiveness also need to be overcome. Some of these problems, such as ohmic heating, are avoided with MR fluids.

The properties of ferrofluids seem now to be well understood, and numerous applications have been found. Unlike ER and MR fluids, particles in ferrofluids have permanent (magnetic) dipoles; thus the particles must be small, around 10 nm, to prevent permanent clumping. If single-domain ferromagnetic particles this small are made, and coated with surfactant to prevent clumping by van der Waals forces, stable ferrofluids can be made whose properties are readily predicted from theory.

REFERENCES

Anderson RA (1992). In *Proceedings of the Conference on Electrorheological Fluids,* Carbondale, IL, Tao R (ed), World Scientific, Singapore, p 81.
Arp PA, Mason SG (1977). *Colloid Polym Sci* 255:566.
Bacri JC, Figueiredo Neto AM (1994). *Phys Rev E* 50:3860.
Bibette J (1993). *J Magn Magn Mater* 122:37.
Block H, Kelly JP, Qin A, Watson T (1990). *Langmuir* 6:6.
Bohon K, Kraus S (1997). Poster presentation, Annual APS meeting.
Bonnecaze RT, Brady JF (1992a). *J Chem Phys* 96:2183.
Bonnecaze RT, Brady JF (1992b). *J Rheol* 36:73.
Brochard F, de Gennes PG (1970). *J Phys (Paris)* 31:691.
Chantrell RW, Bradbury A, Popplewell J, Charles SW (1982). *J Appl Phys* 53:2742.
Chen SH, Amer NM (1983). *Phys Rev Lett* 51:2298.
Chen TJ, Zitter RN, Tao R (1992). *Phys Rev Lett* 68:2555.
Davis LC (1992). *J Appl Phys* 72:1334.

de Gennes PG, Pincus PA (1970). *Phys Kondens Mater* 11:189.

Deinega YF, Vinogradov GV (1984). *Rheol Acta* 23:636.

di Rico L (1963). U.S. Patent 3,109,749.

Fermigier M, Gast AP (1992). *J Colloid Interface Sci* 154:522.

Filisko FE, Radzilowski LH (1990). *J Rheol* 34:539.

Foulc JN, Atten P, Felici N (1994). *J Electrostatics* 33:103.

Gast AP, Zukoski CF (1989). *Adv Colloid Interface Sci* 30:153.

Ginder JM, Elie LD (1992). In *Proceedings of the Conference on Electrorheological Fluids*, Carbondale IL, Tao R (ed), World Scientific, Singapore, p 23.

Ginder JM, Davis LC, Elie LD (1996). In *Proceedings of the 5th International Conference on ER Fluids, MR Suspensions and Associated Technology*, Bullough WA (ed), World Scientific, Singapore, p 505.

Halsey TC (1992). *Science* 258:761.

Halsey TC, Martin JE (1993). *Sci Am* Oct:58.

Halsey TC, Martin JE, Adolf D (1992). *Phys Rev Lett* 68:1519.

Jordan PC (1973). *Mol Phys* 25:961.

Jordan TC, Shaw MT, McLeish TCB (1992). *J Rheol* 36:441.

Kerr RA (1990). *Science* 247:1180.

Klass DL, Martinek TW (1967). *J Appl Phys* 38:67.

Klingenberg DJ (1993). *J Rheol* 37:199.

Klingenberg DJ, Zukoski CF (1990). *Langmuir* 6:15.

Klingenberg DJ, van Swol F, Zukoski CF (1991a). *J Chem Phys* 94:6160.

Klingenberg DJ, van Swol F, Zukoski CF (1991b). *J Chem Phys* 94:6170.

Klingenberg DJ, Zukoski CF, Hill JC (1993). *J Appl Phys* 73:4644.

Kraynik AM (1993). In *Electrorheological (ER) Fluids. A Research Needs Assessment Final Report* DOE/ER/30172, US Department of Energy.

Lemaire E, Meunier A, Bossis G, Liu J, Felt D, Bashtovoi P, Matoussevitch N (1995). *J Rheol* 39:1011.

Marshall L, Goodwin JW, Zukoski CF (1989). *J Chem Soc Faraday I* 85:2785.

Martin JE, Odinek J, Halsey TC (1992). *Phys Rev Lett* 68:1524.

Martin JE, Adolf D, Halsey TC (1994). *J Colloid Interface Sci* 167:437.

Martin JE, Odinek J, Halsey TC, Kamien R (1998). *Phys Rev E* 57:756.

McLeish TCB, Jordan T, Shaw MT (1991). *J Rheol* 35:427.

McTague JP (1969). *J Chem Phys* 51:133.

Melrose JR (1991). *Phys Rev A* 44:R4789.

Melrose JR, Heyes DM (1993). *J Chem Phys* 98:5873.

Parthasarathy M, Klingenberg DJ (1995). *Rheol Acta* 34:430.

Parthasarathy M, Klingenberg DJ (1996). *Mater Sci Eng R* R17:57.

Parthasarathy M, Ahn KH, Belongia BM, Klingenberg DJ (1994). *Int J Mod Phys B* 8:2789.

Parziale AJ, Tilton PD (1950). *AIEE Trans* 69:150.

Promislow JHE, Gast AP, Fermigier M (1995). *J Chem Phys* 102:5492.

Rabinow J (1948). *AIEE Trans* 67:1308.

Rosenzweig RE (1985). *Ferrohydrodynamics*, Cambridge University Press, Cambridge.

Rosenzweig RE (1996). *Science* 271:614.

Saluena C, Rubi JM (1995). *J Chem Phys* 102:3812.

Satoh A, Kamiyama S-I (1995). *J Colloid Interface Sci* 172:37.

Scholten PC, Felius JAP (1990). *J Magn Magn Mater* 85:107.

See H, Doi M (1991). *J Phys Soc Jpn* 60:2778.

See H, Doi M (1992). *J Rheol* 36:1143.

See H, Tamura H, Doi M (1993). *J Phys D Appl Phys* 26:1.

Shliomis MI (1974). *Sov Phys Uspekhi* 17:153.

Stangroom JE (1991). *J Stat Phys* 64:1059.

Tajiri K, Ohta K, Nagaya T, Orihara H, Ishibashi Y, Doi M, Inoue A (1997). *J Rheol* 41:335.

Tao R (1993). *Phys Rev E* 47:423.

Toor WR (1993). *J Colloid Interface Sci* 156:335.

Uejima, H (1972). *J Appl Phys* 11:319.

Vorob'eva TA, Vlodavets IN (1974). *Koll Zh* 36:1154.

Weiss KD, Duclos TG (1994). In *Proceedings of the 4th International Conference on ER Fluids*, Rao R, Roy GD (eds), World Scientific, Singapore, p 43.

Whittle M (1990). *J Non-Newt Fluid Mech* 37:233.

Winslow WM (1947). US Patent 2,417,850.

Winslow WM (1949). *J Appl Phys* 20:1137.

Yang I-K, Shine AD (1992). *J Rheol* 36:1079.

Yang M-C, Scriven LE, Macosko CW (1986). *J Rheol* 30:1015.

Zukoski CF (1993). In *Electrorheological (ER) Fluids. A Research Needs Assessment Final Report* DOE/ER/30172, US Department of Energy.

Chapter 9

FOAMS, EMULSIONS, AND BLENDS

9.1 INTRODUCTION

Milk shakes, shaving cream, mayonnaise, and immiscible polymer blends are all dispersions of one fluid in another. The rheological properties of such dispersions are often crucial: Milk shakes must be "thick," mayonnaise must have just the right yield stress so that it spreads easily and yet doesn't run, and so on. The relationship between the rheological properties of a dispersion and those of the fluids composing it can be complex. For example, a liquid foam, which is typically a dispersion of around 95% air in liquid, acts as an *elastic solid*, with a yield stress and a finite zero-frequency modulus. Although dispersions are often formulated by trial and error, an increased understanding of the microscopic basis of their rheological properties will no doubt help them to be designed more quickly and scientifically.

Just as with dispersed solids, the stability of the dispersed liquid phase against coagulation, or clumping, is a major concern. The tricks that produce stabilization in the former often work also in the latter: (a) electrostatic charging of the interface and (b) steric repulsion using surfactants or polymers that adsorb to the interface (Hunter 1992). For fluid–fluid dispersions, however, coagulation need not be the only mechanism by which the dispersed phase might demix from its host medium. If the dispersed phase is even slightly soluble in the external phase, then large drops or bubbles will grow at the expense of smaller ones even in the absence of coagulation. In foams, this process occurs fairly rapidly as air from smaller bubbles is transported to the larger bubbles by solubilization and diffusion across the liquid films; this allows large bubbles to cannibalize the small ones. Hence, in such systems, only a limited kind of "stability" is attainable.

9.2 EMULSION PREPARATION

Broadly speaking, there are two ways of preparing emulsions: One can *mechanically mix* the separate components, or one can induce phase separation in a homogeneous mixture by a thermal quench or by a chemical reaction. In either case, control of the emulsion morphology is a delicate business, because the emulsified state is at best only kinetically, and not thermodynamically, stable. We first discuss emulsion preparation by phase separation.

388

9.2.1 Phase Separation

The simplest way to produce phase separation is by changing the temperature so that a phase boundary is crossed from a one-phase region of the phase diagram to a region containing two or more phases. Figure 9-1 shows a typical phase diagram of a two-component liquid with an *upper critical solution temperature* (UCST). The UCST is the critical temperature above which the two components are miscible in all proportions, and below which there is a coexistence region of two-phase equilibrium. While mixtures with a UCST are common, some mixtures have a *lower-critical solution temperature* (LCST) in which the diagram of Fig. 9-1 is inverted and phase separation occurs when the temperature is *raised*.

The diagram of Fig. 9-1 is typical of mixtures of two small molecules, or of two polymers of comparable molecular weight and comparable viscosity. Polymer solutions, or blends of two polymers with very different molecular weights, have asymmetric phase diagrams, reflecting the asymmetry of the molecular sizes (see Fig. 9-2).

In a UCST system, when the temperature is reduced to a final value T_f that is below the critical temperature T_c, a mixture with a concentration ϕ not too far from the critical composition ϕ_c will phase separate into two phases whose compositions lie on the opposite sides of the *binodal* envelope line of Fig. 9-1. The dynamics of the separation process of a single phase into these two phases is controlled by T_f, the composition ϕ, the rate of the quench dT/dt, the viscous (or viscoelastic) properties of the phases formed, and the interfacial tension Γ between the two phases. Although a variety of different kinds of behavior can occur, there are two generic types of phase separation, namely, *spinodal decomposition* (SD) and *nucleation and growth* (NG). SD occurs when the mixture is quenched into a part of the phase diagram where the mixture is unstable to small variations in composition, leading to immediate growth of phase-separated domains. When the quenched

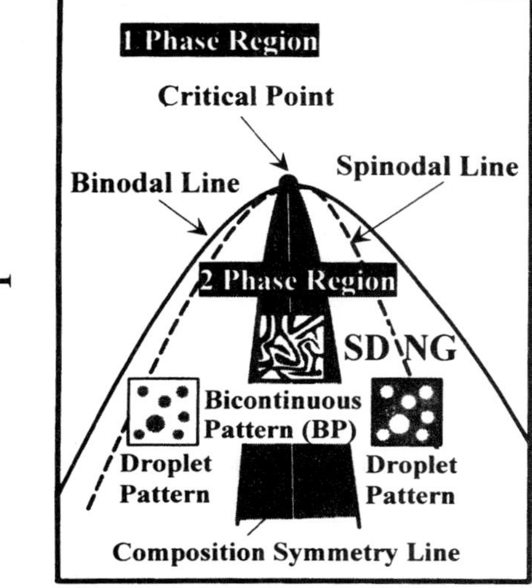

Figure 9-1   Schematic phase diagram of a binary fluid mixture of small molecules. The two-phase region lies under the binodal line, the apex of which defines the critical temperature T_c and critical composition ϕ_c. Between the binodal and the spinodal lines, phase separation is by nucleation and growth (NG), while under the spinodal line it is by spinodal decomposition (SD). Within the region of spinodal decomposition, near the compositional symmetry line, there is a region where the morphology is initially bicontinuous. Outside of this region, one of the phases is a discontinuous droplet phase. Eventually, because of asymmetries, initially bicontinuous structures break apart into a droplet morphologies. (From Tanaka 1995, reprinted with permission from the American Physical Society.)

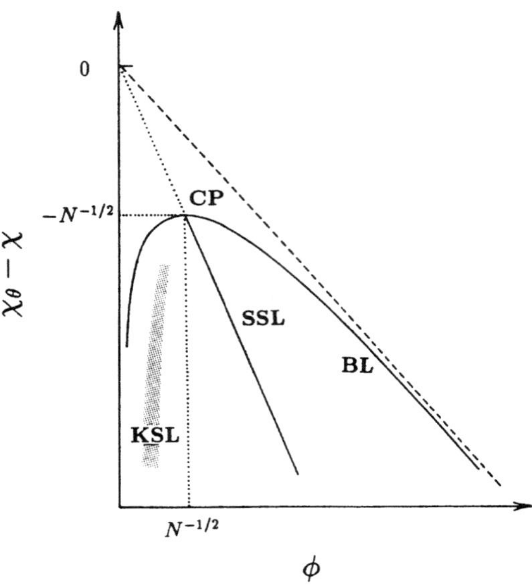

Figure 9.2 Schematic phase diagram of a polymer/solvent mixture, where χ is the Flory "chi" parameter, and $\chi_\theta = 1/2$ is χ at the theta temperature. The quantity $\chi_\theta - \chi$ along the ordinate is a reduced temperature, and ϕ is the polymer volume fraction. CP is the critical point, and BL is the binodal line. SSL and KSL are the static symmetry line and the kinetic symmetry line, respectively. These lines define the phase-inversion boundaries during quenches. In quenches that end at the right of such a line, the polymer-rich phase is the continuous phase, while to the left of the line the solvent-rich phase is the continuous one. SSL applies at long times, after viscoelastic stresses have relaxed, while KSL applies at shorter times before relaxation of viscoelastic stresses. (From Tanaka 1993, reprinted with permission from the American Physical Society.)

mixture is stable to small compositional variations, but not to large ones, then phase separation is suppressed until a nucleus of one phase forms that is large enough to be unstable to further growth; phase separation is then said to occur by NG.

These two modes of phase separation can be explained in terms of a free-energy–composition diagram, such as Fig. 9-3. At $T > T_c$ the homogeneous free energy per unit volume f has a single minimum against composition ϕ, while below T_c it has two minima. If $T < T_c$, and the composition lies between the two minima at ϕ_1 and ϕ_2, the mixture can lower its free energy by phase separating into two phases whose compositions are ϕ_1 and ϕ_2. Thus, all compositions between ϕ_1 and ϕ_2 lie within the *binodal* envelope in Fig. 9-3.

The type of phase separation is, however, dictated by the curvature of the free-energy curve (see the two insets in Fig. 9-3). For $\phi_1 < \phi < \phi_{s1}$, the curvature of $f(\phi)$ is positive; that is, $f''(\phi) \equiv d^2 f/d\phi^2 > 0$. Thus, if patch of initially homogeneous fluid of composition corresponding to point A in Fig. 9-3 suffers a weak compositional fluctuation, thereby producing two patches whose compositions correspond to points B and C, then the average free energy *increases*, as shown. Hence, the fluctuation is thermodynamically unfavorable, and the two patches will remix so that the homogeneous composition, point A, is restored. For $\phi_1 < \phi < \phi_{s1}$, therefore, phase separation will only occur if the fluctuation is large enough to produce a nucleus of composition near ϕ_2, which can grow by selectively drawing toward it the preferred species.

However, for $\phi_{s1} < \phi < \phi_{s2}$, the curvature of $f(\phi)$ is negative, and composition D can demix by a weak fluctuation into patches with composition E and F; this *decreases* the free energy and is thus irreversible. Weak fluctuations therefore grow spontaneously until two distinct phases appear with compositions ϕ_1 and ϕ_2.

This latter case, in which $f'' < 0$, is *spinodal decomposition*. In the former case, with $f'' > 0$, phase separation occurs by *nucleation and growth* NG of finite fluctuations. From

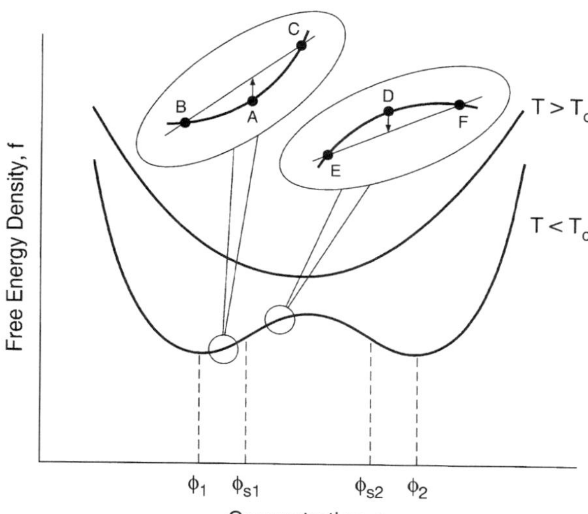

Figure 9.3 Schematic drawing of free energy versus concentration for a binary mixture at temperatures above and below the critical value T_c. The concentrations ϕ_1 and ϕ_2 are the equilibrium binodal concentrations, while ϕ_{s1} and ϕ_{s2} are the spinodal concentrations. As illustrated in the insets, at a homogeneous concentration between ϕ_1 and ϕ_{s1}, a concentration fluctuation leads to an *increase* in free energy, while at a concentration between ϕ_{s1} and ϕ_{s2}, a fluctuation leads to a *decrease* in free energy.

the above discussion and Fig. 9-3, it is clear that $f'' > 0$ only in the flanks of the region under the binodal envelope. Hence, it is in these flanks that NG will occur (see Fig. 9-1). The boundary between the region of NG and that of SD is defined by the condition $f'' = 0$. The critical point is then the temperature T_c and composition ϕ_c at which $f' = f'' = 0$.

Since NG requires the formation of a nucleus of critical size, phase separation might occur only after a (possibly very long) delay period following a quench into the NG region of the phase diagram. Such persistent, but ultimately unstable, states of matter are called *metastable*. The existence of a delay period is not the only way that NG differs from SD. In NG, for example, discrete *droplets* of one phase within another are always formed, while SD can lead to *bicontinuous* structures in which both phases are continuous. (For an example, look ahead to Fig. 9-14.) Compositional, or other, asymmetries can eventually lead to the breakup of one of the continuous phases of the bicontinuous structure into discrete droplets (see Fig. 9-1). This breakup might be very slow to occur if the mixture is a nearly symmetric one.

The stages of spinodal decomposition of two liquids following a sudden drop in temperature are depicted schematically in Fig. 9-4. Figure 9-4 is a one-dimensional depiction of concentration as a function of position r at different stages of time after the temperature quench. In the *early stage*, a small-amplitude sinusoidal composition wave develops against a homogeneous background. The amplitude of this wave grows exponentially in time, while the wavelength stays almost constant. In the *intermediate stage*, the amplitude of the wave continues to grow, but at a less-than-exponential rate, and the wavelength of the pattern begins to increase. In the *transition stage* postulated by Bates and Wiltzius (1989), the wave becomes highly nonsinusoidal; the maximum and minimum concentrations reach their final (i.e., equilibrium) values, but the interfaces between them have not yet thinned to their final widths. In the *late stage*, there are well-defined domains of one phase in the other, with interfaces between them that are sharp compared to the sizes of the domains. In this late stage, neither the interfacial width nor the composition of the domains change

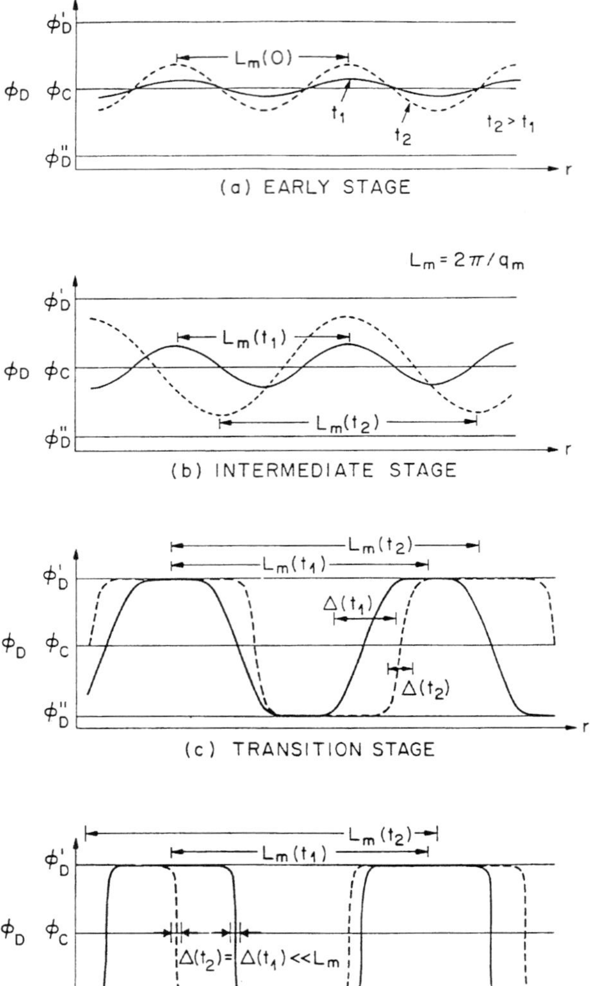

Figure 9.4 Four stages of spinodal decomposition, according to Bates and Wiltzius (1989). In each diagram, the dashed line represents the composition profile at a later time, t_2, than that for the solid line, t_1. The compositions ϕ_D' and ϕ_D'' are the equilibrium compositions in the two phases. In the early stage **(a)**, the wavelength L_m of the sinusoidal pattern remains constant, while the amplitude increases. In the intermediate stage **(b)**, both the amplitude and the wavelength increase with time. In the transitional stage **(c)**, the amplitude is saturated and no longer changes with time, but the wavelength increases, and the width Δ of the interfacial zone decreases, with time. In the final or "late" stage **(d)**, the amplitude is constant and so is the interfacial width Δ, and only the wavelength of the pattern increases with time. (From Bates and Wiltzius, reprinted with permission from J. Chem. Phys. 91:3258, Copyright © 1989, American Institute of Physics.)

much with time, but the domain sizes keep increasing as the pattern coarsens. Finally, if the two components have differing densities, then *gravitational settling*—or its reverse, *creaming*—will eventually occur, leading to stratification of the two components.

9.2.1.1 Early Stage

In the *early* stage of the quench, departures from uniform composition are small, and a linear theory due to Cahn (1965, 1968) is applicable. In this theory, one augments the homogeneous free energy per unit volume by a term that accounts for *gradients* in composition:

$$f_i(\phi, \mathbf{x}) = f(\phi) + \kappa(\nabla\phi)^2 \tag{9-1}$$

Here $f_i(\phi, \mathbf{x})$ is the *inhomogeneous* free energy density, which varies from place to place in the phase-separating sample. It consists of the homogeneous term $f(\phi)$ discussed earlier, plus a term $\kappa(\nabla\phi)^2$ due to the presence of gradients in composition. The coefficient κ is a phenomenological constant.

As discussed earlier, SD occurs whenever $f''(\phi) < 0$. In the absence of the gradient term in Eq. (9-1), one would expect the rate of phase separation by SD to be determined by the distance the molecules must diffuse to grow a compositionally inhomogeneous pattern. One might then expect a pattern with small wavelength to grow faster than one with a large wavelength. However, for a given amplitude of the composition variation, the former has a steeper gradient of composition than the latter, and the gradient penalty term in Eq. (9-1) suppresses its growth. Thus, there is an optimum wavelength with the fastest growth rate. In scattering experiments, such as light scattering, this optimum wavelength manifests itself as a peak scattering intensity at a *wavenumber* q_m, which is equal to 2π divided by the optimum wavelength. (Experimentally, the wavenumber q is related to the scattering angle θ by $q = 4\pi n\lambda^{-1}\sin(\theta/2)$, where n is the average index of refraction of the medium and λ the wavelength of the radiation; see Section 1.6.2.) By solving a linearized form of Eq. (9-1), one finds that

$$q_m = \frac{1}{2}\left(\frac{-f''}{\kappa}\right)^{1/2} \tag{9-2}$$

Thus, the shallower the quench (i.e., the smaller the temperature drop), the smaller the value of $-f''$ and the larger the pattern wavelength $2\pi/q_m$.

The rate of phase separation is also determined by the driving force for phase separation, which increases rapidly (i.e., *exponentially*) during the earliest stage of phase separation. This can be understood by referring again to Fig. 9-3. Suppose the concentration ϕ is initially homogeneous at the point ϕ_{max} where the free energy density f is a maximum on the curve corresponding to $T < T_c$. At this point of unstable equilibrium, phase separation will only commence when a fluctuation causes the concentration to slide slightly to one side or the other of the maximum in f. Once this happens, the driving force for further phase separation is given by the magnitude of the slope of f versus ϕ—that is, by f'. As phase separation occurs and ϕ deviates increasingly from ϕ_{max}, the driving force for continued phase separation increases. If the curvature f'' can be taken as a constant over some range of ϕ, then over this range the rate of phase separation will be an exponential function of time. Hence, the peak scattering intensity I_m, which measures the degree of phase separation, also grows exponentially with some time constant τ_m; that is, $I_m \propto \exp(t/\tau_m)$, where

$$\frac{1}{\tau_m} = \frac{1}{2}\Lambda(-f'')q_m^2 \tag{9-3}$$

and Λ is a phenomenological diffusivity, or "Onsager" coefficient.

Cahn's linear theory for spinodal decomposition is difficult to test in mixtures of simple liquids, because phase separation occurs so rapidly that the range of validity of the linear theory is typically exceeded before data can be acquired. In polymeric mixtures, however, phase separation is much slower, and tests of the theory are easier. For polymers of not-too-dissimilar molecular weight, the Flory–Huggins theory can be used for the free energy f, and Eqs. (9-2) and (9-3) can be written as (de Gennes 1980; Pincus 1981; Binder 1983)

$$q_m^2 = \frac{3}{2} \frac{\varepsilon}{R_g^2} \tag{9-4}$$

$$\frac{1}{\tau_m} = D\varepsilon q_m^2 \tag{9-5}$$

where the reduced *quench depth* ε is defined by

$$\varepsilon \equiv \frac{\chi - \chi_s}{\chi_s} \tag{9-6}$$

R_g is the polymer radius of gyration, χ_s is the value of the χ parameter (see Section 2.3.1) at the spinodal point, and D is the mutual diffusion coefficient of the two polymer components. Bates and Wiltzius (1989) have confirmed the predictions of Eqs. (9-4) and (9-5) for early-time SD of binary blends of perdeuterated and protonated 1,4-polybutadiene. Neutron-scattering studies of SD on a similar system by Jinnai et al. (1993a, 1993b) also confirm the Cahn theory at early times, but the spinodal growth rates deviate somewhat from Eq. (9-5).

If the two polymeric components have very different molecular relaxation times, say $\tau_1 \gg \tau_2$, so that there is *dynamic asymmetry*, then for deep enough quenches ($\varepsilon \gtrsim (\tau_2/\tau_1)^{1/2}$), the fast-moving component will begin to phase separate from the slow-moving one rapidly enough to produce a viscoelastic stress in the slow-moving species. This viscoelastic stress can retard phase separation, even in the linear regime, and the Cahn theory must be modified to account for it (Clarke et al. 1997). In the late stage, the effects of dynamic asymmetry can also be profound (see Section 9.2.1.3).

9.2.1.2 Intermediate Stage

Cahn's linear theory is valid only during the earliest, fastest, stage of SD, in which the compositional variation is small, and phase separation is dominated by sinusoidal composition waves with a dominant wavenumber q_m, given by Eq. (9-2). In the later, *intermediate*, stage, the concentration variations continue to increase in amplitude, producing well-defined "domains" whose composition differs distinctly from that of surrounding material. Simultaneously, the wavelength of the concentration pattern, which can be referred to as the domain size a, begins to increase as well. Theories for the intermediate stage of spinodal decomposition have been summarized by Hashimoto et al. (1986). In such theories, growth rates of the domain size, a, and of the peak scattering intensity, I_m, are expressed as power laws; $a \sim t^\alpha$, and $I_m \sim t^\beta$, with α and β both around 0.33 (Akcasu and Erman 1992). These theories are at best only partially supported by experiment (Bates and Wiltzius 1989).

9.2.1.3 Late Stage

Of greater interest to us is the *late stage*, where the spinodal pattern is described by regions of rather uniform composition separated by *interfaces* whose width Δ is much less than the domain size a (see Fig. 9-4d). In this stage, phase separation proceeds by a *coarsening* process in which the domains get larger while their composition and interfacial width remain nearly constant. Coarsening is driven by the tendency to reduce the interfacial area and thereby reduce the total free energy. If this coarsening process, sometimes called *Ostwald ripening* (Ostwald 1901; Voorhees 1985; Aaronson and Le Gouef 1992), occurs by a diffusive process, then the domain size a increases with time t as $a \sim t^\alpha$, with $\alpha = 1/3$ (Lifshitz and Slyozov 1961). One diffusive process that produces such coarsening is *"evaporation–condensation."* In this process, component A near a highly curved interface with B "evaporates" by dissolving into phase B, and then it diffuses away from that interface. At other locations where the interface is less highly curved, component A condenses back into phase A. This diffusive transport is driven by the curvature-dependence of the chemical potential according to Kelvin's equation (Israelachvili 1991; Voorhees 1985),

$$\mu = \mu_0 + V_m \Gamma H \tag{9-7}$$

where μ_0 is the chemical potential of a molecule at a flat interface, V_m is the molar volume of the diffusing molecule, Γ is the interfacial tension, and $H \propto a^{-1}$ is the mean curvature of the interface.

A second diffusive coarsening process is *diffusion and coalescence* of droplets. That is, the droplets move around by Brownian motion, collide, and occassionally coalesce into larger droplets. This process follows the same slow-diffusion law as evaporation–condensation, namely $a \sim t^{1/3}$ (Vicsek 1989; White and Wiltzius 1995).

If indeed one of the phases has broken up into droplets, then slow diffusive processes are the only coarsening mechanisms possible. However, if both phases are still interconnected and are both fluids, then coarsening can occur by the faster process of *capillary flow*, which is driven by the interfacial tension acting on the curved surfaces between the two fluids (Siggia 1979). The typical pressure drop Δp produced by capillarity is $\Delta p \sim \Gamma/a$. The velocity produced by this pressure drop is $V \sim a\Delta p/\eta \sim \Gamma/\eta$, where η is an average viscosity. The domains grow at a rate $da/dt \sim V \sim \Gamma/\eta$; hence

$$a = B(\Gamma/\eta)t \tag{9-8}$$

with B a constant that has been estimated theoretically to be around 0.04–0.10 (Siggia 1979; San Miguel et al. 1985). In the case of a polymer blend, the interfacial tension near the critical point can be approximated by (Joanny and Leibler 1978; Leibler 1982)

$$\Gamma = \frac{2}{3}k_B T b^{-2} N^{-1/2} \varepsilon^{3/2} \tag{9-9}$$

In Eq. (9-9), b is the statistical segment length, ε is given by Eq. (9-6), and N is the number of monomers in the polymer chain; thus $R_g^2 = b^2 N/6$. The existence of the fast capillary hydrodynamic regime in spinodal quenches of both small-molecule and polymeric blends has been supported by experimental scattering studies, and the experimental values of B, $B \approx 0.075$–0.12 (Bates and Wiltzius 1989; Guenoun et al. 1987), are near the expected range. Thus, the domains grow as $a \sim t^\alpha$ with a power-law exponent $\alpha = 1$ that is much larger than that of diffusive coarsening (for which $\alpha = 1/3$).

Figure 9-5 presents a simplified picture of SD, with three regimes: (1) the linear regime described by the Cahn theory, in which q_m is independent of time t; (2) the slow diffusive coarsening regime in which $q_m \propto t^{-1/3}$; and (3) the fast hydrodynamic regime in which $q_m \propto t^{-1}$.

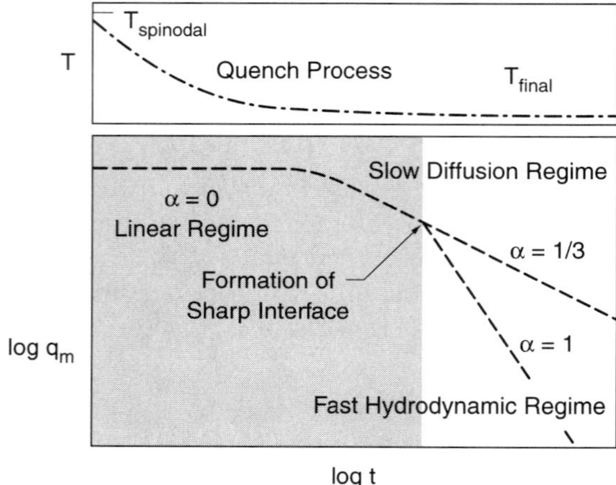

Figure 9.5   Schematic illustration of the phase-separation process after a temperature quench into the spinodal region of the phase diagram. The time dependence of the temperature quench from the spinodal temperature to some final temperature T_{final} is shown in the top diagram. This quench time can be made arbitrarily fast, in which case it has no effect on the time period over which the linear or other regimes persist. The bottom diagram shows the maximum-scattering wavevector q_m of the spinodal pattern as a function of time t, with $q_m \propto t^{-\alpha}$. At first, in the linear regime, q_m is constant, so that $\alpha = 0$; but as the pattern coarsens, q_m decreases, initially as $q_m \propto t^{-1/3}$ due to diffusive Ostwald ripening. Later, when the interfaces are well defined, if the morphology is bicontinuous, there is a crossover to a fast hydrodynamic regime with $q_m \propto t^{-1}$. (From Tanaka 1995, reprinted with permission from the American Physical Society.)

As mentioned above, if one of the phases breaks up into droplets, the fast hydrodynamic regime is excluded and coarsening can only occur by a slow diffusive process. Droplets are avoided when the volume fractions of each of the phases are nearly equal, *assuming that the viscous and elastic properties of the two phases are similar.* If they are not similar, then one expects a change in the criterion for a bicontinuous morphology. For components (labeled 1 and 2) that are mixed by mechanical blending, bicontinuity is achieved under the following empirical condition (Paul and Barlow 1980; Jordhamo et al. 1986; Miles and Zurek 1988):

$$\frac{\phi_1}{\eta_1} \approx \frac{\phi_2}{\eta_2} \qquad \text{(condition for bicontinuity)} \qquad (9\text{-}10)$$

Thus, if $\eta_1 \gg \eta_2$, the more viscous phase will break up into droplets unless its volume fraction is high, $\phi_1 \gg \phi_2$. Onuki derived criterion (9-10) from rather simple considerations, as follows. Since the two phases share an interface, and the shear stress is continuous across an interface, the characteristic shear stress $\eta_1 \dot{\gamma}_1 \sim \eta_2 \dot{\gamma}_2$ must be roughly the same in each phase, where $\dot{\gamma}_i$ is the shear rate in phase i, and $i = 1, 2$. Again because the two phases share interfaces, $a_1/\phi_1 \sim a_2/\phi_2$, where a_i is the typical domain size of phase i. If bicontinuity is to be maintained, Onuki (1994) argued that the velocity should be about the same in each phase; hence $\dot{\gamma}_1 a_1 \sim \dot{\gamma}_2 a_2$. From these relationships, Eq. (9-10) follows. Onuki suggested that Eq. (9-10) should apply not only to mechanically mixed blends, but also to spinodally decomposing ones, in which the shearing flow is produced by interfacial tension rather than by mechanical stirring. However, Eq. (9-10) seems not to have been tested experimentally in spinodally decomposing liquids.

In the case of a polymer–solvent mixture, the phase diagram is highly asymmetric, as depicted in Fig. 9-2, and the critical point is at a low volume concentration of polymer, $\phi_c \propto N^{-1/2}$, where N is the number of segments in the polymer. If a temperature quench, represented by a vertical line in Fig. 9-2, passes through the binodal envelope to the right of the critical point, it will probably then cross over the composition static symmetry line (SSL), because of the skewed shape of the binodal and spinodal envelopes. Thus, as one moves along the quench path, the phase with the highest volume fraction is initially the polymer-rich phase, but in the final composition the polymer-rich phase is in the minority. Thus, one might expect the initially continuous polymer-rich phase to *invert* to become the discrete phase. However, Tanaka (1993, 1994a) has found for mixtures of polystyrene in diethyl malonate that the polymer-rich phase remains the continuous phase long after it has become the minority phase. The minority polymer phase adopts a *network-like* morphology before finally breaking up into droplets (see Fig. 9-6). Tanaka argued that this behavior is caused by *dynamic asymmetry*; that is, the viscoelastic relaxation time of the polymer-rich

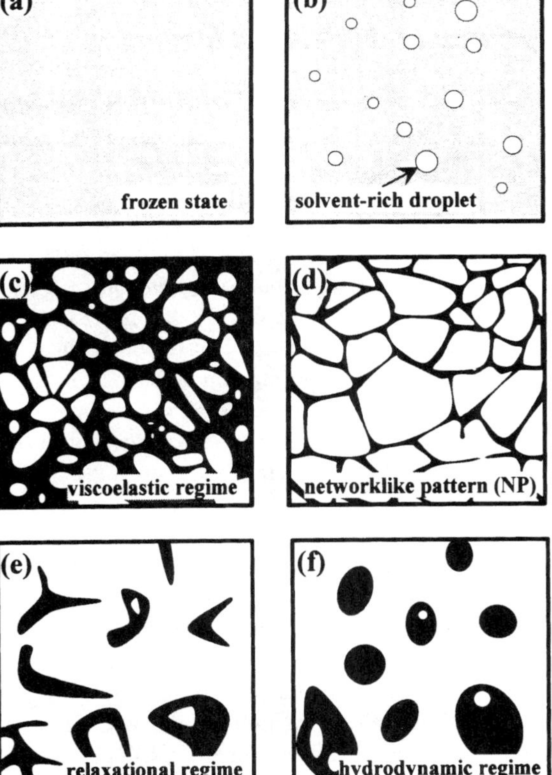

Figure 9.6 Coarsening process during phase separation of a polymer-rich phase (dark) from a solvent-rich phase (light). In the earliest period, image **(b)**, the polymer-rich phase is the majority phase; but eventually, in image **(d)**, the majority phase is the solvent-rich one. The elastic stresses in the polymer-rich phase prevent it from becoming disconnected, however, until times longer than the polymer relaxation time, images **(e)** and **(f)**. (From Tanaka, reprinted with permission from J. Chem. Phys. 100:5323, Copyright © 1994, American Institute of Physics.)

phase is so much longer than that of the solvent-rich phase that the phase separation occurs on a time-scale too fast for the polymer-rich phase to relax. The result is the buildup of elastic stresses in the polymer-rich phase (Taniguchi and Onuki 1996; Onuki and Taniguchi 1997). These elastic stresses retard the breakup of thin polymer strands, thereby preserving the continuity of the polymer phase even when it is in the minority volumetrically. It is well known experimentally and theoretically that elastic stresses hinder the capillary breakup of thin filaments of polymeric liquid (Goldin et al. 1969; Bousfield et al. 1986; Larson 1992a).

Other strange processes have been observed. In quenches of dilute polymer solutions, Tanaka (1992, 1994a) observed the formation of a *moving droplet phase*. In such a quench, described by a line that passes through to the left of the kinetic symmetry line of the phase diagram of Fig. 9-2, droplets of a polymer-rich phase form. These droplets move vigorously by Brownian motion; and they collide frequently with each other but resist coalescence, evidently because of the elasticity of the droplets.

Another peculiar phenomenon is *double phase separation* in which each of the two phases formed during spinodal decomposition of a mixture of "A" and "B" becomes unstable to a second phase separation, in which droplets of B-rich phase appear in the A-rich domain and droplets of A-rich phase appear within the B-rich domains (Tanaka 1994b). This phenomenon is thought to occur when the capillary coarsening process (in which the domain size grows as $a \propto t$) outruns the diffusion process and the A-rich domains are left with a small excess concentration of B over that allowed at bulk equilibrium. This excess of B cannot diffuse to the interface with the A-rich phase as fast as that interface moves away by capillary coarsening. The excess B therefore nucleates into droplets of B-rich phase within the coarsened A domains. The converse occurs in the A-rich domains.

9.2.2 Mechanical Mixing and Droplet Dynamics

Rather than using a thermal quench, a more common way to make an emulsion is by mechanical mixing, or agitation, of two or more liquid components, such as occurs in an old fashioned butter churn. Unless surfactants, or *emulsifiers*, are present, however, when agitation ceases, interfacial tension will drive the two phases back toward separation. This separation occurs by droplet–droplet collision and fusion, if the droplets are Brownian; by sedimentation or creaming, if the droplets are non-Brownian; or by Ostwald ripening, if the droplet phase is soluble in the continuous phase.

It is often of considerable importance that the size distribution of the emulsion droplets be controlled. The size distribution is affected by the shear rate, flow type (e.g., extensional versus shear), surface tension, flow history, viscosity ratio, droplet inertia, fluid composition, non-Newtonian effects, and so on. So many variables can affect droplet sizes that the development of a comprehensive theory would seem to be a quixotic pursuit. Nevertheless, one can gain insight from special cases. The simplest case is that of a suspension of one Newtonian fluid in another in which the droplets are dilute enough that the behavior of one of them is not influenced by the presence of the others. This problem was first considered by G. I. Taylor (1934), who, in a manner typical for him, made the most basic experimental and theoretical discoveries and thus laid the groundwork for all that followed. Taylor's original discoveries have since been confirmed and significantly refined in a series of elegant experimental and theoretical studies by Rumscheidt and Mason (1961), Taylor himself (1964), Acrivos and coworkers (Acrivos and Lo 1978; Barthes-Biesel and Acrivos 1973), and others (Rallison 1980; Torza et al. 1972; Grace 1982; Bentley and Leal 1986).

9.2.2.1 Droplet Breakup in Emulsions

If gravitational settling can be neglected and if the droplet Reynolds number $\mathrm{Re} \equiv \rho_d \dot{\gamma} a^2 / \eta_s$ is small, then the droplet deformation and possible breakup in the flow are controlled by two dimensionless groups, namely the ratio of viscous to capillary forces, or *capillary number*

$$\mathrm{Ca} \equiv \frac{\dot{\gamma} \eta_s a}{\Gamma} \tag{9-11}$$

and the viscosity ratio

$$M \equiv \frac{\eta_d}{\eta_s} \tag{9-12}$$

where η_d is the viscosity of the droplet phase and η_s is the viscosity of the continuous suspending phase. Taylor defined the flow-induced deformation D of the droplet in terms of the lengths L and B of its major and minor axes; thus

$$D \equiv \frac{L - B}{L + B} \tag{9-13}$$

This definition is suitable for axisymmetric droplet shapes or for small deformations. The aspect ratio of the droplet is then $L/B \equiv (1 + D)/(1 - D)$.

Taylor predicted that for small deformations, the droplet deformation D in a steady flow should follow the prediction

$$D = \mathrm{Ca} \cdot f(M) \tag{9-14}$$

where $f(M)$ is near unity, and depends weakly on viscosity ratio M as

$$f(M) \equiv \frac{19M + 16}{16(M + 1)} \tag{9-15}$$

This formula applies to planar extensional flow as well as to shear, if the shear rate $\dot{\gamma}$ in Eq. (9-11) is replaced by $2\dot{\varepsilon}$, where $\dot{\varepsilon}$ is the extension rate. Taylor predicted that droplet breakup should occur when the viscous stresses that deform the droplet overwhelm the surface tension forces that resist deformation; this occurs when D reaches a value D_b given approximately by

$$D_b \approx f(M) \cdot \mathrm{Ca} \approx 0.5 \tag{9-16}$$

Refinements to Taylor's original theory reveal two major exceptions to Eq. (9-16). When the viscosity ratio of the droplet to the medium is low, $M \lesssim 0.05$, the droplets deform greatly before they break. When $M \approx 10^{-3}$, droplets can achieve aspect ratios as high as 18 without breaking in planar extensional flow (see Fig. 9-7) (Bentley and Leal 1986). At low M, highly elongated droplets have pointed ends from which tiny "satellite" droplets are in some cases ejected (Grace 1982) (see Fig. 9-8a). A second exception is the case of viscous droplets, $M \gtrsim 4$, in shearing or nearly shearing flows. For these, droplet deformation remains modest even at high capillary number, and *no breakup is observed* (see Fig. 9-8d).

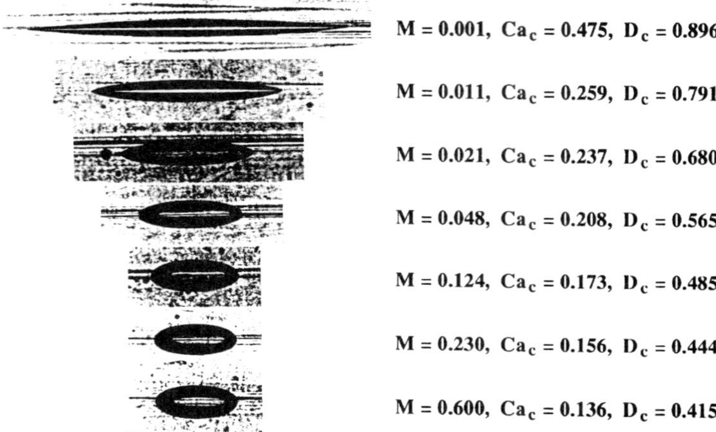

Figure 9.7   Photographs of droplet shapes in planar extensional flow for various viscosity ratios M of the dispersed to the continuous phase. The droplets are viewed in the plane normal to the velocity gradient direction. The critical capillary numbers Ca_c and droplet deformation parameters D_c at breakup are also given. The droplet fluids are silicon oils with viscosities ranging from 5 to 60,000 centistokes, while the continuous fluids are oxidized castor oils; both phases are Newtonian. (From Bentley and Leal 1986, with permission from Cambridge University Press.)

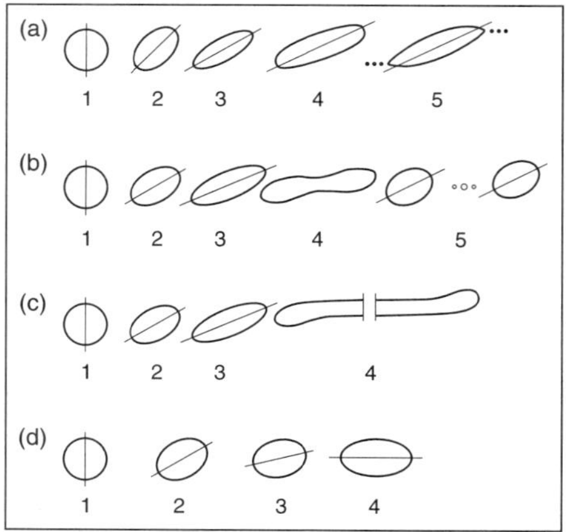

Figure 9.8   Tracings of droplet shapes in shearing flows showing changes of shapes and orientations of droplets as the shear rate increases from left to right. Here the droplets are visualized in the plane containing the velocity and the velocity-gradient directions. **(a)** $M = 2 \times 10^{-4}$: a droplet of water in 52.6 P silicone oil; **(b)** $M = 1$: a droplet of oxidized castor oil in 52.6 P silicone oil; **(c)** $M = 0.7$: a droplet of oxidized castor oil in corn syrup; and **(d)** $M = 6$: a droplet of viscous silicon oil in oxidized castor oil. In **(a)**, picture 5, tiny satellite droplets are spit from the droplet tips. In **(b)**, picture 4, a neck forms which breaks as shown in picture 5, leaving two droplets plus three satellite droplets. In **(c)**, picture 4, the droplet is stretched into a long filament (possibly because of interfacially active ingredients in the corn syrup). In **(d)**, no droplet bursting is seen. (From Rumscheidt and Mason 1961, reprinted with permission from Academic Press.)

Bentley and Leal have measured droplet shapes and critical conditions for droplet breakup over a wide range of capillary numbers, viscosity ratios, and flow types. The "flow type" is conveniently controlled in an apparatus called a "four-roll mill," in which a velocity field is generated by the rotation of four rollers in a container of liquid (see Fig. 1-15). By varying the rotation rate of one pair of rollers relative to that of a second pair, velocity fields ranging from planar extension to nearly simple shear can be produced near the stagnation point.

The velocity gradient in a general two-dimensional linear flow, such as that produced near the stagnation point of a four-roll mill, can be represented as

$$\nabla \mathbf{v} = \frac{1}{2}G \begin{pmatrix} 1+\alpha & 1-\alpha \\ -(1-\alpha) & -(1+\alpha) \end{pmatrix} \tag{9-17}$$

where α is a "flow-type" parameter, which is zero for simple shear and unity for planar extension; G is the shear rate ($\dot{\gamma}$) if the flow is a shearing flow ($\alpha = 0$), and G is the extension rate ($\dot{\varepsilon}$) in a planar extensional flow ($\alpha = 1$). The drop-breakup results of the Bentley–Leal experiments are summarized in Fig. 9-9. In this plot, the capillary number is defined by

$$\mathrm{Ca} \equiv \frac{G\eta_s a}{\Gamma} \tag{9-18}$$

and Ca_c is the value of Ca at which breakup occurs. The predictions of a small-deformation and a large-deformation theory, as well as numerical calculations, are shown to be in good agreement with the experiments for various values of flow-type parameter α. In general, the critical capillary number for droplet breakup is around 0.1–0.5, except for low-viscosity drops or high-viscosity ones with flow types close to that of simple shear. The break-up criterion for low-viscosity droplets has been found to be $\mathrm{Ca}_c M^{1/6} = 0.145\alpha^{-1/2}$ for flows with an extensional component; i.e., $\alpha > 0$ (Bentley and Leal 1986). For simple shearing flow ($\alpha = 0$), a thorough set of experimental data covering $M = 10^{-5}$ to 10 can be found in

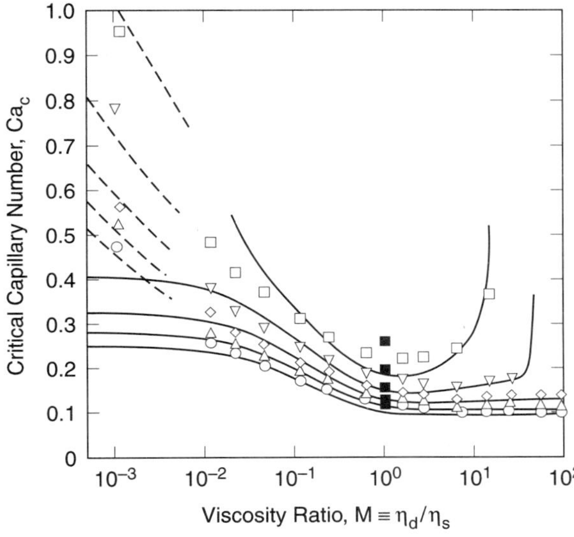

Figure 9.9 Dimensionless critical droplet size for breakup (capillary number $\mathrm{Ca}_c \equiv G\eta_s a/\Gamma$) as a function of viscosity ratio M of the dispersed to the continuous phase for two-dimensional flows in the four-roll mill. The data sets correspond from bottom to top to $\alpha = 1.0\,(\bigcirc), 0.8\,(\triangle), 0.6\,(\Diamond), 0.4\,(\nabla)$, and $0.2\,(\square)$, with α defined in Eq. (9-17). The fluids are those described in Fig. 9-7. The solid lines are the predictions of a small-deformation theory, while the dashed lines are for a large-deformation theory. The closed squares are from Rallison's (1981) numerical solutions (see also Rallison and Acrivos 1978). (From Bentley and Leal 1986, with permission from Cambridge University Press.)

Grace (1982). For $M < 0.01$, these data fit the breakup criterion for low-viscosity droplets, $Ca_c M^{2/3} = 0.054$ (Hinch and Acrivos 1980). For $10^{-3} \leq M \leq 10$, the data can be fitted by an empirical expression of De Bruijn (1989):

$$logCa_c = a + b\, logM + c\, (logM)^2 + \frac{d}{logM - logM_c}$$

with

$$a = -0.506; \quad b = -0.0994; \quad c = 0.124; \quad d = -0.115; \quad M_c = 4.08$$

For a review of work on droplet breakup, see Rallison (1984) and Vinckier (1998).

The results plotted in Fig. 9-9 are for *steady flow*, or flow in which the strain rate is increased very slowly. If, instead, the flow rate is increased in a sudden step, then the capillary number at which droplet breakup occurs is reduced (Bentley and Leal 1986). If the flow rate is abruptly increased so that $Ca \geq 2Ca_c$, the droplets may deform into long cylinders before breaking (Elemans et al. 1993; Vinckier et al. 1997).

9.2.2.2 Droplet Coalescence in Emulsions

In practical situations, mechanical mixing usually involves nondilute droplet concentrations, where both droplet breakup and coalescence of neighboring droplets might occur. Droplet coalescence in shear has been much less studied than breakup. However, Grizzuti and Bifulco (1997) have recently studied coalescence after a step decrease in shear rate in emulsions of Newtonian polyisobutylene (PIB) and polydimethylsiloxane (PDMS) at low droplet volume fractions ($\phi \leq 0.1$). Using an optical microscope to measure the average droplet diameter, they found that the droplet size distribution reaches steady state only very slowly, requiring thousands of strain units to achieve. Furthermore, their steady-state number-averaged diameter d_n, plotted against $\eta_s \dot{\gamma} / \Gamma$ in Fig. 9-10, is less than that predicted by the Taylor theory for breakup, shown by the dotted line. Thus, *coalescence and breakup produce different steady-state droplet sizes*. A similar result was obtained by Minale et al. (1997) who used a rheological method to infer droplet size. The solid line in Fig. 9-10 is the prediction of a theory of shear-induced coalescence by Janssen (1993):

$$d_{coal} = 2M^{-2/5} \left(\frac{4}{\sqrt{3}} h_c \right)^{2/5} \left(\frac{\eta_s \dot{\gamma}}{\Gamma} \right)^{-3/5}$$

This expression is valid for a partially mobile interface; analogous expressions hold for fully mobile and immobile interfaces (Minale et al. 1997). Here, h_c is the "critical thickness" of the liquid gap between droplets at which coalescence occurs. Theoretically, one expects $h_c \simeq (A_H a/8\pi\Gamma)^{1/3}$, where A_H is the Hamaker constant (Chesters 1991). A value $h_c = 0.2\mu$m gives the solid line in Fig. 9-10, this value of h_c is an order of magnitude larger than predicted, no doubt because h_c is used as a fitting parameter to accomodate rough approximations used in the theory.

Figure 9-10 implies that the droplet size distribution can show *hysteresis*. Thus, when the shear rate is increased, one achieves a droplet size controlled by the Taylor theory for breakup, but when the shear rate is then stepped down, the droplet size does not grow back to that predicted by the Taylor theory. However, there is a critical shear rate above which

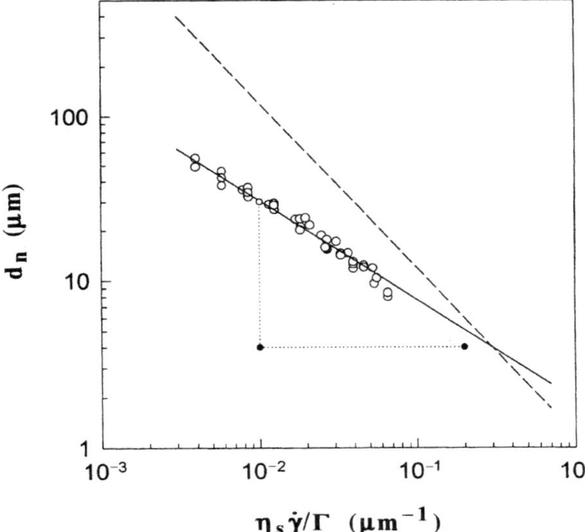

Figure 9.10 Number-average drop diameter as a function of $\eta_s\dot{\gamma}/\Gamma$ for blends of 10% polyisobutylene in polydimethylsiloxane with $M = 0.47$. The solid line is the prediction of the Janssen theory, and the dashed line is for the Taylor theory with $\text{Ca}_c = 0.6$. The dotted line connecting two solid symbols shows a "trajectory" when the shear rate is suddenly reduced. (From Grizzuti and Bifulco 1997, reprinted with permission from Steinkopff Publishers.)

the diameter set by breakup is smaller than that controlled by coalescence. Above this shear rate, where the solid and dashed lines in Fig. 9-10 intersect, we expect hysteresis to disappear, and the steady-state average droplet size should then be set by a dynamic balance of simultaneous breakup and coalescence. Combining the Taylor and Janssen theories, the critical average droplet diameter above which both breakup and coalescence occur is given by

$$d_c = \frac{8}{\sqrt{3}}\text{Ca}_c^{-3/2}M^{-5/3}h_c$$

For $M \approx 1$, this gives $d_c \approx 20h_c$; that is, a few microns. This result is consistent with the measurements of Minale et al. (1997).

The interfacial tension of highly immiscible liquids is typically $\Gamma \geq 10$ dyn/cm. Thus, for low-viscosity emulsions containing, for example, water and light oils ($\eta_s \sim 0.01$ P), the steady-state droplet radius set by the Taylor criterion is $a \sim \Gamma/\eta_s\dot{\gamma} \sim 10^3\dot{\gamma}^{-1}$ cm. This is large, of order millimeters or greater, unless the shear rate is very high, $\dot{\gamma} > 10,000$ sec^{-1}. Smaller droplets are produced by shearing high-viscosity immiscible polymer blends, for which $\eta_s \sim 10^4$ P; in this case, droplets of 1-μm radius can be formed at modest shear rates of around 10 sec^{-1}. Thus, for immiscible polymers blends, we expect coalescence to be more important than it is for low-viscosity liquids under shear.

9.2.2.3 Droplet Dynamics in Immiscible Polymer Blends

Polymers are frequently mixed or *blended* in order to optimize some combination of thermal, mechanical, and other properties without the work and expense of designing and synthesizing new polymers with the desired properties. Unusual combinations of properties—for example, hardness combined with toughness—are most readily achieved by blending rubbery polymers with glassy ones. It is important to understand and control the size and shape of the particles of dispersed phase, because the properties of the blend depend upon them. For example, it has been found that the toughness of blended polymers is often a sensitive function of particle size; in one case, toughness was observed to decrease fourfold when the average particle size increased from 0.7 to 0.8 μm (Wu 1985; Favis and Chalifoux 1987). DuPont workers have shown that if the droplets of the minor phase can somehow be distorted into plate-like shapes, a significant improvement in barrier properties to oxygen and hydrocarbon diffusion can be achieved (Subramanian 1983). Liquid crystalline polymers (LCPs) have been blended into less expensive isotropic polymers to impart to them some of the strength and modulus of the LCPs at a fraction of the cost. In the liquid blend, the LCP phase is prone to form long filaments, especially in extensional flows (Mehta and Isayev 1991; La Mantia et al. 1990). When such a blend is cooled, a fibrous composite results, similar to that obtained by mixing high-modulus solid fibers into a matrix. Thus, the blend is "self-reinforcing" (Collyer and Clegg 1986). Remarkably, the viscosity of such blends is sometimes less than that of either of the components (La Mantia et al. 1990; Lin and Winter 1992; Kim and Jang 1995).

Commercial polymer blends are often formed by melting pellets of the pure components in commercial mixers such as blade-type batch mixers or twin-screw extruders (Wu 1987; Sundararaj and Macosko 1995). The flow fields in such mixers are predominantly shear, but the shear rate varies from place to place, and there are zones where the flow has significant extensional character as well. In addition, the polymer melts are usually non-Newtonian fluids under the conditions of mixing; thus the viscosity is shear-rate-dependent, and there are flow-induced elastic, as well as viscous, forces. Furthermore, droplet coalescence as well as breakup is likely to occur, except at small droplet concentrations. Finally, after blending has occurred at high temperature, there may be a cool-down period in which flow is absent but droplets or domains are still mobile enough to either coalesce, or—if the domains are extended cylinders or sheets—to fragment. Thus, in such blending operations, one would expect the Taylor theory to serve as no more than a very rough guide for predicting the average droplet size.

In several experimental studies, the sizes of droplets formed during polymer blending have been measured. Wu (1987) has found that in extruded polymer blends the dependence of droplet sizes on the viscosity ratio of the dispersed to the matrix phase is qualitatively similar to that of Newtonian liquids shown in Fig. 9-9. In particular, the droplet size a (or, equivalently, the capillary number) is minimized at a viscosity ratio of around unity; Fig. 9-11 is a dimensionless plot of his data. The minimum seems to be rather sharp, producing a V-shaped curve. Similar V-shaped curves have been reported by Favis and Chalifoux (1987), but with the minimum located at a smaller viscosity ratio, $M \approx 0.15$. Wu correlated his data with the expression

$$a = \frac{2\Gamma M^{\pm 0.84}}{\dot{\gamma}\eta_s} \tag{9-19}$$

where the $+$ sign in the exponent applies when $M > 1$ and the $-$ sign applies when $M < 1$. This empirical expression was obtained using blends for which the dispersed phase concentration was 15% by weight and the "effective" shear rate was 100 sec^{-1}.

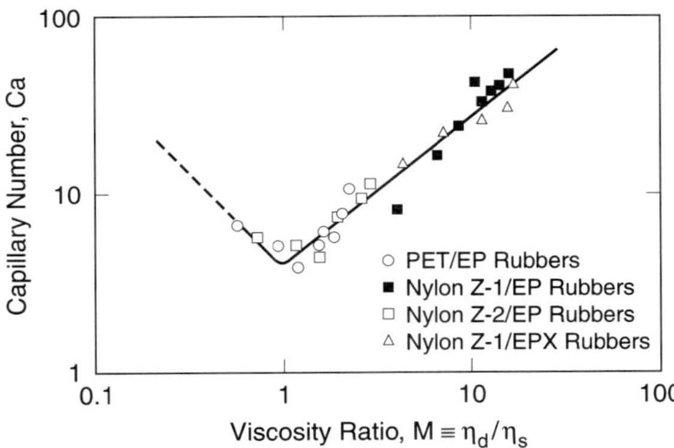

Figure 9.11 Dimensionless droplet size (capillary number) as a function of viscosity ratio M for various blends of 15 wt% ethylene–propylene rubbers in nylon or polyester produced by extrusion at an "effective shear rate" of around 100 sec⁻¹. The components are shear thinning, and the viscosity ratio M is evaluated at the "effective shear rate." (From Wu 1987, reprinted with permission from the Society of Plastics Engineers.)

In drop-breakup experiments with polymers, the shear viscosity is typically shear-rate-dependent. So in plots such as Fig. 9-11, the viscosities are taken to be those of the melts at a nominal shear rate in the mixer. In the experiments of Sundararaj and Macosko, for example, the "nominal" shear rate $\dot{\gamma}$ at a given motor speed is estimated from the linear drag flow that is assumed to exist in the narrowest gap of the mixer.

To use an expression such as Eq. (9-19) for predicting droplet sizes, one also needs a value for the interfacial tension Γ, which for polymer–polymer interfaces are hard to measure. However, the theory of Helfand and Sapse (1975) allows Γ to be estimated from the interfacial width Δ:

$$\Gamma \propto \chi^{1/2}, \qquad \Delta \propto \chi^{-1/2}, \qquad \rightarrow \Gamma \propto \Delta^{-1} \tag{9-20}$$

where χ is the Flory "chi" parameter. An empirical correlation of Wu (1982) is roughly consistent with this relationship between Γ and Δ:

$$\Gamma = 7.6\Delta^{-0.86}, \qquad \Gamma \text{ in } mN/m, \qquad \Delta \text{ in } nm \tag{9-21}$$

From this correlation, Eq. (9-21), the interfacial tension can be estimated if the interfacial width Δ can be measured in an electron micrograph of a blend (Wu 1987).

Notice that in Fig. 9-11 the critical capillary number $\dot{\gamma}\eta_s a/\Gamma$ at which breakup occurs is an order of magnitude higher than predicted by the Taylor theory. Some of this discrepancy is attributable to the effects of droplet–droplet collisions and coalescence that likely occur at the droplet concentration used, namely 15%. Coalescence is not accounted for in Taylor's theory, but it must become increasingly important as ϕ increases. In experiments of Elmendorp and van der Vegt (1986), the average droplet radius approaches the prediction of the Taylor theory at low enough volume fractions, around 0.5% (see Fig. 9-12). A theory to predict collision and coalescence rates at larger ϕ can be found in Elmendorp and van der Vegt (1986). "Population balance equations" for the droplet size distribution have been proposed by Tokita (1979) and others (Sondergaard and Lyngaae-Jorgensen 1995).

The data of Sundararaj and Macosko (1995) (Fig. 9-13), like those of Elmendorp and van der Vegt, show an increase in the average particle diameter d_n with increased concentration, as well as an increase in the width of the size distribution, where one standard deviation in Fig. 9-13 is represented by the "error bars." Sundararaj and Macosko also show

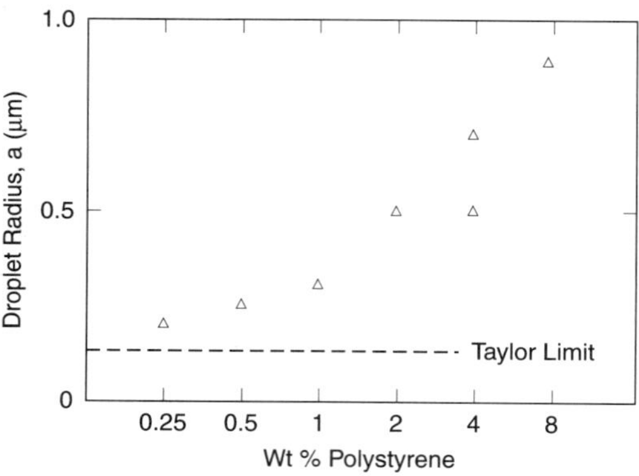

Figure 9.12 Droplet radius versus wt% dispersed polystyrene (PS) in a blend with polypropylene (PP) at an average shear rate of 10 sec^{-1} in a single twin screw extruder at $T = 200°C$. The viscosities of pure PS and PP at these conditions are about 5×10^2 and 2×10^3 Pa·s, respectively; both melts are modestly shear thinning (power-law slopes $\approx -1/3$). The interfacial tension Γ was measured to be 4.9 dyn/cm. The predicted Taylor limit, $a = 0.5 \, \Gamma/\eta_s \dot{\gamma}$, is shown. (From Elmendorp and van der Vegt 1986, reprinted with permission from the Society of Plastics Engineers.)

a decrease in droplet diameter toward a low-ϕ limit at ϕ around 0.5%; but this limiting value exceeds the Taylor value by a factor of four or so. Sundararaj and Macosko attribute this disagreement to the viscoelasticity of the droplet phase. In a systematic study, Mighri et al. (1997) find that an increase in elasticity of the droplet phase reduces its deformation under extensional flow, while the reverse occurs when the elasticity of the matrix phase. It is intuitive that fluid elasticity might act like interfacial tension in suppressing drop breakup. Sundararaj and Macosko suggest a modification of Taylor's criterion that accounts for elastic forces; at breakup, the droplet radius a_c is proposed to be given by

$$a_c = \frac{\Gamma \cdot \mathrm{Ca}_c}{\eta_s \dot{\gamma} - N_1(\dot{\gamma})} \tag{9-22}$$

where $\mathrm{Ca}_c = 0.5$. Equation (9-22) is arrived at by assuming that the viscous stresses deforming a droplet must overcome both interfacial tension and an elastic normal stress N_1 in the droplet phase, which is evaluated at the nominal shear rate $\dot{\gamma}$ of the blender. This formula is crude, and it does not account for differences in shear rates between the droplet and the medium (which are large when the viscosity ratio differs greatly from unity). Nevertheless, because of the shear-rate-dependence of N_1, Eq. (9-22) can predict a *minimum* in droplet size as a function of shear rate that is observed in some cases (Sundararaj and Macosko 1995; Plochocki et al. 1990; Favis and Chalifoux 1987). Viscoelastic forces have indeed been shown to suppress the breakup of thin liquid filaments that would otherwise rapidly occur via Rayleigh's instability (Goldin et al. 1969; Hoyt and Taylor 1977; Bousfield et al. 1986). Elongated *filaments*, for example, are observed in polymer blends (Sondergaard and Lyngaae-Jorgensen 1995).

Sundararaj and Macosko (1995) and Beck Tan et al. (1996) observed that the addition of a block copolymer to the droplet phase before mixing it with the matrix phase had little effect on the resulting droplet size at low droplet volume fraction. Although a block copolymer should reduce the interfacial tension between the two phases, and thereby lead to smaller droplets, the diffusion time of the block copolymer may be too long for it to saturate the new interfacial area that must form rapidly if a droplet is to fragment. However, block copolymers do seem to suppress coalescence, possibly by immobilizing the interface

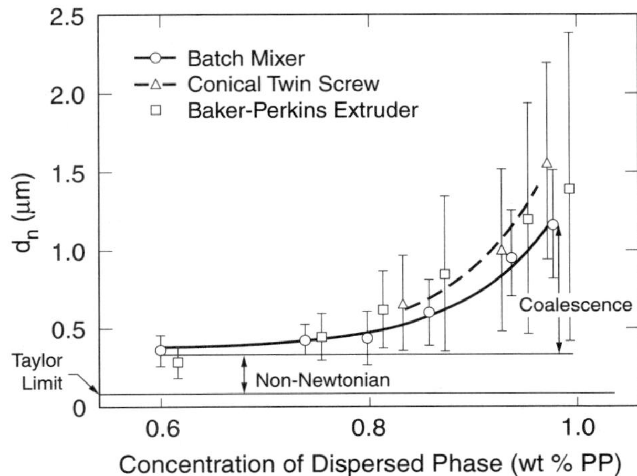

Figure 9.13 Number averaged diameter of droplets d_n of polypropylene ($M_n = 60,000$) in polystyrene ($M_w = 200,000$) as a function of wt% polypropylene mixed in three different mixers at a nominal shear rate of around 65 sec^{-1} and $T = 200°$C. The viscosities of the the PP and PS under these conditions are 840 and 950 Pa·s, respectively. The interfacial tension Γ is 5.0 dyn/cm. The "error bars" represent the distribution of droplet sizes, and they encompass one standard deviation in each direction from the mean. The deviation from the Taylor limit at low concentrations is attributed to non-Newtonian effects, while the increase in droplet size at higher concentrations is attributed to droplet coalescence. Note that similar droplet sizes are obtained in all three different mixers. (Reprinted with permission from Sundararaj and Macosko, Macromolecules 28:2647. Copyright © 1995, American Chemical Society.)

(Elmendorp and van der Vegt 1986). This has been inferred from the observation that the droplet size does not increase as much with increasing droplet volume fraction as it does when the block copolymer is absent (Sundararaj and Macosko 1995; Beck Tan et al. 1996). Milner and Xi (1996) have put forth a theory that explains the suppression of coalescence in terms of an effective entropic repulsive force between droplets with block copolymers at their surfaces.

If the block copolymer is formed *in situ* by chemical reaction of some dispersed-phase molecules with molecules in the matrix, droplet sizes can be much smaller than when such reactions are absent. This process, known as *reactive compatibilization* or *grafting*, is a useful strategy for obtaining fine dispersions of immiscible polymers (Sondergaard and Lyngaae-Jorgensen 1995).

As alluded to in Section 9.2.1.3, when the composition of a binary blend of A and B is varied, the phase continuity *inverts* from droplets of A in a matrix of B at low volume fractions of A, to droplets of B in A at high volume fractions of A. The volume fraction of A at which inversion occurs is of practical importance, since the transport of mass or momentum through a blend is dominated by the properties of the matrix phase. In some cases, it is desirable that *both* components be continuous. If one or both components of the blend are reactive and if each component reacts only with itself, *interpenetrating networks* (IPNs) can be formed, with the network of the second phase occupying the space not occupied by the network of the first phase. Since such structures form intricate labyrinths that can greatly slow, but not stop, diffusion, they might be of value in the manufacture of slow-drug-release membranes (Miles and Zurek 1988).

For polymer blends, the viscosities are typically shear-rate-dependent; Eq. (9-10) for the inversion point has been extended to account for this in an obvious way:

$$\frac{\eta_1(\dot{\gamma})}{\eta_2(\dot{\gamma})} \sim \frac{\phi_1}{\phi_2} \qquad \text{(at the inversion point)} \qquad (9\text{-}23)$$

Miles and Zurek (1988) found that Eq. (9-23) indeed gives the condition for phase inversion for the the three polymer blends they studied, with the largest viscosity contrast being 3:1. A similar range of viscosity ratios was covered by Jordhamo et al. (1986), whose work also supports Eq. (9-23). Interpenetrating structures were found in a modest, but imprecisely determined, band of volume fractions. The effect of the shear-rate-dependent viscosity ratio on phase continuity was confirmed explicitly in a polystyrene/polymethylmethacrylate (PS/PMMA) blend, in which the PMMA was much more shear thinning than the PS. At high shear rates where the PMMA and PS viscosities were similar, a 50/50 blend was bicontinuous (see Fig. 9-14); but at lower shear rates, where the PMMA was much more viscous than the PS, the viscous PMMA phase took the form of discrete droplets (Miles and Zurek 1988). The tendency of the less viscous phase to be continuous is consistent with the observation that any droplets of the low-viscosity phase that might form tend to stretch in the flow to high aspect ratios without breaking (see Section 9.2.1.3). The longer the droplet, the more likely it is to interact and fuse with other droplets. Thus, low-viscosity droplets tend to form a percolated structure, thereby becoming the continuous phase, even at low droplet volume fractions. On the other hand, Newtonian droplets that are more viscous than the matrix tend to resist stretching, and they tend to break if they are forced to stretch. They are therefore unlikely to "percolate" into a continuous phase, unless their volume fraction is high.

One might conjecture that Eq. (9-23) should be significantly modified if one of the phases is much more viscoelastic than the other. As discussed above, an elastic liquid can withstand higher elongations without breaking than a Newtonian liquid, other things being equal. Tanaka's studies of spinodal decomposition (see Section 9.2.1.3) suggest that a viscous, but elastic, liquid might maintain phase continuity when mixed with a less viscous, but Newtonian, one, in disagreement with Eq. (9-23). Systematic blending studies in which viscosity and elasticity are separately controlled will be needed to check this conjecture.

Jordhamo et al. (1986) attempted without much success to apply Eq. (9-23) to the reactive blending process used to form interpenetrating networks or IPNs. In such experiments, the volume fractions and viscosities of both phases change as the two simultaneous polymerization reactions progress. Given the complications of such a system, it is perhaps not surprising that Eq. (9-23) did not predict very well the the condition at which bicontinuity was achieved. Simliar complications likely occur in the formation of *polymer dispersed liquid crystals* (PDLCs). PDLCs are nematic liquid crystalline droplets dispersed in a polymer matrix (Hilsum 1976; Doane 1991). They can be made from a single-phase mixture by a temperature quench, by evaporation of a solvent, or by chemical reaction of a monomer to form a polymer from which the liquid crystal phase separates (West 1990). The liquid crystal is typically the dispersed phase, even when its the final volume fraction is high, around 80%, perhaps because of elastic stresses in the polymer, as discussed by Tanaka (1994a) (see Section 9.2.1.3 and Fig. 9-6). PDLC films act as electrically activated optical

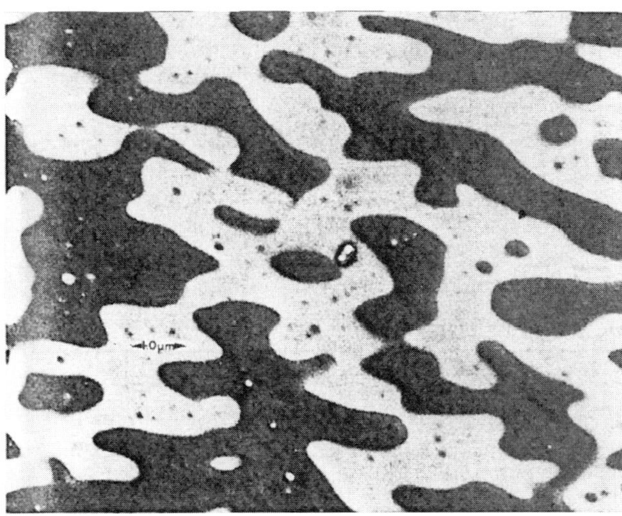

Figure 9.14 Transmission electron micrograph of a section of bicontinuous phase formed by 53 wt% polystyrene and 47 wt% polymethylmethacrylate blended with a Brabender mixer at a rate of 20 rpm, which is roughly equivalent to 90 sec^{-1}, at 200°C. At this shear rate, the two components have about the same viscosity, around 1000 Pa·s. The characteristic domain width is around 1 μm. (From Miles and Zurek 1988, reprinted with permission from the Society of Plastics Engineers.)

switches that do not require polarizers, and hence use light efficiently. Thus, films of PDLC material are contenders for use in optical displays and switchable transmission and reflection gratings (Bunning et al. 1996).

There is such a large body of work on polymer blends that their rheology cannot be reviewed here because of space constraints; the interested reader is referred to books and reviews by Han (1981), Utracki (1990), Wu (1987), Elmendorp and van der Vegt (1991), Sondergaard and Lyngaae-Jorgensen (1995), and Briber et al. (1996). Needed still are studies on polymer blending and its effect on droplet size and size distribution for families of well-characterized polymer blends in which viscosities, elasticities, and other properties are systematically varied. Understanding of the droplet coalescence process in polymer blends seems to be especially underdeveloped. Also undeveloped is the effect of reaction on phase separation, although Glotzer et al. (1995) have recently presented theory and simulations on this topic.

A much-studied topic not addressed at all here is the influence of flow on polymer blends or emulsions that are close to their phase boundaries, where flow-induced mixing or demixing can occur. Reviews of this topic can be found in Rangel-Nafaile et al. (1984) and Larson (1992b).

9.3 RHEOLOGY OF EMULSIONS AND IMMISCIBLE BLENDS

In Sections 9.2.1 and 9.2.2, we considered two basic processes by which emulsions and immiscible polymer blends are formulated—namely, by quenching from a one-phase state and by mechanical mixing. Since blends and emulsions are not at equilibrium, the way they are formulated affects their micromorphology and, thereby, their rheology. Once their micro-morphology is somehow specified, it should be possible in principle to predict their rheological properties. These properties will usually change with time as the blend or emulsion coarsens, and thus the rheological measurements are only well-defined if they are acquired at a rate that is fast compared to the rate of coarsening.

9.3.1 Steady Shear Viscosity and Normal Stresses

The simplest case to consider is steady flow of a dilute suspension of Newtonian drops or bubbles in a Newtonian medium. If the capillary number $\eta_s \dot{\gamma} a / \Gamma$ is small, so that the drops or bubbles do not deform under flow, then at steady state the viscosity of the suspension is given by Taylor's (1932) extension of the Einstein formula for solid spheres:

$$\eta_r \equiv \frac{\eta}{\eta_s} = 1 + \frac{1 + \tfrac{5}{2}M}{1 + M} \phi \tag{9-24}$$

where, as before, η is the viscosity of the suspension and $M \equiv \eta_d / \eta_s$ is the ratio of the viscosities of the dispersed to the suspending fluids. In the limit of a very viscous droplet fluid, $M \to \infty$, the droplets behave like hard spheres, and Einstein's result, $\eta_r = 1 + 5\phi/2$, is recovered. In the opposite limit, $M \to 0$ (which is usually valid for bubbles), the relative viscosity is lower, $\eta_r = 1 + \phi$. The viscosity of a suspension of bubbles is less than that of a suspension of hard spheres at a given volume fraction ϕ, because liquid bounding the surface of a bubble can flow, and hence bubbles (or low viscosity liquids) disturb the flow field of the external fluid less than do hard spheres. Equation (9-24) has been verified experimentally by Nawab and Mason (1958) for $M \equiv \eta_d / \eta_s$ in the range 0.5–5.

For higher concentrations, outside of the dilute regime, Pal (1992) has proposed an empirical equation for the zero-shear viscosity of an emulsion:

$$(\eta_{r,0})^{1/K_1} = \exp\left(\frac{2.5\phi}{1 - \phi/\phi_m}\right) \tag{9-25}$$

Here $\eta_{r,0} \equiv \eta_0 / \eta_s$ is the zero-shear-rate relative viscosity, and

$$K_1 \equiv \left(\frac{0.4 + M}{1 + M}\right) \tag{9-26}$$

The volume fraction at "maximum packing," ϕ_m, produces a divergence in $\eta_{r,0}$ as $\phi \to \phi_m$. At low ϕ, Eq. (9-25) reduces to the Taylor result, Eq. (9-24), while for high droplet viscosity, $M \to \infty$, and arbitrary ϕ, Eq. (9-25) reduces to the so-called Mooney equation for concentrated suspensions of hard spheres (Mooney 1951). The empirical Mooney equation is somewhat similar to the Krieger–Dougherty equation, discussed in Section 6.2.1. Figure 9-15 shows that reasonably good agreement with Eq. (9-25) is obtained for a series of surfactant-stabilized emulsions in which the continuous phase was thickened by addition of polymer, so that $M \equiv \eta_d / \eta_s$ could be varied over a wide range, $M = 0.004$–6.7. The chosen value of ϕ_m is 0.91. In Fig. 9-15, the largest deviations from Eq. (9-25) are observed at high ϕ, where $(\eta_r)^{1/K_1}$ is sensitive to M. This sensitivity is not surprising; for large M, the droplets should behave as undeformable hard spheres and so the viscosity should diverge as $\phi \to \phi_m$

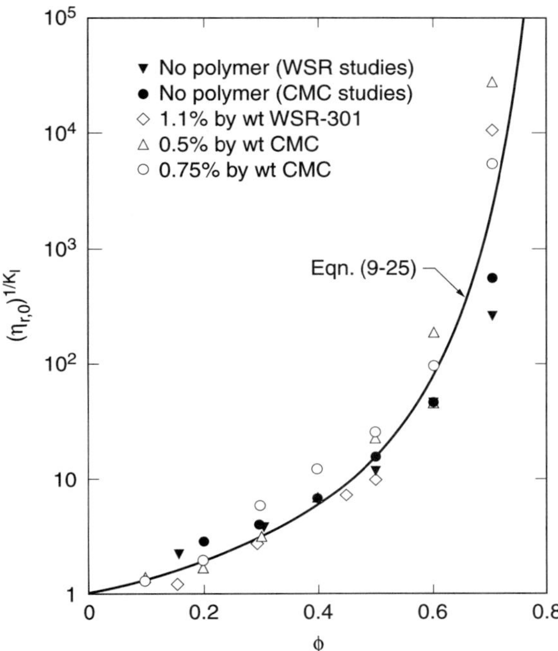

Figure 9.15 Zero-shear relative viscosity $\eta_{r,0}$ to the power $1/K_I$ versus droplet concentration ϕ for emulsions of petroleum oil in aqueous solutions of Triton X-100 surfactant and 1.1% polyethylene oxide (Polyox WSR), or sodium carboxymethyl cellulose (CMC). K_I is given by Eq. (9-26), and the line is the fit using Eq. (9-25) with $\phi_m = 0.91$. (From Pal 1992, with permission from the Journal of Rheology.)

$= 0.63$–0.64, the value for random close packing. For lower M, however, the droplets are deformable and so the apparent divergence of η_r is postponed and ϕ_m is increased.

For higher shear rates, η_r is shear thinning when $\phi \geq 0.6$; the shear thinning becomes more pronounced as ϕ increases. Pal accounts for this simply by adjusting ϕ_m to get the best fit at each shear rate $\dot{\gamma}$. The resulting fits are shown in Fig. 9-16. As $\dot{\gamma}$ increases from 10 to 100 to 1000 sec^{-1}, the best-fit ϕ_m increases from 1.05 to 1.18 to 1.43. Some of the shear thinning in Pal's data arises from shear thinning in the polymer solutions used as the matrix fluids; this was accounted for in Fig. 9-16 by using the value of η_s at the shear rate used in the definition of $M = \eta_d/\eta_s$ which appears in Eq. (9-26).

If the droplet or bubble is deformable (Ca is not negligible), then the emulsion will be *viscoelastic*, even if the two fluids composing it are both Newtonian. The characteristic time constant is

$$\tau \equiv \frac{a\eta_s}{\Gamma} \tag{9-27}$$

For a viscosity ratio M of order unity or less, τ is the relaxation time of the droplet shape and of the resulting viscoelastic stress in the dispersion. Thus, for $\eta_s \approx 1$ cP, $\Gamma \approx 10$ dyn/cm, $a \approx 1$ μm, we obtain $\tau \approx 10^{-7}$ sec, and the stress in the emulsion relaxes almost instantly after cessation of flow. However, when the fluid medium is more viscous and the droplets are bigger, say, $\eta_s \approx 100$ P, $a \approx 10$ μm, with the same Γ, we obtain $\tau \approx 0.01$ sec. Although this latter value of τ corresponds to rather fast relaxation, it can produce appreciable normal stress differences at high shear rates; thus (Choi and Schowalter 1975)

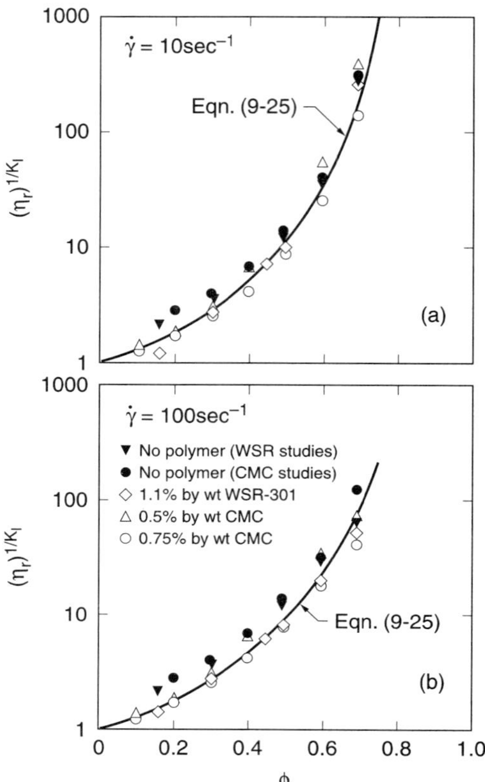

Figure 9.16 Relative viscosity to the power $1/K_I$ versus droplet concentration ϕ at shear rates of 10 sec⁻¹ and 100 sec⁻¹, for emulsions described in the caption to Fig. 9-15. K_I is given by Eq. (9-26), and the lines are the fits using Eq. (9-25) with $\phi_m = 1.05$ for $\dot\gamma = 10$ sec⁻¹ and $\phi_m = 1.18$ for $\dot\gamma = 100$ sec⁻¹. (From Pal 1992, with permission from the Journal of Rheology.)

$$\Psi_1 \equiv \frac{N_1}{\dot\gamma^2} = 2f(M)\eta_s\tau\phi \tag{9-28}$$

$$\Psi_2 \equiv \frac{N_2}{\dot\gamma^2} = \Big[g(M) - f(M)\Big]\eta_s\tau\phi \tag{9-29}$$

where

$$f(M) \equiv \frac{1}{80}\left(\frac{19M+16}{M+1}\right)^2, \quad g(M) \equiv \frac{3(19M+16)\,(25M^2+41M+4)}{560(M+1)^3} \tag{9-30}$$

These equations give the following for the ratio of the second to the first normal stress coefficients:

$$\frac{\Psi_2}{\Psi_1} = \begin{cases} -0.218, & M \to \infty \\ -0.446, & M \to 0 \end{cases} \tag{9-31}$$

Thus, the magnitude of the ratio Ψ_2/Ψ_1 should be higher for emulsions than for single-component entangled polymers, for which Ψ_2/Ψ_1 goes from around -0.25 to near zero as the shear rate increases (see Section 3.7.4.1). This prediction has not yet been tested experimentally.

9.3.2 Linear Viscoelasticity

Oldroyd (1953, 1955) derived expressions for the linear viscoelasticity of suspensions of one Newtonian fluid in another. By using an *effective-medium* approach, he was able to relax the requirement of diluteness. For an ordinary interface whose interfacial tension Γ remains constant during the deformation, Oldroyd's result gives the following for the complex modulus $G^* \equiv G' + iG''$:

$$G^* = i\omega\eta_s \left(\frac{1 + \frac{3}{2}\phi\frac{E}{D}}{1 - \phi\frac{E}{D}} \right) \tag{9-32}$$

where i is the imaginary unit and

$$E \equiv 2i\omega(\eta_d - \eta_s)\,(19\eta_d + 16\eta_s) + \frac{8\Gamma}{a}(5\eta_d + 2\eta_s)$$

$$D \equiv i\omega(2\eta_d + 3\eta_s)\,(19\eta_d + 16\eta_s) + \frac{40\Gamma}{a}(\eta_d + \eta_s) \tag{9-33}$$

At small ϕ, the dilute-suspension result is recovered:

$$G^* = i\omega\eta_s \left(1 + \frac{5}{2}\phi\frac{E}{D}\right) \tag{9-34}$$

In the limit of nondeforming droplets (large Γ/a), $E/D \rightarrow (0.4 + M)/(1 + M)$, and Eq. (9-34) reduces to Taylor's result, Eq. (9-24). Equation (9-32) can be generalized so that the interface supports viscous as well as elastic stresses (i.e., a finite *interfacial viscosity*), and so the interfacial tension can be allowed to change when the interface is undergoing deformation (Palierne 1990).

Equation (9-32) can also be rewritten in the following form:

$$G' = G'_{\text{bulk}} + G'_{\text{interface}}$$

$$G'' = G''_{\text{bulk}} + G''_{\text{interface}} \tag{9-35}$$

where the bulk contributions to G' and G'' are

$$G'_{\text{bulk}} = 0, \qquad G''_{\text{bulk}} = \omega\mu\frac{\tau_2}{\tau_1} \tag{9-36}$$

and the interfacial contributions are

$$G'_{\text{interface}} = \frac{\mu}{\tau_1}\left(1 - \frac{\tau_2}{\tau_1}\right)\frac{\omega^2\tau_1^2}{1 + \omega^2\tau_1^2}$$

$$G''_{\text{interface}} = \frac{\mu}{\tau_1}\left(1 - \frac{\tau_2}{\tau_1}\right)\frac{\omega\tau_1}{1 + \omega^2\tau_1^2} \tag{9-37}$$

The viscosity μ, relaxation time τ_1, and retardation time τ_2 for Oldroyd's model, to low order in ϕ, are given in Table 9-1. The viscosity μ is the shear viscosity of the emulsion in the limit of high interfacial tension.

TABLE 9-1
Viscoelastic Constants for Emulsions

	Oldroyd	Choi and Schowalter
$\dfrac{\mu}{\eta_s}$	$1 + \phi\dfrac{(5M+2)}{2(M+1)}\left[1 + \phi\dfrac{(5M+2)}{5(M+1)}\right]$	$1 + \phi\dfrac{(5M+2)}{2(M+1)}\left[1 + \phi\dfrac{5(5M+2)}{4(M+1)}\right]$
τ_1	$\tau_0\left[1 + \phi\dfrac{(19M+16)}{5(M+1)(2M+3)}\right]$	$\tau_0\left[1 + \phi\dfrac{5(19M+16)}{4(M+1)(2M+3)}\right]$
τ_2	$\tau_0\left[1 - \phi\dfrac{3(19M+16)}{10(M+1)(2M+3)}\right]$	$\tau_0\left[1 + \phi\dfrac{3(19M+16)}{4(M+1)(2M+3)}\right]$

$$\tau_0 \equiv \frac{(19M+16)(2M+3)}{40(M+1)}\frac{\eta_s a}{\Gamma}$$

Oldroyd noted that his coefficient of the ϕ^2 term for μ is much too small. More accurate expressions for μ, τ_1, and τ_2 have been derived by Choi and Schowalter (1975) using a cell model and a perturbation expansion in the concentration ϕ and the droplet deformation. The Choi–Schowalter results can be expressed in the form of Eqs. (9-35)–(9-37), with μ, τ_1, and τ_2 given in Table 9-1.

In both the Oldroyd and the Choi–Schowalter theories, the dispersed and continuous phases are assumed to be *Newtonian*. However, Palierne (1990) has generalized Oldroyd's equations to allow the dispersed and continuous phases to be *viscoelastic*. In the limit of a constant interfacial tension and no interfacial viscosity, his results are analogs of Eqs. (9-32) and (9-33):

$$G^* = G_s^* \left(\frac{1 + \frac{3}{2}\phi\dfrac{E}{D}}{1 - \phi\dfrac{E}{D}}\right) \tag{9-38}$$

and

$$E \equiv 2(G_d^* - G_s^*)(19G_d^* + 16G_s^*) + \frac{8\Gamma}{a}(5G_d^* + 2G_s^*)$$

$$\tag{9-39}$$

$$D \equiv (2G_d^* + 3G_s^*)(19G_d^* + 16G_s^*) + \frac{40\Gamma}{a}(G_d^* + G_s^*)$$

where G_d^* and G_s^* are the frequency-dependent complex moduli of the dispersed fluid and the surrounding matrix fluid, respectively. When, as is usual, the droplet radii are not uniform, the most appropriate average value for a is the volume average, $a_V \equiv (\Sigma\phi_i a_i)/\phi$, where ϕ_i is the volume fraction of droplets with radius a_i (Graebling et al. 1993). Figure 9-17 compares the predictions of the Palierne model for $G'(\omega)$ to measurements on a dispersion of polyisoprene in polydimethylsiloxane, just after pre-shearing the sample at four different shear rates. The volume-average droplet radius a was measured using a microscope and was a decreasing function of the rate of pre-shear, in agreement with the Taylor theory. Note that each curve has two steep portions, separated by a "plateau" region of smaller slope. The high-frequency steep portion is due to the viscoelasticity of the bulk liquids and would be present in a pure, unblended melt. The low-frequency steep portion is due to the interfacial contribution; its height and position along the frequency axis are sensitive to the droplet radius. As the rate of pre-shear increases, the droplet radius decreases, and the total interfacial area increases. The former effect shifts the interfacial contribution to higher frequencies, while the latter increases its modulus, in agreement with Fig. 9-17. The measured droplet radius a scales with prior shear rate $\dot{\gamma}_0$ roughly as $a \propto \dot{\gamma}_0^{-1}$, in agreement with the Taylor theory and the Doi–Ohta theory, discussed in Section 9.3.3 below.

For typical polymer blends, the interfacial tension is often rather low, around 1–10 dyn/cm. For droplets of diameter 1 μm or larger, the interfacial terms are typically negligible except at frequencies low enough that G_s^* and G_d^* are of order 10^3 Pa or less. This is shown in Fig. 9-17, where G' for the two different droplet radii converge at around $G' \sim 3 \times 10^2$ Pa, indicating that the interfacial contribution is negligible at frequencies higher than this. Thus, at high frequency, where the interfacial terms are negligible, we have at low or modest ϕ only the bulk contribution,

$$G_{\text{bulk}}^* = G_s^* + \frac{5}{2}\phi\frac{2(G_d^* - G_s^*)}{2\dfrac{G_d^*}{G_s^*} + 3} \tag{9-40}$$

If the ratio G_d^*/G_s^* is not too far from unity, the denominator of the last term in Eq. (9-40) can be approximated as $2(G_d^*/G_s^*) + 3 \approx 5$, and we obtain simply

$$G_{\text{bulk}}^* \approx (1 - \phi)G_s^* + \phi G_d^* \tag{9-41}$$

which is a linear "mixing rule" for the modulus, in the absence of interfacial terms.

In the range of frequencies low enough that the interfacial terms become important, one can usually assume that both components of the blend are in their terminal regimes, and thus behave as Newtonian liquids. Following this reasoning, Gramespacher and Meissner (1992) divided the linear viscoelastic response of a blend of polystyrene (PS) and poly(methylmethacrylate) (PMMA) into bulk and interfacial terms, as in Eq. (9-35). The interfacial contributions were taken from the Choi–Schowalter theory for Newtonian liquids, Eq. (9-37), with μ, τ_1, and τ_2 from Table 9-1, while the bulk contributions were obtained from Eq. (9-41). With these expressions for G_{bulk}^* and $G_{\text{interface}}^*$, Gramespacher and

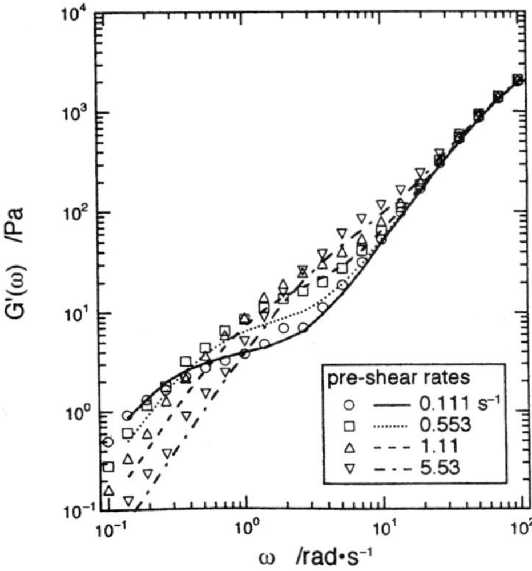

Figure 9.17 Measured storage modulus of 11% polyisoprene ($\eta_0 = 60.9$ Pa·s) in polydimethylsiloxane ($\eta_0 = 73.7$ Pa·s) after pre-shearing at four different shear rates, along with predictions (lines) of the Palierne model, Eqs. (9-38) and (9-39). The interfacial tension, $\Gamma = 3.2$ dyn/cm, and the droplet radii were measured, so there are no fitting parameters. (From Kitade et al. 1997, with permission from the Journal of Rheology.)

Meissner were able to predict without adjustable parameters G' and G'' for the PS/PMMA blend. Graebling et al. (1993) obtained similar agreement experiment and theory using Eq. (9-38). The success of the theory prompted Gramespacher and Meissner to suggest that rheological measurements might be used, along with the theory, as a means of estimating interfacial tensions for polymer–polymer interfaces for which Γ is not known.

9.3.3 Nonlinear Transient Viscoelasticity: The Doi–Ohta Theory

If flow is continuous, or strains are large, droplets or domains deform greatly or burst, and the above expressions for the moduli are inapplicable. Although there have been some experimental and theoretical studies of the droplet size and size distributions in such flows (see Section 9.2.2), there seem to be few theoretical results that allow prediction of the *stresses* in such flows. These flows are immensely complex; the stresses depend not only on the viscoelastic characteristics of the two components of the emulsion, but also on the micromorphology of the phases, which, in turn, depends on the history of the flow.

Doi and Ohta (1991), however, have developed a theory for blends of two immiscible Newtonian fluids of equal viscosity and roughly equal density (to avoid settling). For this special case, the rheology of the fluid differs from that of a single Newtonian fluid only because of the presence of interfaces. The theory originally assumed also a roughly 50:50 blend so that both phases maintain continuity, and coarsen by capillary flow, rather than by slow diffusive processes. However, the theory has been applied, with some success, to blends with compositions other than 50:50, as well as to viscosity ratios M in the range 0.15–2.3 (Vinckier et al. 1996, 1997). For viscosity ratios higher than 4 (and perhaps also for lower ones lower than 0.10), droplets do not break up when the shear rate is increased (see Fig. 9-9), and this leads to failure of the Doi theory (Kitade et al. 1997).

Doi and Ohta define the area of interface (A) per unit volume (V) of sample by $Q \equiv A/V$; Q is inversely proportional to the characteristic domain size a. Q gradually increases in response to an increase in shear rate, and it decreases when the shear rate is reduced.

Shearing not only increases the surface area of interface between the two phases, but also *orients* the interface, so that it becomes more nearly parallel to the shearing surfaces. To quantify the interfacial orientation, Doi and Ohta define an *interface tensor* $\mathbf{q}$:

$$\mathbf{q} \equiv \frac{1}{V} \int_A dA(\mathbf{nn} - \boldsymbol{\delta}) \tag{9-42}$$

where dA is a unit of interfacial area whose orientation is perpendicular to the unit vector $\mathbf{n}$, and the integral is over the whole of the interface A. For the moment, we assume that the viscous stress is large compared to the interfacial stress; then, since both phases have the same viscosity, in a linear flow the velocity gradient is homogeneous, and the interface will therefore *deform affinely*. Affine deformation of the interface implies that Q and $\mathbf{q}$ obey the following equations:

$$\frac{\partial}{\partial t}\mathbf{q} = -\mathbf{q} \cdot \nabla\mathbf{v}^T - \nabla\mathbf{v} \cdot \mathbf{q} + \frac{2}{3}\boldsymbol{\delta}\nabla\mathbf{v} : \mathbf{q} - \frac{Q}{3}(\nabla\mathbf{v} + \nabla\mathbf{v}^T) + \frac{1}{Q}\mathbf{q} : \nabla\mathbf{v}\,\mathbf{q} \tag{9-43}$$

$$\frac{\partial}{\partial t}Q = -\nabla\mathbf{v} : \mathbf{q} \tag{9–44}$$

To derive Eq. (9-44), a *decoupling* approximation was used so that a fourth rank tensor could be replaced by a product of second rank tensors.

The stress tensor of the Doi–Ohta model is given by

$$\sigma = 2\eta_0 \mathbf{D} - \Gamma \mathbf{q} \tag{9-45}$$

where η_0 is the viscosity of each of the two phases.

Equations (9-43) and (9-44) account for the effects of flow on interfacial area and orientation, but omit interfacial shrinkage, or coarsening, which is driven by interfacial tension. Additional terms to account for coarsening must be added to the right sides of Eqs. (9-43) and (9-44). On dimensional grounds, such terms are proportional to Γ/η_0; Doi and Ohta chose arbitrary, but simple, terms consistent with this:

$$\left(\frac{\partial}{\partial t}\mathbf{q}\right)_{\text{relax}} = -\Lambda Q \mathbf{q} \tag{9-46}$$

$$\left(\frac{\partial}{\partial t}Q\right)_{\text{relax}} = -\Lambda \mu Q^2 \tag{9-47}$$

where $\Lambda \propto \Gamma/\eta_0$, and μ is dimensionless.

When these, or other dimensionally correct, relaxation terms are added to the right sides of Eqs. (9-43) and (9-44), distinctive *scaling relationships* can be derived, which can be expressed as follows. Let $\sigma(t, [\dot{\gamma}(t)])$ be the stress tensor at time t during a shearing flow with shear history $[\dot{\gamma}(t)]$. Now choose a new shear history $[\tilde{\dot{\gamma}}(t)] = c[\dot{\gamma}(ct)]$ in which the shear rate at each instant t is a constant c times the shear rate in the old history at time ct. Then, the stress $\tilde{\sigma}$ in the new shear history is given by

$$\tilde{\sigma}(t, [\tilde{\dot{\gamma}}(t)]) = c\sigma(ct, [\dot{\gamma}(t)]) \tag{9-48}$$

This scaling law, Eq. (9-48), implies that all components of the stress tensor are *linear* in the shear rate. Consider for example, a constant-shear-rate experiment. At steady state, not only is the shear stress predicted to be proportional to the shear rate, but so also is the first normal stress difference N_1! This prediction has been nicely confirmed in recent experiments by Takahashi et al. (1994), who studied mixtures of silicon oil and hydrocarbon–formaldehyde resin. Both these fluids are Newtonian, and have the same viscosity, around 10 Pa·s. Figure 9-18 shows that both the shear stress σ and the first normal stress difference $N_1 \equiv \sigma_{11} - \sigma_{22}$ of these mixtures are indeed linear in the shear rate, so that the shear viscosity $\eta \equiv \sigma/\dot{\gamma}$ and the so-called *normal viscosity* $\eta_N \equiv N_1/\dot{\gamma}$ are constants. The first normal stress difference in this mixture must be attributed entirely to the presence of interfaces, since the individual liquids in the mixture have no measurable normal stresses. A portion of the shear stress also comes from the interfacial stress. Figure 9-19 shows that the shear and normal viscosities are both maximized at a component ratio of roughly 50:50. At this component ratio, the interfacial term accounts for roughly half the total shear stress.

The scaling, Eq. (9-48), also implies that time-dependent stresses can be rescaled so that data taken at different shear rates collapse onto a single curve. Consider an example in which the sample is sheared at a rate $\dot{\gamma}_i$ until a steady state is reached, and then the shear rate is suddenly increased by a factor of four to $\dot{\gamma}_f = 4\dot{\gamma}_i$. After this increase in shear rate, the shear stress undergoes an overshoot and an undershoot, while the first normal stress

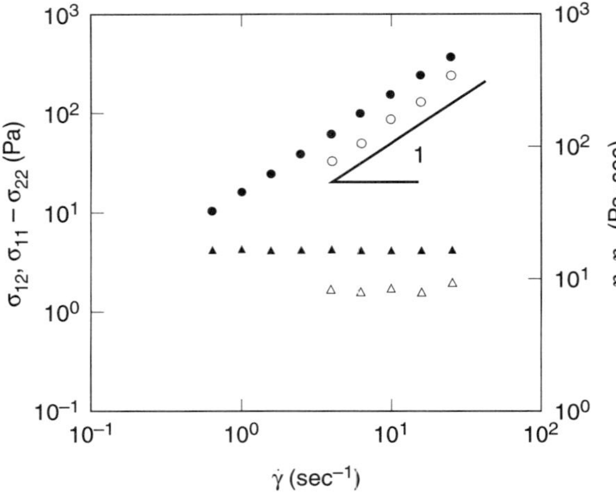

Figure 9.18 Shear-rate dependence of the shear stress (●), first normal stress difference (○), shear viscosity (▲), and normal viscosity (△) for a 1:1 mixture (by weight) of polydimethylsiloxane (viscosity = 9.74 Pa·s), and "Genelite 4050S" (hydrocarbon–formaldehyde resin with viscosity = 10.1 Pa·s), at 25°C. (From Takahashi et al. 1994, with permission from the Journal of Rheology.)

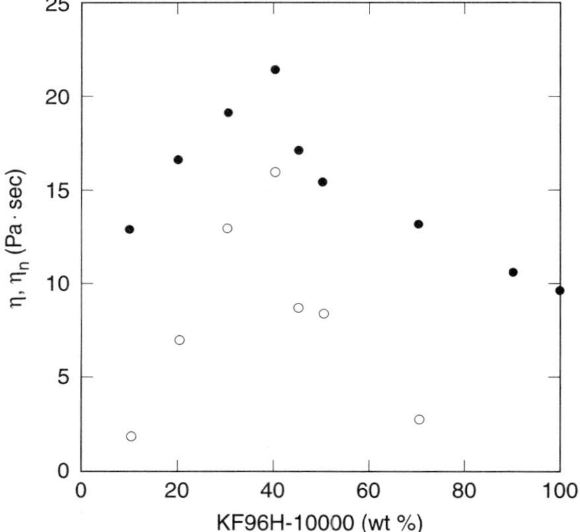

Figure 9.19 Dependence of shear (●) and normal (○) viscosities on weight percent polydimethylsiloxane for blends of the components described in Fig. 9-18. (From Takahashi et al. 1994, with permission from the Journal of Rheology.)

difference undergoes an overshoot only (see Fig. 9-20). For the four curves in Fig. 9-20, the shear rates $(\dot{\gamma}_i,\ \dot{\gamma}_f)$ are (3, 12), (4, 16), (5, 20), and (6, 24) sec^{-1}, from bottom to top. In Fig. 9-21, these data are replotted in the rescaled form $N_1/N_1^{ss}(\dot{\gamma}_i)$ versus strain $\dot{\gamma}_f t$, where $N_1^{ss}(\dot{\gamma}_i)$ is the steady-state first normal stress difference at the initial shear rate $\dot{\gamma}_i$. Note that in this rescaled plot, the data *collapse onto a single curve*. An analogous data collapse can be obtained for the shear stress. Similar scaling behavior has been recently reported by Guenther and Baird (1996) for 25/75 w/w blends of poly(ethylene terephthalate) and nylon 6,6.

   This remarkable *scaling* property, which is shared by some liquid crystalline polymers (see Section 11.3.4), by nematic surfactant solutions (Section 12.4.2), and by some particulate suspensions, is a consequence of the *lack of an intrinsic relaxation time*. In the case

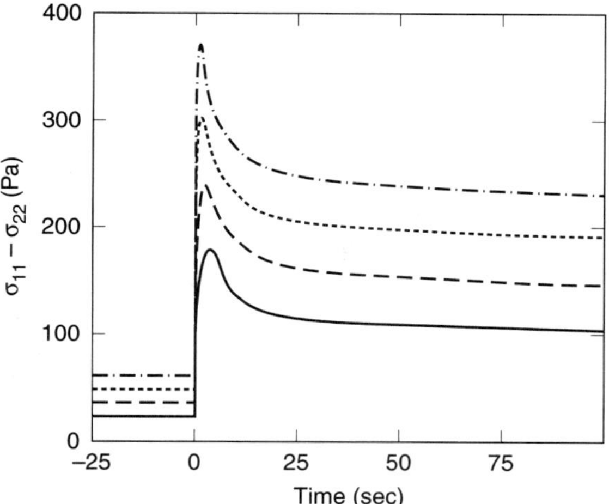

Figure 9.20 First normal stress difference versus time after step increases in shear rate. The shear rates before the step increase, $\dot{\gamma}_i$, and after it, $\dot{\gamma}_f$, are given by $(\dot{\gamma}_i, \dot{\gamma}_f) = (3,12)$, $(4,16)$, $(5,20)$, and $(6,24)$, from the bottom curve to the top one. (From Takahashi et al. 1994, with permission from the Journal of Rheology.)

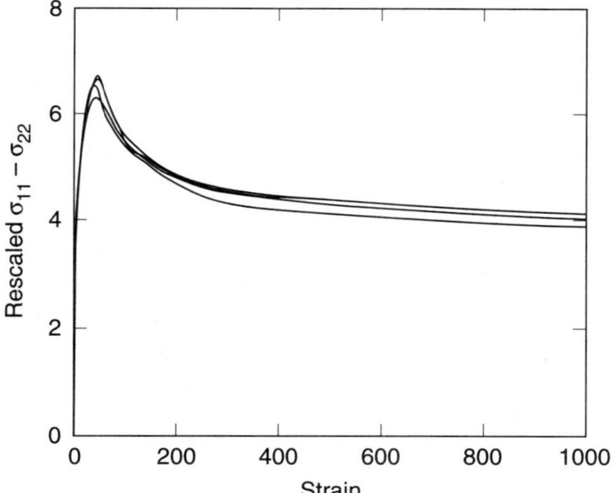

Figure 9.21 The data of Fig. 9-20 replotted as normalized stress $N_1/N_1^{ss}(\dot{\gamma}_i)$ versus strain $\dot{\gamma}_f t$. (From Takahashi et al. 1994, with permission from the Journal of Rheology.)

of blends, the relaxation time τ is proportional to the domain size $a : \tau \approx a\eta_0/\Gamma$, and a, in turn, is set by the shearing rate, $a \sim \Gamma/\eta_0\dot{\gamma}$. Thus, putting these two results together, we obtain $\tau \sim 1/\dot{\gamma}$; that is, the relaxation time is inversely proportional to the shear rate. This leads to the observed scaling properties.

Although this experimentally observed scaling behavior is correctly predicted by the Doi–Ohta theory, the shape of the transient response curve—in particular, the overshoot and undershoot in the shear stress—are *not* predicted. This implies that the relaxation expressions chosen by Doi and Ohta, Eqs. (9-46) and (9-47), are inaccurate. This is not very surprising, since Eqs. (9-46) and (9-47) were chosen rather arbitrarily from many possible forms that satisfy the scaling relationship. Optical microscopy suggests that the overshoot and undershoot are caused by elongation of droplets followed by their breakup (Takahashi and Noda 1995). Vinckier et al. (1997) have shown that the stress growth after start-up or step-up of shearing, up to the first stress maximum, can be predicted nearly quantitatively by combining the Doi–Ohta approach with predictions of an affine-deformation model for droplets.

The Doi–Ohta theory is expected to fail when the two phases have very unequal viscosities, especially when the dispersed phase is the more viscous. In particular, when $M > 4$, the droplets are expected to resist breakup in shear (see Section 9.2.2.1). In addition, the droplets will not deform affinely when $M \neq 1$ (Delaby et al. 1994), and the stress tensor will contain terms other than those in Eq. (9-45) because of the jump in viscosity across the interface (Onuki 1994). In the absence of a detailed theory for the case M deviating greatly from unity, Onuki (1994) made some rough estimates from scaling arguments. His estimates of the droplet radii $a_\parallel$ and $a_\perp$ parallel and perpendicular to the shearing flow direction, and of the blend viscosities and first normal stress differences, are summarized in Table 9-2. In this table, the viscosity of component 1 is taken to be much greater than that of component 2; that is, $\eta_1 \gg \eta_2$. The stress predictions in Table 9-2 seem to be largely untested.

9.3.4 Rheology at High Droplet Volume Fraction

When the volume fraction of droplets is increased systematically, one expects eventually to encounter a point of *phase inversion* where the droplet phase becomes the continuous phase, and vice versa. If, however, surfactant is adsorbed onto the droplet interfaces, the droplets can remain stable even when the droplet phase volume fraction approaches unity. The surfactant also keeps the droplet size distribution from changing with time or when the droplet concentration is changed by addition of more of the continuous phase (Otsubo and Prud'homme 1994). Bibette (1991) has developed a remarkable method, "crystallization fractionation," for preparing emulsions with a monodisperse distribution of droplet sizes. In this method, starting from a polydisperse size distribution, at a high droplet volume fraction a monodisperse droplet fraction forms a close-packed ordered structure which rises to the surface and can be "creamed off." Small (of order microns) monodisperse droplets can also be obtained from an initially very polydisperse distribution of large droplets by large-amplitude oscillatory shearing in thin-gap devices ($\lesssim 200\ \mu$m) (Mason and Bibette 1996). Studies of the rheological properties of such model emulsions have just begun (Mason et al. 1995; Liu et al. 1996).

The droplet concentration in an emulsion can be increased by centrifugation or by dialysis. When the volume fraction ϕ in an emulsion exceeds the random close-packed limit, $\phi \approx 0.64$, the droplets deform, and, as $\phi \to 1$, the continuous phase of the emulsion is confined to thin films between deformed droplets, resembling the liquid films of a foam.

TABLE 9-2
Radii of Droplets and Stresses in Blends with $\eta_1 \gg \eta_2$

	$\phi_1 < \phi_2(\eta_1/\eta_2)$ (phase 1 discrete)	$\phi_1 > \phi_2(\eta_1/\eta_2)$ (phase 1 continuous)
$a_\parallel$	$\Gamma\phi_2/(\eta_2\dot{\gamma})$	$\Gamma/(\eta_2\dot{\gamma})$
$a_\perp$	$a_\parallel$	$(\eta_2/\eta_1)^{1/2}a_\parallel$
$\eta \equiv \sigma_{12}/\dot{\gamma}$	η_2/ϕ_2	η_1
$N_1/\dot{\gamma}$	$\phi_1\eta_2/\phi_2$	$\left(\eta_1^{3/2}/\eta_2^{1/2}\right)\phi_2$

The rheological properties of a dense emulsion with close-packed droplets depends on whether or not the droplets are small enough to be agitated significantly by Brownian motion. If not, because of the high packing density of the droplets, the emulsions should be elastic and have a finite elastic modulus at low frequencies. For liquids with viscosities near that of water, Brownian behavior should dominate for particle radii less than or equal to 1 μm, while non-Brownian behavior occurs when $a \geq 10\ \mu$m.

9.3.4.1 Large Droplets

Princen and Kiss (1986) carried out step-strain measurements on emulsions of paraffin oil droplets of mean radius 10 μm, polydisperse in size, in water containing 20% commercial surfactant as a stabilizer. They showed that the modulus G of these dense emulsions can be represented by the simple formula

$$G \approx 1.77 \frac{\Gamma}{a_{32}} \phi^{1/3}(\phi - \phi_0) \tag{9-49}$$

where $a_{32} = 3V/A$ is the volume(V)-to-surface(A) (or *Sauter* mean) drop radius, and $\phi_0 = 0.71$ is the droplet concentration at which the modulus G collapses to zero, presumably because the droplets no longer press against each other. Equation (9-49) is loosely based on a cell model for foam-like structures, to be described in Section 9.5.1. In the emulsions of Princen and Kiss the stresses show only very slow relaxation (hours) after imposition of the step strain, and thus the emulsions are indeed nearly non-Brownian, as expected. These emulsions also have a *yield stress* σ_y, which can be expressed by (Princen and Kiss 1989)

$$\sigma_y = \frac{\Gamma}{a_{32}} \phi^{1/3} Y(\phi) \tag{9-50}$$

with $Y(\phi)$ an empirical function

$$Y(\phi) = -0.080 - 0.114 \log_{10}(1 - \phi) \tag{9-51}$$

Princen and Kiss (1989) found that in steady shearing the steady-state shear stress σ for their non-Brownian emulsions could be expressed as

$$\sigma = \sigma_y + C(\phi)\eta_s \dot{\gamma} \text{Ca}^{-1/2} \tag{9-52}$$

where $\text{Ca} \equiv \eta_s a_{32} \dot{\gamma}/\Gamma$ is the capillary number, $C(\phi)$ is an empirical expression,

$$C(\phi) = 32(\phi - \phi_0) \tag{9-53}$$

and ϕ_0 was found to be around 0.73. Schwartz and Princen (1987) derived a theoretical expression similar to Eq. (9-52), by assuming that under shear the yield stress is augmented by viscous dissipation due to flow in the *Plateau borders* that connect together thin films between the droplets. However, in the theoretical expression, the dissipative term is proportional to $\text{Ca}^{-2/3}$, while a term proportional to $\text{Ca}^{-1/2}$ fits the experimental data.

9.3.4.2 Small Droplets

While the emulsions studied by Princen and Kiss were solid-like, or elastic, at low frequencies, dense water-in-oil emulsions studied by Pons et al. (1993) were liquid-like at low frequencies, presumably because of the smaller droplet sizes, around 1 μm in radius. Figure 9-22 shows the storage and loss moduli G' and G'' for a water-in-oil emulsion containing 99% water with droplets around 1 μm in radius (Pons et al. 1993). The data in Fig. 9-22 can be fit by a single-relaxation-time Maxwell model, with a high-frequency modulus of order $G_\infty = 216$ Pa and a relaxation time τ of around 3 sec. The dependence of the high-frequency modulus G_∞ on ϕ can be fit to Eq. (9-49), by making adjustments to the prefactor (1.77) by as much as a factor of 3, along with small adjustments to ϕ_0 (see Fig. 9-23).

Pons et al. (1993) found that τ increases rapidly with increasing volume fraction ϕ, and they suggested an equation for this dependence:

$$\tau = \frac{c\eta_s}{G_\infty(1 - \phi)} \tag{9-54}$$

Fits to experimental data could be obtained by choosing rather large fitting values of the empirical coefficient, $c \sim 300$–20,000, depending on the system. So far, no analysis of such flows, and no derivation of a scaling law for τ in such emulsions, seems to be available.

9.3.4.3 Droplets of Intermediate Size

Since Brownian relaxation times scale as the cube of the particle radius (see Section 6.2.2), a fairly abrupt transition from Brownian to non-Brownian behavior might be expected when the droplet radius increases. Surprisingly, however, Otsubo and Prud'homme (1994) observed little dependence on droplet radius over a range of 4–12 μm, for high-density emulsions of oil droplets in an aqueous phase. Figure 9-24 shows G' and G'' for various

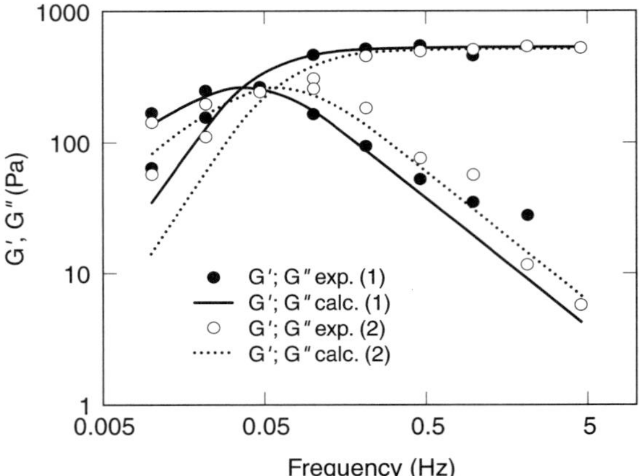

Figure 9.22   Storage and loss moduli G' and G'' for a water-in-oil emulsion containing 99% water, 0.5% $C_{10}H_{22}$ oil, and 0.5% trethylene glycol dodecyl ether ($C_{12}EO_3$) surfactant at 40°C. The droplets are around a micron in size. The closed circles ($\bullet$) are for the first experimental run, while the open circles ($\bigcirc$) are for a repeat run. The solid and dashed lines are fits of the Maxwell model to the first and second runs, respectively. The fitted modulus G_∞ is 216 Pa for both runs, while the fitted relaxation times are 2.7 sec and 4.3 sec. (Reprinted with permission from Pons et al., Journal of Physical Chemistry 97:12320 Copyright © 1993, American Chemical Society.)

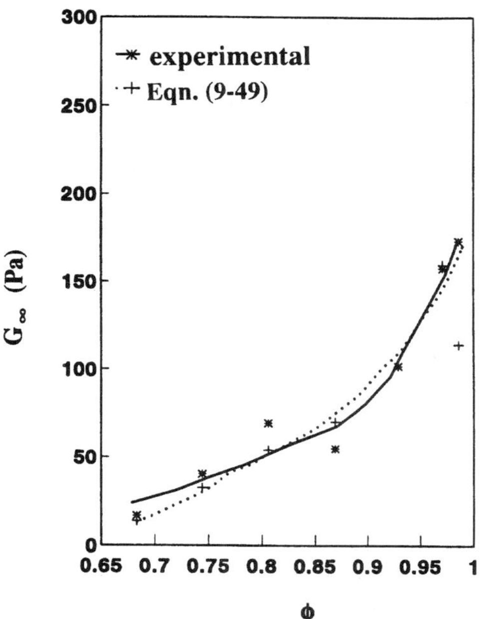

Figure 9.23 High-frequency modulus G_∞ as a function of volume fraction of droplets of water ($a_{32} \approx 1\,\mu$m) in $C_{10}H_{22}$ oil with surfactant $C_{16}EO_4$ at 40°C. The asterisks ($*$) are experimental data and the plus signs ($+$) are the predictions of Eq. (9-49) with the coefficient 1.77 replaced by 0.44, and with $\phi_0 = 0.62$. The interfacial tension is 1.15 dyn/cm. The lines are a guide to the eye. (Reprinted with permission from Pons et al., Journal of Physical Chemistry 97:12320 Copyright © 1993, American Chemical Society.)

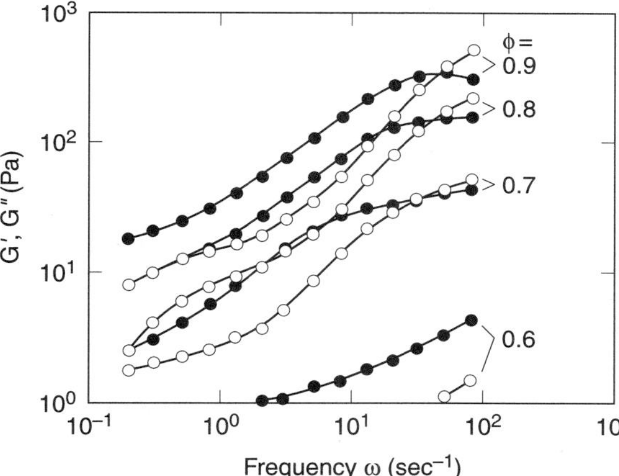

Figure 9.24 Frequency-dependent storage and loss moduli, G' (○) and G'' (●), for emulsions of 4.45-μm-radius oil droplets in an aqueous phase. The oil is a mixture of 24 wt% tritolylphosphate and 76% dioctylphthalate, with a viscosity of 0.065 Pa·s. The aqueous phase is a 20 wt% solution of an anionic surfactant with a viscosity of 0.00265 Pa·s. (From Otsubo and Prud'homme 1994, reprinted with permission from Steinkopff Publishers.)

concentrations of such an emulsion with a ratio $M = 24.7$ of viscosities of dispersed to continuous phase, with a droplet diameter $a = 4.45\,\mu$m and various droplet concentrations ϕ. These data show that the elastic modulus G' of the suspension increases rapidly with ϕ and develops a complex frequency-dependence when ϕ exceeds that for random close packing, $\phi \approx 0.63$. At high frequencies, the modulus G' in Fig. 9-24 appears to be approaching a high-frequency limit, $G' \rightarrow G_\infty$, which can be estimated as twice the value of G' at the frequency where G' and G'' cross, which we define as ω_p. Otsubo and Prud'homme (1994) tabulated values of G_∞ and ω_p for their emulsions with $\phi = 0.90$, drop radii $a \approx 4$–$12\,\mu$m, interfacial tensions $\Gamma = 0.9$–5.9 dyn/cm, and viscosity ratios $M = 8$–25. They found that G_∞ is proportional to Γ/a and is, on average, only about 30% higher than the prediction of Eq. (9-49).

The characteristic relaxation time $\tau = 1/\omega_p$ was found to be insensitive to the droplet size a, weakly dependent on the continuous-phase viscosity, and perhaps weakly dependent on Γ and M also. Although the complete scaling law for τ cannot be deduced from this limited set of data, it is evidently influenced by lubrication flow of liquid in the thin films between the deformed droplets, and perhaps also by the circulatory flow in the viscous droplets.

Figure 9-25 shows the shear viscosity η measured by Otsubo and Prud'homme (1994) for a series of emulsions of Newtonian oil droplets in a Newtonian aqueous phase with a ratio $M = 24.7$ of viscosities of the dispersed to the continuous phase ($M \equiv \eta_d/\eta_s$). At each composition ϕ, data for three average droplet radii are presented: 4.45, 7.7, and 13.15 μm, based on surface-to-volume ratios. The data at $\phi = 0.4$ show a common constant high-shear viscosity, which is *independent of shear rate*. This implies that at high shear rates the emulsion viscosity becomes independent of the droplet surface area, and the surface-tension contributions to the stress become negligible. At lower shear rates, $\dot{\gamma} \leq 30$ sec^{-1}, the viscosity is shear thinning and sensitive to droplet size; thus at these shear rates, the interfaces contribute significantly to the viscosity.

Figure 9-26 is a "master curve" assembled by Otsubo and Prud'homme, consisting of all their data at $\phi = 0.80$. The data seem to collapse when the scaled viscosity $d_{32}\eta/\Gamma\eta_s$ is plotted against $\eta_s\dot{\gamma}$, where $d_{32} = 2a_{32}$. A rearrangement of the Princen and Kiss equations (9-52) and (9-50) gives, however,

$$\frac{\eta a_{32}}{\Gamma \eta_s} = \phi^{1/3} Y(\phi)(\eta_s\dot{\gamma})^{-1} + C(\phi)\left(\frac{\Gamma}{a_{32}}\right)^{-1/2}(\eta_s\dot{\gamma})^{-1/2} \tag{9-55}$$

In this form, the Princen and Kiss equation makes clear that the plot of Otsubo and

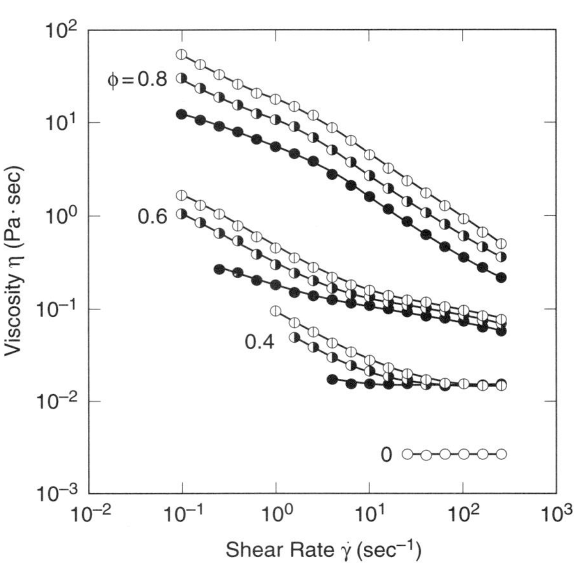

Figure 9.25 Curves of viscosity versus shear rate for emulsions of 4.45 (◑)-, 7.7 (◐)-, and 13.15 (●)-μm-radius oil droplets in an aqueous phase. The oil and water phases were the same as in Fig. 9-24. (From Otsubo and Prud'homme 1994, reprinted with permission from Steinkopff Publishers.)

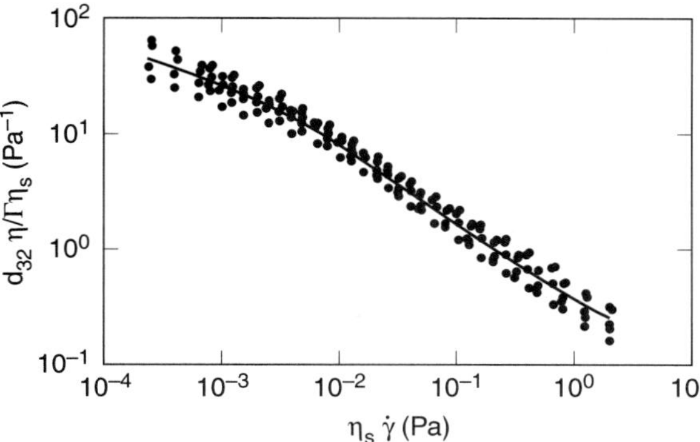

Figure 9.26 "Master curve" of reduced shear viscosity versus reduced shear rate for oil-in-water emulsions with $\phi = 0.80$. The droplet radii are in the range $a = d_{32}/2 \approx 4$–$12\,\mu$m, interfacial tensions $\Gamma = 0.9$–5.9 dyn/cm, and viscosity ratios $M = 8$–25; the ratio Γ/d_{32} is in the range 90–300 Pa. (From Otsubo and Prud'homme 1994, reprinted with permission from Steinkopff Publishers.)

Prud'homme should not be universal, but there should be a variation with the parameter $(\Gamma/a_{32})^{1/2}$. However, in the experiments of Otsubo and Prud'homme, this quantity was varied by a factor of less than two, and deviations from "universal" scaling are therefore within the scatter in Fig. 9-26. In the range $\eta_s\dot{\gamma} = 10^{-2}$–$10^0$ Pa, the data of Otsubo and Prud'homme are about a factor of three larger than the predictions of Eq. (9-55), if $Y(\phi)$ and $C(\phi)$ are taken from Eqs. (9-51) and (9-53), respectively. More significantly, the data of Fig. 9-26 show a trend toward a constant "Newtonian" viscosity plateau at low shear rates, while Eq. (9-55) predicts yield behavior at low shear rates, with a power-law viscosity–shear rate slope of -1. The emulsions of Otsubo and Prud'homme are evidently affected to some extent by Brownian motion, which is not accounted for in Eq. (9-55). Further experimental and theoretical work on emulsion rheology will be required to establish general scaling rules for these complex emulsions.

9.4 STRUCTURE AND COARSENING OF FOAMS

Foams are liquids or elastic solids into which a high volume fraction of bubbles has been introduced, often as high as 90–99%. Examples include foamed rubber, shaving cream, meringue, mousse, soap suds, and the frothy head on a freshly poured glass of beer (Aubert et al. 1986). In keeping with the scope of this book, this section is limited to liquid foams, an example of which is shown in Fig. 9-27. Liquid foams are important as cosmetics, fire-fighting agents, flotation aids, mobility control agent, and so on, and also because they are precursors in the manufacture of solid foams. Foams are intriguing substances: simple to imagine, but rich in physics and mathematics.

Foams can be made by releasing a gas within a foamable liquid. For example, the opening and pouring of a bottle of soda releases pressurized CO_2 gas, producing foam bubbles. Chemical reactions can also generate bubbles within the liquid. Other methods of foam production are to force both liquid and gas through a packed column (Khan et al. 1988), to spray foamable liquid onto a screen on which a fan is blowing (Aubert et al. 1986), or to use the *foam-extrusion* process (Han 1981).

Foams are not stable. They eventually collapse as their liquid films drain and rupture under gravity or as small bubbles are slowly cannibalized by large ones due to diffusion of

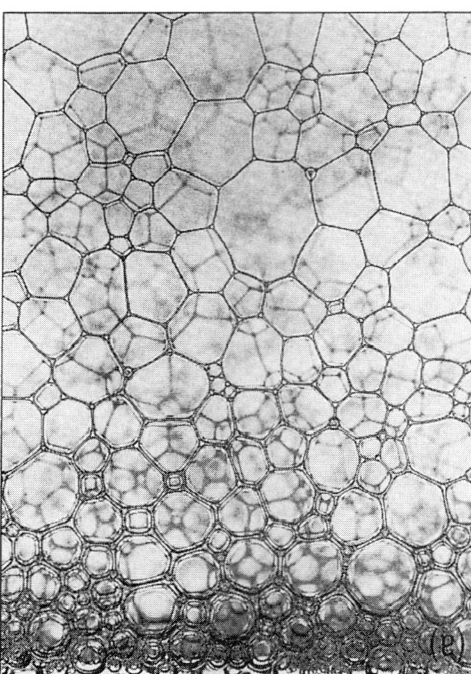

Figure 9.27 Photograph of bubbles in aqueous foams. The foam was made by agitating an aqueous solution of sodium dodecyl sulfate. At the top, the foam is drier and the gas bubbles are polyhedral; near the bottom the foam is wetter and the bubbles are nearly spherical. The bubble size is 3 mm, on average. (From Durian, MRS Bulletin, April 1994, reprinted with permission.)

gas through the liquid films. What stability a liquid froth has is conferred by surfactants, which slow down gas diffusion and help prevent film rupture. The desired longevity of a foam depends on the application; a longer life is desired for shaving cream than for the head on a glass of beer, and thus foam stability must be carefully controlled. The effect of liquid drainage on foam structure is evident in Fig. 9-27; water has drained from the top to the bottom of the froth, resulting in more nearly spherical bubbles there. At the top, the foam is "drier" and the bubbles are polyhedral.

9.4.1 Structure of Dry Foams

Dry foams, for which $\phi \gtrsim 0.95$, can be thought of as *networks of films*. Figure 9-28a shows a model polyhedral foam, in which each film is a face of a polyhedron, specifically a regular tetrakaidecahedron. Polydedral foam models, while oversimplified, capture the essential *topological* features of real foams. These are the meeting of three films at a *line*, called the *Plateau border*, and the joining of four lines at a *point* called a *vertex*. Surface-tension forces acting along each film must balance at the plateau border; this implies that the angle between any two films meeting there is always 120°. Likewise, at each vertex the four Plateau borders form equal tetrahedral angles of 109.47° with each other. These angular constraints, known as *Plateau's laws*, are not satisfied by the regular tetrakaidecahedron in Fig. 9-28a, but are satisfied by a distorted version of it proposed by Kelvin (1887). In the distorted tetrakaidecahedron, the sides of the square faces are curved and the hexagonal

(a)

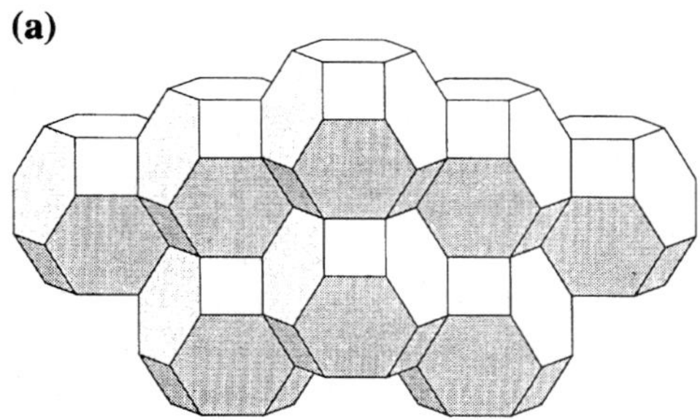

(b)

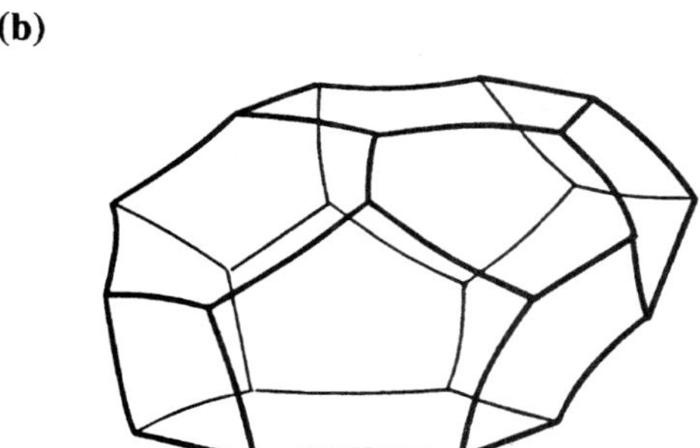

Figure 9.28 Three-dimensional foam models. **(a)** Space-filling regular tetrakaidecahedra with straight edges and flat faces. (From Reinelt 1993, with permission from the Journal of Rheology.) **(b)** The β tetrakaidecahedron with curved edges and bowed faces. (From Gururaj et al. 1995). (Reprinted with permission from Gururaj et al., Langmuir 11:1381. Copyright © 1995, American Chemical Society.)

faces are saddle-shaped. In general, *the polyhedral cells of a dry foam have curved faces and edges.*

While Kelvin's distorted tetrakaidecahedron satisfies Plateau's laws, its faces are squares and hexagons only. Real foams, however, possess a preponderance of pentagonal faces, with a minority of quadrilateral and hexagonal ones. Williams (1968) has therefore proposed the β tetrakaidecahedron, shown in Fig. 9-28b. This has eight pentagonal, two quadrilateral, and four hexagonal faces, and it can be packed tetragonally to fill space. The 14 faces of each β tetrakaidecahedron closely match the average of 13.7 of real froths, and the average number of edges per face, 5.143 for the β tetrakaidecahedron, is also close to the experimental value of 5.196 (Gururaj et al. 1995; Matzke and Nestler 1946). However, the β tetrakaidecahedron does not minimize the surface-to-volume ratio at fixed topology, which is a requirement of real foams (Ross and Prest 1986). A more important limitation of all these models is that they are *periodic*, where all cells are the same, while real froths are *aperiodic*, with polydisperse cell sizes and shapes (see Fig. 9-27).

The films and Plateau borders of real foams have finite thicknesses. However, this thickness can be small, since the air pressure in dry foams presses the surfaces of neighboring bubbles together, driving liquid from the films into the Plateau borders, thereby creating a structure whose cross section is similar to that depicted in Fig. 9-29b. The cross section of the Plateau border thus acquires the shape of a triangle with concave sides. For relatively wet froths ($\phi \approx 0.9$), the triangular plateau borders are thick [Fig. 9-29a (left)], while for very dry foams ($\phi \to 1.0$), the borders collapse to lines whose cross sections in Fig. 9-29a

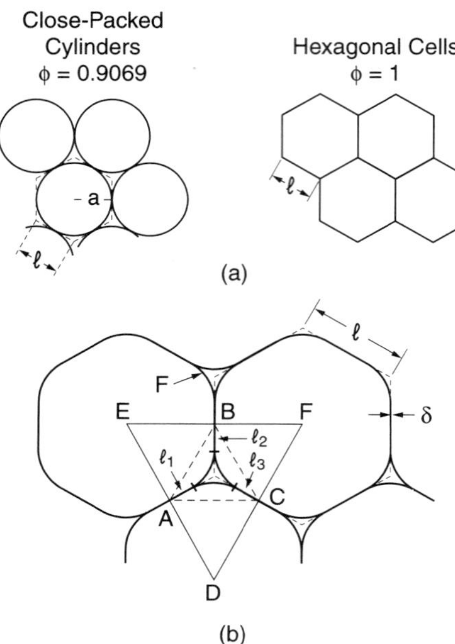

Figure 9.29 Two-dimensional hexagonal foam model. In **(a)**, the foam structure changes from hexagonally close-packed cylinders to regular hexagons as the volume fraction ϕ increases from 0.9069 to unity. In **(b)**, the three films ℓ_1, ℓ_2, and ℓ_3 are shown each with thickness δ. A film splits at each end into two rounded curves each with radius of curvature r. Three such curves define a region known as the Plateau border, which is in the center of the dotted triangle. (From Khan and Armstrong 1989, with permission from the Journal of Rheolgy.)

(right) are shown as points. Drainage of liquid from the films stops when the interfaces on either side of the film are close enough to each other that repulsive surfaces forces—van der Waals, steric, or electrostatic—become high enough to stop further drainage. Quasi-stable film thicknesses are typically around 100 nm in aqueous foams (Durian 1994), which is thin compared to the typical bubble size, $a \sim 10\ \mu$m to 1 cm.

The Plateau borders in a foam form a *network* into which liquid from the films collects, and from which liquid is drained away by gravity. This drainage process, which is important in foam beds used for mass transfer or separation, has been modeled by several groups, starting with Haas and Johnson (1967; see Gururaj et al. 1995 for a recent review).

9.4.2 Coarsening

Once excess liquid has drained and the foam is relatively dry, if film rupture is rare, then the primary coarsening mechanism will be gas diffusion through the liquid films, which allows some bubbles to expand at the expense of others, which shrink and eventually disappear. The chemical potential of the gas in a bubble is proportional to Γ/a, where a is the bubble radius (see Fig. 9-30). Thus, the flux of gas per unit area of bubble surface, which is proportional to the chemical potential, goes as a^{-1}. Since the surface area per bubble, across which mass flux occurs, is proportional to a^2, the rate of change of bubble volume, dV/dt, is proportional to $a^2 \cdot a^{-1} = a$. Thus $da/dt \propto a^{-2} dV/dt \propto a^{-1}$. This argument leads to the prediction that the average bubble radius should grow with time as $t^{1/2}$, which is close to the experimental scaling, $a \propto t^{0.45 \pm 0.05}$ (Durian et al. 1991a, 1991b). The validity of this argument requires that the film thickness not change appreciably during coarsening. If the bubbles were dilute, then the spacing between them would increase during coarsening in

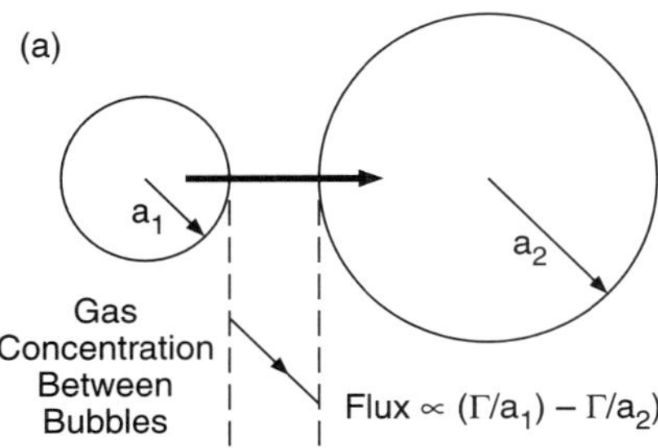

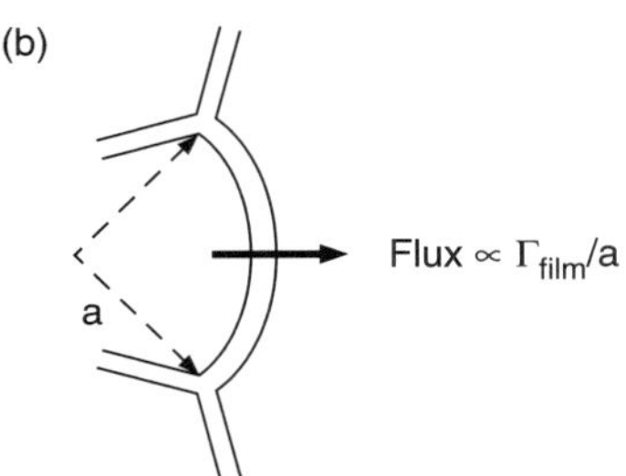

Figure 9.30   In **(a)**, gas diffuses from small bubbles to large ones because of a curvature-induced difference in chemical potential. For a similar reason, in **(b)**, gas diffuses across a curved film. (From Durian, MRS Bulletin, April 1994, reprinted with permission.)

proportion to the bubble size a, and the gas flux per unit area would then scale as $1/a^2$, leading to the the scaling $a \propto t^{1/3}$, the same as in Ostwald ripening (see Section 9.2.1). However, the surface forces that keep the film thickness fixed might also affect the rate of gas diffusion across the interface.

The coarsening process is not smooth. As bubbles change in size, a point is reached when the network topology of the foam becomes unstable, and *bubble rearrangement* occurs (Ashby and Verrall 1973; Weaire and Rivier 1984). In processes of type T1, depicted in Fig. 9-31a, one film disappears and a new one is created, as two neighboring bubbles separate from each other and a new pair of neighbors forms. In process T2, a bubble disappears, thus eliminating three or more films (see Fig. 9-31b). When a T1 or T2 event occurs, films change their positions and curvatures somewhat. This leads to pressure changes within the cells bounding these films, which, in turn, leads to adjustments in all other films bounding those cells. Thus, a cascade of events affecting many bubbles might propagate from a single rearrangement. Hence, foam coarsening proceeds by a slow diffusive growth of some bubbles at the expense of others, punctuated by minor and major rearrangements of foam topology and geometry.

9.4.2.1 Coarsening Theory for Two-Dimensional Foams

The coarsening behavior of two-dimensional dry foams ($\phi \approx 1$) has been studied for many years, originally as a model of grain growth in polycrystalline metals and alloys (Smith 1952). Experimentally, two-dimensional froths can be made by sandwiching a foam between two pieces of plexiglass with a spacing smaller than the cell size and allowing excess liquid to drain off continually (Glazier and Stavans 1989; Stavans 1993). The bubbles of a two-dimensional dry foam are polygons whose sides correspond to films and vertices to

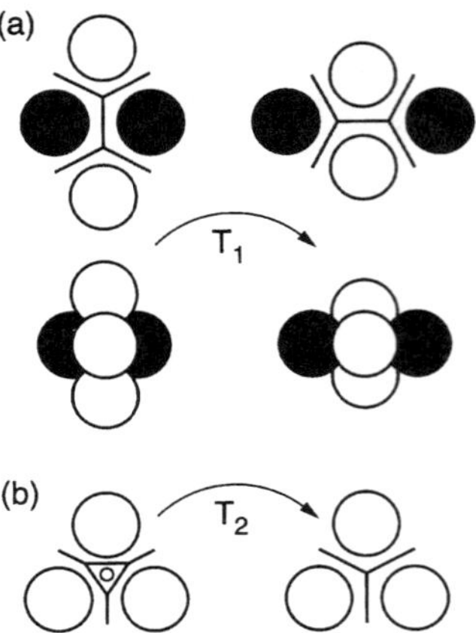

Figure 9.31 Topological changes in foams. In **(a)**, at T1 neighbor-switching process occurs in two and three dimensions. In **(b)**, a three-sided bubble disappears. (From Durian, MRS Bulletin, April 1994, reprinted with permission.)

Plateau borders. Euler's theorem then relates the numbers of polygons (P), sides (S), and vertices (V) by $P - S + V = 2$. This equation, combined with the observation that three sides meet at every vertex, and every side connects two vertices, so that $3V = 2S$, leads to the conclusion that for an infinite two-dimensional foam the average number of sides per polygon is six.

Coarsening is driven by differences in pressure from bubble to bubble. Hence, the sides of the bubbles must in general be curved; bubbles with high pressures are surrounded by convex sides, and ones with low pressure are surrounded by concave sides. The angles at which these sides meet must always be 120° by Plateau's law. This angle is the average interior angle for straight-sided hexagons. Inscribing a straight-sided polygon inside one with convex sides, the interior angles of the straight-sided inscribed polygon are less than those of the convex-sided polygon, and hence on average less than 120°. Convex polygons must therefore have fewer than six sides. This leads to the conclusion that high-pressure, convex bubbles have less than six sides, and low-pressure, concave ones have more than six sides. A (fictitious) foam composed only of hexagons would contain only straight sides, the bubble pressures would then all be equal, and no diffusive coarsening would occur! Real foams coarsen by diffusion of gas from convex bubbles to concave ones, leading to shrinkage of bubbles with fewer than six sides and growth of ones with more than six sides. It was shown by von Neumann (1952) that the rate of growth of the area A_n of a cell with n sides follows a beautifully simple law:

$$\frac{dA_n}{dt} = \kappa(n - 6) \tag{9-56}$$

where κ is a constant proportional to the surface tension and the diffusivity of gas in the liquid film.

This simple law has been the basis of computer simulations and analytic mean-field theories, which predict that the average area per polygon $\bar{A}$ at long times grows as $\bar{A} \propto t$. Hence, the bubble radius grows as $t^{1/2}$, in agreement with experiments (Smith 1952; Glazier and Stavans 1989; Glazier et al. 1990; Stavans 1993). Simulations also show that the distribution of bubble sizes becomes *self-similar* at long times (Anderson et al. 1984);

that is, the mean bubble size changes, but the distribution, normalized by the mean, does not. The distribution of bubble areas is roughly (but not exactly) an exponentially decreasing function of area (Flyvbjerg 1993). The number of polygon sides is distributed about six with a variance of around 1.4–1.5 (Glazier et al. 1990; Weaire and Lei 1990; Flyvbjerg 1993). Polygons with less than four or more than nine sides are rare ($\leq 1\%$).

9.4.2.2 Coarsening Theory for Three-Dimensional Foams

Unfortunately, for three-dimensional foams, no analog of von Neumann's law has been discovered, and theory of coarsening is much less developed. A review of the work on this problem can be found in Glazier and Weaire (1992).

9.5 RHEOLOGY OF FOAMS

Since gas–liquid foams are not stable, rheological characterization must be restricted to the time period after formation of the foam during which changes in foam structure are small. Alternatively, one can use dense liquid–liquid *emulsions* as foam analogs (Princen and Kiss 1986; Otsubo and Prud'homme 1994). Oil-in-water emulsions can be prepared at high volume fractions, $\phi \approx 0.90$ or higher; and since the diffusion coefficient of oil in through water is very small, these emulsions can be kept stable almost indefinitely. In addition, emulsions are almost incompressible under typical flow conditions, while for air–liquid foams, pressures must be kept uniform if complications due to volume changes are to be avoided. Of course, in emulsion droplets there can be viscous dissipation, which is absent in foam bubbles; but the low frequency *elastic* properties of dense emulsions, described in Section 9.3.4, should be qualitatively similar to those of foams.

Measurement of the viscoelastic properties of foams and dense emulsions is also complicated by slip at the rheometer surfaces (Yoshimura and Prud'homme 1988). The liquid in an aqueous foam lubricates flat rheometer fixtures, reducing the strain imposed on the bulk foam. This lubrication is a desirable feature of some foam products such as shaving cream, but it complicates rheological studies. The use of roughened surfaces, such as sandpaper bonded to the rheometer fixtures, seems to be an effective countermeasure (Khan et al. 1988).

Foams usually possess a finite low-frequency elastic modulus, along with static and dynamic yield stresses. These and other aspects of foam flow and rheology can be captured qualitatively and even semiquantitatively by cellular foam models.

9.5.1 Cellular Foam Models

The earliest analysis of the deformation properties of a liquid foam is that of Princen (1983), who modeled two-dimensional foams and dense emulsions at rest by an array of regular hexagons (see Fig. 9-32a). While Princen's model was limited to hexagons with a particular orientation relative to the imposed shearing flow, this restriction was lifted in the work of Khan and Armstrong (1986) and Kraynik and Hansen (1986). Polydisperse cell sizes have also been considered (Weaire et al. 1986; Khan and Armstrong 1987; Weaire and Fu 1988; Kraynik et al. 1991; Okuzono et al. 1993), as well as "wet" foams with

ϕ as low as ≈ 0.9 (Princen 1983; Khan and Armstrong 1987), and three-dimensional foams (Reinelt and Kraynik 1993; Reinelt 1993; Reinelt and Kraynik 1996). Both shear and extensional deformations have been analyzed (Khan and Armstrong 1986; Kraynik and Hansen 1986).

The deformation and stress in a foam are, in general, controlled by the liquid–air surface tension and possibly also by the viscous forces produced by flow of liquid within the liquid films and Plateau borders. The viscous forces are sensitive to shear rate; their importance relative to elastic forces from surface tension is controlled by the capillary number $\mathrm{Ca} \equiv \eta_s a \dot{\gamma} / \Gamma$. By assuming that the viscous dissipation occurs in the thin films separating bubbles, Khan and Armstrong (1987) estimated that at large ϕ viscous stresses should become important when $\mathrm{Ca} \gtrsim 0.01/(1 - \sqrt{\phi})$. For a typical case ($\eta_s = 10^{-2}\mathrm{P}$, $a = 100\ \mu\mathrm{m}$, $\Gamma = 20$ dyn/cm, and $\phi = 0.9$), this implies that viscous stresses can be neglected unless the shear rate exceeds around $10^4\ \mathrm{sec}^{-1}$. Schwartz and Princen (1987), on the other hand, assume that most of the viscous dissipation occurs in the Plateau borders; they find that viscous stresses may become significant at much lower shear rates than estimated by Khan and Armstrong. The magnitude and sources of viscous dissipation in steady and oscillatory flows of foams and dense emulsions are still not well understood, but are under intense investigation (Lequeux and Boltenhagen 1997; Liu et al. 1996), as will be discussed in Section 9.5.2.

Figure 9-32 shows the deformation of an idealized regular hexagonal dry foam in slow shearing flow. The shapes of the deformed cells for $\mathrm{Ca} \rightarrow 0$ are determined by requiring (for reasons of symmetry) an affine motion of the midpoints of the films, and in the deformed state the films meet at $120°$ angles at the Plateau borders. The cells deform according to these rules until some film shrinks to zero area (zero length in two dimensions). At this

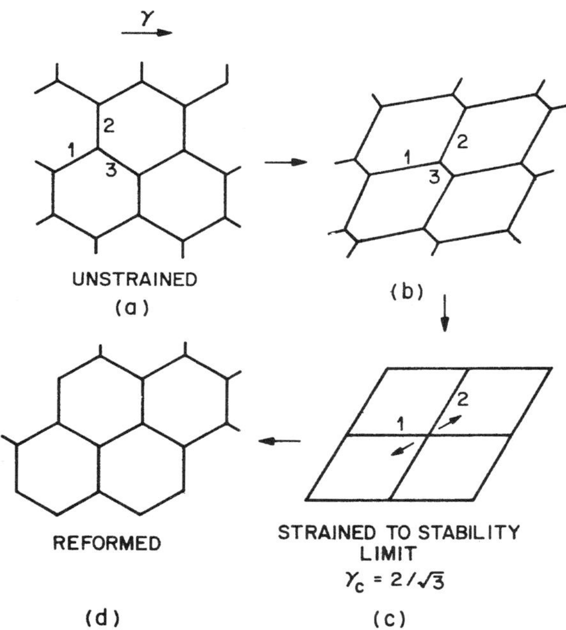

Figure 9.32 (**a–d**) Deformation of a hexagonal foam under shear. Between (**c**) and (**d**), a T1 reorganization occurs, which restores the initial hexagonal foam structure. (Reprinted from J Non-Newt Fluid Mech 22:1, Khan and Armstrong (1986), with kind permission from Elsevier Science - NL, Sara Burgerhartstraat 25, 1055 KV Amsterdam, The Netherlands.)

point the foam is unstable, and a T1 cellular rearrangement must occur. In Fig. 9-32c and 9-32d, a T1 process recreates hexagonal cells with the upper row of cells displaced to the right by one unit cell with respect to the starting configuration. With further straining, the cellular deformation then repeats itself. Different initial foam orientations lead to different deformation sequences. The strain at which the T1 process occurs is a sensitive function of the initial orientation (Khan and Armstrong 1986).

The stress tensor for a two-dimensional hexagonal foam can be computed either from a work-energy argument or by averaging the stresses in the microstructure over a representative volume (or unit cell) of the foam. Both approaches give the following for the shear stress σ_{12} and first normal stress difference N_1:

$$\sigma_{12} = C\frac{2\Gamma}{\sqrt{3}\,\ell}\frac{\gamma}{(\gamma^2 + 4)^{1/2}}, \qquad N_1 = C\frac{2\Gamma}{\sqrt{3}\,\ell}\frac{\gamma^2}{(\gamma^2 + 4)^{1/2}} = \gamma\sigma_{12} \qquad (9\text{-}57a)$$

where γ is the shear strain, Γ is the interfacial tension, ℓ is the length of the film in the hexagonal unit cell, and for this model $C = 1$. These stresses are independent of the initial cell orientation, but the strain at which the foam undergoes a T1 reorganization is not. For the regular three-dimensional tetrakaidecahedron foam model, Reinelt (1993) obtained expressions for σ_{12} and N_1 identical to Eq. (9-57a), except that the prefactor C is $C \approx 0.65$. He also obtained an expression for the second normal stress difference:

$$N_2 = C\frac{2\Gamma}{\sqrt{3}\,\ell}\left(\frac{2 + \gamma^2}{(\gamma^2 + 4)^{1/2}} - 2\right) \qquad (9\text{-}57b)$$

Doi and Ohta (1991) derived very similar expressions for σ_{12}, N_1, N_2 for the elastic stresses of emulsions in step shearing strains. Larson (1997) has shown that Eqs. (9-57a) and (9-57b) can be derived as an approximation from a general phenomenological "film" model for affinely stretching, constant-tension interfaces; thus σ_{12}, N_1, and N_2 can all be represented by the simple tensor expression

$$\boldsymbol{\sigma} = G_0\mathbf{C}^{1/2} \qquad (9\text{-}57\text{-}c)$$

where G_0 is a modulus proportional to the ratio of surface tension to cell or domain size, and $\mathbf{C}$ is the Cauchy tensor, which is the inverse of the Finger tensor $\mathbf{B}$ [see Chapter 1, Eq. (1-16)]. Taking the ratio N_2/N_1 from Eqs. (9-57a) and (9-57b), one finds that the second normal stress difference should be negative and almost as large in magnitude as the first normal stress difference. No experimental test of this prediction for dense emulsions and foams has yet been made.

Figure 9-33a shows the predicted shear stress as a function of strain for the initial foam orientation depicted in Fig. 9-32. The stress grows continuously until at $\gamma = 1.15$ a T1 reorganization occurs which brings the cell structure back to its starting state, and the stress jumps back to zero. Thereafter, the stress history repeats itself. Similar periodic stress patterns and stress jumps have been predicted for the three-dimensional tetrakaidecahedron foam model (Reinelt 1993). If the initial orientation is rotated through an angle of $\pi/12$ with respect to that shown in Fig. 9-32, the stress history also has jumps, but is *aperiodic* (see Fig. 9-33b). Aperiodic behavior is the norm, and periodic stress histories occur only for special initial orientations (Kraynik and Hansen 1986). These unsteady, discontinuous stress

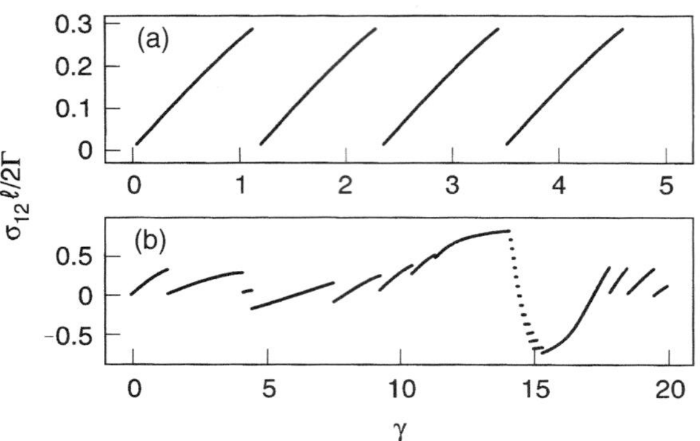

Figure 9.33 Dimensionless shear stress $\sigma_{12}\ell/2\Gamma$ as a function of strain γ at zero capillary number for a hexagonal model foam, with initial orientation: **(a)** as shown in Fig. 9-32, and **(b)** rotated by $\pi/12$. Here Γ is the surface tension and ℓ is the length of a side of the hexagonal foam. (From Kraynik and Hansen 1986, with permission from the Journal of Rheology.

histories are not seen in shearing of real foams, nor are they observed in simulations using irregular foam models (Weaire and Fu 1988; Okuzono et al. 1993), because irregularities average out the stress fluctuations.

Nevertheless, some of the predictions of simple regular foam models are relevant to real foams. One such property is the *linear modulus G_0*, which is the slope of the stress–strain curve at zero strain. From Eq. (9-57a), we obtain

$$G_0 = \frac{\Gamma}{\sqrt{3}\,\ell} = 0.58\frac{\Gamma}{\ell} \qquad \text{(2-D regular hexagonal foam)} \qquad (9\text{-}58)$$

This prediction is analogous to the empirical expression for dense emulsions, Eq. (9-49), in the limit $\phi \to 1$. If the foam is made irregular, with the average area per foam cell held constant, the modulus is predicted to be less than that of Eq. (9-58) by about 14% (Weaire and Fu 1988). For the regular tetrakaidecahedron, Reinelt averaged G_0 over all initial foam orientations and obtained

$$G_0 = 0.52\frac{\Gamma}{a_{32}} \qquad \text{(3-D regular tetrakaidecahedron foam)} \qquad (9\text{-}59)$$

where a_{32} is the surface-to-volume mean bubble radius, which is related to the cell edge length ℓ by $8\sqrt{2}\,\ell^3 = (4/3)\pi a_{32}^3$. For the Kelvin foam, with curved edges and faces, computer simulations with a "surface evolver" give a similar result for the modulus averaged over all orientations, $G_0 = 0.50\Gamma/a_{32}$ (Reinelt and Kraynik 1996). The predictions for two- and three-dimensional foams, Eqs. (9-58) and (9-59), are not too far from each other, since the edge length ℓ for the hexagonal foam can be considered to be roughly the "radius" of a hexagonal bubble. Princen and Kiss (1986) obtained an experimental value close to these estimates, $G_0 \approx 0.509\Gamma/a_{32}$, for dense oil-in-water emulsions when the droplet concentration ϕ was extrapolated to unity. From the data of Khan et al. (1988) for a nearly dry foam ($\phi = 0.97$) with a reported bubble radius $a = 32\mu$m and surface tension $\Gamma = 23$ dyn/cm, one can obtain the result $G_0 \approx 0.15\Gamma/a$, which is about a factor of three to four below the predicted values. Thus, simple cellular foam models give predictions for the modulus of foams that are qualitatively, but not always quantitatively, accurate.

In the dynamic oscillatory data of Khan et al. (1988), G' is a factor of five or so larger than G'' over the measured frequency range. Thus, the dissipation for these foams at small strains is low, as expected. The modulus was found to be a weak function of bubble volume fraction, as expected from the regular cellular foam models (Princen 1983).

The steady-state viscosity of the foam studied by Khan et al. (Fig. 9-34) is very shear thinning, suggesting the existence of a *dynamic yield stress* σ_y. The viscous dissipation in a foam with a dynamic yield stress occurs mainly during the rapid bubble rearrangement events, similar to the dissipation in an electrorheological fluid at low shear rates; for a discussion of dynamic yield stress, see Section 8.2.2.3. In contrast to the modulus, the yield stress increases rapidly with increasing ϕ, in agreement with the predictions of Princen (1983). The differing ϕ dependencies of G_∞ and σ_y imply that the *yield strain* decreases rapidly as ϕ decreases from unity. This occurs because with decreasing ϕ, the Plateau borders grow larger at the expense of the films (compare Figs. 9-29a, left and right). Apparent yield occurs when the area of some fraction of the films shrinks to zero under deformation, which occurs quickly when the plateau borders are small to begin with. The value of the yield stress at high ϕ is about a factor of six lower than that predicted by the hexagonal cell model. Kraynik and coworkers have produced provocative videos of computer simulations of two-dimensional foams under shear and extensional flows. These videos show type T1 events, as well as "shear banding" phenomena that provide insight into the rheological behavior of real foams, and suggest parallels between the flow behavior of foams and that of other complex materials.

9.5.2 Other Models

Durian (1995) has derived a very simple, yet highly instructive, model for relatively wet foams. The model focuses on the bubbles and ignores the detailed shape of the Plateau borders, by assuming that the bubbles press against each other with a force that increases

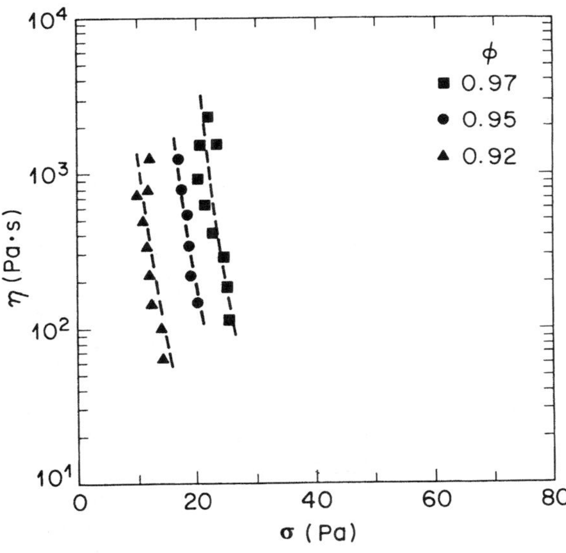

Figure 9.34 Viscosity as a function of shear stress for an aqueous polymer-surfactant foam at the bubble volume fractions ϕ shown. (From Khan et al. 1988, with permission from the Journal of Rheology.)

linearly with the bubble deflection. The bubble deflection is just $1 - r_{ij}/(a_i + a_j)$, where r_{ij} is the distance separating the centers of two adjacent bubbles i and j, and a_i and a_j are the undisturbed radii of those bubbles. This force is zero for separations greater than that at which the undistorted bubbles just touch. Adding together all such forces acting on a bubble, along with a simple Stokes' drag law, produces a set of simple equations for the motion of these bubbles in flow field. For a polydisperse foam, the model predicts a yield stress and a slow-flow regime dominated by intermittent "avalanches," or cascades, of bubble-rearrangement events. Intermittent reorganizations seem to be a hallmark of the flow of disordered solids in general, including foams, glasses, and electrorheological fluids.

Liu et al. (1996) have reported high levels of viscous dissipation in dense emulsions under high-frequency oscillatory shearing deformations. Their data for the complex modulus G^* can be represented empirically by the expression

$$G^* = G' + iG'' = G_p + A(\phi)(i\omega)^{1/2} + \eta_\infty i\omega \tag{9-60}$$

where i is the imaginary unit. This expression contains a low-frequency plateau modulus, G_p, a high-frequency viscosity η_∞ due to the continuous phase in the films, and an anomalous *power-law* contribution $G(i\omega) \propto (i\omega)^{1/2}$ at intermediate frequencies. According to the theory of Liu et al. (1996), the anomalous power-law contribution is caused by slippage of layers of bubbles against each other along randomly oriented "weak planes" where the modulus at that orientation is locally minimum. Averaging over all orientations of the weak planes, a power law $G^* \propto \omega^{1/2}$ is the result. The model of Liu et al. is inspired by the theory of Kawasaki and Onuki (1990) for the low-frequency $\omega^{1/2}$ frequency dependence of G^* for unoriented lamellar block copolymers; this is explained in Section 13.3.1. Thus, there might be a relationship between the rheology of two seemingly disparate complex fluids, namely, lamellar block copolymers and liquid–gas foams!

9.6 SUMMARY

The structural and rheological properties of emulsions, blends, and foams are of great importance in the food, cosmetics, oil-field, and packaging industries. By definition, such fluids are thermodynamically unstable or at best metastable. Hence, conditions of preparation are of extreme importance to both the scientific study and the engineering of these fluids.

Successful theories now exist for the flow and deformation properties of emulsions dilute enough that droplet–droplet interactions are negligible. If the droplet phase is more viscous than the continuous phase, Newtonian droplets deform only slightly before breaking into two or more fragments as the shear or extension rate is increased; for the reverse viscosity ratio, droplets or bubbles become greatly extended before breaking. Under steady-state shearing, the shear stress at which Newtonian droplets break is lowest if their viscosity is equal to that of the matrix fluid. Viscoelastic droplets resist fragmentation, and they are more readily drawn into fine filaments. Because the less viscous phase stretches to a greater extent without breakup than does the more viscous one, the volume-fraction ratios at which emulsions or immiscible blends of Newtonian liquids are co-continuous are shifted toward a high volume fraction of the more viscous phase. Viscoelastic effects can shift this behavior to favor continuity of the more viscoelastic phase even when its volume fraction is low.

In small-amplitude oscillatory shearing, the linear viscoelasticity of emulsions or blends of Newtonian liquids is described by the theories of Oldroyd and of Choi and Schowalter. An extension of Oldroyd's theory to mixtures of viscoelastic liquids by Palierne has been shown to agree well with experimental data. The G' data show a low-frequency "tail" controlled by the average size, interfacial tension, and volume fraction of the droplets. Under steady shearing, a phenomenological theory of Doi–Chta theory accounts for deformation of fluid–fluid interfaces, as well as for the contribution of these interfaces to the elastic stress. The magnitude of the first normal stress difference and the strain-scaling of the transient shear stress after start-up of steady shearing are qualitatively predicted by the Doi–Ohta theory.

Emulsions with a high volume fraction of droplets ($\phi > 0.64$) and foams show solid-like properties such as a yield stress and a low-frequency plateau value of G'. The magnitudes of the yield stress and elastic modulus can be estimated using simple cellular foam models. These and related models show that at low shear rates where the shear stress is close to the yield value, the "flow" occurs by way of intermittent bubble-reorganization events. The dissipative processes that occur during foam and emulsion flows are still under active investigation.

REFERENCES

Aaronson HI, Le Gouef FK (1992). *Metall Trans A* 23:1915.

Acrivos A, Lo TS (1978). *J Fluid Mech* 86:641.

Akcasu AZ, Erman B (1992). *Makromol Chem Macromol Symp* 62:43.

Anderson MP, Srolovitz DJ, Grest GS, Sahni PS (1984). *Acta Metall* 32:783.

Ashby MF, Verrall RA (1973). *Acta Metall* 21:149.

Aubert JH, Kraynik AM, Rand PB (1986). *Sci Am* 254:74.

Barthes-Biesel D, Acrivos A (1973). *In J Multiphase Flow* 1:1.

Bates FS, Wiltzius P (1989). *J Chem Phys* 91:3258.

Beck Tan NC, Tai S-K, Briber RM (1996). *Polymer*, 37:3509.

Bentley BJ, Leal LG (1986). *J Fluid Mech* 167:241.

Bibette J (1991). *J Colloid Interface Sci* 147:474.

Binder K (1983). *J Chem Phys* 79:6387.

Bousfield DW, Keunings R, Marrucci G, Denn MM (1986). *J Non-Newt Fluid Mech* 21:79.

Briber RM, Han CC, Peiffer DG (1996). *Morphological Control in Multiphase Polymer Mixtures*, MRS Symposium Proceedings 461.

Bunning TJ, Natarajan LV, Tondiglia VP, Sutherland RL, Vezie DL, Adams WW (1996). *Polymer* 37:3147.

Cahn JW (1965). *J Chem Phys* 42:93.

Cahn JW (1968). *Trans Am Inst Min Eng* 242:166.

Chesters AK (1991). *Trans I Chem E* 69A:259.

Choi JC, Schowalter WR (1975). *Phys Fluids* 18:420.

Clarke N, McLeish TCB, Pavawongsak S, Higgins JS (1997). *Macromolecules* 30:4459.

Collyer AA, Clegg DW (1986). *High Perform Plast* 4:1.

de Bruijn RA (1989) PhD Thesis, Eindhoven University of Technology, Eindhoven, The Netherlands.

de Gennes P-G (1980). *J Chem Phys* 72:4756.

Delaby I, Ernst B, Germain Y, Muller R (1994). *J Rheol* 38:1705.

Doane JW (1991). *MRS Bull* Jan:22.

Doi M, Ohta T (1991). *J Chem Phys* 95:1242.

Durian DJ (1994). *MRS Bull*, Apr.20.

Durian DJ (1995). *Phys Rev Lett* 75:4780.

Durian DJ, Weitz DA, Pine DJ (1991a). *Science* 252:686.

Durian DJ, Weitz DA, Pine DJ (1991b). *Phys Rev A* 44:R7902.

Elemans EHM, Box HL, Janssen JMH, Meijer HEH (1993). *Chem Eng Sci* 48:267.

Elmendorp JJ, van der Vegt AK (1986). *Polym Eng Sci* 26:1332.

Elmendorp JJ, van der Vegt AK (1991). In *Progress in Polymer Processing*, Vol 2, Utracki LA (ed), Hanser Publishers, Munich.

Favis BD, Chalifoux JP (1987). *Polym Eng Sci* 27:1591.

Flyvbjerg H (1993). *Physica A* 194:298.

Glazier JA, Stavans J (1989). *Phys Rev A* 40:7398.

Glazier JA, Weaire D (1992). *J Phys Condens Mater* 4:1867.

Glazier JA, Anderson MP, Grest GS (1990). *Philos Mag B* 62:615.

Glotzer SC, Di Marzio EA, Muthukumar M (1995). *Phys Rev Lett* 74:2034.

Goldin M, Yerushalmi H, Pfeffer R, Shinnar R (1969). *J Fluid Mech* 38:689.

Grace HP (1982). *Chem Eng Commun* 14:225.

Graebling D, Muller R, Palierne JF (1993). *Macromolecules* 26:320.

Gramespacher H, Meissner J (1992). *J Rheol* 36:1127.

Grizzuti N, Bifulco O (1997). *Rheol Acta* 36:406.

Guenoun P, Gastaud R, Perrot F, Beysens D (1987). *Phys Rev A* 36:4876.

Guenther GK, Baird DG (1996). *J Rheol* 40:1.

Gururaj M, Kumar R, Gandhi KS (1995). *Langmuir* 11:1381.

Haas PA, Johnson HF (1967). *Ind Eng Chem Fundam* 37:1361.

Han CD (1981). *Multiphase Flow in Polymer Processing*, Academic Press, New York.

Hashimoto T, Itakura M, Shimidzu N (1986). *J Chem Phys* 85:6773.

Helfand E, Sapse AM (1975). *J Chem Phys* 62:1327.

Hilsum C (1976). UK Patent 1,442,360.

Hinch EJ, Acrivos A (1980). *J Fluid Mech* 98:305.

Hoyt JW, Taylor JJ (1977). *Phys Fluids* 20:S253.

Hunter RJ (1992). *Foundations of Colloid Science*, Vols I and II, Oxford University Press, New York.

Israelachvili J (1991). *Intermolecular and Surface Forces*, 2nd ed, Academic Press, London.

Janssen JMH (1993). Thesis, University of Eindhoven.

Jinnai H, Hasegawa H, Hashimoto T, Han CC (1993a). *J Chem Phys* 99:4845.

Jinnai H, Hasegawa H, Hashimoto T, Han CC (1993b). *J Chem Phys* 99:8154.

Joanny JF, Leibler L (1978). *J Phys (Paris)* 39:951.

Jordhamo GM, Manson JA, Sperling LH (1986). *Polym Eng Sci* 26:517.

Kawasaki K, Onuki A (1990). *Phys Rev A* 42:3664.

Kelvin, Lord (1887). *Philos Mag* 24:503.

Khan SA, Armstrong RC (1986). *J Non-Newt Fluid Mech* 22:1.

Khan SA, Armstrong RC (1987). *J Non-Newt Fluid Mech* 25:61.

Khan SA, Schnepper CA, Armstrong RC (1988). *J Rheol* 32:69.

Kim BS, Jang SH (1995). *Polym Eng Sci* 32:773.

Kitade S, Ichikawa A, Imura N, Takahashi Y, Noda I (1997). *J Rheol* 41:1039.

Kraynik AM, Hansen MG, (1986). *J Rheol* 30:409.

Kraynik AM, Reinelt DA, Princen HM (1991). *J Rheol* 35:1235.

LaMantia FF, Valenza A, Paci M, Magagnini PL (1990). *Polym Eng Sci* 30:7.

Larson RG (1992a). *Rheol Acta* 31:213.

Larson RG (1992b). *Rheol Acta* 31:497.

Larson RG (1997). *J Rheol* 41:365.

Leibler L (1982). *Macromolecules* 15:1283.

Lequeux F, Boltenhagen P (1997). In *Theoretical Challenges in the Dynamics of Complex Fluids*, McLeish TCB (ed), NATO ASI Series E: Applied Sciences, Vol 339, Kluwer Academic Publishers, London.

Lifshitz IM, Slyozov VV (1961). *J Phys Chem Solids* 19:35.

Lin YG, Winter HH (1992). *Polym Eng Sci* 32:773.

Liu AJ, Ramaswamy S, Mason TG, Gang H, Weitz DA (1996). *Phys Rev Lett* 76:3017.

Mason TG, Bibette J (1996). *Phys Rev Lett* 77:3481.

Mason TG, Bibette J, Weitz DA (1995). *Phys Rev Lett* 75:2051.

Matzke E, Nestler J (1946). *Am J Bot* 33:130.

Mehta A, Isayev AI (1991). *Polym Eng Sci* 31:971.

Mighri F, Ajji A, Carreau PJ (1997). *J Rheol* 41:1183.

Miles IS, Zurek A (1988). *Polym Eng Sci* 28:796.

Milner ST, Xi H (1996). *J Rheol* 40:663.

Minale M, Moldenaers P, Mewis J (1997). *Macromolecules* 30:5470.

Mooney M (1951). *J Colloid Sci* 6:162.

Nawab MA, Mason SG (1958). *J Colloid Sci* 13:179.

Okuzono T, Kawasaki K, Nagai T (1993). *J Rheol* 37:571.

Oldroyd JG (1953). *Proc R Soc A* 218:122.

Oldroyd JG (1955). *Proc R Soc A* 232:567.

Onuki A (1994). *Europhys Lett* 28:175.

Onuki A, Taniguchi T (1997). *J Chem Phys* 106:5761.

Ostwald W (1901). *Z Phys Chem* 37:385.

Otsubo Y, Prud'homme RK (1994). *Rheol Acta* 33:29.

Pal F (1992). *J Rheol* 36:1245.

Palierne JF (1990). *Rheol Acta* 29:204.

Paul DR, Barlow JW (1980). *J Macromol Sci Rev Macromol Chem* C18:109.

Pincus P (1981). *J Chem Phys* 75:1996.

Plochocki AP, Dagli SS, Andrews RD (1990). *Polym Eng Sci* 30:741.

Pons R, Erra P, Solans C, Ravey J-C, Stebe M-J (1993). *J Phys Chem* 97:12320.

Princen HM, (1983). *J Colloid Interface Sci* 91:160.

Princen HM, Kiss AD (1986). *J Colloid Interface Sci* 112:427.

Princen HM, Kiss AD (1989). *J Colloid Interface Sci* 128:176.

Rallison JM (1980). *J Fluid Mech* 98:625.

Rallison JM (1981). *J Fluid Mech* 109:465.

Rallison JM (1984). *Annu Rev Fluid Mech* 16:45.

Rallison JM, Acrivos A (1978). *J Fluid Mech* 89:191.

Rangel-Nafaile C, Metzner AB, Wissbrun KF (1984). *Macromolecules* 17:1187.

Reinelt DA (1993). *J Rheol* 37:1117.

Reinelt DA, Kraynik AM (1993). *J Colloid Interface Sci* 159:460.

Reinelt DA, Kraynik AM (1996). *J Fluid Mech* 311:327.

Ross S, Prest H (1986). *Colloid Interface* 21:179.

Rumscheidt FD, Mason SG (1961). *J Colloid Interface Sci* 16:238.

San Miguel M, Grant M, Gunton JD (1985). *Phys Rev A* 31:1001.

Schwartz LW, Princen HM (1987). *J Colloid Interface Sci* 118:201.

Siggia ED (1979). *Phys Rev A* 20:595.

Smith CS (1952). In *Metal Interfaces*, American Society for Metals, Cleveland, p 65.

Sondergaard K, Lyngaae-Jorgensen J (1995). In *Rheo-Physics of Multiphase Polymer Systems*, Technomic Publishing Co, Lancaster, PA.

Stavans J (1993). *Physica A* 194:307.

Subramanian PM (1983). US Patent 4,410,482.

Sundararaj U, Macosko CW (1995). *Macromolecules* 28:2647.

Takahashi Y, Noda I (1995). In *Flow Induced Structure in Polymers*, Nakatani AI, Dadmun MD (eds), ACS Symposium Series 597, American Chemical Society, Washington, DC.

Takahashi Y, Kurashima N, Noda I, Doi M (1994). *J Rheol* 38:699.

Tanaka H (1992). *Macromolecules* 25:6377.

Tanaka H (1993). *Phys Rev Lett* 71:3158.

Tanaka H (1994a). *J Chem Phys* 100:5323.

Tanaka H (1994b). *Phys Rev Lett* 72:3690.

Tanaka H (1995). *Phys Rev E* 51:1313.

Taniguchi T, Onuki A (1996). *Phys Rev Lett* 77:4910.

Taylor GI (1932). *Proc R Soc Lond A* 138:41.

Taylor GI (1934). *Proc R Soc Lond A* 146:501.

Taylor GI (1964). In *Proceedings of the International Congress on Applied Mechanics*, Munich, p 790.

Tokita N (1979). *Chem Technol* 50:292.

Torza S, Cox RG, Mason SG (1972). *J Colloid Interface Sci* 38:395.

Utracki LA (1990). *Polymer Alloys and Blends, Thermodynamics and Rheology*, Hanser Publishers, Munich.

Vicsek T (1989). *Fractal Growth Phenomena*, World Scientific, Singapore.

Vinckier I (1998). PhD Thesis, Katholieke Universiteit, Leuven, Belgium.

Vinckier I, Moldenaers P, Mewis J (1996). *J Rheol* 40:613.

Vinckier I, Moldenaers P, Mewis J (1997). *J Rheol* 41:705.

von Neumann J (1952). In *Metal Interfaces*, Herring C (ed), American Society for Metals, Cleveland.

Voorhees PW (1985). *J Stat Phys* 38:231.

Weaire D, Fu T-L (1988). *J Rheol* 32:271.

Weaire D, Lei H (1990). *Philos Mag Lett* 62:427.

Weaire D, Rivier N (1984). *Contemp Phys* 25:59.

Weaire D, Fu T-L, Kermode JP (1986). *Philos Mag* 54:L39.

West JL (1990). In *Liquid Crystal Polymers*, ACS Symposium Series 435, American Chemical Society, Washington, DC.

White WR, Wiltzius P (1995). *Phys Rev Lett* 75:3012.

Williams RE (1968). *Science* 161:276.

Wu S (1982). *Polymer Interface and Adhesion*, Marcel Dekker, New York.

Wu S (1985). *Polymer* 26:1855.

Wu S (1987). *Polym Eng Sci* 27:335.

Yoshimura AS, Prud'homme RK (1988). *J Rheol* 32:53.

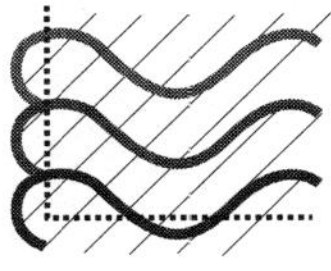

PART IV

LIQUID CRYSTALS AND SELF-ASSEMBLING FLUIDS

Chapter 10

LIQUID CRYSTALS

10.1 INTRODUCTION

In 1888, Friedrich Reinitzer, an Austrian botanist, was investigating the melting transition of an organic compound related to cholesterol, when he saw something very curious (Reinitzer 1888). The solid melted into a cloudy liquid at 145.5°C, but then "melted" again into a clear liquid at 178.5°C! Reinitzer described his observation in a letter to Otto Lehman, a German physicist, who confirmed it using a polarizing microscope to which he had attached a hot stage (Lehmann 1890). Lehmann found other substances with similar behavior, and he and Reinitzer recognized that the cloudy liquid, which was also birefringent, was a new state of matter that Lehmann eventually called a *liquid crystal*. Liquid crystals were so beautiful to look at in a polarizing microscope, and so intriguing in their properties, that although for 100 years they had no practical uses, they were studied intently by several determined scientists, who learned many of their secrets. Since the 1960s, the knowledge they extracted has become very profitable: Liquid crystals are now used in almost all lap-top computers and in many other applications.

Liquid crystals derive both their beauty and their usefulness from their ordered, yet liquid, state. At rest, they possess at least some *orientational order* at rest, but lack the full three-dimensional *positional order* of solid crystals. Orientational order implies that the molecules tend to point in the same direction, while positional order implies that the molecules' centers of mass tend to lie on lattice points. Spherically symmetric molecules, for example, can have positional order, but no orientational order. Thus, when one heats a simple positionally ordered crystalline solid to its melting point, all of its positional order is destroyed in a single step, producing an isotropic liquid. Orientational order is, however, possible for rod-like or disk-like molecules. Crystalline solids made of such molecules can melt into isotropic liquids in multiple steps; at intermediate stages of melting, there can be orientational order with no, or only partial, positional order. That is, the molecules might tend to be aligned in a common direction, but the centers of mass of the molecules take on random positions. Such intermediate phases are liquid crystals.

Molecules that form liquid crystalline phases are often oblate or prolate in shape, with aspect ratios of three or more, and have some rigidity, conferred, for example, by rigid biphenyl or terphenyl groups. Figure 10-1 shows some examples of liquid-crystal formers, namely, N-(p-methoxybenzylidene)-p-butylaniline (MBBA), p-n-hexyloxybenzilidene-p'-aminobenzonitrile (HBAB), and the 4-cyano-4′alkylbiphenyl series, or nCB, where n is the number of carbons (typically 5–12) in the flexible "tail" of the molecule. For example,

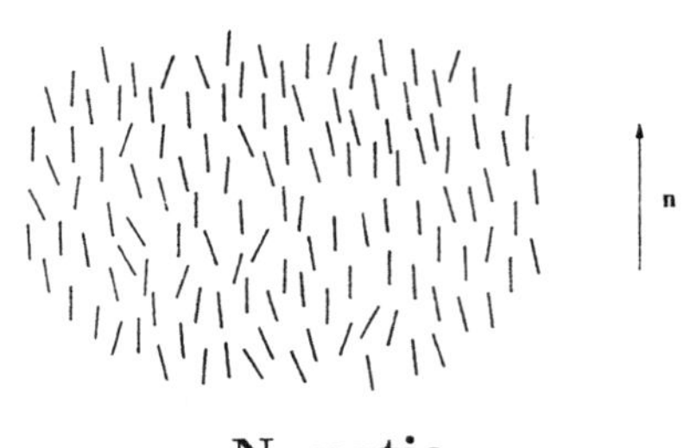

Figure 10.1   Molecules that form nematic and smectic liquid crystals: N-(p-methoxybenzylidene)-p-butylaniline (MBBA), p-n-hexyloxy-benzilidene-p'-aminobenzonitrile (HBAB), and 4-cyano-4'n-biphenyl (nCB), where n is the number of carbons in the flexible "tail." (From Mather 1994, with permission).

5CB is 4-cyano-4'-pentylbiphenyl. Liquid-crystal phases can also form when amphiphilic molecules *associate* to form anisotropic structures which, in turn, spontaneously order into liquid-crystalline phases.

Two of the most basic liquid-crystalline symmetries are depicted in Fig. 10-2; they are *nematic* and *smectic A*. Nematics possess orientational, but no positional, order. The direction of preferred orientation is designated by a unit vector **n** called the "director" (see Section 2.2.2.2). The polarity, or sign, of the vector **n** is of no significance: the "head" of the director is indistinguishable from the "tail," and orientations that are 180° apart are equivalent. Smectics possess both orientational and *one-dimensional*, or layer-like, positional order. All three molecules in Fig. 10-1 form nematics; some of the molecules in the *n*CB series, such as 8CB, also form smectics. There is also the possibility of *columnar* phases that possess *two-dimensional* positional order, as well as orientational order. An example of a columnar phase is the *hexagonal phase* mentioned in Section 2.2.2; it consists of aligned rod-like molecules packed hexagonally in the plane orthogonal to the rod axes. States with three-dimensional positional order are, by definition, fully crystalline. On heating, many liquid-crystal-forming materials pass from the crystalline state, through one or more smectic states, and then to a nematic state, before becoming ordinary isotropic liquids. There are also materials, called *plastic crystals*, which, on heating, lose some orientational order while preserving full three-dimensional positional order.

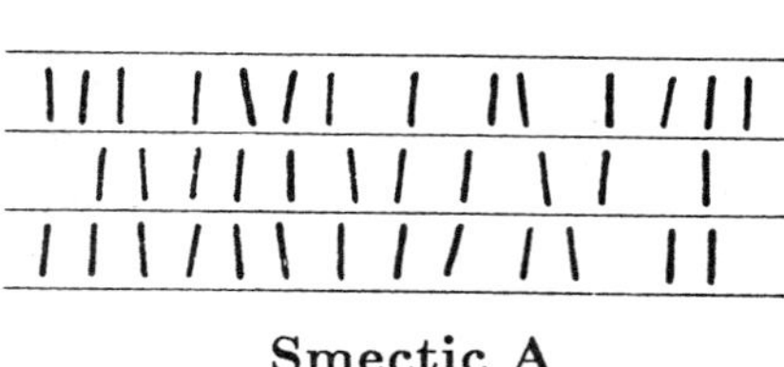

Figure 10.2.   Molecular arrangements in nematic and smectic-A liquid-crystalline phases. (From de Gennes and Prost, Copyright © 1993, by Oxford University Press, Inc. Used by permission of Oxford University Press, Inc.)

Molecules that contain a chiral center can form *chiral* liquid crystalline phases, where the orientation direction rotates in a helical fashion as one moves along the helical axis, which is perpendicular to the locally preferred direction of orientation. Both nematic and smectic phases can be chiral. In a chiral nematic phase, also known as a *cholesteric*, as one moves along the helical axis, the director rotates sinusoidally (see Fig. 10-3). Thus, if z is the helical axis, we obtain

$$n_x = \cos(q_0 z + \phi), \qquad n_y = \sin(q_0 z + \phi), \qquad n_z = 0 \qquad (10\text{-}1)$$

where the chiral *pitch* P is $2\pi/q_0$, or twice the periodicity of the director winding. Often the pitch is quite temperature-sensitive, and in some cholesterics, q_0 even changes sign, indicating a thermally induced transition from right- to left-handed chirality (de Gennes and Prost 1993).

Liquid crystals are finding a variety of important applications. Nematics are commonly used in liquid-crystal *display* devices, such as lap-top computers (Bahudur 1990; Collings 1990). Most, if not all, nematics liquid used in commercial liquid-crystal displays are *mixtures* of several components. The transition to the crystalline state is suppressed in mixtures, which therefore can have a much wider range of temperatures between the isotropic and crystalline states over which the mixture is nematic. Pure components often have a nematic range only 10–25°C wide, which is too narrow for the commercial display market. An early commercial mix, called "E7," is composed of (by weight) 51% 5CB, 25% 7CB, 16% 8OCB, and 8% 5CT, where 8OCB is 4-cyano-4′-octyloxybiphenyl, and 5CT

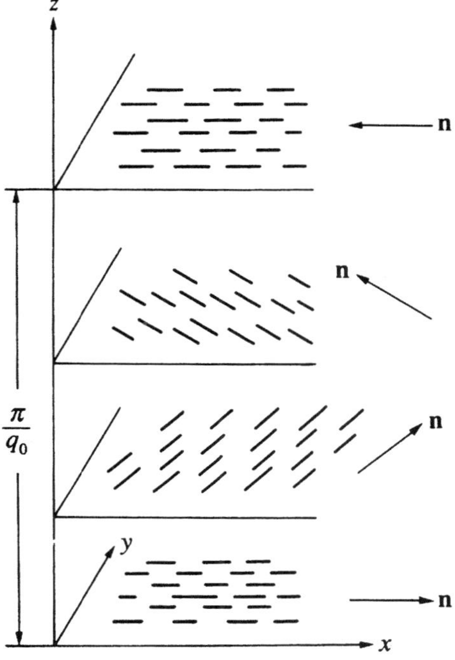

Figure 10.3. The structure of a cholesteric, or twisted nematic, phase. The director rotates through 180° in a distance given by π/q_0, which is half the helical pitch. (From de Gennes and Prost, Copyright © 1993, by Oxford University Press, Inc. Used by permission of Oxford University Press, Inc.)

is 4-cyano-4″-pentyl-*p*-terphenyl. More advanced displays are also under development, using, for example, ferroelectric liquid crystals (which are doped smectic-C liquid crystals), sometimes stabilized by a cross-linked polymer matrix within the liquid crystal (Walba 1995). Cholesterics also have some advantages in display applications (Patel and Lee 1989; Yang et al. 1994; St. John et al. 1995).

Cholesterics are also sometimes also used as temperature indicators, since their pitch is often in the range of visible wavelengths and is strongly temperature-dependent, so that their reflective color can change with temperature when viewed under white light. Cholesteric coatings can thus be used to reveal hot and cold spots, for example, on a human body, thereby assisting medical diagnoses (Maier et al. 1975; Buka 1993). Sometimes high-temperature cholesterics are quenched (cooled rapidly) into a room-temperature solid that retains the optical properties of the cholesteric liquid; the solid can then be fractured into flakes and suspended in paints, producing angle-dependent iridescent colors (Luckhurst and Veracini 1994). Nature discovered the colorful properties of cholesterics under reflected white light long before humans did, and uses them in making the brilliant shells or "cuticles" of beetles (Neville 1975).

Smectics have been explored as possible lubricants, because they can, in principle, slide readily along surfaces (Fuller et al. 1995). Some smectics are formed from soap molecules mixed with water or other solvent. In fact, the name "smectic" is derived from the Greek word for soap; a household example is the gelatinous accretion at the bottom of a soap dish.

Liquid crystals made from soaps or stiff molecules mixed with solvents are called *lyotropic* liquid crystals; their liquid crystalline transitions are driven primarily by changes in concentration. Simulations of hard spherocylinders (cylinders with spherical caps) in an athermal solvent show that transitions from isotropic to nematic to smectic-A phases can be driven by *entropy alone* (Stroobants et al. 1986). Liquid crystals that lack solvent are called *thermotropic*; their phase transitions are driven by temperature changes only. This chapter focuses on small-molecule thermotropic liquid crystals. Surfactant-containing lyotropic liquid crystals are discussed in Chapter 12. Polymeric liquid crystals are discussed in Chapter 11. The interested reader can find more detail on liquid crystals in (a) an introductory text by Collings (1990), (b) a book on phase structure and transitions by Gray and Goodby (1984), and (c) books on physics by Chandrasekhar (1992) and de Gennes and Prost (1993). These volumes also contain spectacularly beautiful photographs of liquid-crystalline microstructures.

10.2 NEMATICS

In the simplest liquid-crystalline phase, namely the *uniaxial nematic,* there is at rest a special direction designated by a unit vector **n** called the director (see Fig. 10-2). In the plane transverse to the director, the fluid is isotropic. The most common nematics are composed of oblong molecules that tend to point in a common direction, which defines the director orientation. Oblate, or disc-like, molecules can also form uniaxial nematics; for these *discotic* nematics, the director is defined by the average orientation of the short axis of the molecule. Lath-like molecules or micelles (shaped like rectangular slabs), in which all three dimensions of the molecule are significantly different from each other, can form *biaxial* nematics (Praefcke et al. 1991; Chandrasekhar 1992; Fiałtkowski 1997). A biaxial nematic must be described by two directors that are mutually perpendicular; the longest axis of the molecule tends to lie parallel to one of the directors, while the second-longest axis is parallel to the other director. The constitutive equation for a biaxial nematic is complex; it has 12 distinct viscosity coefficients and 12 elastic constants (Leslie 1994). Biaxial nematics have been observed in three-component surfactant-containing systems that self-assemble into either prolate or oblate micelles depending on temperature and concentration. Prolate micelles form ordinary uniaxial nematics with the director parallel to the major axis of the micelles; when the micelles are oblate, a uniaxial discotic nematic is obtained with the director parallel to the minor axis. At intermediate conditions, a biaxial nematic can occur (Yu and Saupe 1980). Unless the modifier "biaxial" is used, the term "nematic" usually refers to a uniaxial nematic.

The degree of orientational order in a uniaxial nematic is given by the *order parameter* S, defined by Eq. (2-3). S is zero in the isotropic state, and it approaches unity for hypothetically perfect molecular alignment (i.e., all molecules pointing in the same direction). In single-component small-molecule nematics, such as MBBA, S varies with temperature from $S \approx 0.3$ at T_{NI}, the nematic–isotropic transition temperature, to $S \approx 0.7$ or so at lower temperatures near the crystallization point. Figure 10-4 shows the order parameter measured for MBBA. This variation of S with T, which appears to be typical of small-molecule nematics, is in surprisingly good agreement with the prediction of the Maier–Saupe theory. The Maier–Saupe theory is discussed in Section 2.2.2.2; it contains a potential, Eq. (2-7), with a temperature-independent value of the Maier–Saupe parameter u_{MS}, the value of which is set by the requirement that $u_{\mathrm{MS}}/k_B T_{\mathrm{NI}} = 4.55$. A temperature-independent value of u_{MS} is consistent with the original assumption of Maier and Saupe that the nematic interaction is due entirely to anisotropic van der Waals interactions, but conflicts with enthalpy measurements near the isotropic–nematic transition (de Gennes and Prost 1993). Also, when the order parameter is measured as a function of temperature at *constant volume*,

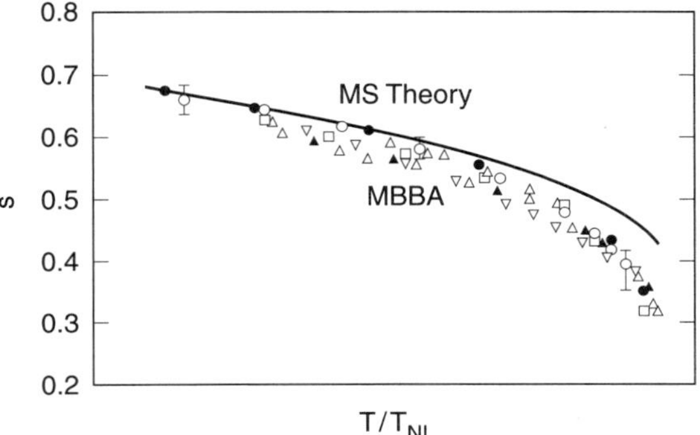

Figure 10.4 Measured values of the order parameter $S = S_2$ versus temperature for MBBA using Raman scattering ($\square$), NMR ($\blacktriangle$), birefringence ($\triangle$), and diamagnetic anisotropy (∇), as described in Deloche et al. (1971). The solid line is the prediction of the Maier–Saupe theory; the dashed line is a modification of the Maier–Saupe theory by Humphries, James, and Luckhurst (1972). (From de Gennes and Prost, Copyright © 1993, by Oxford University Press, Inc. Used by permission of Oxford University Press, Inc.)

S rises significantly less rapidly with decreasing temperature than it does when *S* is measured in the usual way at constant pressure (McColl and Shih 1972). Thus, at least part of the increase in *S* with decreasing T/T_{NI} in Fig. 10-4 is due to increasing molecular packing density. Hence, it is likely that packing entropy, which is neglected by the Maier–Saupe theory, plays an important role in producing the liquid-crystalline state in small-molecule nematics, just as it does in polymeric ones, and the success of the Maier–Saupe theory is somewhat fortuitous.

The flow properties of uniaxial nematics are described by the *Leslie–Ericksen theory* (Leslie 1966, 1968; de Gennes and Prost 1993). The status of the Leslie–Ericksen equation as a constitutive equation for nematics is analogous to that of the Newtonian constitutive equation, Eq. (1-26), as a description of ordinary liquids. The Leslie–Ericksen (LE) constitutive equation is the most general relationships that satisfies the symmetry properties of uniaxial nematic fluids with a single director, for which the stress is *linear* in the local velocity gradient. Generally, the assumption of linearity in the velocity gradient is valid as long as the velocity gradients are small compared to the rate of molecular orientational or positional relaxation. For nonglassy fluids composed of small molecules, molecular relaxation times are typically 10^{-6} sec or less, except very near phase transitions, and the assumption of linearity in the velocity field is usually a very good one. However, for *polymeric* fluids, including polymeric nematics, shear rates can greatly exceed molecular relaxation rates, and the constitutive equation must then accommodate nonlinear dependencies of the stress on the local velocity gradient. Such nonlinearities in polymeric nematics are discussed in Chapter 11.

The LE theory is rather complex since it contains both viscous and elastic stresses. It can best be understood by considering viscous and elastic effects separately. If elastic effects are neglected, the LE equations reduce to *Ericksen's transversely isotropic fluid*, while in the absence of flow the elastic stresses are just those of the Frank–Oseen theory (discussed below in Section 10.2.2).

10.2.1 Ericksen's Transversely Isotropic Fluid

In the absence of elastic stresses, the stress tensor for a flowing nematic is given by (Ericksen 1960, 1961)

$$\boldsymbol{\sigma} = 2\mu\mathbf{D} + 2\mu_1\mathbf{D} : \mathbf{nnnn} + \mu_2(\mathbf{nn} \cdot \mathbf{D} + \mathbf{D} \cdot \mathbf{nn}) \tag{10-2}$$

where μ, μ_1, and μ_2 are constant viscosities, and $\mathbf{D} \equiv \frac{1}{2}(\nabla\mathbf{v} + (\nabla\mathbf{v})^T)$ is the symmetric part of the velocity gradient tensor. Three analogous terms appear in the expression for the stress tensor for suspensions of rigid ellipsoids [see Eq. (6-35)]. (An additional term proportional to $\mu_1\mathbf{nn}$, present in Ericksen's original theory, is here omitted because it produces a yield stress not usually present in liquid nematics.)

Since the director $\mathbf{n}$ can be influenced by the flow, an additional dynamic equation must be provided to couple $\mathbf{n}$ to the velocity gradient. The most general relationship that gives $\partial\mathbf{n}/\partial t$ in terms of $\mathbf{n}$ and $\nabla\mathbf{n}$ that satisfies frame invariance, as well as the constraint that the director is a unit vector ($\mathbf{n} \cdot \mathbf{n} = 1$), and that is linear in $\nabla\mathbf{v}$, is

$$\dot{\mathbf{n}} - \mathbf{n} \cdot \boldsymbol{\omega} - \lambda(\mathbf{n} \cdot \mathbf{D} - \mathbf{nnn} : \mathbf{D}) = 0 \tag{10-3}$$

where $\dot{\mathbf{n}}$ is the substantial time derivative of $\mathbf{n}$, namely $\dot{\mathbf{n}} \equiv \partial\mathbf{n}/\partial t + \mathbf{v}\cdot\nabla\mathbf{n}$, and $\boldsymbol{\omega} \equiv \frac{1}{2}(\nabla\mathbf{v}-(\nabla\mathbf{v})^T)$ is the antisymmetric part of the velocity gradient. The term $-\mathbf{nnn}:\mathbf{D}$ is included in Eq. (10-3) to satisfy the constraint $\mathbf{n}\cdot\mathbf{n}=1$. The coefficient λ is called the *reactive parameter*, or *tumbling parameter*; it controls the rotation of the director in a flow field. For molecules of roughly prolate shape (the usual case), $\lambda > 0$, while for nematics composed of disc-like molecules or particles, one expects $\lambda < 0$.

Consider a simple shearing flow, and define x to be the flow direction, y to be the velocity gradient direction, and z to be the vorticity direction. The director $\mathbf{n}$ can be represented in terms of a polar angle θ and an aximuthal angle ϕ as $\mathbf{n} = (n_x, n_y, n_z) = (\cos\theta\,\cos\phi,\ \sin\theta\,\cos\phi,\ \sin\phi)$ Let us suppose, for the moment, that the director is confined to the x–y plane, which we call the *deformation plane*. Then $\sin\phi = 0$, and $\mathbf{n} = (\cos\theta,\ \sin\theta, 0)$. Now if $|\lambda| > 1$, Eq. (10-3) has a steady-state solution for the director orientation given by

$$\tan\theta = \pm\left(\frac{\lambda - 1}{\lambda + 1}\right)^{1/2} \tag{10-4}$$

There are four such steady-state solutions in the (n_x, n_y) plane, as depicted in the Ericksen diagram shown in Fig. 10-5. Since nematics are nonpolar (i.e., the "head" of the director is indistinguishable from the "tail"), two of the four solutions shown in Fig. 10-5 are redundant; of the remaining two solutions, one is unstable and the other stable. For $\lambda > 1$ and a shear rate $\dot{\gamma}$ that is *positive*, the stable solution is the one with a positive sign in Eq. (10-4). Equation (10-4) tells us that if $\lambda \gg 1$, the stable orientation angle θ approaches $45°$, while if λ is near unity, θ approaches zero; that is, the director becomes parallel with the flow direction.

From Eqs. (10-2) and (10-4), for $\lambda \geq 1$, one finds that the stresses in a shearing flow are

$$\frac{\sigma_{12}}{\dot{\gamma}} = \eta = \mu + \frac{\mu_2}{2} + \frac{\lambda^2 - 1}{2\lambda^2}\mu_1 \tag{10-5a}$$

$$\frac{N_1}{\dot{\gamma}} = \frac{\sqrt{\lambda^2 - 1}}{\lambda^2}\mu_1 \tag{10-5b}$$

$$\frac{N_2}{\dot{\gamma}} = \frac{\sqrt{\lambda^2 - 1}}{2\lambda^2}[\lambda\mu_2 + (\lambda - 1)\mu_1] \tag{10-5c}$$

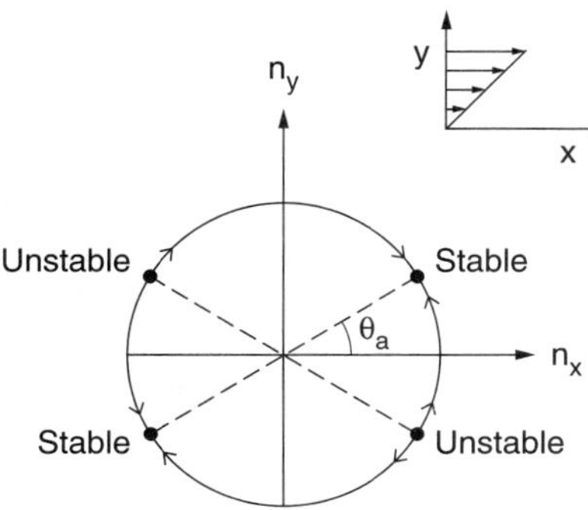

Figure 10.5   Ericksen diagram showing directions of rotation and stable and unstable alignment orientations of a flow-aligning nematic with flow-alignment angle θ_a in a steady shearing flow, with x the flow direction and y the velocity gradient direction.

Note that the normal stress differences N_1 and N_2 are linear in the shear rate $\dot{\gamma}$, as is the shear stress. In contrast, for isotropic viscoelastic fluids, $N_1 \propto N_2 \propto \dot{\gamma}^2$ at low shear rates (see Section 1.4.3).

If $|\lambda| < 1$, Eq. (10-4) has no solutions in the shearing plane. This means that the director *rotates endlessly in the deformation plane*. The period of this rotation—that is, the time it takes to rotate through an angle of π, is

$$P = \frac{2\pi}{\dot{\gamma}\sqrt{1 - \lambda^2}} \tag{10-6}$$

Note that the period is inversely proportional to shear rate $\dot{\gamma}$; hence, the strain period $P\dot{\gamma}$ is independent of shear rate. When $|\lambda| < 1$ the nematic is called a *tumbling* nematic, while when $|\lambda| > 1$, the nematic is *flow-aligning*. As discussed in Sections 10.2.5 and 10.2.6, both cases (tumbling and flow-aligning) can occur in small-molecule liquid crystals.

Equation (10-3) also appears in Section 6.3.1.1 [Eq. (6-26)], where it describes the rotation of axisymmetric non-Brownian ellipsoids in a flow field. For such "spheroids," **n** is the orientation of the axis of symmetry and $\lambda = (p^2 - 1)/(p^2 + 1)$, where p is the aspect ratio of the ellipsoids, L_1/L_2, with L_1 being the length of the symmetry axis, and L_2 the length of each of the other two axes. For finite p, $\lambda < 1$, while for $p \to \infty$, $\lambda \to 1$. In a shear field without Brownian motion, the spheroids therefore *tumble* in Jeffery orbits with a period $\pi(p + p^{-1})/\dot{\gamma}$ (Hinch and Leal 1973). If one considers each molecule in a small-molecule nematic to be a spheroid, one might expect that small-molecule nematics would always tumble. However, in a nematic, tumbling (or flow aligning) refers to the behavior of the *director*, which is the direction of alignment averaged over all molecules in a small region. While each molecule tumbles in a shear field, the *distribution* of molecular orientations often adopts a steady state due to the intermolecular interactions, and so the director often flow aligns (see Section 10.2.4.1).

One can also consider solutions to Eq. (10-3) in which the director is not in the deformation $(x–y)$ plane. In particular, for both tumbling and flow-aligning nematics, Eq. (10-3) has the steady-state solution $\mathbf{n} = (0, 0, 1)$, in which the director is orthogonal to the deformation plane. This orientation is sometimes called the *log-rolling* orientation (Larson and Öttinger 1991). For a flow-aligning nematic with $\lambda > 1$, this orthogonal orientation is unstable to small disturbances, and the director finds its way to the flow alignment angle in the deformation plane, no matter what the initial director orientation is. For a tumbling nematic, the log-rolling orientation is *neutrally stable* to *homogeneous* disturbances; that is, such disturbances neither grow nor decay away, but produce neutrally stable orbits in which the director has both in-plane and out-of-plane components. The log-rolling orientation in a tumbling nematic is, however, unstable to certain periodic *inhomogeneous* disturbances that produce a periodic *roll-cell* pattern (see Section 10.2.6).

The above predictions of the Ericksen theory do not apply rigorously to real nematics, however, because the bounding surfaces or walls influence the director orientation. Often, there is *strong anchoring* at the wall, and the director there is fixed, while away from the wall the director rotates. Thus, shearing flow produces *gradients* in the director field. These gradients or distortions in the director field are energetically unfavorable and lead to elastic stresses known as *Frank stresses*.

10.2.2 Frank–Oseen Theory

The orientation of the director at a wall or bounding surface, the so-called *anchoring* condition, can be controlled by chemical treatment, or rubbing of, the surface (de Gennes and Prost 1993). A director orientation that is orthogonal to the surface is called *homeotropic* alignment. A thin film of lecithin (a surfactant) on glass often produces homeotropic anchoring (Haller and Huggins 1972; Creagh and Kmetz 1972; de Gennes and Prost 1993). Buffing a surface after it has been coated with a thin film, such as polyimide, usually produces nearly parallel orientation in the direction of buffing, although there is often a *2–3° pre-tilt* in the direction normal to the surface. Buffing is used to establish the desired anchoring conditions in liquid-crystal displays.

Director orientations can differ from one surface to another. If the alignments imposed at two different surfaces are not parallel to each other, then the director alignment varies spatially from one surface to the other (see Fig. 10-6). Spatial variations in the director field disrupt somewhat the molecular packing and thus incur a free-energy penalty, the minimization of which determines the equilibrium or static dependence of the director $\mathbf{n}$ on position $\mathbf{x}$. If the change of the director orientation is small over the length a molecule (which for small-molecule nematics is usually the case), then the excess free energy density W_d produced by that gradient can be obtained from the low-order Frank continuum theory (Oseen 1933; Zocher 1933; Frank 1958; Ericksen 1966):

$$2W_d = K_1(\nabla \cdot \mathbf{n})^2 + K_2(\mathbf{n} \cdot \nabla \times \mathbf{n})^2 + K_3(\mathbf{n} \times \nabla \times \mathbf{n})^2$$

$$\qquad\qquad (splay) \qquad\qquad (twist) \qquad\qquad (bend)$$

(10-7)

Here $\nabla \cdot \mathbf{n}$ and $\nabla \times \mathbf{n}$ are the divergence and the curl of $\mathbf{n}$. The three contributions to W_d are associated with the three independent modes of distortion: splay, twist, and bend, depicted in Fig. 10-6. Terms of higher order than quadratic in $\nabla\mathbf{n}$ are only required if spatial distortions become severe. The *Frank constants K_1, K_2, and K_3* are of the order u/a, where a is a molecular length scale [actually the molecule's volume divided by its length

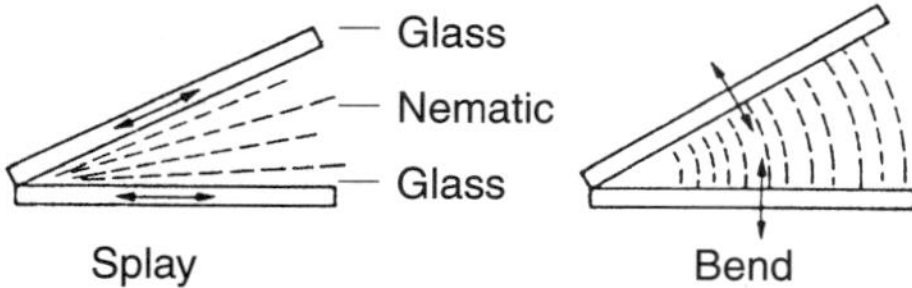

Figure 10.6 Three types of distortion in nematics. (From de Gennes and Prost, Copyright © 1993, by Oxford University Press, Inc. Used by permission of Oxford University Press, Inc.)

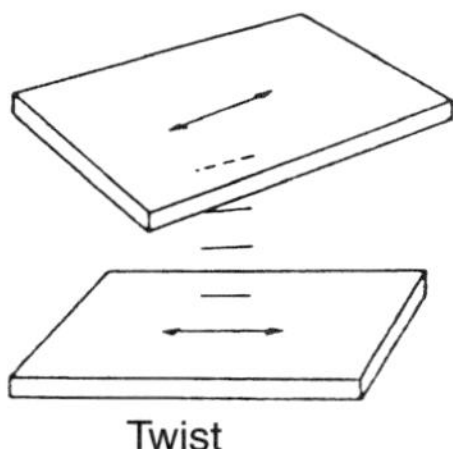

squared; see Eqs. (10-26)] and u is a nematic interaction energy parameter—for example, $u_{MS} = U_{MS}k_B T \sim 5k_B T \sim 2 \times 10^{-13}$ erg. Thus, we estimate that the K values for small molecules should be of magnitude 10^{-6} dyn, which agrees with typical experimental values. For MBBA at 22°C, for example (de Gennes and Prost 1993), we have

$$K_1 = 5.3 \times 10^{-7} \text{ dyn}, \qquad K_2 = 2.2 \times 10^{-7} \text{ dyn}, \qquad K_3 = 7.45 \times 10^{-7} \text{ dyn} \qquad (10\text{-}8)$$

Analyses are simplified by taking all three constants to be equal (the one-constant approximation): $K \equiv K_1 = K_2 = K_3$. It can then be shown that

$$2W_d = K\nabla \mathbf{n} : (\nabla \mathbf{n})^T = Kn_{j,i}n_{j,i} \qquad (10\text{-}9)$$

Here T denotes transpose, and W_d has been expressed in terms of both Gibbs' and Einstein's notation with the conventional summation over repeated indices. The subscript ",i" means partial differentiation with respect to x_i.

The director *anchoring strength* A_a at the wall is defined in terms of an anchoring free energy density $W_a = A_a(\delta\theta)^2$, that is, the energetic cost of an angular deviation $\delta\theta$ from the preferred anchoring direction at the surface (de Gennes and Prost 1993). For MBBA against surfactant-treated glass, which produces homeotropic alignment, A_a has been found to range from 0.065 to 0.013 dyn/cm as the temperature ranges from $T_{NI} - 20°C$ up to T_{NI} (Rosenblatt 1984). Others have found anchoring energies ranging from 10^{-5} to 1 dyn/cm (Jerôme 1991). The surface deviation $\delta\theta$ allows some relaxation of the director distortion in the bulk; the magnitude of $\delta\theta$ can be obtained by minimizing the sum of the bulk distortion energy [Eq. (10-7)] and the anchoring energy. This gives an *anchoring extrapolation length* $b_a = K_i/A_a$, which is the distance from the surface at which the director orientation in the bulk extrapolates to the preferred surface direction. (An analogous extrapolation length was defined for velocity slip boundaries conditions in Section 1.5.2.2.) The surface deviation $\delta\theta$ can therefore be neglected when the extrapolation length is small compared to the distance ξ over which the director in the bulk changes significantly. For the usual strong anchoring, $A_a \geq 0.01$ dyn/cm and $b_a \equiv K_i/A_a \lesssim 1\ \mu$m, which is normally much less than ξ and so $\delta\theta$ is negligible. For *weak anchoring* and steep director gradients, however, b_a can approach the length scale ξ, and $\delta\theta$ is then not negligible. For a review of various types of anchoring, see Jerôme (1991). Molecular models for anchoring and anchoring transitions are discussed by Parsons (1978) and Sonin et al. (1995).

> • Worked Example 10.1 shows how to calculate the Frank–Oseen free energy and use it to predict the response of a liquid crystal to a magnetic or electric field. Such calculations are used to design practical liquid-crystal display devices. They also can be used to determine the values of the elastic constants.

10.2.3 Leslie–Ericksen Theory

In a flowing liquid crystal, both the viscous stresses and Frank elastic stresses are normally important. Thus, the Ericksen theory for the viscous stresses, must somehow be combined with the Frank theory for the elastic stresses. This was accomplished by Leslie, who developed what are now called "the Leslie–Ericksen equations." The Leslie–Ericksen constitutive equation for the viscous stress tensor $\boldsymbol{\sigma}$ is

$$\boldsymbol{\sigma} = \alpha_1 \mathbf{nnnn} : \mathbf{D} + \alpha_2 \mathbf{nN} + \alpha_3 \mathbf{Nn} + \alpha_4 \mathbf{D} + \alpha_5 \mathbf{nn} \cdot \mathbf{D} + \alpha_6 \mathbf{D} \cdot \mathbf{nn} \qquad (10\text{-}10)$$

Equation (10-10) contains six viscosities (the α_i's). Only five of these are independent, however, since Parodi's (1970) relationship expresses one of the viscosities in terms of the others:

$$\alpha_6 = \alpha_2 + \alpha_3 + \alpha_5 \qquad (10\text{-}11)$$

Relationships between the Leslie viscosities (the α's) and Ericksen viscosities (the μ's) can be found in Eq. (11-26).

In Eq. (10-10), $\mathbf{N}$ is defined as

$$\mathbf{N} \equiv \dot{\mathbf{n}} - \mathbf{n} \cdot \boldsymbol{\omega} \qquad (10\text{-}12)$$

$\mathbf{N}$ is the rotation rate of $\mathbf{n}$ relative to that of the background fluid. If the director field is uniform, Eq. (10-3) can be used to express $\mathbf{N}$ in terms of $\mathbf{D}$ and $\mathbf{n}$, and then Eq. (10-10) reduces to Ericksen's equation (10-2).

However, if the director field is not uniform, Frank *distortional stresses* influence the rate of rotation of the director, and a new equation for $\dot{\mathbf{n}}$ (or, equivalently for $\mathbf{N}$) is required to replace Ericksen's equation (10-3). This is obtained from the balance of angular momentum, which gives

$$\mathbf{h} - \mathbf{nn} \cdot \mathbf{h} - \gamma_2 (\mathbf{n} \cdot \mathbf{D} - \mathbf{nnn} : \mathbf{D}) - \gamma_1 \mathbf{N} = 0 \qquad (10\text{-}13)$$

The vector $\mathbf{h}$ is the so-called *molecular field*; it is produced by the same director gradients that produce the distortional energy W_d. Specifically,

$$h_i \equiv \frac{\partial}{\partial x_j} \left(\frac{\partial W_d}{\partial n_{i,j}} \right) - \frac{\partial W_d}{\partial n_i} \qquad (10\text{-}14)$$

The splay, twist, and bend portions of $\mathbf{h}$ are (de Gennes and Prost 1993)

$$\mathbf{h}_S = K_1 \nabla (\nabla \cdot \mathbf{n}) \qquad (10\text{-}14\text{a})$$

$$\mathbf{h}_T = -K_2 [(A \nabla \times \mathbf{n} + \nabla \times (A\mathbf{n})] \qquad (10\text{-}14\text{b})$$

$$\mathbf{h}_B = K_3 [\mathbf{B} \times (\nabla \times \mathbf{n}) + \nabla \times (\mathbf{n} \times \mathbf{B})] \qquad (10\text{-}14\text{c})$$

Here $A \equiv \mathbf{n} \cdot (\nabla \times \mathbf{n})$ and $\mathbf{B} \equiv \mathbf{n} \times (\nabla \times \mathbf{n})$. In the single-constant approximation ($K \equiv K_1 = K_2 = K_3$), $\mathbf{h}$ is given by

$$\mathbf{h} = K \nabla^2 \mathbf{n} \qquad (10\text{-}14\text{d})$$

In Eq. (10-13), γ_1 and γ_2 are given in terms of the Leslie viscosities by

$$\gamma_1 \equiv \alpha_3 - \alpha_2, \qquad \gamma_2 \equiv \alpha_6 - \alpha_5 \qquad (10\text{-}15)$$

For liquid-crystal-display applications, the most important viscosity is the twist viscosity γ_1; values of γ_1 for many different liquid crystals can be found in Yun (1973). The ratio K_i / γ_1, where K_i is one of the Frank constants, has units of length squared per unit time and is thus a diffusivity for orientational order. This ratio divided into the square of the gap between two surfaces then gives the time required for alignment imposed on the surfaces to diffuse throughout the bulk.

If the molecular field $\mathbf{h}$ is zero, then Eq. (10-13) can be reduced to Ericksen's equation (10-3), where the "tumbling parameter" λ is now given by

$$\lambda = -\frac{\gamma_2}{\gamma_1} = -\frac{\alpha_6 - \alpha_5}{\alpha_3 - \alpha_2} = \frac{-(\alpha_3 + \alpha_2)}{\alpha_3 - \alpha_2} \tag{10-16}$$

where the last equality was obtained by using Parodi's relationship in Eq. (10-11).

The total stress tensor in the Leslie–Ericksen theory is the sum of the viscous stress of Eq. (10-10), an isotropic pressure, and the Frank *distortional stress*, given by

$$\sigma_{jk}^{d} \equiv -\left(\frac{\partial W_d}{\partial n_{i,j}}\right) n_{i,k} = -\frac{\partial W_d}{\partial \nabla \mathbf{n}} \cdot (\nabla \mathbf{n})^T \tag{10-17}$$

In the simplest case, where all three Frank constants are equal, this reduces to

$$\sigma^d = -K \nabla \mathbf{n} \cdot (\nabla \mathbf{n})^T \qquad (\text{if } K_1 = K_2 = K_3) \tag{10-17a}$$

In a flowing nematic, the viscous and distortional stresses must satisfy the balance of linear momentum:

$$\nabla \cdot \sigma^d + \nabla \cdot \sigma - \nabla p = 0 \tag{10-18}$$

where p is the pressure. Here we are neglecting inertial and body forces.

The magnitudes of the viscosities (the α_i's) for a single small-molecule nematic can differ from one another by an order of magnitude or more. As a result, the fluid's resistance to flow depends strongly on the directions of the flow and the flow gradient relative to the nematic director. In a shearing flow, the viscosities α_2 and α_3 determine director torques in the orientations shown in Fig. 10-7b and 10-7c. If the director is oriented in the flow direction

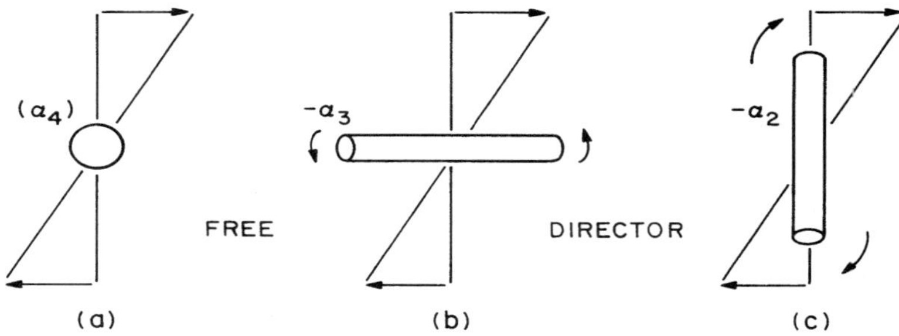

Figure 10.7   (a–c) The Leslie viscosities α_3 and α_2 determine the direction and rate of rotation of the director (represented by the cylinders) in the orientations shown. For negative values of α_3 and α_2 (the usual signs for rod-like nematics), the rotation directions are shown by the arrows. The viscosity α_4 determines the viscosity of the liquid when the director is in the vorticity direction. (Adapted from Skarp et al., reprinted with permission from Mol. Cryst. Liq. Cryst. 60:215, Copyright © 1980, Gordon and Breach Publishers.)

(Fig. 10-7b), then the sign of α_3 determines the direction of rotation; a negative sign for α_3 leads to counterclockwise rotation. If the director is in the flow-gradient direction (Fig. 10-7c), then α_2 determines the torque; a negative value for it leads to clockwise rotation. Thus, if both α_3 and α_2 are negative, the director will rotate clockwise for some orientations and counterclockwise for others, as shown in the diagram in Fig. 10-5. The magnitudes of α_2 and α_3 determine the magnitudes of the torques acting on the director, as well as the range of angles over which clockwise or counterclockwise rotation will occur. The viscosity α_4 is not related to the torque on the director, but instead determines the viscosity of the fluid when the director is oriented in the vorticity direction.

If the director is held in a fixed orientation by a magnetic field strong enough to resist the orienting effects of flow, then shear-rate-independent viscosities can be measured in a simple shearing flow. The three simplest of these, called the *Miesowicz viscosities*, are obtained in each of the three director orientations shown in Fig. 10-8. These viscosities can be related in a simple way to the α_i's, namely,

$$\eta_a \equiv \frac{\alpha_4}{2}, \qquad \eta_b \equiv \frac{\alpha_3 + \alpha_4 + \alpha_6}{2}, \qquad \eta_c \equiv \frac{-\alpha_2 + \alpha_4 + \alpha_5}{2} \qquad (10\text{-}19)$$

Figure 10-9a shows measured values of these three viscosities as functions of temperature for MBBA (Kneppe et al. 1981, 1982). Of course, at temperatures for which MBBA is isotropic, all three viscosities are equal. From the three Miesowicz viscosities (the η's) in Fig. 10-9a, along with the three Leslie viscosities in Fig. 10-9b, the complete set of six Leslie viscosities can be extracted. de Gennes and Prost (1993) give a description of the experimental methods used to measure these viscosities.

10.2.4 Theory for the Leslie Viscosities and Frank Constants

10.2.4.1 The Leslie Viscosities

The lines in Fig. 10-9 are from a molecular theory for these viscosities, based on a solution of the *Smoluchowski equation* for rigid molecules or Brownian particles by Kuzuu and Doi; this is discussed in Section 11.4.2. The theory gives (Kuzuu and Doi 1983; 1984; Larson and Archer 1995).

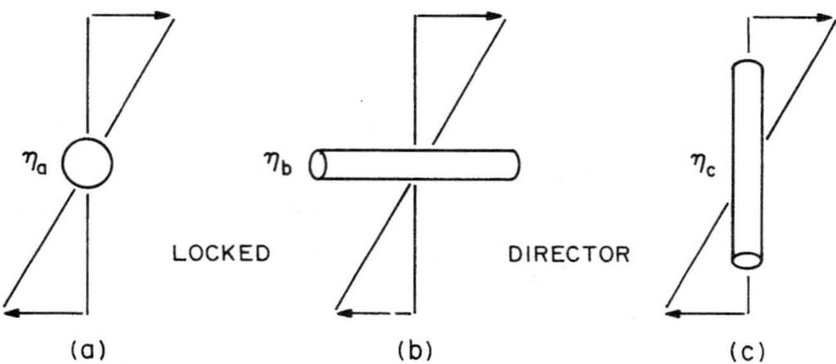

Figure 10.8 (a–c) The Miesowicz viscosities η_a, η_b, and η_c are measured when the director is locked by a strong field in the orientations shown. (Adapted from Skarp et al., reprinted with permission from Mol. Cryst. Liq. Cryst. 60:215, Copyright © 1980, Gordon and Breach Publishers.)

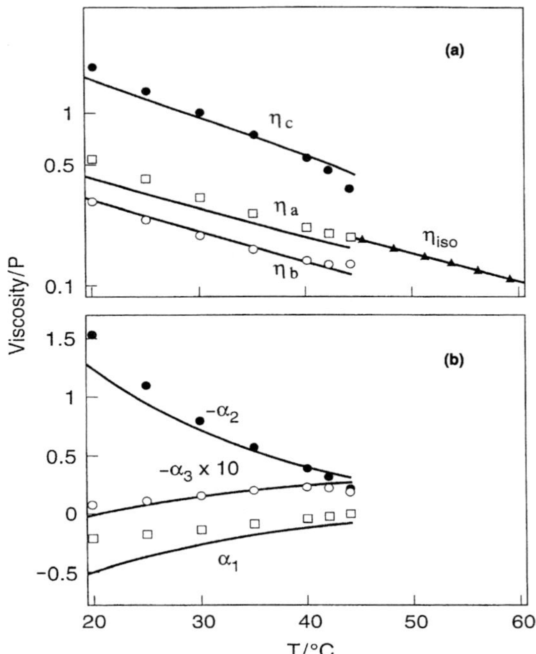

Figure 10.9 (a) The symbols are the Miesowicz viscosities η_a, η_b, and η_c for MBBA as functions or temperature from Kneppe et al. (1982). The lines are the predictions of the modified Kuzuu–Doi molecular theory discussed in the text, with $\alpha_0/\bar{\eta} = 0.6$. (b) As in (a), where here the three Leslie viscosities α_1, α_2, and α_3 are plotted. (From Larson and Archer 1995, by permission of Taylor & Francis.)

$$\alpha_1 = -2\bar{\eta} R^2(p) S_4 \tag{10-20a}$$

$$\alpha_2 = -\bar{\eta} R(p) \left(1 + \frac{1}{\lambda}\right) S_2 \tag{10-20b}$$

$$\alpha_3 = -\bar{\eta} R(p) \left(1 - \frac{1}{\lambda}\right) S_2 \tag{10-20c}$$

$$\alpha_4 = \bar{\eta} R^2(p) \frac{2}{35} (7 - 5S_2 - 2S_4) + \alpha_0 \tag{10-20d}$$

$$\alpha_5 = \bar{\eta} R(p) \left[\frac{1}{7} R(p)(3S_2 + 4S_4) + S_2\right] \tag{10-20e}$$

$$\alpha_6 = \bar{\eta} R(p) \left[\frac{1}{7} R(p)(3S_2 + 4S_4) - S_2\right] \tag{10-20f}$$

In Eqs. (10-20), the six Leslie viscosities are given in terms of the characteristic viscosities $\bar{\eta}$ and α_0 (described below), the "tumbling parameter" λ, the second and fourth moments S_2 and S_4 of the molecular distribution function ψ, and a parameter $R(p)$ that depends on the effective aspect ratio p of the rigid molecules or particles, namely,

$$R(p) \equiv \frac{p^2 - 1}{p^2 + 1} \tag{10-21}$$

For axisymmetric ellipsoids, p is the aspect ratio of their length to diameter, while for cylinders, p is approximately 0.7 times the aspect ratio L/d.

The definitions of the moments S_2 and S_4 of the distribution function are given in terms of averages over the Legendre polynomials $P_2(x)$ and $P_4(x)$:

$$P_2(x) \equiv \frac{3x^2 - 1}{2}, \qquad P_4(x) \equiv \frac{35x^4 - 30x^2 + 3}{8} \tag{10-22}$$

Here x is the cosine of the angle between the orientation of a molecule and that of the director; that is, $x = \mathbf{u} \cdot \mathbf{n}$. The moments S_2 and S_4 are then defined by

$$S_2 \equiv \langle P_2(\mathbf{u} \cdot \mathbf{n}) \rangle, \qquad S_4 \equiv \langle P_4(\mathbf{u} \cdot \mathbf{n}) \rangle \tag{10-23}$$

Except perhaps for α_0, the parameters that enter the equations for the viscosities α_1–α_6 can in principle be determined from molecular characteristics, or can be measured on static samples. The second moment S_2 of the distribution function is just the order parameter S, values of which can be obtained, for example, from optical birefringence experiments (see Section 1.6.3). These show rather good agreement with the simple Maier–Saupe theory (see Fig. 10-4). The fourth moment S_4 is much more difficult to measure; estimates of it from Raman scattering and neutron scattering indicate that for some liquid crystals it is significantly lower than predicted by the Maier–Saupe theory, and possibly even negative in sign for 5CB near T_{NI} (de Gennes and Prost 1993). For MBBA, however, Raman measurements of S_4 are in rather good agreement with the predictions of the Maier–Saupe theory with constant u_{MS}. Therefore, it is convenient to use in Eq. (10-20) the Maier–Saupe values of both S_2 and S_4, given in Table 10-1 as functions of the reduced temperature T / T_{NI}.

> • Worked Example 10.2 shows how the values for S_2 in Table 10-1 can be calculated.

The tumbling parameter λ can be obtained by a numerical solution of the Smoluchowski equation, or analytically using the following simple approximate form (Stepanov 1983; Kröger and Sellers 1995; Archer and Larson 1995):

$$\lambda = R(p)\frac{5S_2 + 16S_4 + 14}{35S_2} \tag{10-24}$$

TABLE 10-1
S_2 and S_4 from the Maier–Saupe Theory

T/T_{NI}	S_2	S_4	T/T_{NI}	S_2	S_4
0.999	0.441	0.127	0.910	0.615	0.256
0.990	0.471	0.145	0.900	0.627	0.268
0.985	0.485	0.154	0.880	0.649	0.290
0.975	0.509	0.171	0.860	0.669	0.311
0.965	0.530	0.186	0.840	0.687	0.331
0.955	0.549	0.200	0.800	0.719	0.370
0.945	0.566	0.213	0.715	0.772	0.446
0.930	0.588	0.232	0.505	0.864	0.619
0.920	0.602	0.244	0.303	0.926	0.775

This approximate expression, using the Maier–Saupe theory for S_2 and S_4 and taking $R(p) \to 1$, agrees reasonably well with measurements of λ for a variety of liquid crystals (see Fig. 10-10), as long as there is no transition to a smectic phase near the temperature range considered. When a smectic-A phase is nearby, as is the case for 8CB, then smectic-like fluctuations of the nematic state can significantly reduce λ. For 8CB, for example, λ drops to around 0.3–0.4 when $T = 34°C$ (Kneppe et al. 1981; Mather et al. 1995), which is around 0.7°C above the transition to the smectic-A phase.

The parameter $R(p) \equiv (p^2 - 1)/(p^2 + 1)$ rapidly approaches unity as p increases, reaching 0.97 for $p = 8$. Hence, for $p \geq 8$, it is reasonable to simply set $R(p) = 1$. For typical small-molecule nematics, estimates of p range from 4 to 8, or so, depending on the molecule and how one estimates the molecular diameter. For 8CB, for example, one can compute the molecule volume v as $M/\rho N_A$, where $M = 278$ is the molecular mass, ρ is the density, and N_A is Avogadro's number. Equating this to the volume of a cylinder $v = \pi d^2 L/4$ and taking the length of 8CB to be 32 Å, one obtains $L/d \approx 8$, $p \approx 0.7(L/d) \approx 5.5$, and $R(p) \approx 0.94$.

According to the molecular theory, the viscosity coefficient $\bar{\eta}$ is given by $vk_B T/2\bar{D}_r$, where $\bar{D}_r$ is the average rotary diffusivity of the molecules. For a typical nematic molecule of molecular weight $M \approx 300$, we have $v = \rho N_A/M \approx 2 \times 10^{21}$ molecules/cm³; and, taking $\bar{D}_r \approx 10^8$ sec⁻¹, we find that $\bar{\eta} \approx 0.4$ P. Note that all contributions to the Leslie viscosities in Eqs. (10-20a)–(10-20f) are proportional to this rotational viscosity $\bar{\eta}$, except for the term α_0 that is added to the expression for α_4. The term α_0 was added to the theory by Larson and Archer (1995) to account phenomenologically for contributions to momentum transport other than those due to rotational motions. These other, translational contributions are expected to be negligible for polymeric nematics, but not for small molecules. The viscosity of liquids composed of spherically symmetric small molecules, for example, is due only to the translational contributions. Larson and Archer assumed that the translational contributions are not orientation-dependent; hence α_0 only appears in the expression (10-20d) for the Leslie viscosity α_4, which is the coefficient of the Newtonian-like term, $\alpha_4 \mathbf{D}$ in Eq. (10-10).

Rather than attempting to compute $\bar{\eta}$ and α_4 from molecular theory, it is more convenient to obtain them from fits to the viscosity data. Consider first the viscosity of the liquid at temperatures above T_{NI}, at which the liquid is in the isotropic state and has no orientational order, so that $S_2 = S_4 = 0$. We then find from Eqs. (10-20a)–(10-20f) that only α_4 is nonzero in the isotropic state; from Eq. (10-10) we find that the Newtonian viscosity η_{iso} of this isotropic liquid is given by $\eta_{\mathrm{iso}} = \alpha_4/2$. From Eq. (10-20d) we then have

$$\eta_{\mathrm{iso}} = \bar{\eta}/5 + \alpha_0/2 = \eta_0 \, \exp(E_a/RT) \tag{10-25}$$

where we have assumed that $\bar{\eta}$ and α_0 have the same, Arrhenius temperature-dependencies. In the isotropic state, the viscosity of MBBA fits this Arrhenius form with $E_a = 32,500$ J mol⁻¹ and $\eta_0 = 1.67 \times 10^{-6}$ P.

Taking $R(p) \approx 1$, this leaves only one temperature-independent constant, $\alpha_0/\bar{\eta}$, left to be obtained by fitting the viscosities in the nematic state below T_{NI}. Since α_0 is an orientation-independent contribution, the ratios of the Miesowicz viscosities $\eta_c:\eta_b:\eta_a$ deviate less from unity as $\alpha_0/\bar{\eta}$ is increased. A value $\alpha_0/\bar{\eta} = 0.6$ fits the Miesowicz viscosities of MBBA reasonably well; the predictions of the theory with $\alpha_0/\bar{\eta} = 0.6$ are given by the lines in Fig. 10-9a. Measurements of the Miesowicz viscosities for other liquid crystals are similar enough to those of MBBA that this theory with $\alpha_0/\bar{\eta} = 0.6$ is likely to work equally well for them. The other independent viscosities of MBBA, in Fig. 10-9b, are also fit fairly well by the theory, including their signs and temperature-dependencies.

Despite this agreement, the theory is probably missing some important molecular physics that are compensated for by use of the fitting parameter. Note in Fig. 10-10 that the tumbling parameter λ is larger than predicted, even when setting $R(p)$ to its large-aspect-ratio asymptote $R(p) = 1$. If the molecular aspect ratios is less than 8 or so, $R(p)$ becomes significantly less than unity, and the gap between the predicted and the measured values of λ becomes greater than that shown in Fig. 10-10.

Baalss and Hess (1988) have proposed a different theory for the viscosities of small-molecule nematics, in which a spherically symmetric interaction potential is affinely transformed into one with ellipsoidal symmetry so that the equipotential surfaces are parallel to the surfaces of the molecules, whose shape is modeled by ellipsoids of revolution. Their theory also assumes that the molecules are perfectly aligned, that is, $S_2 = 1$. The Leslie viscosities predicted by this theory are simply $\alpha_2 = -\alpha_5 = p^2\alpha_3 = p^2\alpha_6 = (1-p^2)\alpha_4/2$; $\alpha_1 = -(p-p^{-1})^2\alpha_4/2$. This rather simple theory predicts that the nematic is always flow-aligning, with $\lambda = 1/R(p)$. This formula for λ is the reciprocal of that predicted by Eq. (10-24), in the limit $S_2 = S_4 = 1$. Thus, the Baalss–Hess theory predicts a greater tendency toward flow-aligning as the particle aspect ratio decreases, while the reverse is predicted by the Kuzuu–Doi theory. Also, the Baalss–Hess theory predicts extreme viscosity anisotropy in the Miesowicz viscosities, $\eta_c/\eta_b = p^4$. Since experimental values of η_c/η_b are of order 10 or less, consistency with the Baalss–Hess theory requires that the molecular aspect ratio be taken to be artificially low, $p < 2$; and if this is done, the tumbling parameter then becomes artificially high, $\lambda \geq 1.67$. Thus, the available data seem to favor the Kuzuu–Doi approach. However, the relative merits of the Kuzuu–Doi versus Baalss–Hess theories would be better assessed if complete sets of temperature-dependent viscosities could be acquired for a homologous series of liquid crystals of varying molecular length. Recently, Fiałkowski (1997) has extended the Kuzuu–Doi theory to the case of biaxial molecules.

10.2.4.2 The Frank Constants

A gradient theory that predicts the Frank constants was proposed by Nehring and Saupe (1972), Priest (1973), and Straley (1973). There have been many modifications and

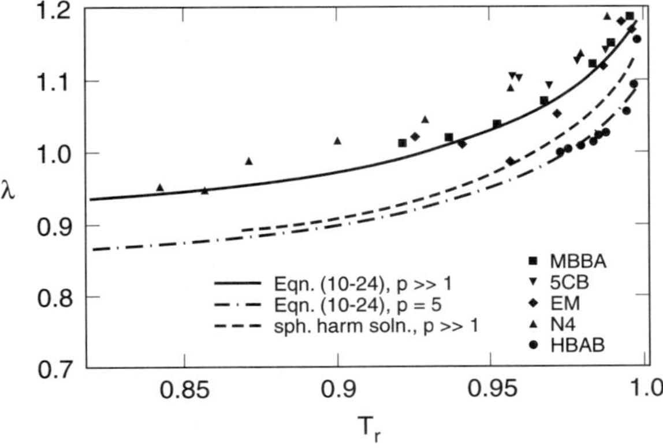

Figure 10.10 "Tumbling parameter" λ versus reduced temperature $T_r \equiv T/T_{\mathrm{NI}}$ for the various liquid crystals listed. The broken line is the exact prediction of the Smoluchowski equation for large aspect ratio p, and the solid and dot-dashed lines are from the approximate expression (10-24) with p large and $p = 5$. (From Archer and Larson, reprinted with permission from J. Chem. Phys. 103:3108, Copyright © 1995, American Institute of Physics.)

extensions of this theory, as referenced by Osipov and Hess (1993) and by Marrucci and Greco (1991). In these gradient theories, a test molecule is assumed to have orientation-dependent interactions with neighboring molecules that lie within an interaction volume around the test molecule. The neighboring molecules are, on average, slightly misaligned with respect to the test molecule, because of an assumed weak gradient in the director, and this produces a gradient free energy from which the Frank constants are computed. For hard rods in a solvent, the rods that must be included in the interaction volume can be determined by geometric considerations (see Section 2.2.2); but for thermotropic nematics in which long-range van der Waals interactions are important, the relevant interaction volume is more ambiguous. Marrucci and Greco assumed that all molecules whose center of mass lie within a distance Λ of the axis of the test rod are included in the interaction volume and thereby obtained

$$K_1 = \frac{\nu k T}{6} U S_2^2 \left[\Lambda^2 + \frac{3L^2}{7} \left(1 - \frac{S_4}{S_2} \right) \right] \tag{10-26a}$$

$$K_2 = \frac{\nu k T}{6} U S_2^2 \left[\Lambda^2 + \frac{L^2}{7} \left(1 - \frac{S_4}{S_2} \right) \right] \tag{10-26b}$$

$$K_3 = \frac{\nu k T}{6} U S_2^2 \left[\Lambda^2 + \frac{3L^2}{7} + \frac{4L^2 S_4}{7 S_2} \right] \tag{10-26c}$$

In the above, $U = 3 U_{\mathrm{MS}}/4$, and L is the molecular length. These expressions contain S_2 and S_4, the second and fourth moments of the orientational distribution function; other expressions also contain the aspect ratio of the molecules (Osipov and Hess 1993). Equation (10-26a)–(10-26c) give the ordering for the elastic constants found in experiments on small-molecule nematics, namely, $K_2 < K_1 < K_3$; and with $\Lambda \approx 1$ nm, roughly the correct magnitudes are obtained as well. As the temperature is lowered from the nematic–isotropic transition, the K's are predicted to increase with the order parameter roughly as S_2^2, which seems to agree, roughly speaking, with measured values of the K's (Leenhouts et al. 1979). Note that in the limit of large Λ, all the K's are equal, while at small Λ (compared to L), $K_1 = 3K_2$. For small-molecule nematics, the choice of a sharply bounded "interaction volume" is, of course, arbitrary, and its size Λ introduces a fitting parameter. Detailed comparisons of predicted and calculated values of the Frank constants for a couple of liquid crystals can be found in Singh (1996). Molecular simulations (Tjipto-Margo et al. 1992; Osipov and Hess 1994) may one day give more rigorous theoretical values of the Frank constants. For polymeric nematics, ratios of Frank constants larger than for small-molecule nematics are both predicted and observed (see Chapter 11).

10.2.4.3 Discotic Nematics

Very few data exist for the viscosities or Frank constants of discotic nematics—that is, nematics composed of disc-like particles or molecules (Chandrasekhar 1992). One can estimate values of the Leslie viscosities from the Kuzuu–Doi equations (10-20) by setting the aspect ratio p equal to the ratio of the thickness to the diameter of the particles; thus $p \to 0$ for highly anisotropic disks. This implies that $R(p) \to -1$, and Eq. (10-20b) implies that the viscosity α_2 is large and *positive* for discoidal nematics, while it is negative for ordinary nematics composed of prolate molecules or particles. If, as expected, α_3 is much smaller in magnitude than α_2, the director (which is orthogonal to the disks) will tend to orient, on average, *perpendicular* to the flow direction, so that one of the long axes of the disks is parallel to the flow direction. A long molecular axis does orient in the flow direction in discotic nematics made of tar pitch in capillary flow (McHugh and Edie 1995). However, the other long axis seems to point in the gradient direction, so that the director orients in the vorticity direction, analogous to a "log-rolling" orientation in prolate nematics. Orientations of the director in either the vorticity or the gradient direction have been found in scattering studies (Hammouda et al. 1995) and from solutions of the Smoluchowski equation (Singh and Rey 1998).

10.2.5 Shearing of Flow-Aligning Nematics

If the director is not held fixed by a magnetic or electric field, and thus is free to rotate in a Couette or Poiseuille shearing flow, then the *overall* stresses are in general *not* proportional to the *overall* shear rate or flow rate, although the *local* stress is linear in the *local* velocity gradient, for a fixed director orientation **n**. Figure 10-11a shows that the effective viscosity measured in Poiseuille, or circular-pipe, flow for *p*-azoxyanisole (PAA) is, in fact, a nonlinear function of the shear rate $\dot{\gamma}$, where $\dot{\gamma}$ is proportional to the volumetric flow rate Q divided by the cube of the tube radius R. The effective viscosity is shear thinning and, when plotted against the shear rate, depends on the tube radius R. Shear thinning is predicted by the Leslie–Ericksen equations (10-10)–(10-18), because they contain nonlinear terms involving gradients of the director field, and these couple to the flow field. Gradients in the director can occur in flowing nematics because the director is influenced by the bounding surfaces of the flow geometry, and the orientation of the director

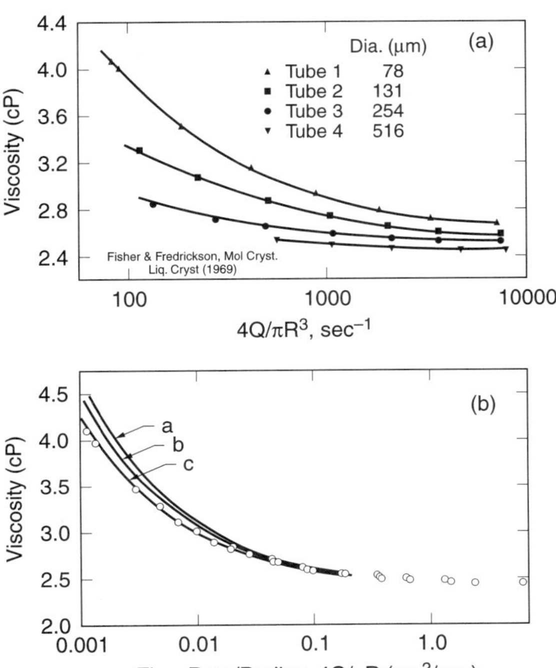

Figure 10.11 (a) Apparent viscosity for Poiseuille flow of *p*-azoxyanisole (PAA) as a function of the effective shear rate $4Q/\pi R^3$, where Q is the volumetric flow rate and R is the tube radius, for homeotropic boundary conditions. (b) The data of (a) are replotted against $4Q/\pi R$, which is proportional to the Ericksen number, defined in Eq. (10-27). The lines are calculated from the Leslie–Ericksen theory using three different estimates of the viscosities. (From Fisher and Fredrickson 1969 and Tseng et al., reprinted with permission from Phys. Fluids 15:1213, Copyright © 1976, American Institute of Physics.)

at boundaries usually differs from that induced in the bulk of the fluid by the viscous flow. In the experiments of Fig. 10-11a, a homeotropic anchoring boundary condition—that is, a perpendicular direction at the tube walls—was used. Near the center of the sample, however, at high flow rates the director orients at the *flow-alignment angle* $\theta_a \lesssim 10°$, defined with respect to the flow direction, with $\tan^2 \theta_a = (\lambda - 1)/(\lambda + 1)$ [see Eq. (10-4)]. Thus, there is a *gradient in the director orientation* between the walls of the tube and the center. As the flow rate increases, an ever greater fraction of the fluid orients at the flow-alignment angle, and the homeotropically aligned fluid is confined to an ever thinner boundary layer at the wall (Wahl and Fischer 1973) (see Fig. 10-12).

The competition between and flow-induced and boundary-induced orientation is quantified by an *Ericksen number*, defined by

$$\mathrm{Er} \equiv \frac{\eta \dot{\gamma}_{\mathrm{eff}}}{K/h^2} = \frac{\eta V h}{K} \tag{10-27}$$

Here η is a typical Leslie viscosity, K is a Frank constant, V is a flow velocity, h is a length scale of the flow geometry, such as the tube diameter in Poiseuille flow, and $\dot{\gamma}_{\mathrm{eff}}$ is the average shear rate. The Ericksen number is the ratio of the flow-induced viscous stress $\eta \dot{\gamma}_{\mathrm{eff}} = \eta V/h$ to the Frank stress K/h^2. The appropriate Leslie viscosity or Frank constant to use in defining the Ericksen number depends on the particular flow geometry. For the tube-flow problem, a reasonable choice is $\eta = \gamma_1$ and $K = K_3$.

In flow through a tube, therefore, the measured *effective* viscosity, which is defined to be proportional to the pressure drop, depends on the Ericksen number. Note that the Ericksen number is proportional to Vh, the velocity times the tube diameter, where we take $h = 2R$. Since the velocity V is proportional to Q/R^2, Vh is proportional to Q/R. Thus, the data for various tube radii in Fig. 10-11a collapse onto a single line when plotted against $4Q/\pi R$ (see Fig. 10-11b). This shows that the effective viscosity is a function of the Ericksen number, which is proportional to the velocity *times* the tube diameter. (For shear-thinning *isotropic* liquids, on the other hand, the viscosity depends on $\dot{\gamma}_{\mathrm{eff}}$, which is the velocity *divided* by the tube diameter.) Because of the orientation-dependence of the viscosity (illustrated in Fig. 10-9a), the wall layer is much more viscous than the core fluid; and since the thickness δ of the wall layer scales as $\delta \sim \sqrt{K/\eta \dot{\gamma}_{\mathrm{eff}}} = \mathrm{Er}^{-1/2} h$, it follows that as Er increases, the more viscous wall layer occupies an ever smaller fraction of the tube, and the average viscosity is then more and more dominated by the less viscous core fluid. Hence, η_{eff} decreases as Er increases, as shown in Fig. 10-11b.

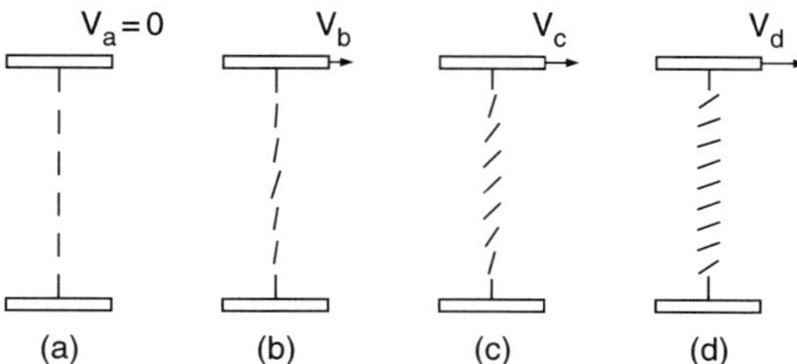

Figure 10.12 (a–d) Schematic illustration of steady-state director configurations of a flow-aligning nematic in a shearing field at flow velocities that increase from left to right, with homeotropic boundary conditions. As the velocity increases, the director in the bulk of the nematic orients toward the flow alignment angle, and the homeotropic orientation is confined to ever thinner boundary layers at the walls. (From Graziano and Mackley, reprinted with permission from Mol. Cryst. Liq. Cryst. 106:103, Copyright © 1984, Gordon and Breach Publishers.)

For small-molecule liquid crystals such as PAA or MBBA, the ratio γ_1/K_3 is around 10^6 sec/μm^2. Since $Vd \approx 4Q/\pi R$, the Ericksen number in Fig. 10-11b is 10^6 times $4Q/\pi R$. Hence, the data in Fig. 10-11b imply that the asymptotic high-shear viscosity in Poiseuille flow is only reached at an Ericksen number in excess of 10^4. The asymptote is approached slowly because the viscosity in the thin boundary layer (η_c) is much higher (at least ten times higher) than than that in the bulk (η_0); hence even a small fraction of it contributes significantly to the overall viscosity. The viscosity in the shear-thinning region should be crudely given by

$$\eta \approx \eta_0 + \eta_c \mathrm{Er}^{-1/2} = \eta_0 + \eta_c \left(\frac{K_3}{h^2\gamma_1\dot{\gamma}}\right)^{1/2} \approx \eta_0 + \gamma_1^{1/2}K_3^{1/2}h^{-1}\dot{\gamma}^{-1/2} \qquad (10\text{-}28)$$

since $\eta_c \approx \gamma_1$. At very low shear rates, the shear thinning should be cut off and the viscosity approaches a high, shear-rate-independent value, η_c.

10.2.6 Shearing of Tumbling Nematics

The steady shear-flow properties of tumbling nematics are very different from those of flow-aligning ones (Gähwiller 1972; Pieranski and Guyon 1974; Cladis and Torza 1975; Carlsson and Skarp 1986). This is shown in rheological studies by Gu et al. (1993) and in microscopic visualization studies by Mather et al. (1996a, 1996b, 1997), with flow-aligning 5CB and tumbling 8CB subjected to torsional shearing flows.

10.2.6.1 Sudden Start-up of Shearing

Consider first a tumbling nematic that is suddenly subjected to a torsional cone-and-plate or plate-and-plate shearing flow with homeotropic boundary conditions at a high Ericksen number, say Er $\gtrsim 10^3$. Since for the tumbling nematic there is no flow-alignment angle at which the viscous torques balance, and the Ericksen number is high enough that Frank elasticity is negligible except at thin wall boundary layers, the director in the bulk is forced to rotate through multiple periods of π. Shear-stress measurements (Gu et al. 1993; Gu and Jamieson 1994) in sudden start-up flow for 8CB in cone-and-plate and plate-and-plate geometries at Er $> 10^3$ show multiple oscillations that slowly dampen with continued shear (see Fig. 10-13). The strain period of the oscillation (about 10 strain units in Fig. 10-13) is independent of the shear rate, but sensitive to temperature.

These multiple stress oscillations are not present at temperatures near T_{NI} where 8CB is a flow-aligning nematic. They are also absent for other nematics, such as 5CB, which are flow-aligners at all temperatures in the nematic range. Instead, flow-aligning nematics with homeotropic anchoring show a single overshoot on start-up of shearing flow (Gu and Jamieson 1994), as expected. Thus, *a pattern of multiple stress oscillations in start-up of shearing flow seems to be a reliable indicator that the nematic is of tumbling character.* Hence, the shear start-up experiment could prove to be a useful diagnostic for some nematics (such as polymeric ones) whose flow-aligning or tumbling character can be difficult to establish any other way. In fact, the first observations of multiple overshoots in shearing flows of nematics were made by Moldenaers and Mewis (1990) for the polymeric nematic PBLG, which was at the time not known to be a tumbling nematic (see Section 11.3.4).

The multiple stress oscillations are produced by the tumbling orbits of the director as it flips end over end, or "winds up," in the shearing flow at high Er. These stress oscillations

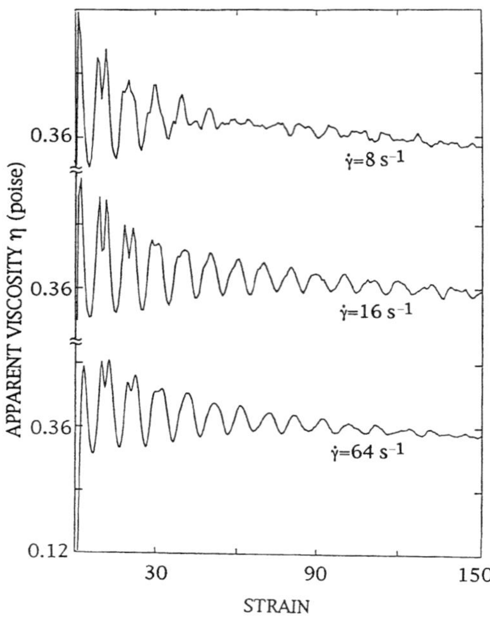

Figure 10.13 Apparent viscosity (shear stress divided by shear rate) versus strain after start-up of cone-and-plate shearing flow of 8CB with homeotropic boundary conditions at 4.8°C below T_{NI} for three different shear rates. (From Gu et al. 1993, with permission from the Journal of Rheology.)

are predicted by the Ericksen theory (which is equivalent to the Leslie–Ericksen theory in the limit of large Er), when $\lambda < 1$. Let us assume that the director is confined to the *deformation plane*, which is the plane containing the flow and the shear-gradient directions, and define the director angle $\theta' = \pi/2 - \theta$ to be the director angle relative to the homeotropic orientation, where θ is the angle with respect to the flow direction. Thus, $\theta' = 0$ corresponds to homeotropic alignment, while $\theta' = \pi/2$ corresponds to alignment in the flow direction. The Ericksen theory then gives

$$\frac{d\theta'}{dt} = \frac{(\alpha_3 \sin^2 \theta' - \alpha_2 \cos^2 \theta')}{\alpha_3 - \alpha_2}\dot{\gamma} \tag{10-29}$$

Note that the director rotation rate is proportional to the shear rate $\dot{\gamma}$. Hence, the period of the stress oscillations is equal to the period P for rotation of the director through an angle of π, given by Eq. (10-6):

$$P = \frac{2\pi}{\dot{\gamma}\sqrt{1 - \lambda^2}} \tag{10-30}$$

As the director rotates, the shearing stress shows periodic maxima and minima. After solving Eq. (10-29) for the strain-dependent director angle $\theta'(\gamma)$, the strain-dependent viscosity $\sigma/\dot{\gamma}$ is given by (Gu and Jamieson 1994)

$$\eta(\dot{\gamma}) = \left[\alpha_1 + \frac{(\alpha_2 + \alpha_3)^2}{\alpha_3 - \alpha_2}\right] \sin^2 \theta' \cos^2 \theta' + \eta_b - \frac{\alpha_3^2}{\alpha_3 - \alpha_2} \tag{10-31}$$

where η_b is a Miesowicz viscosity defined in Eq. (10-19). Thus, by fitting the time-dependence of the shear stress during start-up of shearing to these expressions, one can extract the ratios of some of the viscosities (Gu and Jamieson 1994; Yang and Shine 1993). Figure 10-14 shows that the predicted pattern of stress oscillations agrees well with the experimental one except for a *damping* of the oscillations, which is not predicted by the Ericksen theory.

Figure 10-15a shows that if shear is not continued too long (less than ~ 200 strain units), and then the direction of shearing is suddenly reversed, there is a *reversal* of the damping; that is, the oscillations *increase* in amplitude! The time-dependence of the stress in the reverse shearing is just the opposite of that for the "forward" shearing, and the amplitude of the oscillations in the reverse shearing builds up to that observed at the beginning of the forward shearing when the number of "reverse" strain units equals the number imposed in the "forward" direction. Thus, apparently the reverse shearing corresponds to an "unwinding" of the director pattern produced during the forward shearing. However, if the unidirectional shear is continued for as many as 500 strain units, the damping can no longer be reversed (see Fig. 10-15b).

The damping of the stress oscillations presumably arises from a gradual loss of spatial coherence in the phase of the tumbling orbit across the sample. In a plate-and-plate rheometer, the strain is linearly dependent on the radial distance from the axis of rotation. As a result, the gap-averaged director orientation varies as a function of radial position in the sample. When this source of inhomogeneity in the tumbling orbit is accounted for by integrating the torque contributions predicted by Eq. (10-31) over all radial positions in the sample, the damping of stress oscillations can be predicted (Mather et al. 1997). This radial variation in average director orientation has been visualized directly by optical microscopy of 8CB between parallel glass disks in torsional shearing flow. In this geometry, Mather et al. (1997) discovered that regions of director orientation differing by 180° are separated from each other by axisymmetric *twist walls* that periodically originate at the edge of the sample (where the strain is highest) and migrate radially inward. These twist walls are bands in which the director is twisted out of the local plane of shear, which is the plane defined at each location by the flow direction and the velocity-gradient direction. For a more thorough description of these twist walls, see Mather et al. (1997).

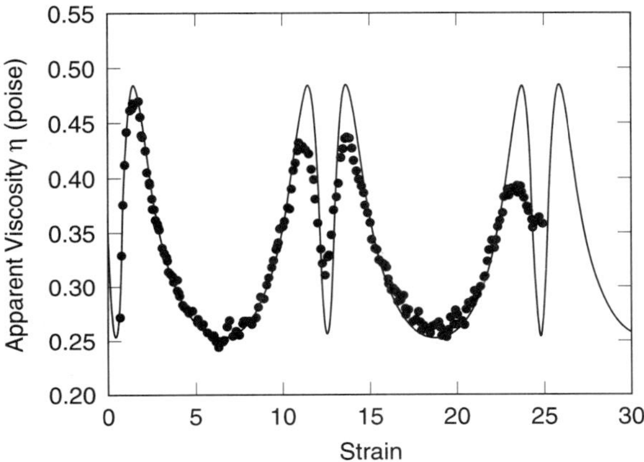

Figure 10.14 Measured shear viscosity η against strain γ for 8CB at 36.6°C (3.2°C below T_{NI}) and a shear rate of $\dot{\gamma} = 16$ sec^{-1}, fit by Eqs. (10-29) and (10-31) of the Ericksen theory with the director confined to the deformation plane. (From Gu and Jamieson, reprinted with permission from J. Rheol. 38:555, Copyright © 1994, American Institute of Physics.)

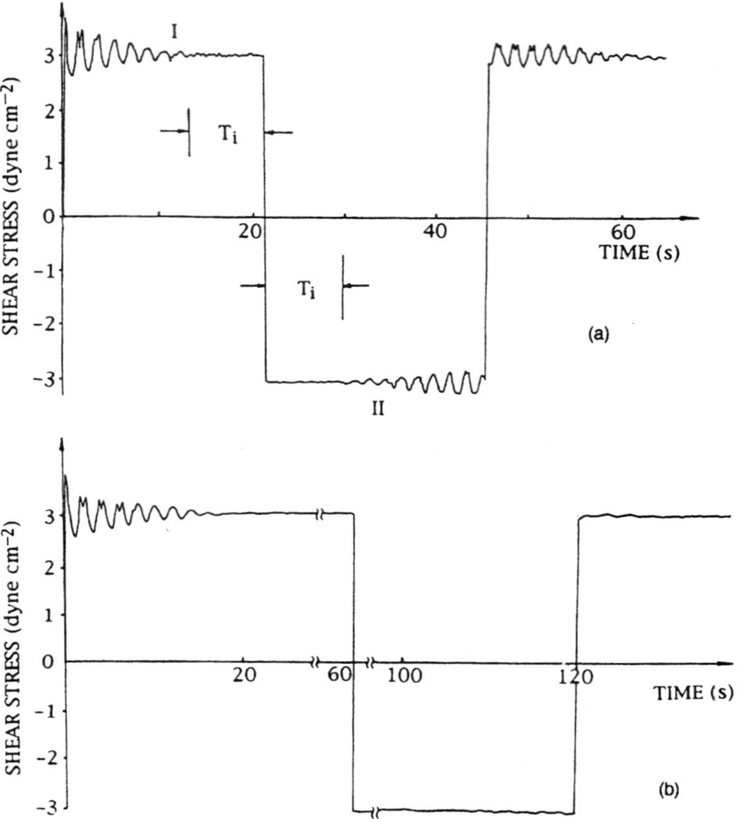

Figure 10.15 Damping of shear stress and reversal of damping as a function of time after start-up of shearing in a cone-and-plate rheometer at $\dot{\gamma} = 8$ sec^{-1} for a tumbling nematic, 8CB. In **(a)** the shearing direction is reversed after imposition of 170 strain units; the damping of the stress oscillations is reversed. In **(b)**, shear reversal occurs after imposition of 480 strain units; the damping is not reversed. (From Gu et al. 1993, with permission from the Journal of Rheology.)

The reversed damping of the shear stress when the flow direction is switched can be explained by unwinding of the inhomogeneous director field. This unwinding is accompanied by reverse propagation of the twist walls, toward the edge of the sample, where they are expelled one after another. Eventually the uniformly homeotropic condition is restored (Mather et al. 1997). If the shearing is continued too long in one direction, however (e.g., $\gamma \approx 500$), *disclinations are nucleated,* and these are not expelled by a reversal in flow direction, but only continue to accumulate. (Disclinations are topological defects described below in Section 10.2.7.) Thus, the damping of stress oscillations can only be reversed when the imposed strain is less than the critical value at which disclinations become abundant.

For cone-and-plate flow, the strain in the sample is uniform, so the damping of the stress oscillations is most likely associated with the nonuniformity of the Ericksen number, which increases as a function of radial position. Not surprisingly, the rate of damping of the stress oscillations in cone-and-plate flow is much more gradual than in plate-and-plate flow; thus the damping is geometry-dependent. Presumably if the Ericksen number were to be increased toward the limit Er $\to \infty$, the damping in the cone-and-plate would become even more gradual.

• Worked Example 10.3 derives Eqs. (10-29) through (10-31).

10.2.6.2 Gradual Increase in Shear Rate

If a homeotropically aligned tumbling nematic is sheared at small Er $\equiv \gamma_1 V h / K_3$, then the director at the midplane of the sample rotates modestly toward the flow direction, until it reaches a steady state where the viscous forces driving the director rotation are balanced by the Frank stresses (see Fig. 10-16 at Er $\sim$ 1). As Er increases in small increments, so that a steady state is attained between each increment in Er, the steady-state director at the midplane is rotated to a greater and greater extent (see Fig. 10-16 at Er $\sim$ 10).

In a parallel-disk device, the shear rate, and hence the Ericksen number, increases linearly with radial position r. Thus, the midplane director angle θ' also increases with r. This produces an increasing *birefringence* which can be observed if the parallel disks are transparent (see Fig. 10-17). The multiple, concentric rings in Fig. 10-17 correspond to

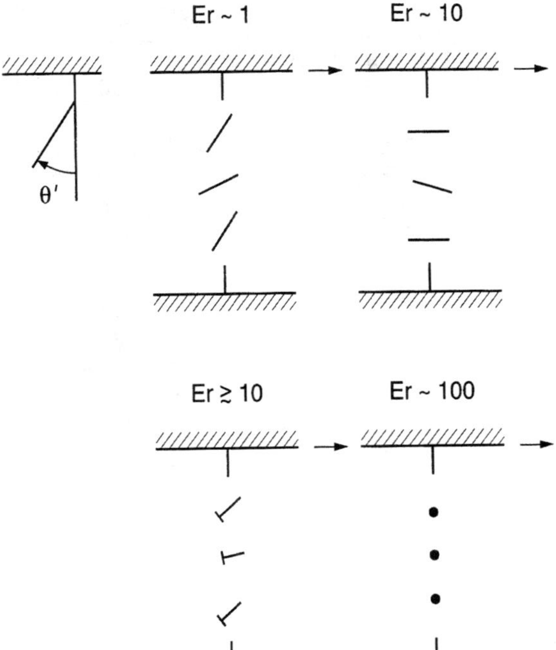

Figure 10.16 Illustration of steady-state director pattern for a tumbling nematic in a simple shearing flow with homeotropic anchoring. At Ericksen number Er $\sim$ 1, the director at the centerplane of the flow rotates modestly toward the flow direction. At higher Er $\sim$ 10, the director at the centerplane rotates through $\theta' = 90°$. When Er becomes still larger (Er $\gtrsim$ 10), the director undergoes an instability that takes it out of the deformation plane as indicated by the "nail" images. For even higher Er ($\sim$100), the director is almost completely normal to the deformation plane, except at the walls.

increasing *orders* of birefringence. A dark *extinction ring* appears each time the birefringence, integrated across the gap, equals π times the wavelength of the monochromatic light used to illuminate the sample. The images in Fig. 10-17 (left) correspond to a low angular velocity Ω of the rotating plate, while those in Fig. 10-17 (right) correspond to a higher value of Ω. The sample gap increases from the top to the bottom of Fig. 10-17. The greater density of rings for higher shear rates means that the director orientation, and hence the birefringence, increases more rapidly with radial position r when Ω is increased.

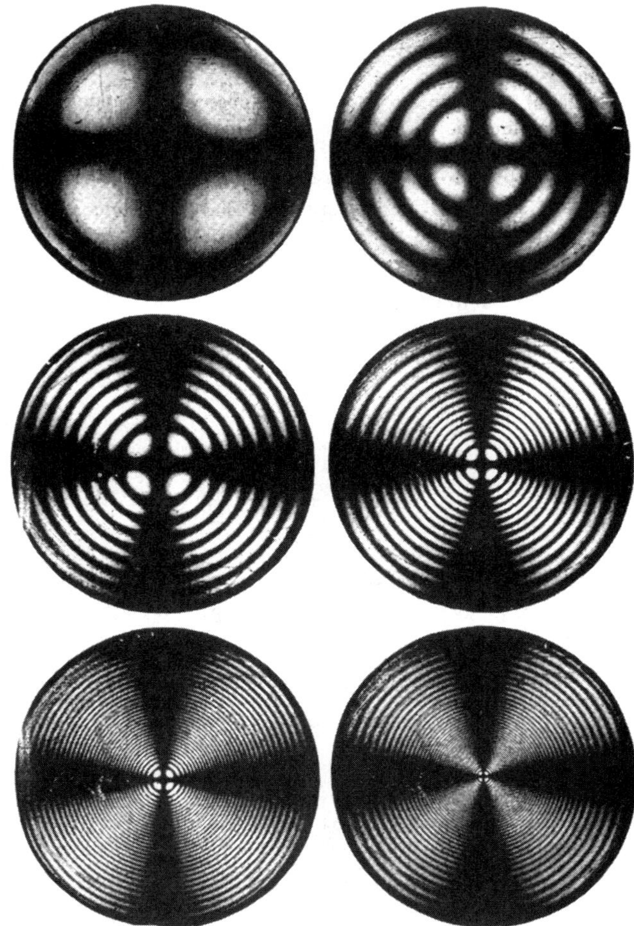

Figure 10.17 Birefringence interference patterns in torsional shearing flow between parallel plates of MBBA, illuminated by monochromatic light and observed between crossed polarizers. The images on the left were obtained at a rotation speed, Ω, of 2.09×10^{-4} sec^{-1}, while those on the right were obtained at 4.18×10^{-4} sec^{-1}. From top to bottom, the gap is 44, 94, and 194 μm. (From Wahl and Fischer, reprinted with permission from Mol. Cryst. Liq. Cryst. 22:359, Copyright © 1973, Gordon and Breach Publishers.)

In a flow-aligning nematic, the director at the midplane eventually approaches the flow-alignment angle, $\theta' = \pi/2 - \theta_a$. But for a tumbling nematic, the director can *pass through the flow direction* so that $\theta' > \pi/2$. At Er $\sim O(10)$ for a tumbling nematic, so that θ' is a few degrees to a few tens of degrees greater than 90°, depending on the values of the Leslie viscosities and Frank constants, calculations with the Leslie–Ericksen equations show that *tumbling instabilities* can ensue, in which θ' at the midplane director abruptly increases (Carlsson 1984; Zuñiga and Leslie 1989). There is also an *out-of-plane* instability (Pieranski et al. 1976; Zuñiga and Leslie 1989; Derfel 1991; Luskin and Pan 1992; Han and Rey 1993; Mather et al. 1996b; Larson and Mead 1993) in which the director begins to rotate toward the vorticity direction, which is the direction perpendicular to both the velocity and the velocity-gradient directions (see Fig. 10-16). In the parallel-disk flow, this manifests itself as a tipping of the birefringence rings (see Fig. 10-18). At Er $\sim O(100)$, the director becomes nearly parallel to the vorticity direction (Mather et al. 1996b; Larson and Mead 1993) (see Fig. 10-16).

As Er is increased further to around 10^3 in 8CB (at 37°C), there is a *roll-cell instability* involving (a) a periodic modulation of the director field in the vorticity direction and (b) a cellular flow. The rolls cells are parallel to the primary flow direction (Pieranski and Guyon 1974) (see Fig. 10-19). These transitions in the director field have been both predicted from the Leslie–Ericksen theory (Manneville and Dubois-Violette 1976; Larson 1993) and

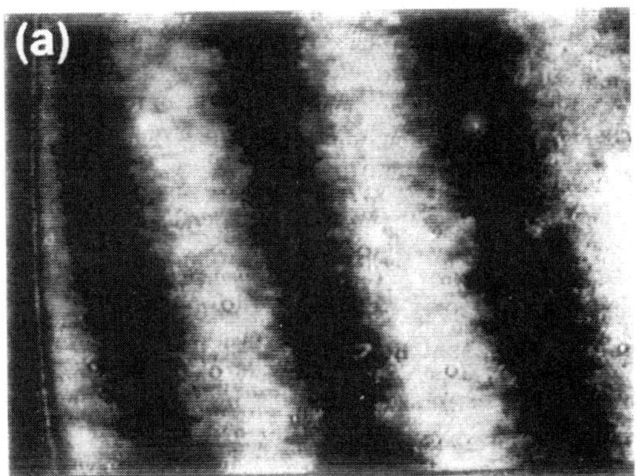

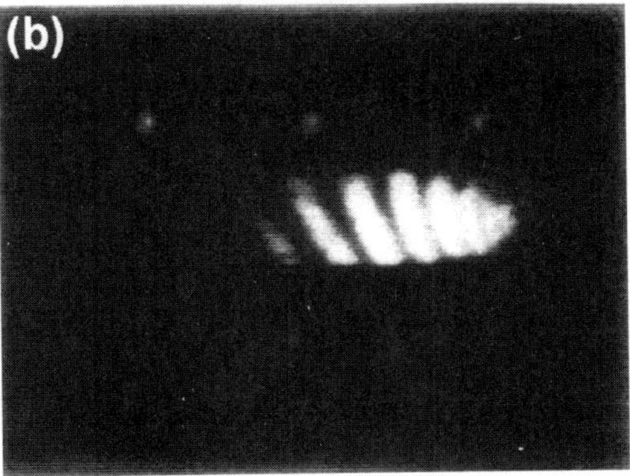

Figure 10.18   Image of 8CB (at 37°C) near the edge of a torsional parallel-plate flow cell with one plate rotating at a rate $\Omega = 0.025$ sec^{-1}. The plates are 43 mm in diameter with a gap of 300 μm. Image **(a)** is magnified with a field of view of 1.8 mm, while **(b)** is a "full slit view" of the sample extending from the axis of rotation nearly to the edge of the sample. The tipping of the birefringent bands away from the vertical (the flow direction) increases with increased radial position, corresponding to an increase in local Ericksen number. The Ericksen number at the sample edge is around 750. (From Mather et al. 1996b, by permission of Taylor & Francis.)

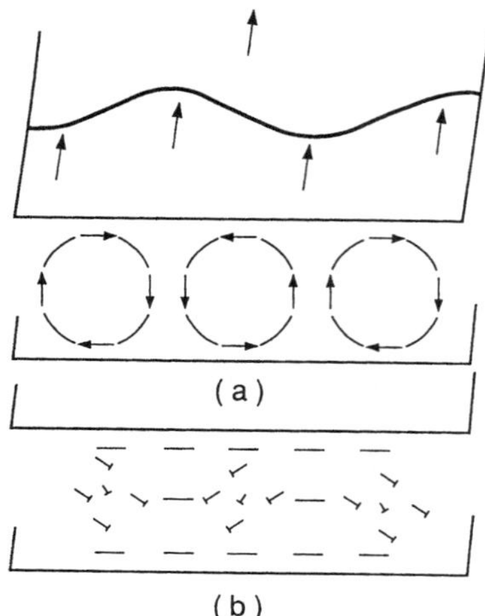

Figure 10.19   Velocity field **(a)** and director pattern **(b)** in roll cells that form in a tumbling nematic initially oriented in the vorticity direction of a shearing flow. (From Larson 1993, with permission from the Journal of Rheology.)

observed in both small-molecule and polymeric tumbling nematics (Guyon et al. 1977; Dubois-Violette et al. 1978; Mather et al. 1996b; Srinivasarao and Berry 1991; Larson and Mead 1993). Once the roll cells form in 8CB, *disclinations* are eventually produced (see Section 10.2.7), and the flow takes on the irregular time-dependent characteristics of *director turbulence* (Gähwiller 1972; Manneville 1981; Cladis and Torza 1976; Mather et al. 1996b) (see Fig. 10-20). Similar transitions exist in polymeric nematics (see Section 11.3.5). Thus, tumbling nematics are especially susceptible to instabilities in shearing flow that produce irregular, disclination-filled samples.

10.2.7 Disclination Dynamics

Nematics frequently contain defects called *disclinations* where the director abruptly changes orientation (Kléman 1989). The most common of these are line disclinations, although point disclinations are also possible. Some of the disclination types, and the director fields surrounding them, are illustrated in Fig. 10-21 (de Gennes and Prost 1993; Stephen and Straley 1974). Within the continuum Frank theory, the disclination line is a singularity, but at the molecular level, where the Frank theory fails, the singularity is replaced by a disclination "core" whose structure is most likely biaxial for $\pm 1/2$ disclinations in small molecule nematics and whose size is roughly that of the molecules (Meiboom et al. 1983; Schopohl and Sluckin 1987; Hudson and Larson 1993). The core structures in polymer nematics are likely to be quite different than those in small molecules, because the ratios of Frank constants differ from unity more in the former than in the latter (Hudson and Larson 1993; Zasadzinski et al. 1986; Mazelet and Kléman 1986; De'Neve et al. 1995). The *strength m* of a disclination (Friedel and Kléman 1970) is determined by the angle $\theta = 2\pi m$ through which the director rotates as one orbits through an angle $\phi = 2\pi$ about the disclination (see Fig. 10-21). The strength m must be a multiple of $\pm 1/2$. The most common disclinations are of strength $\pm 1/2$ or ± 1. A $+1/2$ (or $-1/2$) disclination is a true topological defect, meaning that it can disappear only if it is annihilated by a

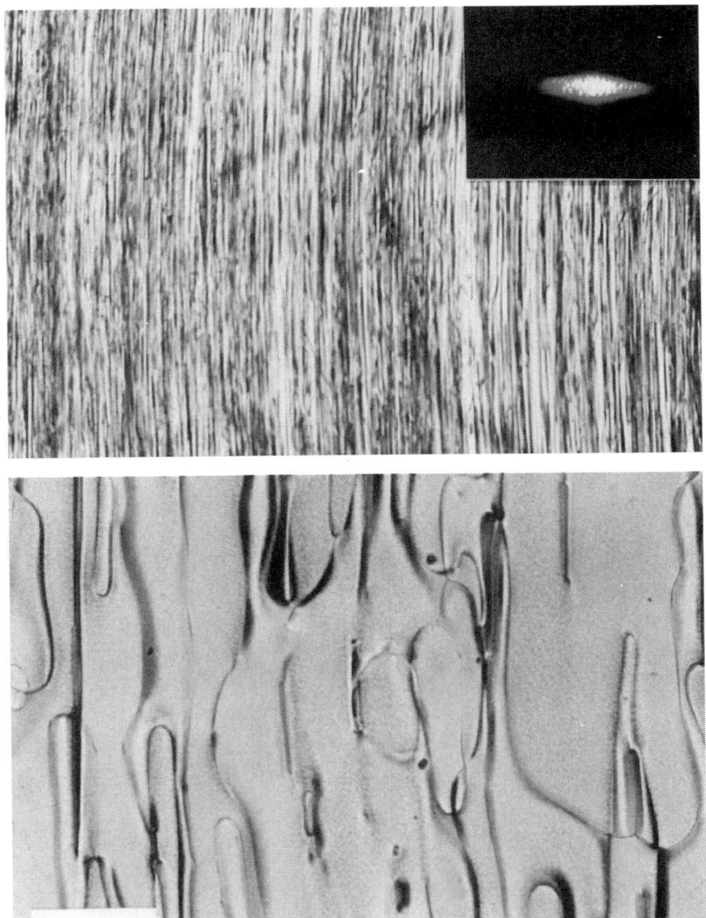

Figure 10.20 Disclination pattern observed in 8CB (37°C) **(top)** during steady-state shearing (the "worm" texture) and **(bottom)** one minute after cessation of shearing with no polarizers. The flow direction is vertical in a torsional plate-plate device with a gap of 300 μm and a rotation rate of 0.6 sec^{-1}; the sample is observed 17 mm from the axis of rotation and Er $\approx 2 \times 10^5$. The bar on the lower figure corresponds to 500 μm. The inset on the upper figure is a laser-light ($\lambda = 633$ nm) scattering pattern; the width of the insert corresponds to about 30 degrees. (From Mather et al. 1996b, by permission of Taylor & Francis.)

$-1/2$ (or $+1/2$) disclination. The ± 1 disclinations are not topological defects; these can disappear without annihilation by "escaping into the third dimension" (Meyer 1973; Cladis and Kléman 1972). In the "escape" process, the apparent disclination "core" thickens; ± 1 disclinations therefore usually appear as thick lines in an optical microscope. Thin lines are $\pm 1/2$ disclinations (Nehring 1973). Thin $\pm 1/2$ lines cannot terminate in the bulk; they either anchor at surfaces or form closed loops (Saupe 1973). The strength of a thin loop is $+1/2$ on one side and $-1/2$ on the other; thus a loop can shrink spontaneously

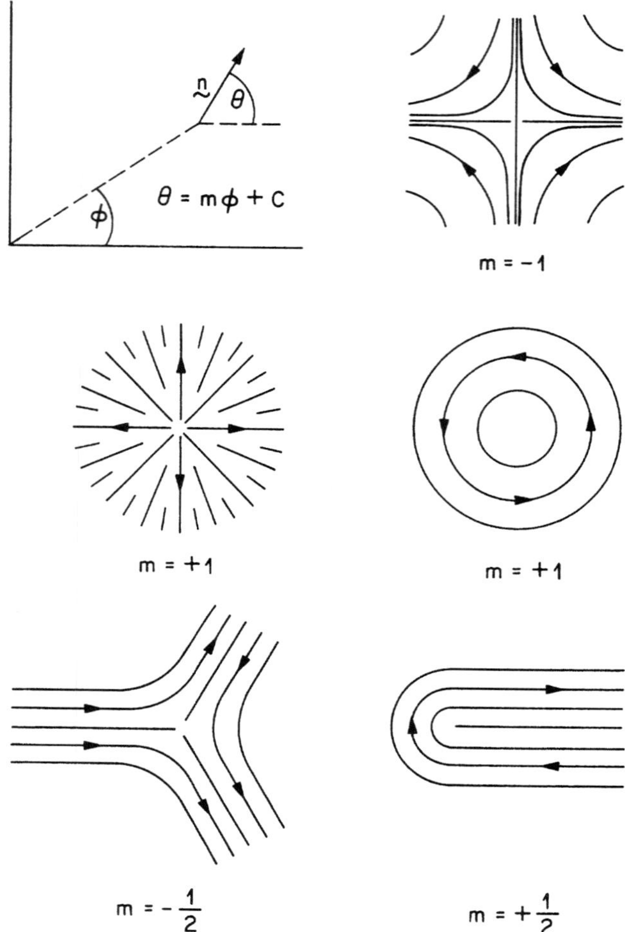

Figure 10.21   Classes of disclinations in a nematic, with the disclination line perpendicular to the plane of the page. [From Larson 1988 (adapted from Chandrasekhar 1992), with permission from Cambridge University Press.]

until it disappears. Further discussion of disclinations can be found in de Gennes and Prost (1993).

As described shortly, disclinations can readily be produced by strong flow fields (Wahl and Fischer 1973; Graziano and Mackley 1984), especially in tumbling nematics (Cladis and Torza 1975; Mather et al. 1996b). A very high density of disclinations is also produced if a material is suddenly quenched from an isotropic to a nematic state by reducing the temperature or increasing the pressure. When such a quench occurs, in each locale the material is suddenly forced to define a direction for the director, and the direction chosen varies from one locale to another. This produces spatial gradients in **n**,

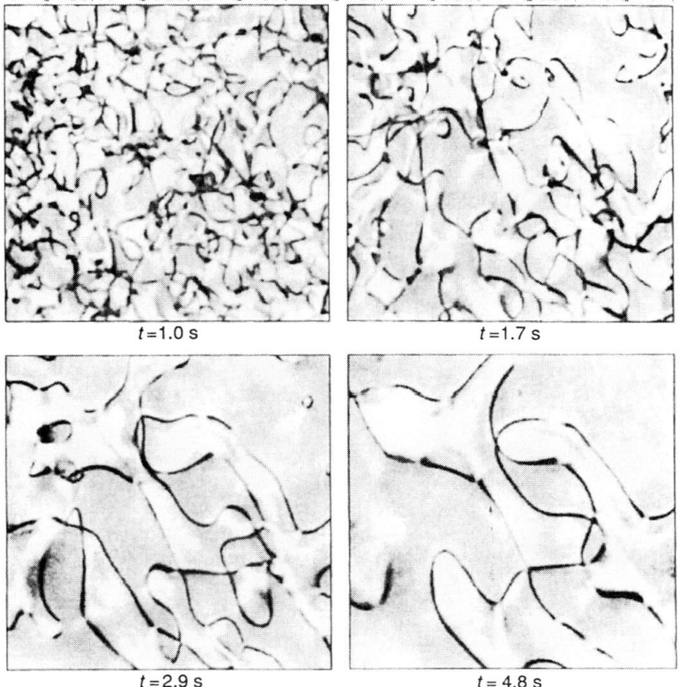

Figure 10.22 A sequence of images of disclinations in 5CB at various times t after a pressure jump $\Delta p = 4.7$ MPa, sufficient to induce a transition from an isotropic state at 3.6 MPa and 33°C to a nematic state. The field of view is 360 μm. (Reprinted with permission from Chuang et al. 1991 Copyright © 1991, American Association for the Advancement of Science.)

along with disclination lines. Figure 10-22 shows a "tangle" of disclination lines produced by a pressure jump across the isotropic–nematic phase boundary for 5CB.

Disclination lines are energetically disfavored because they produce gradients in the director profile and Frank stresses. So if the sample is left alone, the disclination lines and loops spontaneously shrink in length and annihilate one another (see Fig. 10-22) until no disclinations are left—except for any that are pinned by any impurities in the fluid or by wall irregularities, and those trapped because of incompatibilities in anchoring conditions at surfaces (Chuang et al. 1991; Nagaya et al. 1992).

When disclinations are present, the Frank contribution to G' should scale with the average spacing a between disclinations as $G' \sim K/a^2$. For a high density of disclinations ($a \lesssim 10$ μm), G' might exceed 1 Pa.

The rate at which disclinations disappear after a quench follows a simple scaling law,

$$\rho_V \propto \frac{1}{a^2} \propto \frac{\eta}{Kt} \tag{10-32}$$

where, again, η is a typical Leslie–Ericksen viscosity, K is a Frank constant, and ρ_V is the volumetric density of disclinations—that is, the total length of the disclination lines per unit volume of sample. Figure 10-23 shows that this scaling law is obeyed after a sample of 5CB is quenched into the nematic state by a sudden increase in pressure.

As mentioned above, strong flow fields can also generate disclination lines. This may seem surprising, at least for flow-aligning nematics, since one might expect strong flow to align the director to the alignment angle, as well as suppress deviations from this alignment, including the deviations that are associated with disclinations. However, at high Ericksen number, experiments show that torsional shearing tends to produce copious numbers of disclination lines in both flow-aligning and tumbling nematics (Graziano and Mackley 1984; Mather et al. 1996a, 1996b).

For example, Fig. 10-24 shows the steady-state areal density ρ_A of disclinations ($\pm 1/2$'s and ± 1's) as a function of shear rate $\dot\gamma$, measured by microscopic image analysis for flow-aligning 5CB in a torsional shearing flow between glass plates. The areal density ρ_A is the volumetric density ρ_V integrated over the sample gap h. As $\dot\gamma$ is slowly increased by increasing the rotation rate Ω of one of the plates (closed symbols in Fig. 10-24), a high

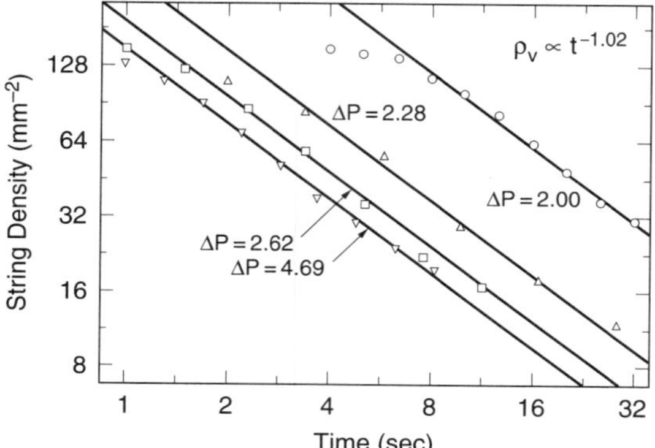

Figure 10.23 The volumetric "string density," or length of disclination line per unit volume ρ_V, is extracted from images such as those in Fig. 10-22 and plotted versus time after the pressure jump Δp (in MPa) into the nematic state. The density ρ_V scales as $\rho_V \propto t^{-1.02}$. (Reprinted with permission from Chuang et al. 1991 Copyright © 1991, American Association for the Advancement of Science.)

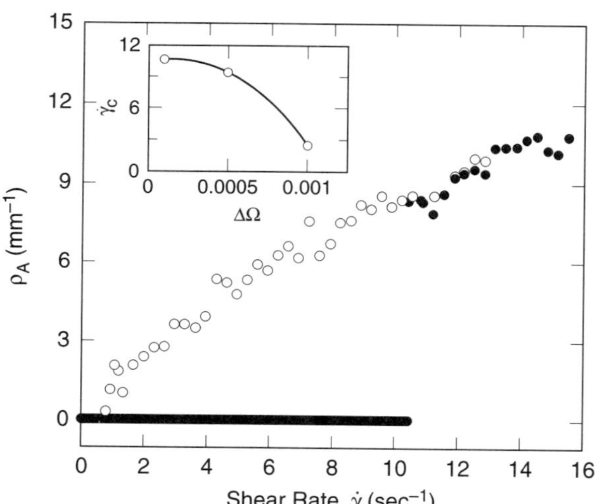

Figure 10.24 Areal density of disclination lines in steady-state shearing of 5CB as a function of shear rate for a gap h of 250 μm. Closed symbols are measurements during increasing shear rate; open symbols are for decreasing shear rate. For increasing shear rate, at a critical shear rate of about 10.5 sec^{-1}, the density of disclinations jumps from zero to a high value. The critical shear rate corresponds to an Ericksen number $Er_c \equiv (\gamma_1 h^2 \dot\gamma / K_3)$ of ~ 4100. The inset shows the critical shear rate as a function of the increment in angular velocity $\Delta\Omega$. (From Mather et al. 1996a, by permission of Taylor & Francis.)

density of disclinations is nucleated at a critical Ericksen number of Er = 4100, where Er $\equiv \gamma_1 \dot{\gamma} h^2 / K_3$. As $\dot{\gamma}$ (or Er) increases further, ρ_A continues to increase. If $\dot{\gamma}$ is then slowly decreased (open symbols in Fig. 10-24), ρ_A decreases as expected, but does not collapse to zero until Er $\propto \dot{\gamma}$ has been decreased nearly to zero.

Thus, there is a large *hysteresis* in the relationship between ρ_A and Er, suggestive of a *subcritical bifurcation*. Evidently, disclination lines are themselves agents for the nucleation of other disclination lines. This inference is supported by direct microscopic observations in shearing flows, in which a single disclination loop is observed to stretch and break into two daughter loops; each of these can then stretch and break again (Mather et al. 1996a). The turbulence generated in the fluid around a disclination loop might also be responsible for the multiplication of loops. How the very first disclination line forms during shearing of a flow-aligning nematic is still a mystery. But it is likely that disclinations are the result of flow instabilities produced by strong nonlinearities at high Ericksen number Er, just as flow instabilities are produced in Newtonian fluid flow by nonlinear effects at high Reynolds number Re (Drazin and Reid 1981). In fact, the irregular disclination-filled flow of some nematics at high Er has been called "director turbulence," because of this analogy to the inertial turbulence of normal isotropic fluids at high Re (Gähwiller 1972; Mannevelle 1981; Cladis and Torza 1976; Cladis and van Saarloos 1992). When the disclination density becomes high enough, disclinations interact and are destroyed by annihilation events, so that a steady-state density of them is eventually established (Mather et al. 1996a).

Tumbling nematics, with $\alpha_3 / \alpha_2 < 0$, seem to be especially susceptible to director turbulence and disclination formation in shearing flows (Cladis and Torza 1975; Mather et al. 1996b). Figure 10-25 shows that at a given Ericksen number, the steady-state density of disclinations is much greater in tumbling 8CB than in flow-aligning 5CB. The greater abundance of disclinations in a sheared tumbling nematic than in a flow-aligning one is not surprising, since in a tumbling nematic there is no preferred alignment angle toward

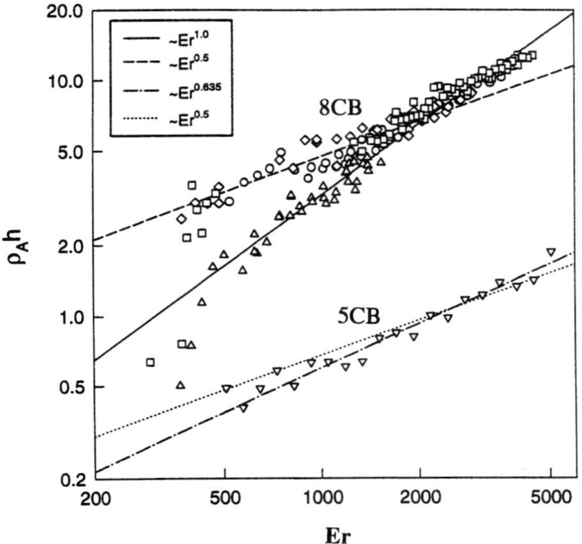

Figure 10.25 Dimensionless areal disclination density versus Ericksen number Er $\equiv \gamma_1 h^2 \dot{\gamma} / K_3$ for 8CB in the tumbling regime ($T = 37°C$) at gaps h of 250 ($\triangle$), 300 ($\diamond$), 350 ($\bigcirc$), and 400 ($\square$) μm. The solid curve is a fit of the data at low Er (except at 250 μm) with an exponent β of 0.5, while the dashed line is a fit with $\beta = 1.0$. Also shown are data for flow-aligning 5CB ($T = 32.5°C$) fit with $\beta = 0.5$ (dotted line) and a best-fit value of 0.635 (dot-dashed line). (From Mather et al. 1996b, by permission of Taylor & Francis.)

which the director in the bulk is attracted. For 8CB at high Ericksen number (Er > 2000), the dimensionless areal disclination density seems to follow a simple scaling law with Ericksen number:

$$\rho_A h \propto \text{Er}^{1.0} = \frac{\gamma_1 \dot{\gamma} h^2}{K_3} \qquad (\text{Er} > 2000) \qquad (10\text{-}33)$$

For lower Ericksen number and at gaps h $\geq$ 300 μm, a different scaling law seems to hold,

$$\rho_A h \propto \text{Er}^{0.5} \qquad (\text{Er} < 2000) \qquad (10\text{-}34)$$

The scaling law in Eq. (10-33) was predicted by Marrucci (1984) by assuming that the disclination density at steady state is set by a balance between the viscous energy density $\eta \dot{\gamma}$ and the Frank elastic energy density K/a^2. Since the areal density ρ_A is proportional to $\rho_V h \propto h/a^2$, this balance is

$$\eta \dot{\gamma} \propto \frac{K}{a^2} \propto \frac{K \rho_A}{h} \qquad (10\text{-}35)$$

Rearranging Eq. (10-35) gives the scaling law in Eq. (10-33), if one chooses the characteristic viscosity η to be γ_1 and chooses the Frank constant K to be K_3.

The scaling law in Eq. (10-34) can be derived by a two-dimensional version of Marrucci's argument: Set $\rho_A \propto 1/a$, and use the first proportionality in Eq. (10-35) to derive Eq. (10-34). The two-dimensional scaling is expected to be valid when the disclination density is low, so that the spacing between disclinations in the plane of the sample is high enough compared to the gap that a typical disclination interacts as strongly with the walls as it does with other disclinations. Significantly, in cases where the two-dimensional scaling applies, the disclinations are found to be confined to a thinner slab near the midplane of the sample than is the case when the three-dimensional scaling works (Mather et al. 1996b). At present, there is no explanation for the absence in thin samples of a crossover from three-dimensional to two-dimensional scaling.

10.3 CHOLESTERICS: CHIRAL NEMATICS

The flow properties of chiral nematics, or *cholesterics*, have been little studied. Because in the cholesteric the director varies spatially in a helical fashion, it is essentially impossible to satisfy *homeotropic* anchoring conditions unless the sample is so thin (of the order of the cholesteric *pitch P*) that the cholestericity is suppressed. For *tangential* anchoring conditions, however, a defect-free cholesteric sample can be prepared between flat plates, if the gap is an integer times one-half the pitch $2\pi/q$. For any thick gap, q can adjust slightly from its equilibrium value q_0 to accommodate this condition. The director field in such a sample is uniform in each plane parallel to the surfaces, but rotates helically as one moves normal to the surfaces, as illustrated in Fig. 10-2. Such director fields are called planar or *Grandjean* textures. Under other anchoring conditions, samples are typically filled with disclinations. The order of magnitude of the pitch is usually a few microns or less, except in some samples for which there is a change in the sign of q_0 as a function of temperature or as a function of concentration in a mixture of right-handed (positive q_0) and left-handed (negative q_0) cholesterics (de Gennes and Prost 1993). Another exceptional case is that of a dilute solution of a cholesteric in a nematic; here the pitch can also be very large.

When the pitch corresponds to an optical wavelength in the visible range, the cholesteric can show brilliant colors on reflection of white light. This, coupled with a strong temperature-sensitivity of P, makes cholesterics convenient thermometers.

The flow properties of cholesterics have scarcely been studied at all. Figure 10-26 shows one of the few sets of measurements of the viscosity of a cholesteric-forming small-molecule material, cholesteryl myristate, as a function of shear rate in flow through a capillary at various temperatures (Sakamoto et al. 1969). As the temperature is lowered, cholesteryl myristate passes through isotropic, chiral nematic, smectic, and crystalline phases. Figure 10-26 shows that the low-shear-rate viscosity jumps by over an order of magnitude when the temperature is decreased through the isotropic-to-cholesteric transition. The cholesteric phase appears to be extremely shear thinning and perhaps has a yield stress.

The viscosity of a typical cholesteric made by doping a nematic with a modest amount of chiral nematic is much lower (around 1 P or so) than that of the typical pure cholesteric. Perhaps this is because the pitch of the doped nematic is higher than that of the typical pure cholesteric, or because the twist elastic constant of the doped nematic is much lower.

A theory by Helfrich (1969) suggests that the viscosity of a chiral nematic phase is high because the cholesteric director is *blocked*; that is, it cannot respond to the flow

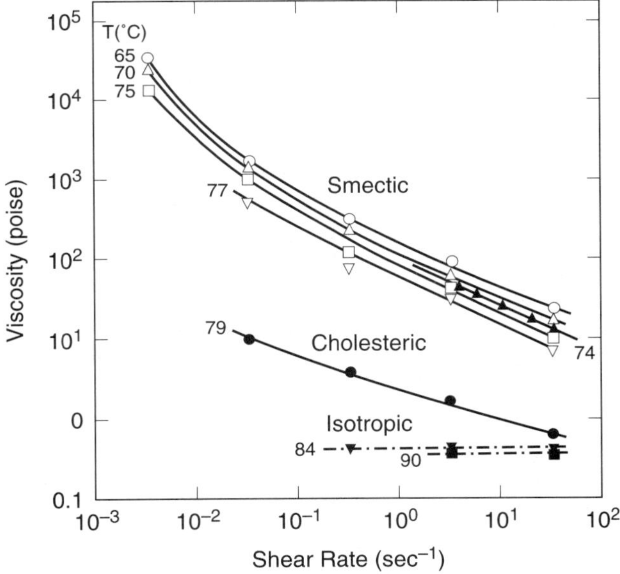

Figure 10.26 Viscosity versus shear rate of cholesteryl myristate in a cone-and-plate rheometer as a function of shear rate. At high temperatures, $T \geq 83°C$, the sample is a low-viscosity, Newtonian isotropic liquid. At intermediate temperatures, $83 \geq T \geq 78°C$, the sample is a shear-thinning cholesteric. At low temperatures the sample is a shear-thinning smectic. (From Sakamoto et al., reprinted with permission from Mol. Cryst. Liq. Cryst. 8:443, Copyright © 1969, Gordon and Breach Publishers.)

gradients, and so the director in a flowing fluid element rotates as it moves to comply with the helical director pattern. The fixed director pattern therefore acts like a mesh, or *porous medium*, through which the fluid must *permeate*. The mesh size is proportional to the helical pitch $P \propto q_0^{-1}$. The permeability of this mesh is then proportional to q_0^{-2}; and in flow through a tube the flow velocity V, by "Darcy's law," is $V \propto (q_0^{-2}/\gamma_1)\, dp/dz$, where dp/dz is the pressure gradient down the axis of the tube. Comparing this result to that for Poiseuille flow of ordinary viscous liquids, one finds that the apparent viscosity of the cholesteric is $\eta_{\mathrm{app}} \propto \gamma_1 (q_0 R)^2$, where R is the tube radius. Thus, the apparent viscosity should depend strongly on tube radius, a prediction, which, surprisingly, has never been checked! The apparent viscosity should also be sensitive to the pitch. The shear thinning in Fig. 10-26 could presumably be explained by flow-induced orientation of the cholesteric superstructure; at high flow rates this structure complies with the velocity gradient and the viscosity drops. If this explanation is correct, the shear-thinning viscosity should depend on the the twist elastic constant, as well as the pitch.

When applied to geometries with moving boundaries, such as the cone-and-plate geometry, the Helfrich argument suggests that the flow should be concentrated in thin zones of width proportional to mesh size, and hence there should be apparent slip.

10.4 SMECTICS

10.4.1 Types of Smectics

Smectics are layered materials. They can be either liquid crystalline or crystalline, depending on the presence or absence of long-range three-dimensional positional order. While the simplest smectic is the smectic-A phase depicted in Fig. 10-27, there are many other kinds of smectics, designated by letters ranging from B to K, that possess less symmetry than smectic A (Gray and Goodby 1984). In some smectics, including smectic C, the direction of average molecular orientation, the director, is tilted with respect to the layers, as represented by the tipped filled ellipses in the side views in Fig. 10-27. The unfilled symbols show the "top" view of the positions of the molecules within each layer. There are also chiral forms of the smectic phases analogous to cholesteric nematic phases (Gray and Goodby 1984; de Gennes and Prost 1993); however, if chirality expresses itself in the smectic-A phase, it generates defect walls (the so-called twist-grain-boundary phase (de Gennes and Prost 1993; Renn and Lubensky 1988), since chirality is incompatible with smectic-A layering.

Within the family of untilted smectics, there is a hierarchy of phases with order ranging from that of smectic A, in which there is no positional order within each layer, to the crystalline smectics B and E, which have long-range positional order within each layer, namely hexagonal and orthorhombic for the B and E phases, respectively. This inplane order, represented by the dots within circles or ellipses of Fig. 10-27, propagates from layer to layer, producing fully three-dimensional crystalline order. The family of tilted smectics includes similar ordered phases: J, G, K, and H.

Within a smectic layer, a state of order called *hexatic* can exist that is intermediate between the liquid order of the smectic A, and the crystalline order of crystalline smectic B. This state of order characterizes the hexatic smectic-B phase. Hexatic order consists of long-range *bond orientational order*, but no long-range positional order. This is illustrated

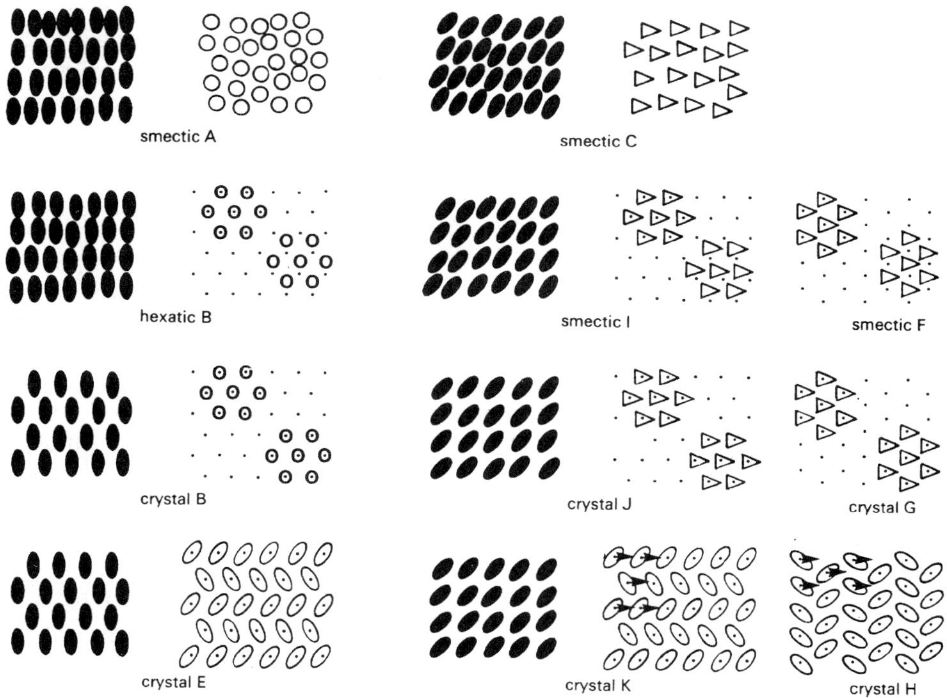

Figure 10.27   Side view and top view of molecular ordering in various smectics. Triangles or arrows are used to represent tilt directions. Hexatic order in hexatic B, smectic I, and smectic F is indicated in the top view by one set of circles or arrows centered on the dots, and the neighboring set not. See text for details. (From Gray and Goodby 1984, with permission.)

in Fig. 10-27 for the hexatic-B phase by the hexagonal arrangement of open circles in the top view, in which the nearest neighbors in the plane are arranged hexagonally around a given molecule. Thus to move from one molecule to a nearest neighboring molecule, one moves in one of six preferred directions. These six preferred directions *do not reorient as one moves from place to place in the layer;* hence imaginary bonds drawn between nearest neighbors have orientations with long-range order. Yet the *positional* correlation between one molecule and another in that plane decays to zero as the second molecule is chosen to be ever farther from the first. As a result, the hexatic phase is noncrystalline.

In the crystalline version of smectic B, there is positional as well as bond-orientational order, and this ordering is called *hexagonal*, as opposed to hexatic. In the tilted version of the hexatic smectic B, there are two different directions the tilt has been found to take with respect to the hexagonal bond orientation, namely toward nearest and toward next nearest neighbors, producing the two different smectics I and F (see Fig. 10-27). Similarly, there are two different tilted crystalline hexagonal phases, smectics J and G. Even in smectic *crystals*, the molecules retain freedom to rotate cooperatively about their axes; hence smectic crystals are not as "solid" as most simple crystals. Nature is wonderfully creative in contriving structures intermediate between simple liquids and crystalline solids! A single material can show many of these intermediate phases. For example, when cooled, terephthalylidene-bis-4-*n*-decylaniline (TBDA) undergoes successive phase transitions from isotropic through nematic, smectic A, smectic C, smectic I, smectic F, and smectic G, and finally (!) to a nonsmectic crystal (Gray and Goodby 1984). For more on the structures of smectics, see Gray and Goodby (1984), Chandrasekhar (1992), and de Gennes and Prost (1993).

10.4.2 Smectic Viscosities

The viscous properties of a smectic A are characterized by the same five independent viscosities that characterize the nematic. As we shall see, however, the elastic properties of the smectic are very different from those of a nematic, and some flows permitted to the nematic are effectively blocked for the smectic. For smectic C, for which the director is tilted with respect to the layers, there are some 20 viscosities needed to characterize the viscous properties (Leslie 1993). Formulas for these, derived using a method analogous to that used for nematics by Kuzuu and Doi (1983, 1984) can be found in Osipov et al. (1995). The smectic phase for which rheological properties are most commonly measured is smectic A, however, and hereafter we will limit our discussion to it.

Since the smectic-A phase is solid-like in one direction and liquid-like in the other two, we expect the rheological properties of a well-ordered smectic-A monodomain to depend drastically on the direction of deformation with respect to the orientation of the layers. The three simplest geometries for shearing flow are illustrated in Fig. 10-28, wherein the normal to the layers is oriented parallel to the vorticity, flow, and gradient directions in Fig. 10-28a, 10-28b, and 10-28c, respectively. Following the convention used for nematics, we shall call these the "a," "b," and "c" orientations, respectively. In smectics and especially block copolymers, these orientations are also given the names (a) "perpendicular," (b) "transverse," and (c) "parallel." In the "parallel" (c) orientation, the smectic layers are parallel to the shearing surfaces. In the "perpendicular" and "transverse" orientations, one direction within each layer is perpendicular to the shearing surfaces.

In the "b" orientation in Fig. 10-28b, uniform shear tends to rotate the layers and change their spacing. Since the layer spacing is a "solid-like" property of the smectic phase, shearing in this orientation should produce a solid-like material response, at least for shearing stresses

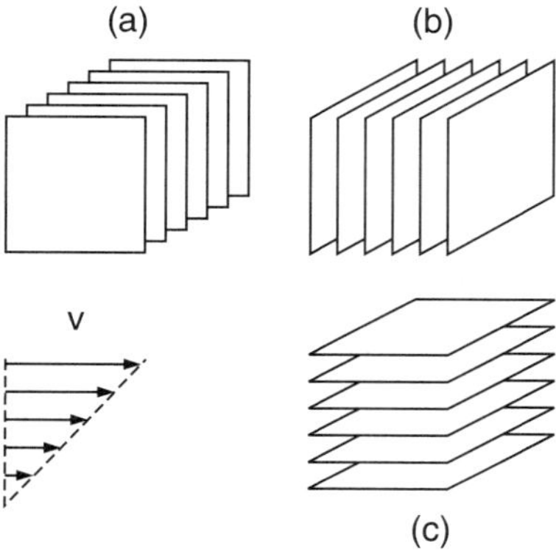

Figure 10.28   (a–c) Alignment directions "a," "b," and "c" of smectic layers in a shearing flow. (Adapted from Horn and Kléman 1978, with permission from EDP Sciences.)

less than some critical value needed to destroy the smectic state. In other orientations, such as the (a) perpendicular and (c) parallel orientations in Fig. 10-28b, shearing does not change the layer spacing, at least as long as the sample remains perfectly well oriented and defect free. For these orientations, the smectic is expected to exhibit a largely viscous response during steady-state shearing, *if the sample remains defect-free.*

10.4.3 Smectic Elasticities

In *thermotropic* (solvent-free) smectic-A phases, two types of distortion are permitted, namely, *splaying* of the director (which corresponds to *bending* of the layers) and layer *compression*. (*Note:* The material itself is assumed to remain incompressible; only the layers compress.) For weak distortions, the free energy cost of these is given by (de Gennes and Prost 1993)

$$W_{d,\text{smectic}} = \frac{1}{2}B\left(\frac{\partial u}{\partial z}\right)^2 + \frac{1}{2}K_1(\nabla \cdot \mathbf{n})^2 \qquad (10\text{-}36)$$

Here u is the position of a layer plane and z is the position coordinate locally parallel to the director $\mathbf{n}$, where $\mathbf{n}$ is parallel to the average molecular axis, which is assumed to remain normal to the layer plane. $\partial u/\partial z \equiv \varepsilon$ is the *compressional* (or *dilational*) *strain*. Thus, layer bending and layer compression are characterized by a *splay* (or layer-bend) modulus K_1 and a *compression* modulus B. Other kinds of distortion present in nematics, such as bend or twisting of the director $\mathbf{n}$, are not compatible with layers that remain nearly parallel, and hence are "forbidden." Equation (10-36) is not invariant to rotations of frame, and its validity is limited to weak distortions; a rotationally invariant expression has been given by Grinstein and Pelcovits (1981).

For small-molecule thermotropic smectic-A phases, typical values of two elastic constants are $K_1 \sim 10^{-7}$ dyn and $B \sim 10^7$ dyn/cm^2 (Ostwald and Allain 1985). For *lyotropic* smectics, such as those made from surfactants in oil or water solvents, the layer compression modulus B can be much lower (see Chapter 12). From B and K_1, a length scale $\lambda \equiv (K_1/B)^{1/2} \approx 1$ nm is defined; it is called the *permeation depth* and its magnitude is roughly given by the smectic layer spacing. The permeation depth λ has nothing to do with the tumbling parameter λ, although we use the same symbol for both quantities. The importance of the permeation depth λ is that it defines the distance scale at which a smectic sample will be induced to exchange layer compression for splay in order to accommodate a macroscopic distortion. Thus, if one bends a stack of smectic layers, the bending might induce a local dilation or compression of the layer spacing only if the radius of curvature of the bend is very small, of order λ.

Alternatively, if one dilates a smectic stack by increasing its thickness by an amount $\delta h > 2\pi\lambda$, then the sample will prefer to bend the layers in an *undulational instability* (Rosenblatt et al. 1977; Ostwald and Allain 1985) in order to restore the lamellar spacing to its preferred value (see Fig. 10-29d). Note that the increase in thickness δh required to produce this instability is independent of the initial thickness h of the stack. Hence for a macroscopic sample of thickness, say, $h = 60\ \mu$m, the *strain* $\delta h/h$ required to induce the undulational instability is extremely small, $\delta h/h \approx 2\pi\lambda/h \approx 10^{-4}$. Thus *smectic monodomains are extremely delicate and can easily be disrupted by mechanical deformation.*

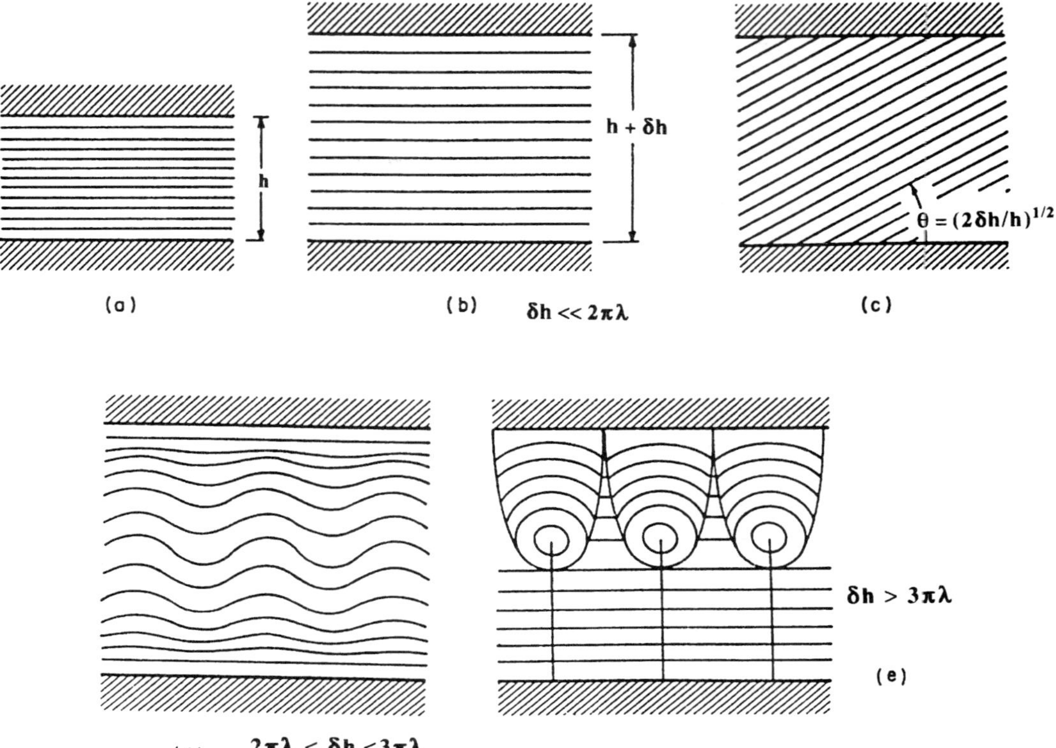

Figure 10.29   Response of an aligned smectic to layer dilation. **(a)** Initial equilibrium sample. **(b)** For a very small dilation $\delta h < 2\pi\lambda$, the layer spacing simply increases. **(c)** A uniform rotation of the layers decreases the spacing toward that of equilibrium, but doesn't satisfy the boundary conditions. **(d)** Hence, the sample undergoes an *undulational instability*, which also narrows the layer spacing while satisfying homeotropic boundary conditions. **(e)** For a large enough dilation, the undulation instability leads to formation of parabolic focal conic defects. (From Rosenblatt et al. 1977, with permission from EDP Sciences.)

10.4.4 Smectic Defects

10.4.4.1 Focal Conics

For a dilational strain only 50% larger than that required to produce an undulational instability, the undulations grow large enough to nucleate topological defects (see Fig. 10-29e). The defects found in this case are *parabolic focal conics,* consisting of a pair of orthogonal parabolic disclination lines with a common focus, as shown in Fig. 10-30b. When the smectic layers are dilated, a grid of these typically forms (see Fig. 10-30a and 10-30b).

Parabolic focal conics are a special case of generic *focal conic defects,* which are composed of layers curved to form toroidal surfaces called *Dupiń cyclides* (see Fig. 10-31). Each such structure contains a pair of disclination lines—one an ellipse and the other

a)

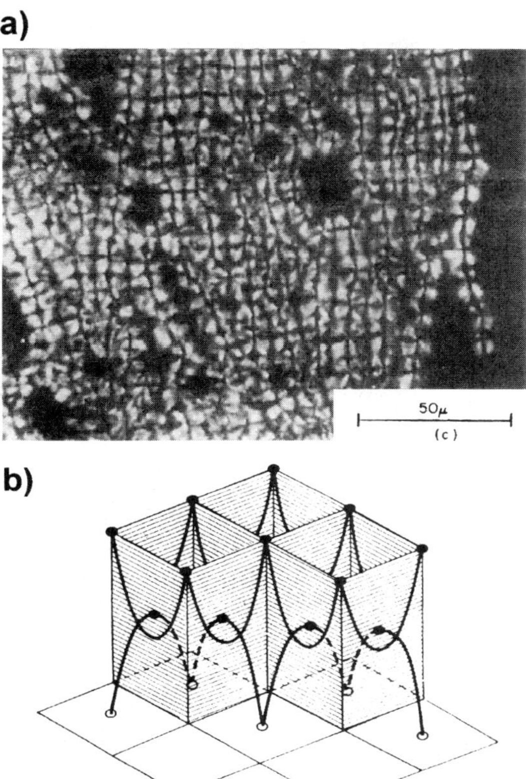

b)

Figure 10.30 (a) An array of parabolic focal conic defects viewed normal to the glass plates separated by a gap of 125 μm. (b) Schematic of (a) showing a three-dimensional "grid" structure of a regular array of parabolic focal conic defects. (From Rosenblatt et al. 1977, with permission from EDP Sciences.)

a hyperbola—that share a common focus. In one degenerate case, the ellipse and the hyperbola both become parabolas; this is the parabolic focal conic shown in Fig. 10-30b. In another a degenerate case, the ellipse becomes a circle and the hyperbola becomes a straight line (see Fig. 10-31b). Focal conic domains are a predominant type of defect in disrupted smectics, because *they maintain a constant layer spacing throughout the domain.* Deviations from constancy in layer spacing only occur when the radius of curvature of a layer becomes as small as the permeation depth λ, which is typically around one layer thickness. Thus, the layers are of constant thickness except within a single layer at the core of a disclination. This means that curved smectic layer lines are analogous to *optical wavefronts*; the layer periodicity mimics the wavelength periodicity of the electric field in a beam of light. Lines that are everywhere perpendicular to the parabolic or hyperbolic smectic layers are analogous to rays of light whose focal points (or lines) correspond to the smectic disclinations; hence, the name *focal conic* defect.

Unaligned, disrupted smectics, such as those prepared by quenching from an elevated temperature into the smectic state, are frequently filled with focal conic domains (Sethna and Kléman 1982; Lavrentovich 1986; Larson and Mather 1997). Parabolic focal conic domains can be packed to fill space as shown in Fig. 10-30a. Another packing, which

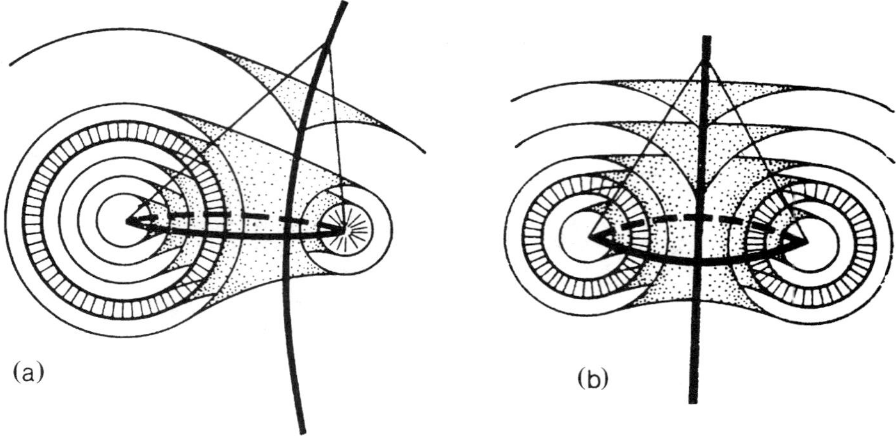

Figure 10.31   Smectic A layers forming Dupiń cyclides. In **(a)**, the disclination lines are an ellipse and a hyperbola; the ellipse passes through the focus of the hyperbola and the hyperbola passes through the focus of the ellipse. In the degenerate case **(b)**, the ellipse becomes a circle and the hyperbola becomes a line. The cone is an isolated confocal domain. (From Lavrentovich, reprinted with permission from Sov. Phys. JETP 64:984, Copyright © 1986, American Institute of Physics.)

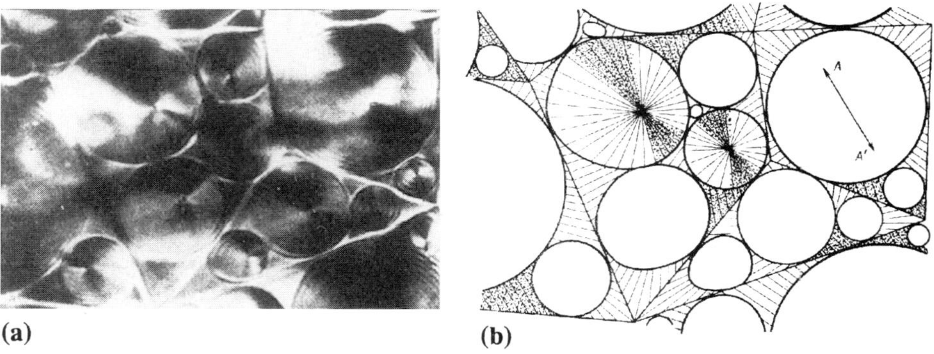

Figure 10.32   **(a)** Polygonal focal conic texture in a 200-μm-thick smectic viewed normal to the glass surfaces using polarized light. **(b)** Sketch of **(a)** showing the polygonal bases of pyramids into which cone-shaped focal domains are packed. (From Lavrentovich 1986, with permission from Sov. Phys. JETP 64:984. Copyright © 1986, American Institute of Physics.)

seems to be prevalent in thick, quenched samples, is the *polygonal focal conic* texture. It is formed by cutting cones from the focal structures shown in Fig. 10-31. These cones are then packed together into a pyramidal region with the cones sharing a common apex. The base of the pyramid is a polygon (see Fig. 10-32b). Small regions between cones are filled in by spherical surfaces that are concentric with the apex of the pyramid. By packing pyramids and inverted pyramids of this kind together, large volumes of space can be filled, as shown in Fig. 10-32b, without dilating or rupturing smectic layers, except at disclination lines.

Besides parabolic and polygonal focal domains, in samples that are reasonably aligned, *isolated* conic domains called *torical focal conic domains* can appear (Lavrentovich et al. 1994). A sketch is shown in Fig. 10-33a, and a *series* of such domains is shown in Fig. 10-33b. If a three-dimensional lattice of such domains forms, the smectic is broken up into discrete multilamellar polyhedra, or deformed "onions." This occurs in lyotropic smectics under flow (see Section 12.4.2.3).

Defect lines in smectics, unlike those in nematics, often do not continuously shrink with time and spontaneously disappear. Instead, there often seems to be a finite energy barrier that must be overcome if a smectic defect is to disappear. This difference between nematics and smectics is a consequence of the layer-spacing constraint that exists in smectics but not nematics. Because of this constraint, topological defects in smectics cannot be removed without ripping layers, and this requires a finite energy.

10.4.4.2 Other Defects

Other, nonfocal defects also occur in smectics. These include *walls*, such as the tilt wall depicted in Fig. 10-34a, as well as *dislocations* such as the *edge* and *screw* dislocations depicted in Fig. 10-34b and 10-34c. The most common defects in small-molecule smectics are those that maintain a constant lamellar spacing, such as focal domains, screw dislocations, and walls. Edge dislocations seem to be more common in lamellar block copolymers, which also have smectic symmetry (see Chapter 13).

(a)

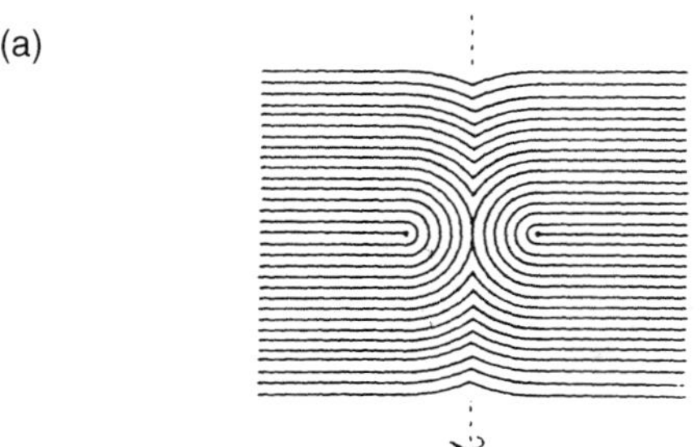

Figure 10.33   (a) Isolated toric focal conic defect. (b) A pair of such defects. (Adapted with permission from Winey et al., Macromolecules 26:2542. Copyright © 1993, American Chemical Society.)

(b)

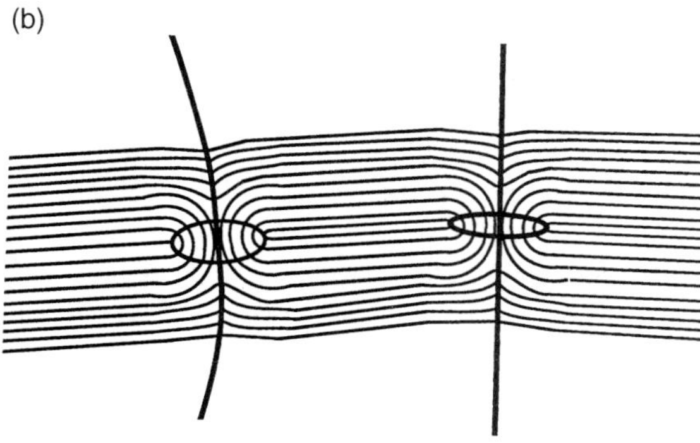

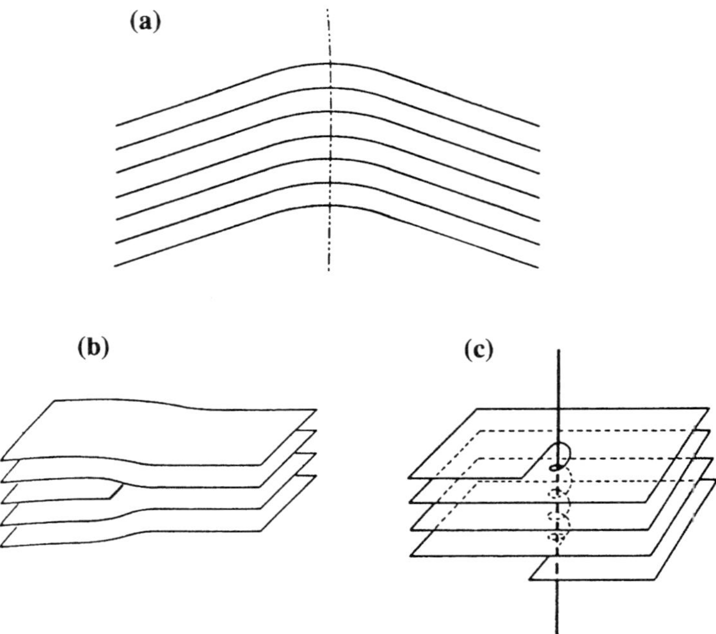

Figure 10.34 Defects in smectic phases: **(a)** a tilt wall, **(b)** an edge dislocation, and **(c)** a screw dislocation. (Adapted from Kléman et al. 1977, by permission of Taylor and Francis.)

While dilation of smectic layers easily produces parabolic focal conic defects (as discussed above), in the opposite case of compression, a stack of layers can recover its preferred spacing by ripping a layer to produce a pair of edge dislocations and then driving the two edge dislocations away from each other toward opposite edges of the sample. In this way the sample expels a layer, thus allowing the remaining layers to increase their thicknesses (Bartolino and Durand 1978; Ostwald and Allain 1985). In samples already containing edge dislocations, no energy is needed to rip the layers, and this process can occur for very small compressions.

10.4.5 Shearing Flow of Aligned Smectics

10.4.5.1 Parallel, or "c," Orientation

Now consider the steady, unidirectional shearing flow of a smectic monodomain with layers parallel to the shearing surfaces, as shown in Fig. 10-28c, with homeotropic anchoring of the molecules. If the layers are perfectly parallel, they can slide smoothly over each other, producing a low-viscosity Newtonian viscous response (Horn and Kléman 1978). However, a dust particle or other imperfection can easily disrupt the layers locally, leaving defects in its wake, as has been beautifully shown by Horn and Kléman (1978). Each moving defect nucleates another, so that a long cascading trail of defects is produced, eventually filling the whole sample with defects. Defects can easily proliferate in thermotropic smectics because of their high layer compression modulus B and small splay-bend constant K_1, which makes the layers more willing to roll up tightly than to compress even in the slightest. Even if dust particles are absent, any nonuniformity in the gap between otherwise "parallel" plates, even one as small as a few layer spacings, is likely to lead to defect formation under steady shear. As we shall see, however (in Section 10.4.8), oscillatory shear can drive out the defects and, in some cases, restore the sample to a state of uniform orientation (Horn and Kléman 1978). The process by which this occurs is not yet understood.

10.4.5.2 Transverse, or "b," Orientation

The effect of shear on monodomains of 8CB in orientation "b" in Fig 10-28b was studied by Marignan et al. (1978). They found that very small shearing strains can introduce defects. This result is not surprising, since shearing in orientation "b" rotates lamellae and thereby changes their spacing. A shear-induced tilt angle of only 10^{-4} radians with respect to the initial orientation is already enough to produce an array of dislocations; at higher tilt angles, focal conic defects appear.

10.4.5.3 Perpendicular, or "a," Orientation

The effect of shear on smectic *monodomains* in orientation "a" seems not to have been studied microscopically. However, the viscosities of a smectic phase magnetically aligned in the "a," "b," and "c" orientations has been measured in a capillary by Bhattacharya and Letcher (1980). The nematic studied was a mixture of 7CB, 8OCB, and 5CT that forms not only a high-temperature nematic phase and a low-temperature smectic-A phase, but also a still-lower-temperature nematic phase referred to as a *reentrant nematic*. Figure 10-35 shows that the viscosities in both nematic regimes follow the typical ordering $\eta_c \gg \eta_a > \eta_b$. In the smectic A phase, however, the ordering changes to $\eta_b \gg \eta_c \gg \eta_a$. The very high viscosity of orientation "b" in the smectic A phase is a consequence of "solid-like" structure in this orientation. But even η_c and η_a increase in the smectic state over the values in the neighboring nematic phases. The viscosity η_a is weakly non-Newtonian, indicating that layer fluctuations might contribute to the viscosity of this phase. It is believed that orientation "a" is more stable than "c" against layer undulation instabilities, since the layer normal in orientation "a" is orthogonal to both the shear and the shear gradient. The relative ordering of the smectic viscosities, $\eta_c > \eta_a$, is consistent with this expectation.

10.4.6 Steady Shearing Flow of Defect-Ridden Smectics

The introduction of defects into a smectic sample destroys its fluidity. This contrasts markedly with nematics, for which the presence of defects hardly alters the fluid's viscosity. Of course, this is because in a smectic the direction locally perpendicular to the layers is solid-like, and when defects are present, all directions acquire some solid-like character. Horn and Kléman (1978) measured shearing stresses in defect-containing smectic samples of 8CB and found that the shear stress was given by the equation of a Bingham plastic:

$$\sigma_{xy} = A + B\dot{\gamma} \tag{10-37}$$

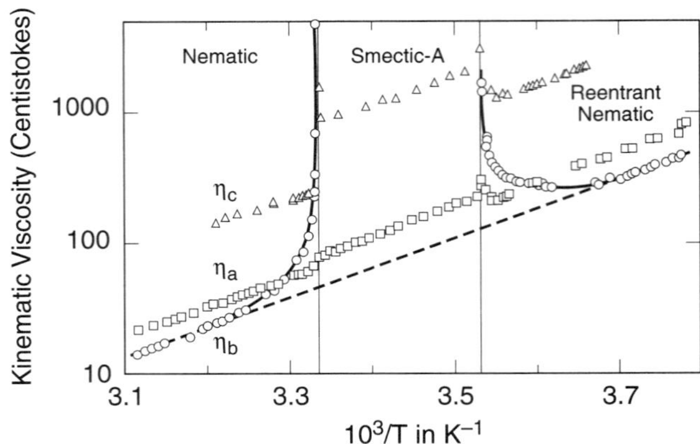

Figure 10.35   Dependence on inverse temperature of the viscosities in each of the orientations depicted in Fig. 10-28 for a ternary mixture of 73% (by weight) 7CB, 18% 8OCB, and 9% 5CT that forms nematic phases at high and low temperature and has an intermediate range of smectic A phase. The dashed line is an Arrhenius temperature-dependence for η_b. (From Bhattacharya and Letcher 1980, reprinted with permission from the American Physical Society.)

When there are no defects in the sample, $A \approx 0$ and B is then the viscosity of the smectic in orientation "b." Both A and B were found to increase as the number of defects in the sample increases. A is an elastic "yield stress" of a defect-containing smectic. The magnitude of A can can be estimated from the density of smectic defects as

$$A \sim \frac{K_1}{a^2} \tag{10-38}$$

where a is the spacing between defects.

For more highly disordered smectics, a power-law behavior, $\eta \propto \dot{\gamma}^n$, rather than Bingham-plastic behavior, has been observed, with $n \approx 0.5$, up to a critical shear rate, above which the behavior is Newtonian (Duke and Chapoy 1976; Panizza et al. 1995; Tamamushi 1974). Figure 10-36 shows the shear stress versus shear rate for 8CB in its smectic A state at a function of shear rate, plotted on a linear and a logarithmic scale. In the low-shear-rate region (I), the shear stress follows a power law with $n = 0.5$; but in the high-shear-rate region, the behavior is Newtonian with a viscosity similar to that of a nematic. The low-shear-rate power-law region shrinks to zero as the transition temperature to the nematic is approached.

10.4.7 Dynamic Modulus

The time-dependent rheological properties of disordered smectics are also peculiar. Figure 10-37 shows apparent G' and G'' data (labeled "MD") measured for 8CB after a quench from the isotropic state. Quenched samples contain a very high density of smectic defects, which produce elastic-like behavior in G' at low frequencies of oscillation. The apparent moduli G' and G'' in Fig. 10-37 are not true linear viscoelastic moduli, since they were

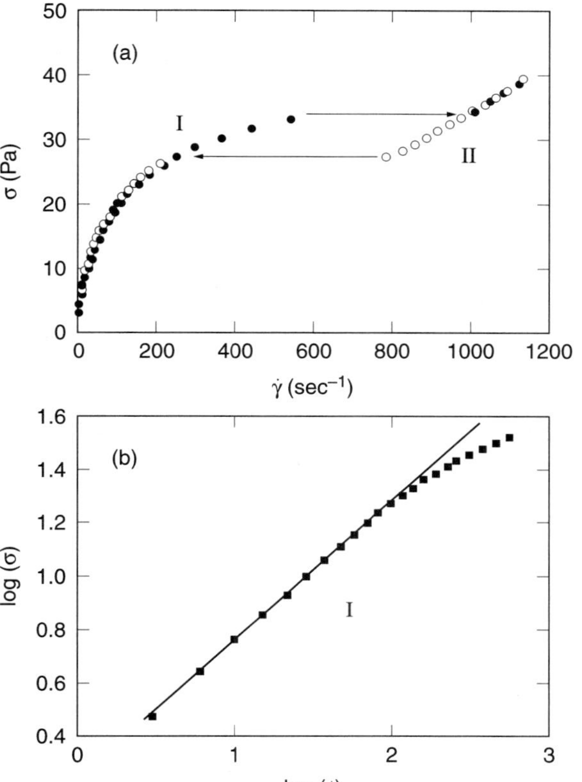

Figure 10.36 Shear stress versus shear rate for 8CB at 31°C in a stress-controlled cone-and-plate plotted **(a)** linearly and **(b)** logarithmically. In **(b)**, only the region I data are plotted. There is hysteresis in the jumps between regions I and II when the stress is increased or decreased. (From Panizza et al. 1995, with permission from EDP Sciences.)

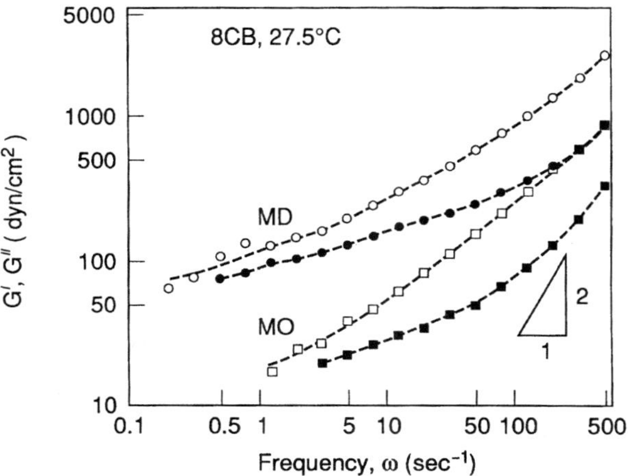

Figure 10.37 Apparent moduli G' (closed symbols) and G'' (open symbols) at a strain amplitude of $\gamma_0 = 0.1$ as functions of frequency for 8CB in the smectic A state. The curves labeled "MD" are for the "macroscopically disordered" smectic formed by a temperature quench into the smectic A state. The curves labeled "MO" are for "macroscopically ordered" smectic formed by large-amplitude ($\gamma_0 = 1$) oscillatory shearing at $\omega = 10$ sec^{-1} for 20 min. (From Larson et al. 1993, reprinted with permission from Steinkopff Publishers.)

measured in a nonlinear viscoelastic regime in which G' and G'' were strain-dependent. In fact, for these defect-ridden smectic materials, no linear regime could be found even for strain amplitudes as low as 1%.

10.4.8 Shear Alignment

As discovered by Horn and Kléman (1978), large-amplitude oscillatory shearing at an amplitude near unity drives out defects and aligns smectic layers. The curves in Fig. 10-37 marked "MO" (macroscopically ordered) correspond to a sample subjected to 20 minutes of oscillatory shearing at an amplitude of unity and a frequency of 10 sec^{-1} in a cone-and-plate instrument. This treatment decreases the low-frequency elastic modulus and increases the sample's fluidity. Continued oscillatory shearing can reduce further the number of these defects, leading to an additional decrease in the low-frequency elastic modulus G'. Thin samples ($\lesssim 300$ μm) sheared between clean glass surfaces that induce homeotropic alignment can be completely freed of defects, and the "b" or "parallel" alignment (the one consistent with homeotropic boundary conditions) is produced (Horn and Kléman 1978; Larson and Mather 1997). Thicker samples never seem to clear themselves, even after hours of oscillatory shearing. Presumably a perfectly aligned sample would show terminal viscoelastic behavior at low frequencies.

Steady shearing flow can also align a smectic, if the shear rate is high and the temperature is close to the temperature T_{AN} at which there is a transition from a smectic A to a nematic phase. Safinya et al. (1993) and Panizza et al. (1995) have shown using x-ray scattering that by shearing within region II of Fig. 10-36, 8CB aligns in the perpendicular, or "a," orientation (see Fig. 10-38). At shear rates within the power-law region I, however, the orientation is a mixture of "a" (perpendicular) and "c" (parallel), and the texture is thought to consist of long "jelly rolls"—that is, smectic layers rolled up into tubes that orient in the flow direction. Since the "a" orientation prevails close to T_{AN} and, indeed, in the nematic state just above the transition at T_{AN} (see Fig. 10-38), it is likely that the coupling of fluctuations to the shearing flow is responsible for orientation in the "a" direction. Intuitively, one expects that in the "c" orientation layer fluctuations will couple to the shearing flow more strongly than they do in the "a" orientation; hence the "a" orientation is favored when such fluctuations are strong, as they are near T_{NI}. Fluctuations are also evidently responsible for a shear-induced "a" orientation in lamellar-forming block copolymers (which have smectic symmetry) at temperatures near the transition to the layered state (see Section 13.3.1.2).

Goulian and Milner (1995) have recently predicted that the "a" orientation should be the stable one in steady shearing of thermotropic smectics even at temperatures well below T_{NA}. However, this analysis ignores defects, which we have seen play a large role in the flow of smectics. Direction "b" seems always to be avoided in small-molecule thermotropic smectics, presumably because since it is incompatible with shearing deformation. In polymeric smectics, however, the long molecules under rapid flow sometimes like to align parallel to the flow direction. It might be for this reason that such samples can under shearing flow become partially oriented in the "b" direction (Alt et al. 1995).

• Problems 10.4 through 10.10 test your practical knowledge about liquid crystals.

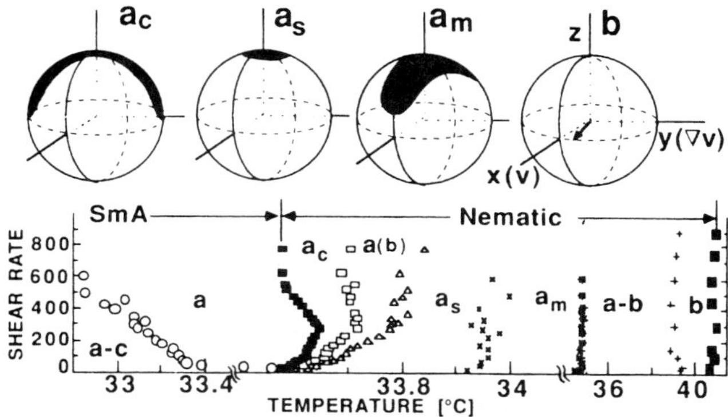

Figure 10.38 Temperature–shear-rate "phase diagram" for 8CB in the vicinity of the transition temperature $T_{AN} = 33.58°C$ from the nematic to the smectic A phase. At temperatures more than 5.5°C above T_{AN}, 8CB is a flow-aligning nematic, and it orients close to the flow direction in orientation "b." As the temperature is lowered, the nematic becomes a tumbler, preferring the "a" orientation, mixed with either "b" or "c." The "a" orientation prevails even in the smectic-A state at temperatures within a degree of T_{AN}, especially at high shear rates. At lower temperatures, the orientation is a mixture of "a" and "c." The globes show the ranges of orientations obtained in each of the states "a_c," "a_s," "a_m," and "b" inferred from x-ray scattering. (From Safinya et al. 1991, reprinted with permission from the American Physical Society.)

10.5 SUMMARY

Liquid crystals are spontaneously ordered states of matter that lack the full three-dimensional positional order of crystals. The flow properties of liquid crystals depend strongly on the direction of flow and of the flow gradient relative to preferred directions of alignment or of long-range positional order in the liquid crystal. The simplest liquid crystals are nematics which have a preferred orientation direction, but no long-range positional order. Small-molecule nematics are generally rather low viscosity fluids, but are highly birefringent and reorient readily under an electric or magnetic field, or under a weak flow. Smectics have one-dimensional positional order, and thus are less fluid than nematics, especially when the velocity gradient is in a direction that tends to change smectic layer spacings.

The fluid mechanics of nematics in simple shearing flows is now reasonably well understood. The Leslie–Ericksen theory in principle provides a complete description of the stresses in a defect-free nematic in an arbitrary flow and is, in principle, nearly as accurate for flows of small-molecule nematics as the Newtonian constitutive equation is for small-molecule isotropic liquids. The six viscosities and three Frank constants of the Leslie–Ericksen theory can be estimated from molecular theory. The shearing flow properties of a nematic depend strongly on whether the nematic is of flow-aligning or tumbling character. The director of a flow-aligning nematic tends to orient at high shear rates toward a preferred alignment direction, while for a tumbling nematic the director has no preferred angle of alignment, but rotates end-over-end when a strong shearing flow is suddenly turned on. Both flow-aligning and tumbling nematics tend to form large numbers of disclination lines under high rates of shear. The motions of these lines in a flow cannot at present be predicted in detail, but some scaling laws seem to describe the dependence of their steady-state density on the shearing rate.

The flow properties of smectic A are much more poorly understood. A smectic A phase has the same six viscosities as the nematic. Twist and bend director distortion modes are "forbidden" in a smectic because of its layered structure, but director splay (layer bend) and modest layer compression or dilation are allowed if a free energy price is paid, the magnitude of which is governed by the corresponding splay and compression/dilation elastic moduli. Smectic A flows freely in a shearing flow only if the layer normal is not parallel to the flow direction, so that the layers either slide over one another or do not move at all. Even when defects are initially absent from a smectic, in most cases steady shearing introduces them by mechanisms that depend on the orientation of the layers relative to the flow. Smectic defects are of various types, the most common of which (focal conics, walls, screws, and edges) preserve smectic layer spacing except near defect cores. Smectics with defects show quasi-solid behavior at low frequencies or shear rates. The defects can to some extent be ironed out of a smectic by imposition of a large-amplitude oscillatory shear flow. The mechanism by which this occurs is not understood.

The flow properties of other liquid crystals, such as chiral nematics (i.e., cholesterics), smectics C, and hexagonal phases, are even more poorly understood.

REFERENCES

Alt DJ, Hudson SD, Garay RO, Fujishiro K (1995). *Macromolecules* 28:1575.

Archer LA, Larson RG (1995). *J Chem Phys* 103:3108.

Baalss D, Hess S (1988). *Z Naturforsch* 43a:662.

Bahudur B (ed) (1990). *Liquid Crystals: Applications and Uses*, Vol 1, World Scientific, Singapore.

Bartolino R, Durand G (1978). *Ann Phys (Paris)* 3.

Bhattacharya S, Letcher SV (1980). *Phys Rev Lett* 44:414.

Buka A (1993). *Modern Topics in Liquid Crystals*, World Scientific, Singapore.

Carlsson T (1984). *Mol Cryst Liq Cryst* 104:307.

Carlsson T, Skarp K (1986). *Liq Cryst* 1:455.

Chandrasekhar S (1992). *Liquid Crystals*, 2nd ed., Cambridge University Press, Cambridge.

Chuang I, Durrer R, Turok N, Yurke B (1991). *Science* 251:1336.

Cladis P, Kléman M (1972). *J Phys (Paris)* 33:591.

Cladis P, Torza S (1975). *Phys Rev Lett* 35:1283.

Cladis P, Torza S (1976). *Colloid Interface Sci* 4:487.

Cladis P, van Saarloos W (1992). In *Solitons in Liquid Crystals*, Lam L and Prost J (eds), Springer-Verlag, Berlin.

Collings PJ (1990). *Liquid Crystals. Nature's Delicate Phase of Matter*, Princeton University Press, Oxford.

Creagh LT, Kmetz AR (1972). *Mol Cryst* 24:15.

de Gennes PG, Prost J (1993). *The Physics of Liquid Crystals*, Oxford University Press, New York.

Deloche B, Cabane B, and Jerôme D (1971). *Mol Cryst Liq Cryst* 15:197.

De'Neve T, Navard P, Kléman M (1995). *Macromolecules,* 28:1541.

Derfel G (1991). *Liq Cryst* 10:647.

Drazin PG, Reid WH (1981). *Hydrodynamic Stability*, Cambridge University Press, Cambridge.

Dubois-Violette E, Durand G, Guyon E, Manneville P, Pieranski P (1978). *Solid State Phys (Suppl)* 14:147.

Duke RW, Chapoy LL (1976). *Rheol Acta* 15:548.

Ericksen JL (1960). *Arch Rat Mech Anal* 24:231.

Ericksen JL (1961). *Trans Soc Rheol* 25:23.

Ericksen JL (1966). *Arch Rat Mech Anal* 23:266.

Fiałkowski M (1997). *Phys Rev E* 55:2902.

Fisher J, Fredrickson AG (1969). *Mol Cryst Liq Cryst* 8:267.

Frank FC (1958). *Disc Faraday Soc* 25:19.

Friedel J, Kléman M (1970). In *Fundamental Aspects of the Theory of Dislocations*, National Bureau of Standards Special Publication no. 317, Vol I, Washington, DC, p 607.

Fuller S, Li Y, Tiddy GJT, Wyn-Jones E, Arnell RD (1995). *Langmuir* 11:1980.

Gähwiller C (1972). *Phys Rev Lett* 28:1554.

Goulian M, Milner ST (1995). *Phys Rev Lett* 74:1775.

Gray GW, Goodby JW (1984). *Smectic Liquid Crystals,* Leonard Hill Publishers, London

Graziano DJ, Mackley MR (1984). *Mol Cryst Liq Cryst* 106:103.

Grinstein G, Pelcovits R (1981). *Phys Rev Lett* 47:856.

Gu D-F, Jamieson AM (1994). *J Rheol* 38:555.

Gu D-F, Jamieson AM, Wang SQ (1993). *J Rheol* 37:985.

Guyon E, Janossy I, Pieranski P, Jonathan JM (1977). *J Optics (Paris)* 8:357.

Haller I, Huggins HA (1972). US Patent 3,656,834.

Hammouda B, Mang J, Kumar S (1995). *Phys Rev E* 51:6285.

Han WH, Rey AD (1993). *J Non-Newt Fluid Mech* 48:181.

Helfrich W (1969). *Phys Rev Lett* 23:372.

Hinch EJ, Leal LG (1973). *J Fluid Mech* 57:753.

Horn RG, Kléman M (1978). *Ann Phys* 3:229.

Hudson SD, Larson RG (1993). *Phys Rev Lett* 70:2916.

Humphries RL, James PG, Luckhurst GR (1972). *Chem Soc Faraday Trans II* 68:1031.

Jerôme B (1991). *Rep Prog Phys* 54:391.

Kléman M (1989). *Rep Prog Phys* 52:555.

Kléman M, Williams CE, Costello MJ, Gulik-Krzywicki T (1977). *Philos Mag* 35:33.

Kneppe H, Schneider F, Sharma NK (1981). *Ber Bunsenges Phys Chem* 85:784

Kneppe H, Schneider F, Sharma NK (1982). *J Chem Phys* 77:3203.

Kröger M, Sellers HS (1995). *J Chem Phys* 103:807.

Kuzuu N, Doi M (1983). *J Phys Soc Jpn* 52:3486.

Kuzuu N, Doi M (1984). *J Phys Soc Jpn* 53:1031.

Larson RG (1988). *Constitutive Equations for Polymer Melts and Solutions*, Butterworths, New York.

Larson RG (1993). *J Rheol* 37:175.

Larson RG, Archer LA (1995). *Liq Cryst* 19:883. (In Figure 2 of this reference, the labels "η_2" and "η_3" are interchanged. Also, the value $E_a = 25,000$ J mol^{-1} cited in the text is incorrect; the correct value of 32,500 J mol^{-1} was used in all calculations.)

Larson RG, Mather PT (1997). In *Theoretical Challenges in the Dynamics of Complex Fluids*, McLeish TCB (ed), NATO ASI Series E: Applied Sciences, Vol 339, Kluwer Academic Publishers, London.

Larson RG, Mead DW (1993). *Liq Cryst* 13:151.

Larson RG, Öttinger HC (1991). *Macromolecules* 24:6270.

Larson RG, Winey KI, Patel SS, Watanabe H, Bruinsma R (1993). *Rheol Acta* 32:245.

Lavrentovich OD (1986). *Sov Phys JETP* 64:984.

Lavrentovich OD, Kléman M, Pergamenshchik VM (1994). *J Phys II (France)* 4:377.

Leenhouts F, Roebers HJ, Dekker AJ, Jonker JJ (1979). *J Phys Colloque C3 (Paris)* 40:C3-291.

Lehmann O (1890). *Z Phys Chem* 5:427.

Leslie FM (1966). *Q J Appl Math* 19:357

Leslie FM (1968). *Arch Rat Mech Anal* 28:265

Leslie FM (1993). *Liq Cryst* 14:121.

Leslie FM (1994). *J Non-Newt Fluid Mech* 54:241 (1994)

Luckhurst GR, Veracini CA (1994). *The Molecular Dynamics of Liquid Crystals*, Kluwer Academic Publishers, Dordrecht.

Luskin M, Pan T-W (1992). *J Non-Newt Fluid Mech* 42:369.

Maier G, Sackmann E, Grabmaier J (1975). *Applications of Liquid Crystals*, Springer-Verlag, Berlin.

Manneville P (1981). *Mol Cryst Liq Cryst* 70:223.

Manneville P, Dubois-Violette E (1976). *J Phys (Paris)* 37:285.

Marignan J, Malet G, Parodi O (1978). *Ann Phys* 3:221

Marrucci G (1984). *Proceedings of the Ninth International Congress on Rheology,* Mena B, Garcia-Rejon A, Rangel-Nafaile C (eds), Acapulco, Mexico, Oct 8–13, 1984.

Marrucci G, Greco F (1991). *Mol Cryst Liq Cryst* 206:17.

Mather PT (1994). PhD Thesis, University of California at Santa Barbara.

Mather PT, Pearson DS, Burghardt WR (1995). *J Rheol* 39:627.

Mather PT, Pearson DS, Larson RG (1996a). *Liq Cryst* 20:527.

Mather PT, Pearson DS, Larson RG (1996b). *Liq Cryst* 20:539.

Mather PT, Pearson DS, Larson RG, Gu D-F, Jamieson A (1997). *Rheol Acta* 36:485.

Mazelet G, Kléman M (1986). *Polymer* 27:714.

McColl JR, Shih CS (1972). *Phys Rev Lett* 29:85.

McHugh JJ, Edie DD (1995). *Liq Cryst* 18:327.

Meiboom S, Sammon M, Brinkman WF (1983). *Phys Rev A* 27:438.

Meyer RB (1973). *Philos Mag* 27:405.

Moldenaers P, Mewis J (1990). *J Non-Newt Fluid Mech* 34:359.

Nagaya T, Hotta H, Orihara H, Ishibashi Y (1992). *J Phys Soc Jpn* 61:3511.

Nehring J, Saupe A (1972). *J Chem Phys* 56:5527.

Nehring J (1973). *Phys Rev A* 7:1737.

Neville AC (1975). *Biology of the Arthropod Cuticle*, Springer-Verlag, New York.

Oseen CW (1933). *Trans Faraday Soc* 29:883

Osipov MA, Hess S (1993). *Molecular Phys* 78:1191.

Osipov MA, Hess S (1994). *Liq Cryst* 16:845.

Osipov MA, Sluckin TJ, Terentjev EM (1995). *Liq Cryst* 19:197.

Ostwald P, Allain M (1985). *J Phys (Paris)* 46:831.

Panizza P, Archambault P, Roux D (1995). *J Phys II (France)* 5:303.

Parodi O (1970). *J Phys (Paris)* 31:581.

Parsons JD (1978). *Phys Rev Lett* 41:877.

Patel JS, Lee SD (1989). *J Appl Phys* 66:1879.

Pieranski P, Guyon E (1974). *Phys Rev Lett* 32:924

Pieranski P, Guyon E, Pikin SA (1976). *J Phys (Paris)* 37,C1:3.

Praefcke K, Kohne B, Gundogan B, Singer D, Demus D, Diele S, Pelzl G, Bakowsky U (1991). *Mol Cryst Liq Cryst* 198:393.

Priest RG (1973). *Phys Rev A* 7:720.

Reinitzer F (1888). *Monatsh Chem* 9:421.

Renn SR, Lubensky TC (1988). *Phys Rev A* 38:2132.

Rosenblatt C (1984). *J Phys (Paris)* 45:1087.

Rosenblatt ChS, Pindak R, Clark NA, Meyer RB (1977). *J Phys (Paris)* 38:1105.

Safinya CR, Sirota EB, Plano RJ (1991). *Phys Rev Lett* 66:1986.

Safinya CR, Sirota EB, Bruinsma RF, Jeppesen C, Plano RJ, Wenzel LJ (1993). *Science* 261:588.

Sakamoto K, Porter R, Johnson J (1969). *Mol Cryst Liq Cryst* 8:443.

Saupe A (1973). *Mol Cryst Liq Cryst* 21:211.

Schopohl N, Sluckin TJ (1987). *Phys Rev Lett* 59:2582.

Sethna JP, Kléman M (1982). *Phys Rev A* 26:3037.

Singh AP, Rey AD (1998). *Rheol Acta* 37:30.

Singh S (1996). *Liq Cryst* 20:797.

Skarp K, Lagerwall ST, Stebler B (1980). *Mol Cryst Liq Cryst* 60:215.

Sonin AA, Yethiraj A, Bechhoefer J, Frisken BJ (1995). *Phys Rev E* 52:6260.

Srinivasarao M, Berry GC (1991). *J Rheol* 35:379.

St. John WD, Lu ZJ, Doane JW (1995). *J Appl Phys* 78:5253.

Stepanov VI (1983). In *Statistical and Dynamical Problems of the Elasticity and Viscoelasticity*,
    Urals Branch of the USSR Academy of Science, Sverdlovsk (in Russian).

Stephen MJ, Straley JP (1974). *Rev Mod Phys* 46:617.

Straley JP (1973). *Phys Rev A* 8:2181.

Stroobants A, Lekkerkerker HNW, Frenkel D (1986). *Phys Rev Lett* 57:1452.

Tamamushi B (1974). *Rheol Acta* 13:247.

Tjipto-Margo B, Evans GT, Allen MP, Frenkel D (1992). *J Phys Chem* 96:3942.

Tseng HC, Silver DL, Finlayson BA (1976). *Phys Fluids* 15:1213.

Wahl J, Fischer F (1973). *Mol Cryst Liq Cryst* 22:359.

Walba DM (1995). *Science* 20:250.

Winey KI, Patel SS, Larson RG, Watanabe H (1993). *Macromolecules* 26:2542.

Yang DK, West JL, Chien LC, Doane JW (1994). *J Appl Phys* 76:1331.

Yang IK, Shine AD (1993). *Macromolecules* 26:1529.

Yu LJ, Saupe A (1980). *Phys Rev Lett* 45:1000.

Yun CK (1973). *Phys Lett* 43A:369.

Zasadzinski JAN, Sammon MJ (1986). *Mol Cryst Liq Cryst* 138:211.

Zocher H (1933). *Trans Faraday Soc* 29 945.

Zuñiga I, Leslie FM (1989). *Liq Cryst* 5:725.

PROBLEMS AND WORKED EXAMPLES

Problem 10.1(a) (Worked Example) Nematic liquids are highly responsive to electric and magnetic fields, making them ideal for information display applications. Consider a layer of liquid crystal between two glass plates, initially uniformly aligned along an easy direction (x) parallel to the glass surfaces. Suppose a magnetic field is applied perpendicular to the plates, leading at high enough fields to a reorientation of the director in the bulk of the sample. This reorientation is called the *Freedericksz transition*. Since the director remains strongly anchored in the easy direction at the glass surfaces, the director field will become distorted, or nonuniform. As a starting step toward calculation of the distorted director field, calculate the Frank–Oseen free energy for a director field that is uniform in the x–y plane but varies spatially in the z direction.

ANSWER:

We represent the director $\mathbf{n}$ by polar and azimuthal angles θ and ϕ, such that

$$\mathbf{n} = (\cos\theta\cos\phi,\ \cos\theta\sin\phi,\ \sin\theta) \equiv (Cc, Cs, S) \tag{A10-1}$$

where we use the shorthand $C \equiv \cos\theta$, $S \equiv \sin\theta$, $c \equiv \cos\phi$, $s \equiv \sin\phi$. We now need to calculate the splay, twist, and bend distortional elasticities in Eq. (10-7). To do so, we remind ourselves of the definitions of the vector dot and cross and of the divergence and the curl operations involving arbitrary vectors $\mathbf{a}$ and $\mathbf{b}$. Using Cartesian coordinates we take $\mathbf{e}_x$, $\mathbf{e}_y$, and $\mathbf{e}_z$ to be unit vectors in directions x, y, and z. Then

$$\mathbf{a}\cdot\mathbf{b} = a_x b_x + a_y b_y + a_z b_z$$

$$\mathbf{a}\times\mathbf{b} = \mathbf{e}_x(a_y b_z - a_z b_y) + \mathbf{e}_y(a_z b_x - a_x b_z) + \mathbf{e}_z(a_x b_y - a_y b_x)$$

$$\nabla\cdot\mathbf{a} = \frac{\partial a_x}{\partial x} + \frac{\partial a_y}{\partial y} + \frac{\partial a_z}{\partial z}$$

$$\nabla\times\mathbf{a} = \mathbf{e}_x\left(\frac{\partial a_z}{\partial y} - \frac{\partial a_y}{\partial z}\right) + \mathbf{e}_y\left(\frac{\partial a_x}{\partial z} - \frac{\partial a_z}{\partial x}\right) + \mathbf{e}_z\left(\frac{\partial a_y}{\partial x} - \frac{\partial a_x}{\partial y}\right)$$

Applying these definitions to the case in which only derivatives with respect to z are retained, and taking the primes to denote differentiation with respect to z, we get

$$\nabla\cdot\mathbf{n} = n_z' = C\theta' \tag{A10-2}$$

$$\nabla\times\mathbf{n} = (-n_y',\ n_x',\ 0)$$

$$\mathbf{n}\cdot\nabla\times\mathbf{n} = -n_x n_y' + n_y n_x'$$

$$= -Cc(-Ss\theta' + Cc\phi') + Cs(-Sc\theta' - Cs\phi') = -C^2\phi' \tag{A10-3}$$

$$\mathbf{n}\times(\nabla\times\mathbf{n}) = (-n_z n_x',\ -n_z n_y',\ n_x n_x' + n_y n_y') \tag{A10-4}$$

Squaring Eqs. (A10-2) and (A10-3) and dotting Eq. (A10-4) into itself give the splay, twist, and bend free energy densities, which, using Eq. (10-7), add up to

$$2W_d(z) = K_1(\nabla\cdot\mathbf{n})^2 + K_2(\mathbf{n}\cdot\nabla\times\mathbf{n})^2 + K_3(\mathbf{n}\times\nabla\times\mathbf{n})^2$$

$$= (K_1\cos^2\theta + K_3\sin^2\theta)\left(\frac{d\theta}{dz}\right)^2 + \cos^2\theta(K_2\cos^2\theta + K_3\sin^2\theta)\left(\frac{d\phi}{dz}\right)^2 \tag{A10-5}$$

Now, since the field is applied in the z direction and anchoring is in the x direction, we expect the component n_y to remain zero, reducing Eq. (A10-5) to

$$2W_d(z) = (K_1\cos^2\theta + K_3\sin^2\theta)\left(\frac{d\theta}{dz}\right)^2 \tag{A10-6}$$

Problem 10.1(b) (Worked Example) Calculate the critical magnetic field required to induce the Freedericksz transition described in part (a), where d is the gap between the plates.

ANSWER:

The interaction of a magnetic field **H** with a nematic liquid produces a magnetization that depends on the orientation of the field relative to the nematic director. The magnetization **M** is $\chi_{\parallel}\mathbf{H}$ if **H** is parallel to **n** and $\chi_{\perp}\mathbf{H}$ if **H** is orthogonal to **n**, where the χ's are diamagnetic susceptibilities. The magnetic free energy density is then given by (de Gennes and Prost 1993):

$$W_M = -\int_0^H \mathbf{M} \cdot d\mathbf{H} = \tfrac{1}{2}\chi_\perp \mathbf{H}^2 - \tfrac{1}{2}\chi_a (\mathbf{n} \cdot \mathbf{H})^2 \tag{A10-7}$$

where $\chi_a \equiv \chi_{\parallel} - \chi_\perp$ is the anisotropy in the susceptibility. Since χ_a is usually positive, the free energy is lowered when the director rotates toward the field direction. But because of strong anchoring at the glass surfaces, this rotation distorts the director field, which increases the distortional free energy. From Eqs. (A10-6) and (A10-7), the total free energy density is

$$W(z) = \tfrac{1}{2}(K_1 \cos^2\theta + K_3 \sin^2\theta)\left(\frac{d\theta}{dz}\right)^2 - \tfrac{1}{2}\chi_a \sin\theta^2 \mathrm{H}^2 \tag{A10-8}$$

where the orientation-independent part of the magnetic free energy is of no consequence and has been dropped. The steady-state director profile under a magnetic field can now be calculated by finding the function $\theta(z)$ that minimizes the integral over the sample of Eq. (A10-8) under strong anchoring boundary conditions. This minimizing function is a solution of an Euler equation derived from Eq. (A10-8).

Here, we shall content ourselves with obtaining an analytic result for the minimum field H_c necessary to induce the first slight reorientation away from uniform alignment. If the director field is only very slightly distorted, then the angle θ is small, and Eq. (A10-8) reduces to

$$W(z) = \tfrac{1}{2}K_1 \left(\frac{d\theta}{dz}\right)^2 - \tfrac{1}{2}\chi_a \mathrm{H}^2\theta^2 \tag{A10-9}$$

The boundary conditions are $\theta = 0$ at $z = 0$ and at $z = d$. The function that minimizes W can be represented as a series of sine functions:

$$\theta = \sum_{i=1}^{\infty} \sin\left(\frac{\pi i z}{d}\right) \tag{A10-10}$$

The total free energy per unit area of sample is the integral of W over z from 0 to d. When Eq. (A10-10) is substituted into Eq. (A10-9) and the integral is carried out, the orthogonality properties of the sine functions guarantee that each mode i contributes independently to the total free energy. Thus, the lowest free energy state corresponds to one that contains the least distortion—that is, the lowest mode, $i = 1$. Keeping only this mode reduces Eq. (A10-9) to

$$W(z) = \tfrac{1}{2}K_1 \left(\frac{\pi}{d}\right)^2 \cos^2\left(\frac{\pi z}{d}\right) - \tfrac{1}{2}\chi_a \mathrm{H}^2 \sin^2\left(\frac{\pi z}{d}\right) \tag{A10-11}$$

When this is integrated over the gap, both the cosine squared and sine squared terms integrate to just $\pi d/2$. Thus,

$$\int_0^d W(z)\, dz = \left\{ \tfrac{1}{2}K_1 \left(\frac{\pi}{d}\right)^2 - \tfrac{1}{2}\chi_a \mathrm{H}^2 \right\}\left(\frac{\pi d}{2}\right) \tag{A10-12}$$

For very weak fields, the positive distortion term in the curly brackets will be larger in magnititude than the negative magnetic term, and the integrated free energy will be positive, unless θ is zero everywhere. Thus, unless the field is strong enough that the right side of Eq. (A10-12) is negative, there will be no reorientation. The critical field H_c required to induce the Freedericksz transition is therefore that for which the magnetic and distortional terms in the curly brackets sum to zero. Taking this to be the case, and rearranging, gives the critical field:

$$H_c = \left(\frac{\pi}{d}\right)\left(\frac{K_1}{\chi_a}\right)^{1/2}$$

Measurement of the critical field for this transition is therefore a convenient method for measuring the splay elastic constant K_1.

Problem 10.1(c) Compute the critical magnetic field H_c for a Freedericksz transition in which the director is initially oriented in the z direction, *perpendicular* to the plates, and strongly anchored in this orientation at the plates, and the field is imposed in the x direction, *parallel* to the plates. Which Frank constant controls this Freedericksz transition? Can you think of a configuration that would permit measurement of the twist elastic constant K_2?

Problem 10.2 (Worked Example) Using the Maier–Saupe theory described in Section 2.2.2.2, derive an implicit equation relating the order parameter S as a function to the Maier–Saupe interaction strength u_{MS}.

ANSWER:

Let us define the z direction to be the orientation direction of the director **n**. Then the order parameter tensor **S** is

$$\mathbf{S} = S\begin{pmatrix} 0 & 0 & 0 \\ 0 & 0 & 0 \\ 0 & 0 & 1 \end{pmatrix} \tag{A10-13}$$

where S is the scalar order parameter. If **u** is a unit vector in the orientation direction of a molecule, and we define the angle θ as the polar angle of **u** with respect to the z direction, then $\cos\theta$ is the projection of **u** against the z direction. Then using Eq. (A10-13), we obtain

$$\mathbf{uu} : \mathbf{S} = S\cos^2\theta \tag{A10-14}$$

From Eq. (2-7),

$$V_{\text{nem}} = \text{const} - \frac{3}{2}u_{MS}\mathbf{uu} : \mathbf{S} = \text{const} - \frac{3}{2}Su_{MS}\cos^2\theta \tag{A10-15}$$

Equation (2-6) then gives

$$\psi(\mathbf{u}) = C\exp\left(\frac{-\text{const}}{k_BT} + \frac{3}{2}S\frac{u_{MS}}{k_BT}\cos^2\theta\right) \tag{A10-16}$$

where C is the normalization constant. From the definition of the scalar order parameter S in Eq. (2-3), we obtain

$$S = \frac{3}{2}\langle\cos^2\theta\rangle - \frac{1}{2}$$

$$= \frac{3}{2}\int_0^\pi\int_0^{2\pi} C_1\exp\left(\frac{3}{2}SU_{MS}\cos^2\theta\right)\cos^2\phi\,d\phi\,\sin\theta\,d\theta - \frac{1}{2} \tag{A10-17}$$

where we have combined constants into $C_1 \equiv C\exp(-\text{const}/k_BT)$, and we have defined $U_{MS} \equiv u_{MS}/k_BT$. The integral over ϕ just gives a multiplier of 2π. If we set $x \equiv \cos\theta$, then we get $dx = -\sin\theta\,d\theta$. Equation (A10-17) then becomes

$$S = 3\pi \int_0^1 C_1 \exp\left(\frac{3}{2} SU_{MS}x^2\right) x^2 \, dx - \frac{1}{2} \tag{A10-18}$$

To obtain the constant C_1, we apply the normalization condition:

$$1 = \int_0^\pi \int_0^{2\pi} \psi \, d\phi \sin\theta \, d\theta = 2\pi C_1 \int_0^1 \exp\left(\frac{3}{2} SU_{MS}x^2\right) dx$$

This gives

$$C_1 = \left\{ 2\pi \int_0^1 \exp\left(\frac{3}{2} SU_{MS}x^2\right) dx \right\}^{-1} \tag{A10-19}$$

Combining Eq. (A10-18) and (A10-19), we get the final result

$$S = \frac{3}{2} \frac{\int_0^1 \exp[(3/2)SU_{MS}x^2]x^2 \, dx}{\int_0^1 \exp[(3/2)SU_{MS}x^2] \, dx} - \frac{1}{2} \tag{A10-20}$$

Since the order parameter S appears on both sides of Eq. (A10–20), a numerical solution is required to extract the relationship between S and U_{MS}. Note that there is always the trivial solution $S = 0$; but for U_{MS} around 4 or greater, there are also solutions for which $S > 0$. As an exercise, solve Eq. (A10-20); you should get the results given in Table 10-1, where $T/T_{NI} = U_{MS,NI}/U_{MS}$ with $U_{MS,NI} = 4.55$ and $S_2 = S$.

Problem 10.3(a) (Worked Example) From the Leslie–Ericksen equations at high shear rate, derive Eq. (10-29) for the director angle θ' in a shearing flow of a tumbling nematic.

ANSWER:

We start with Eq. (10-3) for the time evolution of the director **n** in the absence of Frank elastic stresses:

$$\dot{\mathbf{n}} - \mathbf{n} \cdot \boldsymbol{\omega} - \lambda(\mathbf{n} \cdot \mathbf{D} - \mathbf{nnn} : \mathbf{D}) = 0 \tag{A10-21}$$

where we can express the director **n** in the deformation plane as

$$\mathbf{n} = (\cos\theta, \ \sin\theta) = (c, s) \tag{A10-22}$$

We are here using the angle θ measured in the counterclockwise direction with the flow direction corresponding to $\theta = 0$. (Later, we will switch to the angle $\theta' \equiv \pi/2 - \theta$.) In Eq. (A10-22), we use the shorthand $c \equiv \cos\theta$; $s \equiv \sin\theta$. Now we evaluate the terms in Eq. (A10-21) using Eq. (A10-22):

$$\dot{\mathbf{n}} = (-s, c)\,\dot{\theta} \tag{A10-23}$$

$$\mathbf{n} \cdot \boldsymbol{\omega} = (c, s) \cdot \frac{1}{2}\begin{pmatrix} 0 & -1 \\ 1 & 0 \end{pmatrix} \dot{\gamma} = \frac{1}{2}(s, -c)\dot{\gamma} \tag{A10-24}$$

$$\mathbf{n} \cdot \mathbf{D} = (c, s) \cdot \frac{1}{2}\begin{pmatrix} 0 & 1 \\ 1 & 0 \end{pmatrix} \dot{\gamma} = \frac{1}{2}(s, c)\dot{\gamma} \tag{A10-25}$$

$$\mathbf{nn} : \mathbf{D} = \frac{1}{2}(s, c) \cdot (c, s)\dot{\gamma} = cs\dot{\gamma} \tag{A10-26}$$

$$\mathbf{nnn} : \mathbf{D} = (c, s)sc\dot{\gamma} \tag{A10-27}$$

From these expressions, we can assemble the first component of the vector equation, Eq. (A10-21):

$$- s\dot{\theta} = -\tfrac{1}{2}\dot{\gamma}s - \lambda(\tfrac{1}{2}s\dot{\gamma} - sc^2\dot{\gamma}) = 0 \tag{A10-28}$$

which implies that

$$- \dot{\theta} = \tfrac{1}{2}\dot{\gamma} + \lambda(\tfrac{1}{2} - c^2)\dot{\gamma} = [\tfrac{1}{2} + \lambda(\tfrac{1}{2} - c^2)]\dot{\gamma} \tag{A10-29}$$

We now use Eq. (10-16) in Chapter 10 to write λ in terms of the Leslie–Ericksen viscosities:

$$\lambda = \frac{-(\alpha_3 + \alpha_2)}{\alpha_3 - \alpha_2} \tag{A10-30}$$

When inserted into Eq. (A10-29), this gives

$$\begin{aligned}
-\dot{\theta} &= \left[\frac{1}{2} - \frac{\alpha_3 + \alpha_2}{\alpha_3 - \alpha_2}\left(\frac{1}{2} - c^2\right)\right]\dot{\gamma} \\
&= \left[\frac{1}{2}(\alpha_3 - \alpha_2) - (\alpha_3 + \alpha_2)\left(\frac{1}{2} - c^2\right)\right]\frac{\dot{\gamma}}{(\alpha_3 - \alpha_2)} \\
&= [-\alpha_2 + c^2\alpha_2 + c^2\alpha_3]\frac{\dot{\gamma}}{\alpha_3 - \alpha_2} \tag{A10-31}
\end{aligned}$$

We now switch to an angle θ' measured clockwise from the velocity-gradient direction; that is, $\theta = \pi/2 - \theta'$. Using trigonometric identities, $\cos\theta = \sin\theta'$ and $\sin\theta = \cos\theta'$. Also, $-\dot{\theta} = \dot{\theta}'$. Thus, switching from θ to θ' in Eq. (A10-31), and rearranging a bit, we get the desired result:

$$\dot{\theta}' = \frac{(\alpha_3 \sin^2\theta' - \alpha_2 \cos^2\theta')}{\alpha_3 - \alpha_2}\dot{\gamma} \tag{A10-32}$$

Problem 10.3(b) (Worked Example) Derive Eq. (10-31), the time- or strain-dependent shear viscosity in the absence of Frank elastic stresses.

ANSWER:

Again, we express **n** in terms of the angle θ using Eq. (A10-22). We need the xy component of the stress tensor given by the Leslie–Ericksen equation, Eq. (10-10). We consider, in turn, the xy component of each term of this equation. We use Eq. (A10-26) to evaluate the first term of Eq. (10-10):

$$(\mathbf{nnnn} : \mathbf{D})_{xy} = n_x n_y \, \mathbf{nn} : \mathbf{D} = cs(cs\dot{\gamma}) = c^2 s^2 \dot{\gamma} \tag{A10-33}$$

To evaluate the second term of Eq. (10-10), we need to obtain **N**. Using the definition of **N** in Eq. (10-12) along with Eq. (10-3), which is valid in the absence of Frank elastic stresses, we get

$$\mathbf{N} \equiv \dot{\mathbf{n}} - \mathbf{n} \cdot \boldsymbol{\omega} = \lambda(\mathbf{n} \cdot \mathbf{D} - \mathbf{nnn} : \mathbf{D})$$

Using Eqs. (A10-25) and (A10-27), we obtain

$$N_x = \lambda\dot{\gamma}(\tfrac{1}{2}s - c^2 s), \qquad N_y = \lambda\dot{\gamma}(\tfrac{1}{2}c - s^2 c)$$

We then find that

$$(\mathbf{nN})_{xy} = n_x N_y = \lambda\dot{\gamma}c^2(\tfrac{1}{2} - s^2) \tag{A10-34}$$

and

$$(\mathbf{Nn})_{xy} = N_x n_y = \lambda \dot{\gamma} s^2 (1/2 - c^2) \tag{A10-35}$$

We follow similar procedures to obtain

$$D_{xy} = 1/2\dot{\gamma}; \quad (\mathbf{nn} \cdot \mathbf{D})_{xy} = 1/2 c^2 \dot{\gamma}, \qquad (\mathbf{D} \cdot \mathbf{nn})_{xy} = 1/2 s^2 \dot{\gamma} \tag{A10-36}$$

We now combine the results of Eqs. (A10-33) through (A10-36) to evaluate the shear stress σ_{xy} from Eq. (10-10):

$$\eta = \frac{\sigma_{xy}}{\dot{\gamma}} = \alpha_1 c^2 s^2 + \alpha_2 \lambda c^2 (1/2 - s^2) + \alpha_3 \lambda s^2 (1/2 - c^2)$$

$$+ 1/2 \alpha_4 \dot{\gamma} + 1/2 \alpha_5 c^2 + 1/2 \alpha_6 s^2$$

This can be reorganized as

$$\eta = [\alpha_1 - (\alpha_2 \lambda + \alpha_3 \lambda)] s^2 c^2 + 1/2 \alpha_2 \lambda c^2 + 1/2 \alpha_3 \lambda s^2$$

$$+ 1/2 \alpha_4 + 1/2 \alpha_5 c^2 + 1/2 \alpha_6 s^2 \tag{A10-37}$$

Using Parodi's relationship, Eq. (10-11) to substitute for $\alpha_5 = \alpha_6 - \alpha_2 - \alpha_3$, and rearranging, we get

$$\eta = [\alpha_1 - (\alpha_2 \lambda + \alpha_3 \lambda)] s^2 c^2$$

$$+ 1/2 \alpha_2 c^2 (\lambda - 1) + 1/2 \alpha_3 (\lambda s^2 - c^2) + 1/2 \alpha_4 + 1/2 \alpha_6 \tag{A10-38}$$

We now using the identity $s^2 = 1 - c^2$ to restructure the third-to-the last term in the above:

$$1/2 \alpha_3 (\lambda s^2 - c^2) = 1/2 \alpha_3 (\lambda - \lambda c^2 - c^2) = -1/2 \alpha_3 c^2 (\lambda + 1) + 1/2 \lambda \alpha_3 \tag{A10-39}$$

Using Eq. (10-16), a couple of other identities can be established:

$$\lambda - 1 = \frac{-2\alpha_3}{\alpha_3 - \alpha_2}, \qquad \lambda + 1 = \frac{-2\alpha_2}{\alpha_3 - \alpha_2} \tag{A10-40}$$

Then we can use Eqs. (A10-39) and (A10-40) to simplify two terms of Eq. (A10-38):

$$1/2 \alpha_2 c^2 (\lambda - 1) + 1/2 \alpha_3 (\lambda s^2 - c^2) = 1/2 c^2 [\alpha_2 (\lambda - 1) - \alpha_3 (\lambda + 1)] + 1/2 \alpha_3 \lambda$$

$$= c^2 \left(\frac{-2\alpha_2 \alpha_3}{\alpha_3 - \alpha_2} + \frac{2\alpha_3 \alpha_2}{\alpha_3 - \alpha_2} \right) + 1/2 \lambda \alpha_3 = 0 + 1/2 \lambda \alpha_3 \tag{A10-41}$$

With this simplification, Eq. (A10-38) reduces to

$$\eta = [\alpha_1 - (\alpha_2 \lambda + \alpha_3 \lambda)] s^2 c^2 + 1/2 \lambda \alpha_3 + 1/2 \alpha_4 + 1/2 \alpha_6 \tag{A10-42}$$

We now write $1/2 \lambda \alpha_3$ as $1/2 \alpha_3 + 1/2 (\lambda - 1) \alpha_3$. We can then replace $\alpha_3 + \alpha_4 + \alpha_6$ in Eq. (A10-42) by η_b, according to Eq. (10-19). Then, using Eq. (A10-40) for $\lambda - 1$, and Eq. (10-16) for λ, Eq. (A10-42) finally becomes

$$\eta = \left[\alpha_1 + \frac{(\alpha_2 + \alpha_3)^2}{\alpha_3 - \alpha_2} \right] \sin^2 \theta' \cos^2 \theta' + \eta_b - \frac{\alpha_3^2}{\alpha_3 - \alpha_2}$$

where we have used the identities $\sin^2 \theta = \cos^2 \theta'$ and $\cos^2 \theta = \sin^2 \theta'$ to replace the angle θ by θ'.

Problem 10.3(c) (Worked Example) Derive Eq. (10-30).

ANSWER:

The equation for the rotation of the director, Eq. (10-3), is identical to that for the rotation of a rigid spheroid, Eq. (6-26) in Chapter 6, if we identify λ with $(p^2 - 1)/(p^2 + 1)$. Inverting this to obtain p^2 in terms of λ, we get

$$p^2 = \frac{1+\lambda}{1-\lambda} \tag{A10-43}$$

The period P for the rotation of the ellipsoid through an angle of π is given by Eq. (6-28), which can be rearranged using Eq. (A10-43):

$$P = \frac{\pi}{\dot{\gamma}} \left(p + \frac{1}{p} \right) = \frac{2\pi}{\dot{\gamma}(1 - \lambda^2)^{1/2}}$$

which is the desired result.

Problem 10.4 How does a smectic liquid-crystalline phase differ from a nematic?

Problem 10.5 Why are nematic liquid crystals used in displays such as those for lap-top computers?

Problem 10.6 Name two applications of cholesteric liquid crystals.

Problem 10.7 Derive Eq. (10.4) for the alignment angle θ of the director in shearing flow of a flow-aligning nematic.

Problem 10.8 You are measuring the elasticities and viscosities of a room-temperature nematic at reduced temperatures and you find that below about 10°C the twist and bend constants K_2 and K_3 become very large, while the splay constant K_1 retains a modest value. Also, the Miesowicz viscosity η_b becomes enormous while η_a goes up only modestly. What could explain this behavior?

Problem 10.9 Why are focal conic defects so common in smectics?

Problem 10.10 You would like to reduce the density of defects present in a smectic A liquid. Which method would be more likely to succeed: steady shearing or large-amplitude oscillatory shearing?

Chapter **11**

LIQUID-CRYSTALLINE POLYMERS

11.1 INTRODUCTION

In 1923, Vorlander attempted to synthesize long liquid-crystal-forming molecules by connecting benzene rings together at their para positions using esther linkages (Vorlander 1923). He noted that the melting point of these materials increases rapidly with increasing molecular length. This work culminated in the synthesis of poly(p-benzamide); its melting point, however, was so high that it charred before melting. The problem of achieving a lower melting point for such materials was finally overcome by Jackson and coworkers at Eastman Kodak, by using random copolymers of aromatic polyesters (Jackson and Kuhfuss 1976). Meanwhile, starting in the 1930s, it was noticed that when viruses, such as tobacco mosaic, cucumber, and potato viruses, were dissolved at high concentration in solution, they formed birefringent phases (Bawden and Pirie 1937; Donald and Windle 1992). After this discovery, the possibility of using solutions of rigid molecules to make artificial silk was recognized; and in England, polyglutamates—in particular poly(γ-benzyl-L-glutamate) (PBLG)—was synthesized. Its solution properties were studied in detail by Robinson et al. (1958). The first commercially successful lyotropic (solvent-based) liquid-crystalline polymer (LCP) was synthesized at the Du Pont Laboratories by Kwolek (1971), as well as in a parallel effort at Akzo Chemical. These efforts, combined with the publication of Jackson and Kuhfuss' seminal paper on thermotropic (solvent-free) LCPs in 1976, led to an explosion of scientific and engineering interest in LCP's that continues to this day.

LCPs are commercially interesting because of their unusual bulk properties and their processability, resulting from chain stiffness and high molecular orientation. Solidified LCP fibers can have tensile moduli in the tens or hundreds of gigapascals and can have fracture strengths of order 1–4 GPa (Donald and Windle 1992; Prevorsek 1982). In addition, LCP materials are unusually resistant to solvent attack, moisture uptake, and thermal expansion, softening, or decomposition. Their processability is enhanced by the low viscosity they have in the nematic melt state. Commercial uses range from fibers for bulletproof vests, to packaging for electrical and optical components and for strong, lightweight aircraft parts (Collyer 1992; Covault 1996), where the high cost of LCPs is not an overriding issue. Thermotropic LCPs are particularly attractive for injection molding, because shrinkage is low, injection and clamping pressures are low (because the melt viscosity is low), and cycle times are fast (because thermal conductivity is high and heat of fusion is low) (Naitove

1986). In the future, there may be applications in the area of nonlinear optical components and, for side-group LCPs, in field-orientation devices (Donald and Windle, 1992).

Solidified LCPs often have a "hierarchial structure," consisting of fibrils and microfibrils in fiber products and layered "skin/core" morphologies in injection-molded parts (Sawyer and Jaffe 1986; Weng et al. 1986). Such morphologies are evidently produced by the response to flow of the unique molecular structures present in LCPs. For a review and thorough bibliography covering the chemistry, morphologies, and applications of LCPs, see Cox (1987).

11.2 MOLECULAR CHARACTERISTICS OF LIQUID-CRYSTALLINE POLYMERS

Polymer molecules that are capable of producing liquid-crystalline solutions or melts are generally semirigid or contain rigid units, called *mesogens*. The distribution of orientations of the molecules, or at least of the mesogens, is anisotropic in the liquid crystalline state. In *main-chain* LCPs the rigid units are contained in the backbone, whereas in *side-chain* LCPs the mesogens are attached to the backbone by flexible spacers (see Fig. 11-1). Polymers with mesogens both in the backbone and in sidegroups are also possible. This chapter deals with the rheological properties of main-chain nematic LCPs. The interested reader can learn about side-chain LCPs in Kannan et al. (1993, 1994a, 1994b), Gu et al. (1993), and Colby et al. (1993).

As with small-molecule materials, liquid crystallinity in polymers generally occurs at conditions intermediate between those for which isotropic liquid and crystalline solid states

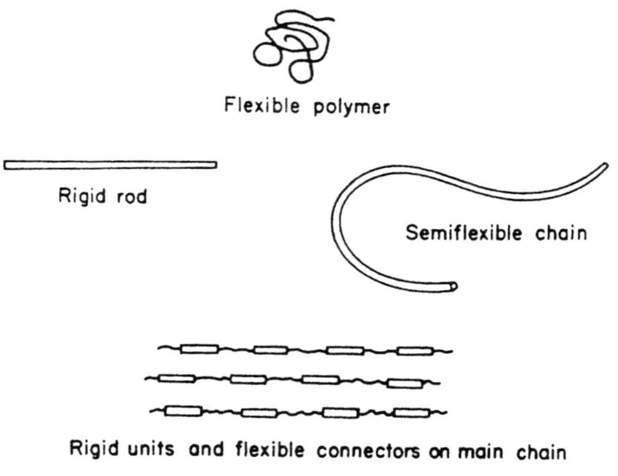

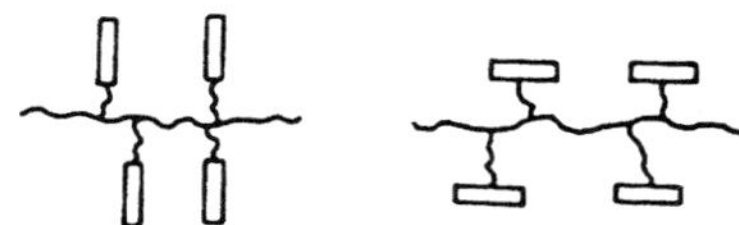

Figure 11.1   Schematic illustration of different kinds of polymers, with rigid units (rectangles) and flexible ones (lines) along the backbone or in side branches. (From Jackson and Shaw 1991, with permission from the Institute of Metals.)

exist. In single-component bulk materials, called *thermotropes*, the liquid-crystalline state is accessed from the crystalline state by increasing the temperature, or from the isotropic state by decreasing it. In *lyotropic* polymers, on the other hand, a solvent is added to produce liquid crystallinity. We will designate the mass fraction of polymer in a lyotropic solution by C, the volume fraction by ϕ, and the number of molecules per unit volume by v. For cylindrical rod-like polymers with length L and diameter d, $\phi = \pi d^2 L v / 4$. The symmetry of an LCP phase can be nematic, smectic, or columnar. As with small molecules, a single polymeric species can exhibit multiple-liquid crystalline phases, depending on temperature or concentration.

Some main-chain liquid-crystal-forming molecules are depicted in Fig. 11-2. These molecules differ considerably in stiffness and in the conditions under which they form nematic or other liquid-crystalline phases. As discussed in Section 2.2.4, the tendency to form a nematic state is controlled mainly by the *axial ratio*, b_K/d, where b_K is the Kuhn length, which is twice the persistence length λ_p, and d, the effective diameter of the molecule. In general, the stiffer the molecule, the more prone it is to form liquid-crystalline, rather than isotropic, phases. However, stiffness also promotes crystallinity, which must be suppressed if the liquid crystalline state is to assert itself (Donald and Windle, 1992). Since the addition of a solvent is typically very effective in destroying crystallinity, in lyotropic systems liquid crystallinity is compatible with high molecular stiffness, such as that of PBG or PPTA (trade name: Kevlar). (We use the abbreviation PBG to designate either the enantiomorph PBLG or its racemic mixture with PBDG.) For thermotropic materials, however, molecular kinks or flexible spacers must typically be introduced into the polymer backbone to suppress crystallinity, and these usually also reduce molecular stiffness. Thus, thermotropic LCPs are usually composed of *semiflexible* molecules whose persistence length λ_p is much shorter than the molecular length L; that is, $L \gg \lambda_p$.

In main-chain LCPs, molecular flexibility can be distributed more-or-less uniformly along the chain, as is the case for PBLG, HPC, or Vectra A, or it can be concentrated in flexible spacers, as in OQO(phenylsulfonyl)10 (see Fig. 11-2). The former are called *persistently flexible* molecules, and are often modeled by the *worm-like chain*, with a uniform bending modulus, while for the latter, a reasonable model might be the freely jointed chain (see Fig. 11-3 and Section 2.2.3.2). For a recent discussion of the phase behavior and dynamics of worm-like chains, see Sato and Teramoto (1996).

Some stiff or semiflexible polymer molecules are *chiral*; that is, they lack mirror symmetry. Poly(γ-benzyl-L-glutamate) (PBLG) is a left-handed molecule that forms a left-handed chiral nematic, or cholesteric, phase. The right-handed enantiomer is poly(γ-benzyl-D-glutamate) (PBDG). Nematics without chirality can be formed from chiral molecules by making a *racemic mixture*—that is, a mixture of equal proportions of the L and D enantiomers. For polymeric nematics, chirality often has little effect on rheological properties except perhaps at low flow rates. This is because the chiral structures in polymers are usually weak and easily suppressed by flow, except, perhaps, in the so-called Region I of the flow curve (see below). Therefore, in this chapter the term "nematic" will usually include both chiral and achiral nematics, unless noted otherwise.

Although uniaxial nematics possess rotational symmetry about the nematic axis, the molecules composing them are usually not axisymmetric (Emsley et al. 1981; Bunning et al. 1986), but the axial asymmetry of the molecules is averaged out by the rotations of the molecules about their backbones. In some nematics, however, molecular interactions

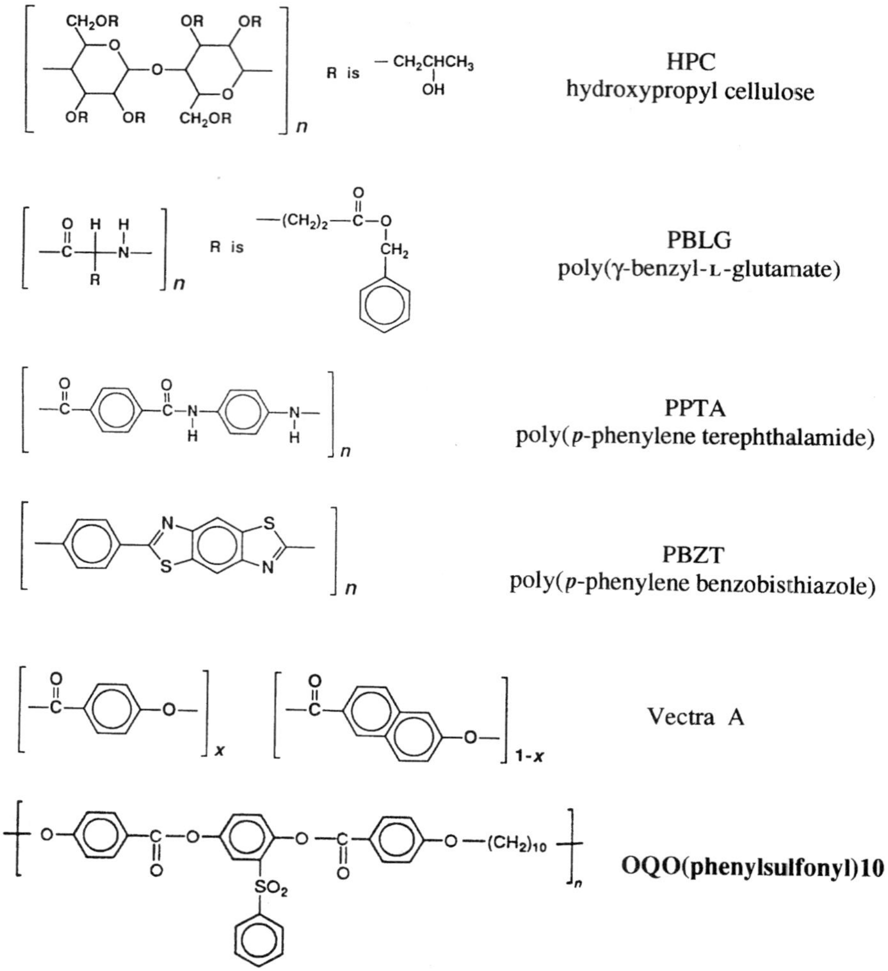

Figure 11.2 Chemical structures of some well-studied LCPs. The top four structures are usually used as lyotropic polymers, and the bottom two are thermotropes. Estimates of the persistence lengths (λ_p) and effective diameters (**d**) of these molecules are $(\lambda_p, d) = (12, 1.04)$ nm for HPC, $(90, 1.5)$ nm for PBLG, $(29, 0.6)$ nm for PPTA, $(20, 0.6)$ nm for PBZT, $(12, 0.6)$ nm for Vectra[R] A, and $(6, 0.5)$ nm for OQO(phenylsulfonyl)10. These values, listed in Baek et al. (1994), Brelsford and Krigbaum (1991), and Farmer et al. (1993), are both solvent- and temperature-dependent, and thus should be used with caution. (From Donald and Windle 1992, with permission from Cambridge University Press.)

restrict these rotations enough to break the axisymmetry of the nematic phase. The result is a *biaxial nematic*, for which there are two independent nematic axes. A biaxial nematic has no rotational symmetry, although it remains positionally disordered. At least one commercial nematic polymer, Vectra B950, has been discovered to be biaxial by an examination of its disclination lines (De'Neve et al. 1992). Vectra B950 is a thermotropic copolyesteramide

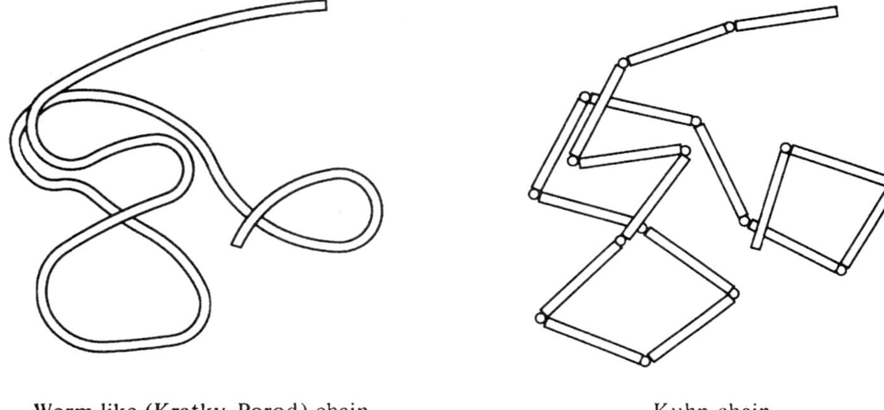

Figure 11.3 Schematic drawings of a worm-like polymer chain (with continuous flexibility) and a Kuhn chain (with rigid links joined at hinges that allow free rotation about the angle between the links). The length of the links has been chosen so that the contour length and mean-square end-to-end length of the Kuhn chain are the same as those of the worm-like chain. (From Donald and Windle 1992, with permission from Cambridge University Press.)

based on 60/20/20 2,6-hydroxynaphthoic acid/hydroxybenzoic acid/aminophenol. Apart from its apparently higher value of the Frank constant K_1 (De'Neve et al. 1994), it is not known to what extent the rheological properties of biaxial nematics differ from those of uniaxial nematics.

11.3 FLOW PROPERTIES OF NEMATIC LCPs

In this chapter, we discuss the rheological properties of main-chain nematic LCPs. Rheological data for side-chain nematics can be found in Zentel and Wu (1986), Gu et al. (1993), Colby et al. (1993), Kannan et al. (1993, 1994), and Rubin et al. (1995). Other, excellent reviews of main-chain LCP rheology are those of Wissbrun (1981), Marrucci and Greco (1993), Srinivasarao (1995), Mewis and Moldenaers (1996), and Burghardt (1998).

The flow properties of nematic LCPs are often extraordinary and are only partially understood. Furthermore, these properties vary considerably from one LCP to another, and even a single LCP sample may behave very differently in different regimes of shear rate. Despite their complexity, there are some typical features, described in the next sections.

11.3.1 Dependence of Viscosity on Concentration or Temperature

The transition from the isotropic to the nematic state is marked by a *decrease* in the shear viscosity (Hermans 1962; Gillmor et al. 1994) (see Figs. 11-4a and 11-5). This is not surprising, since molecules can slide past each other more readily in the nematic state than in the isotropic state. However, since the nematic is accessed from the isotropic state by an

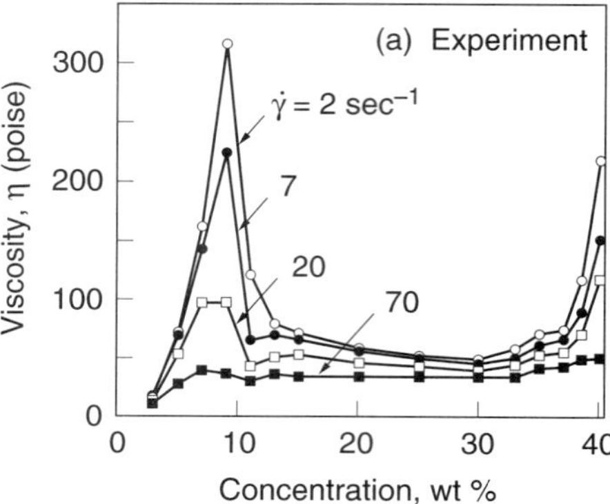

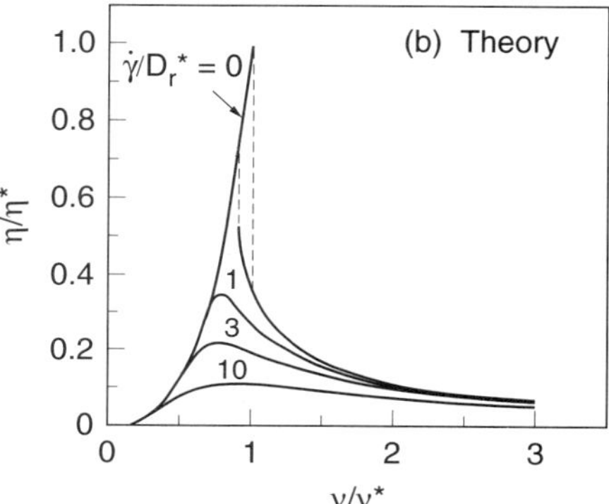

Figure 11.4 **(a)** Measured shear viscosity versus PBLG concentration (molecular weight = 238,000) in *m*-cresol at various shear rates. The viscosity drops abruptly at the transition from the isotropic to nematic state, which occurs at a concentration of around 10 wt% for this polymer. **(b)** Dimensionless viscosity versus normalized concentration at various reduced shear rates predicted by Doi's original molecular theory (see discussion in Section 11.4.3). (From Larson et al. 1993, reprinted with permission from Springer Verlag.)

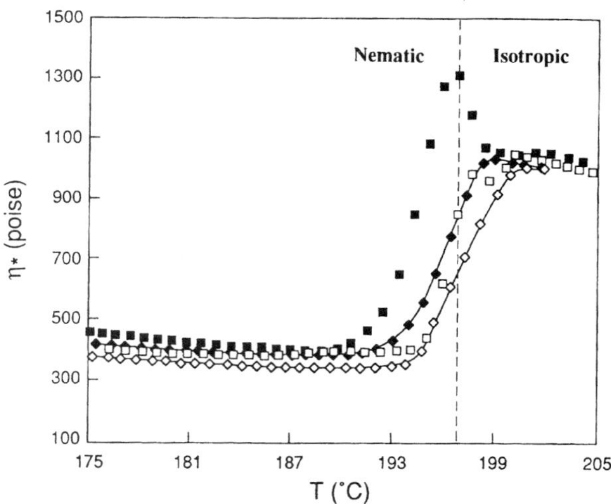

Figure 11.5 Complex dynamic viscosity as a function of temperature for a main-chain polyether consisting of a methyl stilbene mesogen and a mixture of seven- and nine-carbon aliphatic spacers. The polymer has a molecular weight of 36,000. The diamonds and squares are for temperature ramp rates of 0.1°C and 2.0°C/min, respectively; the open and closed symbols are for heating and cooling, respectively. The dashed line marks the isotropic–nematic transition. (From Gillmor et al. 1994, with permission from the Journal of Rheology.)

increase in concentration or by a decrease in temperature, the lower viscosity of the nematic compared to the isotropic phase implies a reversal in the usual trend of increasing viscosity with increasing polymer concentration or decreasing temperature. Hence, the incidence of an isotropic-to-nematic transition is announced by an easily noticed rheological anomaly. Nevertheless, it is prudent to confirm the existence of such a transition by inspecting the sample under a polarizing microscope.

For lyotropic LCPs, there is a *biphasic window* of concentrations over which nematic and isotropic phases coexist. The polymer concentrations in the coexisting isotropic and nematic phases are designated by C_1 (or $\phi_1 = \pi d^2 L \nu_1/4$) and C_2 (or $\phi_2 = \pi d^2 L \nu_2/4$), respectively. There is also a theoretical concentration C^*, at which the isotropic phase becomes unstable to orientational fluctuations. According to the Onsager theory, $\phi_2/\phi^* = 1.047$ and $\phi_2/\phi_1 = 1.27$ (see Section 2.2.2). *Thermotropic* LCPs often have a biphasic window of *temperatures* over which isotropic and nematic phases coexist. This biphasic window exists in nominally "single-component" thermotropics because of polydispersity; the nematic phase is typically enriched in the longer molecules relative to the coexisting isotropic phase (D'Allest et al. 1986).

11.3.2 Region I Shear Thinning

A useful scheme for categorizing the viscosity-versus-shear rate behavior of main-chain LCPs is the "three-region" shear-viscosity curve of Onogi and Asada (see Fig. 11-6). In Fig. 11-6, Region II is a "Newtonian plateau," and Regions I and III are, respectively, a high-shear-rate and a low-shear-rate shear-thinning regime. Figures 11-7 and 11-8 show these three regimes for solutions of hydroxypropylcellulose (HPC) in water and for solutions of poly(1,4-phenylene-2,6-benzobisthiazole) (PBZT; also called PBT) in methane sulfonic acid. (Only the open symbols in Fig. 11-8 correspond to fully nematic liquids.) Region I has also been observed in PBG solutions in *m*-cresol (Asada et al. 1985; Larson et al. 1993; Walker et al. 1995) and poly(para-benzamide) solutions in dimethyl acetamide (Papkov et al. 1974). For isotropic solutions of flexible or rigid polymers, regions analogous to II and III are common. In Fig. 11-8, for example, isotropic solutions of PBZT (the filled symbols) have a low-shear-rate Newtonian plateau (or Region II) and a high-shear-rate shear-thinning region (or Region III), but no Region I. For both isotropic and nematic polymer solutions, the Newtonian plateau typically exists over ranges of shear rate for which the distribution

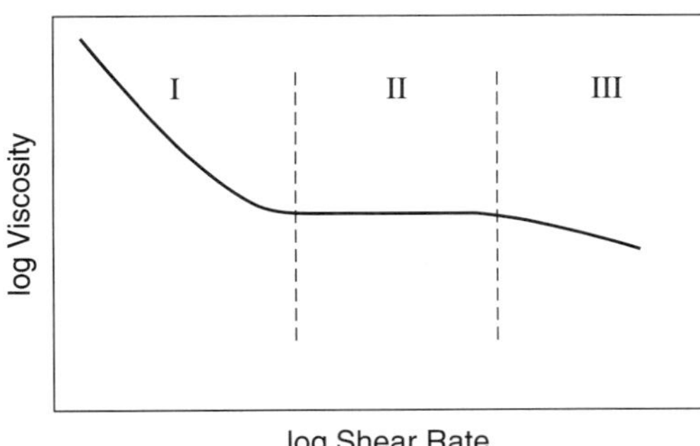

Figure 11.6 Onogi and Asada's three-region flow curve (solid line). For some LCPs, the low-shear-rate Region I is not found. (Adapted from Onogi and Asada 1980, by permission of Plenum Publishing Corp.)

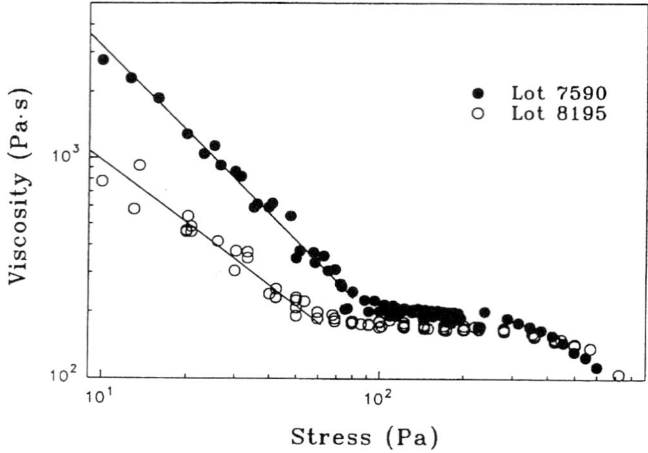

Figure 11.7   Viscosity versus shear-rate curve for 60% hydroxypropylcellulose (HPC) in water, made from two different batches of of Klucel E. The power-law slopes at low shear rate are −0.51 for Lot 8195 and −0.56 for Lot 7590. (From Walker and Wagner 1994, with permission from the Journal of Rheology.)

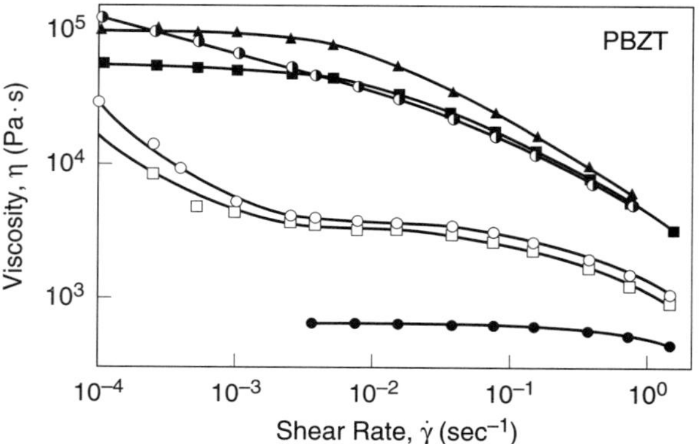

Figure 11.8   The steady-state viscosity for solutions of poly(1,4-phenylene-2,6-benzobisthiazole) (PBZT) with viscosity-averaged molecular weight 37,400 (molecular length 170 nm), in methane sulfonic acid at concentrations of 1.5 (●), 2.52(■), 3.0(▲), 3.43(◐), 6.11(○), and 8.2(□)% (by weight) at 60°C. The solid symbols are isotropic solutions, the half-tone symbol is biphasic, and the open symbols are fully liquid crystalline. (From Einaga et al. 1985, with permission from the Society of Polymer Science, Japan.)

of molecular orientations and conformations is not significantly distorted by the flow (see Section 3.7.5.1 and Fig. 3-32). The high-shear-rate shear-thinning region occurs at shear rate where the molecular distribution function is disturbed by flow. Thus, compared to isotropic liquids, the most distinguishing new feature in the flow curve of LCPs is the appearance (in some cases) of a shear thinning region at low shear rates, the so-called Region I. This feature must be attributed to the liquid-crystalline character of these fluids.

Region I shear thinning has not been observed for all LCPs; in fact, it is perhaps absent as often as it is present in experimentally reported flow curves for LCPs. However, since all rheometers have a shear stress below which they cannot measure stresses reliably, the existence of a Region I at lower shear rates than those accessed experimentally is always at least a theoretical possibility.

In lyotropic LCPs, the shear-rate range over which Region I exists is sensitive to polymer concentration (Walker and Wagner 1994). In one batch of PBLG, Region I appears suddenly as the polymer concentration is increased above about 37% (see Fig. 11-9). In other PBLG solutions, Region I behavior has been reported at much lower concentrations (Asada et al. 1985).

Some thermotropic LCPs show indications of a Region I in their flow curves. Figures 11-10 and 11-11 show the shear viscosity curves for Vectra A900 and Vectra B950. Of the two, Vectra A shows a clearer sign of Region I. Nevertheless, for either material, the range of shear rates is too small for definite identification of a Region I. Also, in some thermotropes, apparent Region I behavior can be eliminated by shear aligning the nematic (Kim and Han 1993).

11.3.3 Negative First Normal Stress Differences

Another unusual property of some, but not all, nematic LCPs is the existence of a shear-rate range over which the steady-state first normal stress difference N_1 is *negative* (Kiss and Porter 1978; Moldenaers and Mewis 1986; Grizzuti et al. 1990; Baek et al. 1993a). Figure

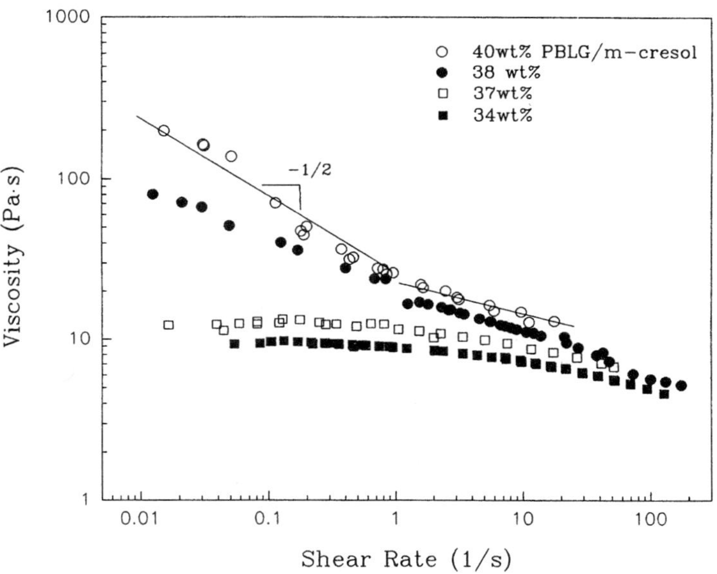

Figure 11.9 Steady-state shear viscosity versus shear rate for PBLG solutions (molecular weight = 238,000) in *m*-cresol for several concentrations. The 38 wt% and 40 wt% samples show Region I behavior. (From Walker et al. 1995, with permission of the Journal of Rheology.)

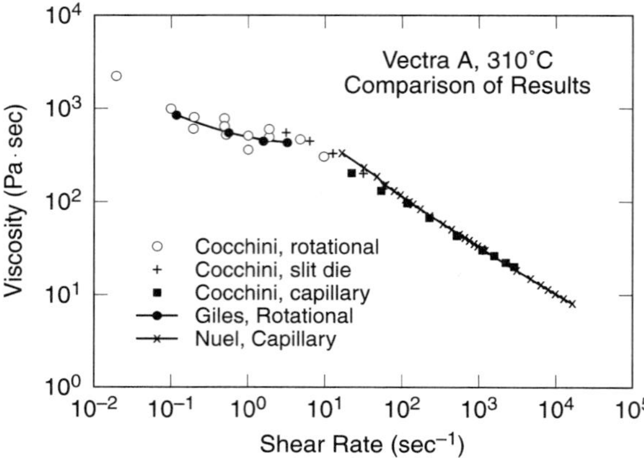

Figure 11.10   The viscosity of Vectra^R A900 versus shear rate at 310°C measured on various rotational and capillary shear instruments. (From Giles and Denn 1994, with permission of the Journal of Rheology.)

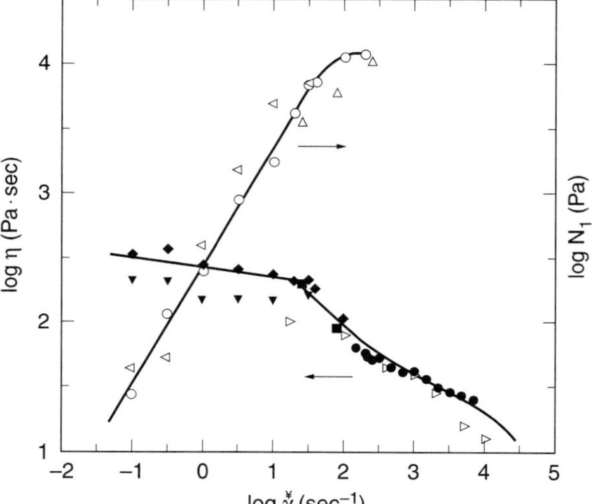

Figure 11.11   Shear viscosity η and first normal stress difference N_1 as functions of shear rate $\dot\gamma$ for Vectra^R B-950 at 300°C. The different symbols are for data acquired on different cone-and-plate and capillary rheometers. (From De'Neve et al. 1993a, with permission of the Journal of Rheology.)

11-12 shows N_1 and the viscosity η versus shear rate $\dot\gamma$ for a PBLG solution, measured by Kiss and Porter (1980a). There are two sign changes: N_1 is positive at the lowest shear rates, negative over about a decade of intermediate shear rates, and positive again at the highest shear rates.

For a given polymer, the range of shear rates over which N_1 is negative depends strongly on concentration and molecular weight. For some LCPs, only positive values of N_1 have been reported at steady state. Although the lack of a measurable range of negative N_1 does not preclude the possibility that N_1 is negative outside the experimental shear-rate range, all undisputedly negative N_1 values reported so far have been found near the onset of Region III. Thus, LCPs that don't show negative N_1 values near the onset of Region III might well lack any region of negative N_1. Figure 11-13, for example, shows η and N_1 values for the thermotropic polyester nematic poly(phenylsulfonyl)-p-phenylene 1,10-decamethylenebis(4-oxybenzoate), sometimes called PSHQ10, but which we shall call

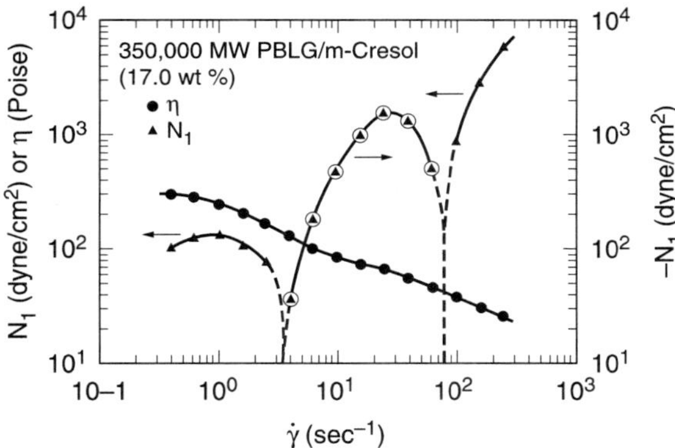

Figure 11.12 Shear viscosity and first normal stress difference versus shear rate for 17% PBLG (molecular weight 350,000) in *m*-cresol. The circled triangles are negative N_1 values. (From Kiss and Porter, reprinted with permission from Mol. Cryst. Liq. Cryst. 60:267, Copyright © 1980, Gordon and Breach Publishers.)

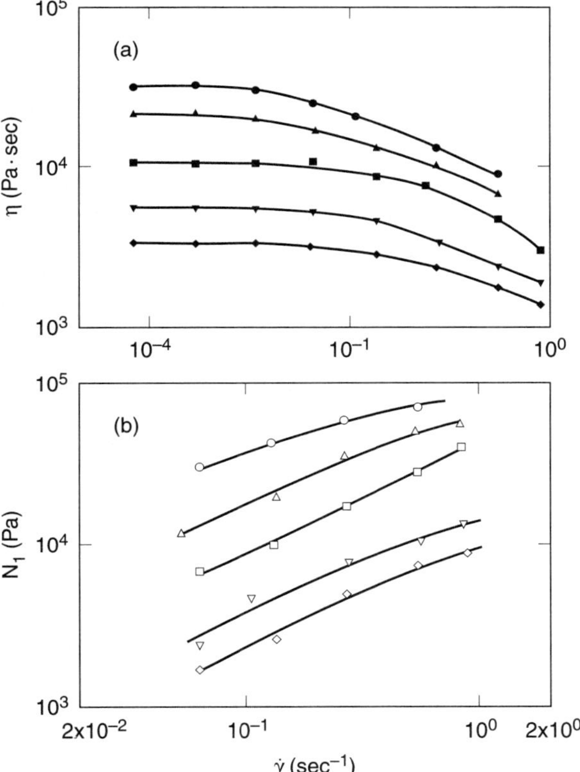

Figure 11.13 Viscosity **(a)** and first normal stress difference **(b)** versus shear rate for OQO(phenylsulfonyl)10 thermotropic polyester (called PSHQ10 by Kim and Han), at 140°C for six molecular weights: 35,000, 37,600, 38,900, 45,200, 49,500, and 53,400. The curves for the viscosities and normal stress differences shift upward monotonically with increasing molecular weight. These data were measured after the sample had been shear aligned by previous shearing. (Reprinted with permission from Kim and Han, Macromolecules 26:6633. Copyright © 1993, American Chemical Society.)

OQO(phenylsulfonyl)10. For this polymeric nematic, N_1 tends to level off at high shear rates, but remains positive even at shear rates well beyond that at which η shows significant shear thinning.

Steady-state negative N_1 values have been reported for some thermotropic LCPs, but none of these has shown all three N_1 regimes: positive, then negative, and then positive again, as the shear rate increases. Also, in thermotropic LCPs, spuriously negative N_1 values can apparently be produced (Cocchini et al. 1992), and it is thus prudent to regard reports of negative N_1 in thermotropes with caution. *Transient* negative N_1 values after start-up of steady shearing are often reported, even in shear rate ranges where N_1 eventually becomes positive as steady state is approached.

11.3.4 Transient Shear Stresses

For LCPs, complex transient shear stresses are often generated when shearing flow is started, the shear direction is reversed, or the rate of shear is suddenly changed. Often there are multiple overshoots and undershoots in stress as a function of time, and frequently 50–100 strain units or even more must be imposed before a steady state is reached (Moldenaers et al. 1990; Vermant et al. 1994a; Picken et al. 1991). In Region II, characteristic viscoelastic response times are found to scale inversely with the shear rate $\dot{\gamma}$ (Moldenaers and Mewis 1986). As a result, curves of transient stress versus time collapse onto a single line when plotted against time multiplied by the imposed shear rate (Moldenaers et al. 1989, 1990; Picken et al. 1991; Mortier et al. 1996). This *scaling* behavior is often beautifully precise. Figure 11-14, for example, shows the normalized shear stress after a sudden reversal in the direction of shearing of a PBLG solution, with no change in the magnitude of the shear rate. The results measured for shear rates varying by a factor of 50 superimpose almost perfectly, if the stresses are normalized with respect to the final steady-state stress and *if stresses are plotted against strain*, $\gamma = \dot{\gamma}t$. This superposition implies that the time period P between maxima in the stress response is inversely proportional to the shear rate $\dot{\gamma}$. Similar multiple stress overshoots that scale with strain are observed in *tumbling* small-molecule nematics (see Sections 10.2.1 and 10.2.6). Their existence suggests that PBG solutions are of the tumbling, rather than flow-aligning, type, an inference which has been confirmed in the experiments discussed below.

Analogous scaling behavior is also observed in other transient shearing flows. Figure 11-15, for example, shows that after shearing an HPC solution to steady state at some shear rate $\dot{\gamma}_0$ and then removing the shearing torque, the rate at which shearing strain is recovered

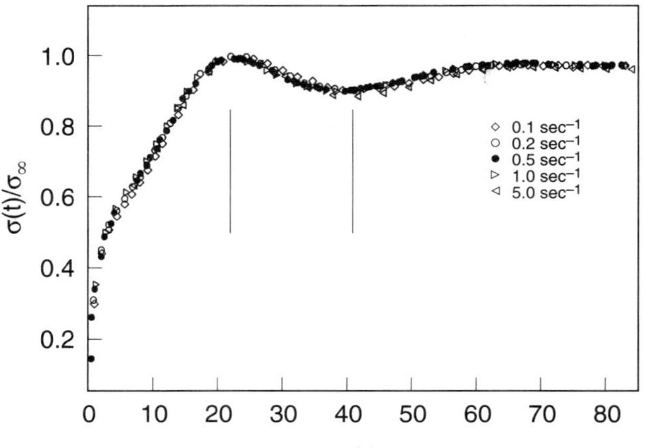

Figure 11.14 Reduced shear stress (transient stress divided by steady-state stress) plotted against strain for a 37% PBLG solution (molecular weight = 238,000) in *m*-cresol after reversal of the shearing direction for various rates of shear. The bars mark a half-period of the oscillation. (From Walker et al. 1995, with permission of the Journal of Rheology.)

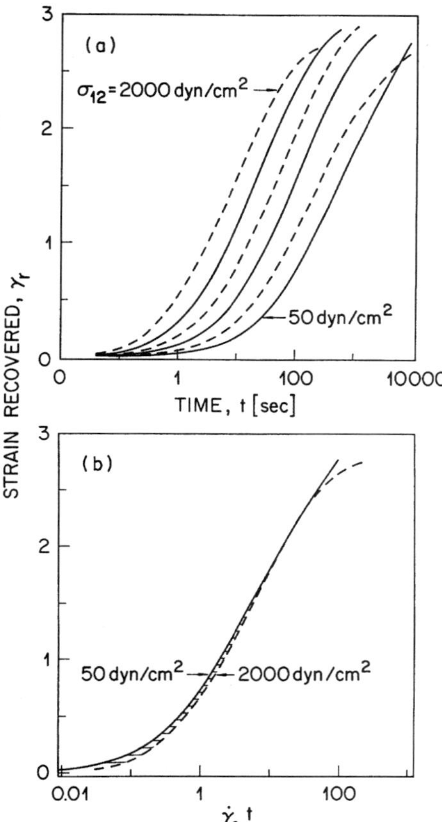

Figure 11.15 (a) Recoverable strain γ_r versus time for a 52.8% liquid-crystalline hydroxypropylcellulose solution in water at room temperature, 22°C. The curves from left to right were obtained by imposing stresses of 50, 125, 250, 500, 1000, and 2000 dyn/cm² before recovery. These imposed shear stresses correspond to various values of the prior shear rate $\dot{\gamma}_0$. (b) The same plotted against $\dot{\gamma}_0 t$; all curves fall within the indicated region between the curves for 50 and 2000 dyn/cm². (From Larson and Mead 1989, with permission from the Journal of Rheology.)

increases as the prior shear rate $\dot{\gamma}_0$ is increased. When the data are plotted against the product of prior shear rate $\dot{\gamma}_0$ and time t, all curves collapse onto a common line, showing that the time of recovery is inversely proportional to $\dot{\gamma}_0$. This indicates that shearing flow in the LCP creates an elastic microstructure with a "texture" relaxation time τ_{tex} that scales as $\tau_{\text{tex}} \propto \dot{\gamma}_0^{-1}$. Similar scaling has been observed for strain recovery in PBLG solutions (Larson and Mead 1989) and in a thermotropic LCP, Vectra A900 (Giles and Denn 1994). Early observations of such scaling laws were reported by Moldenaers and Mewis (1986) in their studies of (a) the evolution with time of the linear moduli G' and G'' after cessation of shearing and (b) the long-time stress relaxation after cessation of shearing (Moldenaers and Mewis 1990). These results suggest that the elastic energy stored in the shear-induced distortions of the nematic texture is proportional to the shear rate $\dot{\gamma}$.

11.3.5 Textures of Sheared LCPs

The transient phenomena discussed above are no doubt produced by distortions of the nematic director field under shear; these distortions, known as "textures," can be visualized with an optical microscope. In addition, measurements of scattering dichroism during shearing show optical transients that are analogous to those of the shear stresses (Moldenaers et al. 1989). Since scattering dichroism is sensitive to structures with sizes near the wavelength of light, micron-size "domains" are evidently involved in the peculiar transient rheological behavior described above. In small-angle light-scattering experiments under shearing flow, multilobe scattering patterns are observed that also reflect textures whose size scale is sensitive to the shear rate (Ernst et al. 1990; Vermant et al. 1994a).

Textures exist in small-molecule nematics, but are much more persistent in LCPs because of the high viscosities of the latter. One can estimate that the annealing time t_a for spontaneous disappearance of a disclination texture from a nematic layer of thickness h is

$$t_a \sim \frac{\eta h^2}{K}$$

where K is a characteristic Frank constant. For small-molecule nematics the effective "director diffusivity" K/η is around $10^{-6} \text{cm}^2/\text{sec}$, while for polymeric nematics we have $K/\eta \sim 10^{-8}$–10^{-10} cm/sec. Thus, a 250-μm-thick sample of a small-molecule nematic might free itself of disclinations in a few minutes; for a polymeric nematic, days would be required. Application of magnetic or electric fields can greatly accelerate the process, however. Because of the very long times required for a polymeric nematic to spontaneously free itself of disclinations, the rest state of a textured LCP is often ill-defined or dependent on the previous flow history. For this reason, to obtain reliable transient rheological data, LCPs are often *pre-sheared* until a steady state is reached, which can then serve as a well-defined starting state for rheological tests. This protocol was used to obtain the transient strain reversal and recoverable shear data reported in Section 11.3.4. Another way to achieve a well-defined starting state is to quench the sample from an isotropic state into the nematic state and then begin rheological testing a well-defined time period after the quench (Kim and Han 1993).

Some textures in LCPs are fairly regular; one example is the so-called "band" pattern, which consists of birefringent bands oriented perpendicular to the shearing direction (Elliott and Ambrose 1950; Kiss and Porter 1980a; Viney et al. 1983; Navard 1986; Srinivasarao and Berry 1991; Gleeson et al. 1992; Donald and Windle 1992; Vermant et al. 1994b). These bands have been observed most frequently after cessation of steady shearing, but they can also appear briefly after start-up of shear, if the sample is well-aligned in the flow direction before the start of flow (Larson and Mead 1992; Yan and Labes 1994). Bands have been observed in virtually all LCPs, both lyotropic and thermotropic, and have sometimes been seen in elongational flow as well as in shear (Srinivasarao 1995). Similar bands have even been observed in nematics formed by oriented "worm-like" micelles (Roux et al. 1995), such as those discussed in Section 12.4.2.1. Whether these bands appear after or during shear, they do not persist indefinitely, but disappear eventually. After cessation of shear, the time required for the bands to appear seems to be roughly inversely proportional to the prior shear rate $\dot{\gamma}_0$, which is the same scaling as has been observed in rheological measurements discussed in Section 11.3.4. A numerical simulation of the Leslie–Ericksen equation by Han and Rey (1995) has recently shown that bands during start-up of shearing can be explained by an instability of the director to a periodic out-of-plane director rotation.

Other quasi-periodic textures have been reported, including birefringent "stripes" parallel to the flow direction (Gleeson et al. 1992). A low-shear-rate stripe texture observed in PBG solutions can be attributed to *roll cells* (Larson and Mead 1993), similar to those that occur in small-molecule tumbling nematics (see Section 10.2.6). Examples of roll cells and other quasi-one-dimensional textures induced during or after shearing are shown in Fig. 11-16.

There are also very irregular textures, such as the "thread" and "worm" textures (Alderman and Mackley 1985; De'Neve et al. 1993a, 1993b), and the "tight" texture (Romo-Uribe et al. 1997). The "thread" texture is an irregular mass of disclination lines in incompletely relaxed nematics under quiescent conditions or at low shear rates (see Fig. 10-22 and the lower half of Fig. 10-20, for example). If allowed to relax, the disclination lines shrink in length and annihilate each other, eventually leaving a uniform *monodomain*, if the sample is thin and the bounding surfaces have been suitably cleaned or treated to produce a uniform anchoring condition for the director. In the absence of flow, the threaded texture is typically much more persistent in polymeric than in small-molecule nematics, because of the high viscosities of the former, $\eta \sim 10^2$–10^4 P.

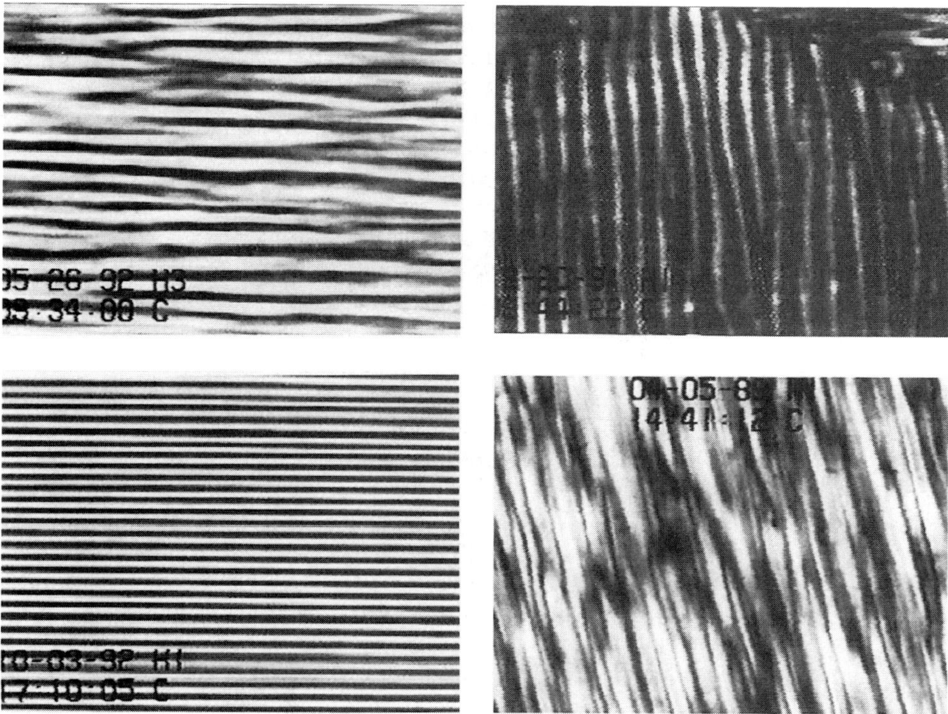

Figure 11.16 "Stripe" and "band" patterns produced by shearing PBG solutions between glass plates under crossed polaroids in a microscope. The field of view is 890 μm, and the flow direction is horizontal. The two "stripe" patterns form at steady state; the more irregular of the two (**upper left**) is produced by roll cells at a low shear rate (around 0.07 sec^{-1}), while the regular stripe pattern (**lower left**) occurs at high shear rate, 25 sec^{-1}. The perpendicular "band" patterns are transients that occur either during start-up of shearing (**upper right**), or after cessation of shearing (**lower right**). The detailed conditions under which these patterns are formed are discussed by Larson (1994).

Although for small-molecule nematics the "threaded" texture can persist under a weak shearing flow, at higher shear rates the disclination density usually increases dramatically, producing the "worm" texture as discussed by De'Neve et al. (1993a). In the "worm" texture, the densely packed disclinations are generally aligned along the flow direction, and there is also a net molecular orientation in the flow direction. The worm texture is also seen in the small-molecule tumbling nematic 8CB (see Section 10.2.6 and Fig. 10-20). In LCPs the worm texture can be very fine-grained (of order a micron or so) and is sometimes called a "speckled" texture (Vermant et al. 1994a), but the existence of orientational anisotropy and relaxation of this texture after cessation of flow distinguish it from the "tight" texture described below. At shear rates above those that produce the worm texture, uniform monodomains and "ordered textures" are sometimes created (Larson and Mead 1993; De'Neve et al. 1993a).

The "tight" texture is a very fine texture evidently associated with a very high disclination density that so randomizes the director that it produces, on average, orientational isotropy when the material is at rest (Romo-Uribe et al. 1997). Melts or solutions containing the tight texture can be oriented to some extent by shear, but they spring back to an isotropic tight texture when the flow is stopped. This texture seems to be associated with the most viscous LCPs; and it occurs, for example, in Vectra-like melts when the molecular weight exceeds 10,000 or so. At lower molecular weights, these melts show the threaded texture at rest (Romo-Uribe et al. 1997). The tight texture seems to occur in melts whose flow curves possess a Region I (Walker et al. 1995, 1997). For a review of the textures of LCPs with and without flow, see Srinivasarao (1995).

11.3.6 The Role of Microcrystallites and Chemical Reactions

It is an unpleasant fact that many of the thermotropic LCPs of greatest interest for commercial application are difficult to study rheologically because of restricted thermal stability at the melt temperature and because of slow structural changes in the melt state that limit the reproducibility of experimental data. These difficulties are to some extent byproducts of the desirable features of LCPs. Applications of LCPs often take advantage of the high melting points that are achievable; a melting temperature near or even above 300°C is desirable for some electronic applications. But when such a material does melt, its stability to oxidation or transesterification reactions may by so marginal that little time is available for careful rheological studies. For LCPs with such high melt temperatures, the isotropization temperature T_{NI} is usually so far above the temperature range at which the material has even marginal thermal stability that any effort to access T_{NI} to achieve a well-defined starting state for rheological studies is hopeless.

Thermotropic LCPs are potentially useful in precision injection molding because, when cooled, the crystallization transition causes only slight changes in volume. This occurs because the state of molecular order in the nematic differs not so much from that in the crystal. Apparently the similar molecular packing in the nematic and crystalline states allows such LCPs to form small microcrystallites slowly even in the "melt" nematic state. The formation of such microcrystallites, even to a slight degree, produces effective "cross-link" points that have a profound effect on rheological properties, leading to quasi-solid elasticity in the melt, similar to that of a physical gel (see Section 5-4). In addition, the slow, temperature-dependent rate of microcrystallite formation leads to (a) a thermal-history-dependence of the rheological properties and (b) sensitivity to sample-preparation protocols.

To avoid these problems, "model" thermotropic LCPs are often studied whose melting temperatures are low enough to access both the nematic and isotropic states without chemical reactions, and which have side groups that suppress crystallite formation (Kim and Han 1993; Chang and Han 1997).

11.4 MOLECULAR DYNAMICS OF POLYMERIC NEMATICS

11.4.1 Molecular and Gradient Elasticities

Because nematic liquid-crystalline polymers by definition are both anisotropic and polymeric, they show elastic effects of at least two different kinds. They have director *gradient* elasticity because they are nematic, and they have *molecular elasticity* because they are polymeric. As discussed in Section 10.2.2, Frank gradient elastic forces are produced when flow creates inhomogeneities or gradients in the continuum director field. Molecular elasticity, on the other hand, is generated when the flow is strong enough to shift the molecular order parameter $S \equiv S_2$ from its equilibrium value S^{eq}. (Microcrystallites, if present, can produce a third type of elasticity; see Section 11.3.6.)

Significant shifts in S are not expected to occur in small-molecule nematics unless the shear rate is extraordinarily high. For polymeric nematics, however, molecular relaxation times τ are typically 0.001–10 sec, or even higher, and therefore molecular elastic effects are produced at shear rates $\dot{\gamma} \sim \tau^{-1} = 0.1$–$1000\,\text{sec}^{-1}$. Thus, the order parameter S is significantly distorted away from that of equilibrium when the *Deborah number* De (discussed in Section 3.6.2.1.1) is of order unity or greater, where

$$\text{De} \equiv \tau\dot{\gamma} \tag{11-1}$$

The relaxation time τ is given by $\tau \equiv 1/6\overline{D}_r$, where $\overline{D}_r$ is the rotational diffusivity of the molecules in the nematic phase.

The dimensionless *Ericksen number* that characterizes the strength of the viscous forces compared to the Frank gradient elastic forces is defined in Section 10.2.5 [Eq. (10-27)] as

$$\text{Er} \equiv \frac{\eta V h}{K} \tag{11-2}$$

Here η is a characteristic Leslie viscosity, K is a Frank constant, V is a flow velocity, and h is a length scale of the flow geometry, such as the gap between shearing surfaces. Naively, one might expect Frank elastic forces to be negligible whenever $\text{Er} \gg 1$, but this is not always so. The reason is that as Er increases, and viscous forces become more important, the distortion of the director field also tends to increase, at least locally near solid walls or disclination lines, leading to a commensurate increase in the strength of the local Frank elasticity. At high shear rates, gradients in the director can become so steep that significant variations in **n** occur over the length of a single polymer molecule. When this occurs, the Frank theory of gradient elasticity fails, and gradient elasticity is expected to saturate (Marrucci 1991), so that it does not increase in strength with further increases in $\dot{\gamma}$.

Gradient elasticity is expected to saturate at a shear rate high enough that molecular elasticity becomes important. This is because flow-induced director gradients become steep on the molecular length scale when the shearing torques exerted on individual molecules are high enough to disturb the molecular order parameter. This suggests that two ranges of shear rate should be considered: De $\ll 1$, where molecular elasticity is negligible and Frank elasticity is important, and De $\gtrsim 1$, where molecular elasticity is important and Frank elasticity is, for the most part, negligible. The overlap region in which both gradient and molecular elasticities might be important is difficult to consider theoretically (see, however, Edwards et al. 1990).

When De $\ll 1$ and molecular elasticity is negligible, the flow properties of polymeric nematics can, in principle, be described by the Leslie–Ericksen equations (see Section 10.2.3). However, at moderate and high De, the Leslie–Ericksen continuum theory fails, and a molecular theory is required to describe the effect of flow on the distribution of molecular orientations.

11.4.2 The Smoluchowski Equation

A molecular theory for the effect of flow on the orientation of ideal monodisperse rigid rod-like polymers has been developed by Doi (1980) and by Hess (1976). In this theory, a *Smoluchowski equation* is derived for the probability $\psi(\mathbf{u})$ that a rod-like molecule is oriented parallel to a unit vector $\mathbf{u}$:

$$\frac{\partial \psi}{\partial t} + \frac{\partial}{\partial \mathbf{u}} \cdot [(\mathbf{u} \cdot \nabla \mathbf{v} - \mathbf{uuu} : \nabla \mathbf{v})\psi] - \overline{D}_r \frac{\partial}{\partial \mathbf{u}} \cdot \left[\frac{\partial \psi}{\partial \mathbf{u}} + \psi \frac{\partial}{\partial \mathbf{u}} \left(\frac{V_{\text{nem}}}{k_B T} \right) \right] = 0 \qquad (11\text{-}3)$$

In Eq. (11-3), $\nabla \mathbf{v}$ is the velocity gradient, $\partial/\partial \mathbf{u}$ is the gradient operator on the unit sphere, and V_{nem} is a nematic potential, such as that of Onsager [Eq. (2-5)], or of Maier and Saupe [Eq. (2-7)]. $\overline{D}_r$ is a preaveraged molecular rotary diffusivity in the nematic phase:

$$\frac{1}{\overline{D}_r} \equiv D_r^{-1} \left(\frac{4}{\pi} \right)^2 [\int \psi(\mathbf{u})\psi(\mathbf{u}') \sin(\mathbf{u}, \mathbf{u}') \, d^2u \, d^2u']^2 \qquad (11\text{-}4)$$

Here D_r is the rotary diffusivity predicted for a rod in an isotropic solution of like rods from the theory of Doi and Edwards (1986) for semidilute solutions, namely,

$$D_r = \beta D_{r0}(\nu L^3)^{-2} \qquad (11\text{-}5)$$

where D_{r0} is the rotary diffusivity for the rod in dilute solution, and β is a dimensionless coefficient, which has been computed to be about 1350 for perfectly rigid rods (Teraoka and Hayakawa 1989; Bitsanis et al. 1990). Fits to experimental data sometimes give much larger values than this, $\beta \approx 10^4$, probably because of polydispersity and flexibility effects (Chow et al. 1985). According to dilute-solution theory for high-aspect ratio ellipsoidal particles (Doi and Edwards 1986; Kirkwood and Auer 1951), D_{r0} is given by Eq. (6-32b) from Chapter 6:

$$D_{r0} = \frac{3k_B T \, (\ln(L/d) - \gamma)}{\pi \eta_s L^3} \qquad (11\text{-}6)$$

where η_s is the solvent viscosity, d is the rod diameter, and the small constant γ can be taken to be 0.8, or neglected altogether. Equation (11-6) is in good agreement with D_{r0} measured in dilute solutions of PBLG (Ookubo et al. 1976; Warren et al. 1973).

In Eq. (11-4), the rotary diffusivity is preaveraged over the orientation distribution, so that it could be taken outside the derivative $\partial/\partial\mathbf{u}$ in Eq. (11-3); a more exact expression for the rotary diffusivity is given in Doi (1980). Although the averaging of the rotary diffusivity removes its dependence on $\mathbf{u}$, a dependence on the average degree of molecular orientation is retained. The higher the degree of orientation, the less the molecules hinder each other's rotary motion, and the higher $\overline{D}_r$ becomes.

Once ψ is computed by somehow solving Eq. (11-3), the birefringence tensor and the stress tensor can be obtained. The birefringence tensor is proportional to $\nu\mathbf{S}$, where $\mathbf{S}$ is the *order parameter tensor,*

$$\mathbf{S} \equiv \left\langle \mathbf{uu} - \frac{1}{3}\delta \right\rangle \tag{11-7}$$

Here δ is the unit tensor, and $\langle\cdots\rangle$ denotes the average over the distribution function ψ [see Eq. (2-4)]. The scalar order parameter S is related to $\mathbf{S}$ by

$$S \equiv \frac{3}{2}(\mathbf{S} : \mathbf{S})^{1/2} \tag{11-8}$$

S lies in the range 0–1, with $S = 0$ for a completely isotropic distribution of orientations and $S = 1$ for rods that are all aligned in the same direction.

The stress tensor is given by (Doi and Edwards 1986)

$$\sigma = 3\nu k_B T\mathbf{S} + \nu \int \psi \left(\frac{\partial}{\partial\mathbf{u}} V_{\text{nem}} \right) \mathbf{u}d^2u + \nu\zeta_{\text{str}}\mathbf{D} : \langle\mathbf{uuuu}\rangle \tag{11-9}$$

The first two terms in this expression for the stress tensor are elastic terms due to Brownian motion and the nematic potential, respectively. The last term is a purely viscous term produced by the drag of solvent as it flows past the rod-like molecules [see Eq. (6-36)]. ζ_{str} is a drag coefficient, which for modestly concentrated solutions is predicted to follow the dilute-solution formula (Doi and Edwards 1986; see Section 6.3.1.4):

$$\zeta_{\text{str}} = \frac{\pi \eta_s L^3}{6[\ell n(L/d) - \gamma]} = \frac{k_B T}{2D_{r0}} \tag{11-10}$$

The ratio of the viscous to the elastic stress contribution is controlled by the dimensionless ratio

$$\beta_V^* \equiv \frac{\zeta_{\text{str}} D_r^*}{k_B T} \tag{11-11}$$

where D_r^* is the value of D_r at the concentration ν^*. Since at fixed polymer volume fraction ϕ, D_r decreases strongly with polymer length L as $D_r \propto L^{-7}$, while ζ_{str} only increases as L^3, theoretically the viscous contribution should become small compared to the elastic contributions for very long, rigid rods. However, this limit is probably not reached for real liquid-crystalline solutions, especially since the effective value of ζ_{str} may exceed its dilute-solution value at higher polymer concentrations (Maffettone et al. 1994). Also, since at high shear rates the elastic contribution to the shear stress decreases with increasing shear rate $\dot{\gamma}$ while the viscous contribution grows roughly linearly with $\dot{\gamma}$, the viscous contribution to the shear stress is dominant at high Deborah number, even when $\beta_V \equiv \zeta_{str} D_r / k_B T$ is as small as 0.02 (Maffettone et al. 1994; Baek et al. 1994).

11.4.3 Doi's Approximate Solution of the Smoluchowski Equation

If Eq. (11-3) is multiplied by **uu** and integrated over the unit sphere, one obtains an evolution equation for the second moment tensor **S** (Doi 1980; Doi and Edwards 1986). In this evolution equation, the fourth moment tensor $\langle \mathbf{uuuu} \rangle$ appears, but no higher moments, if one uses the Maier–Saupe potential to describe the nematic interactions. Doi suggested using a closure approximation, in which $\langle \mathbf{uuuu} \rangle$ is replaced by $\langle \mathbf{uu} \rangle \langle \mathbf{uu} \rangle$, thereby yielding a closed-form equation for **S**, namely,

$$\frac{\partial}{\partial t}\mathbf{S} = \mathbf{F} + \mathbf{G} \tag{11-12}$$

with

$$\mathbf{F} = -6\overline{D}_r \left[\left(1 - \frac{U}{3} \right) \mathbf{S} - U(\mathbf{S} \cdot \mathbf{S} - \tfrac{1}{3}\mathbf{S} : \mathbf{S}\delta) + U\mathbf{SS} : \mathbf{S} \right] \tag{11-13a}$$

$$\mathbf{G} = \tfrac{2}{3}\mathbf{D} + \nabla\mathbf{v}^T \cdot \mathbf{S} + \mathbf{S} \cdot \nabla\mathbf{v} - \tfrac{2}{3}\mathbf{D} : \mathbf{S}\delta - 2\mathbf{D} : \mathbf{SS} \tag{11-13b}$$

where U is proportional to the strength of the nematic interactions, and hence is proportional to the concentration v. Within this closure approximation, $U = U^* = 3$ corresponds to the concentration $v = v^*$.

A closure approximation must also be invoked to express the rotary diffusivity $\overline{D}_r$ in terms of **S**. Doi chose the following approximation:

$$\overline{D}_r \approx D_r \left[1 - \frac{3}{2}\mathbf{S} : \mathbf{S} \right]^{-2} \tag{11-14}$$

The stress tensor is given by

$$\sigma = -\frac{v k_B T}{2\overline{D}_r}\mathbf{F} \tag{11-15}$$

This expression neglects the viscous term, $v\zeta_{str}\mathbf{D} : \langle \mathbf{uuuu} \rangle$.

From the above equations, predictions of the stresses in arbitrary flow histories can readily be computed. Figure 11-4b shows the shear viscosity as a function of reduced concentration, U/U^*, for various dimensionless shear rates, predicted by this theory. These predictions are in reasonable qualitative agreement with experimental data (Fig. 11-4a) for PBLG solutions, except at high concentrations (>30%). Figure 11-17 shows the predicted shear stress and first normal stress difference as functions of dimensionless shear rate for the approximate Doi theory with $U = 6$. The first normal stress difference N_1 is predicted to be positive and to increase monotonically with shear rate. However, as discussed above, many LCPs, at least lyotropic ones, show regions of *negative* N_1. As Marrucci and Maffettone (1989) have shown, negative N_1 is a consequence of *director tumbling*. The closure approximation used to obtain Eq. (11-12) destroys the prediction of tumbling inherent in the original equation (11-3) and thus leads to predictions of positive, rather than negative, N_1. Despite this serious limitation, the approximate theory is probably qualitatively valid for non-shear-flow histories, such as extensional flow, or mixed shear and extension. It may even provide qualitatively reasonable predictions in simple shearing flows of *flow-aligning* nematic polymers, perhaps including some thermotropic LCPs (Baek et al. 1994).

11.4.4 Tumbling and Flow-Aligning

At low shear rates, molecular elasticity should be negligible, and polymeric nematics should, in principle, obey the Leslie–Ericksen constitutive equation, Eqs. (10-10)–(10-18), just as isotropic polymeric liquids obey the Newtonian constitutive equation at low shear rates. Thus, the flow properties of polymeric nematics should be similar to those of small-molecule nematics, when the shear rate is low enough that the molecular order parameter is not appreciably disturbed by the flow and when Region I is absent. In Section 10.2.5 and 10.2.6, where the flow properties of small-molecule nematics are described, the differences between "flow-aligning" and "tumbling" nematics are detailed. When sheared, the director in a flow-aligning nematic tends toward the flow-aligning angle; this angle is usually a few degrees from the flow direction, in the deformation plane. Tumbling nematics, on the other hand, have no preferred alignment angle; instead, at any orientation angle in the deformation plane, the director experiences a viscous torque tending to rotate it.

From slow-shear-rate solutions of the Smoluchowski equation, Eq. (11-3), with the Onsager potential, Semenov (1987) and Kuzuu and Doi (1983, 1984) computed the theoretical Leslie–Ericksen viscosities. They predicted that $\alpha_3/\alpha_2 < 0$ (i.e., tumbling behavior) for all concentrations in the nematic state. The ratio α_3/α_2 is directly related to the *tumbling parameter* λ by $\lambda \equiv (1 + \alpha_3/\alpha_2)/(1 - \alpha_3/\alpha_2)$. (*Note:* the tumbling parameter "λ" is not to be confused with the persistence length λ_p.) Thus, $\lambda < 1$ whenever $\alpha_3/\alpha_2 < 0$. As discussed in Section 10.2.4.1, an approximate solution of Eq. (11-3) predicts that for long, thin, stiff molecules, λ is related to the second and fourth moments S_2 and S_4 of the molecular orientational distribution function (Stepanov 1983; Kröger and Sellers 1995; Archer and Larson 1995);

$$\lambda = \frac{5S_2 + 16S_4 + 14}{35S_2} \tag{11-16}$$

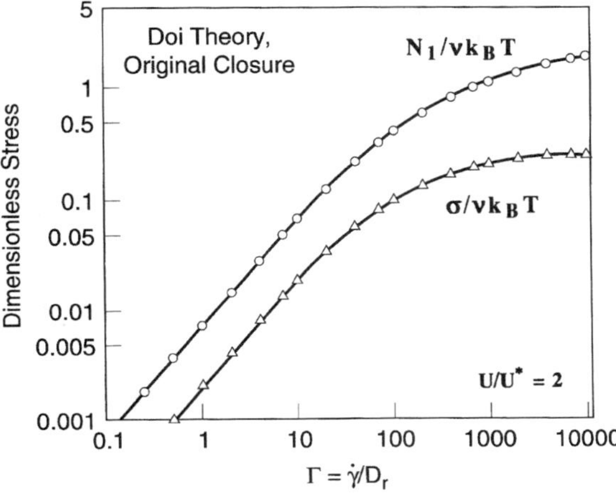

Figure 11.17 Dimensionless shear stress and first normal stress difference as functions of dimensionless shear rate predicted by the Doi with the closure approximation for the fourth moment, $\langle \mathbf{uuuu} \rangle = \langle \mathbf{uu} \rangle \langle \mathbf{uu} \rangle$ and $U = 2U^* = 6$.

Nearly exact numerical solutions of the Smoluchowski equation show that for the Maier-Saupe potential, $\lambda \leq 1$ when $S \equiv S_2 > 0.524$. For the Onsager potential, $\lambda < 1$ for all values of the order parameter within the nematic range. Values of λ for the Onsager potential are plotted in Fig. 11-18.

The Smoluchowski equation (11-3) is valid for *rigid-rod* polymers. Most liquid-crystal-forming polymers, however, are *semiflexible*; typically $L/\lambda_p > 1$. Thus, the Smoluchowski equation for rigid rods might be inaccurate for many LCPs. Although a general theory for the effect of flexibility on the ratio α_3/α_2 is yet to be developed, a couple of limits have been examined. Semenov (1987; Semenov and Khokhlov 1988) derived a generalized Smoluchowski equation for the limit $L/\lambda_p \to \infty$, and found α_3/α_2 to be positive. In the limit $L/\lambda_p \to \infty$, the nematic is therefore predicted to be *flow-aligning* for all values of the order parameter in the nematic state. Subbotin (1993) considered the opposite case of slightly flexible rods in which $0 < L/\lambda_p \ll 1$, and he found that λ is less than predicted for perfectly rigid rods (i.e., the tumbling effect is stronger in slightly flexible molecules than it is in perfectly stiff ones) except at high order parameter. Thus, combining the results of Subbotin and of Semenov, we infer that for moderate S, an increase in L/λ_p at first produces a decrease in λ, but eventually λ rises above unity when L/λ_p becomes large. The value of L/λ_p at which λ crosses unity (α_3/α_2 crosses zero) as a function of S is unknown. Unfortunately, there is as yet no general theory valid for L/λ_p in the typical experimental range, $1 \lesssim L/\lambda_p \lesssim 10$.

Direct experimental determinations of the sign of α_2/α_3 in LCPs by optical conoscopy or similar direct methods are rare, mainly because the optical observations must be made

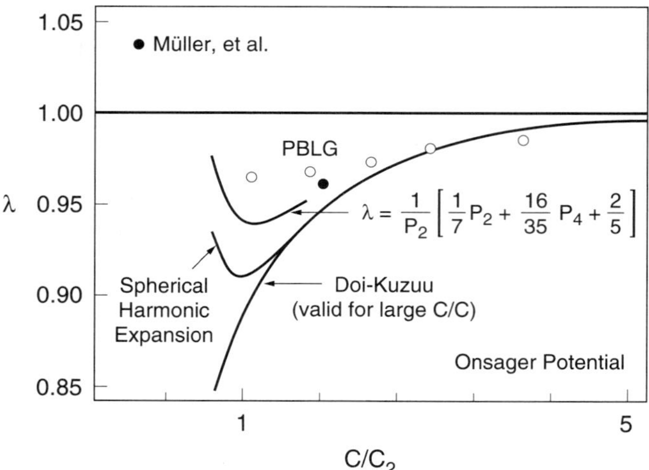

Figure 11.18   Predictions of the tumbling parameter λ as a function of reduced concentration C/C_2 from the Smoluchowski equation for hard rods with the Onsager potential. The "exact" result from the spherical-harmonic expansion is shown, compared to approximate results from an analytic formula and from the perturbation expansion of Kuzuu and Doi. The open circles ($\bigcirc$) are estimates from the periods of shear stress oscillations in transient shearing flows for PBG solutions (see Walker et al. 1995), and the closed circle ($\bullet$) is from a direct conoscopic measurement of Müller et al. (1994).

on untextured monodomain samples, and for polymeric LCPs these are hard to prepare. Nevertheless, the sign of α_2/α_3 has been determined directly for at least three LCPs so far: lyotropic PBG and PBZT (also called PBT) (Burghardt and Fuller 1991; Srinivasarao and Berry 1991) and a thermotropic material, OQO(OEtOEt)10, which is the same as OQO(phenylsulfonyl)10 depicted in Fig. 11-2, except for the (ethoxy)ethoxy side branch that replaces the phenylsulfonyl (Srinivasarao et al. 1992). In these OQO materials, rigid mesogens alternate with 10-carbon flexible spacers. Tumbling ($\alpha_3/\alpha_2 < 0$) was observed in both of the the lyotropic samples, PBG and PBZT. For PBG, Müller et al. (1994) reported a value $\alpha_3/\alpha_2 = -1/44$ by observation of the movement of conoscopic fringes under shearing flow. In thermotropic OQO(OEtOEt)10, tumbling was reported at low temperatures near T_{NA}, the transition temperature from nematic to a smectic A phase, and flow-aligning was reported at higher temperatures near T_{NI}, the temperature of the transition to an isotropic phase. This behavior is similar to that of small-molecule thermotropic nematics, such as 8CB, which also shows a transition from flow-aligning to tumbling as T approaches T_{NA}.

Although the sign of the viscosity ratio, $\mathrm{sgn}(\alpha_2/\alpha_3)$, has been determined by direct optical methods in only these three LCPs, $\mathrm{sgn}(\alpha_2/\alpha_3)$ can also be inferred from rheological measurements on monodomains or even textured samples. In Section 11.3.4, we discussed the multiple overshoots in shear stress that occur after start-up of steady shearing of small-molecule tumbling nematics, in contrast to the single overshoot for homeotropically aligned, flow-aligning, small-molecule nematics. Overshoots, and the strains at which they occur, have been used in combination with the Leslie–Ericksen theory by Yang and Shine (1993) to infer the existence of tumbling and to estimate values of some of the Leslie viscosities in a nematic solution of poly(hexylisocyanate) (PHIC). For textured LCPs that tumble, λ can be inferred from the time period P between the shear stress overshoots in transient shear experiments, if P is assumed to be equal to the period P for rotation of the director through an angle of π in a small "domain." With this assumption, P can be related to λ by

$$P = \frac{2\pi}{\dot{\gamma}\sqrt{1 - \lambda^2}} \tag{11-17}$$

For PBG, Eq. (11-17) gives the values of λ plotted in Fig. 11-18. These values are in reasonable agreement with the predictions of the approximate solution of the Smoluchowski equation.

Another rheological test for tumbling is the measurement of the first normal stress difference N_1 as a function of shear rate $\dot{\gamma}$. As discussed below in Section 11.5.2, the Smoluchowski equation (11-3) for rigid rods predicts the appearance of a region of *negative* N_1, bounded at both lower and higher shear rates by regions of positive N_1, whenever $\alpha_2/\alpha_3 < 0$. Hence a region of negative N_1 is a signal that $\alpha_2/\alpha_3 < 0$. Where comparisons have been made, the implications of the rheological tests have been consistent with direct optical measures of director tumbling.

Rheological signatures of tumbling have been observed in most lyotropic LCPs, including PBG, HPC, PHIC, and PPTA. Thus, either direct or indirect evidence supports the existence of director tumbling in lyotropic LCPs spanning a wide range of molecular flexibilities from very flexible (HPC: persistence length $\lambda_p \approx 9$–12 nm), to moderately flexible (PPTA: $\lambda_p \approx 20$–30 nm), to relatively stiff (PBG: $\lambda_p \approx 90$–150 nm). These results suggest that tumbling may be the norm for most lyotropic LCPs.

For thermotropic LCPs, the situation is more ambiguous. With the exception of the polymer OQO(OEtOEt)10 studied by Srinivasarao et al. (1992), there have been no direct optical determinations of $\mathrm{sgn}(\alpha_2/\alpha_3)$ for thermotropics. And rheological data for most thermotropics are not decisive. To date, no thermotropic LCP has shown the two transitions in N_1 found in lyotropics, namely N_1 changing from positive to negative, and from negative back to positive, as $\dot{\gamma}$ increases. Hydroquinone (OQO) polymers similar to that studied by Srinivasarao et al. show only positive values of N_1 at steady state (see Fig. 11-13), and shear-aligned samples show no shear stress oscillations on start-up of steady shearing. Although a couple of thermotropes have been reported to have low-shear-rate regions of negative N_1 (Prasadarao et al. 1982; Gotsis and Baird 1986; Guskey and Winter 1991), the experimental accuracy of these reports has been questioned (Cocchini et al. 1992). Nevertheless, it is difficult, especially with thermotropes, to cover a range of shear rates that is wide enough to be sure that N_1 does not change sign at shear rates outside of the experimental range. Thus, the sign of α_2/α_3 for thermotropic polymers is either positive or unknown.

11.5 MOLECULAR THEORY FOR THE RHEOLOGY OF POLYMERIC NEMATICS

11.5.1 Leslie Viscosities and Frank Constants of Monodomains

At low enough shear rates, polymeric nematics ought to obey the same Leslie–Ericksen continuum theory that describes so well the behavior of small-molecule nematics. The main difference is that polymers have a much higher molecular aspect ratio than do small molecules, which leads to greater inequalities in the the numerical values of the various viscosities and Frank constants and to much higher viscosities.

Because of the difficulty with which polymeric nematic monodomains are prepared, there are few measurements of Leslie viscosities and Frank constants for LCPs reported in the literature. The most complete data sets are for PBG solutions, reported by Lee and Meyer (1990), who dissolved the polymer in a mixed solvent of 18% dioxane and 82% dichloromethane with a few percent added dimethylformamide. Some of these data, measured by light scattering and by the response of the nematic director to an applied magnetic field, are shown in Figs. 11-19 and 11-20 and in Table 11-1. While the twist constant has a value of around $K_2 \approx 0.6 \times 10^{-7}$ dyn, which is believed to be roughly independent of concentration and molecular weight, the splay and bend constants K_1 and K_3 are sensitive to concentration and molecular weight.

These trends can be compared with the theoretical predictions, plotted in Fig. 11-21, obtained from a gradient expansion of the Onsager excluded-volume potential for hard cylindrical rods. A quantitative comparison of the theory with experiments can be made by noting that the diameter of PBG is around 1.5 nm, yielding $d/k_B T \approx 3.75 \times 10^6$ dyn^{-1}. For the lowest concentration ($C = C_2$) above the biphasic window, $\pi L^2 d\nu/4$ in Fig. 11-21 is equal to 4.20, and the hard-rod theory predicts $K_1 = 2.2 \times 10^{-7}$ dyn, $K_2 = 0.7 \times 10^{-7}$ dyn, and $K_3 = 14 \times 10^{-7}$ dyn at room temperature, which are of the right order of magnitude but are not in quantitative agreement with the theory. The hard-rod theory predicts that $K_1/K_2 = 3$ for all rod concentrations, and $K_3/K_2 = 20\text{--}100$ for C ranging from C_2 to

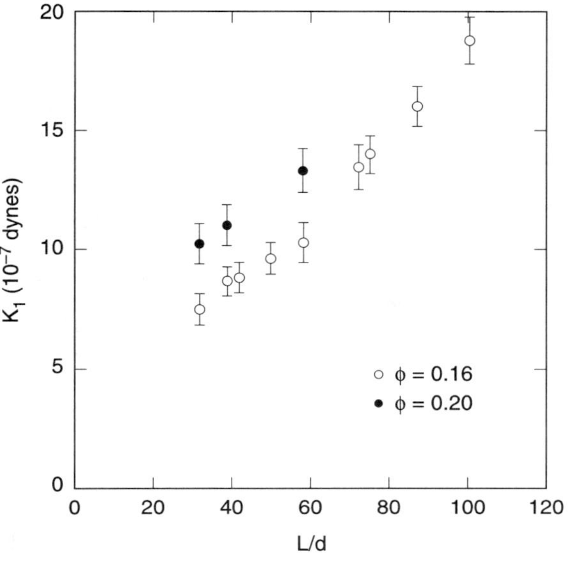

Figure 11.19 Splay constant K_1 as a function of molecular aspect ratio L/d at polymer volume fractions of 0.16 and 0.20, obtained from light-scattering data, taking $K_2 = 0.6 \times 10^{-7}$ dynes, in solutions of PBG in a mixed solvent composed of 18% dioxane and 82% dichloromethane. An aspect ratio of 100 corresponds to a molecular weight of 210,000. (From Lee and Meyer 1990, by permission of Taylor & Francis.)

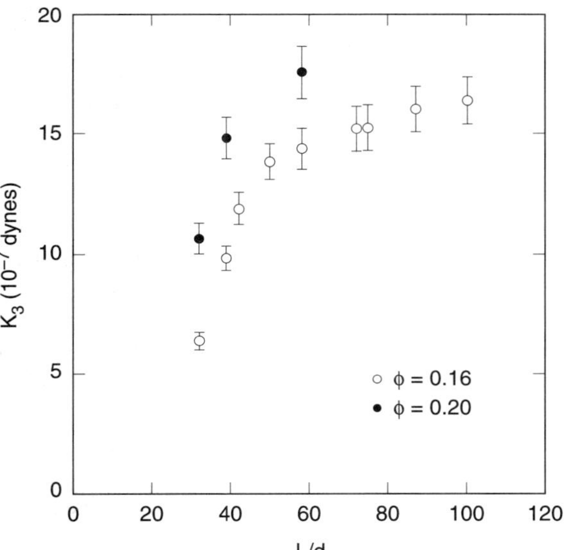

Figure 11.20 The bend constant K_3 as a function of molecular aspect ratio, obtained for the same solutions as in Fig. 11-19. (From Lee and Meyer 1990, by permission of Taylor & Francis.)

$2 \times C_2$. Although these ratios are not consistent with data for PBG solutions, they are consistent with measurements of the Frank constants of suspensions of the tobacco mosaic virus (Lee and Meyer 1986), which is an extremely rigid particle. Thus, discrepancies between theory and measurements for PBG are most likely due to the partial flexibility of PBG molecules.

A Landau–de Gennes theory for the Frank constants of long semiflexible worm-like chains at low order parameter S gives (Shimada et al. 1988) $K_1 = K_3 = 10K_2 =$

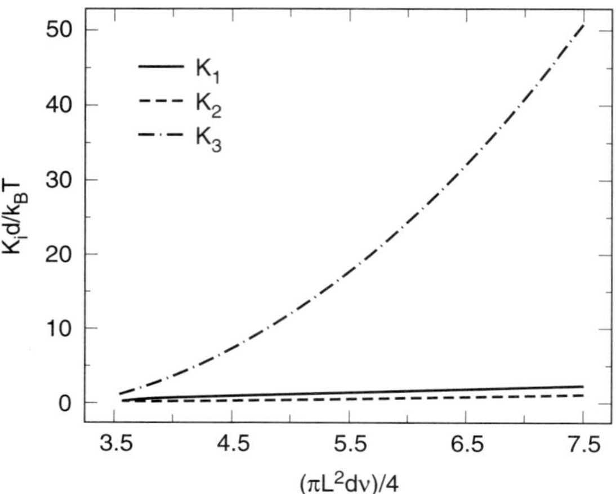

Figure 11.21 Predictions of the dimensionless Frank constants for hard rods as functions of reduced concentration $\pi L^2 dv/4$, using a numerical solution of the Onsager equation. At the concentration $v = v^*$, the abscissa has the value $\pi L^2 dv/4 = 4$. (From Lee and Meyer, reprinted with permission from J. Chem. Phys. 84:3443, Copyright © 1986, American Institute of Physics.)

$(25/14)L\lambda_p vk_B T S^2$. These predictions yield $K_1 = K_3 = 10K_2 \approx 20 \times 10^{-7}$ dyn for PBG molecules of typical length $L \approx \lambda_p \approx 100$ nm, of concentration such that $vk_B T \approx 2 \times 10^4$ dyn/cm^2, and of order parameter $S \approx 0.8$. Although the Landau–de Gennes theory is, in principle, only valid for large L/λ_p and small S, these predictions for the K's are in better accord with Lee and Meyer's PBG data than those obtained from the hard-rod theory. Thus only for highly rigid molecules or particles (such as the tobacco mosaic virus) is the bend constant K_3 much larger than both the splay and twist constants, while for semiflexible molecules, $K_1 \approx K_3 \gg K_2$.

Measurements of the twist viscosity $\gamma_1 \equiv \alpha_3 - \alpha_2$ for PBG solutions are reported in Fig. 11-22; the dependence on molecular weight is steep. The shear viscosity of textured PBLG solutions in m-cresol is also a steep function of molecular weight, $\eta \propto M^{5.5}$, with η ranging from 3 to 2000 Pa s as M varies from 150,000 to 510,000, for a fixed concentration of 20% (Asada et al. 1985). This power law is close to the theoretical one, $\eta \propto M^6$, from Doi's theory for rigid rods. Ratios of viscosities measured by light scattering are reported in Table 11-1 for PBG solutions with average molecular weights of 70,000 and 130,000 and concentrations of around 15% and 20% by volume (Lee and Meyer 1990). For these molecular weights, ϕ^* (the concentration at which the isotropic solution becomes unstable to formulation of the nematic) can be estimated from the data of Robinson et al. (1958), to be around 15% and 12% for the lower- and higher-molecular-weight samples, respectively.

The viscosities γ_1, η_{splay}, and η_{bend} in Table 11-1 are related to the Leslie α's and to the Miesowicz viscosities η_b and η_c, defined in Eq. (10-19), by

$$\gamma_1 = \alpha_3 - \alpha_2, \qquad \eta_{\text{splay}} = \gamma_1 - \frac{\alpha_3^2}{\eta_b}, \qquad \eta_{\text{bend}} = \gamma_1 - \frac{\alpha_2^2}{\eta_c} \qquad (11\text{-}18)$$

Formulas for the Leslie viscosities, in turn, were derived from the Smoluchowski equation for hard rods by Kuzuu and Doi (1983, 1984; Semenov 1987), and are given in Eqs. (10-20) with $\alpha_0 = 0$. These formulas require as inputs values of S_2, S_4, λ, and $\overline{D}_r$, which are functions of polymer concentration C. Reasonably reliable analytic functions for these dependencies were obtained by Kuzuu and Doi using a perturbation expansion for large order parameter, yielding

TABLE 11-1
Viscosity Ratios for PBG Compared to Hard-Rod Theory

	Experimental	Theory $\beta_V^* = 0.03$	Theory $\beta_V^* = 0$
$M = 70{,}000$; $\phi = 0.159$; $\phi/\phi^* = 1.06$			
$\eta_{\text{splay}}/\gamma_1$	0.85	0.76	0.59
$\gamma_1/\eta_{\text{bend}}$	108	102	222
η_c/η_b	92	78	132
$M = 70{,}000$; $\phi = 0.203$; $\phi/\phi^* = 1.35$			
$\eta_{\text{splay}}/\gamma_1$	0.92	0.84	0.66
$\gamma_1/\eta_{\text{bend}}$	166	193	513
η_c/η_b	153	161	339
$M = 130{,}000$; $\phi = 0.156$; $\phi/\phi^* = 1.3$			
$\eta_{\text{splay}}/\gamma_1$	0.97	0.83	0.65
$\gamma_1/\eta_{\text{bend}}$	148	175	448
η_c/η_b	144	145	293
$M = 130{,}000$; $\phi = 0.202$; $\phi/\phi^* = 1.7$			
$\eta_{\text{splay}}/\gamma_1$	1.00	0.89	0.73
$\gamma_1/\eta_{\text{bend}}$	254	347	1180
η_c/η_b	254	308	830

$$S_2 = 1 - 0.1473(\nu/\nu^*)^{-2} - 3.344 \times 10^{-2}(\nu/\nu^*)^{-4} - 1.49 \times 10^{-2}(\nu/\nu^*)^{-6}$$

$$S_4 = 1 - 0.4909(\nu/\nu^*)^{-2} - 2.713 \times 10^{-2}(\nu/\nu^*)^{-4} - 1.56 \times 10^{-2}(\nu/\nu^*)^{-6} \qquad \text{(11-19a)}$$

$$\frac{\alpha_i^E}{\eta^*} = a_{i1}(\nu/\nu^*) + a_{i2}(\nu/\nu^*)^{-1} + a_{i3}(\nu/\nu^*)^{-3} \qquad \text{(11-19b)}$$

with the coefficients a_{ij} given in Table 11-2.

Here $\eta^* \equiv \nu^* k_B T/10 D_r^*$ is the viscosity of the (hypothetical) isotropic phase at C^*.

The viscous stress, $10(\nu/\nu^*)\eta^*\beta_V^* \mathbf{D} : \langle \mathbf{uuuu} \rangle$, contributes to the Leslie viscosities, as follows (Larson 1996):

$$\alpha_1^V = 10\left(\frac{\nu}{\nu^*}\right)\eta^*\beta_V^* S_4, \qquad \alpha_4^V = \frac{4}{21}\left(\frac{\nu}{\nu^*}\right)\eta^*\beta_V^*(7 - 10S_2 + 3S_4)$$

$$\alpha_2^V = \alpha_3^V = 0, \qquad \alpha_5^V = \alpha_6^V = \frac{20}{7}\left(\frac{\nu}{\nu^*}\right)\eta^*\beta_V^*(S_2 - S_4) \qquad \text{(11-20)}$$

These contributions should be added to the "elastic" viscosities [the α_i^E's in Eq. (11-19b)] calculated by Kuzuu and Doi, who neglected the viscous stress. Table 11-1 shows values of ratios of viscosities computed from the hard-rod theory both without the viscous term, and with a viscous term with $\beta_V^* = 0.03$, computed from Eqs. (11-19a) and (11-20), for PBG of molecular weights $M = 70{,}000$ and $130{,}000$. The molecular lengths are given by $L = M/M_L = 48$ nm and 89 nm, where $M_L = 1460$ nm^{-1} for PBG molecules. The

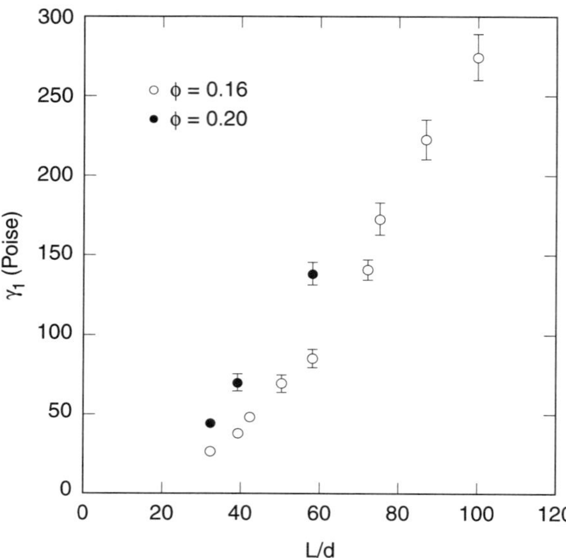

Figure 11.22   Twist viscosity γ_1 as a function of molecular aspect ratio L/d, obtained for the same solutions as in Fig. 11-19. (From Lee and Meyer 1990, by permission of Taylor & Francis.)

predicted viscosity ratios are in reasonably good agreement with the hard-rod theory if we take $\beta_V^* = 0.03$ (see Table 11-1). Note in Table 11-1 that when the viscous term is dropped, the predicted viscosity ratios are much larger than the measured ones. Thus, at least for these PBG solutions, the viscous stress contribution cannot be neglected.

According to Eqs. (11-5), (11-10), and (11-11), β_V^* is proportional to $(vL^3)^{-2} \propto \phi^{-2}L^{-4}$. Thus, as the molecular weight increases, β_V^* is predicted to decrease dramatically. For PBG solutions of molecular weight 130,000, one predicts from Eq. (11-11) that β_V^* should be only around 0.004, and the viscous stress should become negligible at high molecular weight. However, Table 11-1 shows that the very large viscosity ratios predicted for negligible β_V^* are not seen in the experiments. Better agreement with the experiments is obtained if β_V^* is held fixed at 0.03, even for the higher-molecular-weight samples. This effect might be explained by molecular flexibility. Flexibility increases the value of β, which leads to a correspondingly larger value of β_V^*, according to Eqs. (11-5) and (11-11). Thus, even for high-molecular-weight samples, the viscous stress cannot always be neglected.

Few other sets of viscosities exist for polymeric nematics. Yang and Shine (1993) obtained three of the Leslie viscosities for monodomains of poly(n-hexyl isocyanate) (PHIC) from rheological measurements in the presence of an electric field, and they obtained values reasonably consistent with the predictions of the Kuzuu–Doi expressions. From monodomains of the polyion PBZT, poly(1,4-phenylene-2,6-benzobisthiazole) in methane sulfonic acid, some of the Leslie–Ericksen parameters have been extracted via light-scattering and magnetic-field-reorientation studies (Berry 1988; Srinivasarao and Berry 1992). As with PBG, the ratios K_1/K_2 and K_3/K_2 are roughly 10, but the ratio $\gamma_1/\eta_{\text{bend}}$ was found to be only around 7, much less than for PBG. The behavior of these PBZT solutions deviates from theoretical expectations in other respects as well. Although molecular modeling indicates that the PBZT molecule should be quite flexible, with λ_p around 21.5 nm (Farmer et al. 1993), both intrinsic viscosity measurements (Berry 1989) and

TABLE 11-2
Coefficients of Eq. (11-19b), Calculated by Kuzuu and Doi

i	a_{i1}	a_{i2}	a_{i3}
1	$-5/2$	0.767	0.182
2	$-5/2$	-0.215	-0.011
3	0	0.123	0.051
4	0	0.123	0.038
5	$5/2$	-0.153	-0.081
6	0	-0.245	-0.041

the low value of the critical volume fraction required to form a wholly nematic state, namely $\phi_2 \lesssim 0.04$ (Einaga et al. 1985), suggest that the PBZT molecule is effectively very stiff, with $\lambda_{p,\text{eff}}$ around 200 nm or higher. In addition, in the nematic state, the order parameter S is very high, $S \approx 0.94$ (Mattoussi et al. 1992), even when ϕ is not much above ϕ_2. Furthermore, although the displacement of conoscopic images under shear indicates that PBZT solutions are tumbling nematics, the displacement does not seem to follow the Leslie–Ericksen equations, even at low shear rates (Srinivasarao and Berry 1991). While roll cells are observed, as expected for a tumbling nematic when the shearing flow is orthogonal to the initial director orientation (see Section 10.2.6), the width of these rolls does not seem to follow theoretical predictions (Larson 1993). Finally, these PBZT solutions exhibit a very pronounced Region I-type shear thinning, even at concentrations close to ϕ_2 (see Fig. 11-8). It is possible that the charges distributed along the molecule contribute to its effective stiffness, and perhaps are responsible for some of the other anomalies as well (Berry 1989).

For thermotropic LCPs, viscosity data for carefully prepared monodomains are rare. However, Klein et al. (1989) have aligned thermotropic samples of the polymer DDA9, poly(oxy(3-methyl-1,4-phenylene)-azoxy(2-methyl-1,4-phenylene)oxy-(1,12-dioxo-1,12-dodecanediyl), in the magnetic field of a nuclear magnetic resonance (NMR) spectrometer. These aligned samples were then rotated with respect to the field, so the twist viscosity γ_1 could be inferred from the rotational response time of the director to the magnetic field. These viscosities show a remarkably strong dependence on molecular weight, $\gamma_1 \propto M^6$, over the molecular weight range 2250–8400, with γ_1 nearly equal to 10^6 P for the sample of highest molecular weight at $T = 119°C$. Kim and Han (1993) reported a similarly steep molecular-weight dependence, $\eta \propto M^6$, for the shear viscosity of a thermotrope, OQO(phenylsulfonyl)10, that had been prealigned by a strong shearing flow, with $\eta = 3 \times 10^5$ P for $M = 53,000$ at 140°C. A similar power law, $\eta \propto M^{6.5}$, was reported for the high-temperature *isotropic* phase of this polymer. While the range of molecular weights in this study was narrow, $M = 35,000$–53,000, it was wide enough to observe a decade increase in viscosity. Furthermore, a very similar thermotrope, OQO(OEt)10, differing only in the side branch, showed a two-decade increase in the viscosity (Baek et al. 1994) and relaxation time (Hudson et al. 1993b) in both the nematic and isotropic phases when the molecular weight was increased from 5000 to 30,000. A power-law exponent of 4.1 was recently reported for a Vectra-like series of nematic polyesters with no flexible spacers (Romo-Uribe and Windle 1995). Thus, in both the nematic and isotropic states, the dependence of the viscosity on molecular weight of these semiflexible thermotropic polymers is steeper than that of flexible polymers. Although the Doi theory for rigid-rod polymers predicts a steep molecular-weight dependence of the viscosity similar to that observed, $\eta \propto M^6$ (Kim and Han 1993), one would not expect the theory to be relevant to semiflexible polymers. A possible explanation is the molecular-weight dependence of the glass transition observed in these polymers (Hudson et al. 1993b).

The Frank constants for thermotropic polymers are rarely measured. Zheng-Min and Kléman (1984) report values of $K_1 \approx 30 \times 10^{-7}$ dyn, $K_2 \approx K_3 \approx 3 \times 10^{-7}$ dyn for a semiflexible thermotropic polyester with molecular weight of around 8500. The high value of the splay constant, relative to those of twist and bend, in this bulk, semiflexible, thermotropic nematic, is consistent with an argument of de Gennes (1977) that if hairpin conformations can be neglected, splay deformation requires either density variations or a high concentration of chain ends, and both are suppressed in bulk nematics composed of long molecules. By observing the director pattern around a 1/2 disclination line via electron microscopy, Hudson et al. (1993a) estimated the ratio K_1/K_3 for two polyesters with differing rigidity. They found that K_3 and K_1 are within a factor of 2–3 of each other, but that K_1/K_3 varies somewhat as one approaches within 1 μm of the core. The variation of K_1/K_3 with distance from the core indicates a failure of the Frank theory at small distances from the core. The more flexible polyester tended to have a higher ratio K_1/K_3, as expected. For a soluble semiflexible polydiacetylene adopting a "worm-like" configuration, Wang et al. (1994) used a similar method to obtain $K_1/K_3 \approx 3$. There is clearly a great need for further measurements of the elastic and viscous constants for polymeric nematics, especially thermotropic ones.

11.5.2 Rheology in Region III: The Effect of Molecular Elasticity

In Region III, the shear rate is high enough to distort significantly the degree of molecular order—that is, to change the scalar order parameter S. Hence, in this region the Leslie–Ericksen equations do not apply. One can still predict rheological properties, however, by solving numerically the Smoluchowski equation (11-3) for the molecular orientation distribution ψ outside of the Leslie–Ericksen limit. The most accurate solutions presently available for this equation are obtained using a high-order expansion in spherical harmonic functions (Larson 1990; Larson and Öttinger 1991), but qualitatively similar predictions were first obtained for orientations in the deformation plane with simpler, two-dimensional solutions (Marrucci and Maffettone 1989). The stress tensor can be calculated from the orientation distribution ψ using Eq. (11-9). In the numerical solutions, Frank elasticity is neglected; hence, for polydomains, Eq. (11-3) is assumed to describe each domain, with negligible interdomain interactions.

The term "director" is usually only defined in the limit of vanishing shear rate where the order parameter tensor has uniaxial symmetry. Even at finite shear rates, however, the order parameter tensor $\mathbf{S}$ remains well-defined; it can be represented pictorially by an ellipsoidal shape, whose major axis is the largest eigenvalue of $\mathbf{S}$. If, for example, the major axis of $\mathbf{S}$ lies in the x–y plane, then its angle θ measured clockwise with respect to the x direction is given by

$$\frac{1}{2}\tan(2\theta) = \frac{S_{xy}}{S_{xx} - S_{yy}} \tag{11-21}$$

If we now generalize the definition of "director" to be the unit vector parallel to the major axis of the order parameter tensor, we find that at vanishingly low shear rates, where the order parameter tensor is nearly uniaxial, this definition reduces to the usual meaning of the term "director."

As the shear rate increases, the numerical solutions of the Smoluchowski equation (11-3) begin to show deviations from the predictions of the simple Ericksen theory. In particular, the scalar order parameter S begins to oscillate during the tumbling motion of the director (for a discussion of tumbling, see Sections 11.4.4 and 10.2.6). The maxima in the order parameter occur when the director is in the first and third quadrants of the deformation plane; i.e., $0 < \theta - n\pi < \pi/2$, where n is an integer. Minima of S occur in the second and fourth quadrants. The amplitude of the oscillations in S increases as $\dot{\gamma}$ increases, until S is reduced to only 0.25 or so over part of the tumbling cycle.

As $\dot{\gamma}$ is increased still further, the tumbling behavior of the director is replaced by a "wagging" motion wherein the director oscillates back and forth about the flow direction. The frequency of this wagging motion increases with further increases in $\dot{\gamma}$; and after another transitional shear rate is exceeded, the wagging motion damps out and a steady-state is attained in both S and in the director orientation (Larson 1990).

The instantaneous stresses computed from the molecular theory are time oscillatory in the tumbling and wagging regimes. Such oscillatory stresses could be observed experimentally only if one could prepare a monodomain and shear it without introducing orientational inhomogeneities or defects. But most experiments on polymeric nematics are carried out on polydomain samples that are riddled with disclinations. Nevertheless, if the Frank gradient stresses are small compared to viscous stresses, the average stresses in the polydomain might be estimated simply by averaging the stress tensor in a single uniformly oriented domain over a period of its oscillation. This assumes that, after prolonged steady shearing, each "domain" settles into a tumbling or wagging orbit that is identical to that of neighboring domains except for a phase shift. The orbit of each domain, it is assumed, is affected by interactions with other domains only in that a steady-state distribution of domain orientations is produced. This calculation neglects variations in the shear rate from one domain to another.

With these rather drastic approximations, the steady-state shear stress σ and first and second normal stress differences, N_1 and N_2, can be calculated from the time-dependent numerical solutions of the Smoluchowski equation. Figure 11-23a plots N_1 and N_2 for the Onsager potential with $U = 10.67$, which corresponds to the lowest concentration in the wholly nematic state. These results are in amazingly good agreement with the measured normal stress differences for PBLG solutions (see Fig. 11-23b). Similar qualitative agreement with the theory is obtained for the normal stress differences of HPC solutions at modest concentrations (Baek et al. 1993a).

Not only are the signs of N_1 and N_2 correct in each of the three ranges of shear rate, but so are the relative locations of the intersection points of N_1 with N_2, as well as the general shapes of the curves. For PBLG, even the magnitudes of the predicted stresses are found to be close to the experimental ones, if one notes that $\nu k_B T \approx 10^3$ dyn/cm^2 for the concentration considered. The transition from tumbling to wagging, which produces the first sign change in N_1, also causes an increase in average birefringence Δn (see Fig. 11-24). At the shear rate at which N_1 first changes sign, the birefringence begins to increase. The birefringence levels off at high shear rates, as the steady-state flow regime is entered.

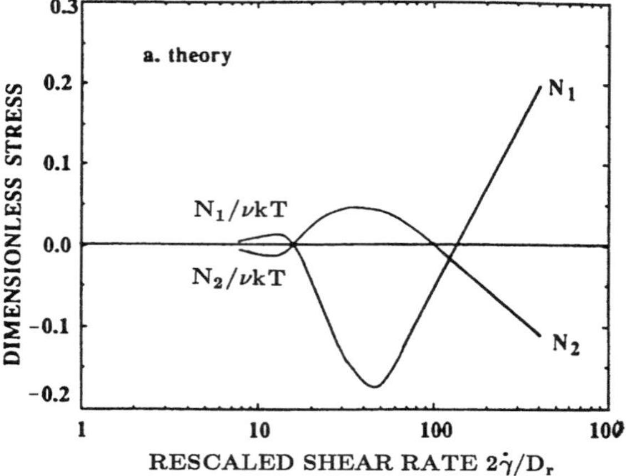

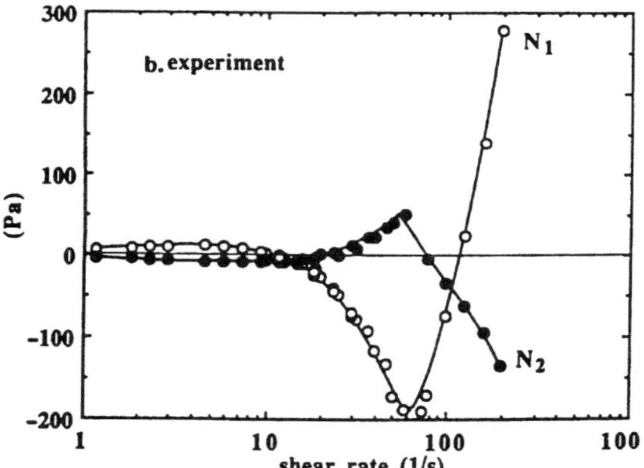

Figure 11.23   Comparison of theoretical and experimental first and second normal stress differences N_1 and N_2. The theoretical results **(a)** were calculated from the Smoluchowski equation (11-3) using the Onsager potential with $U = 10.67$, the minimum value for a fully nematic state. $\dot{\gamma}/D_r$ is the dimensionless shear rate (or Deborah number), where D_r is the rotary diffusivity of a hypothetical isotropic fluid at the same concentration. Only the molecular–elastic contribution to the stress tensor was considered. The experimental results **(b)** are for 12.5% (by weight) PBLG (molecular weight $= 238,000$) in *m*-cresol. (Reprinted with permission from Magda et al., Macromolecules 24:4460. Copyright © 1991, American Chemical Society.)

This increase in birefringence is qualitatively predicted by the molecular theory; but the magnitude of the observed increase is considerably greater than predicted (Hongladarom et al. 1993).

In Fig. 11-23, the most significant deviations between theory and experiment occur in Region II, the constant-viscosity region at shear rates below the positive maximum in N_1. (Region I is absent in this material; the lowest rates shown are in Region II.) Note in Fig. 11-23 that as the shear rate decreases, the normal stress differences decrease toward zero more rapidly in the theory than in the experiment. In Region II, N_1 and N_2 are predicted by the Smoluchowski equation to be quadratic in shear rate, $N_1 \propto -N_2 \propto \dot{\gamma}^2$, while experiments show instead that $N_1 \propto -N_2 \propto \dot{\gamma}$, a scaling consistent with that predicted by the mesoscopic theory described in the next section, 11.5.3. In this low-shear-rate region (Region II), disturbances to the domain orbits from Frank elastic stresses (which are proportional to $\dot{\gamma}$) cannot be neglected.

However, in Region III—which includes shear rates above the first, positive maximum in N_1—the agreement between theory and experiment for PBLG solutions is very good,

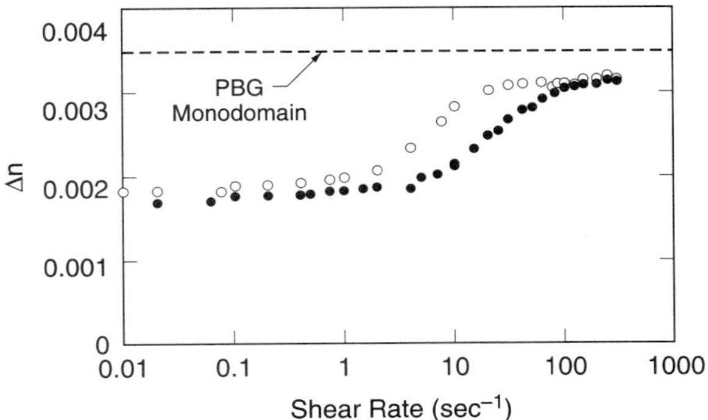

Figure 11.24   Steady-state birefringence as a function of shear rate for PBG solutions in 13.5% (by weight) m-cresol, for molecular weights of ($\bigcirc$) 301,000 and ($\bullet$) 238,000. The dashed line gives the birefringence of monodomains of the solutions with a molecular weight of 301,000 [Hongladarom et al. (1993); see also the thorough review of Burghardt (1997)]. (Reprinted with permission from Hongladarom et al., Macromolecules 26:772. Copyright © 1993, American Chemical Society.)

at least when the concentration C is not too much higher than C_2. The only adjustable parameter in the theory is the rotary diffusivity D_r in Eq. (11-4), which sets the scale for the shear rate. In principle, D_r can be computed from molecular theory using Eq. (11-5). By comparing theory and experiment in Fig. 11-23, we find that the best-fit value, $\beta \approx 15,000$, is about a factor of 10 larger than the theoretical value $\beta = 1350$. A similar discrepancy between the theoretical and best-fit values of β for high-molecular-weight PBG samples was discussed in Section 11.5.1. The higher "best-fit" value of β can be relationized, at least to some extent, by the nonideal characteristics of the PBLG solutions, such as molecular flexibility and polydispersity.

Once β has been obtained by fits to data for one molecular weight M and concentration C, Eq. (11-5) can be used to predict the rotary diffusivity D_r at other values of concentration and molecular weight. A convenient way to test the rheological consequences of these predictions is to compare the predicted value of $\dot{\gamma}_{max}$, the shear rate at which N_1 reaches its positive maximum, with experimental measurements of $\dot{\gamma}_{max}$. At a fixed value of C/C_2, $\dot{\gamma}_{max}$ is predicted to decrease very steeply with molecular weight, as $\dot{\gamma}_{max} \propto M^{-5}$, which is roughly what is observed in PBLG solutions with C/C_2 slightly higher than unity, with M ranging from 86,000 to 238,000 (Gleeson et al. 1992). The predicted dependence of $\dot{\gamma}_{max}$ on C/C^* at fixed M is also steep and is shown in Fig. 11-25a, along with the prediction for $N_{1,max}$, the first normal stress difference at $\dot{\gamma}_{max}$. The predicted monotonic increases in these quantities are observed experimentally for PBLG solutions for concentrations up to around 40% by weight (see Fig. 11-25b). Similar experimental results can be found in Kiss and Porter (1980b). At concentrations above about 40%, PBLG in m-cresol forms a viscous gel-like phase at room temperature, an occurrence that seems to be common in solutions of rigid polymers.

For LCPs other than PBG, trends with molecular weight and concentration have been much less thoroughly studied. For HPC solutions in m-cresol, $\dot{\gamma}_{max}$ decreases much less steeply with M than for PBLG, no doubt because HPC molecules are much more flexible than PBG molecules (Baek et al. 1994).

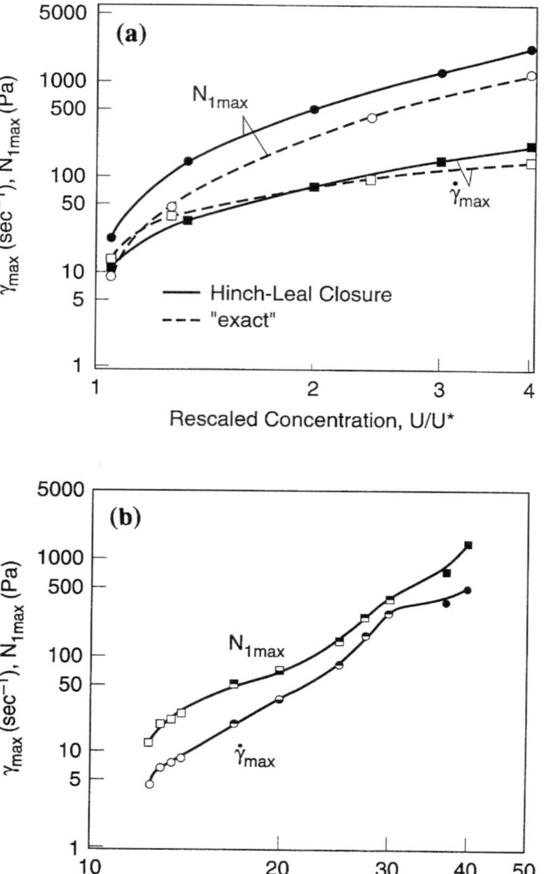

Figure 11.25 **(a)** Shear rate $\dot{\gamma}_{max}$ at which the first normal stress difference reaches its positive maximum $N_{1\,max}$, versus reduced concentration U/U^*, predicted by the Smoluchowski equation using an "exact" spherical harmonic solution and using the "Hinch–Leal" approximate closure (see Baek et al. 1993b). These predictions use the estimates $D_r^* = 6$ sec^{-1} and $v^*kT = 10^3$ Pa. **(b)** Measured values of $\dot{\gamma}_{max}$ and $N_{1\,max}$ for PBLG (molecular weight = 238,000) as a function of concentration in m-cresol. The different symbols denote measurements using different cone-and-plate fixtures. (From Baek et al. 1993, with permission of the Journal of Rheology.)

For concentrations of HPC or PBG greater than about 35–40%, the rheological behavior begins to diverge drastically from the predictions of the Smoluchowski equation (Baek et al. 1994). For example, for HPC solutions in m-cresol with $C \geq 35$wt%, $\dot{\gamma}_{max}$ decreases rather than increases with increasing C; and the activation energy of the solution, inferred from the shift of rheological data with temperature, begins to exceed greatly that of the solvent (Baek et al. 1994). Large activation energies are also found for HPC in water at high concentrations (Walker and Wagner 1994). In addition, at high concentrations of polymer, the shear viscosity shows shear thinning at low shear rates, consistent with so-called Region I behavior, discussed in Section 11.5.4. At $C \gtrsim 55$ wt% HPC, the region of negative N_1 shifts upward to become a local *positive* minimum (Baek et al. 1994).

Most of the discussion in this section on Region III rheology has focused on HPC and PBG solutions, because for these LCPs, Region III behavior can clearly be identified. Many other LCPs, lyotropic and thermotropic, show no steady-state negative N_1, perhaps because they are not tumbling nematics or because for them Region III is not reached even at the highest shear rates attained in the experiments. In thermotropic Vectra-like LCPs of modest molecular weight ($M_w < 10,000$), a low-shear-rate "log-rolling" regime is found in which the steady-state director orientation is in the vorticity direction (Romo-Uribe and Windle 1996). Such a regime has been predicted from solutions of the Smoluchowski equation (Larson and Öttinger 1991), but in the Vectra-like LCP the occurrence of this regime might instead be caused by a transition to a smectic phase.

11.5.3 Rheology in Region II: The Effect of Texture Elasticity

One might think that at low shear rates, the structure of a nematic LCP would be only slightly disturbed from equilibrium, and hence its flow properties ought to be predictable by assuming that disturbances from equilibrium are small. This assumption is only true in part. At the molecular level, a polymeric nematic is nearly at equilibrium when $\dot\gamma$ (or really $De \equiv \tau\dot\gamma$) is small. But on a larger, supermolecular scale, the director pattern remains near equilibrium only when the Ericksen number $\mathrm{Er} \equiv \eta\dot\gamma h^2/K$ is small, and this is rarely the case for polymeric nematics. Instead, because of their high viscosity, the Ericksen number of polymeric nematics in shearing flow is almost always high, even if the Deborah number is low. At high Ericksen number, both flow-aligning and tumbling nematics are unstable to the production of disclinations that lead to a disordered flowing state called *director turbulence* (see Section 10.2.6). Even starting from a monodomain, under shearing flow, polymeric monodomains can only be kept free of disclinations under special conditions, such as very low shear rate (e.g., $\dot\gamma \lesssim 10^{-3}\ \mathrm{sec}^{-1}$, so that $\mathrm{Er} \lesssim 10^3$), a small shearing strain ($\gamma \lesssim 5$), or the application of a strong electric or magnetic field. Therefore, practical theories of polymer nematics in shearing flow, even at $\mathrm{De} \ll 1$, must account for the irregular disclination-filled textures typical of these fluids. Some preliminary steps have been taken toward development of microscopic theories of the dynamics of disclination lines in shearing flows (Marrucci and Maffettone 1993; Rey 1993; Windle et al. 1994), but experimental studies of disclination dynamics are required to guide their development. Thus, in what follows, we consider a phenomenological approach to understanding the texture and rheological properties of LCPs in Region II.

Microscopic observations (Alderman and Mackley 1985; Larson and Mead 1992; Vermant et al. 1994a; Mather et al. 1996) and polarimetry studies (Burghardt and Hongladarom 1994) show that the textures of flowing polydomain nematics and nematic polymers become finer as the shear rate $\dot\gamma$ increases, at least for samples showing no Region I behavior. (LCPs with Region I behavior seem to have a very fine, submicron texture that is insensitive to shear; see Section 11.5.4.) Marrucci (1984) suggests that the characteristic distance a between disclinations at steady state in Region II should follow a simple scaling law

$$a \propto \left(\frac{\eta\dot\gamma}{K}\right)^{-1/2} \tag{11-22a}$$

where η and K are a characteristic viscosity and Frank constant, respectively. Using microscopic image analysis, Mather et al. (1996) measured the volumetric disclination density ρ_V (length of disclination lines per unit volume) for the tumbling small-molecule nematic 8CB and found that at high Ericksen number ($\mathrm{Er} \gtrsim 2000$) this scaling law seems to be obeyed (see Section 10.2.7). Using the twist viscosity γ_1 for η and K_3 for K, they found for 8CB

$$a \equiv \rho_V^{-1/2} \approx 20 \left(\frac{\gamma_1 \dot{\gamma}}{K_3} \right)^{-1/2} \qquad (11\text{-}22b)$$

Since neither γ_1 nor K_3 were varied in these experiments, the appropriateness of these choices for the characteristic viscosity and Frank constant was not tested. For nematic polymers, no direct microscopic measurement of ρ_V has been attempted yet, but indirect measurements using scattering (Walker et al. 1997) and polarimetry (Burghardt and Hongladarom 1994) indicate that a scaling law similar to Eq. (11-22b) applies, with a somewhat weaker exponent (around -0.2 to -0.4). For rod-like nematic polymers such as PBG, the value of γ_1 estimated from hard-rod theory is two orders of magnitude greater than the zero-shear viscosity η_0. Taking $K_3 \approx 10^{-6}$ dyn and assuming that Eq. (11-22b) applies to PBG and HPC solutions, we estimate that

$$a \equiv \rho_V^{-1/2} \approx 20 \, (\eta_0 \dot{\gamma})^{-1/2} = 20 \, \sigma^{-1/2} \qquad \text{for PBG or HPC solutions} \quad (11\text{-}22c)$$

where σ is the shear stress in units of dyn/cm^2, and the texture size a is in microns. For large shear stresses, $\sigma \gtrsim 400$ dyn/cm, this formula predicts that the texture size a will be submicron, in agreement with values of a inferred from light-scattering measurements for HPC solutions (Walker and Wagner 1994). For lower shear stresses in the range 1–100 dyn/cm^2, a is predicted to be in the range 2–20 μm, which is much greater than the microscopic molecular length $L \approx 0.1\mu$m, but much smaller than the macroscopic rheometer gap $h \sim 1$ mm. In this regime ($L \ll a \ll h$), bulk properties, such as birefringence or stress, are therefore averages over a statistically large number of regions whose size is a.

In this lower range of shear stresses, $\sigma = 1$–100 dyn/cm^2, it is useful to define a nematic "domain" as a region whose linear size is of order a, which is small enough that within it the director orientation is nearly constant, yet is large enough compared to the molecular size that continuum properties, such as birefringence and stress, are well-defined. (Within disclination cores, molecular orientations vary on length scales comparable to the molecules themselves; hence these cores must be excluded from "domains" defined as above.) Within a "domain," we may use continuum theory, specifically the Leslie–Ericksen equations (10-10)–(10-18), to describe orientation and stress. A property such as birefringence can then be averaged over enough domains that a well-defined average, called a *mesoscopic* average, is obtained. The *mesoscopic length scale A*, which is the size of the region over which the averaging occurs, must therefore be larger than the domain size a, perhaps by a factor of ten, so that $\sim 10^3$ domains are contained in the averaging volume.

In typical rheometers, the imposed shear rate and shear stress are usually macroscopically uniform throughout the fluid, so that mesoscopic averaging is appropriate as long as $A \approx 10 \times a \leq h$, the rheometer gap. However, in commercial polymer processing equipment, such as extruders and mold cavities, the imposed flow field or stress varies from place to place, and mesoscopic averaging is then useful only if the mesoscopic scale A is smaller than the distances over which such variations occur. This was the case in the experiments of Bedford and Burghardt (1995), who found that the use of mesoscopic averages allows accurate prediction of the birefringence in a nonhomogeneous slit-flow of an LCP.

Inspired by a mesoscopic theory for mixtures of immiscible liquids (Doi and Ohta 1991; see also Section 9.3.3), Larson and Doi (1991) have derived mesoscopic equations for polydomain nematics by multiplying the Leslie–Ericksen equation (10-13) by the director orientation vector $\mathbf{n}$ and averaging Eqs. (10-10), (10-13), and (10-17) over a mesoscopic length scale encompassing many domains. Neglecting spatial variations in the velocity gradient $\nabla\mathbf{v}$ caused by variations in $\mathbf{n}$ (this is a crude approximation), one obtains

$$\langle[\boldsymbol{\sigma}_{\text{tot}}]\rangle = 2\mu\mathbf{D} + 2\mu_1\mathbf{D}:\overline{\mathbf{S}}\left(\overline{\mathbf{S}} + \frac{1}{3}\boldsymbol{\delta}\right) + \frac{2}{3}\mu_2\mathbf{D} + \mu_2(\overline{\mathbf{S}}\cdot\mathbf{D} + \mathbf{D}\cdot\overline{\mathbf{S}})$$

$$+ \frac{\alpha_2}{\gamma_1}\langle[\mathbf{n}\hat{\mathbf{h}}]\rangle + \frac{\alpha_3}{\gamma_1}\langle[\hat{\mathbf{h}}\mathbf{n}]\rangle + \langle[\boldsymbol{\sigma}^d]\rangle \tag{11-23}$$

and

$$\frac{d}{dt}\overline{\mathbf{S}} = \boldsymbol{\omega}^T\cdot\overline{\mathbf{S}} + \overline{\mathbf{S}}\cdot\boldsymbol{\omega} + \lambda\left(\frac{2}{3}\mathbf{D} + \mathbf{D}\cdot\overline{\mathbf{S}} + \overline{\mathbf{S}}\cdot\mathbf{D} - 2\overline{\mathbf{S}}:\mathbf{D}\left(\overline{\mathbf{S}} + \frac{1}{3}\boldsymbol{\delta}\right)\right)$$

$$+ \frac{1}{\gamma_1}(\langle[\hat{\mathbf{h}}\mathbf{n}]\rangle + \langle[\mathbf{n}\hat{\mathbf{h}}]\rangle) \tag{11-24}$$

Here $\boldsymbol{\sigma}_{\text{tot}}$ is the sum of the viscous stress $\boldsymbol{\sigma}$ from Eq. (10-10) and the Frank distortional stress $\boldsymbol{\sigma}^d$ from Eq. (10-17), and (Kawaguchi 1996)

$$\hat{\mathbf{h}} \equiv \mathbf{h} - \mathbf{n}\mathbf{n}\cdot\mathbf{h}$$

The mesoscopic averages are designated by enclosure in the brackets $\langle[\cdot]\rangle$. The "mesoscopically averaged" order parameter $\overline{\mathbf{S}}$ is defined as

$$\overline{\mathbf{S}} \equiv \langle[\mathbf{n}\mathbf{n}]\rangle - \frac{1}{3}\boldsymbol{\delta} \tag{11-25}$$

The Ericksen viscosities μ, μ_1, and μ_2 in Eq. (11-23) are given by

$$\mu = \frac{\alpha_4}{2}, \qquad \mu_1 = \frac{\alpha_1}{2} - \frac{(\alpha_2 + \alpha_3)(\alpha_6 - \alpha_5)}{2(\alpha_2 - \alpha_3)}, \qquad \mu_2 = \frac{(\alpha_2\alpha_6 - \alpha_3\alpha_5)}{(\alpha_2 - \alpha_3)} \tag{11-26}$$

In the derivation of Eqs. (11-23) and (11-24), the following closure approximation was used:

$$\langle[\mathbf{nnnn}]\rangle \approx \langle[\mathbf{nn}]\rangle\langle[\mathbf{nn}]\rangle \tag{11-27}$$

Additional closure approximations are needed to eliminate the unknown quantities $\langle[\hat{\mathbf{h}}\mathbf{n}]\rangle$ and $\langle[\boldsymbol{\sigma}^d]\rangle$. Clever experiments by Moldenaers and Mewis (1993), in which steady and oscillatory shearing flows were superimposed, indicate that the Frank contribution $\langle[\boldsymbol{\sigma}^d]\rangle$ to the stress tensor is negligible compared to the viscous contribution $\langle[\boldsymbol{\sigma}]\rangle$ in Region II. Thus, we drop $\langle[\boldsymbol{\sigma}^d]\rangle$ from the above, leaving only the indirect contribution of Frank elasticity, $\langle[\mathbf{n}\hat{\mathbf{h}}]\rangle \approx \langle[\hat{\mathbf{h}}\mathbf{n}]\rangle$, to be determined. This latter term, while not contributing directly to the stress, contributes indirectly by modifying the tumbling orbits of the domains. The molecular field $\mathbf{h}$, defined in Eq. (10-14), involves gradients of the director field $\mathbf{n}(\mathbf{x})$.

To understand the role of the terms $\langle[\mathbf{n}\hat{\mathbf{h}}]\rangle$ and $\langle[\hat{\mathbf{h}}\mathbf{n}]\rangle$, let us for the moment drop them from Eq. (11-24); thus

$$\frac{d}{dt}\overline{\mathbf{S}} = \boldsymbol{\omega}^T\cdot\overline{\mathbf{S}} + \overline{\mathbf{S}}\cdot\boldsymbol{\omega} + \lambda\left(\frac{2}{3}\mathbf{D} + \mathbf{D}\cdot\overline{\mathbf{S}} + \overline{\mathbf{S}}\cdot\mathbf{D} - 2\overline{\mathbf{S}}:\mathbf{D}(\overline{\mathbf{S}} + \frac{1}{3}\boldsymbol{\delta})\right) \tag{11-28}$$

which is just the mesoscopic average of $\mathbf{n}$ times the evolution equation for the director:

$$\frac{\partial}{\partial t}\mathbf{n} = \mathbf{n} \cdot \boldsymbol{\omega} - \lambda(\mathbf{n} \cdot \mathbf{D} - \mathbf{D} : \mathbf{nnn}) \tag{11-29}$$

For $\lambda < 1$, Eq. (11-29) describes the tumbling orbits of individual domains, and Eq. (11-28) describes the collective effect of the tumbling of many such domains. The important point is that *neither Eq. (11-28) nor (11-29) has a steady state.* Each domain tumbles endlessly, with period P given by Eq. (11-17), and even though the domain orientations are out of phase with each other, they all have the same period and so the initial distribution of domain orientations is revisited over and over again with periodicity P. Thus, without some source of *dispersion*, the distribution of domain orientations cannot evolve toward a steady state. An analogous result is obtained in suspensions of ellipsoidal particles, when interactions among the particles are neglected (see Section 6.3.1.1). In the case of suspensions, hydrodynamic interactions cause the Jeffreys orbit of one particle to be disturbed slightly by its neighbors. This produces an effective dispersive force that eventually damps out oscillations in the particle orientation distribution function and produces a steady state (see Section 6.3.2.2). Presumably, an analogous effect occurs in polydomain liquid crystals. But for liquid crystals, the source of the interaction must be sought in the Frank gradient elasticity, which couples the director rotation of one domain to that of its neighbors and is represented in Eq. (11-24) by the terms containing $\langle[\mathbf{n\hat{h}}]\rangle$ and $\langle[\mathbf{\hat{h}n}]\rangle$.

The magnitude of the dispersion term $\langle[\mathbf{n\hat{h}}]\rangle \approx \langle[\mathbf{\hat{h}n}]\rangle$ can be estimated using the scaling relationship in Eq. (11-22a), which is supported by the transient experiments on lyotropics and some thermotropics, as discussed earlier. From Eq. (10-14), $\mathbf{h}$ is proportional to the second spatial derivative of the director field, and hence the magnitude of $\mathbf{h}$ and of $\hat{\mathbf{h}}$ should be proportional to a^{-2}, the inverse square of the characteristic domain size. In steady-state shearing flow in Region II, we expect a to be given by Eq. (11-22a). On the other hand, in the absence of flow, a textured sample of nematic liquid (produced from the isotropic state by a jump in temperature or pressure) coarsens according to the scaling law $d\rho_V/dt \propto -\rho_V^2$, as found in scaling arguments and the experiments of Chuang et al. (see Section 10.2.7) on small-molecule nematics. Linearly combining these two scaling laws gives

$$\frac{d}{dt}\rho_V = \zeta\dot{\gamma}\rho_V - \rho_V^2 \tag{11-30}$$

where the coefficient of the ρ_V^2 term has been absorbed into the definition of ρ_V, leaving one dimensionless phenomenonological constant ζ.

Although we expect for dimensional reasons that the average magnitude of $\hat{\mathbf{h}}$, and hence the magnitude of $\langle[\mathbf{n\hat{h}}]\rangle$, will be proportional to ρ_V, the tensorial form of $\langle[\mathbf{n\hat{h}}]\rangle$ is unknown. To obtain $\langle[\mathbf{n\hat{h}}]\rangle$, without having to revert back to (an almost impossible) microscopic calculation of the director field, Larson and Doi (1991; Kawaguchi 1996) assumed that $\langle[\mathbf{n\hat{h}}]\rangle$ is a function of the mesoscopic order parameter $\overline{\mathbf{S}}$—that is, that $\langle[\mathbf{n\hat{h}}]\rangle = Ka^{-2}\,\mathbf{f}(\overline{\mathbf{S}})$. Dimensional reasoning then leads to the ansatz that

$$\langle[\mathbf{n\hat{h}}]\rangle = -\frac{1}{2}\varepsilon\gamma_1\rho_V\overline{\mathbf{S}} \tag{11-31}$$

where ε is a dimensionless constant.

Equations (11-24) and (11-31) imply that in the absence of flow, $\overline{\mathbf{S}}$ will tend to evolve toward the null tensor at long times; that is, alignment will tend to disappear. This indeed is the case in some LCPs, such as HPC solutions (Hongladarom et al. 1994), but for others, such as PBLG solutions, the orientation increases toward that of a monodomain when flow ceases (Hongladarom and Burghardt 1993; Burghardt 1998). The latter behavior is perhaps more to be expected than the former, since the ground state of the LCP is an aligned monodomain. In any event, the most important effect of the term $\langle[\mathbf{n}\hat{\mathbf{h}}]\rangle$ is to dampen oscillations in average properties, so that a steady state is eventually reached.

Larson and Doi simplified Eqs. (11-23), (11-24), and (11-30) by setting $\langle[\sigma^d]\rangle = 0$ (for reasons already discussed) and taking $\zeta = 1$. Thus, the equations of the *mesoscopic theory* simplify to

$$\langle[\sigma_{\text{tot}}]\rangle = 2\mu\mathbf{D} + 2\mu_1\mathbf{D} : \overline{\mathbf{S}}(\overline{\mathbf{S}} + \tfrac{1}{3}\delta) + \tfrac{2}{3}\mu_2\mathbf{D} + \mu_2(\overline{\mathbf{S}}\cdot\mathbf{D} + \mathbf{D}\cdot\overline{\mathbf{S}})$$
$$- \tfrac{1}{2}\varepsilon(\alpha_2 + \alpha_3)\rho_V\overline{\mathbf{S}} \tag{11-32}$$

and

$$\frac{d}{dt}\overline{\mathbf{S}} = \omega^T\cdot\overline{\mathbf{S}} + \overline{\mathbf{S}}\cdot\omega + \lambda(\tfrac{2}{3}\mathbf{D} + \mathbf{D}\cdot\overline{\mathbf{S}} + \overline{\mathbf{S}}\cdot\mathbf{D} - 2\overline{\mathbf{S}} : \mathbf{D}(\overline{\mathbf{S}} + \tfrac{1}{3}\delta)) - \varepsilon\rho_V\overline{\mathbf{S}} \tag{11-33}$$

Only the single parameter ε needs to be specified; λ and the viscosities μ, μ_1, and μ_2 in Eqs. (11-32)–(11-33) can be obtained from Eq. (11-26) and α's computed using the Kuzuu–Doi theory (1983, 1984; see Section 11.5.1). In the limit $\varepsilon \ll 1$, one can solve Eqs. (11-30), (11-32), and (11-33) perturbatively to give

$$\overline{S}_{12} = \frac{\lambda}{3(1-\lambda^2)}\varepsilon$$

$$\overline{S}_{11} = \frac{1+\lambda/3}{2\lambda + \varepsilon/\overline{S}_{12}}, \qquad \overline{S}_{22} = \frac{-1+\lambda/3}{2\lambda + \varepsilon/\overline{S}_{12}}, \qquad \overline{S}_{33} = \frac{-2\lambda/3}{2\lambda + \varepsilon/\overline{S}_{12}} \tag{11-34}$$

Since $1 - \lambda$ is small, around 0.05–0.10, these expressions yield the simple approximate results $\overline{S}_{12} \approx \lambda\varepsilon/(6(1-\lambda))$, $\overline{S}_{11} - \overline{S}_{33} \approx 1 - 2.5(1-\lambda)$, and $\overline{S}_{22} - \overline{S}_{33} = (\lambda - 1)/2 = -\alpha_3/(-\alpha_2 - \alpha_3)$. Note that the above components of the mesoscopic birefringence tensor $\overline{S}_{ij}$ are predicted to be *independent of shear rate* at shear rates low enough to be in Region II.

For a concentration just high enough to form a wholly nematic solution, $C/C^* = 1.122$; and taking $\varepsilon = 0.03$ (Larson and Doi 1991), the exact solution to the mesoscopic equations gives for the birefringence tensor (Hongladarom and Burghardt 1994; Burghardt 1998).

$$\langle[\mathbf{nn}]\rangle_{\text{mesoscopic}} = \begin{pmatrix} 0.866 & 0.048 & 0 \\ 0.048 & 0.048 & 0 \\ 0 & 0 & 0.085 \end{pmatrix}$$

Hongladarom and Burghardt (1994) measured the components of this tensor for a 13.5% solution of PBDG using polarimetry and found that, in accord with the mesoscopic theory, the birefringence is indeed independent of shear rate in Region II. The experimental birefringence tensor was found to be

$$\langle[\mathbf{nn}]\rangle_{\text{experiment}} = \begin{pmatrix} 0.727 & 0.100 & 0 \\ 0.100 & 0.0968 & 0 \\ 0 & 0 & 0.177 \end{pmatrix}$$

These predicted and measured values above should be compared to each other and to those for an ideal monodomain oriented in the flow direction:

$$\langle[\mathbf{nn}]\rangle_{\text{monodomain}} = \begin{pmatrix} 1 & 0 & 0 \\ 0 & 0 & 0 \\ 0 & 0 & 0 \end{pmatrix}$$

Comparing these results, we find that the observed shear-induced disruption in orientation, relative to that of a monodomain, is roughly twice as large as predicted by the mesoscopic theory.

Stresses can be predicted from the mesoscopic theory by inserting Eqs. (11-34) for the mesoscopic birefringence into the equation for the stress tensor, Eq. (11-32), to give the following approximate expressions for the stresses:

$$\frac{\sigma}{\dot{\gamma}} = \mu + \tfrac{1}{2}\mu_2\lambda + O(\varepsilon^2), \qquad N_1/\dot{\gamma} = O(\varepsilon), \qquad N_2/\dot{\gamma} = O(\varepsilon) \qquad (11\text{-}35)$$

The shear stress σ and the normal stress differences N_1 and N_2 are all predicted to be linear in the shear rate $\dot{\gamma}$. This scaling has indeed been observed for LCPs in Region II (Kiss and Porter 1980b; Moldenaers and Mewis 1986; Baek et al. 1993b). In contrast, for isotropic polymeric materials, N_1 is proportional to $\dot{\gamma}^2$ at low shear rates. As noted earlier, if ε is identically zero, then there are no domain interactions, and hence the stresses never reach a steady state. However, in the singular limit that ε approaches zero without reaching it, from Eq. (11-35) we find that $\sigma \rightarrow \mu + \mu_2\lambda/2$ and $N_1 \rightarrow N_2 \rightarrow 0$; hence only σ is nonzero as this limit is approached.

The concentration-dependence of the shear viscosity predicted by the mesoscopic theory is also in qualitative agreement with experiments on PBG solutions (Larson 1996). The zero-shear viscosity is predicted to decrease at first with increasing concentration in the nematic phase, but then show an *upturn* due to the increasing importance of viscous and texture stresses at high concentration. The upturn is in qualitative agreement with experimental data on PBG, HPC, and other solutions (Suto et al. 1989; Baek et al. 1993b; Aharoni 1980) (see Fig. 11-4a). The upturn is not predicted by the original Doi theory, which neglects texture stresses (see Fig. 11-4b). The texture stresses are proportional to ε; a value $\varepsilon \approx 0.03$–0.1 is consistent not only with the viscosity upturn, but also with stress relaxation measurements in PBG solutions after cessation of steady shearing. These show that some 30–55% of the stress relaxes slowly with a characteristic time constant τ_{tex} that obeys the usual texture scaling law $\tau_{\text{tex}} \propto \dot{\gamma}_0^{-1}$ (Moldenaers and Mewis 1990; Walker et al. 1995; Larson 1996; Walker et al. 1996).

For transient shearing flows, Larson and Doi solved Eqs. (11-30), (11-32), and (11-33) numerically to obtain predictions of the mesoscopic theory for step-up, flow-reversal, and recoverable-strain histories. Equation (11-30) ensures that the experimental scaling behavior is captured by the theory; for example, the recoverable strain and the shear stress in flow reversal experiments are functions of $\dot{\gamma}t$, and not $\dot{\gamma}$ and t separately. The magnitude of the predicted recoverable strain agrees with experiments for PBLG solutions for $\varepsilon = 0.03$. Having obtained ε by comparing the predictions of the theory with recoverable-strain experiments, the predictions for the shear stress in step-up and flow reversal are shown in Fig. 11-26 for various dimensionless concentrations C/C^*. The agreement with

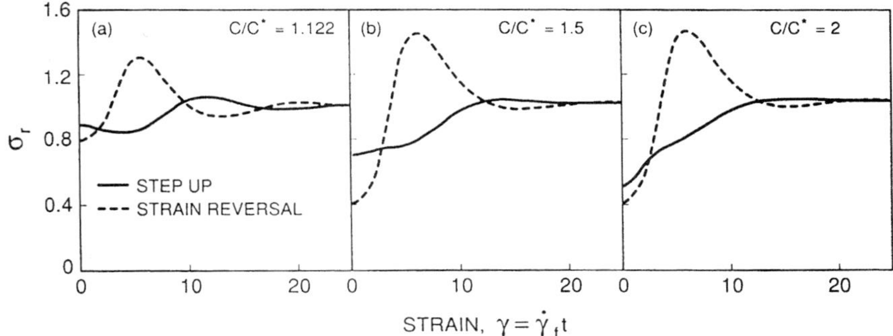

Figure 11.26   Reduced shear stress $\equiv \sigma/\sigma_\infty$ versus shear strain $\gamma \equiv \dot{\gamma}t$, after a step up in shear rate from $\dot{\gamma}_i$ to $\dot{\gamma}_f$ with $\dot{\gamma}_i/\dot{\gamma}_f = 0.1$, and after a reversal in shear direction, predicted by the mesoscopic theory with $\varepsilon = 0.03$ and $C/C^* = $ **(a)** 1.122, **(b)** 1.5, and **(c)** 2.0. (From Larson and Doi 1991, with permission of the Journal of Rheology.)

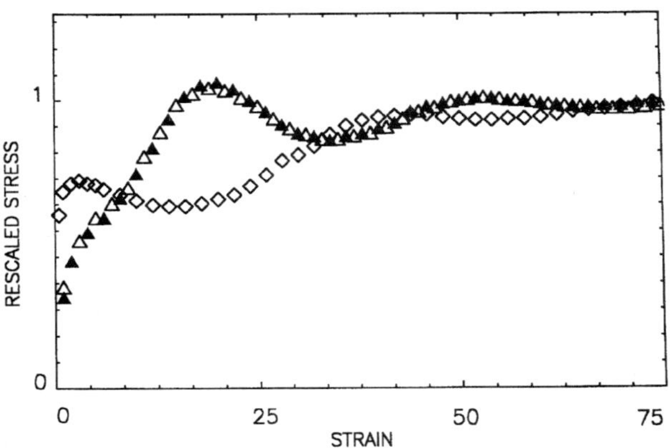

Figure 11.27 Reduced shear stress $\sigma_r \equiv \sigma/\sigma_\infty$ versus shear strain $\gamma \equiv \dot{\gamma}t$, after a step up in shear rate from $\dot{\gamma}_i$ to $\dot{\gamma}_f$ with $\dot{\gamma}_i/\dot{\gamma}_f = 0.1$ and $\dot{\gamma}_f = 1\ \text{sec}^{-1}$ ($\diamond$) and after a reversal in shear direction, with $\dot{\gamma} = 0.4$ ($\triangle$) and 1 sec^{-1} ($\blacktriangle$) sec^{-1} for 12% PBLG ($M = 250{,}000$) in *m*-cresol). (From Moldenaers et al. 1990, reprinted with permission from Moldenaers et al., in *Liquid Crystalline Polymers*, ACS Symp Series 435. Copyright © 1990, American Chemical Society.)

measurements for PBLG solutions is qualitatively good (see Fig. 11-27). However, the transient birefringence predicted by the theory deviates significantly from experimental measurements of it (Hongladarom and Burghardt 1993).

A couple of other important limitations of the theory must be acknowledged. First, as noted earlier, on cessation of flow, some LCPs show a gradual increase in orientation rather than the predicted decrease. This suggests that the tensorial form assumed for $\langle[\mathbf{n\hat{h}}]\rangle$, namely $\langle[\mathbf{n\hat{h}}]\rangle \propto \overline{\mathbf{S}}$, might not be completely appropriate. A second limitation of the theory is that Eq. (11-30) for the disclination density is not appropriate for small-amplitude oscillatory shearing, because it implies that small-strain-amplitude high-frequency oscillatory shearing leads to an increase in the disclination density; and this is not observed. A more generally valid mesoscopic constitutive equation for textured LCPs is not yet in hand.

11.5.4 Rheology in Region I

Finally, we come to one of the most interesting, and perplexing, phenomena in LCPs. This is the existence in some LCPs of Region I shear thinning at low shear rates (see Fig. 11-6 and the discussion in Section 11.3.2). Region I cannot be interpreted as classical yield (as described in Section 1.5.3), since the scaling of viscosity with shear rate $\dot{\gamma}$ is a power law $\eta \propto \dot{\gamma}^{-n}$, with $n < 1$; typically $n \approx 0.5$.

Very recently, Burghardt and coworkers (Ugaz et al. 1997, 1998) have found that for PBG solutions very near the concentration at which Region I first occurs, Bragg peaks appear in x-ray scattering that are consistent with the formation of a hexagonal phase, probably coexistent with the nematic. Since a hexagonal phase is solid-like in two dimensions, it is no surprise that the appearance of such a phase would dramatically increase the viscosity and alter its shear-rate dependence. (An analogous phenomenon occurs with smectics; see Section 10.4.6.) The appearance of grains of hard, hexagonal phase in a matrix of nematic might pin the director in the nematic and thereby provide a source of "trapped disclinations" in the Marrucci argument described below.

It has also been suggested that Region I in HPC solutions is associated with cholestericity (Hongladarom et al. 1994). At low shear rates in the Region I regime, HPC solutions show diffraction characteristics of the cholesteric phase. At higher shear rates, in Region II, the cholestericity is broken down by the shear (Hongladarom and Burghardt 1998). High viscosities and pronounced shear thinning are indeed observed in small-molecule cholesterics (see Section 10.3).

Since Region I behavior was first reported by Onogi and Asada (1980), it has been associated with a particular texture observed over a range of shear rates coincident with Region I. This texture, originally called a "piled polydomain," is reported not to vary much as a function of shear rate, within Region I. Indeed, Hongladarom et al. (1994) and Walker et al. (1997) show that in PBLG and HPC solutions that show a Region I, the steady-state degree of orientation under shearing flow becomes nearly isotropic as the shear rate decreases into Region I, and the texture size approaches a constant.

Region I behavior seems to be observed only in *high-viscosity* nematics, where the shear viscosity in Region II exceeds approximately 100 P. For such viscous nematics, the characteristic texture length scale, a, estimated from Eq. (11-22c), is very small, even at modest shear rates. Indeed, according to Eq. (11-22c), for $\eta = 100$ P, $a \sim 0.6\ \mu$m at a shear rate of only $10\ \sec^{-1}$. Light-scattering and neutron-scattering measurements in HPC and PBLG samples showing Region I behavior confirm that a is very small, $a \approx 1\ \mu$m (Walker and Wagner 1994). Thus, the characteristic texture scale can be driven down nearly to molecular length scales in very viscous nematic liquids. Clearly, by this point, if not before, Frank theory, not to mention Leslie–Ericksen theory, must fail, and a new kind of dynamics must emerge.

In addition, measurements of orientation after cessation of shearing show that flow-induced orientation *decreases* after cessation of shearing toward an *isotropic* state of orientation, and the dynamic moduli increase with time (Hongladarom et al. 1994). This behavior is seen in PBLG solutions only under conditions where a pronounced Region I is seen; for lower concentrations where only Region II is present, the orientation *increases* after cessation of shearing (Walker et al. 1995).

Region I can be *hysteretic*. Figure 11-28 shows the shear viscosity versus shear rate for a 40% PBLG solution after a fresh sample has been loaded in the rheometer. While evidence of a weak "Region I" appears initially at low shear rate, a much more prominent Region I appears after the sample has been sheared at a rate exceeding a critical value of around 3 sec^{-1}—that is, a shear stress of around 500 dyn/cm^2. At this shear stress, the texture size predicted from Eq. (11-22b) is very small, around 0.9 μm, consistent with microscopic observations (Walker et al. 1995). Hysteresis is not observed with other LCPs showing Region I, but it is possible that these other fluids, which are more viscous, can never escape the "upper branch."

These results suggest that for high-viscosity nematics, defects can be packed together so tightly under even modest shearing that a defect-saturated "tight texture" is achieved, from which the material has difficulty relaxing. At rest, the annealing of these densely packed defects, and therefore coarsening of the defect structure, requires flow of the nematic fluid around a given defect, which may be exceedingly slow. If so, the texture length scale a does not change as a function of shear rate, but is "pinned" at some molecular value a_0, perhaps set in some cases by the presence of hexagonal domains or cholesteric pitch. Flow induces orientation, but when flow ceases, the director rotates back toward an isotropic distribution dictated by the presence of the trapped disclinations. This inference is consistent with measurements of orientation relaxation after cessation of shearing for PBLG in Region I.

Under steady shearing, these "trapped disclinations" should play the role of an anchoring condition, much like the role solid walls play in the flow properties of small-molecule nematics. A scaling analysis of this problem in Section 10.2.5 gives an equation, (10-28), for the steady-state shear viscosity for flow between surfaces with strong, homeotropic anchoring:

$$\eta \sim \eta_0 + \gamma_1^{1/2} K_3^{1/2} a_0^{-1} \dot{\gamma}^{-1/2} \tag{11-36}$$

where the domain size a_0 has been substituted for the gap width h. For 40% PBLG, we estimate $\gamma_1 \approx 10^4$ P (about 30–100 times the Region II shear viscosity), and $K_3 \approx 10^{-5}$ dyn. Taking $a_0 \approx 0.5\mu$m, we find in Region I that $\eta \sim 600\dot{\gamma}^{-1/2}$, in reasonable agreement with the experimental results shown in Fig. 11-28.

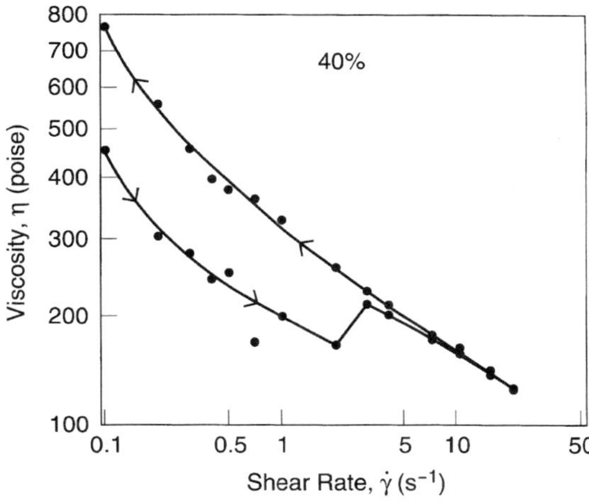

Figure 11.28 Viscosity as a function of shear rate for 40% PBLG (molecular weight = 238,000) in *m*-cresol, showing a hysteresis loop. The lower branch was obtained for a fresh sample; the viscosities on the upper branch were obtained after pre-shearing at a rate in excess of 3 sec⁻¹. (From Larson et al. 1993, reprinted with permission from Springer Verlag.)

The above explanation for Region I is inspired by (and conceptually similar to) an early model for Region I by Marrucci (1984), who also assumed an "intrinsic" domain spacing a_0. In Marrucci's model, while the domain spacing is fixed, the domain size shrinks as a result of shearing stresses. The domains themselves are taken to be impenetrable, so that fluid flows only in the gaps between them that open up to an extent that depends on the shearing stress. The domain size follows the same boundary-layer scaling used to derive Eq. (11-36), so it is no surprise that the viscosity is predicted to scale as $\dot{\gamma}^{-1/2}$ in Region I.

11.6 SUMMARY

The rheology of liquid-crystalline polymers is rich with interesting details. Even a defect-free nematic polymer is characterized by five independent viscosities and three elastic constants. In addition, under continuous shearing flow, a dense, defect-ridden texture is generated, which greatly complicates the behavior of these liquids. Furthermore, some LCPs show extremely unusual rheological properties, including negative first normal stress differences, time-shear rate scaling laws, and Region I shear thinning at the lowest accessible shear rates. Despite their complexity, the behavior of several solvent-based lyotropic LCPs can now be explained at least qualitatively. The shear-flow rheology of these lyotropics is dominated by *director tumbling*, in which viscous stresses drive the director to rotate continuously. From continuum theory, director tumbling can be shown to produce instabilities and secondary flows, giving rise to defects. From molecular theory, the negative first normal stress difference is shown to be a consequence of director tumbling, or, more specifically, a consequence of the frustration of tumbling that occurs when shearing is strong enough to distort the molecular order parameter. The density of disclination defects generated in a shearing flow can be estimated using scaling theory. This, combined with the Leslie–Ericksen equation averaged over regions large compared to the texture size scale (a so-called mesoscopic average), allows one to predict the time-shear rate scaling laws frequently observed.

Despite these successes, there are numerous areas for future work. First, many more rheological data need to be obtained for thermotropics LCPs, especially for monodomain samples. Very few values of the Leslie viscosities and Frank constants have been reported for thermotropic LCPs, and the trends with molecular weight and chain stiffness can only be guessed at. These data are especially needed, not only to test molecular theories, but also because even the simplest efforts to tailor rheological properties to suit applications require an understanding of the influence of molecular weight on flow properties, such as viscosity. Even the sign of α_2/α_3, which controls whether the LCP tumbles or flow aligns, is unknown for many thermotropic LCPs. In addition, the rheological properties of some important commercial lyotropic LCPs, such as poly(1,4-phenylene-2,6-benzobisthiazole) or PBZT, seem not to follow theoretical predictions very well, and an explanation of this is needed. Finally, the origin of the mysterious "Region I" behavior is still an unsettled issue.

REFERENCES

Aharoni SM (1980). *Polymer* 21:1413.

Alderman NJ, Mackley MR (1985). *Faraday Discuss Chem Soc* 79:149.

Archer LA, Larson RG (1995). *J Chem Phys* 103:3108.

Asada T, Tanaka T, Onogi S (1985). *J Appl Polym Sci Appl Polym Symp* 41:229.

Baek S.-G., Magda JJ, Cementwala (1993a). *J Rheol* 37:935.

Baek S.-G., Magda JJ, Larson RG (1993b). *J Rheol* 37:1201.

Baek S.-G., Magda JJ, Larson RG (1994). *J Rheol* 38:1473.

Bawden FC, Pirie NW (1937). *Proc R Soc Lond Ser A* 234:66.

Bedford BD, Burghardt WR (1995). *J Rheol* 38:1657.

Berry GC (1988). *Mol Cryst Liq Cryst* 165:333.

Berry GC (1989). In *The Materials Science and Engineering of Rigid Rod Polymers,* Adams WW, Eby RK, McLemore DE (eds), *Proc Mater Res Soc,* Pittsburgh, p 181.

Bitsanis I, Davis HT, Tirrell M (1990). *Macromolecules* 23:1157.

Brelsford GL, Krigbaum WR (1991). In *Liquid Crystallinity in Polymers*, Ciferri A (ed), VCH Publishers, New York.

Bunning JD, Crellin DA, and Faber TE (1986). *Liq Cryst* 1:37.

Burghardt WR (1998). *Macromol Chem Phys,* 199:471.

Burghardt WR, Fuller GG (1991). *Macromolecules* 24:2546.

Burghardt WR, Hongladarom K (1994). *Macromolecules* 27:2327.

Burghardt WR, Hongladarom K (1994). *Macromolecules* 27:2327.

Chang S, Han CD (1997). *Macromolecules* 30:2021.

Chow AW, Fuller GG, Wallace DG, Madri JA (1985). *Macromolecules* 18:905.

Cocchini R, Nobile MR, Acierno D (1992). *J Rheol* 36:1307.

Colby RH, Gillmor JR, Galli G, Laus M, Ober CK, Hall E (1993). *Liq Cryst* 13:233.

Collyer AA (1992). *Liquid Crystal Polymers: From Structures to Application*, Elsevier, London.

Covault C (1996). *Aviation Week & Space Technol,* March 25:289.

Cox MK (1987). *Liquid Crystal Polymers*, Rapra Review Reports, Vol 1, No 2, Pergamon Press, New York.

D'Allest, JF, Wu PP, Blumstein A, Blumstein RB (1986). *Mol Cryst Crys Lett* 3:103.

de Gennes PG (1977). *Mol Cryst Liq Cryst Lett* 34:177.

De'Neve T, Kléman M, Navard P (1992). *J Phys II (France)* 2:187.

De'Neve T, Navard P, Kléman M (1993a). *J Rheol* 37:515.

De'Neve T, Navard P, Kléman M (1993b). *C R Acad Sci Paris Ser II* 316:1037.

De'Neve T, Navard P, Kléman M (1994). *Liq Cryst* 18:67.

Doi M (1980). *Ferroelectrics* 30:247.

Doi M, Edwards SF (1986). *The Theory of Polymer Dynamics*, Oxford University Press, New York.

Doi M, Ohta T (1991). *J Chem Phys* 95:1242.

Donald AM, Windle AH (1992). *Liquid Crystalline Polymers,* Cambridge University Press, New York.

Edwards BJ, Beris AN, Grmela M (1990). *J Non-Newt Fluid Mech* 35:51.

Einaga Y, Berry GC, Chu SG (1985). *Polym J* 17:239.

Elliott A, Ambrose EJ (1950). *Discuss Faraday Soc* 9:246.

Emsley JW, Luckhurst GR, Stockley CP (1981). *Mol Phys* 44:565.

Ernst B, Navard P, Hashimoto T, Takebe T (1990). *Macromolecules* 23:1370.

Farmer BL, Chapman BR, Dudis DS, Adams WW (1993). *Polymer* 34:1588.

Giles DW, Denn MM (1994). *J Rheol* 38:617.

Gillmor JR, Colby RH, Hall E, Ober CK (1994). *J Rheol* 38:1623.

Gleeson JT, Larson RG, Kiss G, Cladis PE (1992). *Liq Cryst* 11:341.

Gotsis AD, Baird DG (1986). *Rheol Acta* 25:275.

Grizzuti N, Cavella NS, Cicarelli P (1990). *J Rheol* 34:1293.

Gu D, Smith SR, Jamieson AM, Lee M, Percec V (1993). *J Phys II (France)* 3:937.

Guskey SM, Winter HH (1991). *J Rheol* 35:1191.

Han WH, Rey AD (1995). *Macromolecules* 28:8401.

Hermans J (1962). *J Colloid Sci* 17:638.

Hess S (1976). *Z Naturforsch* 31A:1034.

Hongladarom K, Burghardt WR (1993). *Macromolecules* 26:785.

Hongladarom K, Burghardt WR (1994). *Macromolecules* 27:483.

Hongladarom K, Burghardt WR, Baek S-G, Cementwala S, Magda JJ (1993). *Macromolecules* 26:772.

Hongladarom K, Burghardt WR (1998). *Rheol Acta* 37:46.

Hongladarom K, Secakusuma V, Burghardt WR (1994). *J Rheol* 38:1505.

Hudson SD, Fleming JW, Gholz E, Thomas EL (1993a). *Macromolecules* 26:1270.

Hudson SD, Lovinger AJ, Larson RG, Davis DD, Garay RO, Fujishiro K (1993b). *Macromolecules* 26:5643.

Jackson CL, Shaw MT (1991). *Int Mater Rev* 36:165.

Jackson WL, Kuhfuss HF (1976). *J Polym Sci Polym Chem Ed* 14:2043.

Kannan RM, Kornfield JA, Schwenk N, Boeffel C (1993). *Macromolecules* 26:2050.

Kannan RM, Rubin SF, Kornfield JA, Boeffel C (1994). *J Rheol* 38:1609.

Kawaguchi M (1996). Private communication of an error in Larson and Doi (1991).

Kim SS, Han CD (1993). *Macromolecules* 26:6633.

Kirkwood JG, Auer PL (1951). *J Chem Phys* 19:281.

Kiss G, Porter RS (1978). *J Polym Sci Polym Symp* 65:193.

Kiss G, Porter RS (1980a). *Mol Cryst Liq Cryst* 60:267.

Kiss G, Porter RS (1980b). *J Polym Sci Polym Phys Ed* 18:361.

Klein T, Jun HX, Esnault P, Blumstein A, Volino F (1989). *Macromolecules* 22:3731.

Kröger M, Sellers HS (1995). *J Chem Phys* 103:3108.

Kuzuu N, Doi M (1983). *J Phys Soc Jpn* 52:3486.

Kuzuu N, Doi M (1984). *J Phys Soc Jpn* 53:1031.

Kwolek SL (1971). Du Pont, US Patent 3,600,350.

Larson RG (1990). *Macromolecules* 23:3983.

Larson RG (1993). *J Rheol* 37:175.

Larson, RG (1994). In *Spatio-Temporal Patterns*, Cladis PE, Palffy-Muhoray P (eds), SFI Studies in the Sciences of Complexity, Proceedings, Vol XXI, Addison-Wesley, Reading, MA.

Larson RG (1996). *Rheol Acta* 35:150.

Larson RG, Doi M (1991). *J Rheol* 35:539.

Larson RG, Mead DW (1989). *J Rheol* 33:185.

Larson RG, Mead DW (1992). *Liq Cryst* 12:751.

Larson RG, Mead DW (1993). *Liq Cryst* 13:151; corrigendum (1995).

Larson RG, Öttinger HC (1991). *Macromolecules* 24:6270.

Larson, RG, Promislow J, Baek S-G, Magda JJ (1993). *Ordering in Macromolecular Systems*, A. Teramoto (ed), Proceedings of Osaka University Macromolecular Symposium, Osaka, Japan, June 3–6, 1993, Springer-Verlag, New York.

Lee S-D, Meyer RB (1986). *J Chem Phys* 84:3443.

Lee S-D, Meyer RB (1990). *Liq Cryst* 7:15.

Maffettone PL, Marrucci G, Mortier M, Moldenaers P, Mewis J (1994). *J Chem Phys* 100:7736.

Magda JJ, Baek S-G, DeVries KL, Larson RG (1991). *Macromolecules* 24:4460.

Marrucci G (1984). *Proceedings of the Ninth International Congress on Rheology,* Mena B, Garcia-Rejon A, Rangel-Nafaile C (eds), Acapulco, Mexico, Oct. 8–13.

Marrucci G (1991). *Macromolecules* 24:4176.

Marrucci G, Greco F (1993). In *Advances in Chemical Physics*, Vol 86, Prigogine I, Rice SA (eds), Wiley, New York, p 331.

Marrucci G, Maffettone PL (1989). *Macromolecules* 22:4076.

Marrucci G, Maffettone PL (1993). *Proceedings of the International Workshop on Liquid Crystal Polymers,* WLCP 93, Capri, Italy, June 1–4.

Mather PT, Pearson DS, Larson RG (1996). *Liq Cryst* 20:539.

Mattoussi H, Srinivasarao M, Kaatz PG, Berry GC (1992). *Macromolecules* 25:2860.

Mewis J, Moldenaers P (1996) *Curr Opin Colloid Interface Sci* 1:466.

Moldenaers P, Fuller G, Mewis J (1989). *Macromolecules* 22:960.

Moldenaers P, Mewis J (1986). *J Rheol* 30:567.

Moldenaers P, Mewis J (1990). *J Non-Newt Fluid Mech* 34:359.

Moldenaers P, Mewis J (1993). *J Rheol* 37:367.

Moldenaers P, Yanase H, Mewis J (1990). In *Liquid-Crystalline Polymers*, Weiss RA, Ober CK (eds), ACS Symposium Series 435, American Chemical Society, Washington, DC.

Mortier M, Moldenaers P, Mewis J (1996). *Rheol Acta* 35:57.

Müller JA, Stein RS, Winter HH (1994). *Rheol Acta* 33:473.

Naitove MH (1986). *Plast Technol* 32, Feb. p23.

Navard P (1986). *J Polym Sci Polym Phys Ed* 24:435.

Onogi S, Asada T (1980). In *Rheology*, Proceedings of the Eighth International Congress on Rheology, Naples, Italy 1980, Plenum, New York.

Ookubo N, Komatsubara M, Nakajima H, Wada Y (1976). *Biopolymers* 15:929.

Papkov, SP, Kulichikhin VG, Kalmykova VD, Malkin AYa (1974). *J Polym Sci Polym Phys Ed* 12:1753.

Picken SJ, Aerts J, Doppert HL, Reuvers AJ, Northolt MG (1991). *Macromolecules* 24:1366.

Prasadarao M, Pearce EM, Han CD (1982). *J Appl Polym Sci* 27:1343.

Prevorsek DC (1982). In *Polymer Liquid Crystals,* Ciferri A, Krigbaum WR, Meyer RB (eds), Academic Press, New York.

Rey AD (1993). *Mol Cryst Liq Cryst* 225:313.

Robinson C, Ward JC, Beevers RB (1958). *Discuss Faraday Soc* 25:29.

Romo-Uribe A, Windle AH (1995). *Macromolecules* 28:7085.

Romo-Uribe A, Windle AH (1996). *Macromolecules* 29:6246.

Romo-Uribe A, Lemmon TJ, Windle AH (1997). *J Rheol* 41:1.

Roux DC, Berret J-F, Porte G, Peuvrel-Disdier E, Lindner P, (1995). *Macromolecules* 28:1681.

Rubin SF, Kannan RM, Kornfield JA, Boeffel C (1995). *Macromolecules* 28:3521.

Sawyer LC, Jaffe M (1986). *J Mater Sci* 21:1897.

Sato T, Teramoto A (1996). *Adv Polym Sci* 126:85.

Semenov AN (1987). *Sov Phys JETP* 58:321.

Semenov AN, Khokhlov AR (1988). *Sov Phys Usp* 31:988.

Shimada T, Doi M, Okano K (1988). *J Phys Soc Jpn* 57:2432.

Srinivasarao M (1995). *Int J Mod Phys B* 9:2515.

Srinivasarao M, Berry GC (1991). *J Rheol* 35:379.

Srinivasarao M, Berry GC (1992). *Mol Cryst Liq Cryst* 223:99.

Srinivasarao M, Garay RO, Winter HH, Stein RS (1992). *Mol Cryst Liq Cryst* 223:29.

Stepanov VI (1983). In *The Statistical and Dynamical Problems of Elasticity and Viscoelasticity,* Urals Branch of the USSR Academy of Science, Sverdlovsk (in Russian).

Subbotin A (1993). *Macromolecules* 26:2562.

Suto S, Hiromasa H, Nishibori W, Karasawa M (1989). *J Appl Polym Sci* 37:1147.

Teraoka I, Hayakawa R (1989). *J Chem Phys* 91:2643.
Ugaz VM, Cinader DK Jr, Burghardt WR (1997). *Macromolecules* 30:1527
Ugaz VM, Cinander DK Jr, Burghardt WR (1998). *J Rheol* 42:379.
Vermant J, Moldenaers P, Picken SJ, Mewis J (1994a). *J Non-Newt Fluid Mech* 53:1.
Vermant J, Moldenaers P, Mewis J, Picken SJ (1994b). *J Rheol* 38:1571.
Viney C, Donald AM, Windle AH (1983). *J Mater Sci* 18:1136.
Vorlander D (1923). *Z Phys Chem* 105:211.
Walker LM, Wagner NJ (1994). *J Rheol* 38:1525.
Walker LM, Wagner NJ, Larson RG, Mirau PA, Moldenaers P (1995). *J Rheol* 39:925.
Walker LM, Mortier M, Moldenaers P (1996). *J Rheol* 40:967.
Walker LM, Kernick WA III, Wagner NJ (1997). *Macromolecules* 30:508.
Wang W, Hashimoto T, Lieser G, Wegner G (1994). *J Polym Sci Polym Phys Ed* 32:2171.
Warren TC, Schrag JL, Ferry JD (1973). *Biopolymers* 12:1905.
Weng T, Hiltner A, Baer E (1986). *J Mater Sci* 21:744.
Windle AH, Assender HE, Lavine MS (1994). *Philos Trans R Soc Lond A* 348:73.
Wissbrun KF (1981). *J Rheol* 25:619.
Yan NX, Labes MM (1994). *Macromolecules* 27:7843.
Yang IK, Shine AD (1993). *Macromolecules* 26:1529.
Zentel R, Wu J (1986). *Makromol Chem* 187:1727.
Zheng-Min S, Kléman M (1984). *Mol Cryst Liq Cryst* 111:321.

Chapter **12**

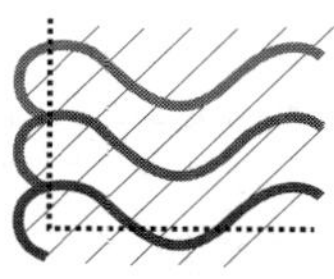

SURFACTANT SOLUTIONS

12.1 INTRODUCTION

Surfactant molecules are *amphiphilic*; that is, they possess *hydrophilic* (water-loving) groups chemically bonded to *lyophilic* (oil-loving) groups. A typical lyophilic group is a slender hydrocarbon chain of length 8–20 carbon atoms, frequently called a surfactant "tail." The hydrophilic group is usually short and bulky and is often called the "head." Some examples of amphiphiles are listed in Table 12-1; the head groups are underlined. Table 12-1 includes anionic, cationic, and nonionic surfactants, and the last entry in Table 12-1 is a two-tailed surfactant. In a watery milieu, the tails huddle together to minimize their exposure to water. The reverse occurs in oil.

Thus, in solutions, amphiphiles *self-assemble* into aggregates such as spheroidal and worm-like micelles, vesicles, and bilayers (see Fig. 12-1). At high concentrations, such aggregates can form ordered phases of nematic, hexagonal, smectic, cubic, or other symmetry (Seddon 1990; Berret et al. 1995). Amphiphiles are not only of great interest scientifically, but are of immense practical value as detergents, emulsifiers, encapsulants, lubricants, and so on. Long, worm-like micelles act like "living polymers"—that is, polymers that can reassemble themselves from their constituent surfactant molecules even after being torn to pieces, say, by a strong flow field. Such self-assembling "polymers" have possible applications as viscosifiers or as drag reducing agents that suppress turbulence in flows that might irreversibly fragment ordinary drag-reducing polymers (Zakin et al. 1996). Bilayers composed of phospholipids—which are fatty-acid-derived surfactants—serve biological functions as membranes in animal and plant cells. The double-tailed phospholipids are prevalent in animal membranes, and glycolipids are common in plant membranes (Israelachvili 1992; Seddon 1996). Some biological amphiphiles have as many as seven tails, and tail lengths can range up to 90 carbons (Seddon 1996)!

TABLE 12-1
Examples of Some Surfactants

SDS	Sodium dodecyl sulfate	$C_{12}H_{25} - \underline{O - SO_3^-} \; Na^+$
CTAB	Cetyltrimethylammonium bromide	$C_{16}H_{33} - \underline{N^+(CH_3)_3}Br^-$
$C_{12}EO_6$	Hexaoxyethylene dodecyl ether	$C_{12}H_{25} - \underline{(O - CH_2CH_2)_6 - OH}$
DDAB	Didodecyldimethylammonium bromide	$(C_{12}H_{25})_2 - \underline{N^+(CH_3)_2}Br^-$

Lipid	Critical packing parameter $v/a_0 l_c$	Critical packing shape	Structures formed
Single-chained lipids (surfactants) with large head-group areas: *SDS in low salt*	< 1/3	Cone 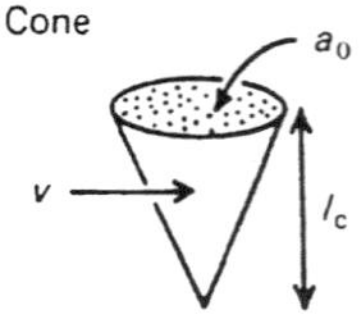	Spherical micelles 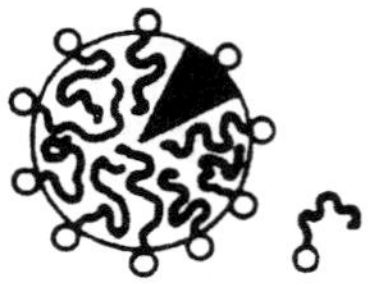
Single-chained lipids with small head-group areas: *SDS and CTAB in high salt, nonionic lipids*	1/3-1/2	Truncated cone 	Cylindrical micelles 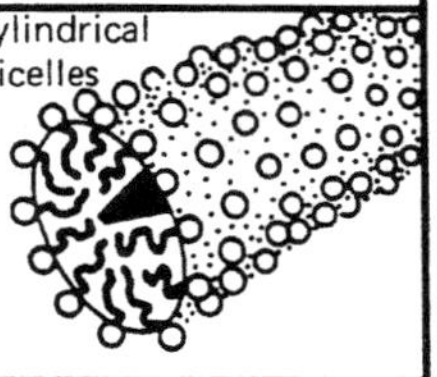
Double-chained lipids with large head-group areas, fluid chains: *Phosphatidyl choline (lecithin), phosphatidyl serine, phosphatidyl glycerol, phosphatidyl inositol, phosphatidic acid, sphingomyelin, DGDG[a], dihexadecyl phosphate, dialkyl dimethyl ammonium salts*	1/2-1	Truncated cone 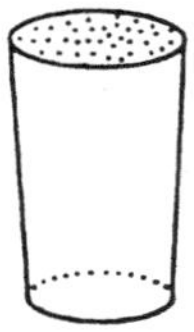	Flexible bilayers, vesicles 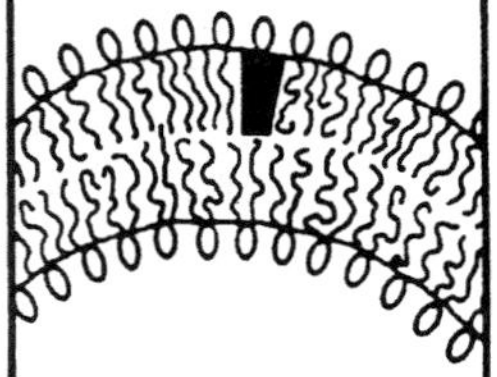
Double-chained lipids with small head-group areas, anionic lipids in high salt, saturated frozen chains: *phosphatidyl ethanolamine, phosphatidyl serine + Ca^{2+}*	~1	Cylinder 	Planar bilayers 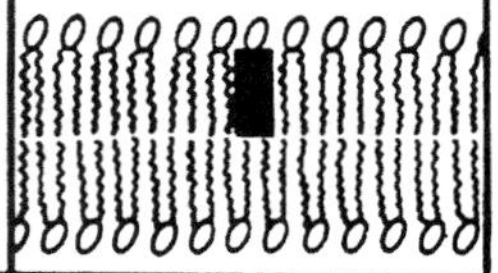
Double-chained lipids with small head-group areas, nonionic lipids, poly (cis) unsaturated chains, high *T*: *unsat. phosphatidyl ethanolamine, cardiolipin + Ca^{2+} phosphatidic acid + Ca^{2+} cholesterol, MGDG[b]*	> 1	Inverted truncated cone or wedge 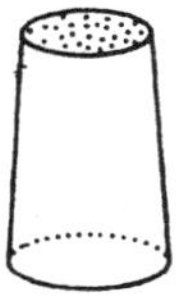	Inverted micelles 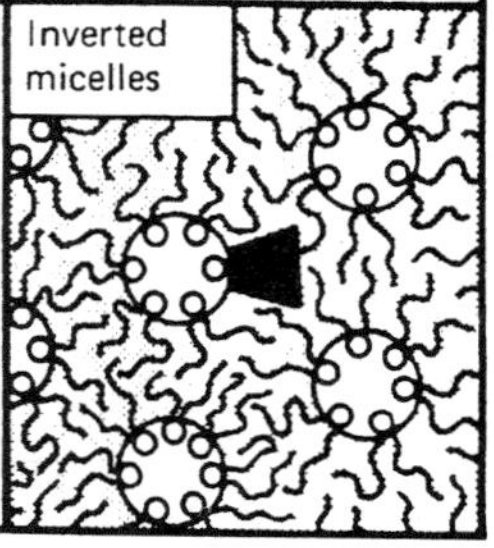

[a] DGDG, digalactosyl diglyceride, diglucosyl diglyceride.
[b] MGDG, monogalactosyl diglyceride, monoglucosyl diglyceride.

Figure 12.1 Packing shapes of surfactants and the structures they form. (From Israelachvili 1992, reprinted with permission from Academic Press.)

Surfactants and phospholipids are *polymorphic*; that is, they form many different structures, depending on concentration, temperature, and salinity (Luzzati and Spegt 1967; Luzzati et al. 1968; Kekicheff and Cabane 1987; Turner et al. 1992; Ström and Anderson 1992; Seddon 1996). Diverse microstructures and phase behaviors are exhibited by surfactant solutions; a comprehensive review can be found in Laughlin's (1994) book. Of course, one would like to be able to predict the microstructures possible for a given surfactant structure, as well as predict the dependence of microstructure on the amounts and types of solvents present, on temperature, on salinity, and so on. One method for doing so is the *molecular packing* approach; another is the surfactant *surface-curvature* approach, and a third relies on molecular *simulations*. There are also composition-field methods for predicting microstructures in surfactant solutions, analogous to those discussed in Section 1.7.4. The next section is a brief synopsis of some of these methods; for a more thorough review, see Gompper and Schick (1994).

12.2 METHODS OF PREDICTING MICROSTRUCTURES

12.2.1 Packing Argument

As discussed by Israelachvili (1992), the shapes of surfactant aggregates can, to a first approximation, be anticipated based on the *packing* of simple molecular shapes (Tanford 1980). Figure 12-1 from Israelachvili illustrates this principle: Conical molecules with bulky head groups attached to slender tails form spherical micelles; cylindrical molecules with heads and tails of equal bulkiness form bilayers; and wedge-shaped molecules with tails bulkier than their heads form inverted micelles containing the heads in their interiors. A simple dimensionless molecular parameter that controls the shape of the aggregates is the molecular *shape parameter* $v/\ell_c a_0$; here v is the volume occupied by the hydrocarbon tail, ℓ_c is the tail's "maximum effective length," and a_0 is the area on the surface of the aggregate that the head group occupies (Israelachvili et al. 1976). When $v/\ell_c a_0 \leq 1/3$, spherical micelles are expected; when $1/3 \leq v/\ell_c a_0 \leq 1/2$, cylinders are expected; for $v/\ell_c a_0 \approx 1$, bilayers should form; and for $v/\ell_c a_0 > 1$, one expects inverse structures (see Fig. 12-1). More exotic cubic "strut" phases are expected when $1/2 < v/a_0\ell_c < 2/3$ or when $1 < v/a_0\ell_c < 3/2$; these are discussed in Section 12.4.1.2.

The maximum effective tail length ℓ_c and the tail volume v of a saturated hydrocarbon chain of n_c carbon atoms are estimated as (Tanford 1980)

$$\ell_c \approx (1.54 + 1.265 n_c) \text{ Å}; \qquad v \approx (27.4 + 26.9 n_c) \text{ Å}^3 \qquad (12\text{-}1)$$

The head-group area a_0 is harder to estimate; it is not a constant for a given molecule, but depends on the solvent environment. For example, at low ionic strength, the head groups of ionic surfactants repel each other more strongly and therefore occupy a larger effective area on the aggregate's surface than they do at high ionic strength. Thus, as the ionic strength is decreased, a_0 increases; as a result, $v/\ell_c a_0$ decreases, and transitions from cylindrical to spherical micelles can occur, as discussed in Section 12.3.1.2. Values of a_0 can be obtained, for example, from measurements of surface tension versus surfactant concentration, using the Gibbs adsorption equation (Herb et al. 1994).

12.2.2 Surface-Curvature Argument

An alternative approach for predicting surfactant microstructures is the Helfrich *curvature expansion* (Helfrich 1973; Safran 1994; Gelbart and Ben-Shaul 1996). This expansion is most useful when the surfactant is relatively dilute and forms a monolayer separating oil and water regions. The theory can also be used to predict the shapes and shape fluctuations of bilayers in water or oil. If the surfactant layer is thin compared to the separation between neighboring layers, the layer can be thought of as a two-dimensional *film*, characterized only by its local principal radii of curvature, R_1 and R_2. The two radii of curvature are often rearranged into the *mean curvature*, $H \equiv \frac{1}{2}(R_1^{-1} + R_2^{-1})$, and the *Gaussian curvature*, $K \equiv (R_1 R_2)^{-1}$. If these curvatures are small enough, the free energy of a unit area of surfactant surface is

$$f = 2k(H - H_0)^2 + \bar{k}K \tag{12-2}$$

where k and $\bar{k}$ are elastic bending moduli and H_0 is the *spontaneous curvature*—that is, the curvature the surfactant-laden interface would locally adopt if free from global constraints. In general, H and K vary from place to place on the surface. The total contribution of the monolayer or bilayer to the solution free energy is then obtained by integrating Eq. (12-2) over the entire surfactant film. For the simple spherical, cylindrical, and planar geometries, H and K are constants.

The surface free energy, Eq. (12-2), is used as the basis of the *film models* of surfactant microstructure and phase behavior (Talmon and Prager 1978; de Gennes and Taupin 1982; Widom 1984; Andelman et al. 1987; Huse and Leibler 1988; Golubović and Lubensky 1990; Safran 1994). In such models, aggregate shapes and shape transitions are predicted by minimizing the free energy with respect to the curvature of the film, holding composition or chemical potential fixed. For example, if $\bar{k}$ is negative, then the shape that minimizes the free energy f in Eq. (12-2) is that of monodisperse spheres with radius $R_{opt} = [1 + (\bar{k}/2k)]/H_0$. However, the volume fraction of surfactant ϕ_A, together with the thickness δ of the surfactant film, fix the surface area between oil and water, and the volume fraction of water ϕ_W fixes the total volume of water; the two together fix both the number and radius of the spheres of water that might form in oil. This radius is $3\delta\phi_W/\phi_A$, which need not equal R_{opt}. Within the constraints of fixed area and fixed volume of water, shapes other than spheres might therefore have lower free energy than that of spheres. Figure 12-2 shows the shapes of water-containing objects in oil predicted to have lowest free energy as a function of composition and parameters k, $\bar{k}$, and H_0. As the volume fraction of water ϕ_W increases at fixed surfactant volume fraction ϕ_A, transitions occur from lamellae to cylinders to spheres, or directly from lamellae to spheres, if the ratio $-\bar{k}/k$ is large. Finally, when so much water is added that the radius $3\delta(\phi_W/\phi_A)$ of the spheres equals the preferred value R_{opt}, any additional water is ejected into a second phase, leading to emulsification failure. These predictions of the film model are valid only for small ϕ_A. Furthermore, H_0, k, and $\bar{k}$ often cannot readily be inferred from molecular structure and solvent environment, and they must be estimated by matching the observed phase behavior and microstructures to those predicted by the model.

12.2.3 Molecular Simulations

Although simple arguments based on molecular packing or film curvature can rationalize the structures observed in many experimental systems, one would like to calculate aggregate

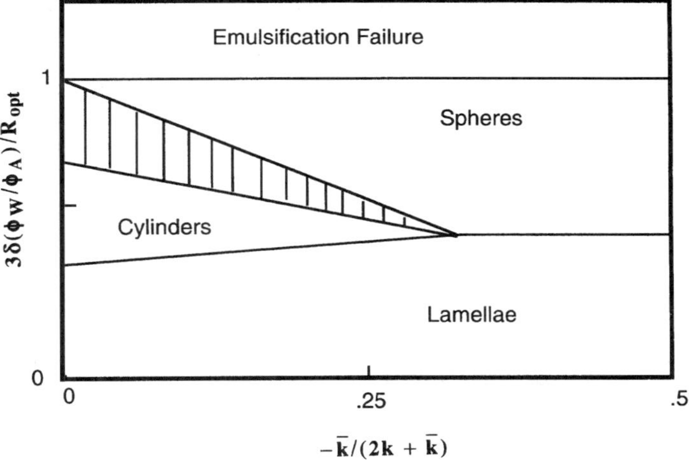

Figure 12.2   Regions of stability of spheres, cylinders, and lamellae in oil predicted by a mean-field "film" theory using Eq. (12-2). The hatched region has coexisting spheres and cylinders. Above the region of spheres, there is a region of "emulsification failure," where the sphere phase coexists with an excess water phase. In the ordinate, δ is the thickness of the surfactant film, which is roughly the surfactant molecular length, ϕ_A and ϕ_W are the volume fractions of surfactant (or amphiphile) and water, and $R_{\mathrm{opt}} = (1 + 0.5\bar{k}/k)/H_0$. (From Safran 1994, with permission from Addison-Wesley Publishing Company, Copyright © 1994.)

shapes and sizes using a more detailed description of the surfactant molecule. In principle, with molecular dynamics or Monte Carlo computer simulations (see Section 1.7), one could determine equilibrium structures and even dynamics of surfactant aggregates composed of molecules whose structure and interactions are defined at the atomic or near-atomic level (Owenson and Pratt 1984; Watanabe et al. 1988). Practically, however, the large spatial scales (tens of nanometers) and long equilibration times (milliseconds to hours) of real surfactant solutions preclude realistic simulations (Larson 1997).

Fortunately, it is possible to devise "coarse-grained" computer models, which, while not realistic in detail, include more molecular and structural detail than can be captured by simple packing or film models (Larson et al. 1985, 1996, 1997; Widom 1986; Schick and Shih 1986; Care 1987; Gompper and Schick 1989; Smit et al. 1990; Mackie et al. 1997). In a coarse-grained model, the surfactant tail might be represented by three or four contiguous units on a cubic lattice, for example, and the head group might be represented by one or more such units. Solvent molecules are often just single sites on the lattice, and interactions among the units are usually represented by pairwise additive nearest-neighbor attractive and repulsive energies. These models are useful for testing and verifying simple packing and curvature arguments that predict aggregate size and shape distributions.

12.3 DISORDERED MICELLAR SOLUTIONS

First we consider surfactant phases containing aggregates without positional or orientational order.

12.3.1 Micellar Shapes and Shape Transitions

The aggregation of surfactants in solution is governed by the ordinary laws of chemical thermodynamics, or *mass action*, which dictate that at equilibrium the chemical potential μ_n of aggregates containing n surfactant molecules is the same for all n, including $n = 1$—that is, isolated surfactants, or unimers. Thus (Tanford 1980; Israelachvili 1992)

$$\mu_1 = \mu_n \qquad \text{for all } n \tag{12-3}$$

For a solution of aggregates dilute enough to be ideal,

$$\mu_n = \mu_n^0 + \frac{k_B T}{n} \ln\left(\frac{X_n}{n}\right) \tag{12-4}$$

where X_n is the mole fraction of surfactant in aggregates containing n molecules, and μ_n^0 is the standard chemical potential for formation of an aggregate of size n. Equating μ_1 to μ_n and using Eq. (12-4), we find that

$$X_n = n\left[X_1 \exp\left((\mu_1^0 - \mu_n^0)/k_B T\right)\right]^n \tag{12-5}$$

Now let n_0 be the value of n that minimizes μ_n^0, and assume that $n_0 \gg 1$. As we shall soon see, we can also assume that $\mu_n^0 - \mu_{n_0}^0$ is greater than $k_B T$ for n significantly less than n_0. Aggregates of size n_0 or greater will then tend to be favored over aggregates of other sizes. Then we define

$$X_1^* \equiv \exp\left(-(\mu_0^1 - \mu_{n_0}^0)/k_B T\right) < 1 \tag{12-6}$$

From Eq. (12-5), we then find that $X_{n_0} = n_0(X_1/X_1^*)^{n_0}$. Since n_0 is large, it follows that as X_1 increases, X_{n_0} will remain very small until X_1 approaches X_1^*, and then X_{n_0} will grow rapidly. Since X_{n_0} is a mole fraction, it cannot exceed unity, and therefore X_1 must remain less than X_1^*. Hence, as the total inventory of surfactant increases from zero, most of it remains isolated as unimer until the mole fraction of surfactant approaches X_1^*, and abruptly thereafter almost all additional surfactant will aggregate into micelles of size n_0 or so (see Fig. 12-3). The concentration at which most additional surfactant forms large aggregates is called the *critical micelle concentration*, or *CMC*, and its value is close to X_1^*. The detailed distribution of micelle sizes is controlled by the dependence of μ_n^0 on n.

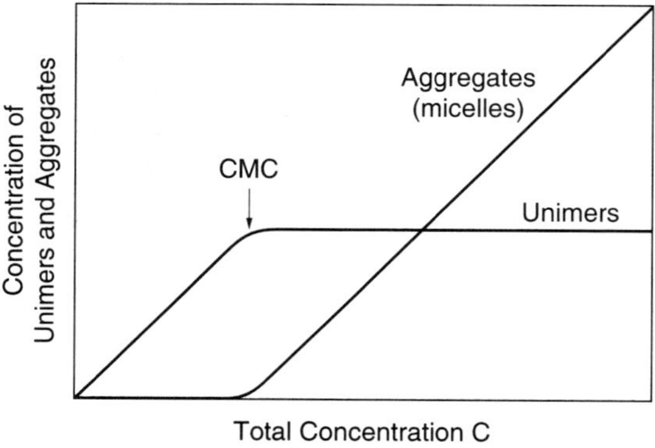

Figure 12.3 Schematic illustration of unimer and micelle concentrations as functions of the total concentration of surfactant. (Adapted from Israelachvili 1992, reprinted with permission from Academic Press.)

12.3.1.1 Spherical Micelles

For spherical micelles, Israelachvili (1992) has estimated the dependence of μ_n^0 on n from the surface area a occupied by the average head group on the surface of the aggregate:

$$\mu_n^0 \approx \Gamma a + \frac{k}{a} \qquad (12\text{-}7)$$

Here Γ is the surface energy per unit area of contact between the solvent (e.g., water) and the hydrocarbon tails, and k is a phenomenological constant. Thus, the first term on the right side of (12-7) is just a surface-tension contribution. Since $1/a$ is the number of head groups per unit area on the surface of the aggregate, the second term, k/a, is a penalty for crowding together of the head groups. Equation (12-7) is a crude approximation; other suggested expressions for μ_n^0, along with comparisons to computer simulations, can be found in Desplat and Care (1996).

Equation (12-7) can be rewritten as

$$\mu_n^0 = 2\Gamma a_0 + \frac{\Gamma}{a}(a - a_0)^2 \qquad (12\text{-}8)$$

where $a_0 = \sqrt{k/\Gamma}$ is the value of a that minimizes μ_n^0.

For a spherical micelle whose hydrocarbon core has radius R_c, n can be computed from either the core volume ($4\pi R_c^3/3 = vn$) or its surface area ($4\pi R_c^2 = an$). Eliminating R_c between these two expressions, one finds that

$$n = \frac{36\pi v^2}{a^3} \qquad \text{(spherical micelles)} \qquad (12\text{-}9)$$

Then, since there is an optimal area per head group a_0, there is also an optimal aggregation number n_0 and an optimal radius $R_{c,0}$ at which μ_n^0 is a minimum:

$$n_0 = \frac{36\pi v^2}{a_0^3}, \qquad R_{c,0} = \frac{3v}{a_0} \qquad (12\text{-}10)$$

If n deviates from this optimal value slightly—that is, if $\Delta n \equiv n - n_0 \ll n_0$—then Eqs. (12-8) and (12-9) imply that

$$\mu_n^0 \approx \mu_{n_0}^0 + \frac{\Gamma a_0}{9n_0^2}(\Delta n)^2 \qquad (12\text{-}11)$$

This result, combined with Eq. (12-4) and the condition of equality of chemical potentials, $\mu_n = \mu_{n_0}$, implies that

$$X_n = n\left\{ \frac{X_{n_0}}{n_0} \exp\left[\frac{-\Gamma a_0}{9n_0 k_B T}(\Delta n)^2 \right] \right\}^{n/n_0}$$

$$\approx X_{n_0} \exp\left[\frac{-\Gamma a_0}{9n_0 k_B T}(\Delta n)^2 \right] \qquad \text{for } \frac{\Delta n}{n_0} \ll 1 \qquad (12\text{-}12)$$

This is a *Gaussian* distribution of micelle sizes, with a standard deviation σ of

$$\sigma = \left(\frac{9n_0 k_B T}{2\Gamma a_0} \right)^{1/2} \qquad (12\text{-}13)$$

For hydrocarbon tails and polar or ionic head groups, $\Gamma \approx 30$ dyn/cm and $a_0 \approx 0.60$ nm^2, and therefore (Israelachvili 1992)

$$\sigma \approx \sqrt{n_0} \qquad (12\text{-}14)$$

This result is consistent with data on micelles of sodium alkyl sulfate (Aniansson et al. 1976), namely, $\sigma/\sqrt{n_0} \approx 1$–$2$. Lattice simulations of spheroidal micelles also show a result roughly consistent with this (Larson 1992), namely, $\sigma/\sqrt{n_0}$ increasing from 1.4 to 2.3 as the surfactant volume fraction increases from 0.08 to 0.20. Figure 12-4 shows the micelle size distribution from lattice simulations for an idealized surfactant whose head and tail groups each occupy four sites on the lattice. As the amphiphile volume fraction ϕ_A increases from 0.04 to 0.20, the average micelle size increases somewhat, from 68 to 84, as is expected from the law of mass action. At a volume fraction of 0.20, the micelle size distribution seems to develop a *second peak* at an aggregation of around 140. This second peak corresponds to large "supermicelles," which are usually prolate in shape (Mackie et al. 1997) *and are the precursors to a change to cylindrical micellar shape that occurs at higher concentration and/or lower temperature.*

12.3.1.2 Micellar Growth and Transition to Cylindrical Shapes

If the optimal micelle core radius $R_{c,0}$ from Eq. (12-10) exceeds the maximum effective tail length ℓ_c (i.e., if $R_{c,0}/\ell_c = 3v/a_0\ell_c > 1$), then optimally sized spherical micelles are not possible, and either the surface area per head group must deviate from its optimal value a_0, or there will be a transition to a shape with a higher surface-to-volume ratio, such as prolate ellipsoids or cylinders. Since the highest possible core radius is $R_{c,0} \approx \ell_c$, we estimate from Eq. (12-10) that the highest aggregation number of a spherical micelle is

$$n_{\max} = \frac{4\pi \ell_c^3}{3v} \qquad (12\text{-}15)$$

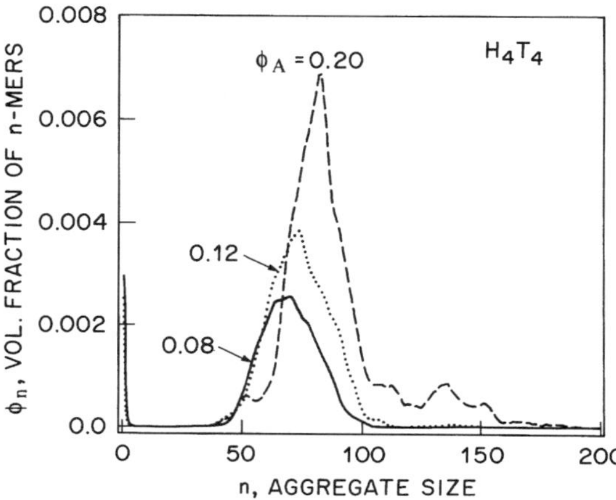

Figure 12.4 Volume fraction ϕ_n of surfactant in aggregates with aggregation number n for various total surfactant volume fractions ϕ_A according to Monte Carlo lattice simulations for a model "surfactant" H$_4$T$_4$ containing four head unit and four tail units, with each unit occupying a single site on the lattice. (From Larson, reprinted with permission from J. Chem. Phys. 96:7904, Copyright © 1992, American Institute of Physics.)

For alkyl surfactants with n_c carbons in their tails, Eqs. (12-1) and (12-15) yield

$$n_{\max} = \frac{4\pi}{3} \frac{(1.54 + 1.265 n_c)^3}{(27.4 + 26.9 n_c)} \tag{12-16}$$

$$\approx 0.4 n_c^2 \qquad \text{for } n_c = 8\text{--}16 \tag{12-17}$$

Thus, for alkyl surfactants, the maximum aggregation number for spherical micelles should increase from 26 to 58, as n_c increases from 8 to 12 (Missel et al. 1983).

At low ionic strength, the experimental aggregation numbers inferred from light-scattering studies of micelles of sodium alkyl sulfate agree almost exactly with the maximum aggregation number predicted by Eq. (12-17) (see Fig. 12-5). However, as the ionic strength of the micellar solution is increased, the aggregation number begins to exceed the predicted maximum for spherical micelles, especially for surfactants with longer tails (see Fig. 12-5).

This finding implies that for the alkyl sulfate surfactants, the ratio $v/\ell_c a_0$ must be close to 1/3 at low ionic strength, so that the spherical micelles are close to the maximum size allowed by the length of their tail groups. As the ionic strength is increased, the ionic head groups can be expected to repel each other less strongly, and a_0 decreases. Thus, as the ionic strength is increased, $v/\ell_c a_0$ should begin rise above 1/3, and a transition to nonspherical micelles can be expected. This shape change is evidently responsible for the observed increase in the aggregation number with increasing ionic strength in Fig. 12-5.

Further increases in ionic strength lead to an explosive growth in average micelle size. Figure 12-6 shows the values of the hydrodynamic radii $\bar{R}_h$ for micelles of alkyl surfactants at three high ionic strengths as functions of temperature. These hydrodynamic radii were inferred from micelle diffusion rates obtained by dynamic light scattering (discussed in Section 1.6.2). Near room temperature, $\bar{R}_h$ can exceed 200 Å, which is much higher than the

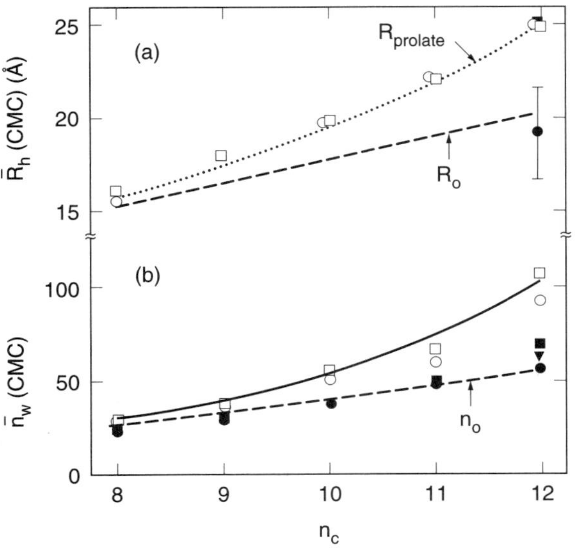

Figure 12.5 Hydrodynamic radius $\bar{R}_h$ (**a**) and average aggregation number $\bar{n}_w$ (**b**) at the CMC as functions of the number of carbons n_c in the hydrocarbon tail, for sodium alkyl sulfate micelles at low concentration near the CMC, deduced from light-scattering data. (●) Pure H_2O; (▼) 0.01 M NaCl; (■) 0.03 M NaCl; (○) 0.1 M NaCl; (□) 0.2 M NaCl. The dashed curves give the predictions for spherical micelles, for which $\bar{n}_w \approx 0.4 n_c^2$. (Reprinted with permission from Missel et al., Journal of Physical Chemistry 87:1264 Copyright © 1983, American Chemical Society.)

largest possible radius of spherical micelles of these surfactants. Thus, at low temperatures and high ionic strengths, alkyl sulfates form *cylindrical micelles.*

12.3.1.3 Cylindrical Micelles

If n is the number of surfactant molecules in a long cylindrical aggregate of core radius R_c and length L, then $n = \pi R_c^2 L/v = 2\pi R_c L/a$, which implies that $R_c = 2v/a$. For a long cylinder, the radius R_c will presumably adjust itself so that a equals its optimal value over all but the ends of the cylinder; hence $R_c = 2v/a_0$. On the ends of the cylinder, however, a cannot equal the optimal value; and one finds that the area per head group for hemispherical end caps is $a_s = 3a_0/2$. Since the area per surfactant molecule is at its optimal value everywhere except on the end caps, the chemical potential μ_n^0 for a surfactant in a cylindrical aggregate varies with aggregation number *only because of the end caps* (Israelachvili 1992). For a cylindrical micelle of aggregation number n and optimal head group area a_0, the number n_s of surfactant molecules that make up the end caps is $n_s = 4\pi R_c^3/3v \approx 0.4n_c^2$ for nearly fully stretched alkyl tails [see Eqs. (12-16) and (12-17)]. For a micelle of aggregation number n, the free energy of micellization $F_{\mathrm{mic}}(n)$ is then

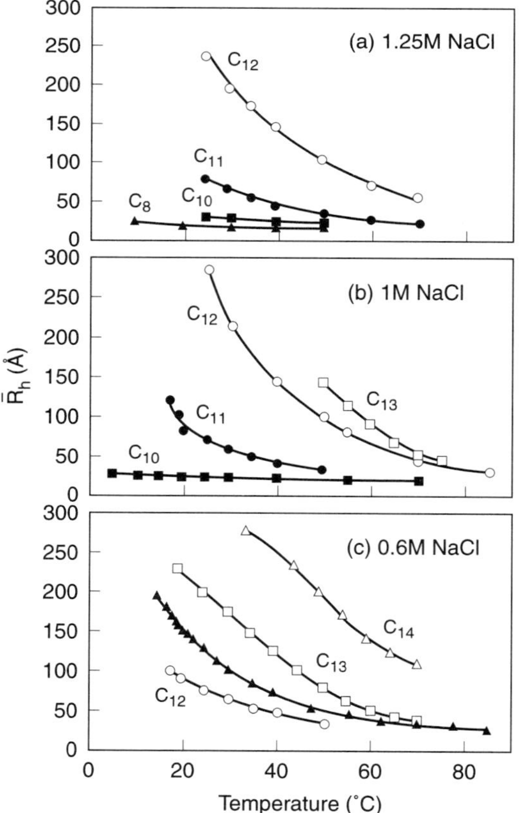

Figure 12.6   The hydrodynamic radius $\bar{R}_h$ as a function of temperature for alkyl sulfate micelles of varying number of carbon atoms at NaCl concentrations of **(a)** 1.25 M, **(b)** 1 M, and **(c)** 0.6 M. The surfactant concentrations are in the range 0.05–2.0 g/dL. (Reprinted with permission from Missel et al., Journal of Physical Chemistry 87:1264 Copyright © 1983, American Chemical Society.)

$$F_{\text{mic}}(n) = n_s \mu_s^0 + (n - n_s)\mu_r^0 = n\mu_r^0 + n_s(\mu_s^0 - \mu_r^0) \tag{12-18}$$

where μ_s^0 is the chemical potential of a surfactant molecule in the spherical end cap, and μ_r^0 is the chemical potential in the remaining rod-like aggregate. Hence, the overall chemical potential of a surfactant molecule is

$$\mu_n^0 = \frac{1}{n} F_{\text{mic}}(n) = \mu_r^0 + \frac{n_s}{n}(\mu_s - \mu_r^0) \tag{12-19}$$

Thus the difference in chemical potential between n-mers and unimers has the following simple dependence on n:

$$\mu_n^0 - \mu_1^0 = \mu_r^0 - \mu_1^0 + \frac{n_s}{n}(\mu_s^0 - \mu_r^0) \tag{12-20}$$

$$= E_\infty + \frac{E_{\text{sciss}}}{n} \tag{12-21}$$

where $E_\infty \equiv \mu_r^0 - \mu_1^0$ is the free energy, per surfactant molecule, for formation of an infinite cylinder, and $E_{\text{sciss}} \equiv n_s(\mu_s^0 - \mu_r^0)$ is the free energy for formation of two end caps; hence it is the free energy for *scission* of a cylindrical micelle into two smaller ones.

Using a chemical potential with this n dependence [Eq. (12-20)], along with the law of mass action, Eq. (12-4), Israelachvili (1992) has shown that the micellar size distribution well above the critical micelle concentration (CMC) (where the CMC is roughly equal to $\exp(E_\infty/k_B T)$) is approximated by

$$X_n = n[1 - (X e^{E_{\text{sciss}}/k_B T})^{-1/2}]^n e^{E_{\text{sciss}}/k_B T} \approx n e^{-n/\sqrt{X e^{E_{\text{sciss}}/k_B T}}} e^{-E_{\text{sciss}}/k_B T} \tag{12-22}$$

where X is the total mole fraction of surfactant. This distribution of micelle sizes is broad, with a maximum at a value of $n = n_0$ given by setting $dX_n/dn = 0$:

$$n_0 = X^{1/2} e^{E_{\text{sciss}}/2k_B T} \tag{12-23}$$

and a mean value twice as large as this:

$$\langle n \rangle \approx \frac{1}{X} \int_0^\infty n X_n dn = 2X^{1/2} e^{E_{\text{sciss}}/2k_B T} = 2n_0 \tag{12-24}$$

Equations (12-22)–(12-24) tell us that cylindrical micelles differ from spherical ones in that cylindrical micelles have a broad distribution of aggregation numbers, and their mean aggregation number $\langle n \rangle$ grows relatively rapidly with surfactant concentration X, namely as $\sqrt{X}$. The greater breadth of the distribution of cylindrical micelles, compared with that of spherical micelles, and the greater dependence of average size on surfactant concentration derive from the differences in the n dependence of the chemical potential [compare Eq. (12-20) with Eq. (12-7)]. According to Eq. (12-24), $\langle n \rangle$ for cylindrical micelles is also sensitive to changes in E_{sciss} caused by variations in temperature or salinity.

Indeed, Missel et al. (1983) have shown that the size of alkyl sulfate cylindrical micelles depends on temperature, NaCl concentration, and alkyl chain length and that these dependences can be fit amazingly well by a simple semiempirical expression for E_{sciss}:

$$\frac{E_{\text{sciss}}}{k_B T} = n_s \frac{(\mu_s - \mu_r)}{k_B T} = 0.4 n_c^2 \left(\frac{\alpha'}{T} + \beta' + \Gamma' n_c + \Delta' \ln[\text{NaCl}] \right) \qquad (12\text{-}25)$$

where [NaCl] is the salt concentration in molarity, and the constants α', β', Γ', and Δ' have the best-fit values

$$\alpha' = 224 \pm 4 \; K, \qquad \beta' = -0.814 \pm 0.135$$

$$\Gamma' = 0.040 \pm 0.011, \qquad \Delta' = 0.215 \pm 0.022 \qquad (12\text{-}26)$$

Equation (12-25) can be rationalized to some extent. The multiplier $0.4 n_c^2$ is just n_s, the number of surfactant molecules in the end caps. The term α'/T can be attributed to the surface energy for contact between the micellar core and the surrounding aqueous medium. The constant α' can be estimated using the reasonable values $a_0 \approx 60 \; \text{A}^2$ and $\Gamma \approx 30$ dyn/cm. From these, we estimate that $\alpha' \approx a_0 \Gamma / 2 k_B \approx 228 \; K$, in excellent agreement with Eq. (12-26). Thus, the surface energy term is about $0.75 k_B T$ per surfactant molecule. The term $\Delta' \ell \, n \, [\text{NaCl}]$ evidently arises from electrostatic contributions to the free energy; both a counterion binding model (Porte and Appell 1981) and a Gouy–Chapman double-layer model produce a term of this form. The latter model yields a value $\Delta' = 0.15$, not far from the best-fit value in Eq. (12-26). β' is thought to arise from a combination of electrostatic effects and the temperature-dependence of the surface tension.

Finally, the term $\Gamma' n_c$ is best rationalized as an entropy penalty for stretching hydrocarbon tails. To become fully stretched, long-tailed surfactants must give up more entropy than do short-tailed surfactants; hence long tails are more prone to curl up a bit. Thus for long-tailed surfactants, the "maximum effective length" ℓ_c does not grow linearly with the tail volume v, and the Israelachvili ratio $v/\ell_c a_0$ then grows somewhat with increasing tail length. Hence, a transition to cylindrical micelles can occur as the tail length is increased. This entropic effect becomes large in the case of block copolymer micelles; lengthening of the block contained within a spherical block-copolymer micelle induces a transition to cylindrical geometry, thereby reducing the stretching of the interior block (see Section 13.2.3).

12.3.2 Rheology of Disordered Solutions of Spherical Micelles

Disordered solutions of spherical micelles are not particularly viscoelastic, or even viscous, unless the volume fraction of micelles becomes high, greater than 30% by volume. Figure 12-7, for example, shows the relative viscosity (the viscosity divided by the solvent viscosity) as a function of micellar volume fraction for a solution of hydrated micelles of lithium dodecyl sulfate in water. Qualitatively, these data are reminiscent of the viscosity–volume-fraction relationship for suspensions of hard spheres, shown as a dashed line (see Section 6.2.1). The micellar viscosity is higher than that of hard-sphere suspensions because of micellar ellipsoidal shape fluctuations and electrostatic repulsions.

12.3.3 Rheology of Dilute Wormy Micellar Solutions

As cylindrical micelles grow, their total length can eventually exceed their persistence length (defined in Section 2.2.4), and they then become semiflexible "wormy" micelles.

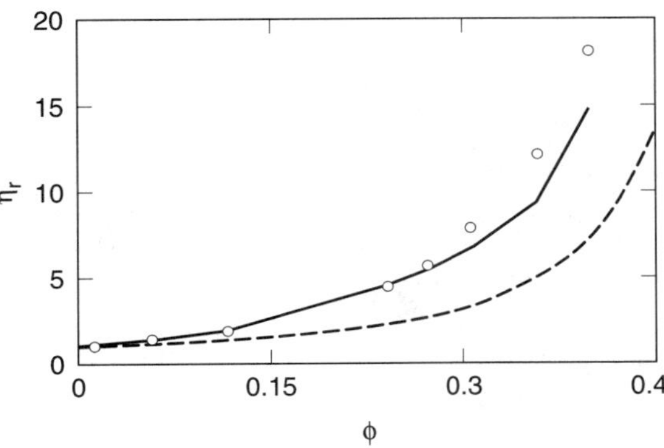

Figure 12.7 Relative viscosity $\eta_r = \eta/\eta_s$ versus hydrated micellar volume fraction ϕ for lithium dodecyl sulfate in water (symbols). The dashed line is the prediction for hard spheres, and the solid line is a theoretical prediction using a measured neutron-scattering structure factor to account for shape fluctuations and electrostatic interactions. (From Liu and Sheu 1996, reprinted with permission from the American Physical Society.)

Using quasi-elastic light scattering and magnetic birefringence, Porte et al. (1980) have inferred a persistence length of around $\lambda_p \approx 20$ nm for cetylpyridinium bromide (CPyBr) micelles. Shikata et al. (1994) found from birefringence measurements that $\lambda_p = 26$ nm for CTAB/NaSal micelles. Other estimates are summarized in Shikata et al. (1994). Figure 12-8 shows electron micrographs of fast-frozen masses of "wormy" micelles 2000 nm or longer in length, composed of a similar surfactant, cetyltrimethylammonium chloride with added sodium salicylate (CTAC/NaSal). These wormy micelles only form when the ratio of the molarity of NaSal to that of CTAC exceeds 0.20; below this concentration of NaSal, the micelles are spherical. As [NaSal]/[CTAC] increases above 0.20, the solution is transformed from a low-viscosity Newtonian liquid to a highly viscoelastic solution.

Although the micrographs in Fig. 12-8 suggest that long wormy micelles readily overlap and "entangle" with one another, at low enough surfactant concentrations the wormy micelles are presumably in a "dilute" state in which entanglements are rare. The existence of a dilute wormy micellar regime can be inferred from rheological experiments in which the shear viscosity of low-concentration ($\sim$1–5 mM; $\phi \lesssim 0.002$) surfactant solutions only modestly exceeds that of the solvent, but the birefringence and normal stresses are large, suggesting the presence of elongated objects, i.e., wormy micelles (Hu et al. 1993). One might naively expect the rheology of "dilute" solutions of wormy micelles to be quite simple. Nothing could be further from the truth! When sheared, such solutions often exhibit a perplexing *shear-thickening* phenomenon (Rehage and Hoffman 1982; Hu et al. 1993; Wang et al. 1994; Hu and Matthys 1995). The shear thickening typically occurs after an inception period that can last thousands of strain units before the shear stress and first normal stress difference abruptly increase and thereafter fluctuate irregularly. An apparently analogous phenomenon has been observed in extensional flow, but with a lower critical strain rate than in shear. It has been suggested that this slow viscosity buildup is caused by flow-induced micellar growth (Wang et al. 1994), as predicted by recent theories (Cates and Turner 1990; Bruinsma et al. 1992). Pine and coworkers (Liu and Pine 1996; Boltenhagen et al. 1997) have shown using direct imaging that the shear thickening and stress fluctuations are accompanied by the formation of highly elastic, time-dependent, gel-like structures, also known as "shear-induced structures" (SIS). The possibilities of an elastically driven hydrodynamic instability, along with "shear-induced structures," in these solutions have also been suggested (Huang et al. 1996). Shear-induced structures are also thought to form

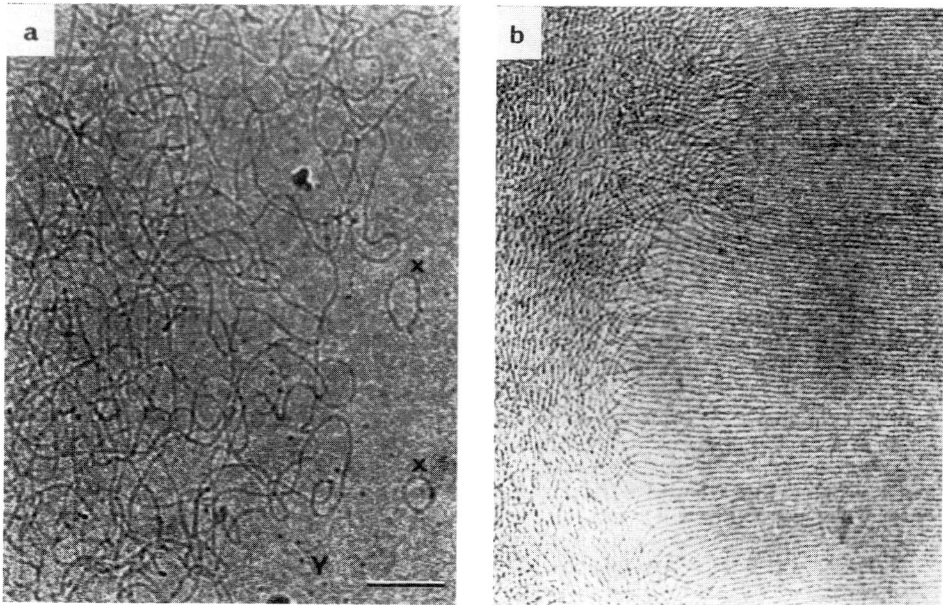

Figure 12.8   Cryo-TEM micrographs of 5-nm-diameter wormy micelles in 0.05 M CTAC with added **(a)** 0.05 M NaSal and **(b)** 0.10 M NaSal. All samples were squeezed to form thin films. Orientation induced by squeezing was allowed time to relax in **(a)**, but not in **(b)**. Bar = 100 nm. (Reprinted with permission from Clausen et al., Journal of Physical Chemistry 96:474 Copyright © 1992, American Chemical Society.)

in more concentrated solutions of entangled worm-like micelles, as evidenced by light scattering and by shear thinning that is less pronounced than one would expect from the linear viscoelastic response according to the so-called Cox–Merz rule (Kodoma et al. 1997; see Section 1.3.1.5). Much work is still necessary to sort out and understand the bizarre phenomena occurring in these solutions.

12.3.4 Rheology of Entangled Wormy Micellar Solutions

Figure 12-8 suggests that at high enough concentrations, these micelles can entangle with each other, like the entanglements in concentrated solutions of long synthetic polymers. These giant "wormy" micelles are sometimes called "equilibrium polymers" or "living polymers" (Cates 1990), because their length distribution is not fixed by chemical synthesis, but can vary reversibly in response to changes in concentration, salinity, temperature, and even flow.

Entangled solutions of "wormy" micelles, like solutions of ordinary polymers, show pronounced viscoelastic effects. Figure 12-9 shows the storage and loss moduli for a solution of 15 mM cetylpyridinium salicylate (CPySal) with 11 mM added sodium salicylate. For this solution, the storage modulus is found to be comparable to the loss modulus over a significant range of frequencies. In contrast, solutions of spherical micelles at comparable volume fractions are Newtonian with a viscosity only slightly larger than that of the solvent, as one might expect by applying the Stokes–Einstein law for suspensions of spherical particles. Even more remarkably, when the added salt concentration is increased from 11 to 12.5 mM, the linear relaxation spectrum changes from that shown in Fig. 12-9 to that shown in Fig. 12-10 (Rehage and Hoffmann 1988). The moduli in Fig. 12-10 are close to that predicted by a one-relaxation-time Maxwell model, for which

$$G' = \frac{G_0\omega^2\tau^2}{1 + \omega^2\tau^2}, \qquad G'' = \frac{G_0\omega\tau}{1 + \omega^2\tau^2} \qquad (12\text{-}27)$$

At this concentration of salt and of surfactant, a nearly single-relaxation-time description holds over four or more decades of frequency, even to values of ω high enough that $G''/G' < 0.01$! Such nearly perfect monoexponential relaxation is rare and has only been found for a few other complex fluids, as discussed in Section 5.4.1.

A single-relaxation-time response is also observed for this fluid in other flow histories, including start-up and cessation of steady shearing, if the shear rate $\dot{\gamma}$ is low enough. At higher shear rates, the viscoelastic response is more complex. Figure 12-11 shows the time-dependence of the shear viscosity η after start-up of steady shearing for a solution

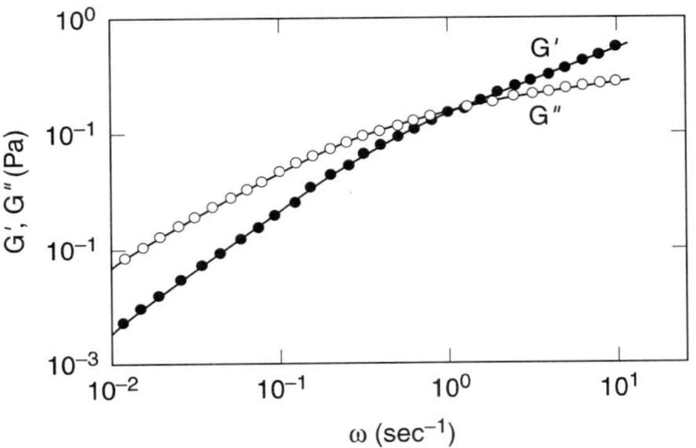

Figure 12.9 Storage and loss moduli G' and G'' as functions of frequency ω for a solution of 15 mM CPyCl and 11 mM NaSal at $T = 20°C$. (Reprinted with permission from Rehage and Hoffmann, Journal of Physical Chemistry 92:4712 Copyright © 1988, American Chemical Society.)

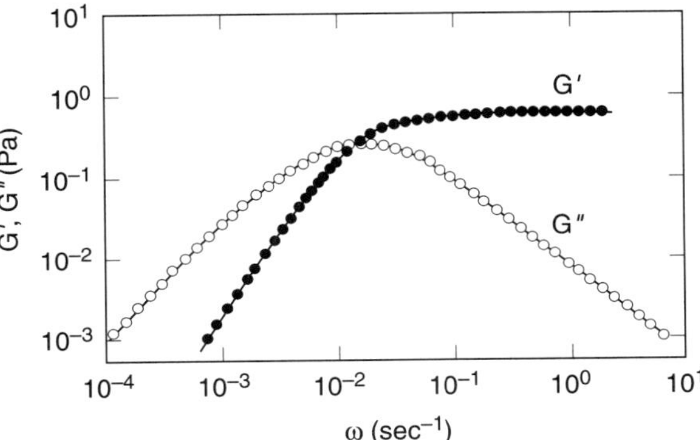

Figure 12.10 Storage and loss moduli G' and G'' as functions of frequency ω for a solution of 15 mM CPyCl and 12.5 mM NaSal at $T = 20°C$. The data are similar to the predictions of the single-relaxation-time Maxwell model. (Reprinted with permission from Rehage and Hoffmann, Journal of Physical Chemistry 92:4712 Copyright © 1988, American Chemical Society.)

of CTAB/NaSal at various shear rates (Shikata et al. 1988b). At the lowest rate ($\dot{\gamma} =$ 0.12 sec^{-1}), the stress response is monotonic and is well-described by a single Maxwell element. At higher shear rates, there is a stress overshoot, and the steady-state viscosity decreases strongly with increased shear rate. At even higher shear rates, peculiar multiple over- and undershoots are observed after start-up of steady shearing. Electron microscopy of wormy micelles fast-frozen immediately after being subjected to flow show clearly that the micelles are highly aligned by the flow (see Fig. 12-8b, for example).

The behavior shown in Fig. 12-11 (except for the multiple oscillations) is reminiscent of entangled flexible polymers. Such similarities in rheological properties, along with the obvious analogy between entangled "wormy" micelles and entangled polymers, suggests that theories for relaxation of entangled polymers, such as reptation, might be applicable to solutions of "wormy" micelles (Cates 1987, 1990, 1997). However, to apply a theory for the dynamics and rheology of entangled polymer molecules to "wormy" micelles, one must somehow account for the reversible assembly and disassembly of these micelles in solution.

Long wormy micelles break and reconnect fairly rapidly. At equilibrium, the typical time for breakage of a micelle must equal the time for reconnection. This typical time can be measured by "temperature-jump" experiments in which the temperature of a small volume of solution is suddenly raised slightly, and the time for the distribution of micelle sizes to adjust to this change is monitored by some optical property, such as scattered light (Lang and Zana 1987). In this way, the breakage (or equivalently, reconnection) time τ_{br} is inferred to be 0.1–1.0 sec for CTAB micelles over a range of surfactant concentrations in 0.25 M KBr (Candau et al. 1990; Cates and Candau 1990). Breakage times can also be extracted from rheological measurements, as discussed at the end of Section 12.3.4.1. Breakage and

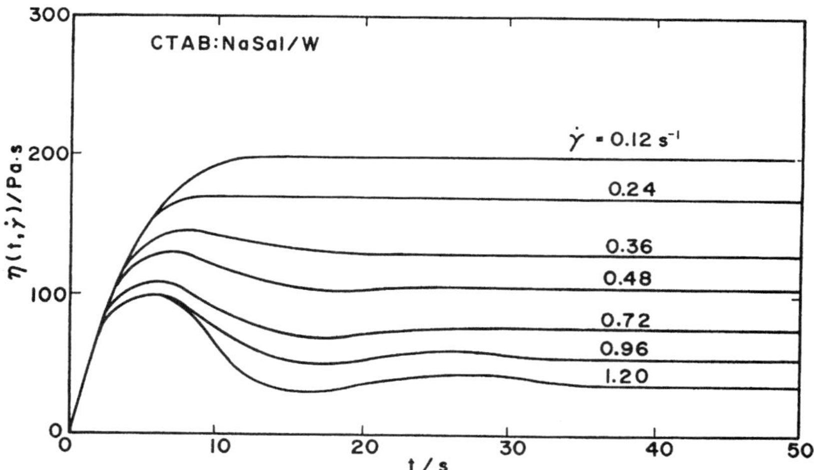

Figure 12.11 Viscosity (stress divided by shear rate) after start-up of steady shearing at various shear rates for a solution of 0.1 M CTAB and 0.4 M NaSal. (Reprinted from J Non-Newt Fluid Mech 28:171, Shikata et al. (1988), with kind permission from Elsevier Science - NL, Sara Burgerhartstraat 25, 1055 KV Amsterdam, The Netherlands.)

reconnection endow wormy micelles with a Houdini-like ability to escape entanglements that is not available to ordinary synthetic polymers. Nature has bestowed a similar ability on cellular DNA molecules, which can break and reconnect via an enzyme, Topoisomerase II, which permits the DNA to disentangle during cellular mitosis, or cell division (Holm 1994).

The kinetic and structural details of micelle breakage and reconnection depend on surfactant type and salinity. Figure 12-12 illustrates some simple possibilities (Candau et al. 1993; Khatory et al. 1993). In one limit, Fig. 12-12a, the micelles are nonintersecting; they break somewhere along their contour by forming end-caps, yielding two shorter wormy micelles. The reverse process can also occur, in which two such micelles join end-to-end by elimination of two end-caps. In another limit, Fig. 12-12c, the micelles interconnect to form a network; network junctions break and re-form. Of course, a behavior intermediate between these extremes is also possible (see Fig. 12-12b). The limit in Fig. 12-12a predominates at low electrolyte concentrations, presumably because electrostatic repulsions keep the micelles from intersecting each other. At high salt concentration, both indirect and direct fast-freezing electron microscopy indicate that the micelles become branched and intersect each other. Figure 12-13 shows a generic "phase diagram," with regimes of concentration and salinity where intersecting and nonintersecting worm-like micelles, as well as liquid-crystalline phases, typically occur. It has been suggested (Shikata et al. 1988a) that branched junction points might be short-lived intermediate states that allow otherwise unbranched micelles to pass through each other, thus relaxing stress. Theories of the viscoelasticity of

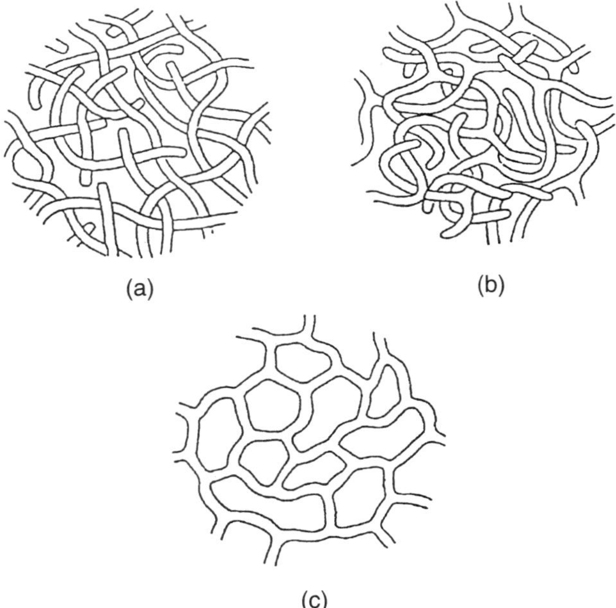

(a)   (b)

(c)

Figure 12.12   Structures of wormy micelles. (**a**) Entangled linear micelles, (**b**) intermediate case, and (**c**) network of interconnected micelles. (From Candau et al. 1993, with permission from EDP Sciences.)

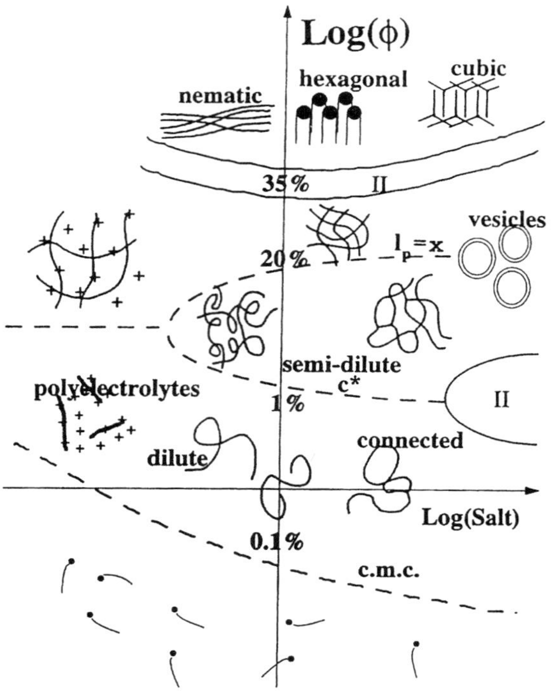

Figure 12.13 Schematic diagram illustrating typical micellar structures found as functions of surfactant concentration ϕ and salinity. At very low ϕ, surfactants are isolated unimers. At higher concentration, worm-like micelles can form, which become entangled as the ϕ continues to increase. If the surfactants are anionic or cationic and the salinity is low, the micelles repel each other electrostatically, forming the analog of "polyelectrolytes"; at high salinity, electrostatic screening allows the micelles to intersect. At still higher ϕ, ordered phases (nematic, hexagonal, cubic) occur. The regions labeled "II" are coexistence regions. (From Lequeux and Candau, figure 4, with kind permission of Kluwer Academic Publishers, Copyright 1997.)

solutions of wormy micelles depend to some extent on which model is applicable to the given solution.

12.3.4.1 Cates' Reptation Model

Cates (1987, 1990) has described in detail a model for the viscoelasticity of wormy micelles. Cates' original model is based on the assumption that micelles do not form branched states; that is, they are described by the case envisaged in Fig. 12-12a. His theory rests on the following three premises: (i) Chain breakage is a unimolecular process; each segment of unit length on any chain has the same probability of breaking as any other; (ii) successive breakage and recombination events are uncorrelated; and (iii) each chain relaxes by reptation. The first assumption implies that the time for breakage of a micellar chain is inversely proportional to the length of the chain. Thus, for a chain with length equal to the average length $\bar{L} = \langle L \rangle$, the breakage time is

$$\tau_{\rm br} = \frac{1}{k\bar{L}} \qquad (12\text{-}28)$$

where k is a constant.

Within these three assumptions, one can identify at least two important limiting cases. In the first, the micelle breakage time $\tau_{\rm br}$ is long enough that the typical micelle can relax by reptation before it breaks. In this case, one obtains the linear relaxation modulus $G(t)$

simply by averaging the linear modulus for pure reptation, Eq. (3-67), over the equilibrium distribution of micelle lengths, Eq. (12-22). In so doing, one uses the prediction of reptation theory that the reptation time for a micelle of length L is proportional to L^3. Because of this strong, cubic power law, the broad distribution of micelle lengths described by Eq. (12-22) yields an even broader distribution of relaxation times, and the linear modulus can be approximated by

$$G(t) \propto \exp[-(t/\tau_{\mathrm{rep}})^\alpha], \qquad \text{with } \alpha = 1/4 \tag{12-29}$$

where τ_{rep} is the reptation time (or "disengagement" time, $\tau_{\mathrm{rep}} = \tau_d$) of a micelle of length $\bar{L}$. Equation (12-29) is a *stretched exponential* or Kohlrausch–Williams–Watts (KWW) relaxation modulus, often used to describe the relaxation of glasses (see Section 4.2). The small value of the exponent, $\alpha = 1/4$, implies that the relaxation deviates strongly from single exponential behavior (for which α would be unity).

The linear relaxation moduli of some wormy micelle solutions can indeed be fit to a stretched exponential. The exponent α has been found to increase from 0.3 to unity as the ratio of salt to surfactant concentration increases in solutions of CPyCl/NaSal (see Fig. 12-14). An exponent of $\alpha = 1$ corresponds to single-exponential relaxation, similar to that shown in Fig. 12-10.

The narrowing of the viscoelastic spectrum as salinity increases is analogous to "motional" narrowing of line widths in NMR spectroscopy. The narrowing process occurs when the breakage time τ_{br} becomes smaller than the reptation time τ_{rep}. In the limit $\tau_{\mathrm{br}} \ll \tau_{\mathrm{rep}}$, the Cates model predicts that $G(t)$ is *monoexponential*, with a time constant τ given by

$$\tau = (\tau_{\mathrm{br}} \, \tau_{\mathrm{rep}})^{1/2} \tag{12-30}$$

This remarkable result can be understood as follows (Cates 1990). Consider a test segment of the "tube" confining the reptating chain. This segment can only relax its orientation when a chain end passes through it. The chain end nearest this segment only lasts a time τ_{br} before it is reconnected with another chain. Thus, a chain end will only pass through the test segment if it is close enough to the test segment that it can reptate through it before reconnecting with another chain end. Since the whole chain of length $\bar{L}$ escapes its tube in a time τ_{rep}, in the more limited time τ_{br} only a fraction of the chain $\ell/\bar{L} \sim (\tau_{\mathrm{br}}/\tau_{\mathrm{rep}})^{1/2}$ can diffuse out of its tube. (The square root appears because diffusion

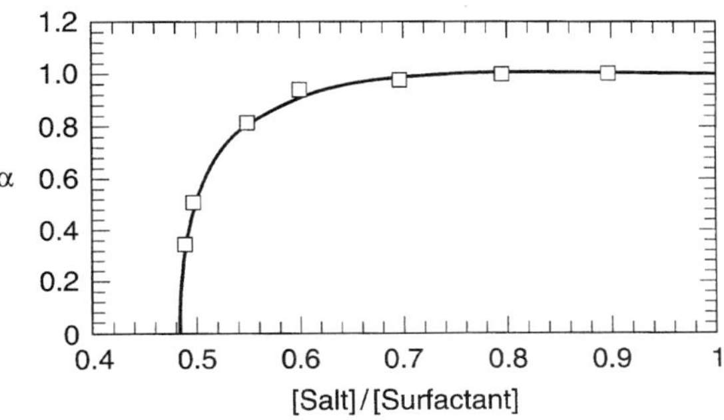

Figure 12.14 The stretched-exponential exponent α as a function of the concentration ratio [NaSal]/[CPyCl] for 0.1 M CPyCl at 20°C. (From Rehage and Hoffmann 1991, by permission of Taylor & Francis.)

distances are proportional to the square root of the time available.) This means that the tube segment can only relax if it is within a curvilinear distance ℓ from the end of the tube. If $\ell \ll \bar{L}$, this is unlikely the first time the chain breaks, but becomes likely after the chain has broken $N = \bar{L}/\ell$ times. Thus, the relaxation time is given by

$$\tau \approx N\tau_{\text{br}} = (\bar{L}/\ell)\tau_{\text{br}} = (\tau_{\text{rep}}\tau_{\text{br}})^{1/2} \tag{12-31}$$

The relaxation modulus is a single exponential decay because a typical tube segment relaxes only after it has been passed off from one chain to another many times. Hence, before relaxing, each tube segment samples chains of many lengths as well as many different distances from a chain end; this produces "motional averaging," which results in nearly monoexponential relaxation.

Although motional averaging might occur in ways other than that envisioned by Cates, temperature-jump experiments have yielded values of τ_{br} that indicate $\tau_{\text{br}} < \tau_{\text{rep}}$ in the region where the relaxation is nearly monoexponential, in agreement with Cates' theory. In addition, Cates' theory offers distinctive predictions for the concentration-dependencies of the viscoelastic behavior; these allow the theory to be tested rather stringently. To obtain these predictions, we note that in the semi-dilute regime, the mean-field reptation time is $\tau_{\text{rep}} \sim \bar{L}^3\phi^2$, where ϕ is the volume fraction of surfactant. Hence, from Eqs. (12-31) and (12-28), one obtains the mean-field predictions (Cates and Candau 1990):

$$\tau \sim (\tau_{\text{br}}\tau_{\text{rep}})^{1/2} \sim (\bar{L}^2\phi^2)^{1/2} \sim \phi^{1.5}$$
$$\eta_0 \sim G_0\tau \sim \phi^{3.5} \tag{12-32}$$

where we have used the result $\bar{L} \propto \langle n \rangle \propto \sqrt{X} \propto \phi^{1/2}$ from Eq. (12-24), along with the fact that the plateau modulus G_0 is proportional to ϕ^2. These exponents, $\alpha_\tau = 1.5$ and $\alpha_\eta = 3.5$, as well as the concentration exponent for self-diffusion, are in good agreement with measured values of these exponents for CTAB/KBr micelles at a salt concentration [KBr] = 0.25 M (Candau et al. 1989), as well as for CPyCl/NaSal micelles in 0.5 M NaCl (Berret et al. 1994).

Granek and Cates (1992) have recently shown that high-frequency deviations from monoexponential behavior in G' and G'' can be accounted for by extending Cates' model to allow for "primitive-path fluctuations" (see Section 3.7.2.1) and Rouse relaxation modes. These become more important as the number of entanglements per molecule decreases. By fitting the extended theory to G' and G'' data, one can extract separately the two time constants τ_{br} and τ_{rep} (Kern et al. 1994); τ_{br} is roughly the frequency at which G'' deviates from monoexponential Maxwellian behavior. A convenient way of exposing these deviations from monoexponential behavior is to plot G'' versus G', a format known as a "Cole–Cole" plot (see Fig. 12-15). Monoexponential behavior corresponds to a semicircular shape. Deviations from this shape at high frequency allow one to extract the ratio $\zeta \equiv \tau_{\text{br}}/\tau_{\text{rep}}$ (Kern et al. 1994). The ratio G''/G_0 at the high-frequency local minimum is roughly the inverse of the average number of entanglements per worm-like micelle.

12.3.4.2 The Effect of Added Salts

At lower ionic strength, 0.1 M, Candau et al. (1988) found a steeper power law than predicted by Cates' theory, namely $\eta \sim \phi^5$. This deviation has been attributed to incompletely

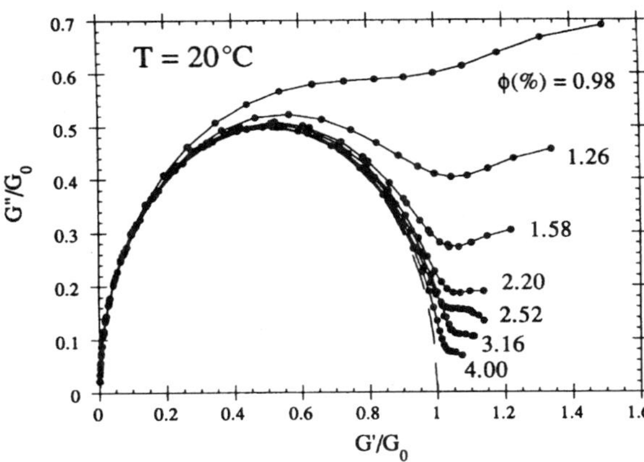

Figure 12.15 Cole–Cole plot for CPyCl/NaSal at various total surfactant volume percentages ϕ. Here G_0 is the plateau modulus for G', which is used to normalize the data. The dashed line is a semicircle (distorted into an ellipse because of the differing scales of abscissa and ordinate), which represents ideal monoexponential behavior. (From Berret et al. Langmuir 9:2851, Copyright © 1993, American Chemical Society.)

screened electrostatic interactions that lead to smaller micelles at lower surfactant concentrations (Candau et al. 1993). Deviations from Cates' theory are also observed at high electrolyte concentrations. Figure 12-16 shows that the viscosities of solutions of CTAB and $CPyClO_3$ have *maxima* as functions of added electrolyte concentration. At salinities greater than that for which η_0 is a maximum, Candau et al. (1993) found that the viscosity exponents decrease to as low as 1.0, much smaller than 3.5, which is that for Cates' theory. The relaxation remains nearly monoexponential, however. Porte et al. (1980) and Candau et al. (1993) have proposed that at high salinities, wormy micelles interconnect to form a dynamic three-dimensional network (see Fig. 12-12c) because of the high screening of electrostatic repulsions. Such a network could also obey a monoexponential relaxation law, but might follow concentration scaling laws that differ from those of Cates' theory.

The existence of interconnected cylindrical micelles has been predicted by Monte Carlo lattice simulations (Larson 1992), but not yet observed by direct experiments for single-tail surfactants. For multi-tail surfactants, such interconnected micelles have also been both simulated (Karaborni et al. 1994) and observed by fast-freeze electron microscopy (Danino et al. 1995). These interconnections are expected to be more likely when the head groups are small and not too repulsive, so that head-group crowding at the joint between connected micelles is not severe. Since these joints are saddle-shaped surfaces, surfactants might be expected to form such interconnections readily if they don't strongly resist negative Gaussian curvature—that is, if $\bar{k}$ in Eq. (12-2) is positive, or negative but small in magnitude. Turner and Cates (1992) have extended the Cates' theory by considering the kinetics of intermicelle reactions ignored in the original theory, including the formation of threefold or fourfold interconnected intermediates, such as those depicted in Fig. 12-13. In the Turner–Cates theory these intermediates are assumed to be rare, but nevertheless to dominate the kinetics. The case of a threefold intermediate gives a relaxation time similar to that of reversible scission, namely $\tau \propto \tau_{br}^{1/2}\tau_{rep}^{1/2}$, while for a fourfold intermediate we have $\tau \propto \tau_{br}^{1/3}\tau_{rep}^{2/3}$ when $\tau_{br} \ll \tau_{rep}$.

For some surfactant/salt combinations, the dependence of viscosity on added salt concentration is even more peculiar than that shown in Fig. 12-16. Rehage and Hoffmann (1988) found that the zero-shear viscosity of CPyCl has two maxima in viscosity as a function of the concentration of added NaSal. The peculiar behavior of this and similar systems is related to the specific binding of the Sal^- ion to the micellar surface (Rehage and Hoffmann 1991; Olsson et al. 1986). Thus, the shape of the viscosity-salt concentration curve is highly sensitive to the particular added ion (Rehage and Hoffmann 1991). Counterions that contain a hydrophobic group, such as salicylate, 4-ethylbenzoate, and 4-*n*-propyl benzoate, are especially prone to bind to the micellar surface, as shown by nuclear magnetic resonance measurements (Shikata et al. 1988a; Rehage and Hoffmann 1991). The specific binding can be strong enough that at high counterion concentration the micellar surface charge can change sign from positive to negative. This sign change occurs near the viscosity minimum between the two viscosity maxima. In some systems, near the minimum in viscosity, there is a phase separation into micelle-rich and micelle-poor phases (Rehage and Hoffmann 1991). Even systems affected by counterion binding effects are often governed by a single relaxation time, whose dependence on salt concentration can be complex.

Figure 12-17 shows the storage and loss moduli for CTAB/NaSal solutions at various ratios of NaSal molarity (C_S) to CTAB molarity (C_D). At the lowest concentrations of added NaSal, below those shown in Fig. 12-17, the solution is a low-viscosity Newtonian liquid. As the concentration of added salt is increased, G' and G'' become measurable (see Fig. 12-17), but the longest relaxation time τ is short, and the frequency-dependence shows a broad distribution of relaxation times, similar to that predicted by the Rouse theory (Shikata and Kotaka 1991; Shikata et al. 1987, 1988a, 1989). At higher salinities, $C_S/C_D \geq 0.7$, the relaxation spectrum has only a single relaxation time, and τ has a peculiar nonmonotonic dependence on the concentration difference $C_S - C_D$, for $C_S^* \equiv C_S - C_D > 0$. For $C_S \gg C_D$, τ decreases as $(C_S^*)^{-5} C_D^0$. The modulus scales as $C_D^{2.2} C_S^0$. The $C_D^{2.2}$ scaling is similar to that found in entangled solutions of ordinary polymers.

The observed concentration-dependence of τ is due to the specific binding of Sal^- to the micelles, which changes the shape of the micellar surface. At low $[Sal^-]$ the micelle surfaces are highly charged, and electron microscopy shows that the micelles are spherical (Clausen et al. 1992). As $[Sal^-]$ increases, the surface charges are neutralized, so that the surface area per head group a_0 decreases, and the micelles begin to grow; thus τ increases. As $[Sal^-]$ increases still more, electron microscopy shows that the micelles remain very

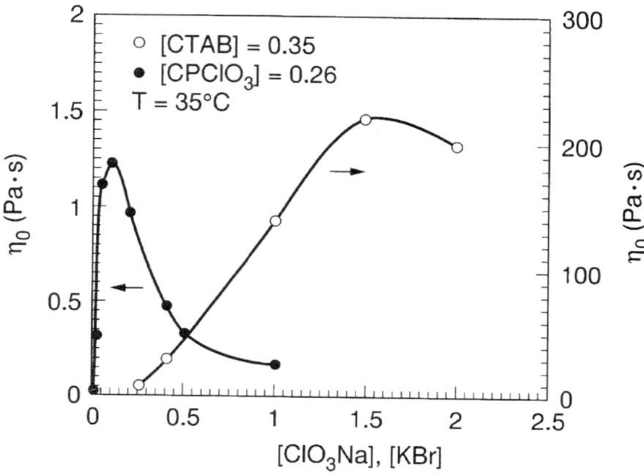

Figure 12.16 Zero-shear viscosities of CPClO₃/ClO₃Na and CTAB/KBR as functions of salt concentration. (From Candau et al. 1993, with permission from EDP Sciences.)

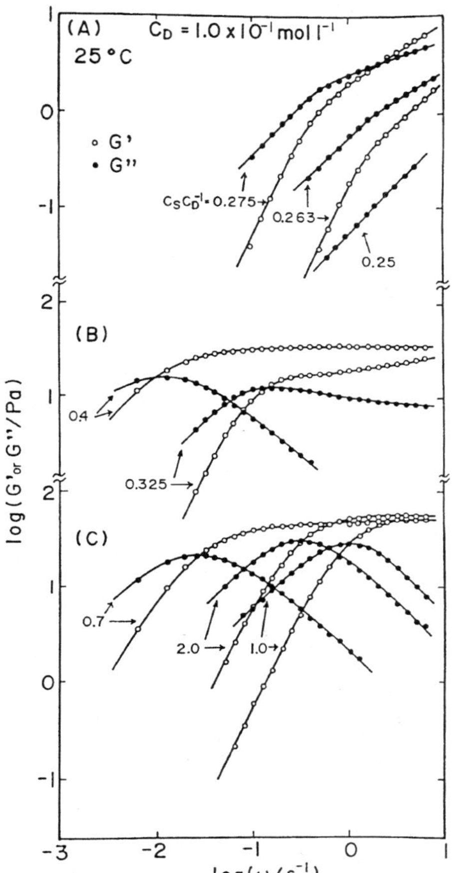

Figure 12.17 Dependence of the storage and loss moduli G' and G'' on frequency for CTAB/NaSal solutions with 0.1 M CTAB, and various ratios of the molarity of salt (C_S) to surfactant (C_D) (**A**), (**B**), and (**C**) correspond to successively higher ranges of C_S/C_D. (Reprinted from J. Non-Cryst. Solids 131-133:831, Shikata and Kotaka 1991, with kind permission from Elsevier Science - NL, Sara Burgerhartstraat 25, 1055 KV Amsterdam, The Netherlands.)

long even though τ begins to decrease. This can be explained if excess Sal$^-$ ions accelerate the kinetics of micellar breakage and reconnection or interconnection, thus allowing more rapid relaxation of long micelles.

12.3.4.3 The Effect of Temperature

For CTAB/KBr solutions (and presumably many others), the experimental temperature-dependencies of the stress relaxation time τ and the breakage time τ_{br} follow Arrhenius laws (Cates and Candau 1990). For τ, one finds the dimensionless activation energy $E_\tau/k_B T = 42$ at 300 K; and for τ_{br}, it was found that $E_{\mathrm{br}}/k_B T = 48$. From the formula $\tau = (\tau_{\mathrm{br}}\tau_{\mathrm{rep}})^{1/2}$, the activation energy for τ should be $E_\tau = (E_{\mathrm{br}} + E_{\mathrm{rep}})/2$, where E_{br} and E_{rep} are the activation energies for τ_{br} and τ_{rep}. Then, from $\tau_{\mathrm{rep}} \propto \bar{L}^3$ and Eq. (12-24), which gives the activation energy for $\bar{L} \propto \langle n \rangle$ as $E_{\mathrm{sciss}}/2$, we infer that

$$E_\tau = \frac{1}{2}\left(E_{\mathrm{br}} + \frac{3E}{2}\right) \tag{12-33}$$

where E is the temperature-independent part of E_{sciss}. From the measured values of E_τ and E_{br}, we then obtain $E/k_BT = 24$ at room temperature. From Eq. (12-25), using $n_c = 16$ for CTAB, we estimate $E/k_BT = 0.4n_c^2\alpha'/T = 76$ at 300 K, about three times as large as the experimental value of 24. This discrepancy is as yet unexplained.

12.3.4.4 Nonlinear Rheology

As remarked earlier, the nonlinear viscoelastic behavior of entangled wormy micellar solutions is similar to that of entangled flexible polymer molecules. Cates and coworkers (Cates 1990; Spenley et al. 1993, 1996) derived a full constitutive equation for entangled wormy micellar solutions, based on suitably modified reptation ideas. The stress tensor obtained from this theory is (Spenley et al. 1993)

$$\sigma = \frac{15}{4}G_0\left(\mathbf{W} - \frac{1}{3}\delta\right) \tag{12-34}$$

where δ is the unit tensor and the tensor $\mathbf{W}$ is

$$\mathbf{W}(t) = \int_0^\infty \frac{1}{\tau}\exp\left[-\int_{t'}^t \left(\frac{1}{\tau} + v(t'')\right)\right]\tilde{\mathbf{Q}}(t - t')\,dt' \tag{12-35}$$

where $v \equiv \mathbf{W} : \nabla\mathbf{v} \equiv \text{trace}(\mathbf{W} \cdot \nabla\mathbf{v})$, and

$$\tilde{\mathbf{Q}} \equiv \frac{1}{4\pi}\int \frac{(\mathbf{u} \cdot \mathbf{E})\,(\mathbf{u} \cdot \mathbf{E})}{|\,\mathbf{u} \cdot \mathbf{E}\,|}d^2\mathbf{u} \tag{12-36}$$

and $\mathbf{E}$ is the inverse deformation gradient tensor (see Section 1.4.2.1).

The relationship between stress and shear rate predicted by this theory is similar to that of the Doi–Edwards reptation theory for flexible polymers (see Section 3.7.4.4). Figure 12-18a shows that the predicted stress for wormy micellar solutions exhibits a maximum $\sigma_{\max}$ at a shear rate $\dot{\gamma}_{\max} \approx 2.6\tau^{-1}$. High-frequency relaxation processes not accounted for by this simple theory are added to the theory to produce a stress upturn at high shear rates, shown by the long-dashed line in Fig. 12-18a. Fluids with a flow curve of this shape are expected to be unstable to the formation of zones of differing shear rate when the overall shear rate lies somewhere along the horizontal short-dashed line in Fig. 12-18a. For any overall shear rate on this line, the fluid stratifies into at least two layers, one with a low shear rate given by the value at the left end of the short-dashed horizontal line (where it intersects the solid line) and the other with a high rate given by the value at the right end of the short-dashed line (where it intersects the long-dashed line). The worm-like micelles in the low-shear-rate zone are only modestly oriented, while those in the high-shear-rate zone are highly oriented. As the overall shear rate increases, the proportional of fluid sheared at the high rate increases, so that the average of the shear rate over both zones equals the imposed value. Using birefringence, the predicted flow stratification in a circular Couette device has been directly observed in these solutions (Decruppe et al. 1995; Makhloufi et al. 1995). However, because these solutions often have a nematic phase at a higher concentration, it is possible to interpret the observed stratified state as the coexistence of an isotropic with a shear-induced nematic state, as discussed in Berret et al. (1997).

In the stratified-flow regime, the shear stress remains constant, equal to the value $\sigma_{\max}$, in both zones. Thus, the shear stress measured in such a flow should be independent of

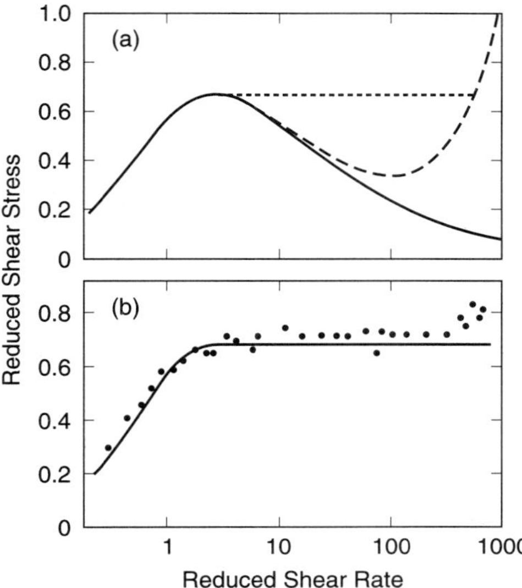

Figure 12.18   **(a)** The solid line is the reduced shear stress σ/G_0 as a function of reduced shear rate $\dot{\gamma}\tau$ predicted by Eqs. (12-34)–(12-36). The long-dashed line is a hypothetical modification obtained by considering high-frequency "Rouse" modes. The short-dashed line is the stress predicted by flow stratification. **(b)** Comparison of the theory with flow-stratification to experimental data of Rehage and Hoffmann (1991). (From Spenley et al. 1993, reprinted with permission from the American Physical Society.)

the macroscopic shear rate for all rates between the terminal points in Fig. 12-18a. This predicted behavior has been observed in wormy micellar solutions (see Fig. 12-18b). The measured shear stress along the horizontal plateau for this solution is $0.67G_0$, as predicted, where G_0 is the linear viscoelastic plateau modulus. The first normal stress difference in the region of a shear stress plateau is a linear function of shear rate and becomes very large at high shear rates within the plateau; this behavior is not yet completely understood. Furthermore, as the concentration is varied within the semidilute regime, the dimensionless plateau shear stress can become higher than predicted by Eqs. (12-34)–(12-36), namely around $0.9G_0$, and the value of $\dot{\gamma}$ at the beginning of the plateau can be a factor of two or more lower than the predicted value of $2.6/\tau$ (Aït-Ali and Makhloufi 1997; Porte et al. 1997). For some concentrations, Fischer et al. (1997) have shown that both linear and nonlinear rheological properties of a semidilute CTAB/NaSal solution are described almost unbelievably accurately by a simple single-mode Giesekus model with $\alpha = 1/2$. The Giesekus constitutive equation, given by Eqs. (3-80) and (3-81b), was derived from an ad hoc model intended for entangled polymer melts and solutions; why it should be so spectacularly successful for semidilute wormy micelles is a mystery. Additional discussion of stratified flows can be found in Section 3.7.5.3.

As discussed in Section 12.3.3, unusual time- and shear-rate-dependencies have been reported for some wormy micellar solutions at dilute concentrations—for example, 1–5 mTAB/NaSal. At higher concentrations, 7–250 mM, of a similar surfactant, tetradecyltrimethyammonium bromide in NaSal, the extensional viscosity increases with increasing extension rate until a maximum is reached, and extension thinning then follows (Prud'homme and Warr 1994). Prud'homme and Warr interpret the maximum as the critical extension rate for fracture of micelles; a force balance at this extension rate then leads to an estimate 0.7–2.0 μm for the micelle length in these solutions. Flow instabilities observed at high extension rates lead to "flare" birefringence patterns analogous to those seen in extensional flows of dilute solutions (Odell et al. 1988).

12.4 SURFACTANT LIQUID CRYSTALS

If the concentration of surfactant becomes high enough, surfactant structures often develop long-range order, and hence they become *liquid crystalline*. They are *lyotropic* liquid crystals, because the transition to the liquid-crystalline state is induced by concentration changes. Surfactant solutions can form nematic and smectic-A liquid-crystalline phases analogous to those discussed in Chapter 10. In addition, *hexagonal* and *cubic* phases are common in surfactant solutions.

12.4.1 Phase Behavior

12.4.1.1 Nematic, Hexagonal, and Lamellar Phases

The phase diagrams of surfactant-containing liquids are extremely diverse, and they depend on the ratios of head to tail length of the surfactant, the presence of multiple tail or head groups, salinity, the presence and concentrations of "cosurfactants," and so on. However, amid the extreme diversity, one finds some recurring patterns. Figure 12-19 shows the composition–temperature phase diagrams of four aqueous solutions of the following single-tailed surfactants: a nonionic surfactant (hexaoxyethyleneglycol mono *n*-dodecyl ether), an anionic surfactant (lithium perfluorooctanoate), a cationic surfactant (dodecyltrimethylammonium chloride), and a zwitterionic phospholipid (1-oleoyl-*sn*-glycero-3-phosphocholine). Despite the great differences among these one-tailed surfactants, their phase diagrams show remarkable similarities. In each case, there are, with increasing surfactant concentration, transitions from a disordered solution of spherical micelles, to hexagonal packings of cylinders, to a bicontinuous cubic phase (discussed below), to a lamellar phase. The term "bicontinuous" means that continuous sample-spanning passages can be found that reside entirely within the aqueous fluid; and likewise for the oleic fluid. In some systems, there is also a cubic packing of spheres wedged between the hexagonal cylinders and the disordered spheres (Seddon 1996; Larson 1994), of space group Pm$\bar{3}$n or Fd3m. A hexagonal sphere packing has also been found in this region (Clerc 1996). These symmetries are quite different from the simple BCC packing (space group Im$\bar{3}$m) often found in block copolymer systems (see Section 13.2.2). The ordered sphere packings tend to occur when the surfactant tail is short and not as bulky as the head group; in these cases, spherical micelles resist a transition to cylinders until after their concentration becomes high enough to induce positional ordering. Note, also, in each case in Fig. 12-19 the boundary between disordered (isotropic) solutions and ordered liquid crystalline solutions occurs at the highest temperature for the lamellar phase, at the next highest temperature for the hexagonal cylindrical phase, and at the lowest temperature for the cubic phases. Presumably this pattern is dictated by entropic considerations, which at higher temperatures disfavor highly ordered, low-symmetry phases.

The similarities among the phase diagrams for disparate surfactant molecules suggest that the phase transitions are driven by rather basic physical considerations. And, indeed, similar phase transitions are found in simulations with a simple lattice model (see Fig. 12-20). The agreement between experimental phase diagrams and that of the lattice model indicates that these transitions are driven by simple space-filling constraints. That is, the

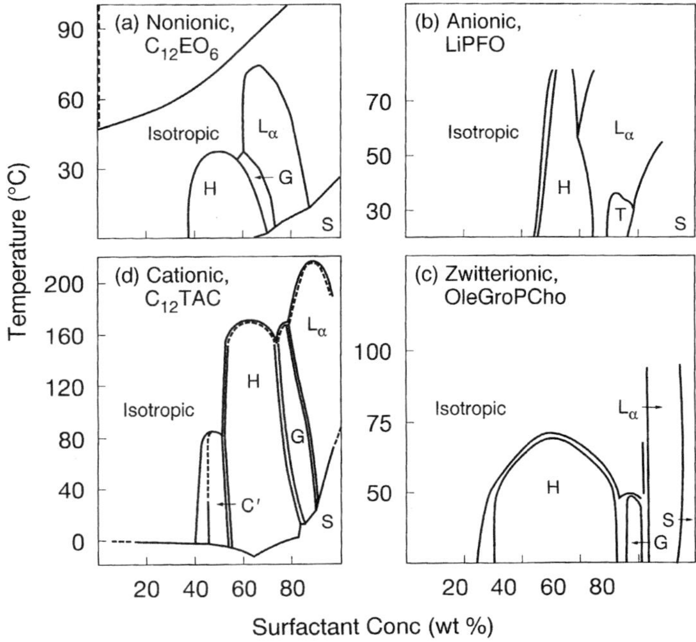

Figure 12.19 Phase diagrams of four surfactants in water. H and L_α stand for hexagonal and lamellar, respectively; G and T are cubic gyroid and tetragonal mesh phases, respectively; C' is a cubic packing of spherical micelles; and S stands for crystalline solid. **(a)** Hexaoxyethylene dodecyl ether ($C_{12}EO_6$) (Clerc et al. 1990). **(b)** Lithium perfluorooctanoate (Kekicheff and Tiddy 1989). **(c)** 1-Oleoyl-*sn*-glycero-3-phosphocholine (Arvidson et al. 1985). **(d)** Dodecyltrimethylammonium chloride. (From Balmbra et al., reprinted with permission from Nature, 222:1159, Copyright © 1969, MacMillan Magazines, Ltd.)

compactness of the lyophilic domains must decrease when their volume fraction increases; hence, the progression: spheres, cylinders, and lamellae, with cubic gyroid phases as a compromise between cylinders and lamellae. Comparing Fig. 12-19 to 12-20, we note also that modest differences among the surfactant phase diagrams of Fig. 12-19 can be accounted for, to some extent, by differences in the effective ratio of head size to tail size, where this ratio for the lattice model in Fig. 12-20 varies from 4/7 to 6/4. Thus, the Israelachvili concept of a "shape parameter" provides some guidance in interpreting these phase transitions.

For more extreme values of the shape parameter than those covered in Figs. 12-19 or 12-20, the phase diagram can be considerably different. For example, when the tail group becomes much larger than the head group, at even low concentrations of surfactant in water, cylindrical, rather than spherical, micelles form. If these micelles are stiff enough, then at concentrations lower than those needed to form a hexagonal phase these worm-like micelles can become nematically ordered. For example, Fig. 12-21 shows the phase diagram for CPyCl/hexanol in 0.2 M NaCl (Roux et al. 1995). This diagram shows a transition from a disordered phase of worm-like micelles to an ordered nematic phase for a total surfactant + hexanol concentration of around $\phi = 0.36$. Since the diameter d of

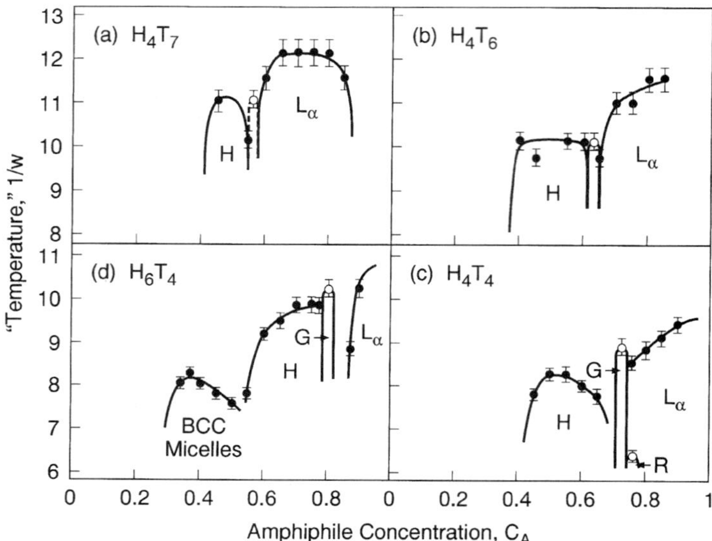

Figure 12.20   Simulated phase diagrams of **(a)** H_4T_7, **(b)** H_4T_6, **(c)** H_4T_4, and **(d)** H_6T_4 in single-site "water" on a cubic lattice. The nomenclature is the same as in Fig. 12-19; R stands for a rhombohedral-like mesh phase. The gyroid phase in **(a)** may be metastable. (From Larson 1996, with permission from EDP Sciences.)

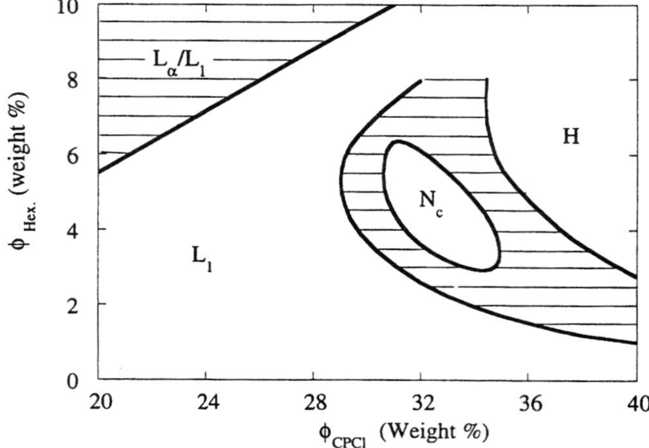

Figure 12.21   Phase diagram for the surfactants cetylpyridinium chloride and hexanol in 0.2 M NaCl at 30°C. L_1 is a disordered state of wormy micelles, N_c is the nematic state of oriented micelles, H is a state of hexagonally ordered cylindrical micelles, and L_α is a lamellar state. The hatched areas are coexistence regions. (Reprinted with permission from Roux et al., Macromolecules 28:1681. Copyright © 1995, American Chemical Society.)

the wormy micelles is around 4 nm and the persistence length λ_p is around 20 nm, the axial ratio ζ is about $2\lambda_p/d = 10$, and thus the transition to the nematic phase occurs at a value of $\phi\zeta = (0.36) \cdot 10 \approx 3.6$—considerably lower than the values of 11.4 predicted for worm-like chains, and 4.86 predicted for freely jointed ones (see Section 2.2.4).

12.4.1.2 Strut, Mesh, and Ribbon Phases

Exotic phases are found at compositions between lamellae and hexagonal cylinders (see Figs. 12-19 and 12-20). Some examples of the morphologies of these phases are shown in Fig. 12-22; these include cubic "strut" phases, tetragonal and rhombohedral "mesh" phases, and rectangular "ribbon" phases.

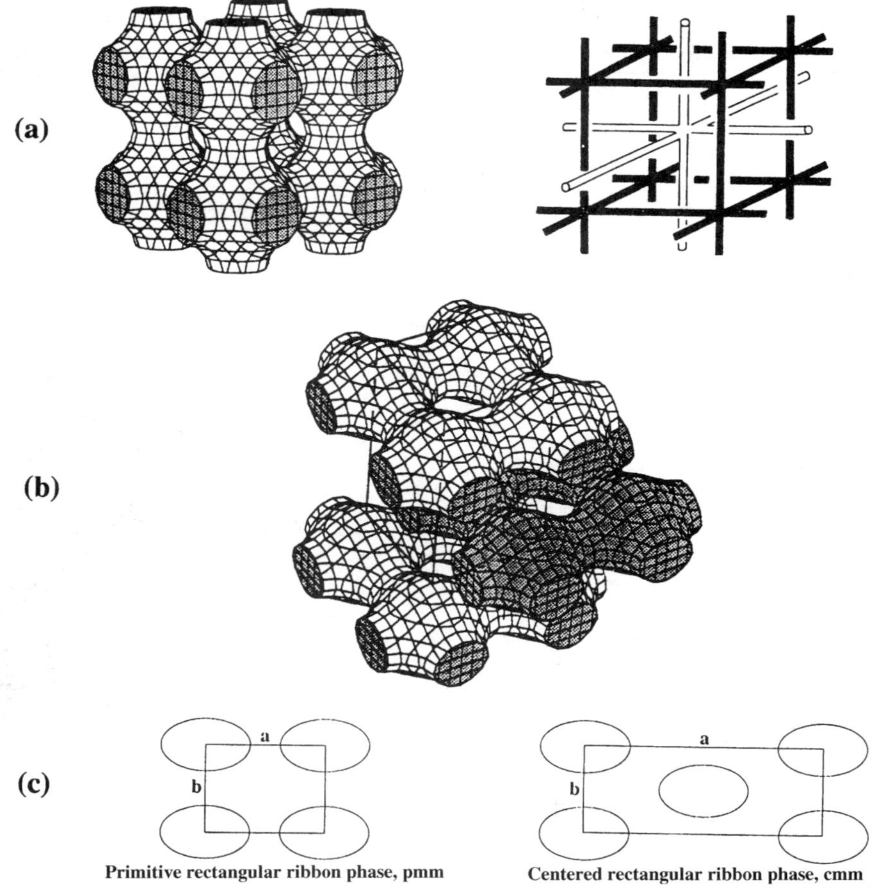

Figure 12.22 Types of phases that occur between hexagonal and lamellar phases: **(a)** *Strut phase*; the left image is the minimal "P" surface over which surfactant is draped, and the right is the topology of the two strut networks on either side of the minimal surface; **(b)** tetragonal *mesh phase*; **(c)** *ribbon phases* containing cylinders with ellipsoidal cross sections.

In the remarkable bicontinuous "strut" phases, the surface that separates hydrophilic from hydrophobic material is triply periodic and multiply connected; the material on each side of this surface forms a regular network of struts, and the two networks do not intersect each other (see Fig. 12-22a). The "struts" composing these networks are not necessarily cylindrically shaped (Ström and Anderson 1992). Figure 12-23 compares the topologies of the three most common strut phases: the P phase with space-group symmetry $Im\bar{3}m$, the D or "double diamond" phase with symmetry $Pn\bar{3}m$, and the G or "gyroid" phase with symmetry $Ia\bar{3}d$. Each of these mesophases possesses two nonintersecting networks of struts, with six struts meeting at each node in the P phase, four to a node in the D phase, and three to a node in the G phase. In type I strut phases, both strut networks are channels of hydrophobic material (tails + oil), while in type II (or inverse phases) the struts are hydrophilic material (heads + water). The surfactant, then, forms a bilayer centered on the multiply connected surface separating the networks of struts, such as, for example, the surface depicted in Fig. 12-22a. For type I phases, opposing layers of head groups are sandwiched in the interior of the bilayer with the periodic surface passing between

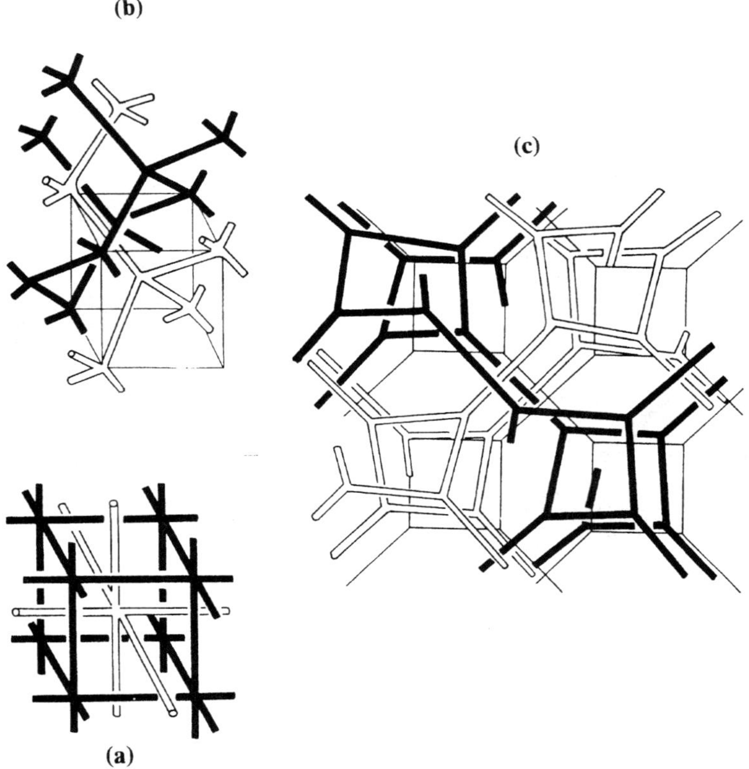

Figure 12.23   Topologies of **(a)** P phase; **(b)** D phase; and **(c)** G phase. (From Charvolin and Sadoc 1987, with permission from EDP Sciences.)

the opposing head layers, and each layer of tail groups is then embedded in one of the hydrophobic strut networks. Type II phases are the reverse of this; the tail groups are inside the bilayer next to the periodic surface, and the exterior head groups, along with any water present, comprise the hydrophilic strut networks. Thus, the networks in Fig. 12-23 could contain either hydrophobic or hydrophilic material, depending on whether they represent type I or type II phases. The periodic surfaces are thought to be *minimal surfaces*, which means that their mean curvature $H \equiv (R_1^{-1} + R_2^{-1})/2$ is zero. This is possible when $R_1 = -R_2$; that is, the surfaces are everywhere saddle-shaped!

The other phases are less exotic. The mesh phases consists of lamellae with ordered holes, while ribbon phases are deformed cylinders on a rectangular lattice (see Fig. 12-22). These phases can are usually type I phases with the tails inside the deformed cylinders or inside the hole-filled lamellae, but they could also be inverse, type II, phases. Type II mesh and ribbon phases seem not to have been reported much; type II strut phases are common for two-tailed lipids, such as those in cell membranes. In fact, type II strut phases evidently serve biological functions, since they have been found to exist in cellular structures such as the endoplasmic reticulum and the mitochondrion (Seddon 1996).

The phases that exist between hexagonal and lamellar vary from one surfactant to the next. For example, a mesh intermediate phase is formed by the perfluorooctanoate surfactant (and also occurs in the lattice simulation), while cubic phases are produced by the other three surfactants in Fig. 12-19. Also, there are surfactants whose phase behavior in water differs greatly from those of Fig. 12-19. Figure 12-24, for example, shows the phase

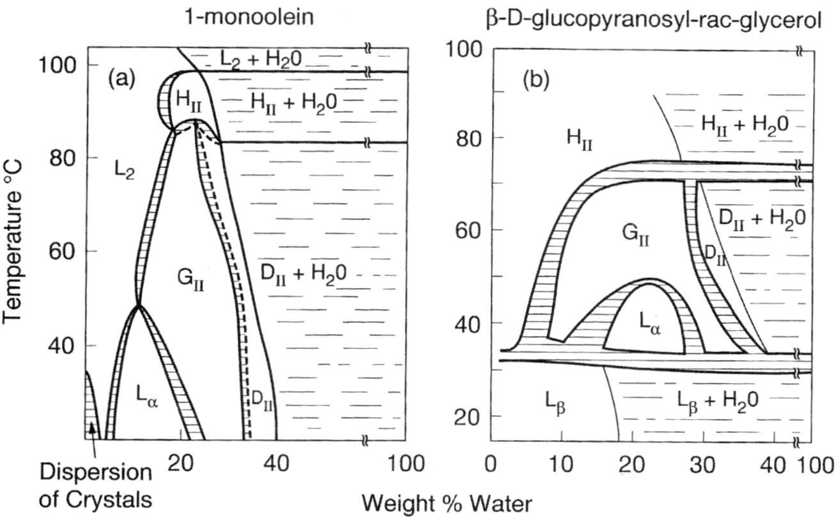

Figure 12.24 Phase diagrams for **(a)** 1-monoolein in water and **(b)** di-dodecyl alkyl-β-D-glucopyranosyl-rac-glycerol in water. "H_{II}" is the inverse hexagonal phase, G_{II} is the inverse gyroid Ia$\bar{3}$d, and "D_{II}" is the inverse double-diamond Pn$\bar{3}$m phase. In the inverse phases, the aqueous phase is inside the channels. [Part (a): Reprinted with permission from Larsson et al., Journal of Physical Chemistry 93:7304 Copyright © 1989, American Chemical Society. Part (b): Reprinted with permission from EDP Sciences.]

diagram for the single-tailed lipid 1-monoolein and the doubled-tailed lipid di-dodecyl alkyl-β-D-glucopyranosyl-rac-glycerol in water. For these surfactants, there are two type II cubic phases, the gyroid and the double-diamond, and the transitions among the phases as temperature or composition is changed do not fit the pattern of Fig. 12-19. Even more remarkable is the phase diagram for the two-tailed surfactant didodecyldimethylammonium bromide in water and styrene at 20°C (see Fig. 12-25). This system contains five distinct bicontinuous cubic phases, spread over a huge range of water concentrations, 9% to 80%! Four of them have been identified as the type II phases: G, D, P, and the Neovius phase C(P). The extent of the region of bicontinuous ordered phases is sensitive to the hydrocarbon solvent; smaller solvents that more strongly penetrate the hydrocarbon tails of the two-tailed surfactant enlarge the bicontinuous region (Maddaford and Toprakcioglu 1993; Toprakcioglu 1997).

Cubic strut phases are common in the phase diagrams of two-tailed surfactants. These surfactants have a relatively high value of the $v/a_0\ell_c$ parameter, because the volume-to-length ratio v/ℓ_c of the double tail is twice that of a single tail. A high value of $v/a_0\ell_c$ is consistent with the formation of type II bicontinuous and other inverse phases, such as the inverse hexagonal phase in Fig. 12-24.

The values of $v/a_0\ell_c$ for which one should expect bicontinuous cubic phases can be derived by an extension of the packing argument of Israelachvili et al. (Hyde 1990, 1992; Ström and Anderson 1992). Consider a small patch of area $a(0)$ on a curved surface. If this patch is displaced a small distance ξ in a direction normal to the surface, then the area $a(\xi)$ of the displaced surface becomes

$$a(\xi) = a(0)[1 + 2H\xi + K\xi^2] \tag{12-37}$$

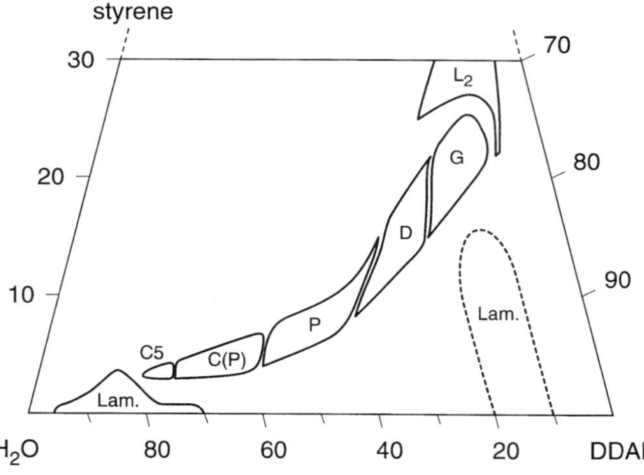

Figure 12.25 Phase diagram of didodecyldimethylammonium bromide (DDAB) in water and styrene at 20°C. The phases include an oil-rich isotropic phase L_2, lamellar phases, and five distinct cubic strut phases, including the G, D, P, C(P), and an unknown phase "C5." Above the cubic phases are regions of two- and three-phase coexistence. (From Ström and Anderson 1992, reprinted with permission from Langmuir 8:691. Copyright © 1992, American Chemical Society.)

where H and K are the mean and Gaussian curvatures of the original patch. If we integrate this expression to obtain the volume v between the original and the displaced patch, we obtain

$$v = \int_0^{\ell_c} a(\xi)\, d\xi = a(0)\ell_c[1 + H\ell_c + \frac{1}{3}K\ell_c^2] \tag{12-38}$$

Now let the original surface be the dividing surface separating head and tail groups, and let $a(0) = a_0$ be the area occupied by a head group on that surface, ℓ_c the tail length, and v the tail volume. Then Eq. (12-38) implies that

$$\frac{v}{a_0\ell_c} = 1 + H\ell_c + \frac{1}{3}K\ell_c^2 \tag{12-39}$$

From this expression, one finds that for tails enclosed in spherical micelles of radius $R = \ell_c$, one obtains $H = -1/\ell_c$, $K = 1/\ell_c^2$, and therefore $v/a_0\ell_c = 1/3$. Similarly, for cylindrical micelles, Eq. (12-39) yields $v/a_0\ell_c = 1/2$, and for lamellae it yields $v/a_0\ell_c = 1$. Thus from Eq. (12-39), the packing criteria for simple geometries, described in Section 12.2.1, are recovered.

However, Eq. (12-39) is general and can be applied to other surfaces, including minimal surfaces for which $H = 0$ (Hyde 1990, 1992). For type II minimal surfaces, the ends of the tails reside on the minimal surface, and the junctions between heads and tails lie on a surface displaced a distance ℓ_c away from the minimal surface. Hence, from Eq. (12-37) we obtain

$$a_0 = a(\ell_c) = a(0)\left[1 + K\ell_c^2\right] \tag{12-40}$$

and from Eq. (12-38) we have

$$\frac{v}{a(0)\ell_c} = 1 + \frac{1}{3}K\ell_c^2 \tag{12-41}$$

There is a remarkable relationship between the average Gaussian curvature of a surface and its *topology* as quantified by the *genus,* which is the number of holes in a multiply connected surface. The relationship, the *Gauss–Bonnet theorem,* when applied to a surface of constant Gaussian curvature, is

$$g = 1 - \frac{KA}{4\pi} \tag{12-42}$$

where g is the genus of a surface of area A with Gaussian curvature K. Let A be the area of a minimal surface contained in a single unit cell of a cubic phase, and define a dimensionless surface area of the minimal surface, $\sigma \equiv A/V^{2/3}$, where V is the volume of the unit cell of the cubic phase. Now A equals the number n of surfactant molecules in a unit cell times $a(0)/2$, the area on the minimal surface per surfactant. (The factor of two enters because the minimal surface is enclosed in a *bilayer*; that is, the surface is bounded by two monolayers, and the patch $a(0)$ is therefore shared by two surfactant molecules.) Similarly, V equals n times the volume v/Φ occupied by a surfactant molecule, where Φ is the volume fraction of the unit cell occupied by tails, and v is the tail volume, as usual. Thus

$$\frac{V}{A} = \frac{2v}{a(0)\Phi} \tag{12-43}$$

Combining Eqs. (12-40)–(12-43), Hyde derived an equation relating the volume fraction of tails Φ to the shape parameter $v/a_0\ell_c$ for type II cubic phases:

$$\Phi = 2\frac{v}{a_0\ell_c}\sqrt{\frac{12\sigma^3}{2\pi(2g-2)}\frac{\frac{v}{a_0\ell_c}-1}{\left(3\frac{v}{a_0\ell_c}-1\right)^3}} \tag{12-44}$$

[Equation (12-44) corrects a factor-of-two omission in Hyde's formula.] Engblom and Hyde (1995) then define the *homogeneity index* HI as

$$\text{HI} \equiv \sqrt{\frac{\sigma^3}{2\pi(2g-2)}} \tag{12-45}$$

The quantity $\chi \equiv 2-2g$ is known as the *Euler characteristic*. Unlike the genus, the Euler characteristic is proportional to the volume of the unit cell chosen. Theoretical values for the genus, Euler characteristic, and homogeneity index for the conventional unit cells of some periodic minimal surfaces are given in Table 12-2. Note that the homogeneity indices for the low-genus unit cells are quite close to the value 3/4. This value would be obtained exactly if the Gaussian curvature K of the minimal surface were completely uniform over the whole minimal surface (Engblom and Hyde 1995).

Since the G phase has the highest homogeneity index, it can accommodate the highest surface area per unit volume on which the surfactant layers can assemble. Hence, the G phase is predicted to have the highest tail volume fraction for a given value of $v/a_0\ell_c$, followed by the D phase, the I-WP phase, the P phase, and the C(P) phase. This progression is consistent with experimental observations: The G phase generally appears at higher surfactant concentrations than do the D, P, and C(P) phases (see Figs. 12-24 and 12-25, for example). For apparently unknown reasons, the I-WP phase has not been observed experimentally. Free energy functions have been proposed for the various cubic strut phases by Ström and Anderson (1992) and Turner et al. (1992). While these theories differ in detail, they both predict the progression G $\to$ D $\to$ P with increasing water concentration. Indeed, this progression is insensitive to the details of the free energy model, since it arises from the relative magnitudes of the "homogeneity index," which dictates the relative ability of each phase to accommodate added water.

For single-tailed surfactants, the parameter $v/a_0\ell_c$ is often less than unity, and type I

TABLE 12-2
Characteristics of Periodic Minimal Surfaces (Conventional Unit Cell)

Minimal Surface	Space Group	Genus g	Euler Characteristic χ	Homogeneity Index HI
G	Ia$\bar{3}$d	5	-8	0.7665
D	Pn$\bar{3}$m	2	-2	0.7498
I-WP	Im$\bar{3}$m	4	-6	0.7425
P	Im$\bar{3}$m	3	-4	0.7163
C(P)	Im$\bar{3}$m	9	-16	0.6640

cubic phases are then expected. Arguments by Hyde analogous to those used to obtain Eq. (12-44) yield the analogous formula for type I cubic phases:

$$\Phi = 2\frac{v}{a_0 \ell_c}\, \text{HI}\, \frac{\frac{1}{3}\left(-1 + 2\frac{v}{a_0 \ell_c}\right)^2}{\left(\frac{-1}{3} + \frac{v}{a_0 \ell_c}\right)^3} \tag{12-46}$$

which corrects Eq. (28) of Hyde (1990), and HI is defined in Eq. (12-45). This formula predicts that type I cubic phases will appear for $1/2 < v/a_0\ell_c \lesssim 2/3$; i.e., between cylinders and lamellae, as observed. If we approximate HI by its approximate value 3/4, then Eq. (12-46) can be inverted to give (Engblom and Hyde 1995)

$$\frac{v}{a\ell_c} = \frac{2 - n - n^2}{3(1 - n^2)} \tag{12-47}$$

where

$$n \equiv \cos\,\Delta - \sqrt{3}\sin\,\Delta, \qquad \Delta \equiv \cos^{-1}(1 - \Phi)$$

An equation analogous to Eq. (12-46) for mesh phases predicts that they can occur in this same range of values for $v/a_0\ell_c$ as the strut phases, namely $v/a_0\ell_c = \frac{1}{2}-\frac{2}{3}$ (Hyde 1990). Mesh phases have tetragonal or rhombohedral symmetry, and hence for each there are two independent spacings that define their unit cells, these being determined by (a) the spacing between holes within a layer and (b) the interlayer spacing. Because of this extra degree of freedom, mesh phases can, in principle, accommodate a wider range of surfactant volume fractions for a given value of $v/a_0\ell_c$ than can the cubic phases. The free energy trade-offs involved in determining whether mesh or strut phases should be observed for a given surfactant have not yet been explored.

12.4.2 Rheology

12.4.2.1 Nematic Phases

In the nematic phase (N_c shown in Fig. 12-21), Roux and coworkers (Roux et al. 1995; Berret et al. 1995) have observed rheological behavior and microscopic flow-induced "band" textures that are qualitatively identical to that of tumbling lyotropic nematic polymers! Figure 12-26, for example, shows that the time-dependent normalized shear stress after shear-flow reversal plotted against shear strain shows multiple oscillations and strain scaling qualitatively identical to that observed in nematic polymers (compare Fig. 12-26 with Fig. 11-14). From the period of the oscillation, one finds the "tumbling parameter" to be $\lambda = 0.99$, near the boundary to flow-aligning behavior. It would appear that the rheology and flow-induced textures of lyotropic surfactant-based nematics can be understood using theories similar to those described in Chapter 11.

12.4.2.2 Hexagonal Phases

Richtering and coworkers (Richtering et al. 1994; Linemann et al. 1995) have have carried out light-scattering studies and rheological studies of hexagonal surfactant phases formed

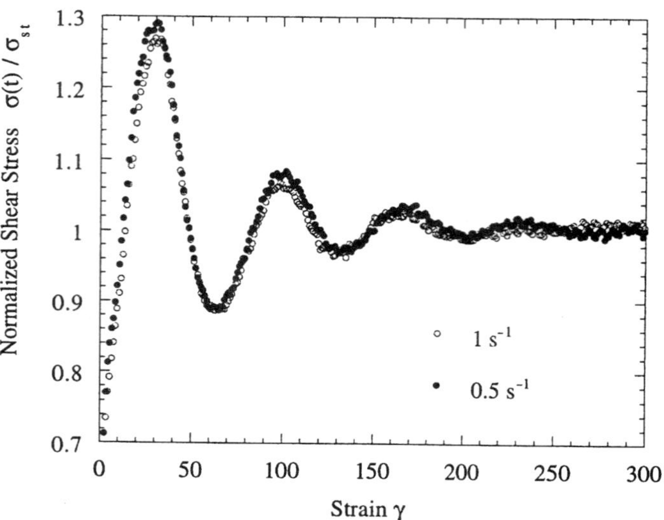

Figure 12.26   Dependence of the shear stress normalized by its steady-state value on the shear strain at two different shear rates for a nematic solution of wormlike micelles of CPyCl and hexanol at a volume fraction of CpyCl + hexanol of 0.36 in 0.2 M NaCl. (From Berret et al. 1995, with permission from EDP Sciences.)

by a nonionic surfactant with a branched head group. During start-up of shearing they have observed unusual rheological and orientational transients, including the appearance of what they interpret to be a "log-rolling" orientation of the cylinders in the vorticity direction (see Section 11.5.2), after which the cylinders apparently orient toward the flow direction. The final orientation appears to be similar to that observed in hexagonal phases of block copolymers. The dynamic moduli G' and G'' show nonterminal power-law behavior at low frequencies, also similar to that observed in block copolymers (see Chapter 13).

12.4.2.3 Smectic Phases

In principle, the rheological properties of lyotropic smectic (or lamellar) phases should be similar to those of small-molecule thermotropic smectics, discussed in Chapter 10. However, there are some important differences. First, if the volume fraction of solvent is high, then the splay elastic constant K_1, and especially the layer compression modulus B, should be much lower than in thermotropic systems, since the former have many fewer layers per unit sample thickness than the latter, for a given molecular length. In addition, since lyotropic smectics contain at least two components (surfactant and solvent), there are in principle a couple of other elastic constants, associated with gradients in composition and with coupling of these gradients to layer compression (Brochard and de Gennes 1975; Nallet et al. 1989; Lu and Cates 1994).

The rheology of lyotropic smectic phases under shear has some similarities, as well as some differences, with respect to that of thermotropic smectics discussed in Section 10.4. In particular, the shear viscosity η shows a power-law shear-thinning region as a function of shear rate or shear stress (Roux et al. 1993; Bergenholtz and Wagner 1996), followed in at least some cases by a jump to a high shear rate state in which the viscosity is independent of shear rate (see Fig. 12-27). In the shear-thinning regime, the power-law index of the shear stress $\sigma \propto \dot{\gamma}^n$ is around $n \approx 0.2$ for the system sodium dodecyl sulfate/pentanol/dodecane/water studied by Roux et al. (1993). While the power-law regime and the high-shear-rate Newtonian regime are also observed in thermotropic smectics (see Section 10.4.6), the power-law exponent is larger ($n \approx 0.5$), in thermotropic 8CB, and the lyotropic system has a low-shear-rate Newtonian region that seems to be absent in the thermotropic smectics (compare Fig. 12-27 with Fig. 10-36).

Diat et al. (1993) have determined the structure of the lyotropic smectic phase in the three flow regimes shown in Fig. 12-27 using optical microscopy and neutron scattering. The orientation diagram is shown in Fig. 12-28. At low shear rates, there is a partially disordered, presumably defect-ridden, structure in which the lamellae orient predominantly in the "c" orientation (see Section 10.4.5) with the lamellar layers parallel to the shearing surfaces. At intermediate shear rates, in the shear-thinning region, they observe the breakdown of the lamellae into "onions"—that is, multilamellar vesicles (MLVs) (Roux et al. 1993; Roux 1997). This contrasts with the thermotropic smectic 8CB, for which the structure in the shear-thinning region appears to be multilamellar cylinders, or "jelly rolls" (see Section 10.4.8). At high shear rates, where the viscosity is again Newtonian, the onions disappear and are replaced by layers with orientation "c," the same as at low shear rates, except that at high shear rates the lamellar are much better aligned. In the surfactant system AOT/brine, orientation "a" forms rather than "c" in the high shear rate "Newtonian" regime (Bergenholtz and Wagner 1996), which is similar to the behavior of the thermotropic smectic 8CB.

The radius a of the onions in the intermediate shear-rate regime of lyotropic smectics depends on shear rate, scaling roughly as a $\sim \dot{\gamma}^{-1/2}$. A similar "texture size" scaling rule is found in nematics (see Section 10.2.7); there it reflects a balance of shear stress $\eta\dot{\gamma}$ against Frank elastic stress. In smectics, the two important elastic constants B and K_1 have differing dimensional units, and the viscosity is strongly shear thinning, so it is not obvious how the appropriate scaling law should be derived. Once formed, these onions are stable for months

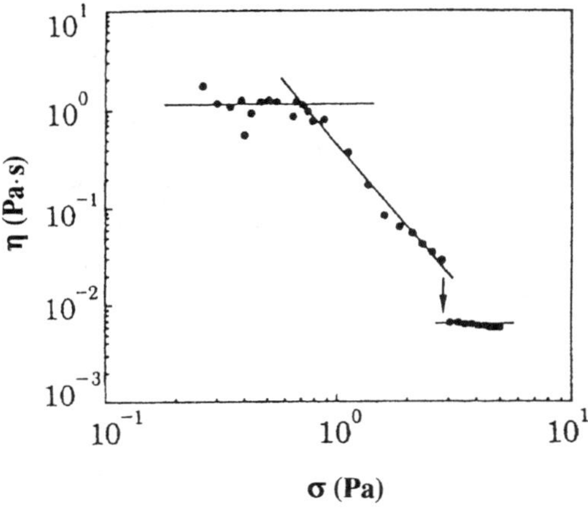

Figure 12.27   Viscosity as a function of shear stress σ for a lamellar phase of sodium dodecyl sulfate/water/pentanol mixed with 71 vol% dodecane in a stress-controlled Couette cell. (From Roux et al. 1993, with permission).

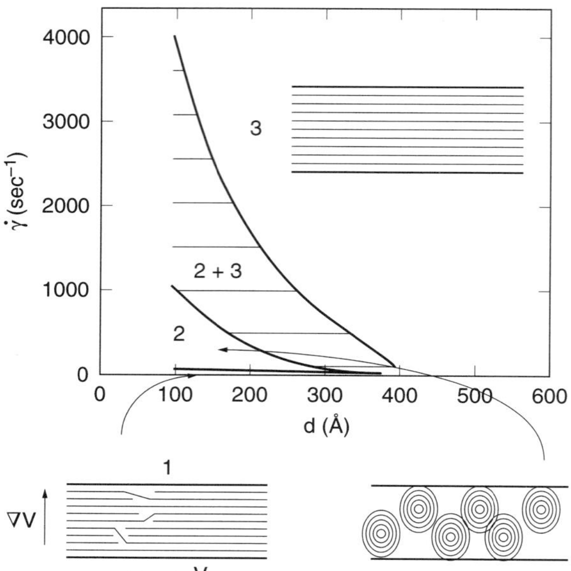

Figure 12.28 Orientation diagram of shear rate $\dot{\gamma}$ versus lamellar layer thickness d, which is controlled by the surfactant concentration, for water and sodium dodecyl-sulfate in the mass ratio 1.55 to 1 mixed with a solution of dodecane and pentanol. At low shear rates (region 1) defect-ridden lamellae appear, predominantly oriented parallel to the shearing surfaces (orientation "c"). At intermediate shear rates (region 2), "onions" form, while at high shear rates (region 3), lamellae in orientation "c" again form, but with few or no defects. (From Roux et al. 1993, with permission from EDP Sciences.)

if the flow is stopped. If the flow rate is merely decreased, however, the onions can increase gradually in size, eventually reaching a size that is independent of the shear-rate history.

The onion radius a is also strongly dependent on surfactant concentration, scaling with lamellar layer spacing d as $a \propto d^{-2}$. This might be a partly consequence of the dependence of the elastic constants on layer spacing. A larger layer spacing implies smaller elastic constants; hence at a given shear rate, the onion size at which the elastic stresses balance the viscous ones becomes smaller with larger d. At low shear rates the onion size depends significantly on the rheometer gap; this remains unexplained.

At high shear rates in some systems, the onions become large and very monodisperse in size, and they then *order* into a *macrocrystalline* packing. At rest, it is clear that the onions are not spherical, but polyhedral, because they must fill space. In the perfectly ordered macrocrystalline state, the typical shape of the space-filling onions appears to be that of the Kelvin tetrakaidecahedron, which is a model structure for liquid *foams* (see Section 9.5.1). These well-defined MLVs might be important as encapsulants in the pharmaceutical or cosmetics industries (Roux and Diat 1992).

The orientation in the high-shear-rate region 3 is nearly perfect; there are few or no defects present, while in region 1, the sample is highly defective. The highly defective state is to be expected in ordinary smectics because small eccentricities or imperfections in the gap spacing should, under shear, lead to layer dilation, and hence to the undulation instability discussed in Section 10.4.4.1. This instability readily produces masses of defects. Thus, the absence of defects in region 3 suggests that at these shear rates the shearing stresses are high enough to offset the compressional stresses produced by shearing with an inevitably somewhat uneven gap (Diat et al. 1993). If so, one can estimate that the boundary to region 3 is given by the condition that $\eta\dot{\gamma} \sim B\varepsilon$, where ε is a characteristic layer compression, given by $\varepsilon \sim \delta/h$, where δ is the unevenness in rheometer gap and h is the gap thickness.

This argument implies that as the gap thickness h decreases with fixed error δ in the gap, the boundary to region III should shift upward in proportion to $1/h$. This seems to agree with the experiments. If this argument is correct, then from the concentration-dependence of this boundary one can infer the concentration-dependence of B; it must then scale as $B \propto \phi^5$.

12.4.2.4 Cubic Strut Phases

Very few rheological data have been reported for ordered cubic phases of surfactant solutions. However, Radiman et al. (1994) have reported dynamic oscillatory data for solutions of didodecyldimethylammonium bromide (DDAB) in deuterated water and octane. Their data for type II cubic P and D phases are shown in Fig. 12-29. They report that the samples "have the texture and appearance of clear, highly viscous, gels," which is not surprising, considering their solid, network-like structure. Note that at the highest frequencies in Fig. 12-29, G' exceeds G'' by a decade or so, implying highly elastic behavior at these frequencies. Remarkably, when tapped, these gels "ring" like a bell. Ringing has also been reported for surfactant/oil/water mixtures that are apparently organized into closely packed, three-dimensionally ordered droplets (Gradzielski et al. 1990). Ringing implies very low viscous damping at the frequency of ringing, around 10^3 Hz (or $2\pi \times 10^3$ rad/sec). Thus, G' must exceed G'' at this frequency by perhaps two or three decades, which is consistent with

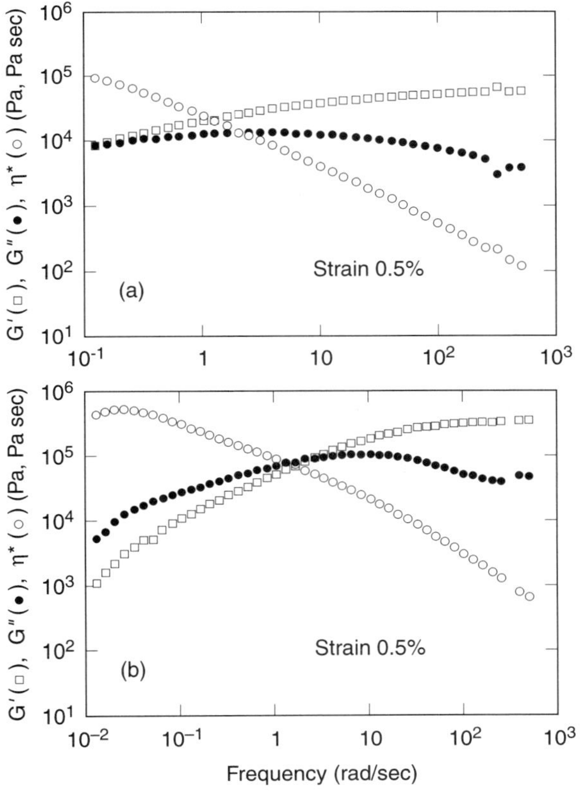

Figure 12.29 Frequency-dependence of G' ($\square$), G'' ($\bullet$), and η^* ($\bigcirc$) for **(a)** the P phase and **(b)** the D phase of a mixture of didodecyldimethyl-lammonium bromide (DDAB) in D$_2$O and octane. (Reprinted with permission from Radiman et al., Langmuir 10:61. Copyright © 1994, American Chemical Society.)

extrapolations of the data of Fig. 12-29 to higher frequencies. Since the P and D phases are ordered, one expects elastic behavior to dominate at low frequencies also; this low-frequency elastic behavior presumably exists below the range covered in Fig. 12-29. In the frequency range 10^{-2}–10^{2} sec^{-1}, considerable dissipation occurs, which Radiman et al. attribute to transport of surfactant molecules along the bilayer, allowing partial relaxation of shear-induced deformation of the cubic-phase structure.

If these cubic-phase gels are heated, they melt into isotropic liquid-like phases. As the melting point is approached, the storage and loss moduli become equal to each other and follow a power law in frequency, reminiscent of critical gels (see Section 5.3). This suggests that the melting transition for these cubic phases is not far from being second order.

At strains γ_0 above about 1–5% the moduli decrease as nonlinearities set in, and at high γ_0 the moduli are inversely proportional to γ_0. Thus, the stress is roughly constant above a critical stress σ_c. This suggests a "yield" phenomenon, in which the cubic phase ruptures at high strains and deformation is concentrated along slip planes. A model along these lines has been proposed by Jones and McLeish (1995). The possibility of aligning these phases by using steady or large-amplitude oscillatory shearing seems not yet to have been considered (see Section 13.3).

12.5 SUMMARY

The structural units of surfactant-containing liquids are self-assembled aggregates, such as spherical or cylindrical micelles, or bilayers. These supramolecular structural units can then further self assemble into ordered phases, with cubic, hexagonal, smectic, or other symmetry. Consequently, the structural and flow properties of such liquids are amazingly rich. Laws of mass action, combined with geometric packing arguments, allow rationalization, if not prediction, of the phase behavior of many surfactant solutions.

The flow properties of disordered micellar phases are now reasonably well understood. For spherical micelles the viscosity can be estimated from modified hard-sphere-suspension theories, while for disordered "semidilute" cylindrical micelles the Cates theory of entangled "living polymers" provides at least a good starting point, and in some cases nearly quantitative prediction of rheological properties.

Theories for the rheology of ordered surfactant fluids are less developed. For surfactant *nematics*, the flow and flow-orientation properties seem to be very similar to those of lyotropic (solvent-containing) tumbling polymer nematics. Hence, for low shear rates, the Leslie–Ericksen theory should apply to them. The rheology of surfactant *smectics* is much less understood; some of the behavior is analogous to that of thermotropic small-molecule smectics, but the former also show some different phenomena, such as a shear-induced "onion" phase. More highly ordered phases, such as bicontinuous cubic phases, have dissipative properties at modest frequencies and high-modulus "gel-like" properties at high frequencies. Theories for the flow behavior of such fluids are just beginning to be formulated.

REFERENCES

Aït-Ali A, Makhloufi R (1997). *J Rheol* 41:307.
Andelman D, Cates ME, Roux D, Safran SA (1987). *J Chem Phys* 87:7229.

Aniansson EAG, Wall SN, Almgren M, Hoffmann H, Kielmann I, Ulbricht W, Zana R, Lang J, Tondre C (1976). *J Phys Chem* 80:905.

Arvidson G, Brentel I, Khan A, Lindblom G, Fontell K (1985). *Eur J Biochem* 152:753.

Balmbra RR, Clunie JS, Goodman JF (1969). *Nature* 222:1159.

Bergenholtz J, Wagner NJ (1996). *Langmuir* 12:3122.

Berret J-F, Appell J, Porte G (1993). *Langmuir* 9:2851.

Berret J-F, Porte G, Decruppe (1997). *Phys Rev E* 55:1668.

Berret J-F, Roux DC, Porte G (1994). *J Phys II (France)* 4:1261.

Berret J-F, Roux DC, Porte G Lindner P (1995). *Europhys Lett* 32:137.

Boltenhagen P, Hu Y, Matthys EF, Pine DJ (1997). *Phys Rev Lett* 79:2359.

Brochard F, de Gennes P-G (1975). *Pramana* Suppl 1:1.

Bruinsma R, Gelbart WM, Ben-Shaul A (1992). *J Chem Phys* 96:7710.

Candau SJ, Hirsch E, Zana R, Adam M (1988). *J Colloid Interface Sci* 122:430.

Candau SJ, Hirsch E, Zana R, Delsanti M (1989). *Langmuir* 5:1225.

Candau SJ, Merikhi F, Waton G, Lemaréchal P (1990). *J Phys (Paris)* 51:977.

Candau SJ, Khatory A, Lequeux F, Kern F (1993). *J Phys IV Colloq C1* 3:197.

Care CM (1987). *J Chem Soc Faraday Trans I* 83:2905.

Cates ME (1987). *Macromolecules* 20:2289.

Cates ME (1990). *J Phys Chem* 94:371.

Cates ME (1997). In *Theoretical Challenges in the Dynamics of Complex Fluids*, McLeish TCB (ed), NATO ASI Series E: Applied Sciences, Vol 339, Kluwer, London, p 257.

Cates ME, Candau SJ (1990). *J Phys Condens Matter* 2:6869.

Cates ME, Turner MS (1990). *Europhys Lett* 11:681.

Charvolin J, Sadoc JF (1987). *J Phys (Paris)* 48:1559.

Clausen TM, Vinson PK, Minter JR, Davis HT, Talmon Y, Miller WG, (1992). *J Phys Chem* 96:474.

Clerc M (1996). *J Phys II (France)* 6:961.

Clerc M, Levelut AM, Sadoc JF (1990). *Colloid Phys C7* 51:97.

Danino D, Talmon Y, Levy H, Beinert G, Zana R (1995). *Science* 269:1420.

Decruppe JP, Cressely R, Makhloufi R, Cappelaere E (1995). *Colloid Polym Sci* 273:346.

de Gennes PG, Taupin C (1982). *J Phys Chem* 86:2294.

Desplat J-C, Care CM (1996). *Mol Phys* 87:441.

Diat O, Roux D, Nallet F (1993). *J Phys II (France)* 3:1427.

Engblom J, Hyde ST (1995). *J Phys II (France)* 5:171.

Fischer P, Holz T, Rehage H (1997). *Rheol Acta* 36:12.

Gelbart WM, Ben-Shaul A (1996). *J Phys Chem* 100:13169.

Golubović L, and Lubensky TC (1990). *Phys Rev A* 41:4343.

Gompper M, Schick M (1989). *Phys Rev Lett* 62:1647.

Gompper M, Schick M (1994). *Self-Assembling Amphiphilic Systems*, Academic Press, New York.

Gradzielski M, Hoffmann H, Oetter G (1990). *Colloid Polym Sci* 268:167.

Granek R, Cates ME (1992). *J Chem Phys* 96:4758.

Helfrich W (1973). *Z Naturforsch, Teil C* 28:693.

Herb CA, Chen LB, Sun WM (1994). In *Structure and Flow in Surfactant Solutions*, Herb CA, Prud'homme RK (eds), ACS Symposium Series 578; American Chemical Society, Washington, DC, Chapter 10.

Holm C (1994). *Cell* 77:955.

Hu Y, Matthys EF (1995). *Rheol Acta* 34:450.

Hu Y, Wang SQ, Jamieson AM (1993). *J Rheol* 37:531.

Huang C-M, Magda JJ, Larson RG, Pine D, Liu C-H (1996). *Proceedings of the XIIth*

International Congress on Rheology, Ait-Kadi A, Dealy JM, James DF, Williams MC (eds), Quebec City.

Huse DA, Leibler S, (1988). *J Phys (Paris)* 49:605.

Hyde ST (1990). *Colloid Phys C7* 51:209.

Hyde ST (1992). *Pure Appl Chem* 64:1617.

Israelachvili J (1992). *Intermolecular and Surface Forces*, 2nd ed, Academic Press, London.

Israelachvili J, Mitchell DJ, Ninham BW (1976). *J Chem Soc Faraday Trans I* 72:1525.

Jones JL, McLeish TCB (1995). *Langmuir* 11:785.

Karaborni S, Esselink K, Hilbers PAJ, Smit B, Karthäuser J, van Os NM, Zana R (1994). *Science* 266:254.

Kekicheff P, Cabane B (1987). *J Phys (Paris)* 48:1571.

Kekicheff P, Tiddy GJT (1989). *J Phys Chem* 93:2520.

Kern F, Lequeux F, Zana R, Candau SJ (1994). *Langmuir* 10:1714.

Khatory A, Lequeux F, Kern F, Candau SJ (1993). *Langmuir* 9:1456.

Kodoma IA, Ylitalo C, van Egmond JW (1997). *Rheol Acta* 36:1.

Lang J, Zana R (1987). *Surfactant Solutions*, Zana R (ed), Marcel Dekker, New York p 405 and references herein.

Larson RG (1992). *J Chem Phys* 96:7904.

Larson RG (1994). *Chem Eng Sci* 49:2833.

Larson RG (1996). *J Phys II (France)* 6:1441.

Larson RG (1997). *Curr Opin Colloid Interface Sci* 2:361.

Larson RG, Scriven LE, Davis HT (1985). *J Chem Phys* 83:2411.

Larsson K (1989). *J Phys Chem* 93:7304.

Laughlin RG (1994). *The Aqueous Phase Behavior of Surfactants*, Academic Press, New York.

Lequeux F, Candau SJ (1997). In *Theoretical Challenges in the Dynamics of Complex Fluids*, McLeish TCB (ed), NATO ASI Series E: Applied Sciences, Vol 339, Kluwer, London, p 181.

Linemann R, Läuger J, Schmidt G, Kratzat K, Richtering W (1995). *Rheol Acta* 34:440.

Liu C, Pine DJ (1996). *Phys Rev Lett* 77:2121.

Liu Y-C, Sheu EY (1996). *Phys Rev Lett* 76:700.

Lu C-YD, Cates ME (1994). *J Chem Phys* 101:5219.

Luzzati V, Spegt PA (1967). *Nature* 215:701.

Luzzati V, Tardieu A, Gulik-Krzywicki T (1968). *Nature* 217:1028.

Mackie AD, Panagiotopoulos AZ, Szleifer I (1997). *Langmuir* 13:5022.

Maddaford PJ, Toprakcioglu C (1993). *Langmuir* 9:2868.

Makhloufi R, Decruppe JP, Aït-Ali A, Cressely R (1995). *Europhys Lett* 32:253.

Missel PJ, Mazer NA, Benedek GB, Carey MC (1983). *J Phys Chem* 87:1264.

Nallet F, Roux D, Prost J (1989). *J Phys (Paris)* 50:3147.

Odell JA, Müller AJ, Keller A (1988). *Polymer* 29:1179.

Olsson U, Söderman O, Guering P (1986). *J Phys Chem* 90:5223.

Owenson B, Pratt L (1984). *J Phys Chem* 88:2905.

Porte G, Appell J (1981). *J Phys Chem* 85:2511.

Porte G, Appell J, Poggi Y (1980). *J Phys Chem* 84:3105.

Porte G, Berret J-F, Harden JL (1997). *J Phys II (France)* 7:459.

Prud'homme RK, Warr GG (1994). *Langmuir* 10:3419.

Radiman S, Toprakcioglu C, McLeish T (1994). *Langmuir* 10:61.

Rehage H, Hoffmann H (1982). *Rheol Acta* 21:561.

Rehage H, Hoffmann H (1988). *J Phys Chem* 92:4712.

Rehage H, Hoffmann H (1991). *Mol Phys* 74:933.

Richtering W, Läuger J, Linemann R (1994). *Langmuir* 10:4374.

Roux D (1997). In *Theoretical Challenges in the Dynamics of Complex Fluids*, McLeish TCB (ed), NATO ASI Series E: Applied Sciences, Vol 339, Kluwer, London, p 203.

Roux D, Diat O (1992). French Patent 92-04108.

Roux D, Nallet F, Diat O (1993). *Europhys Lett* 24:53.

Roux DC, Berret J-F, Porte G, Peuvrel-Disdier E, Lindner P (1995). *Macromolecules* 28:1681.

Safran SA (1994). *Statistical Thermodynamics of Surfaces, Interfaces, and Membranes*, Addison-Wesley, Reading, MA. In the caption to Fig. 8.3 of this reference, the expression for r inadvertently exchanges ϕ_S and ϕ_W.

Schick M, Shih WH (1986). *Phys Rev B* 34:1797.

Seddon JM (1990). *Biochim Biophys Acta* 1031:1.

Seddon JM (1996). *Ber Bunsenges Phys Chem* 100:380.

Shikata T, Kotaka T (1991). *J Non-Cryst Solids* 131-133:831.

Shikata T, Hirata H, Kotaka T (1987). *Langmuir* 3:1081.

Shikata T, Hirata H, Kotaka T (1988a). *Langmuir* 4:354.

Shikata T, Hirata H, Takatori E, Kotaka T (1988b). *J Non-Newt Fluid Mech* 28:171.

Shikata T, Hirata H, Kotaka T (1989). *Langmuir* 5:398.

Shikata T, Dahman SJ, Pearson DS (1994). *Langmuir* 10:3470.

Smit B, Hilbers PAJ, Esselink K, Rupert LAM, van Os NM, Schlijper AG (1990). *Nature* 348:624.

Spenley NA, Cates ME, McLeish TCB (1993). *Phys Rev Lett* 71:939.

Spenley NA, Yuan XF, Cates ME (1996). *J Phys II (France)* 6:551.

Ström P, Anderson DM (1992). *Langmuir* 8:691.

Talmon Y, Prager S (1978). *J Chem Phys* 69:2984.

Tanford C (1980). *The Hydrophobic Effect*, Wiley, New York.

Toprakcioglu C (1997). In *Theoretical Challenges in the Dynamics of Complex Fluids*, McLeish TCB (ed), NATO ASI Series E: Applied Sciences, Vol 339, Kluwer, London.

Turner DC, Wang Z-G, Gruner SM, Mannock DA, McElhaney RN (1992). *J Phys II (France)* 2:2039.

Turner MS, Cates ME (1992). *J Phys II (France)* 2:503.

Wang SQ, Hu YT, Jamieson AM (1994). In *Rheology of Surfactant Solutions*, Herb CA, Prud'homme RK, (eds), ACS Symposium Series 578, American Chemical Society, Washington, DC, p 278.

Watanabe K, Ferrario M, Klein ML (1988). *J Phys Chem* 92:819.

Widom B (1984). *J Chem Phys* 81:1030.

Widom B (1986). *J Chem Phys* 84:6943.

Zakin JL, Myska J, Chara Z (1996). *AIChE J* 42:3544.

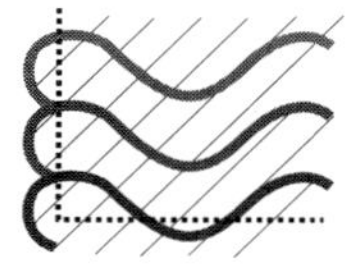

Chapter **13**

BLOCK COPOLYMERS

13.1 INTRODUCTION

Block copolymers are made of two or more chemically distinct sequences, or polymeric "blocks" (see Fig. 13-1) (Molau 1971; Aggarwal 1976; Folkes 1985; Bates and Fredrickson 1990) The simplest type of block copolymer is the *diblock*, which contains a sequence of type "A" monomers covalently bonded to a sequence of type "B" monomers. A *triblock* copolymer contains three blocks. Two of these blocks can be chemically the same, as in an ABA triblock, or all three can be different, as in an ABC triblock copolymer. Obviously, the number of possible block sequences increases rapidly with the number of blocks and the number of different types of block in the chain. One can also synthesize block copolymers with branched architecture, such as star-branched block copolymers, in which each of the arms of the star contains either the same or different block sequences (see Fig. 13-1). One or more of the blocks could also be stiff or liquid crystalline (Chiellini et al. 1994; Chen et al. 1996; Radzilowski et al. 1997; Jenekhe and Chen 1998). For a given type of block copolymer, the degree of polymerization N of the whole molecule, or the degree of polymerization N_i of one or more of the blocks, can be varied. Thus, the number of different types of block copolymers is practically endless.

Interest in block copolymers, both commercial and scientific, is stimulated by the fact that under suitable conditions, the different blocks *microseparate from each other to form microstructured materials with one or more types of domain*. Thus, block copolymers *self-assemble* into *composite* materials, with properties that can be controlled by varying the types of blocks, their volume fractions, molecular weights, and so on. A simple, but very useful, composite of this kind is formed when a block that forms a rubbery matrix (like isoprene or butadiene) is attached to a block (like styrene) that forms hard glassy domains at room temperature. The resulting microseparated structure is both *rigid*, with a high modulus, and *tough*, with a high fracture stress (Argon and Cohen 1990). The fracture toughness is believed to be especially enhanced in triblock copolymers in which there are two minority glassy blocks, one on each end of the molecule, with the rubbery block in the center. With this architecture, the rubbery middle blocks, which reside in the matrix, act as *bridges* between neighboring glassy domains, thus making a *network-like* structure (see Fig. 13-2). This design yields a rather stiff, but tough, material, useful for high-impact plastics. At elevated temperatures, the glassy domains melt, and the material can be processed and shaped into a useful form, much like other thermoplastic polymers (Folkes 1985).

(A – B)$_n$ Block Copolymer Architectures

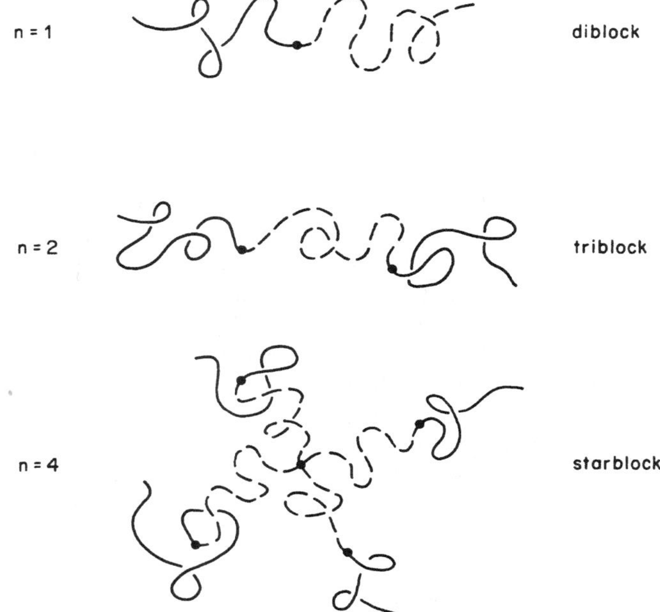

Figure 13.1. Block-copolymer architectures: diblock, triblock, and starblock, where the solid and dashed lines represent chemically different polymers that have been tethered together. (From Bates and Fredrickson 1990, with permission from the Annual Review of Physical Chemistry, Volume 41, Copyright © 1990, by Annual Reviews, Inc.)

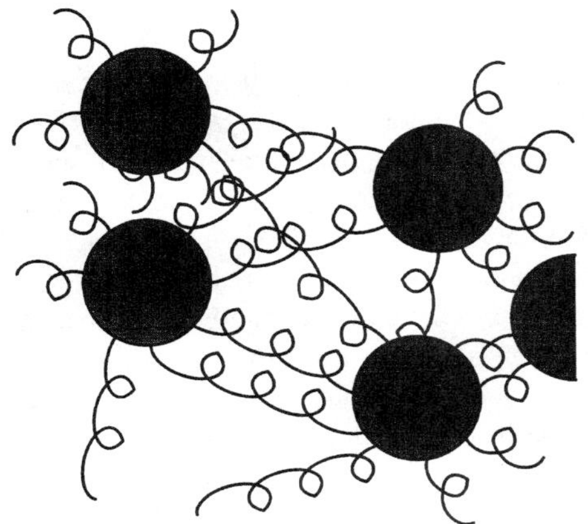

Figure 13.2 Illustration of the network morphology of a microphase-separated triblock copolymer with the glassy end blocks in hard spherical domains bridged by the rubbery center blocks.

Block copolymers are also used as *compatibilizers*: An A–B block copolymer, when added to a blend of A and B homopolymers, reduces the tendency of the two homopolymers to phase separate, and thus it stabilizes the blend. It does so by migrating to the interface between A and B regions, where the A block can happily mix in the A domain while the B block can mix in the B domain.

Block copolymers containing *water-soluble blocks* such as poly(ethylene oxide) (PEO) or polyethylene glycol (PEG) are of great interest in the biomedical–pharmaceutical arena. PEO, in particular, has good biocompatibility. A diblock or triblock of PEO and poly(propylene oxide) (PPO) will absorb onto hydrophobic surfaces in water, forming a "brush" with PEO extended into water; such block copolymers can protect surfaces from attact by the body's immune system, for example (Alexandridis 1996). Micelles formed by such block copolymers can be used for drug delivery or in cosmetics.

An exotic application of block copolymers is as optical waveguides; when the domain size is comparable to its wavelength, light is preferentially channeled into the high-index domains of a well-aligned block copolymer (Chen et al. 1995). Rod-coil block copolymers form large vesicle-like containers that can extract and purify fullerenes (Jenekhe and Chen 1998). Diblock copolymers can also be used to create fine-scale surface patterns for lithograph processing of semiconductors (Park et al. 1997).

This chapter deals almost exclusively with *neat*, or pure, diblock copolymer melts. Polymer blends are discussed in Chapter 9, micellar solutions in Chapter 12, and stabilized suspensions in Chapter 6. In the following, Section 13.2 briefly reviews the thermodynamics of block copolymers, and Section 13.3 describes the rheological properties and flow alignment of lamellae, cylinders, and sphere-forming mesophases of block copolymers. More thorough reviews of the thermodynamics and dynamics of block copolymers in the liquid state have been written by Bates and Fredrickson (1990; Fredrickson and Bates 1996). The processing of block copolymers and mechanical properties of the solid-state structures formed by them are covered in Folkes (1985). Biological applications are discussed in Alexandridis (1996).

13.2 THERMODYNAMICS OF BLOCK COPOLYMERS

13.2.1 The Flory χ Parameter

The tendency of differing blocks to microseparate from each other is quantified by Flory's "chi parameter" χ, introduced in Chapter 2. An increasing, positive value of χ implies an increasing tendency for the two chemically dissimilar species to segregate from each other. As discussed in Section 2.3.1.2, for a *blend* of two different homopolymers (A and B) of equal degree of polymerization $N_A = N_B$ at a 50/50 composition, the Flory–Huggins theory predicts that phase separation should occur at a critical value of $\chi_c = 2/N_A$. For a block copolymer, the different components are tethered together, and so it cannot *macroscopically* phase separate. Nevertheless, for a symmetric diblock copolymer, when χ reaches a value close to that at which the two blocks would phase separate if they weren't bonded together, namely $\chi \gtrsim 2/N_A = 4/N$, neutron scattering (Almdal et al. 1990; Rosedale et al. 1995; Stühn et al. 1992) and computer simulations (Fried and Binder 1991; Binder 1994; Larson 1994a) show that *fluctuating* A-rich and B-rich domains begin to appear, and the A and B blocks begin to stretch out so that their respective radii of gyration exceed their equilibrium values. Nevertheless, the diblock melt remains *disordered*, until a larger value of χ is reached, around $\chi = \chi_{ODT} = 5.25/N_A = 10.5/N$, according to the mean-field theory of Leibler (1980; see below). Thus, for a given temperature, the molecular weight at which a symmetric diblock A–B copolymer orders should be roughly five times larger than the molecular weight at which A and B homopolymers of equal molecular weight phase separate. This prediction has been confirmed in recent experiments of Lohse et al. (1993) for mixtures and diblocks of poly(ethylene-propylene) and atactic polypropylene. For these diblock copolymers, the molecular weight required for an ordering transition is rather high, around 200,000. For lower molecular weights, χ_{ODT} should be shifted to values higher than predicted by the Leibler theory (see below).

Typically, χ can be increased by lowering the temperature. The temperature-dependence of χ is often represented as

$$\chi = \frac{A}{T} + B \tag{13-1}$$

where T is expressed in degrees Kelvin and A and B are material-dependent constants. In the Flory–Huggins theory, $\chi k_B T$ is the interaction "energy" per monomer when chains A and B are mixed together. The tendency of two polymeric species to phase separate—or to microseparate if they are tethered together in a block copolymer—is then proportional to χN, where N is the degree of polymerization of the polymer molecule. Since the value of χ is typically obtained by fits to experimental data, all thermodynamic effects not accounted for by the ideal entropy-of-mixing term in the Flory–Huggins theory [see Eq. (2-33)] get lumped into χ. Thus, χ usually reflects entropic as well as energetic interactions (see Section 2.3.1).

The temperature dependence of χ varies considerably from one pair of monomers to another. For example, for the pair styrene/isoprene, one estimate of χ is (Lin et al. 1994)

$$\chi = \frac{26.5}{T} + 1.18 \times 10^{-2} \tag{13-1a}$$

while for ethylene-propylene/ethylethylene, Bates and coworkers (Rosedale et al. 1995) estimate

$$\chi = \frac{4.46}{T} + 2.5 \times 10^{-2} \tag{13-1b}$$

For the pair styrene/isoprene, various other dependences of χ on temperature have been reported in the literature, which give values of χ that at 100°C differ by almost a factor of two, from 0.06 to around 0.14 (see Fig. 13-3) (Lin et al. 1994; Han et al. 1995). These estimates of χ are not obtained by direct calorimetric measurements (the energetic effects are too minute to measure), but by fitting the predictions of thermodynamic theories (such as that of Leibler or Fredrickson and Helfand; see below) to x-ray or neutron scattering data for diblock copolymers or blends. The values of χ thereby obtained are only as accurate as the theories to which the data are fitted.

Furthermore, in defining χ, one must deal with the complication that the monomer volumes of the two blocks usually differ, and so one must arbitrarily choose a *reference volume, v*, of a reference chain segment. This could, for example, be the volume of one

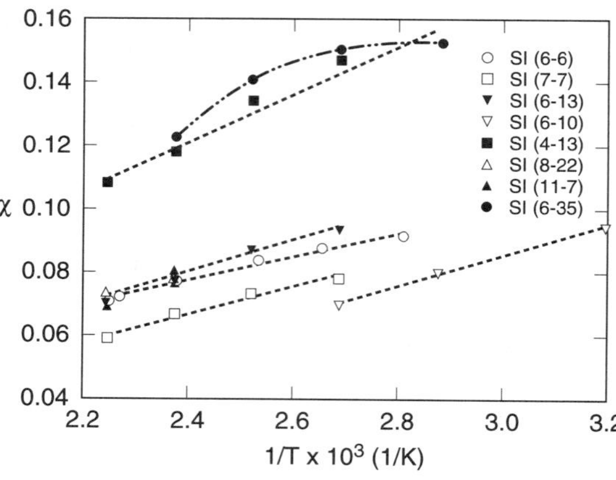

Figure 13.3 Dependence of the apparent χ parameter on inverse temperature, extracted from fits of the Fredrickson–Helfand fluctuation theory to neutron scattering data for the polystyrene–polyisoprene diblock copolymers listed. The reference volume v is 1.503×10^{-22} cm^3. Note that for most of the samples, a linear relationship between χ and $1/T$ is observed, consistent with Eq. (13-1). The apparent χ values for highly asymmetric diblocks are higher than those for more symmetric ones. (Reprinted with permission from Lin et al., Macromolecules 27:7769. Copyright © 1994, American Chemical Society.)

of the monomers, or a geometric mean of two different monomers. Equation (13-1a)—for example, for PS-PI—is based on a reference volume $v = (v_{PI} v_{PS})^{1/2} = 1.503 \times 10^{-22}$ cm^3, where v_{PI} and v_{PS} are the estimated volumes of polyisoprene and polystyrene monomers, respectively, at 80°C. To be consistent with this, the number of segments N in the whole chain must be chosen so that the volume of the whole molecule is Nv. For PEP-PEE, χ in Eq. (13-1b) was obtained by fits to the Fredrickson–Helfand theory (see below) using a temperature-dependent segment volume $v = 108 \exp[6.85 \times 10^{-4}(T - 25°C)]$. In comparing measured values of χ and N from different sources, one must therefore be careful to convert reported values to a common reference volume.

Despite this ambiguity and some differences in reported χ values for the same A/B pair, when one compares the χ values for different pairs of monomers, large variations in χ and in its temperature-dependence still stand out. These differences imply that for a given degree of polymerization, the temperature at which microseparation occurs varies enormously from one pair of monomers to another. Or, if one wishes to synthesize a block copolymer that microseparates within a desired temperature range, the degree of polymerization required will depend greatly on the chemical composition of the blocks.

There are A/B pairs whose χ parameter does not follow the simple form given in Eq. (13-1). For example, block copolymers of deuterated polystyrene/poly(n-butyl methacrylate) studied by Karis et al. (1995) form ordered, microseparated phases at both high temperatures and at low ones and are disordered at intermediate temperatures, implying that χ has a *minimum* at intermediate temperatures and that its temperature-dependence is *nonmonotonic*.

13.2.2 Microphase Separation and Ordering

In even the simplest pure diblock and triblock copolymer melts, numerous distinct microphase-separated morphologies have been observed, as depicted at the top of Fig. 13-4 (Winey et al. 1992; Förster et al. 1994; Hadjuk et al. 1994; Schulz and Bates 1996). The simplest types of domain shapes are *spheres*, *cylinders*, and *lamellae*. For a pure AB diblock

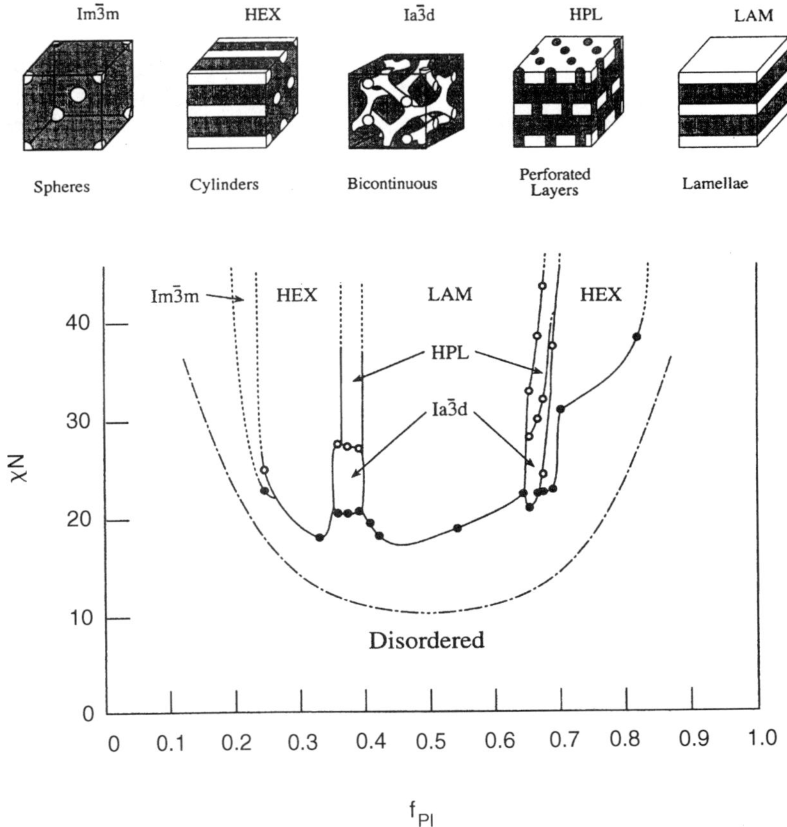

Figure 13.4   Phase diagram of a PS–PI diblock copolymer showing regions of BCC spheres (Im$\bar{3}$m), hexagonal cylinders (HEX), Ia$\bar{3}$d gyroid, hexagonally perforated lamellae (HPL), lamellar (LAM), and disordered phases; f_{PI} is the volume fraction of polyisoprene. The dot–dash line represents the mean-field order–disorder transition based on the formula $\chi = 71.4/T - 0.0857$ with reference segment volume $v = 144$ Å^3. (Reprinted with permission from Khandpur et al., Macromolecules 28:8796. Copyright © 1995, American Chemical Society.)

or ABA triblock, the domain microstructure is most strongly affected by the total volume fraction f_A of blocks A. If the A blocks are greatly in the minority volumetrically—that is, if $f_A \lesssim 0.18$–0.23—then spheres of A typically form in a matrix of B. If the volume fraction of A is somewhat greater than this, up to around 0.3–0.35, then cylinders of A usually form. For roughly equal amounts of A and B, one obtains lamellae (Hasegawa et al. 1987; Vavasour and Whitmore 1993). For still higher f_A, the inverse phases, cylinders or spheres of B in A, appear. In addition to these, there is also the simple *homogeneous state* in which the different blocks are mixed together. The homogeneous state is most often observed at high temperature when the blocks are not too long or are not too chemically distinct from one another, so that the blocks do not segregate from each other (see Fig. 13-5).

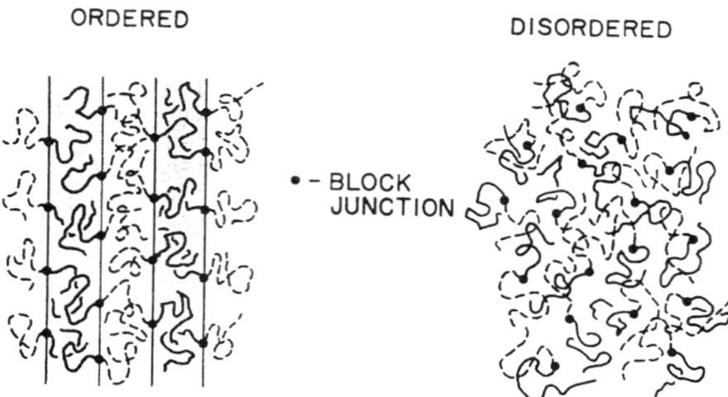

Figure 13.5 Diagram of an ordered inhomogeneous lamellar phase and a disordered homogeneous phase of a diblock copolymer. (From Bates and Fredrickson 1990, with permission.

The domains in a microphase-separated block copolymer can be spatially ordered, as depicted in Fig. 13-4, or disordered. Disordered microdomains—usually spheres, or in blends sometimes flexible cylinders ("worms") or vesicles—are common when the volume fraction of these domains is small, less than 0.10 or so (Kinning et al. 1988). Phases that are either homogeneous or contain disordered microdomains are *isotropic* states of matter. On the other hand, if the concentration of one of the blocks is greater than 10% by volume and the temperature is not too high, then *ordered*, phases typically form. Similar ordering transitions occur for similar reasons in surfactant-containing phases (see Chapter 12). The most common ordered phases are lamellae with *smectic* or one-dimensional layer-like ordering, cylinders with *hexagonal columnar* ordering, and spheres with *cubic macrocrystalline* ordering. Figure 13-6 shows transmission electron micrographs of the lamellar and hexagonal cylindrical phases for polystyrene-polyisoprene block copolymers.

The cubic packing symmetry of the spherical domains is often body-centered cubic (BCC) (Thomas et al. 1987; Almdal et al. 1993; Okamoto et al. 1994a; Chu et al. 1995; Adams et al. 1996) and, when solvent is present, sometimes face-centered cubic (FCC) (McConnell et al. 1993). The BCC packing is characteristic of spheres that interact via

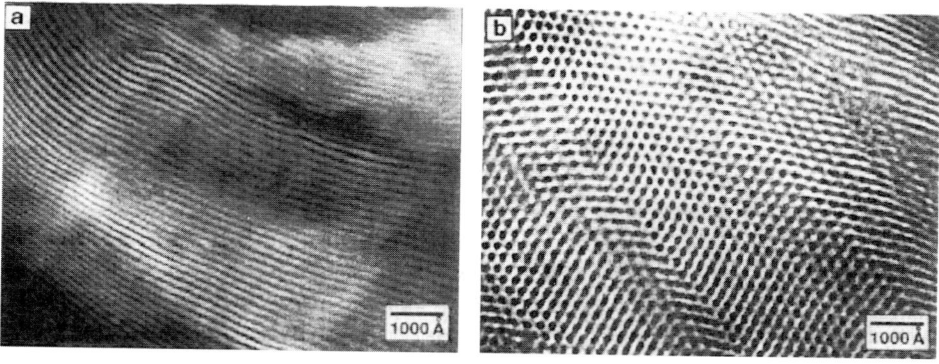

Figure 13.6 Transmission electron micrographs of thin-sectioned PS–PI block copolymer phases: **(a)** lamellar phase, **(b)** hexagonal cylindrical phase. (Reprinted with permission from Förster et al., Macromolecules 27:6922. Copyright © 1994, American Chemical Society.)

a "soft" potential with a long tail; spheres with a short-ranged interaction (such as hard spheres) order into an FCC packing. If one varies the temperature, there can be a transition at a thermodynamically precise temperature from an ordered state to a disordered one, or vice versa. Such a transition is called an *order–disorder transition*, and the temperature at which it occurs is labeled T_{ODT}. There are also transitions from one kind of ordered structure to another (Förster et al. 1994); each such transition occurs at a distinct *order–order* transition temperature, T_{OOT}.

A somewhat different nomenclature, "T_{MST}," is sometimes used instead of "T_{ODT}," where "MST" refers to "microseparation transition." This nomenclature can be misleading, however, since microseparation of one type of block from another occurs gradually as the temperature is lowered, and it is distinct from the ordering transition at which symmetry is broken and an ordered phase appears at a sharply defined temperature. This is particularly true when the composition of one of the blocks is low ($\lesssim 0.15$), so that well-segregated, but disordered, spheres (or polymeric "micelles"), and sometimes "worms" and vesicles, appear. Thus, we prefer to avoid the potentially misleading nomenclature "T_{MST}."

Often, for compositions between those for cylinders and those for lamellae (i.e., $0.35 \lesssim f_A \lesssim 0.40$), exotic ordered phases form, such as block-copolymer versions of the bicontinuous *strut* or *mesh* phases discussed in Section 12.4.1.2 (Thomas et al. 1988; Hasegawa et al. 1987; Hashimoto et al. 1992; Hamley et al. 1994; Förster et al. 1994; Hadjuk et al. 1994). The compositional ranges over which these phases occur can vary by a few percent from one type of block copolymer to another. The most common strut phase for both surfactants and diblock copolymers appears to be the "gyroid" (G) phase with Ia$\bar{3}$d symmetry, depicted in Fig. 12-23. Transmission electron-micrograph images of the gyroid phase formed by a polystyrene–polyisoprene diblock copolymer are shown in Fig. 13-7. The *double diamond* (D) phase, depicted in Fig. 12-23, might also occur if a homopolymer A is added to an AB diblock (Matsen 1995), but it is easy to mistake the G for the D phase, unless careful x-ray diffraction data are obtained (Förster et al. 1994). The P phase seems

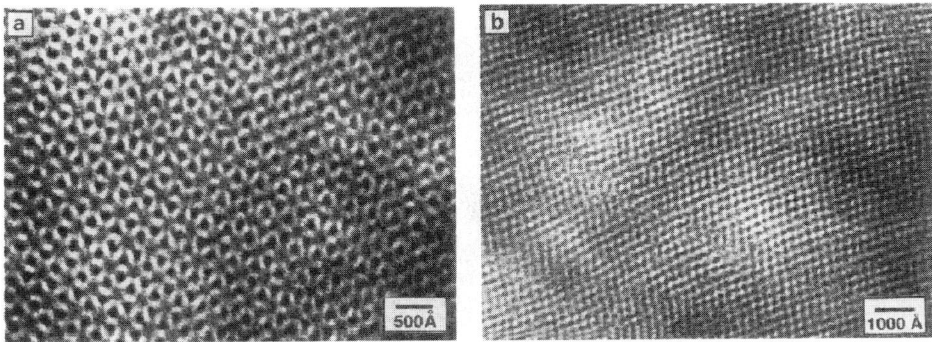

Figure 13.7   Transmission electron micrographs of a thin-sectioned PS–PI block copolymer gyroid phase: **(a)** sectioned at an orientation showing the "wagon-wheel" three-fold projection of the cubic structure; **(b)** sectioned at an orientation showing the four-fold projection. (Reprinted with permission from Förster et al., Macromolecules 27:6922. Copyright © 1994, American Chemical Society.)

not to have been observed yet in block copolymers. Mesh phases consisting of hexagonally perforated lamellae (HPL) (forming a phase with rhombohedral symmetry) have also been found in diblock copolymer melts (Hamley et al. 1994; Förster et al. 1994), including PS–PI (see Fig. 13-4). These mesh phases in block copolymers may be only metastable, however (Qi and Wang 1997). Strut and mesh phases in polystyrene–polyisoprene block copolymers are confined to the composition window $0.36 \leq f_{PI} \leq 0.39$ and the complementary window $0.65 \leq f_{PI} \leq 0.68$.

ABC-type *triblock* copolymers, in which A, B, and C are all chemically distinct from each other, can form a variety of additional structures, such as a cylindrical phase ordered on a square lattice, a tricontinuous diamond phase (Mogi et al. 1992; Nakazawa and Ohta 1993), and some exquisitely delicate structures, such as cylinders decorated by rings (see Fig. 13-8) or by strands winding around them in helical fashion (Stadler et al. 1995; Krappe et al. 1995). Recent findings have greatly expanded the known types of ordered block copolymer phases.

When an ordered phase is prepared from a disordered one by thermal quenching or by solvent casting, the long-range order expected at equilibrium is usually disrupted by *defects* (points, lines, and walls) analogous to those in liquid crystals (see Chapter 10). As a result, domains or *grains* of uniform order are usually only of order a micron or so in size (Gido et al. 1993; Hudson et al. 1995), and, on the scale of the whole sample, there is little macroscopic orientation. However, as discussed in Section 13.3, application of shearing deformation can induce macroscopic order.

13.2.3 Theories for Block-Copolymer Phase Behavior

Theories for the phase behavior and morphology of block copolymers and their mixtures with solvent or other polymers are available for the simpler cases. As reviewed by Bates and Fredrickson (1990), two general types of theory are available. *Strong-segregation theories* apply when $\chi N \gtrsim 50$, and the chemically distinct blocks are highly segregated so that the interfaces separating them are sharp compared to the domain width. *Weak-segregation theories* apply when $\chi N \lesssim 10$–15, so that compositional variations are small or modest and can be approximated by a low-amplitude sinusoidal variation (see Fig. 13-9). An analogous

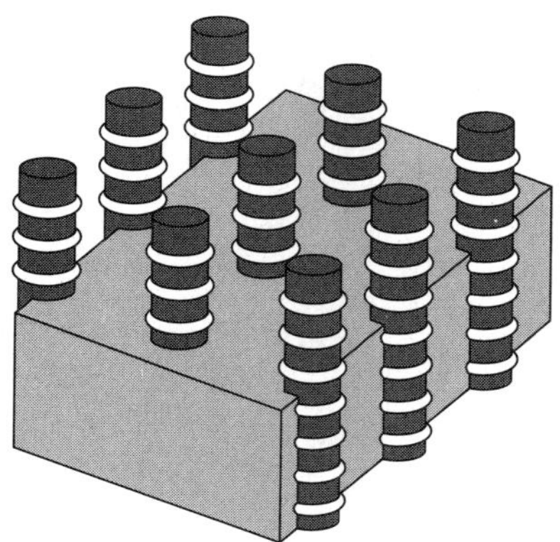

Figure 13.8 Microstructure of an ABC triblock copolymer, polystyrene–poly(ethylene-butylene)–poly(methyl methacrylate) in which the center "B" block (representing the white domains) is more incompatible with the end blocks (the black and the gray) than the end blocks are with each other. In this case, the center block forms rings that decorate the dark cylinders. In other cases, the center block forms a thin cord that winds helically around the dark cylinder. (Reprinted with permission from Stadler et al., Macromolecules 28:3080. Copyright © 1995, American Chemical Society.)

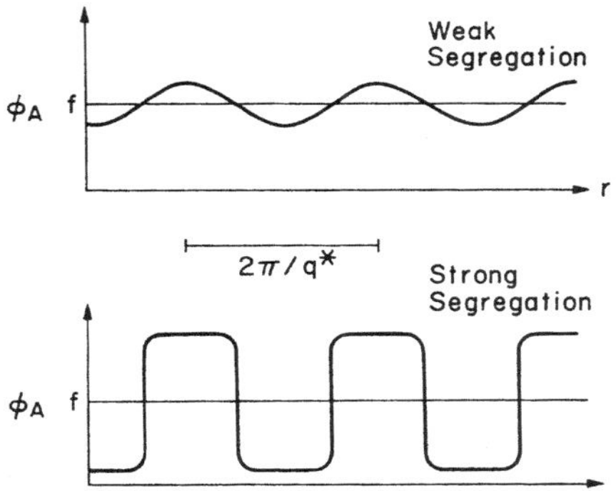

Figure 13.9 Compositional profiles of block "A" of a block copolymer whose overall volume fraction is f in the weak and strong segregation limits. (From Bates and Fredrickson 1990, with permission from the Annual Review of Physical Chemistry, Volume 41, Copyright © 1990, by Annual Reviews, Inc.)

distinction between regimes of segregation occurs in theories for spinodal decomposition, where early-time dynamics are governed by weak sinusoidal composition variations, and late-time dynamics are governed by structures with well-defined interfaces (see Fig. 9-4).

13.2.3.1 Strong Segregation

In the strong segregation limit, fluctuations are not important, and mean-field theories are accurate. Such theories for block-copolymer phase behavior were developed by Meier (1969), Leary and Williams (1970), and especially Helfand and Wasserman (1976, 1982) and Semenov (1985). Helfand and Wasserman (1982), for example, predicted that, for a polystyrene–polyisoprene diblock, the boundaries corresponding to the sequence "spheres, cylinders, lamellae, inverse cylinders, inverse spheres" would occur at polystyrene weight fractions of 0.1, 0.3, 0.65, and 0.85, in reasonable agreement with the experimental values of 0.17, 0.30, 0.65, and 0.77, respectively (Hasegawa et al. 1987). The asymmetry of this sequence reflects differences in statistical segment length and density between the styrene and isoprene blocks. It appears that cubic bicontinuous phases, such as the gyroid, do not exist in the strong segregation limit (Matsen and Bates 1996).

Strong-segregation theories predict for the domain size D, the interfacial tension Γ, and the interfacial width Δ (Helfand and Wasserman 1982; Bates and Fredrickson 1990; Wang 1994):

$$D = b^{2/3} v^{1/3} N^{2/3} \left(\frac{\Gamma}{k_B T} \right)^{1/3} \tag{13-2a}$$

$$\Gamma \sim \chi^{1/2} b v^{-1} k_B T \tag{13-2b}$$

$$\Delta = \sqrt{2/3}\, b \chi^{-1/2} \tag{13-2c}$$

where b is the statistical segment length [see Eq. (2-15)], which is assumed to be the same for the two blocks. The above formulae were derived for lamellar patterns, but the scalings with N and χ apply for other domain types. The scaling $D \propto N^{2/3}$, in particular, implies that each block is stretched beyond its random-flight size, which scales as $N^{1/2}$. This *chain stretching* in the strong-segregation limit is driven by the tendency toward a reduced interfacial area between A domains and B domains, at the expense of decreased configurational entropy.

The predictions of the strong segregation theories have been verified experimentally for diblocks when $\chi N \gtrsim 50$.

In the strong-segregation limit, the phase behavior is controlled mainly by f_A, the fraction of component A in the diblock, and, to a lesser extent, by a *conformational asymmetry parameter*, which is the ratio (Hamley et al. 1994)

$$\varepsilon = \frac{b_A{}^2/v_A}{b_B{}^2/v_B} \tag{13-3}$$

where b_i is the statistical segment length of a monomer of block i and v_i is the volume it occupies in the bulk (Förster et al. 1994; Hamley et al. 1994). In the case $\varepsilon = 1$, the blocks are *conformationally symmetric*. If $\varepsilon > 1$, then block A prefers a more expanded conformation than does block B. Hence, when plotted with f_A on the abscissa, an increasing value of ε shifts the phase diagram toward the right. For example, the lamellar window of compositions is predicted to shift from around $f_A = 0.35$–0.65 for $\varepsilon = 0$ to $f_A = 0.4$–0.7 for $\varepsilon = 1.7$ at high χN (Vavasour and Whitmore 1993; Matsen and Schick 1994). The lamellar phase at the higher values of f_A contains thin B layers and thick A ones, which is favored when $\varepsilon > 1$, because this gives the A chains more room to stretch than the B ones. This shifting with ε is analogous to the role the *shape parameter* plays in the phase behavior of surfactant solutions (see Chapter 12). For polyisoprene–polystyrene diblock copolymers, for example, polyisoprene has the more expanded conformation and $\varepsilon = (b_{PI}^2/v_{PI})/(b_{PS}^2/v_{PS}) = 1.5$ (Khandpur et al. 1995).

Using strong-segregation theory, Semenov (1985), followed by Milner et al. (1988), have predicted the typical configuration of a polymer block as it extends from the interface into the interior of a lamellar domain. The calculation method can be applied to other interesting problems such as the statistical mechanics of a "brush" of polymer grafted onto a flat or slightly curved surface. Such problems arise when grafted polymers are used for steric stabilization of colloidal suspensions (see Chapter 7).

13.2.3.2 Weak Segregation

In the weak-segregation limit, the phase behavior is influenced not only by f_A and by the conformational asymmetry parameter ε, but also by χN, and, when fluctuation effects are accounted for, by N. The simplest theory for the weak-segregation limit is the *mean-field theory* of Leibler (1980), which neglects fluctuation effects on the free energy. A mean-field theory that encompasses both the weak and strong segregation regimes has been developed by Vavasour and Whitmore (1993) and by Matsen and Schick (1994). It reduces to the Helfand–Wasserman theory at high χN and to the Leibler theory at low χN. Figure 13-10 shows the phase diagram predicted from this theory by Matsen and Bates (1996) for a conformationally symmetric diblock, $\varepsilon = 1$. Within this mean-field theory, the phase behavior is controlled by f and the product χN. The ordered phases appear at values of $\chi N > (\chi N)_s = 10.5$, where $(\chi N)_s$ is the *spinodal* value of χN when $f = 0.5$. Note that the lamellar phase is predicted to occupy a range of compositions f that shrinks to a point as χN approaches 10.5. Flanking the lamellar range are zones of gyroid and of hexagonally ordered cylinders. Flanking these are zones of BCC spherical domains. The transition from the homogeneous phase to the lamellar phase with increasing χN at $f = 0.5$ and $\chi N = 10.5$ in the mean-field theory is a *second-order transition*. Thus, the

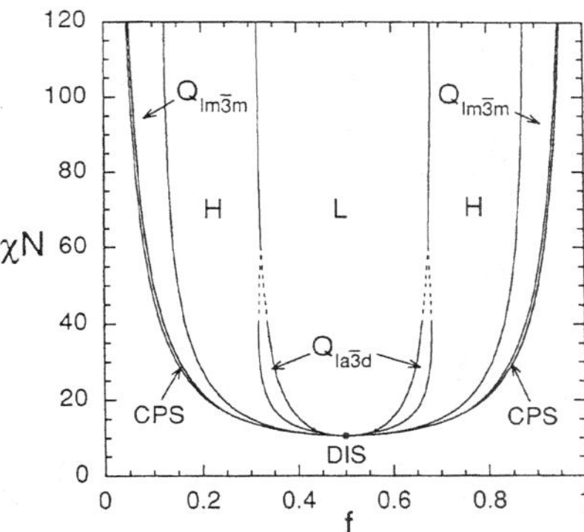

Figure 13.10 Phase diagram for a conformationally symmetric diblock copolymer predicted by the mean-field theory, showing the regions where the equilibrium phases are disordered (DIS), lamellar (L), gyroid ($Q_{Ia\bar{3}d}$), hexagonal cylindrical (H), BCC cubic ($Q_{Im\bar{3}m}$), and close-packed spheres (CPS, which is either face-centered cubic or hexagonally close-packed). (Reprinted with permission from Matsen and Bates, Macromolecules 29:1091. Copyright © 1996, American Chemical Society.)

length scale of *critical fluctuations* is predicted to grow without bound as this transition point is approached, and the scattering function $S(k)$ is predicted to become singular.

Although the gross shape of the phase envelope predicted by the mean-field theory, as well as the regions of lamellar and hexagonal phases, are more or less in agreement with experiments on diblock copolymers (see Fig. 13-4), the predictions of the theory near the critical point at $f = 0.5$, $\chi N = 10.5$ are incorrect. Fredrickson and Helfand (1987) showed that the second-order transition predicted by the mean-field theory is corrected to a *first-order* transition when the effects of fluctuations on the free energy are accounted for using a so-called Brazovskii Hamiltonian (Brazovskii 1975).

The phase behavior predicted by the Fredrickson–Helfand (FH) fluctuation theory depends not only on f and χN, but also on the molecular weight of the block copolymer. This molecular-weight dependence enters the theory in the form of a parameter $\overline{N} \equiv Nb^6v^{-2}$, where again N is the number of statistical segments, each of length b and volume v, in the chain. If small-molecule solvent is present, its volume fraction multiplies the expression for $\overline{N}$. Thus, $\overline{N}^{1/2}$ is roughly the volume *pervaded* by the molecule ($N^{3/2}b^3$) divided by the volume occupied by it (Nv); in a dense melt it is roughly the number of other molecules that can fit into the open spaces in the volume pervaded by a single coil, and therefore the number that are in contact with it. The greater the value of $\overline{N}$, the greater the number of other molecules that a single molecule contacts, and the closer the behavior of the melt is to mean-field predictions. Using Eqs. (3-64a) and (3-64b) from Chapter 3, along with $N_e = M_e/M_0$, $\langle R^2 \rangle_0 = b^2N$, and $\rho = M_0/vN_A$, we find that $\overline{N}_e = N_eb^6v^{-2} = 375$. Thus $\overline{N}/375$ is the number of entanglements per molecule. Hence, the effects of fluctuations and those of entanglements in a block copolymer are linked together, and the more mean-field-like the thermodynamics are, the more the dynamics are dominated by entanglement effects. For polystyrene–polyisoprene block copolymers, for example, $\overline{N} = 10^4$ corresponds to a molecular weight of around 60,000 (using $v = 1.44 \times 10^{-22}$cm^3, $b \approx 8$Å, and a segment mass of around 74). As a rough rule of thumb, for flexible hydrocarbon melts, $\overline{N}$ is roughly a factor of 10 or so greater than the number of statistical segments N, and a factor of roughly 10 less than the molecular weight M in daltons.

The weak-segregation phase diagram predicted by the FH theory for $\overline{N} = 10^4$ is compared to the Leibler mean-field diagram in Fig. 13-11. In the FH theory, at $f = 0.5$, the ordering transition to a lamellar phase occurs when

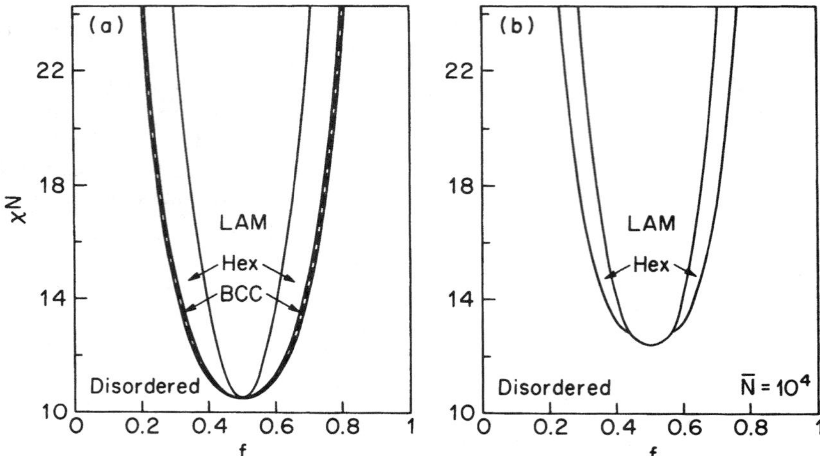

Figure 13.11   Phase diagram for a diblock copolymer in the weak-segregation limit predicted by **(a)** the Leibler mean-field theory and **(b)** the Fredrickson–Helfand fluctuation theory. (From Bates et al., reprinted with permission from J. Chem. Phys. 92:6255, Copyright © 1990, American Institute of Physics.)

$$(\chi N)_{\mathrm{ODT}} = 10.5 + 41\overline{N}^{-1/3} \tag{13-4}$$

Thus, there is an increasing upward shift in $(\chi N)_{\mathrm{ODT}}$ from the Leibler value of 10.5 as the molecular weight *decreases*; the Leibler result is recovered in the limit of infinite molecular weight. Since the FH theory is a perturbative one that assumes the shift in $(\chi N)_{\mathrm{ODT}}$ to be small, the theory should work best for high-molecular-weight polymers ($\overline{N} \gtrsim 10^4$) and small values of χ ($\chi \lesssim 0.01$).

The FH theory predicts that there is a window of compositions in which there is a direct transition from the disordered state to the lamellar state, rather than just the single point at $f = 0.5$ in the Leibler theory. The FH theory also predicts that the transition to BCC spheres only occurs at highly asymmetric values of f and high values of $\overline{N}(> 10^4)$, and the thin zone of BCC phase that in the Leibler theory exists even for f near 0.5 is suppressed in the FH theory. The theory also predicts that fluctuations will produce deviations in the scattering function $S(k)$ from the mean-field theory so that $S(k)$ is non-singular near the transition. All these features of the FH theory are in accord with experiments. Rosedale et al. (1995) show using three different polyolefin block copolymers that the range of values of χN over which non-mean-field fluctuation effects appear in the scattering (and in the rheology; see below) shrinks with increasing $\overline{N}$, in accord with theory. It appears that for $f \approx 0.5$, these fluctuation corrections to the mean-field theory for the disordered state become pronounced when χN exceeds the Leibler critical value, 10.5. Thus, fluctuations have profound effects when χN is in the range (Rosedale et al. 1995)

$$10.5 \lesssim \chi N \lesssim (\chi N)_{\mathrm{ODT}} \qquad \text{(fluctuation--dominated regime)} \tag{13-5}$$

As $\overline{N}$ decreases, $(\chi N)_{\mathrm{ODT}}$ exceeds 10.5 to a greater degree, and there is an increasing range of χ values (and hence of temperatures) over which fluctuation effects are important. This range of χ values corresponds to a range of temperatures that can be as small as $\sim 10°C$ to as large as $\sim 75°C$, depending on the polymer.

The phase behavior of symmetric ABA *triblock* copolymers is similar to that of diblocks, except that the value of χN at which ordering occurs is a little less than twice that for the corresponding diblock. This can be understood very simply by considering the symmetric triblock to be two AB diblocks (each of length N/2) stitched together "tail to tail" at their "B" ends. The phase behavior and morphology is almost the same as it would be if the chains weren't stitched together except for a bit of lost entropy that makes the disordered state a little less stable than it would be if the chains weren't stitched together.

13.2.4 Simulations of Block-Copolymer Phase Behavior

The phase behavior and ordering transitions of diblock copolymers have been investigated to a limited extent by Monte Carlo lattice simulations. Predictions of scattering and chain stretching effects in the disordered state near the transition seem to be in reasonable agreement with experiments (Minchau et al. 1990; Binder et al. 1994; Binder 1994). Figure 13-12 shows images from such simulations of a symmetric block copolymer 24 lattice sites long, with $\overline{N} \approx 78$ (Larson 1994b) in (a) the well-ordered lamellar state, (b) the ordered state near the disordering transition, and (c) the fluctuating state on the disordered side of the transition. At high values of χN the lamellae are well-defined (see Fig. 13-12a), while at lower χN values the fluctuations create "rips" in the lamellar layers (Fig. 13-12b), which at still lower χN become so numerous that the lamellar pattern is torn apart, leaving an isotropic pattern of fluctuations (Fig. 13-12c). The dependence of $(\chi N)_{\text{ODT}}$ on $\overline{N}$ obtained from the simulations agree qualitatively with weak-segregation theories that include the effects of compositional fluctuations and chain stretching (Larson 1994b; Barrat and Fredrickson 1991). However, the maximum values of $\overline{N}$ reached in the simulations are about 5–10 times smaller than that for the shortest block copolymer studied experimentally. Hence, relative to experiments, the simulations present an exaggerated view of the effects of fluctuations.

13.3 RHEOLOGY AND SHEAR-ALIGNING OF BLOCK COPOLYMERS

The rheological properties of block copolymers depend on a large number of factors: the size and shape of the domains, the strength of the segregation, the rheological properties of the material in the domains and in the matrix, the number of blocks in the molecule, the degree of macroscopic or global alignment of any ordered structure, and others. However, a general characteristic of block copolymers with ordered domains is *nonterminal* behavior at low frequency or low shear stress, while the behavior at high frequency or shear stress remains similar to that of homopolymers. The ordered superstructure is responsible for the low-frequency nonterminal behavior, which is that of a quasi-solid even at very slow deformations. An exceptional case is that of *well-aligned lamellar* copolymers, whose low-frequency response can be almost liquid-like, as discussed below. The behavior of block copolymers at low frequencies is similar to that of the analogous small-molecule liquid-crystalline phases, discussed in Section 10.4.8.

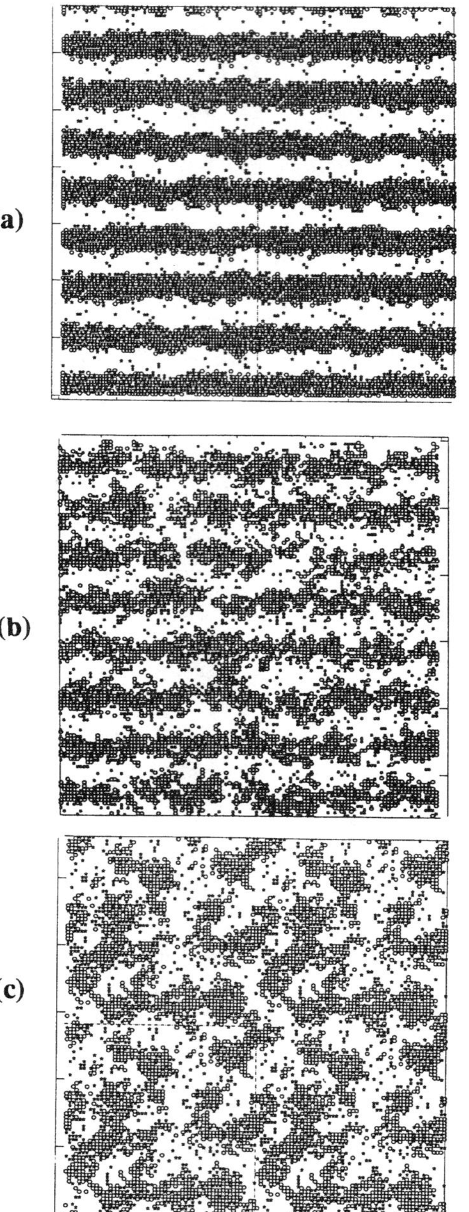

Figure 13.12 Simulated composition patterns of a lamellae-forming symmetric diblock copolymer of length 24 units on a cubic lattice **(a)** in the well-ordered state at $\chi N \approx 34$; **(b)** at $\chi N \approx 16$ just above $(\chi N)_{ODT}$; and **(c)** at $\chi N \approx 15$ just below $(\chi N)_{ODT}$ in the disordered state. The lamellar spacing in **(a)** is 31% greater than in **(b)** because of chain stretching at high χN, but the size of image **(a)** has been rescaled so that the spacing looks the same as in **(b)**. (From Larson, reprinted with permission from Molecular Simulation 13:321, Copyright © 1994, Gordon and Breach Publishers.)

The earliest rheological studies on block copolymers were carried out on triblocks with hexagonally ordered *cylindrical* microdomains (Chung and Gale 1976; Gouinlock and Porter 1977). In these triblocks, the end blocks were much smaller than the center

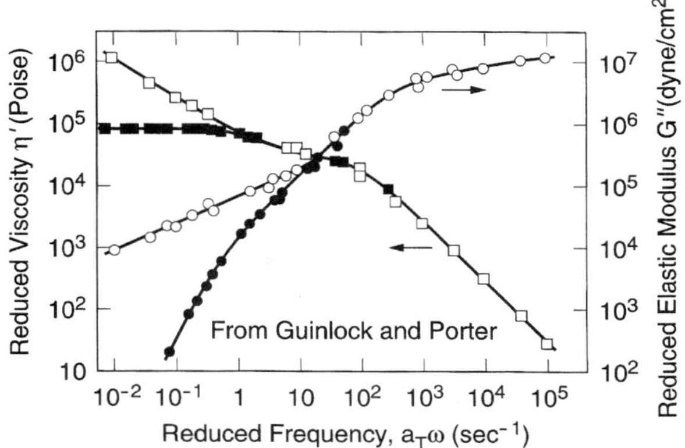

Figure 13.13 Reduced storage modulus G' and dynamic viscosity $\eta' \equiv G''/\omega$ as functions of reduced frequency $a_T\omega$ for a cylinder-forming polystyrene–polybutadiene–polystyrene triblock copolymer with block molecular weights of 7000–43,000–7000. The curves are time–temperature-shifted to a reference temperature of 138°C; the open symbols were obtained in the low-temperature ordered state; the closed symbols were obtained in the high-temperature disordered state. (From Gouinlock and Porter 1977, reprinted with permission from the Society of Plastics Engineers.)

block and were contained within the cylindrical domains. Figure 13-13, for example, shows G' and $\eta' \equiv G''/\omega$ for a cylinder-forming styrene–butadiene–styrene triblock copolymer, S7–B43–S7, both above and below the order–disorder transition temperature (Gouinlock and Porter 1977). In this figure, time–temperature superposition was used to shift the *high-frequency* data onto master curves, as discussed for homopolymers in Section 3.4.2. However, the low-frequency data for G' (and for η') do not superimpose, but fall onto two branches. One branch contains the data for the sample in the disordered state at high temperature, and the other branch contains the low-temperature data in the hexagonally ordered state. The high-temperature response of the disordered phase is similar to that of ordinary homopolymers; that is, G' and η' approach "terminal" behavior at low frequencies: $G' \propto \omega^2$ and $\eta' = $ constant. However, for the low-temperature ordered phase, neither G' nor η' show terminal behavior. Instead, both seem to follow an approximate power law corresponding to $G' \propto G'' \propto \omega^{0.4}$ (Gouinlock and Porter 1977; Morrison et al. 1990). This behavior is intermediate between that of a true viscoelastic liquid, for which $G' \propto \omega^2$, $G'' \propto \omega$, and that of a true elastic solid, for which $G' = $ const.

Behavior that is intermediate between that of a solid and that of a liquid is perhaps not surprising for a block copolymer with hexagonally ordered cylinders, since such a material has solid-like order in the two directions perpendicular to the cylinders and liquid-like order parallel to the cylinders. Similar behavior is observed in lamellar block copolymers, which has solid-like order in the direction normal to the lamellae and has liquid-like order in the other two directions. For lamellar block copolymers, solid-like behavior at low frequencies typically arises from the disrupting effect of *defects*, such as those present in smectic liquid crystals (see Section 10.4.8).

High densities of defects are frequently present in ordered block-copolymer phases, especially when the ordered state is reached by thermal quenching or solution-casting from an initially disordered state. This occurs because in the absence of an aligning field, when initially isotropic material elements break symmetry to become anisotropic, their orientations vary willy-nilly throughout the sample, yielding many defects. Thus, anisotropic order is confined to submicron-sized *domains* or *grains*, while globally the sample might be only weakly anisotropic, or even isotropic. Keller and coworkers (Keller et al. 1970; Keller and Odell 1985) showed, however, that global order can be produced by a shearing flow—in particular, in *extrusion* through a capillary. It was later shown that similar global ordering could be also induced by steady or large-amplitude oscillatory shearing between two parallel surfaces (Hadziioannou et al. 1979, 1982; Morrison et al. 1990). An advantage of this latter method is that progress toward a well-aligned state can be monitored by measuring the shearing stress during the oscillatory shearing. Another way to align block copolymers is by *roll casting*, in which a shearing flow produced by a pair of concentric rollers is imposed during solvent evaporation (Albalak and Thomas 1993).

Shear alignment of block copolymers might prove commercially useful as a means of producing maximally anisotropic properties. For example, Keller and Odell (1985) have shown that shear alignment of a polystyrene–polyisoprene–polystyrene (SIS) block copolymer with cylindrical domains produces a huge anisotropy in mechanical properties. The low-frequency modulus parallel to the cylinders can be 100 times higher than that perpendicular to them! A sheet of such material would easily roll up in one direction into a tube whose rigidity to bending would be high.

As discussed below, the quality of the alignment (and even its direction in the case of lamellar morphology), is influenced by temperature, as well as the frequency and strain amplitude of the aligning shear field. No general theory for the alignment of block-copolymer phases has yet been developed. However, studies of a number of different block-copolymer systems show that in ordered states with cylindrical domains, shear orients the cylinders *parallel* to the flow, while for lamallar microdomains, two different shear-induced orientations are commonly found, depending on alignment conditions; in both of these orientations, the flow direction lies in the plane of the lamellae.

The steady-state flow properties of block copolymers are often hard to measure. In steady shear, the shear stress often does not reach a clear steady-state value (Lyngaae-Jorgensen 1985). In cone-and-plate rheometers, steady shearing of an ordered block copolymer can result in edge fracture and flow irregularities, as might be expected when one forces a quasi-solid structure to flow (Winey et al. 1993a).

13.3.1 Lamellar Microdomains

The rheology and shear alignment of lamellar block copolymers have been studied most intensively for two chemically distinct types of diblock copolymers, namely a fully saturated polyolefin block copolymer poly(ethylenepropylene)–poly(ethylethylene) (PEP–PEE) (Koppi et al. 1992; Kannan and Kornfield 1994) and polystyrene–polyisoprene or (PS-PI) diblocks (Winey et al. 1993a, 1993b; Larson et al. 1993; Patel et al. 1995; Gupta et al. 1996; Zhang et al. 1995; Zhang and Wiesner 1995; Han et al. 1995; Pinheiro et al. 1996). PEP–PEE–PEP and PS–PI–PS *triblock* copolymers have also been investigated. (Tepe et al. 1997; Riise et al. 1995). The PEP–PEE diblock has a rather low χ parameter [see Eq. (13-1b)], and therefore ordered phases at accessible temperatures are only achieved when the molecular weight is high ($M \approx 50,000$) and when molecular entanglements are present. The two blocks of PEP–PEE have fairly similar glass transition temperatures, and one expects the rheological properties of the two blocks to be fairly similar. Thus, these blocks have a low *mechanical contrast* between them. PS–PI diblocks, on the other hand, are characterized by rather large χ values [see Eq. (13-1a)]; and even when the molecular weight is at or below the entanglement threshold ($M \approx 20,000$), ordered lamellar phases can form at accessible temperatures. Because polystyrene's glass transition temperature ($T_g \approx 100°C$) is much higher than that of polyisoprene ($T_g \approx -70°C$), the polystyrene block of PS–PI is much more viscous than the polyisoprene block, and the layers have a *high mechanical contrast*.

13.3.1.1 The Linear Modulus

Despite these significant differences between PEP–PEE and PS–PI block copolymers, the frequency dependencies of the linear moduli of the two materials are similar at low frequencies. In particular, both materials in the unaligned state show *nonterminal* behavior at low frequency. Figure 13-14 shows G' for a roughly symmetric PEP–PEE of molecular weight 50,000 at a series of temperatures both above and below the lamellar order-disorder temperature $T_{ODT} = 96°C$. As usual, the data in Fig. 13-14 have been shifted along the frequency axis to bring the high-frequency data into superposition. At high temperature, in the disordered state well above T_{ODT}, G' shows the ordinary terminal behavior at low frequency, namely $G' \propto \omega^2$. As T is reduced toward T_{ODT}, an anomalous low-frequency contribution to G' appears, which becomes more pronounced as T_{ODT} is approached more closely. This "anomaly" will be discussed in more detailed below. With further cooling, the temperature drops below T_{ODT} and there is a discontinuous jump in the modulus and a change in shape of the low-frequency behavior of G' versus ω. In the ordered, but *unaligned*, state, terminal behavior is absent even at the lowest frequencies accessed experimentally. This result for the lamellar morphology is qualitatively similar to that observed for cylindrical microdomains of S7–B43–S7 in Fig. 13-13. Creep tests show that terminal behavior is lacking in lamellar PEP–PEE, even at an extraordinarily low shear rate of 10^{-7} sec^{-1} (Koppi et al. 1992). Thus, for all practical purposes, these materials have no terminal behavior.

The behavior of PS–PI is similar in some respects to that of PEP–PEE. Figure 13-15 shows G' as a function of frequency for a range of temperatures that includes $T_{ODT} = 152°C$, for a PS–PI sample studied by Winey et al. (1993a, 1993b). Similar behavior has been observed by others (Gupta et al. 1996; Han et al. 1995; Pinheiro et al. 1996). As with PEP–PEE, terminal behavior is observed for $T > T_{ODT}$, and nonterminal behavior is obtained when T drops below this temperature. For PS–PI, one does not find the anomalous low-frequency behavior observed in the disordered state near T_{ODT} for PEP–PEE.

In the ordered state, lamellar block copolymers frequently show departures from linear viscoelastic behavior at low strain amplitudes of around 1% (Rosedale and Bates 1990; Winey et al. 1993a). Homogeneous polymers, on the other hand, typically show departures from linear behavior only when the strain amplitude exceeds 10% or so.

Both Figs. 13-14 and 13-15 indicate that there is a critical reduced frequency ω_c above which the storage moduli are similar in the ordered and disordered states (see also Fig. 13-13). Hence, for $\omega > \omega_c$, the modulus is controlled by *molecular* relaxations, while

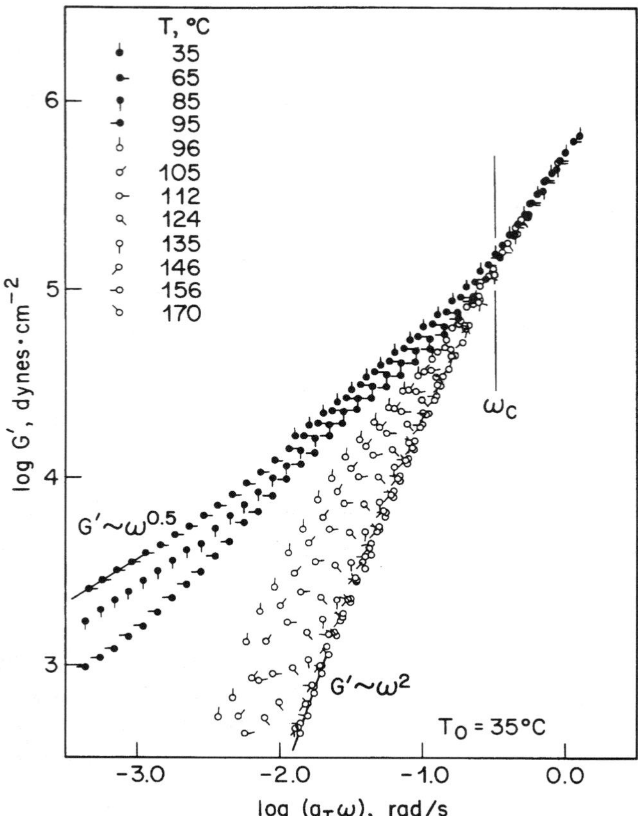

Figure 13.14 Storage modulus G' as a function of reduced frequency for a symmetric PEP–PEE sample at temperatures above $T_{ODT} = 96°C$ (open symbols), and below it (filled symbols), time–temperature shifted to obtain superposition at $\omega > \omega_c$ with a reference temperature of 35°C. (Reprinted with permission from Rosedale and Bates, Macromolecules 23:2329. Copyright © 1990, American Chemical Society.)

below it, deformation of the microdomains dominates the modulus. Kawasaki and Onuki (1990) have developed a theory for orientationally disordered lamellae that predicts a low-frequency regime in which $G' \propto G'' \propto \omega^{0.5}$. A modification of this theory by Milner and Fredrickson (1996) suggests that at very low frequencies the power law should terminate in a low-frequency elastic plateau given by $G' \sim 0.02B$, where B is the layer compression modulus.

In the disordered state near the ordering transition, the storage moduli for the PEP–PEE diblock copolymer, shown by open symbols in Fig. 13-14, fail to superpose at low frequency, and there is a low-frequency anomaly. This anomaly results in a shift of the terminal region (where $G' \propto \omega^2$) toward a lower frequency. It is believed that the anomaly is produced by an extra viscoelastic contribution from lamellar-like composition fluctuations that are coherent over a spatial range longer than the lamellar spacing, when T is slightly larger than T_{ODT}

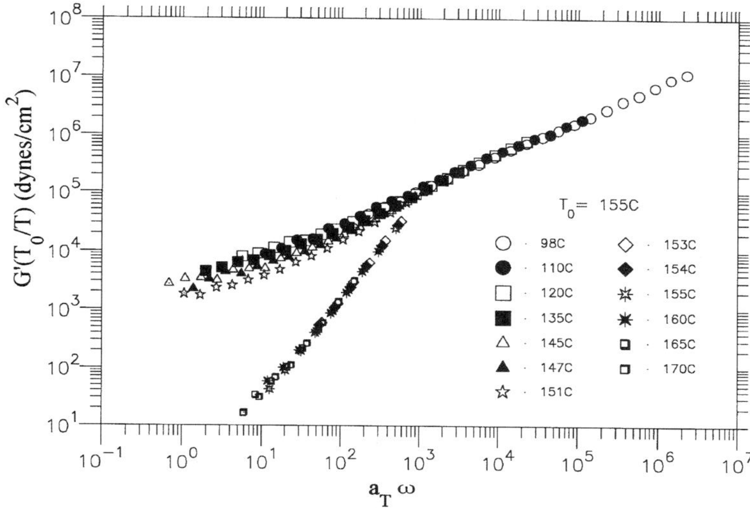

Figure 13.15   Reduced storage modulus versus reduced frequency $a_T \omega$ for a lamellae-forming polystyrene–polyisoprene diblock copolymer ($M = 22{,}000$) at temperatures above the order–disorder transition temperature $T_{ODT} = 152°C$, and quenched to temperatures below it. The disordered samples show terminal behavior, and the ordered (but unoriented) ones show nonterminal behavior. (Reprinted with permission from Patel et al., Macromolecules 28:4313. Copyright © 1995, American Chemical Society.)

(Bates 1984; Fredrickson and Larson 1987; Fredrickson and Helfand 1988). The relaxation of these fluctuations involves collective motion of many molecules, and thus it is slower than the relaxation time of individual molecules. In small-amplitude oscillatory shearing, the fluctuation waveform is deformed, producing a slowly relaxing stress. Presumably, this accounts for (a) the anomalous contribution to G' and (b) a similar, but smaller, contribution to G'' (Rosedale and Bates 1990; Jin and Lodge 1997). (Similar anomalies are observed in polymer blends.) An asymmetric version of this PEP–PEE polymer that forms cylindrical domains shows an even larger low-frequency anomaly (Almdal et al. 1992).

The fluctuation contribution to the modulus is smaller or not even visible in other block copolymers (Rosedale et al. 1995; Han et al. 1995). Bates and coworkers (Rosedale et al. 1995) have shown that in polyolefin block copolymers, the fluctuation contributions to the rheology occur within the range of χN values for which the structure factor $S(k)$ deviates from mean-field theory, namely the range $10.5 \lesssim (\chi N) \lesssim (\chi N)_{ODT}$. This range of χN values becomes larger with decreasing polymer molecular weight [$\overline{N}$ in Eq. (13-4)]. The temperature range over which the fluctuations are important also depends on the sensitivity of χ to temperature, controlled by the "A" term in Eq. (13-1) (Han et al. 1995). The high temperature sensitivity of χ in PS–PI block copolymers may partly explain why fluctuation effects are not seen in the PS–PI rheology.

Within the fluctuation-dominated region of the disordered phase of a diblock near T_{ODT}, *shearing can apparently induce a transition to the ordered state* (Koppi et al. 1993). Cates and Milner (1989) predicted such a phenomenon, based on a shear-induced suppression of fluctuations, leading to a shift of $(\chi N)_{ODT}$ toward its mean-field value, which is 10.5 for $f = 0.5$. The lamellae produced by shearing the disordered state near T_{ODT} align in the perpendicular orientation, defined below. Analogous shear-induced transitions to hexagonal cylindrical phases have been predicted (Marques and Cates 1990) and observed (Bates et al. 1994; Nakatani et al. 1996; Fredrickson and Bates 1996).

13.3.1.2 Alignment

13.3.1.2.1 OBSERVATIONS. In both PS-PI and PEP-PEE, large-amplitude oscillatory shearing aligns lamellae and renders the moduli more nearly terminal at low frequencies. Figure 13-16 is a typical plot of G', G'', B', and B'', the in-phase and out-of-phase components of the stress and birefringence, both divided by the shear strain amplitude, versus time, during large-amplitude oscillatory shearing of a PEP–PEE sample (Kannan and Kornfield 1994). The birefringence in some block copolymers, such as PS–PI, is almost entirely *form birefringence*—that is, birefringence due to the differences in the indices of refraction between the two blocks (Folkes and Keller 1971). However, for PEP–PEE, the form effect is small, because the indices of refraction of the PEP and PEE blocks are almost the same, and for this diblock, *intrinsic birefringence* is thought to dominate. Intrinsic birefringence is produced by orientation of polymer molecules (see Section 1.6.3). Notice in Fig. 13-16 that the shear-induced improvement in the degree of alignment, as indicated by the reductions in modulus and birefringence amplitudes, is rapid at first, but that very slow reductions continue even up to 25,000 sec, corresponding to 100 cycles of strain. In general, the shear-induced alignment can be divided into a "fast" and a "slow" process (Chen et al. 1997a). The degree of alignment, as well as its direction, can be inferred not only from birefringence, but more commonly from neutron or x-ray scattering. The alignment is never perfect; residual misorientation seems to persist no matter how long the large-amplitude shearing is prolonged.

There are in principle three possible simple orientation directions of the lamellar layers with respect to the shearing flow, as shown in Fig. 13-17. These are referred to as follows: the *parallel* or "c" orientation, in which the layers are parallel to the shearing surfaces or rheometer tools so that the normal to the layers is parallel to the shear-gradient direction; the *perpendicular* or "a" orientation, in which the normal to the layers is parallel to the vorticity axis; and the *transverse* or "b" orientation, in which the normal to the layers is parallel to the flow direction. (See also the discussion on smectics in Section 10.4.5.) The PEP–PEE and PS–PI samples both align in either the parallel or the perpendicular directions, depending on the temperature and frequency of oscillation. For higher-molecular-weight PS–PI samples, a state of *mixed parallel and transverse* alignment has also been observed (see below).

The reduction in moduli produced by the aligning process can be explained, in part, by considering the orientation-dependence of the moduli of lamellar monodomains. Koppi et al. (1992) measured the moduli G' and G'' in each of the three orientations depicted in Fig. 13-17, where the shear amplitude γ_0 was kept low enough that the orientation direction was not changed during the measurements. Figure 13-18 shows that the low-frequency modulus is *lower in the perpendicular and parallel orientation than in the transverse orientation* at frequencies below ω_c. Since the initial, globally unaligned sample contains regions whose local lamellar orientations are in the high-modulus transverse direction, the

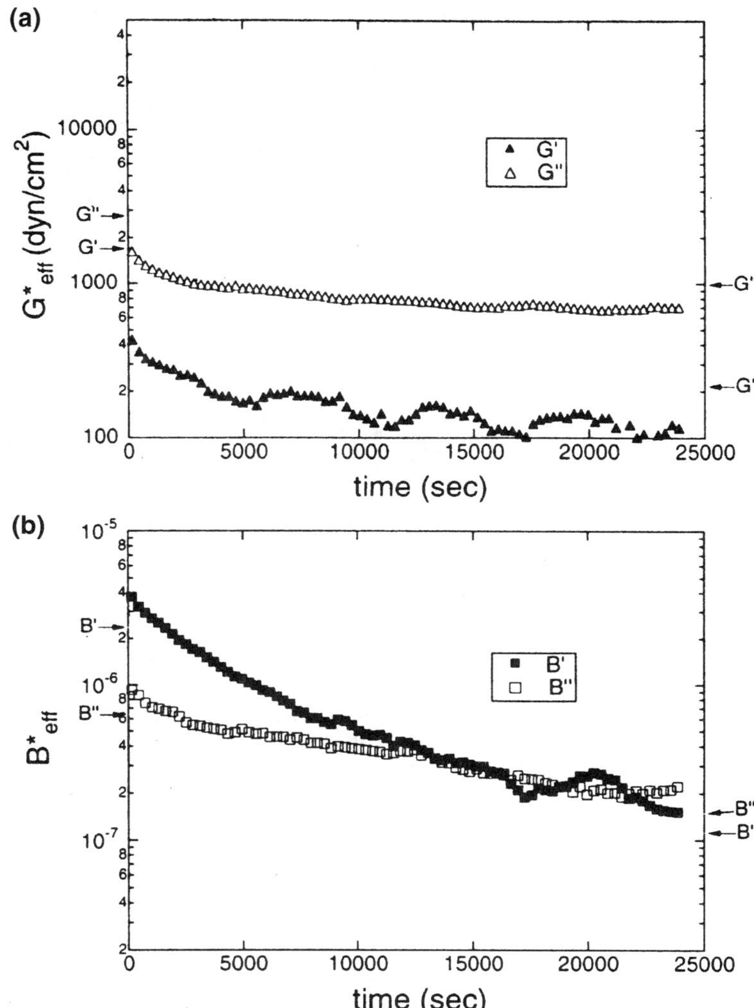

Figure 13.16 (a) Effective storage and loss moduli as functions of time during large-amplitude oscillatory shear alignment of PEP–PEE lamellar diblock copolymer at a strain amplitude of 0.9 and a frequency of 0.02 sec^{-1}. The initial state of the sample was a random orientation produced by quenching from the disordered state. (b) The in-phase (B′) and out-of-phase (B″) components of the oscillating shear (12) component of the birefringence measured during the alignment process. The continued decrease in birefringence after the moduli have almost stopped changing shows that slow improvement in alignment continues even up to 25,000 sec, or 100 cycles of strain. The arrows at the left and right show the low-amplitude values of the quantities respectively before and after alignment. (Reprinted with permission from Kannan and Kornfield, Macromolecules 27:1177. Copyright © 1994, American Chemical Society.)

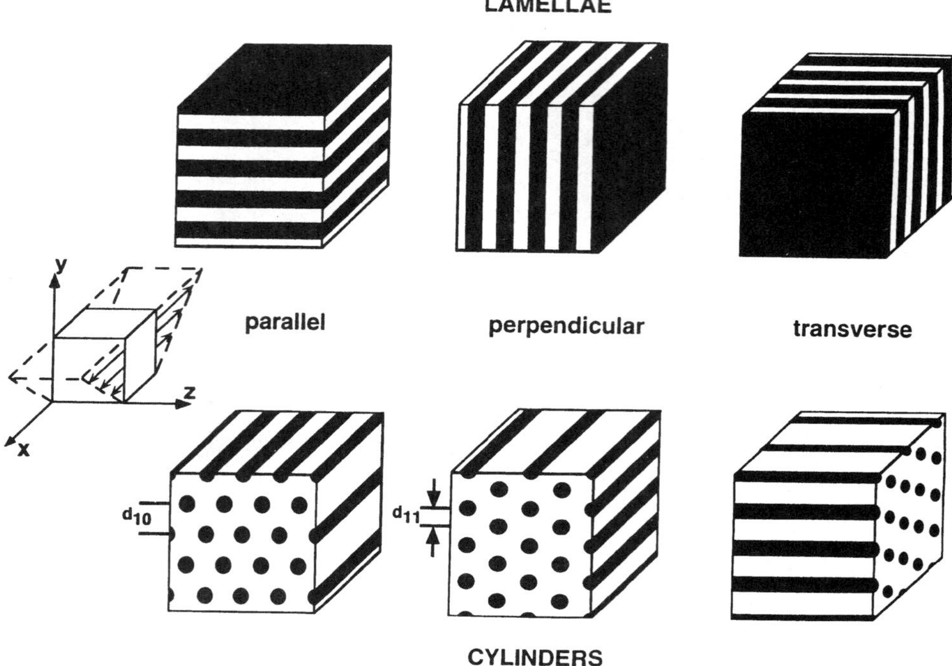

Figure 13.17 Definitions of "parallel," "perpendicular," and "transverse" orientations of lamellae and hexagonal cylinders, with respect to the axes of the shearing flow. (From Fredrickson and Bates 1996, with permission from Annual Reviews.)

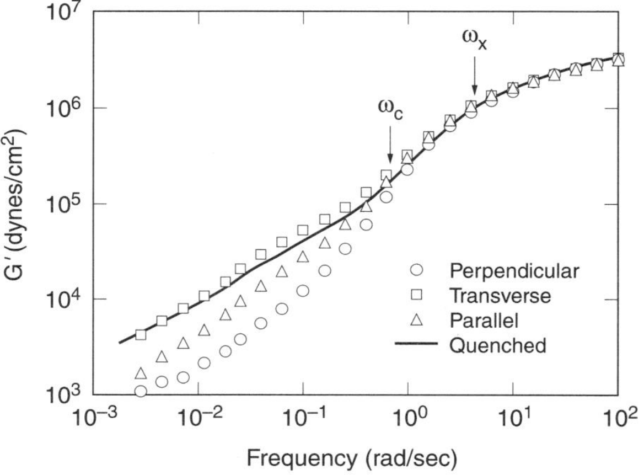

Figure 13.18 Storage shear modulus G' for lamellar samples of PEP–PEE ($M_n \approx 50{,}000$) at 40°C. The solid line is for an unaligned sample quenched from the disordered state ($T_{\mathrm{ODT}} = 93$°C), while the symbols are for a shear-aligned sample oriented in the parallel ($\triangle$), perpendicular ($\bigcirc$), and transverse ($\square$) orientations. ω_x is the frequency at which G' and G'' cross; ω_c is the frequency above which the molecular relaxations dominate the modulus. (From Koppi et al. 1992, reprinted with permission from the American Physical Society.)

modulus of an unaligned sample is decreased when these regions are realigned into either the parallel or perpendicular orientation.

For PEP–PEE, two regimes of alignment have been observed. Parallel alignment is obtained at low frequencies, and perpendicular alignment is found in a window at high frequency and high temperatures near T_{ODT} (Koppi et al. 1993). For low-molecular-weight PS–PI, on the other hand, as many as three regimes of alignment have been observed (see Fig. 13-19). As with PEP–PEE, parallel alignment is observed at low frequencies while perpendicular alignment is observed at higher frequencies (Zhang et al. 1995, 1996; Wiesner 1997). But, in addition, at still higher frequencies there is another regime of parallel alignment (Patel et al. 1995; Gupta et al. 1996; Zhang et al. 1995). In PS–PI, the low-frequency regime of parallel alignment is only observed if the sample is annealed well below T_{ODT} (but above T_g for the polystyrene) for hours before shear alignment. If, instead, the sample is quenched from the high-temperature disordered isotropic state down to the temperature of interest and shear-aligned soon thereafter, the perpendicular alignment rather than the parallel one is obtained at the lowest frequencies (Patel et al. 1995; Wiesner 1997). This thermal-history dependence of the alignment is an important clue to consider when evaluating proposed alignment mechanisms (see below). Another important clue is the observation, shown in Fig. 13-19, that a critical strain seems to be necessary to induce the perpendicular alignment.

For higher-molecular-weight, strongly segregated materials, such as PS–PI with $M_w >$ 50,000 and polystyrene-poly(ethylene propylene) with $M_w = 130,000$, a shear-induced *mixed state* consisting of partial alignment in *both* the parallel and transverse directions has also been observed, either as a transient en route to a final parallel alignment (Zhang and Wiesner 1995) or as the final state (Okamoto et al. 1994b, Pinheiro et al. 1996).

Alignment is also sensitive to block architecture. In a *triblock* PS–PI–PS block copoly-

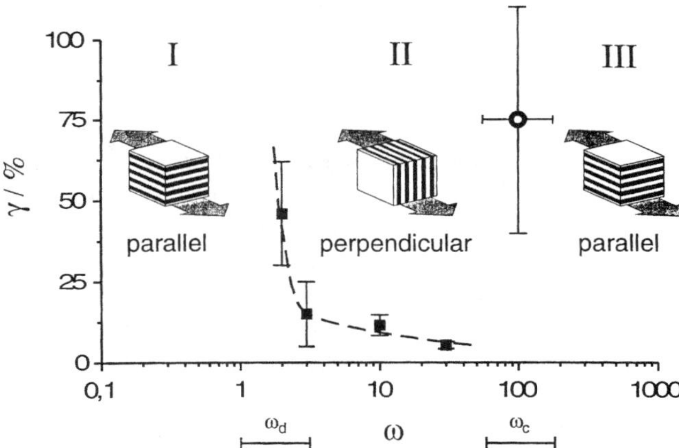

Figure 13.19   Orientation diagram for PS–PI diblock copolymer at $T = 136°C$, just below T_{ODT}. The frequencies ω_c and ω_d were obtained from linear viscoelastic data; they mark the transitions from relatively elastic behavior at high and low frequencies to more dissipative behavior at intermediate frequencies (from Wiesner 1997, reprinted with permission from Hüthig & Wepf Verlag.)

mer (Riise et al. 1995), *no transition* from parallel to perpendicular alignment with decreasing alignment frequency was observed; perpendicular orientation was dominant at all frequencies and temperatures studied. However, for a PEP–PEE–PEP triblock, Tepe et al. (1997) observed only the opposite: parallel alignment at all frequencies and temperatures!

The shear-alignment behavior of PS–PI block copolymer *solutions*, where a solvent mixes with both blocks, has been studied by Balsara and coworkers (Balsara and Hammouda 1994; Balsara et al. 1994b). Mostly parallel alignment was observed, apparently with some rippling of the layers. Little else is known about the alignment behavior of solvated block copolymers.

13.3.1.2.2 EXPLANATIONS. The aligning behavior of block copolymers under shear is complicated and is sensitive to block architecture, mechanical contrast between the layers, proximity of the ODT, frequency, and strain. No general theory seems to predict the alignment direction under all conditions. Theories and rationalizations are available, however, to explain some of the trends.

Focusing first on *diblock* copolymers, consider the transition between regions II and III on the alignment diagram of Weisner (Fig. 13-19) for low-molecular-weight PS–PI. This transition occurs at ω_c, the frequency at which the time–temperature-shifted moduli for disordered samples merge with those for ordered samples (see Figs. 13-14, 13-15, and 13-18). Above ω_c the *molecules* are unable to remain completely relaxed during the deformation, while below ω_c the molecules are relaxed and the stress is predominately due to deformation of the lamellar *pattern*. Thus, the high-frequency parallel alignment must be produced by a *molecular* deformation mechanism, while the lower-frequency perpendicular alignment is produced by a *pattern* deformation mechanism. This interpretation is confirmed by studies of alignment direction as a function of both frequency and temperature. Figure 13-20 shows that when the frequencies for attaining high-frequency parallel and perpendicular alignment at various temperatures are time–temperature-shifted using the shift factor a_T used to superpose the rheological data, perpendicular alignment is found for reduced frequencies $a_T\omega$ less than $a_T\omega_c$, and parallel alignment at reduced frequencies greater than $a_T\omega_c$.

At frequencies higher than $a_T\omega_c$, the molecular characteristics of the block copolymer influence the stresses, and presumably also the alignment direction. As noted, the polystyrene block in PS–PI is much more viscous that the polyisoprene block; that is, the blocks in PS-PI have a large *mechanical contrast*. Thus, the molecular stresses generated by deformation of styrene domains will be higher than those in the isoprene domains, and above $a_T\omega_c$ these stresses will be large enough to influence the lamellar orientation. In particular, we might expect the lamellae to orient so that the high stresses in the viscous styrene block are reduced by redistributing the imposed strain preferentially into the less viscous isoprene blocks (Patel et al. 1995). This redistribution of the local strain can more easily occur in lamellae aligned in the parallel orientation than in the perpendicularly orientation, since in the former the viscous styrene blocks can slide over, and be "lubricated" by, the less viscous isoprene ones. For perpendicular alignment, each styrene lamella spans the gap between the shearing surfaces and must experience the entire macroscopic shearing strain. This interpretation is supported by Fig. 13-20, which shows the frequency-dependent complex modulus G^* for PS–PI S12–I9 aligned in both parallel and perpendicular orientations. Note that there is a crossover; the sample with parallel alignment

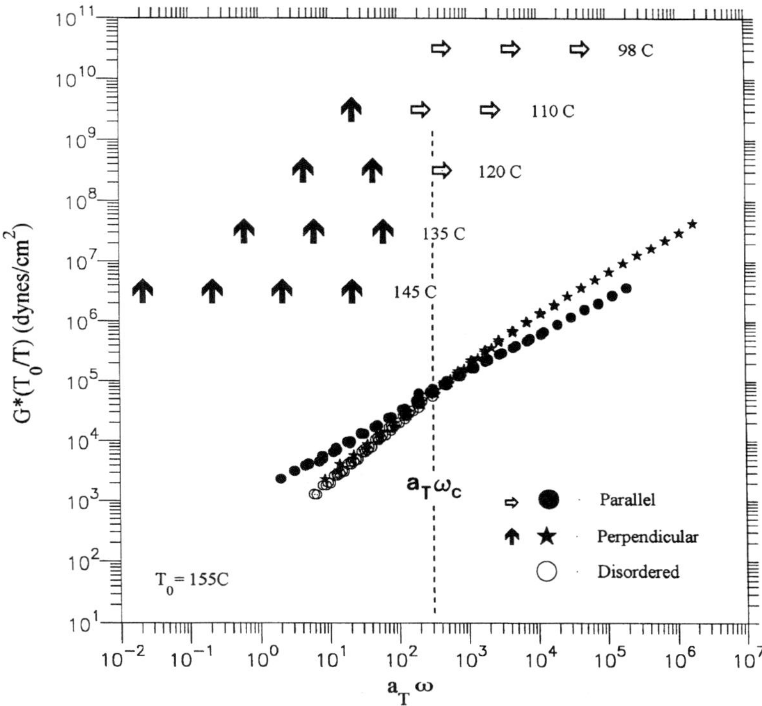

Figure 13.20 Master curves of $G^*(T_0/T^*)$ versus reduced frequency $a_T\omega$ for the PS–PI block copolymer described in Fig. 13-15 in the high-temperature disordered state and in the low-temperature ordered state flow-aligned into either the parallel or perpendicular directions. The arrows show the time–temperature-shifted frequencies at which the parallel (sideways arrow) and perpendicular (up arrow) orientations are achieved by large-amplitude shearing. The frequency $a_T\omega_c \approx 300$ sec^{-1} is the reduced frequency below which G' in the disordered and quenched states no longer superpose. (Reprinted with permission from Patel et al., Macromolecules 28:4313. Copyright © 1995, American Chemical Society.)

has the lower modulus at high frequency (where viscoelastic stresses in the polystyrene layer are large), while the sample with perpendicular alignment has lower modulus at low frequency. For this PS–PI sample at least, alignment is in the direction that produces the lower modulus.

For PEP–PEE, there is little mechanical contrast; both blocks have similar glass transitions and entanglement densities, and hence similar viscosities. The lack of mechanical contrast explains why the moduli superpose for different alignment conditions above ω_c in Fig. 13-18; contrast this with the behavior of PS–PI shown in Fig. 13-20. This lack of mechanical contrast in PEP–PEE may well explain why no high-frequency parallel-alignment regime has been found for this block copolymer. Studies on other block copolymers, with varying degrees of mechanical contrast, should allow testing of this proposed mechanism for high-frequency alignment.

Having found a credible mechanism for parallel alignment above ω_c, it remains to explain alignment regimes I and II in Fig. 13-19. Since the symmetry of a block-copolymer lamellar phase is identical to that of a *smectic-A liquid crystal,* one expects the flow properties of the former to be similar to those of the latter *at frequencies low enough that polymer modes of relaxation are not significant*—that is, at frequencies less than ω_c. At these low frequencies, the stresses in the block copolymer should be dominated by the layered microstructure and its associated defects, and not by molecular deformations within the individual blocks. Similarity in the rheological behavior of block copolymers and smectic liquid crystals is illustrated in Fig. 13-21. This figure shows that the moduli G' and G'' measured for a PS–PI diblock copolymer before and after alignment by large-amplitude oscillatory shearing are very similar to the moduli for the small-molecule smectic 8CB measured before and after alignment under the same conditions of frequency, strain, and time (Larson et al. 1993).

As discussed in Section 10.4.8, small-molecule *thermotropic* smectics, such as 8CB, show two flow alignment regimes under steady shearing (see the left half of Fig. 10-38 labeled "SmA"). At high shear rates and high temperatures, 8CB orients preferentially in the "a," or perpendicular, direction under steady shear, the same as is seen in block copolymers in region II. Recent theory indicates that layer fluctuations destabilize the "c,"

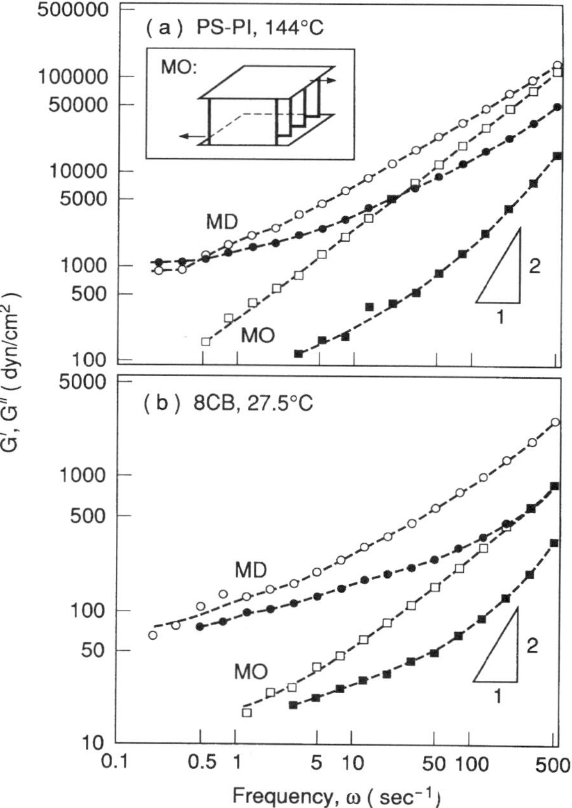

Figure 13.21 Apparent moduli G' (closed symbols) and G'' (open symbols) as functions of frequency at a strain amplitude of 10% for **(a)** a PS–PI lamellar block copolymer and **(b)** a smectic-A liquid crystal, 8CB. In both **(a)** and **(b)**, the curves labeled "MD" are "macroscopically disordered," or unaligned, states produced by a quench from the disordered state. "MO" refers to "macroscopically ordered" samples aligned by large-amplitude oscillatory shearing at an amplitude of unity and a frequency of 10 sec[-1] for 20 min. (From Larson et al. 1993, reprinted with permission from Steinkopff Publishers.)

or parallel, orientation and lead to orientation in the perpendicular state (Goulian and Milner 1995). Intuitively, when the lamellae are in the parallel orientation, wave-like distortions or defects in the layers such as shown in Fig. 13-12b could easily couple to the vorticity of a shearing flow and lead to rotation of the layers, thus destabilizing the parallel orientation (Koppi et al. 1992). In the perpendicular orientation, fluctuations are likely to couple less strongly to shear. Such considerations may explain shear-induced perpendicular alignment in block copolymers and smectics at deformation rates that are high (but not so high as to induce molecular stresses, i.e., to induce a transition from region II to region III).

A *weak-segregation theory* has also been developed by Fredrickson (1994) to explain the alignment behavior of diblock copolymers. In this theory, perpendicular alignment is predicted near T_{ODT} at frequencies that are high compared to rates of fluctuations (but low compared to molecular time scales) as a result of coupling of composition fluctuations to the shearing field. Well below T_{ODT}, these fluctuations are small, and mechanical contrast between the blocks, however small, dominates and favors parallel alignment.

At the lowest shear rates, under *steady flow*, 8CB adopts a mixture of parallel and perpendicular alignment (see Fig. 10-38). Block copolymers, under large-amplitude *oscillatory flow*, adopt nearly pure parallel orientation. Obviously, the difference between these results could well arise from the differences between oscillatory and steady flow, a hypothesis that should be tested by performing oscillatory shear alignment of 8CB. Alignment in this low-frequency region I evidently emerges because of the role of *smectic defects*. In 8CB sheared at low rates, the observed alignment is associated with "jelly-roll" defects discussed in Section 10.4.8. In block copolymers, defects are implicated by the observation that region I is not obtained if shear alignment commences shortly after a quench into the lamellar state, presumably before well-defined defects characteristic of a smectic phase have had a chance to form. How and why such defects lead to parallel alignment is a matter of speculation. Apparently, shearing flow induces migration of defects in such a way that parallel alignment is favored.

Thus, three mechanisms of alignment—one for frequencies and temperatures at which molecular stresses dominate, and the other two at conditions where stresses from the lamellar pattern dominate—can rationalize the alignment behavior of the low-molecular-weight PS–PI diblock copolymers. The alignment characteristics of PEP–PEE are qualitatively similar to those of PS–PI except for the absence of region III, evidently because of a lack of mechanical contrast.

We noted the dominance of the *perpendicular* orientation at all aligning frequencies for PS–PI–PS triblock copolymers. This might be explained by *bridging* configurations in which the two end styrene blocks on each PS–PI–PS molecule reside in different styrene lamellae, with the isoprene block spanning the intervening isoprene layer. The alternative configuration for a such a triblock molecule is an isoprene *loop*, with the two styrene blocks both residing in the same lamella. For $f = 0.5$, bridging and looping configurations are estimated to be equally probable. In the parallel orientation, bridging configurations across the isoprene layers should tend to reduce the fraction of strain that gets allocated to the isoprene layers. However, in diblocks, as noted earlier, at high frequencies the parallel orientation is believed to be favored because it allows the shear to be concentrated in the isoprene layers, rather than in the more viscous styrene layers. For the triblock, this advantage of the parallel orientation is nullified by the presence of bridges across the isoprene layers, which impedes deformation of these layers. This might explain the dominance of the perpendicular orientation at all frequencies in triblock PS–PI–PS.

It would be interesting to test this explanation by determining the shear alignment direction of the "inverted" triblock, PI–PS–PI, which has a styrene blocks in the middle. Presumably, if the above explanation for the prevalence of perpendicular alignment in PS–PI–PS is correct, then the inverted triblock PI–PS–PI should show parallel alignment at high frequencies.

For multiblock copolymers, all layers should be "stitched together," making it impossible for them to slide over one another in continuous shear in the parallel orientation and therefore suppressing this orientation. Although multiblock copolymers with more than three blocks seem not to have been studied as yet, a smectic liquid-crystalline polymer composed of alternating flexible and stiff monomer units has been flow-aligned under oscillatory shear (Alt et al. 1995). Such a material can be considered to be an analog of a multiblock copolymer, if a "block" is taken to be one monomer unit in length and one of the block types is stiff. Consistent with the behavior expected of multiblocks, oscillatory shearing did not produce parallel alignment in this material. Instead, at high strains the expected perpendicular orientation was obtained, while at low strains ($\lesssim 10\%$) the transverse orientation, or a combined perpendicular/transverse orientation, was found (Alt et al. 1995).

As mentioned earlier, for some strongly segregated, entangled *diblock* copolymers sheared at high frequency, a *bimodal mixture of parallel and transverse alignment* is induced (Pinheiro et al. 1996; Zhang and Wiesner 1995). The existence of a significant component of transverse alignment in any of these materials might seem surprising, since transverse lamellae rotate under shear and change their layer spacing, and this orientation should therefore be strongly disfavored. However, in the strong segregation limit, thermodynamic forces preferentially orient the chains in the direction orthogonal to the lamellar layers, and if the shear rate is high enough to exert significant stresses on individual molecules, one might expect the molecules to align in the flow direction. This molecular orientation is compatible with lamellar layering only if the layers are in the transverse orientation.

Kornfield and coworkers (Kannan et al. 1993; Kannan and Kornfield 1994; Gupta et al. 1996; Chen et al. 1997a, 1997b) have used used transient birefringence measurements, x-ray scattering, and transmission electron microscopy to monitor the orientation kinetics of lamellae under large-amplitude oscillatory shear. They and other authors have suggested at least three different mechanisms of alignment. In one, the oscillatory shear introduces a "rocking" motion of grains of misaligned material; asymmetry in the rocking motion leads to a gradual drift toward the favored alignment; birefringence measurements support this mechanism. A second possibility is that misaligned regions are locally *melted* into the disordered state by strain energy. Obviously, this mechanism is more probable at temperatures close to T_{ODT} where the energy required for melting is low (Larson 1994a; Kodama and Doi 1996). A third possibility is that stress drives defects of opposite sign together so that they annihilate. Material through which a defect passes reorients preferentially into the favored direction. Recent electron-microscopy studies on quenched samples provides some hints regarding the relative importance of these and possibly other mechanisms of alignment (Chen et al. 1997a).

13.3.1.3 Elastic Constants and Defects

When the layers of a lamellar block copolymer are distorted, the free energy density is augmented by a distortional term that can, like the smectic-A phase, be described as the sum of layer compression/dilation and layer-bending energies:

$$W_d = \tfrac{1}{2}B\varepsilon^2 + K_1(\nabla \cdot \mathbf{n})^2 \tag{13-6}$$

where ε is the dilational strain and $\mathbf{n}$ is the unit vector locally normal to the layers (see Section 10.4.3). The compressional and bending (or director-splay) constants B and K_1 can be estimated from block-copolymer thermodynamic theories. In the strong segregation limit, Wang (1994) found for a symmetric diblock that

$$B \approx 3\left(\frac{\Gamma}{k_B T}\right)^{2/3} N^{-2/3} b^{-2/3} v^{-1/3} k_B T \tag{13-7a}$$

$$K_1 \approx \frac{3}{16}\left(\frac{\Gamma}{k_B T}\right)^{4/3} N^{2/3} b^{2/3} v^{1/3} k_B T \tag{13-7b}$$

where Γ is the interfacial tension given in terms of χ by Eq. (13-2b), and b and v are the statistical segment length and volume. The *permeation depth* is $\sqrt{K_1/B} = D/4$, where D is the thickness of a layer of A (or B), given by Eq. (13-2a). Hence, the permeation length is of order the molecular size, as is the case with smectic-A phases.

In the weak-segregation limit, Amundson and Helfand (1993) estimated

$$B = 55A_0^2 v k_B T, \qquad K_1 = 3.6A_0^2 R_g^2 v k_B T \tag{13-8}$$

where v is the number of polymer chains per unit volume, R_g is the polymer radius of gyration, and A_0 is one-half the amplitude of the sinusoidal compositional waveform; it has a maximum value of 0.25 when the composition varies sinusoidally from pure A to pure B. In the weak-segregation limit, $A_0 \leq 0.1$.

From Eqs. (13-7) and (13-8), one can estimate that typical values for B and K_1 for weak to moderately segregated diblocks should be $B \sim 10^6$–10^7 dyn/cm^2 and $K_1 \sim 10^{-7}$–10^{-6} dyn. By compressing shear-aligned diblocks, Hudson et al. (1995) have extracted experimental estimates of B that are around a decade or so lower than these values.

The compressive modulus predicted for a triblock ($f = 0.5$) in the strong-segregation limit is theoretically the same as for the diblock formed by cutting the triblock in the middle. The bend modulus for the triblock is predicted to be about twice as large as that of the diblock because of the presence of "bridging" configurations (Turner 1995).

The bend and especially the compressive moduli of lamellar block copolymers are therefore typically lower, or at least no higher, than those of small-molecule smectic-A liquid crystals, while the viscosities of the former are usually much larger than the latter. By analogy with nematics, for layered materials one can define characteristic *Ericksen numbers* as $\mathrm{Er}_K \equiv \eta\dot{\gamma}h^2/K_1$ and $\mathrm{Er}_B \equiv \eta\dot{\gamma}/B$, where η is a characteristic viscosity and h the sample thickness [see Eq. (10-27)]. For $\dot{\gamma} \sim 1\,\mathrm{sec}^{-1}$, and $h \approx 0.1$ cm, one obtains $\mathrm{Er}_K \sim 10^4$ for typical small-molecule smectics with $\eta \sim 1$ P, and $\mathrm{Er}_K \sim 10^8$ for block copolymers with $\eta \sim 10^4$ P. Hence, for a given shear rate, bend elastic forces are even more more easily overwhelmed by viscous forces in block copolymers than in small-molecule smectics, and defects can easily be generated and multiplied. The spacing a between defects in a *nonaligning* flow in a block copolymer is estimated to be $a \sim \sqrt{K/\eta\dot{\gamma}} \sim 0.1\,\mu$m, close to the molecular length scale. Hence, flows could easily pulverize the lamellae to such an extent that individual defects are no longer discernible in the rubble. However, flows that *align* the block copolymer layers, such as moderate-amplitude oscillatory shearing flows ($\gamma_0 \sim 1$), can eliminate defects and thus increase a.

When defects are identifiable in block copolymers, they are of the type expected for smectics, including focal conics and walls (Hudson et al. 1995). However, screw dislocations seem to be rare in sheared lamellar block copolymers, while edge dislocations are abundant (Hudson et al. 1995); the reverse is true in small-molecule smectics. The relative scarcity of edge dislocations in small-molecule smectics is understood to be a consequence of their high compressive modulus and the unavoidable change in local layer spacing produced by an edge. In small-molecule smectics, if edge dislocations were to appear, they would quickly cluster together to form lower-energy compound dislocations and, ultimately, disclination lines, which typically organize into focal domains. In block copolymers, however, viscosities are high and defect mobilities are low; hence clustering is very slow. This, along with high viscoelastic forces generated during flow of block copolymers, might help explain the abundance of edge dislocations in lamellar block copolymers (Hudson et al. 1995).

13.3.2 Cylindrical Microdomains

Some of the features of the thermodynamics, rheology, and flow alignment of cylinder-forming block copolymers are similar to lamellae-forming ones; but there are notable differences as well. Like lamellae-forming samples, cylinder-forming block copolymers show a first-order transition from a disordered state to an ordered state when χN reaches a critical value that depends on f_A, on $\overline{N}$, and on the "conformational asymmetry parameter" ε. For diblocks, the transition to cylinders occurs when f_A is in the range 0.2–0.35 or 0.65–0.8, and $\chi N \gtrsim 15$. The transition between the disordered state and the hexagonal cylindrical state is much more hysteretic than the transitions between disordered and lamellar states for similar materials (Almdal et al. 1992; Winter et al. 1993). Thus, curves of G' versus temperature at fixed, low frequency differ over a range of around 20°C between heating and cooling cycles (Winter et al. 1993; Almdal et al. 1992), and the apparent disordering temperature on heating can be 10–20°C higher than the apparent ordering temperature on cooling. After a shallow quench, the growth of large domains of hexagonal material ($1\,\mu$m or larger) can be extremely slow—that is, can require many hours (Winter et al. 1993). Using depolarized light scattering, Dai et al. (1996) observed two growth regimes: (1) a faster initial regime in which dilute ordered domains grow at the expense of disordered material and (2) a slower long-time regime in which crowded ordered domains grow apparently by defect annihilation.

Balsara et al. (1994a) observed that a defect-filled "grainy" hexagonal cylindrical ordered state disorders on heating at a temperature 13°C lower than does a shear-aligned "monodomain" of the same material! Thus, the presence of the defects significantly destabilizes the hexagonally ordered state. Winter et al. (1993) have observed that when a cylinder-forming PS–PI–PS triblock copolymer is sheared at a temperature of around 10°C below

T_{ODT}, the hexagonal phase is suppressed. These peculiar observations suggest that near T_{ODT} the hexagonal phase is especially delicate, and its free energy is only very slightly less than that of the disordered state.

As with lamellae-forming samples, when the hexagonal phase forms after a quench from the disordered state, terminal behavior in G' and G'' is replaced by nonterminal behavior (see Fig. 13-13). For hexagonal cylinders, the low-frequency behavior is generally "more elastic" than for lamellae; the apparent power-law exponent in $G' \propto \omega^n$ is usually somewhat lower ($n \sim 0.0$–0.4) for the former than for the latter ($n \sim 0.5$) (Gouinlock and Porter 1977; Morrison et al. 1990; Winter et al. 1993; Almdal et al. 1992). This is perhaps not surprising, since ordered hexagonal cylinders are "solid-like" (i.e., positionally ordered) in two dimensions and lamellae are positionally ordered in only one.

As with lamellae, large-amplitude oscillatory shearing, or steady shearing, can align defect-filled "grainy" samples of hexagonal cylinders. *The cylinders seem always to align in the flow direction* (Morrison et al. 1990; Winter et al. 1993 Jackson et al. 1995) with the (10) plane of the hexagonal unit cell parallel to the x–y, or deformation, plane. In this orientation, the most densely packed layers of cylinders are parallel to the shearing surfaces. This orientation is called the "parallel" orientation, to distinguish it from the alternative "perpendicular" orientation *of the unit cell* in which the (11) plane is parallel to the x–y plane (see Fig. 13-17). The "transverse" orientation, in which the cylinders themselves are orthogonal to the flow direction, seems never to be produced; and if it is artificially chosen as the starting orientation, the shearing flow rotates the cylinders toward the flow direction within around 10 strain units (Scott et al. 1992). Once aligned, high strains can fragment the cylinders (Odell and Keller 1977; Scott et al. 1992). As with the lamellar phase, shear alignment can drastically reduce the elastic modulus (Morrison et al. 1990), in some cases nearly restoring terminal behavior (Almdal et al. 1992).

Jackson et al. (1995) have shown that prolonged shearing of a cylinder-forming SBS block copolymer not far below T_{ODT} reduces the diameter of some of the aligned cylinders (from 21 to 12 nm) and changes their packing symmetry; regions containing the thinner cylinders coexist with regions of the thicker ones. This shear-induced phase transition seems to be analogous to *Martensitic* transitions in shape-memory metals (Olson 1992). A smaller shear-induced reduction in spacing of a *lamellar* block copolymer has been recently reported by Pinheiro et al. (1996). Such an effect of shear on lamellar spacing has been predicted theoretically by Williams and MacKintosh (1994).

13.3.3 Spherical Microdomains

As with cylinder- and lamellae-forming block copolymers, the rheological behavior of block copolymers that form spherical domains depends on whether or not the domains possess macrocrystalline order. If the domains are disordered, then the low-frequency moduli show terminal behavior typical of ordinary viscoelastic liquids; that is, G' and G'' fall off steeply as the frequency becomes small (Watanabe and Kotaka 1983, 1984; Kotaka and Watanabe 1987). When the spherical domains are ordered, however, elastic behavior is observed at low frequency; that is, G' approaches a constant at low frequency, and a yield stress is observed in steady shearing.

13.3.3.1 Disordered Microdomains

Block copolymers that form disordered spherical microdomains are liquid-like under slow flows, but show viscoelastic behavior under flows fast enough to distort either the polymer chains or the equilibrium disordered arrangement of the spherical domains. These contributions to the viscoelastic stress have been illuminated in a beautiful set of experiments by Watanabe et al. (1997). They studied polystyrene–polyisoprene diblocks in which the styrene formed hard glassy spheres at low temperature in a matrix composed of a polyisoprene "corona" mixed with a low-molecular-weight polyisoprene homopolymer. Both the linear and nonlinear viscoelastic properties of this mixture could be decomposed into (a) a fast part associated with relaxation of the polyisoprene strands composing the corona and (b) a slow part associated with translational diffusion of the polystyrene "hard spheres." The contributions to the stress after a step shear strain strain γ could be represented by a sum, $G_f(t)h_f(\gamma) + G_s(t)h_s(\gamma)$, containing linear moduli G_f and G_s for the fast and slow relaxations and corresponding nonlinear "damping functions" $h_f(\gamma)$ and $h_s(\gamma)$. The "fast" damping function for the corona chains, shown in Fig. 13-22, is very similar to that for entangled linear and star polymers (which obey the Doi–Edwards theory; see Section 3.7.4.2), and the "slow" damping function for the polystyrene spheres is similar to that for a hard-sphere suspension of silica particles! Thus the nonlinear properties of the disordered block-copolymer micelles can be explained by combining the rheological features of polymer chains and of hard-sphere suspensions. Even more remarkably, once these linear and nonlinear functions are specified, the rheological behavior of the diblock copolymer solution in steady shear and start-up of steady shearing are well-predicted by the appropriate K–BKZ equation (an equation given in Section 3.7.4.4). This equation also works for hard-sphere suspensions, but only up to the shear rate at which shear thickening occurs (see Section 6.2.6). Disordered spherical micelles lack a shear-thickening regime, which indicates that the block-copolymer coronas suppress the clustering phenomenon which is evidently necessary for shear thickening.

13.3.3.2 Ordered Microdomains

As with lamellar and hexagonal block-copolymer phases, steady or large-amplitude oscil-

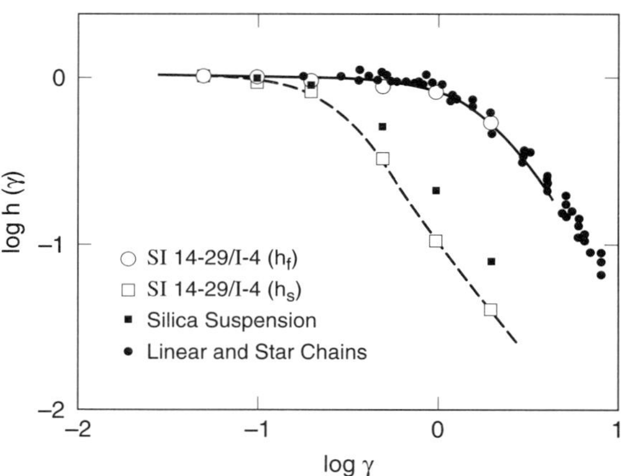

Figure 13.22 Damping functions $h_f(\gamma)$ and $h_s(\gamma)$ for the fast and slow relaxation processes of a 15 wt% solution of a micelle-forming polystyrene–polyisoprene diblock copolymer (molecular weights, respectively, of 14,000 and 29,000) in a low-molecular-weight ($M = 4,000$) polyisoprene. Damping functions for linear and star polymers and for silica dispersion are shown for comparison. (From Watanabe et al. 1997, with permission from Macromolecules 30:5905. Copyright © 1997, American Chemical Society.)

latory shear can align the symmetry axes of ordered micellar phases. The shear alignment directions of BCC and FCC micellar phases are similar to those for the corresponding BCC and FCC colloidal crystalline phases discussed in Section 6.4.3. For a BCC phase, the (111) direction of the unit cell orients parallel to the flow direction, with the (110) planes slipping over each other, parallel to the rheometer tools, just as in charged colloidal suspensions (Koppi et al. 1994; McConnell et al. 1995; Ackerson et al. 1986). For an FCC micellar phase (formed when solvent is present), the 111 layers slide over each other, also like colloidal crystals (McConnell et al. 1995). At low shear rates, an unaligned polycrystalline state has been observed, while at very high shear rates, the ordered crystalline states formed at moderate shear rates melt into disorder.

In contrast to lamellar microdomains, where removal of defects and alignment of domains can, in principle, restore terminal behavior to G' and G'', even after alignment, G' for three-dimensionally ordered spherical domains remains that of a soft solid, albeit with a much lower modulus than that of the unoriented material (Koppi et al. 1994).

The source of elasticity in block copolymers containing well-ordered spherical domains is analogous to that for simple three-dimensional crystalline solids. When a mechanical deformation displaces spherical domains from their equilibrium lattice positions, the domains are pulled back by the thermodynamic forces that are responsible for the existence of the macrocrystalline order in the static sample. Similar forces exist when the domains are cylindrical or lamellar, but these latter morphologies, if well-ordered and oriented appropriately, can sustain shearing deformations without displacement of the domains from their equilibrium positions, since they are not ordered in all dimensions as spherical domains are.

Of course the modulus of a block copolymer with ordered spherical microdomains is much lower than that of a crystalline solid. Near the disordering transition, the potential energy holding each domain or atom in place is of order $k_B T$, and the modulus is roughly $\nu k_B T$, where ν is the number of domains or atoms. This gives an elastic modulus 10^3–10^5 dyn/cm^2 for typical block copolymers with spherical domains, as opposed to 10^9–10^{10} dyn/cm^2 for atomic crystals. Ordered spherical diblock copolymers are therefore *soft solids*. They deflect under an imposed shear stress, but do not flow continuously unless that stress exceeds a critical value, the *yield stress* (Watanabe and Kotaka 1984).

Unusual stress waveforms are observed during oscillatory deformation on such materials when the strain amplitude is much greater than the yield strain. Watanabe and Kotaka (1984) plotted large-amplitude oscillatory shear data in the form of Lissajous figures—that is, plots of periodic stress versus periodic strain. In the nonlinear regime, above the yield strain, unusual rhombic Lissajous figures were obtained at low frequencies, and bent ellipses were obtained at higher frequencies. Qualitatively, the waveforms are similar to those depicted in Fig. 13-23.

These shapes are similar to those predicted by Doi and coworkers (Ohta et al. 1993; Doi et al. 1993) using a simple two-dimensional model for the deformation of materials with ordered domains. They considered the case of a two-dimensional hexagonally well-ordered macrolattice subjected to shearing deformations with the shear velocity parallel to a crystallographic direction of the macrolattice, so that rows of circular domains slide past each other; as illustrated in Fig. 1-21. The strain in a layer, say layer i, can be defined as $\gamma_i \equiv (x_{i+1} - x_i)/h$, where h is the distance between layers, and x_j is the displacement in the flow direction of layer j. The elastic stress generated in layer i is a simple periodic function of the strain in that layer:

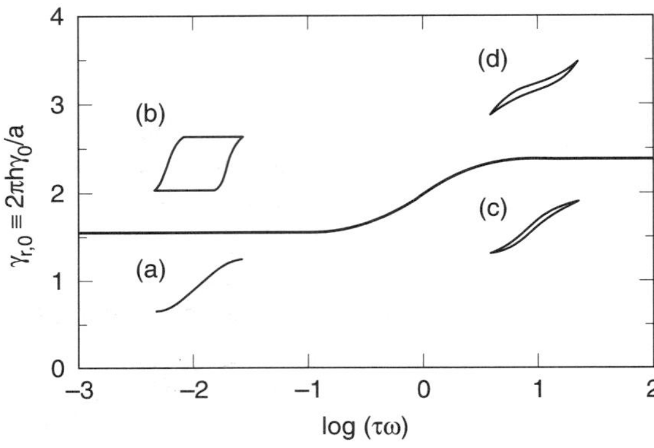

Figure 13.23 **(a–d)** Sketch of Lissajous figures of stress versus strain in oscillatory shearing at low and high dimensionless frequency ($\tau\omega$) and low and high strain amplitude (γ_0). These figures are qualitatively similar to those observed by Watanabe et al. (1994) in BCC sphere-forming polystyrene–polybutadiene diblock copolymers in a solvent, and are predicted by a theory of Doi et al. described in the text. (Reprinted with permission from Doi et al., Macromolecules 26:4935. Copyright © 1993, American Chemical Society.)

$$\sigma_i^e = G_e \sin \gamma_{r,i} \tag{13-9}$$

where the reduced strain $\gamma_{r,i}$ is related to the layer strain γ_i by

$$\gamma_{r,i} \equiv 2\pi \frac{h}{a} \gamma_i \tag{13-10}$$

and a is the periodicity of the lattice structure, and $h = (\sqrt{3}/2)a$. To this elastic stress component with modulus G_e a viscoelastic term with modulus G_{ve} is added that accounts for stresses generated by distortions of the domain shapes that occur at higher rates of deformation. Thus, the total stress in layer i is given by

$$\sigma_i = G_e \sin \gamma_{r,i} + G_{\text{ve}} \int_{-\infty}^{t} dt' \exp[-(t-t')/\tau]\dot{\gamma}_{r,i}(t') \tag{13-11}$$

The shear stress must be the same in all layers; thus $\sigma_i = \sigma$. Eq. (13-11) can also be written in a simple differential form:

$$\tau(G_{ve} + G_e \cos \gamma_{r,i})\dot{\gamma}_{r,i} + G_e \sin \gamma_{r,i} = \sigma + \tau\dot{\sigma} \tag{13-12}$$

A constraint must be imposed that the average over all layers of the strain equals the imposed strain:

$$\gamma_r = \frac{1}{N} \sum_{i=1}^{N} \gamma_{r,i} \tag{13-13}$$

where $\gamma_r \equiv 2\pi (h/a)\gamma$, and γ is the true macroscopic strain.

Equation (13-12) and (13-13) plus a small noise term—which introduces small imperfections into the macrolattice—complete the description of the the model. Because of the periodic nature of the elastic stress [Eq. (13-9)] in large-strain shearing, instabilities occur that allow layers to slip with respect to each other, and the strain γ_i then varies from layer to layer. Thus flow occurs by *yielding* a mechanism similar to that discussed in Section 1.5.3.

The predictions of the theory of Doi et al. (1993) are in good qualitative agreement with experiments on diblock copolymers with ordered spherical domains. In particular, the theory predicts that peculiar Lissajous figures (Fig. 13-23) observed by Watanabe and Kotaka

(1984) in large-amplitude shearing flow. The rhombus-shaped Lissajous figure shown in Fig. 13-23 is obtained at frequencies low enough that only the elastic stress contributes significantly to the total stress, while the more ellipsoidally shaped figure is produced at higher frequencies where the viscoelastic term in Eq. (13-11) is larger.

13.4 SUMMARY

The rheological and flow properties of ordered block copolymers are extraordinarily complex; these materials are well-deserving of the apellation "complex fluids." Like the liquid-crystalline polymers described in Chapter 11, block copolymers combine the complexities of small-molecule liquid crystals with those of polymeric liquids. Hence, at low frequencies or shear rates, the rheology and flow-alignment characteristics of block copolymers are in some respects similar to those of small-molecule liquid crystals, while at high shear rates or frequencies, "polymeric modes" of behavior are more important.

Lamellar block copolymers usually (but not always) align under shearing flow in either the "parallel" or "perpendicular" orientations, with normals to the lamellae respectively in the gradient or vorticity directions, depending on the conditions of shear. A high degree of mechanical contrast between the blocks seems to favor the parallel alignment at high frequencies where "polymer modes" of relaxation dominate. Multiblock architecture, on the other hand, seems to favor the perpendicular orientation, with the exception of triblocks with little viscoelastic contrast. Cylinders align in the flow direction, with the hexagonal unit cell preferentially in the "parallel" orientation, so that the most densely packed planes of cylinders are parallel to the shearing surfaces. Phases of BCC spheres align with the most densely packed 111 direction parallel to flow and the most densely packed planes, the 110 planes, parallel to the shearing surfaces.

Microscopically ordered, but macroscopically unaligned or poorly aligned, samples show strong departures from terminal viscoelastic behavior at low frequencies. Alignment by shearing flow brings the low-frequency behavior of samples with lamellar or hexagonal cylindrical domains closer to terminal. Oscillatory shearing flow seems to produce better alignment than steady flow, although optimum alignment may require hundreds of cycles of oscillation. Strain amplitudes of order unity are often the most effective; however, substantial alignment at amplitudes of only 5% or less have been reported under some conditions.

Beyond these remarks, few statements regarding the rheological behavior of block copolymers have general validity; the detailed rheology and flow-alignment characteristics depend on the type of domain formed (lamellae, cylinders, spheres, etc.), the block architecture (diblock, triblock, etc.), the strength of the microseparation (strong versus weak), the degree of "mechanical contrast," the presence of solvents or homopolymers, and other factors. So far, theoretical work has provided guidance only in restricted situations. Thus, the rheological behavior of block copolymers is perhaps the least understood of all categories of complex fluids considered in this book.

REFERENCES

Ackerson BJ, Hayter JB, Clark NA, Cotter L (1986). *J Chem Phys* 84:2344.

Adams JL, Quiram DJ, Graessley WW, Register RA, Marchand GR (1996). *Macromolecules* 29:2929.

Aggarwal SL (1976). *Polymer* 17:938.

Albalak RJ, Thomas EL (1993). *J Polym Sci Polym Phys Ed* 31:37.

Alexandridis P (1996). *Curr Opin Colloid Interface Sci* 1:490.

Almdal K, Rosedale JH, Bates FS, Wignall GD, Fredrickson GH (1990). *Phys Rev Lett* 65:1112.

Almdal K, Bates FS, Mortensen K (1992). *J Chem Phys* 96:9122.

Almdal K, Koppi KA, Bates FS (1993). *Macromolecules* 26:4058.

Alt DJ, Hudson SD, Garay RO, Fujishiro K (1995). *Macromolecules* 28:1575.

Amundson K, Helfand E (1993). *Macromolecules* 26:1324.

Argon AS, Cohen RE (1990). *Adv Polym Sci* 91-92:301.

Balsara NP, Hammouda B (1994). *Phys Rev Lett* 72:360.

Balsara NP, Dai HJ, Kesani PK, Garetz BA, Hammouda B (1994a). *Macromolecules* 27:7406.

Balsara NP, Hammouda B, Kesani PK, Jonnalagadda SV, Straty GC (1994b). *Macromolecules* 27:2566.

Barrat J-L, Fredrickson GH (1991). *J Chem Phys* 95:1281.

Bates FS (1984). *Macromolecules* 17:2607.

Bates FS, Fredrickson GH (1990). *Annu Rev Phys Chem* 41:525.

Bates FS, Rosedale JH, Fredrickson GH (1990). *J Chem Phys* 92:6255.

Bates FS, Koppi KA, Tirrell M, Almdal K, Mortensen K (1994). *Macromolecules* 27:5934.

Binder K (1994). *Adv Polym Sci* 112:181.

Binder K, Deutsch H-P, Müller M, Fried H, Kikuchi M (1994). *Il Nuovo Cimento* 16:653.

Brazovskii SA (1975). *Sov Phys JEPT* 41:85.

Cates ME, Milner ST (1989). *Phys Rev Lett* 62:1856.

Chen JT, Thomas EL, Ober CK, Mao G-P (1996). *Science* 273:343.

Chen JT, Thomas EL, Zimba CG, Rabolt JF (1995). *Macromolecules* 28:5811.

Chen Z-R, Issaian A, Kornfield JA, Smith SD, Grothaus JT, Satkowski MM (1997a). *Macromolecules,* 30:7096.

Chen Z-R, Kornfield JA, Smith SD, Grothaus JT, Satkowski MM (1997b). *Science* 277:1248.

Chiellini E, Galli G, Angeloni S, Laus M (1994). *Trends Polym Sci* 2:244.

Chu JH, Rangarajan P, Adams JL, Register RA (1995). *Polymer* 36:1569.

Chung CI, Gale JC (1976). *J Polym Sci Polym Phys Ed* 14:1149.

Dai HJ, Balsara NP, Garetz BA, Newstein MC (1996). *Phys Rev Lett* 77:3677.

Doi M, Harden JL, Ohta T (1993). *Macromolecules* 26:4935.

Folkes MJ (ed) (1985). *Processing, Structure and Properties of Block Copolymers*, Elsevier, New York.

Folkes MJ, Keller A (1971). *Polymer* 12:222.

Förster S, Khandpur AK, Zhao J, Bates FS, Hamley IW, Ryan AJ, Bras W (1994). *Macromolecules* 27:6922.

Fredrickson GH (1994). *J Rheol* 38:1045.

Fredrickson GH Bates FS (1996). *Annu Rev Mater Sci* 26:501.

Fredrickson GH, Helfand E (1987). *J Chem Phys* 87:697.

Fredrickson GH, Helfand E (1988). *J Chem Phys* 89:5890.

Fredrickson GH, Larson RG (1987). *J Chem Phys* 86:1553.

Fried H, Binder K (1991). *Europhys Lett* 16:237.

Gido SP, Gunther J, Thomas EL, Hoffman D (1993). *Macromolecules* 26:4506.

Gouinlock EV, Porter RS (1977). *Polym Eng Sci* 17:535.

Goulian M, Milner ST (1995). *Phys Rev Lett* 74:1775.

Gupta VK, Krishnamoorti R, Kornfield JA, Smith (1996). *Macromolecules* 29:875.

Hadjuk DA, Harper PE, Gruner SM, Homeker CC, Kim G, Thomas EL, Fetters LJ (1994). *Macromolecules* 27:4063.

Hadziioannou, G, Mathis A, Skoulios A (1979). *Colloid Polym Sci* 257:136.

Hadziioannou G, Picot C, Skoulios A, Ionescu M-L, Mathis A, Duplessix R Gallot Y, Lingelser J-P (1982). *Macromolecules* 15:263.

Hamley IW, Gehlsen MD, Khandpur AK, Koppi KA, Rosedale JH, Schulz MF, Bates FS, Almdal K, Mortensen K (1994). *J Phys II (France)* 4:2161.

Han CD, Baek DM, Kim JK, Ogawa T, Sakamoto N, Hashimoto T (1995). *Macromolecules* 28:5043.

Hasegawa H, Tanaka H, Yamasaki K, Hashimoto T (1987). *Macromolecules* 20:1651.

Hashimoto T, Koizumi S, Hasegawa H, Izumitani T, Hyde ST (1992). *Macromolecules* 25:1433.

Helfand E, Wasserman ZR (1976). *Macromolecules* 9:879.

Helfand E, Wasserman ZR (1982). In *Developments in Block Copolymers*, Vol I, Goodman I (ed), Elsevier, New York.

Hudson SD, Amundson KR, Jeon HG, Smith SD (1995). *MRS Bull* Sept:42.

Jackson CL, Barnes KA, Morrison FA, Mays JW, Nakatani AI, Han CC (1995). *Macromolecules* 28:713.

Jenekhe SA, Chen XL (1998). *Science* 279:1903.

Jin X, Lodge TP (1997). *Rheol Acta* 36:229.

Kannan RM, Kornfield JA (1994). *Macromolecules* 27:1177.

Kannan RM, Kornfield JA, Schwenk N, Boeffel C (1993). *Macromolecules* 26:2050.

Karis TE, Russell TP, Gallot Y, Mayes AM (1995). *Macromolecules* 28:1129.

Kawasaki K, Onuki A (1990). *Phys Rev A* 42:3664.

Keller A, Odell JA (1985). In *Processing Structure and Properties of Block Copolymers*, Folkes MJ (ed), Elsevier, New York.

Keller A, Pedemonte E, Willmouth FM (1970). *Kolloid-Z Z Polym* 238:385.

Khandpur AK, Förster S, Bates FS, Hamley AI, Ryan AJ, Bras W, Almdal K, Mortensen K (1995). *Macromolecules* 28:8796.

Kinning DJ, Winey KI, Thomas EL (1988). *Macromolecules* 21:3502.

Kodama H, Doi M (1996). *Macromolecules* 29:2652.

Koppi KA, Tirrell M, Bates FS, Almdal K, Colby RH (1992). *J Phys II (France)* 2:1941.

Koppi KA, Tirrell M, Bates FS (1993). *Phys Rev Lett* 70:1449.

Koppi KA, Tirrell M, Bates FS, Almdal K, Mortensen K (1994). *J Rheol* 38:999.

Kotaka T, Watanabe H (1987). In *Current Topics in Polymer Science*, Vol II, Ottenbrite RM, Utracki LA, Inoue S (eds), Hanser Publishers, New York.

Krappe U, Stadler R, Voigt-Martin I (1995). *Macromolecules* 28:4558.

Larson RG (1994a). *Mol Sim* 13:321.

Larson RG (1994b). *Macromolecules* 27:4198.

Larson RG, Winey KI, Patel SS, Watanabe H, Bruinsma R (1993). *Rheol Acta* 32:245.

Leary DF, Williams MC (1970). *J Polym Sci B* 8:335.

Leibler L (1980). *Macromolecules* 13:1602.

Lin CC, Jonnalagadda SV, Kesani PK, Dai HJ, Balsara NP (1994). *Macromolecules* 27:7769.

Lohse DJ, Fetters LJ, Doyle MJ, Wang H-C, Kow C (1993). *Macromolecules* 26:3444.

Lyngaae-Jorgensen J (1985). In *Processing Structure and Properties of Block Copolymers*, edited by MJ Folkes (ed), Applied Science, New York.

Marques CM, Cates ME (1990). *J Phys (France)* 51:1733.

Matsen MW (1995). *Macromolecules* 28:5765.

Matsen MW, Bates FS (1996). *Macromolecules* 29:1091.

Matsen MW, Schick M (1994). *Macromolecules* 27:7157.

McConnell GA, Gast AP, Huang JS, Smith SD (1993). *Phys Rev Lett* 71:2102.

McConnell GA, Lin MY, Gast AP (1995). *Macromolecules* 28:6754.

Meier DJ (1969). *J Polym Sci C* 26:81.

Milner ST, Fredrickson GH (1996), unpublished.

Milner ST, Witten TA, Cates ME (1988). *Europhys Lett* 5:413; *Macromolecules* 21:2610.

Minchau B, Dünweg B, Binder K (1990). *Polym Commun* 31:348.

Mogi Y, Kotsuzi H, Kaneko Y, Mori K, Matsushita Y, Noda I (1992). *Macromolecules* 25:5408.

Molau GE (ed) (1971). *Colloidal and Morphological Behavior of Block and Graft Copolymers*, Plenum Press, New York.

Morrison FA, Winter HH, Gronski W, Barnes JD (1990). *Macromolecules* 23:4200.

Nakatani AI, Morrison FA, Douglas JF, Mays JW, Jackson CL, Muthukumar M, Han CC (1996). *J Chem Phys* 104:1589.

Nakazawa H, Ohta T (1993). *Macromolecules* 26:5503.

Odell JA, Keller A (1977). *Polym Eng Sci* 17:544.

Ohta T, Enomoto Y, Harden JL, Doi M (1993). *Macromolecules* 26:4928.

Okamoto S, Saijo K, Hashimoto T (1994a). *Macromolecules* 27:3753.

Okamoto S, Saijo K, Hashimoto T (1994b). *Macromolecules* 27:5547.

Olson GB (1992). In *Martensite*, ASM International; Metals Park, OH, Chapter 1.

Park M, Harrison C, Chaikin PM, Register RA, Adamson DH (1997) *Science* 276:1401.

Patel SS, Larson RG, Winey KI, Watanabe H (1995). *Macromolecules* 28:4313.

Pinheiro BS, Hajduk DA, Gruner SM, Winey KI (1996). *Macromolecules* 29:1482.

Qi S, Wang Z-G (1997). *Phys Rev E* 55:1682.

Radzilowski LH, Carragher BO, Stupp SI (1997). *Macromolecules* 30:2110.

Riise BL, Fredrickson GH, Larson RG, Pearson DS (1995). *Macromolecules* 28:7653.

Rosedale JH, Bates FS (1990). *Macromolecules* 23:2329.

Rosedale JH, Bates FS, Almdal K, Mortensen K, Wignall GD (1995). *Macromolecules* 28:1429.

Schulz MF, Bates FS (1996). In *Physical Properties of Polymers Handbook*, Mark JE (ed), AIP Press, New York.

Scott DB, Waddon AJ, Lin Y-G, Karasz FE, Winter HH (1992). *Macromolecules* 25:4175.

Semenov AN (1985). *Sov Phys JETP* 61:733.

Stadler R, Auschra C, Beckmann J, Krappe U, Voigt-Martin I, Leibler L (1995). *Macromolecules* 28:3080.

Stühn B, Mutter R, Albrecht T (1992). *Europhys Lett* 18:427.

Tepe T, Schulz MF, Zhao J, Tirrell M, Bates FS, Mortensen K, Almdal K, (1995). *Macromolecules* 28:3008.

Tepe T, Hillmayer MA, Weimann PA, Tirrell M, Bates FS, Almdal K, Mortensen K (1997). *J Rheol* 41:1147.

Thomas EL, Kinning DJ, Alward DB, Henkee CS (1987). *Macromolecules* 20:2934.

Thomas EL, Anderson DM, Henkee CS, Hoffman D (1988). *Nature* 334:598.

Turner MS (1995). *Macromolecules* 28:6878.

Vavasour JD, Whitmore MD (1993). *Macromolecules* 26:7070.

Wang Z-G (1994). *J Chem Phys* 100:2298.

Watanabe H, Kotaka T (1983). *J Rheol* 27:223.

Watanabe H, Kotaka T (1984). *Polym Eng Rev* 4:73.

Watanabe H, Yao M-L, Sato T, Osaki K (1997). *Macromolecules* 30:5905.

Wiesner U (1997). *Macromol Chem Phys,* submitted.

Williams DRM, MacKintosh FC (1994). *Macromolecules* 27:7677.

Winey KI, Thomas EL, Fetters LJ (1992). *Macromolecules* 25:2645.

Winey KI, Patel SS, Larson RG, Watanabe H (1993a). *Macromolecules* 26:2542.

Winey KI, Patel SS, Larson RG, Watanabe H (1993b). *Macromolecules* 26:4373.

Winter HH, Scott DB, Gronski W, Okamoto S, Hashimoto T (1993). *Macromolecules* 26:7236.

Zhang Y, Wiesner U (1995). *J Chem Phys* 103:4784.

Zhang Y, Wiesner U, Spiess HW (1995). *Macromolecules* 28:778.

Zhang Y, Wiesner U, Yang Y, Pakula T, Spiess HW (1996). *Macromolecules* 29:5427.

Appendix
MOMENTUM-BALANCE EQUATIONS IN THE ABSENCE OF INERTIA

Rectangular Coordinates

$$\frac{\partial}{\partial x}\sigma_{xx} + \frac{\partial}{\partial y}\sigma_{yx} + \frac{\partial}{\partial z}\sigma_{zx} - \frac{\partial p}{\partial x} + \rho g_x = 0 \tag{A-1}$$

$$\frac{\partial}{\partial x}\sigma_{xy} + \frac{\partial}{\partial y}\sigma_{yy} + \frac{\partial}{\partial z}\sigma_{zy} - \frac{\partial p}{\partial y} + \rho g_y = 0 \tag{A-2}$$

$$\frac{\partial}{\partial x}\sigma_{xz} + \frac{\partial}{\partial y}\sigma_{yz} + \frac{\partial}{\partial z}\sigma_{zz} - \frac{\partial p}{\partial z} + \rho g_z = 0 \tag{A-3}$$

Cylindrical Coordinates

$$\frac{1}{r}\frac{\partial}{\partial r}(r\sigma_{rr}) + \frac{1}{r}\frac{\partial}{\partial \theta}\sigma_{\theta r} + \frac{\partial}{\partial z}\sigma_{zr} - \frac{\sigma_{\theta\theta}}{r} - \frac{\partial p}{\partial r} + \rho g_r = 0 \tag{A-4}$$

$$\frac{1}{r^2}\frac{\partial}{\partial r}(r^2\sigma_{r\theta}) + \frac{1}{r}\frac{\partial}{\partial \theta}\sigma_{\theta\theta} + \frac{\partial}{\partial z}\sigma_{z\theta} + \frac{\sigma_{\theta r} - \sigma_{r\theta}}{r} - \frac{1}{r}\frac{\partial p}{\partial \theta} + \rho g_\theta = 0 \tag{A-5}$$

$$\frac{1}{r}\frac{\partial}{\partial r}(r\sigma_{rz}) + \frac{1}{r}\frac{\partial}{\partial \theta}\sigma_{\theta z} + \frac{\partial}{\partial z}\sigma_{zz} - \frac{\partial p}{\partial z} + \rho g_z = 0 \tag{A-6}$$

Spherical Coordinates

$$\frac{1}{r^2}\frac{\partial}{\partial r}(r^2\sigma_{rr}) + \frac{1}{r\sin\theta}\frac{\partial}{\partial \theta}(\sigma_{\theta r}\sin\theta) + \frac{1}{r\sin\theta}\frac{\partial}{\partial \phi}\sigma_{\phi r} - \frac{\sigma_{\theta\theta} + \sigma_{\phi\phi}}{r}$$
$$- \frac{\partial p}{\partial r} + \rho g_r = 0 \tag{A-7}$$

$$\frac{1}{r^3}\frac{\partial}{\partial r}(r^3\sigma_{r\theta}) + \frac{1}{r\sin\theta}\frac{\partial}{\partial\theta}(\sigma_{\theta\theta}\sin\theta) + \frac{1}{r\sin\theta}\frac{\partial}{\partial\phi}\sigma_{\phi\theta} + \frac{\sigma_{\theta r} - \sigma_{r\theta} - \sigma_{\phi\phi}\cot\theta}{r}$$

$$-\frac{1}{r}\frac{\partial p}{\partial\theta} + \rho g_\theta = 0 \tag{A-8}$$

$$\frac{1}{r^3}\frac{\partial}{\partial r}(r^3\sigma_{r\phi}) + \frac{1}{r\sin\theta}\frac{\partial}{\partial\theta}(\sigma_{\theta\phi}\sin\theta) + \frac{1}{r\sin\theta}\frac{\partial}{\partial\phi}\sigma_{\phi\phi}$$

$$+\frac{\sigma_{\phi r} - \sigma_{r\phi} - \sigma_{\phi\theta}\cot\theta}{r} - \frac{1}{r\sin\theta}\frac{\partial p}{\partial\phi} + \rho g_\phi = 0 \tag{A-9}$$

Common Notation

A	area, cross-sectional area
A_H	Hamaker constant
B	$2\beta^2 L^2 = 3L^2/\langle R^2\rangle_0$, polymer extensibility
B	smectic-layer compressive modulus
$\mathbf{B}$	$\mathbf{E}^T \cdot \mathbf{E}$, Finger strain tensor
$\mathbf{C}$	$\mathbf{B}^{-1}$, Cauchy strain tensor
Ca	$\dot\gamma \eta_s a/\Gamma$, capillary number
C_∞	characteristic ratio, defined by $\langle R^2\rangle_0 = C_\infty n b_n^2$
D	translational diffusivity
D	closest separation distance between surfaces
D	dimensionality
D_f	fractal dimension
$\mathbf{D}$	$\frac{1}{2}[\nabla\mathbf{v} + (\nabla\mathbf{v})^T]$, rate of strain tensor
De	τ/t_f, Deborah number
D_r	rotational diffusivity
D_{r0}	rotational diffusivity in dilute solution
D_s	short-time diffusivity
E	energy
$\mathbf{E}$	electric field
$\mathbf{E}$	$\partial\mathbf{x}/\partial\mathbf{x}'$, inverse deformation gradient tensor
Er	$\eta\dot\gamma h/K$, Ericksen number
E_{sciss}	energy of scission of a worm-like micelle
F	force, free energy
$\mathbf{F}$	force vector
$\mathbf{F}$	$\mathbf{E}^{-1}$, deformation-gradient tensor
G	modulus
G'	storage modulus
G''	loss modulus
G^*	$G' + iG''$ or $[(G')^2 + (G'')^2]^{1/2}$, complex modulus
G_N^0	entanglement plateau modulus
H	enthalpy
H	$\frac{1}{2}(R_1^{-1} + R_2^{-1})$, mean curvature of surface (Chapter 12)
$\mathbf{H}$	magnetic field (Chapter 8)
I	scattering intensity
K	$R_1 R_2$, Gaussian curvature of surface (Chapter 12)
$\mathbf{K}$	$(\nabla\mathbf{v})^T$, transpose of velocity gradient
K_1	Frank splay constant

K_2	Frank twist constant
K_3	Frank bend constant
L	length of molecule or worm-like micelle
M	molecular weight
M	η_d/η_s, viscosity ratio of dispersed to continuous phases
$\mathbf{M}$	magnetization of medium
$\mathbf{M}_d$	magnetization of particles
Ma	$\eta_s\dot{\gamma}/(2\varepsilon_0\varepsilon_s\beta^2 E^2)$, Mason number
M_c	threshold molecular weight for entanglement
M_e	entanglement molecular weight
M_n	number-averaged molecular weight
M_w	weight-averaged molecular weight
M_0	monomer molecular weight
N	number of monomers in a polymer chain
N	number of surfactant molecules in a micelle (Chapter 12)
$\overline{N}$	Nb^6v^{-2}, Eq. (13-4)
N_A	Avogadro's number, 6.023×10^{23}
N_K	number of Kuhn steps in a polymer chain [Eqs. (2-17) and (2-18)]
N_s	number of springs
N_1	$\sigma_{11} - \sigma_{22}$, first normal stress difference
N_2	$\sigma_{22} - \sigma_{33}$, second normal stress difference
P	probability
$\mathbf{P}$	electric polarization
Pe	$\eta_s\dot{\gamma}a^3/k_BT$, Peclet number
$\mathbf{Q}$	Doi–Edwards "universal" strain tensor [Eq. (3-70)]
R	gas constant, 1.98 cal/(mol K)
$\mathbf{R}$	end-to-end vector of polymer molecule or strand
Re	Reynolds number
R_g	polymer radius of gyration
$R(p)$	see Eq. (10-21)
R_1	first principal radius of curvature of a surface
R_2	second principal radius of curvature of a surface
S	S_2, second-moment orientational order parameter [Eq. (10-23)]
S	entropy
$\mathbf{S}$	nematic tensor order parameter
S_c	configurational entropy
S_4	fourth-moment orientational order parameter [Eq. (10-23)]
$S(\mathbf{q})$	$\langle\rho_\mathbf{q}\rho_{-\mathbf{q}}\rangle$, structure factor
T	temperature
$\mathbf{T}$	$\sigma - p\delta$, state-of-stress tensor
T_{AN}	smectic-A–nematic transition temperature
T_{NI}	nematic–isotropic transition temperature
T_{ODT}	order–disorder transition temperature
T_g	glass transition temperature
T_f	fictive temperature (see Fig. 4-20)
T_0	temperature at which $\tau \to \infty$ [Eq. (4-4)]

V	scalar velocity
V_s	slip velocity
W	free energy or potential
W_d	distortional free energy in a liquid crystal
Wi	$\tau\dot{\gamma}$, Weissenberg number
X_n	mole fraction of aggregates containing n surfactant molecules (Chapter 12)
a	radius of sphere
a	entanglement spacing (Chapter 3)
a	area occupied by surfactant head group (Chapter 12)
a	domain or texture size
a_T	temperature shift factor
a_0	optimal area occupied by head group (Chapter 12)
b	slip extrapolation length
b	$\equiv b_N$, length of random-walk step in a chain of N steps
b_K	length of Kuhn step [see Eqs. (2-17) and (2-18)]
b_n	length of equivalent random-walk step in a chain of n steps
c	mass concentration
c^*	polymer–polymer overlap concentration
d	diameter of particle, molecule, or polymer chain
det	determinant
e	charge on an electron, 1.6×10^{-19}
f	free energy per unit volume
f	v_f/v_g, fractional free volume (Chapter 4)
f	volume fraction of block in block copolymer (Chapter 13)
g	gravitational constant, 980 cm/sec^2
$g(r)$	radial distribution function
h	gap between surfaces, as in a rheometer
$h(\gamma)$	damping function in step shear
i	$\sqrt{-1}$, imaginary unit
k_B	Boltzmann's constant, 1.38×10^{-16} erg/°C
log	base 10 logarithm
ℓ	carbon–carbon bond length, 1.53 Å
ℓ_c	effective length of surfactant tail
ln	natural logarithm
m	mass
$\mathbf{m}$	magnetic dipole moment
n	index of refraction
n	number of carbon–carbon bonds in a polymer chain
n	number of surfactant molecules in a micelle (Chapter 12)
$\mathbf{n}$	nematic director
$\mathbf{n}$	index of refraction tensor
n_c	number of carbons in a surfactant tail
n_i	number density of ions of type i
p	pressure
p	L/d, aspect ratio of length to diameter of particle or molecule
p	fraction of allowed bonds (Chapter 5)

p_c	fraction of allowed bonds at percolation transition (Chapter 5)
q	electric charge
$\mathbf{q}$	scattering vector
q_m	wavenumber at maximum scattering intensity
r	distance of center-of-mass separation
t	time
t'	past time
t_f	characteristic flow time
tr	trace
u	energetic parameter
$\mathbf{u}$	unit vector
$\mathbf{u}$	electric dipole moment
v	specific volume, $1/\rho$
v	volume of surfactant tail (Chapter 12)
v_f	free volume per unit mass (Chapter 4)
v_o	occupied volume per unit mass, $v_o = v - v_f$
$\mathbf{x}$	position vector
$\mathbf{x}'$	position at past time t' of the fluid particle that resides at position $\mathbf{x}$ at time t
z	charge valence

GREEK

Γ	surface tension, interfacial tension
Δ	difference, or layer thickness
ΔS	$S_{\text{liq}} - S_{\text{cryst}}$, excess entropy
Δ_d	depletion layer thickness
Θ	theta temperature
Ψ_1	$N_1/\dot{\gamma}^2$, first normal stress coefficient
Ψ_2	$N_2/\dot{\gamma}^2$, second normal stress coefficient
Ω	rotation rate in torsional shearing
Ω_0	molecular collision rate, $\sim 10^{13}\,\text{sec}^{-1}$
α	flow type parameter in Eqs. (1-11) and (3-55)
α	polarizability
α	$d\ln v/dT$, thermal expansion coefficient
β	$3/(2N_K b_K^2) = 3/(2\langle R^2\rangle_0)$
β	$(\varepsilon_r - 1)/(\varepsilon_r + 2)$ [see also Eq. (8-2)]
β^*	$\tau_0\dot{\gamma}$
γ	shear strain
γ_0	shear-strain amplitude
γ_1	rotational viscosity of nematic [Eq. (10-15)]
$\dot{\gamma}$	shear rate
$\boldsymbol{\delta}$	unit tensor
ε	dielectric constant
ε_0	permittivity of free space, $8.85 \times 10^{-12}\,\text{C}^2\,\text{J}^{-1}\,\text{m}^{-1}$
ε'	real part of complex dielectric constant
ε''	imaginary part of complex dielectric constant

ε^*	$\varepsilon' + i\varepsilon''$, complex dielectric constant
ε_r	$\varepsilon_p/\varepsilon_s$, ratio of particle to solvent dielectric constant
$\dot{\varepsilon}$	extension rate
ζ	drag coefficient
ζ	zeta potential (Chapter 7)
ζ_0	monomeric friction coefficient
ζ_{str}	viscous drag coefficient [Eqs. (6-37), (6-51), or (6-52)]
η	$\sigma_{12}/\dot{\gamma}$, viscosity
η'	G''/ω
η''	G'/ω
η_d	viscosity of droplet phase
η_r	η/η_s, relative viscosity
η_0	zero-shear viscosity
$\overline{\eta}_u$	uniaxial extensional viscosity
θ	angle
θ_a	flow-alignment angle in a nematic (Fig. 10-4)
κ	inverse of Debye double-layer distance [Eq. (2-48)]
λ	stretch ratio
λ	wavelength of radiaion
λ	$-W_{e,min}/(2k_BT)$ or $W_{m,min}/(2k_BT)$ (Chapter 8)
λ	nematic tumbling/flow-aligning parameter [Eq. (10-3)]
λ	$(K_1/B)^{1/2}$, permeation depth (Chapter 10)
λ_p	persistence length of semi-flexible molecule [Eq. (2-24a)]
μ	chemical potential
ν	number of molecules, particles, or entanglements per unit volume
$\nu^{(R)}$	number of polymer molecules per unit particle-free volume
ρ	mass density
ρ_A	disclination density per unit area
ρ_V	disclination density per unit volume
ρ_q	Fourier transform of $\rho(\mathbf{x})$
σ	σ_{12} shear stress
σ	surface-charge density
$\overline{\sigma}$	extensional stress
$\boldsymbol{\sigma}$	extra-stress tensor
τ	relaxation time
τ_B	Baxter "sticky-sphere" interaction parameter [Eq. (7-20)]
τ_{br}	breakage time of a worm-like micelle
τ_d	reptation, or "disengagement time" of a polymer molecule
τ_r	retraction time
τ_{rep}	reptation time of a worm-like micelle (Chapter 12)
τ_v	volumetric relaxation time
τ_0	characteristic relaxation time, $[\eta]_0 M\eta_s/N_A k_B T$
τ_1	longest relaxation time
ϕ	volume fraction
ϕ_c	critical volume fraction for phase separation
ϕ_m	maximum-packing volume fraction

χ	Flory "chi" energetic parameter
ψ	orientational distribution function
ψ	electrostatic potential
ω	frequency
$\boldsymbol{\omega}$	$\frac{1}{2}[\nabla\mathbf{v} - (\nabla\mathbf{v})^T]$, vorticity tensor
∇	gradient, $(\partial/\partial x_1, \partial/\partial x_2, \partial/\partial x_3)$
$[\cdot]$	molarity, or dilute-solution intrinsic property
$\langle\cdot\rangle$	ensemble average

SUPERSCRIPTS

T	transpose
b	Brownian
d	drag
d	Frank distortional
depl	depletion
e	elastic
hyd	hydration
p	polymer or particle contribution
s	solvent, or spring
v	viscous
vdw	van der Waals

SUBSCRIPTS

ODT	order–disorder
c	critical
d	drag, droplet phase
depl	depletion
e	electrostatic
g	glass property
hyd	hydration
i	relaxation mode number
n	number averaged
p	particle property
r	relative
s	solvent, or suspending fluid, or surface property
vdw	van der Waals
w	weight averaged
x	flow dirction
y	yield, or velocity-gradient direction
z	vorticity direction
0	zero-shear property, as in η_0, $\Psi_{1,0}$, etc., or dilute-solution property
1	flow direction
2	flow-gradient direction
3	vorticity direction
∞	high-frequency or high-shear-rate property

AUTHOR INDEX

SUBJECT INDEX